**Handbook of RAFT Polymerization**

*Edited by*
*Christopher Barner-Kowollik*

*Related Titles*

Soares, J. B. P., McKenna, T. F. L.

**Polyolefin Reaction Engineering**

2008
ISBN: 978-3-527-31710-3

Matyjaszewski, K., Gnanou, Y., Leibler, L. (eds.)

**Macromolecular Engineering**

**Precise Synthesis, Materials Properties, Applications**

2007
ISBN: 978-3-527-31446-1

Tadmor, Z

**Principles of Polymer Processing 2e**

2006
E-Book
ISBN: 978-0-471-79276-5

Shaw, M. T., MacKnight, W. J.

**Introduction to Polymer Viscoelasticity**

2005
ISBN: 978-0-471-74045-2

Meyer, T., Keurentjes, J. (eds.)

**Handbook of Polymer Reaction Engineering**

2005
ISBN: 978-3-527-31014-2

Advincula, R. C., Brittain, W. J., Caster, K. C., Rühe, J.(eds.)

**Polymer Brushes**

**Synthesis, Characterization, Applications**

2004
ISBN: 978-3-527-31033-3

Odian, G.

**Principles of Polymerization**

2004
ISBN: 978-0-471-27400-1

Goodwin, J.

**Colloids and Interfaces with Surfactants and Polymers - An Introduction**

2004
ISBN: 978-0-470-84142-6

Scheirs, J., Long, T. E. (eds.)

**Modern Polyesters**

**Chemistry and Technology of Polyesters and Copolyesters**

2003
ISBN: 978-0-471-49856-8

Elias, H.-G.

**An Introduction to Plastics**

2003
ISBN: 978-3-527-29602-6

# Handbook of RAFT Polymerization

*Edited by*
*Christopher Barner-Kowollik*

WILEY-VCH

WILEY-VCH Verlag GmbH & Co. KGaA

**The Editor**

***Professor Barner-Kowollik***
Center for Advanced Macromolcular Design
University of New South Wales
P.O. Box
Sydney, NSW 2052
Australien



**Library of Congress Card No.:** applied for

**British Library Cataloguing-in-Publication Data**
A catalogue record for this book is available from the British Library.

**Bibliographic information published by the Deutsche Nationalbibliothek**
Die Deutsche Nationalbibliothek lists this publication in the Deutsche Nationalbibliografie; detailed bibliographic data are available in the Internet at http://dnb.d-nb.de.



**Composition** Aptara, New Delhi, India
**Printing** Betz Druck GmbH, Darmstadt
**Bookbinding** Litges & Dopf GmbH, Heppenheim
**Cover Design** Adam Design, Weinheim

Printed in the Federal Republic of Germany
Printed on acid-free paper

**ISBN** 978-3-527-31924-4

# Contents



*Handbook of RAFT Polymerization.* Edited by Christopher Barner-Kowollik

ISBN: 978-3-527-31924-4

# List of Contributors

***Leonie Barner***
Centre for Advanced Macromolecular Design
School of Chemical Sciences and Engineering
The University of New South Wales
NSW 2052
Australia

***Christopher Barner-Kowollik***
Centre for Advanced Macromolecular Design
School of Chemical Sciences and Engineering
The University of New South Wales
NSW 2052
Australia

***Brian C. Benicewicz***
NYS Center for Polymer Synthesis
Dept. of Chemistry & Chemical Biology
Cogswell Laboratory
Rensselaer Polytechnic Institute
Troy, NY 12180-3580
USA

***Marie-Thérèse Charreyre***
Unité Mixte CNRS-bioMérieux
École Normale Supérieure de Lyon
46 Allée d'Italie
69364 Lyon Cedex 07
France

***Michelle L. Coote***
Research School of Chemistry
Australian National University
Canberra ACT 0200
Australia

***Matthias Destarac***
Université Paul Sabatier
Laboratoire Hétérochimie Fondamentale et Appliquée
UMR 5069
118, route de Narbonne
31062 Toulouse Cedex 9
France

***Arnaud Favier***
Unité Mixte CNRS-bioMérieux
École Normale Supérieure de Lyon
46 Allée d'Italie
69364 Lyon Cedex 07
France

***Ekaterina I. Izgorodina***
Research School of Chemistry
Australian National University
Canberra ACT 0200
Australia

*Handbook of RAFT Polymerization.* Edited by Christopher Barner-Kowollik

ISBN: 978-3-527-31924-4

**Thomas Junkers**
Centre for Advanced Macromolecular Design
School of Chemical Sciences and Engineering
The University of New South Wales
NSW 2052
Australia

**Elizabeth H. Krenske**
Research School of Chemistry
Australian National University
Canberra ACT 0200
Australia

**Bertrand de Lambert**
Unité Mixte CNRS-bioMérieux
École Normale Supérieure de Lyon
46 Allée d'Italie
69364 Lyon Cedex 07
France

**Yu Li**
Dept. of Chemistry & Chemical Biology
Cogswell Laboratory
Rensselaer Polytechnic Institute
Troy, NY 12180-3580
USA

**Tara Lovestead**
Centre for Advanced Macromolecular Design
School of Chemical Sciences and Engineering
The University of New South Wales
NSW 2052
Australia

**Andrew B. Lowe**
University of Southern Mississippi
Department of Chemistry and Biochemistry
118 College Drive #5043
Hattiesburg, MS 39406-0001
USA

**Charles L. McCormick**
University of Southern Mississippi
School of Polymers and High Performance Materials
Polymer Science Building, Room 314
118 College Drive #10076
Hattiesburg, MS 39406-0001
USA

**Graeme Moad**
CSIRO Molecular and Health Technologies
Bag 10, Clayton South, VIC 3169
Australia

**Michael J. Monteiro**
Australian Institute of Bioengineering and Nanotechnology
School of Molecular and Microbial Sciences
University of Queensland
Brisbane QLD 4072
Australia

**Carl N. Urbani**
Australian Institute of Bioengineering and Nanotechnology
School of Molecular and Microbial Sciences
University of Queensland
Brisbane QLD 4072
Australia

**Sebastien Perrier**
Key Centre for Polymer Colloids
School of Chemistry
The University of Sydney
NSW 2006
Australia

**Ezio Rizzardo**
CSIRO Molecular and Health Technologies
Bag 10, Clayton South, VIC 3169
Australia

***Linda S. Schadler***
Dept. of Chemistry & Chemical Biology
Cogswell Laboratory
Rensselaer Polytechnic Institute
Troy, NY 12180-3580
USA

***Martina H. Stenzel***
Centre for Advanced Macromolecular Design
School of Chemical Sciences and Engineering
The University of New South Wales
NSW 2052
Australia

***Daniel Taton***
Laboratoire de Chimie des Polymères Organiques (UMR 5629), ENSCPB
CNRS-Université Bordeaux 1
16, Av. Pey Berland
33607 PESSAC Cedex
France

***San Thang***
CSIRO Molecular and Health Technologies
Bag 10, Clayton South, VIC 3169
Australia

***Samir Z. Zard***
Laboratoire de Synthèse Organique
Ecole Polytechnique
91128 Palaiseau
France

# 1
# Introduction

*Christopher Barner-Kowollik*

The *reversible addition–fragmentation chain transfer* (RAFT) process is the most recent of the living/controlled free radical methodologies that have revolutionized the field of free radical polymerization. It was invented in 1998 by the Commonwealth Scientific and Industrial Research Organization (CSIRO) in Melbourne, Australia, by a team of several researchers [1]. Almost simultaneously, a group of researchers in France patented a process they termed *macromolecular design via the interchange of xanthates* (MADIX) [2], which employed xanthates as controlling agents but proceeds by an identical mechanism as the CSIRO-reported RAFT process. Both processes are based on earlier developed chemistries, such as the small radical reactions pioneered by Zard and coworkers [3]. Equally, polymerizations employing reversible addition–fragmentation chain transfer principles, which showed some of the characteristics of living polymerization, were first reported in 1995 by the CSIRO group [4]. The RAFT process employs a fundamentally different conceptual approach compared to nitroxide-mediated polymerization (NMP) and atom transfer radical polymerization (ATRP) in that it relies on a degenerative chain transfer process and does not make use of a persistent radical effect to establish control. Such an approach has the important consequence that the RAFT process features quasi-identical rates of polymerization – apart from deviations caused by the chain length dependence of some rate coefficients – as the respective conventional free radical polymerization processes. Among the other unique features of the RAFT process is its high tolerance to functional monomers – such as vinyl acetate and acrylic acid – which can be polymerized with living characteristics with ease. The RAFT process is an equally powerful tool for the construction of complex macromolecular architectures via variable approaches – i.e. the so-called Z- and R-group designs – that allow for almost limitless possibilities in the synthetic protocols in terms of the low molecular weight scaffolds that support the complex structure.

The popularity of the RAFT process has steadily increased since the first report in 1998. Figure 1.1 gives the number of RAFT-related papers as function of the year (up to April 2007). It is clear that the research interest in the RAFT process continues to be very strong. The present book aims at collating the current state of the art

*Handbook of RAFT Polymerization.* Edited by Christopher Barner-Kowollik

ISBN: 978-3-527-31924-4

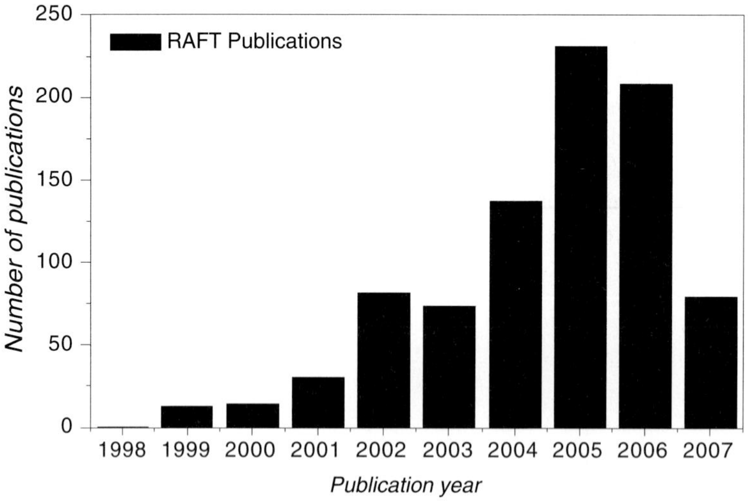


**Fig. 1.1** RAFT publications as function of publication year (ISI Web of Science Database, 24.04.2007, search for 'reversible addition fragmentation chain transfer polymerization').

in RAFT and MADIX. It assembles a cross-section of the world's leading experts on the RAFT/MADIX process, with the view of providing an encompassing picture. It ranges from the underpinning fundamental rate coefficients and equilibrium constants obtained via high-level quantum chemical approaches (Chapter 2) as well as the – sometimes complex – kinetics and mechanism of RAFT and its employment as a powerful kinetic tool (Chapters 3 and 4) to the uses of the process for the formation of simple, yet well-defined polymers in bulk (Chapter 6) to the synthesis of highly complex star and block copolymer structures (Chapter 9). As the MADIX process – although essentially identical to RAFT from a mechanistic perspective – was developed simultaneously, it seems apt to cover its particular strengths and applications in a separate chapter (Chapter 10) accompanied by a contribution on the small radical chemistry of thiocarbonylthio compounds (Chapter 5). The rationale for including a chapter solely on small radical chemistry early in the book is to provide synthetic depths and insight into thiocarbonylthio chemistry from the perspective of an organic chemist. Further chapters highlight the strengths and limitations of RAFT in the context of emulsion and miniemulsion polymerization (Chapter 8) and in homogeneous aqueous systems (Chapter 7). The RAFT process has also been frequently employed to successfully functionalize surfaces as well as micro- and nanoparticles, and Chapter 11 is dedicated to this topic. The controlled functionalization of surfaces is of high importance for a range of applications, including diagnostic kits and tissue engineering. Chapter 12 examines the functional variety that is achievable in RAFT-controlled end groups as well as the stability of RAFT-made polymers. One of the most outward looking chapters aims to examine the materials that have been prepared via the RAFT process for applications ranging from drug delivery to optoelectronics (Chapter 13).

The reader may ask: where to from here? Synthetically, further improvements can be made, especially with a view to generating (complex) block copolymers of monomers with very disparate reactivities, such as styrene and vinyl acetate. While theoretical and some practical advances have been made toward a universal RAFT agent, more research is required in this area. Alternatively, the fusion of RAFT chemistry with highly orthogonal pericyclic reactions (*click* chemistry) is only now emerging and further combination and clever exploitation of these concepts will allow for the construction of extremely well-defined complex polymer systems. Kinetically, the ongoing fruitful discussions about the mechanism of the RAFT process – stimulated by the formation of an IUPAC working party on its mechanism – have led to some clarifications, yet important questions remain to be more fully investigated: How does the size of the equilibrium constant vary with chain length and substitution? If cross-terminations occur, are they reversible and what is the value of the associated termination rate coefficient? Are such side reactions a consequence of slowly fragmenting intermediates or do short-lived adduct radicals react with large termination rate coefficients? Some of the answers to these questions are emerging now and it will be fascinating to follow the progress in this field. However, it is important to note that complex kinetic situations – including the often-quoted rate retardation and inhibition effects – do not occur in the majority of RAFT processes. Ultimately, the application of the RAFT process in materials that benefit society has yet to come to full fruition. We are seeing encouraging trends where RAFT chemistry is employed in variable areas including drug and gene delivery, diagnostic applications, tissue engineering and regenerative medicine, membrane science, bioconjugation as well as the preparation polymers with optoelectronic properties. Yet, more work is required to enhance the existing materials and enable RAFT chemistry to deliver tangible benefits for society.

It is an interesting observation that the RAFT process – although it is in many aspects an extremely powerful living/controlled free radical technique – has still not attracted the overall popularity of living/controlled polymerization based on atom transfer concepts. It is a sometimes advanced notion that this is correlated with the fact that RAFT/MADIX agents are (as of April 2007) not commercially available (except one) and have to be synthesized prior to polymerization. Although this is largely correct, read in Chapter 3 of this book: "While a wide variety of RAFT agents have been made featuring variable Z and R groups, it is important to point out that in principle it should be possible to control polymerization of all monomers using just two RAFT agent structures (e.g. a cyanoisopropyl dithiobenzoate) for styrenics, acrylates, methacrylates, acrylamides and a xanthate or dithiocarbamate (e.g. *O*-ethyl-*S*-cyanomethyl xanthate) for vinyl acetate, *N*-vinyl pyrrolidone and similar monomers."

The current compilation of fascinating RAFT chemistry will further widen its use as a versatile tool in the toolbox of living/controlled polymerization and enhance further collaborative efforts between theoreticians, organic chemists, kineticists, polymer chemists, physical chemists, material scientists and industry. Concomitantly, the current book aims at advancing applications of RAFT for the generation of advanced materials that will positively impact on the lives of all of us.

## References

1. J. Chiefari, Y. K. Chong, F. Ercole, J. Krstina, J. Jeffery, T. P. Le, R. T. A. Mayadunne, G. F. Meijs, C. L. Moad, G. Moad, E. Rizzardo, S. H. Thang, *Macromolecules* **1998**, *31*, 5559–5562; T. P. Le, G. Moad, E. Rizzardo, S. H. Thang, Int. Pat. 9801478 **1998** [*Chem. Abstr.* **1998**, *128*, 115390].
2. P. Corpart, D. Charmot, T. Biadatti, S. Z. Zard, D. Michelet, WO 9858974 **1998** [*Chem. Abstr.* **1999** *130*, 82018].
3. P. Delduc, C. Tailhan, S. Z. Zard, *J. Chem. Soc. Chem. Commun.* **1988**, 308–310.
4. J. Krstina, G. Moad, E. Rizzardo, C. L. Winzor, C. T. Berge, M. Fryd, *Macromolecules* **1995**, *28*, 5381–5385.

# 2
# Quantum-Chemical Studies of RAFT Polymerization: Methodology, Structure-Reactivity Correlations and Kinetic Implications

*Michelle L. Coote, Elizabeth H. Krenske, and Ekaterina I. Izgorodina*

## 2.1
## Introduction

Controlled/living free-radical polymerization techniques such as reversible addition–fragmentation chain transfer (RAFT) [1, 2] rely upon a kinetic strategy for controlling the molecular weight and architecture of the resulting polymer. Rather than eliminating the radical–radical terminations entirely, their frequency is instead minimized with respect to the number of growing polymer chains through the reversible trapping of the growing polymeric radical as a dormant species. In the RAFT process this is achieved using dithioester compounds, known as RAFT agents. The propagating radical adds to the thiocarbonyl sulfur center of the dithioester to produce an intermediate carbon-centered radical. This carbon-centered radical can then undergo β-scission, either to re-form the propagating radical or to liberate a new carbon-centered radical (the 'leaving group') (equation 2.1).

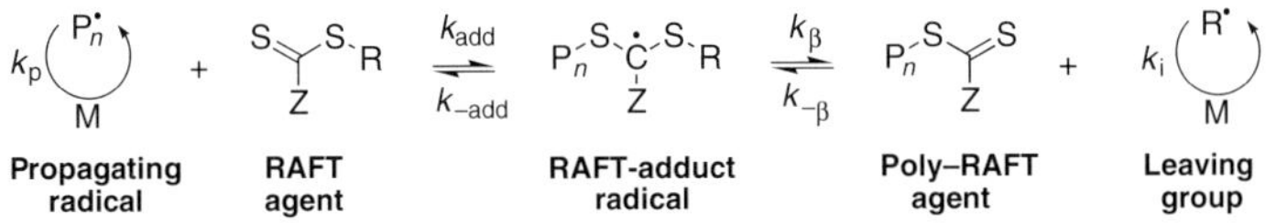

(2.1)

The R group of the RAFT agent is chosen so that it undergoes β-scission from the RAFT-adduct radical in preference to the propagating species, but is still capable of reinitiating polymerization. As a result, the initial RAFT agent (S=C(Z)SR) is rapidly converted to the poly-RAFT agent, R• is converted to more propagating species, and eventually there is a symmetrical equilibrium established between the propagating radical and dormant poly-RAFT agent (equation 2.2).

*Handbook of RAFT Polymerization.* Edited by Christopher Barner-Kowollik

ISBN: 978-3-527-31924-4

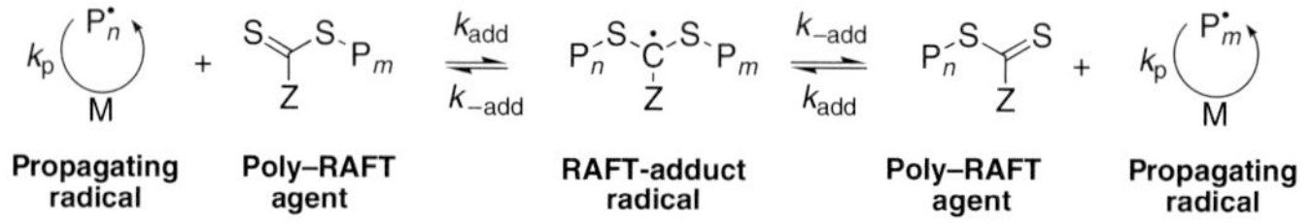


(2.2)

To achieve control, a delicate balance of the rates of these various reactions is required, so as to ensure that the dormant species is orders of magnitude greater in concentration than the active species, but the exchange between the two forms is rapid. The reactivity of the RAFT agent must be tailored to match the reactivity and stability of the polymeric propagating radical; information on the mechanism, kinetics and thermodynamics of these individual steps can greatly assist in the selection of appropriate RAFT agents and reaction conditions.

As in any complex multistep process, the kinetics and thermodynamics of the individual reactions are difficult to study via experimental approaches without recourse to kinetic model-based assumptions. This is because the experimentally observable properties of the process are not the rates and equilibrium constants of the individual reactions, but rather the overall polymerization rate, the average molecular weight distribution of the resulting polymer and the concentrations of some of the major species. To infer the individual rate and equilibrium constants from these measured quantities, one has to assume a kinetic scheme and often make additional simplifying assumptions (such as the steady-state assumption). This problem is exacerbated in controlled radical polymerization processes such as RAFT because there are tens or even hundreds of reactions that are potentially kinetically distinct and should thus be considered in a complete kinetic model of the process. For practical kinetic models it becomes necessary to restrict the number of adjustable parameters via further simplifying assumptions such as the neglect of chain length effects and side reactions. However, while some such simplifications are probably justifiable under certain circumstances, they can also be a potentially large source of systematic error. For example, depending upon the type of data measured and the associated model-based assumptions used, alternative *experimental* values for the equilibrium constant in cumyl dithiobenzoate- (CDB-) mediated polymerization of styrene at 60 °C differ by 6 orders of magnitude [3–5].

Computational quantum chemistry offers an attractive solution to this problem, as it allows the individual reactions to be studied without recourse to kinetic model-based assumptions. Using ab initio molecular orbital theory, it is possible to predict the kinetics and thermodynamics of chemical reactions from first principles, assuming only the laws of quantum mechanics and a few fundamental physical constants (such as the masses and charges of the electron, proton and neutron). Moreover, such calculations yield a range of additional properties such as the geometries and vibrational frequencies of the reactants, products and transition structures, and the distribution of charge and spin density within them. Indeed, as noted by Paul Dirac in 1929, "the underlying physical laws necessary for the mathematical theory of a large part of physics and the whole of chemistry are thus completely known, and the difficulty is only that the exact application of these laws leads to equations much too complicated to be soluble" [6]. Therein also lies the problem: there is no analytical solution to the many-electron Schrödinger equation;

instead, numerical simplifications and approximations must be made. A variety of methods exist, differing principally in the size of their basis set and their treatment of electron correlation. The most accurate methods reliably deliver 'sub-kcal accuracy' but require enormous computational resources, their computational cost scaling exponentially with the size of the system; cheaper methods can be used to study much larger systems but are much less reliable.

To apply quantum-chemical methods to reactions of relevance for free-radical polymerization processes necessarily involves a compromise in which, on the one hand, we select the most computationally efficient methods that still deliver acceptable accuracy, and, on the other, we design small model systems that still mimic the kinetic behavior of the polymeric reactions. Recently we have shown that a successful compromise is now possible for the prediction of propagation rate coefficients in free radical polymerization [7] and, on the basis of extensive assessment studies [7–12], we have designed a methodological approach that delivers 'chemical accuracy' for a variety of radical reactions including radical addition to dithioester compounds. Already, computational calculations have helped to provide an insight into structure–reactivity patterns in RAFT polymerization [3, 9, 13–22] and have led to the design of a new class of RAFT agent [23, 24]. Moreover, computational chemistry is also making a contribution to our understanding of other controlled radical polymerization processes, such as atom transfer radical polymerization [20, 25–27] and nitroxide-mediated polymerization [28, 29], as well as conventional radical polymerization [30–41]. With continuing advances in computer power, more applications are anticipated and computational chemistry will take its place along side experimental methods as a practical kinetic tool.

In this chapter, which updates an earlier review article [42], we show how quantum chemistry can be used to study the RAFT polymerization process. We begin with a description of the types of methods that are used, and discuss their accuracy and their outstanding problems. We then show how computational chemistry can be used to clarify and improve models for reaction kinetics, and indeed how computational calculations can be used directly in kinetic simulations to predict the macroscopic outcome of a process. Finally, we examine the RAFT process at a deeper mechanistic level, using computational data to model and explain structure–reactivity trends, and show how these insights can be applied to the practical problem of RAFT agent design.

## 2.2
## Methodology

To select reliable yet cost-effective theoretical procedures, assessment studies are performed. In essence, one takes a small prototypical example of the class of chemical reaction under study and calculates the geometries, frequencies, barriers, enthalpies, rate coefficients and other properties at a variety of levels of theory, ranging from the extremely accurate but highly computationally intensive down to those that are computationally inexpensive but potentially subject to very large errors. For each type of property, one compares the results at the lower levels of theory with the highest-level results and (where possible) also with reliable gas-phase experimental data. For the case of addition–fragmentation processes, assessment studies have been performed for the prototypical reactions $\bullet CH_3 + S{=}C(R)R' \rightarrow CH_3SC\bullet(R)R'$ (R, R′ = H, $CH_3$) [8] and also the more RAFT-related systems, $R'SC(Z){=}S + \bullet R \rightarrow R'SC\bullet(Z)SR$ (various combinations of R, R′ = $CH_3$, $CH_2CH_3$, $CH_2CN$, $C(CH_3)_2CN$, $CH_2COOCH_3$, $CH(CH_3)COOCH_3$, $CH_2OCOCH_3$, $CH_2Ph$, $CH(CH_3)Ph$, and Z = $CH_3$, H, Cl, CN, $CF_3$, $NH_2$, Ph, $CH_2Ph$, $OCH_3$, $OCH_2CH_3$, $OCH(CH_3)_2$, $OC(CH_3)_3$, F) [10]. In addition, the accuracy of the harmonic oscillator approximation [9, 18] and the applicability of standard (rather than variational) transition state theory have been explored [9]. On the basis of these studies, the following guidelines for performing theoretical calculations are suggested.

## 2.2.1 Electronic-Structure Calculations

In general, it is possible to conduct reliable geometry optimizations and frequency calculations for the stationary species (i.e. the reactants and products) in the RAFT process at relatively low levels of theory, such as B3-LYP/6-31G(d) or HF/6-31G(d) [8]. For example, provided the energies are calculated at a consistent level of theory, the reaction enthalpies for $\bullet CH_3 + S{=}CRR'$ (R, R′ = H, $CH_3$) vary by less than 1 $kJ{\cdot}mol^{-1}$, regardless of whether low levels such as B3-LYP/6-31G(d) or HF/6-31G(d) or higher levels such as CCSD(T)/6-311+G(d,p) are used for the geometry optimizations [8]. Likewise, provided the recommended scale factors are used [43], the zero-point vibrational energy (and hence the frequency calculations) at these lower levels of theory agrees to within 1–2 $kJ{\cdot}mol^{-1}$ of the CCSD(T)/6-311+G(d,p) calculations [8]. However, for transition structures, extra precautions are required. In particular, density functional theory (DFT) methods such as B3-LYP significantly overestimate the length of the forming bond in the transition state, finding transition structures that are too early. HF fares somewhat better, but does underestimate the forming bond length, finding transition structures that are too late. To address this problem, the transition structures should be corrected to higher levels of theory via IRCmax [8]. In the IRCmax method [44, 45], one calculates the minimum energy path of the chemical reaction at a low level of theory and then calculates single-point energies along this reaction path at a higher level of theory, such as RMP2/6-311+G(3df,2p). The IRCmax transition structure is then identified as that species corresponding to the maximum point of the minimum energy path, as calculated at the *higher* level of theory. In essence, one optimizes the most sensitive part of the geometry optimization (i.e. the reaction coordinate) at the higher level of theory, at a fraction of the cost of a full geometry optimization at that level. For the case of radical addition to C=S bonds, the IRCmax transition structures have forming bond lengths within less than 0.05 Å, and provide reaction barriers within 1 $kJ{\cdot}mol^{-1}$, of those obtained using full geometry optimizations at the CCSD(T)/6-311+G(d,p) level [8].

In contrast to geometries and frequencies, accurate calculations of the energetics of these types of chemical reactions require very high levels of theory. Ideally, such calculations should be performed using W1 theory or better. W1 aims to approximate coupled cluster energies [URCCSD(T)] with an infinite basis set via extrapolation, and includes corrections for core correlation and relativistic effects [46]. It has been shown to deliver 'kJ accuracy' when assessed against a large test set of gas-phase experimental data [46], and has performed well in assessment studies for the prototypical system $\bullet CH_3 + S{=}CH_2$ [8] as well as other types of radical reaction such as radical addition to C=C bonds [11] and hydrogen atom abstraction [12]. Unfortunately, such calculations are very computationally intensive and currently only practicable for up to approximately 5 nonhydrogen atoms. The G3 family of methods provides a lower-cost alternative to W1. Like W1, they attempt to approximate coupled cluster energies with a large basis set, but achieve this via additivity approximations at the MP2 and/or MP4 levels of theory. As a result, they are less expensive than W1 but also less reliable. They have been shown to deliver 'kcal accuracy' when assessed against a large test set of experimental thermochemical data [47, 48] and to provide good agreement with W1 for a variety of radical reactions [11, 12]. For the specific problem of radical addition to C=S double bonds, the G3 methods provide excellent agreement with W1 for the reaction barriers; for reaction enthalpies, the errors are slightly larger (ca. 10 $kJ \cdot mol^{-1}$) but are likely to be reasonably systematic for a class of reactions and therefore suitable for studying substituent effects [8]. Nonetheless, they are also currently too computationally intensive for all but the simplest RAFT systems, being currently only feasible for systems of up to approximately 17 nonhydrogen atoms.

For practical RAFT systems, lower-cost procedures are necessary and unfortunately these procedures can be subject to large errors. In particular, current DFT methods fail comprehensively to model the effects of substituents in addition–fragmentation processes [10]. Moreover, this failing has also been observed for other radical reactions such as R—X bond-dissociation reactions (R = Me, Et, *i*-Pr, *t*-Bu; X = H, $CH_3$, $OCH_3$, OH, F) [49], propagation rate coefficients [7] and even simple closed shell systems such as the cyclization energies of alkenes [50] and alkane isomerization energies [51, 52]. Restricted open-shell second-order Møller–Plesset perturbation theory (RMP2), which is slightly more expensive than DFT but can still be practically applied to relatively large systems, has been shown to fare much better than DFT. It generally provides reasonable absolute values (within 10 $kJ \cdot mol^{-1}$) and excellent relative values (within 4 $kJ \cdot mol^{-1}$ or better) for the barriers and enthalpies of the addition–fragmentation processes [8, 10] as well as a variety of other radical reactions [7, 11, 12, 49]. However, it breaks down when the attacking radical is substituted with a group (such as phenyl or CN) that delocalizes the unpaired electron, with errors of over 15 $kJ \cdot mol^{-1}$ being reported in some cases [9, 10]. Since such radicals are common in practical RAFT systems, this severely compromises the utility of RMP2.

Given these problems, we have designed an alternative approach that is based on the ONIOM procedure [53]. In the ONIOM method, one first defines a 'core' section of the reaction that, in the very least, contains all forming and breaking bonds and would preferably include the principal substituents attached to them. In forming the core system, deleted substituents are replaced with 'link atoms' (typically hydrogens) chosen so that the core system provides a good chemical model of the reaction center. One then calculates the core system at both a high level of theory and also a lower level; the 'full system' is calculated at only the lower level. The full system at the high level of theory is then approximated as the sum of (a) the core system at the high level and (b) the substituent effect measured at the low level of theory. The approximation is valid if the low level of theory measures the substituent effect accurately; this in turn depends upon the level of theory chosen and the way in which the core system is defined.

For the RAFT systems, we know from above that W1 provides accurate absolute values of barriers and enthalpies, and G3(MP2)-RAD provides excellent relative values. For small systems, such as reaction 2.3, the W1 enthalpy could thus be approximated as the sum of the W1 enthalpy for the core reaction 2.4, and the difference in the G3(MP2)-RAD enthalpies for 2.3 and 2.4.

$$\bullet CH_2Ph + S{=}C(CH_3)SCH_3 \rightarrow PhCH_2SC\bullet(CH_3)SCH_3 \quad (2.3)$$

$$\bullet CH_3 + S{=}CH_2 \rightarrow CH_3SCH_2\bullet \quad (2.4)$$

For larger systems, where G3(MP2)-RAD calculations are not currently feasible, we add an additional 'ONIOM layer' in which the full system is calculated using RMP2/6-311+G(3df,2p), the core system is calculated using G3(MP2)-RAD and the 'inner core' is studied at W1. For example, one could approximate the W1 enthalpy for reaction 2.5 as the sum of the W1 value for the inner core (2.4), the G3(MP2)-RAD difference for the core (2.3) and inner core (2.4), and the RMP2/6-311+G(3df,2p) difference for the full (2.5) and core (2.3) systems.

$$\begin{aligned} &\bullet CH(Ph)CH_2C(CH_3)2CN + S{=}C(CH_2Ph)SC(CH_3)_2Ph \\ &\quad \rightarrow (CH_3)_2C(CN)CH_2CH(Ph)SC\bullet(CH_2Ph)SC(CH_3)_2Ph \end{aligned} \quad (2.5)$$

The ONIOM procedure provides an accurate alternative to high-level calculations on the full system, provided that the lower level of theory measures the substituent effects accurately. While the RMP2/6-311+G(3df,2p) method normally provides excellent relative values of barriers and enthalpies, it does break down in situations where the unpaired electron is highly delocalized. It is therefore extremely important to partition the full and core systems carefully, such that the delocalized radical is treated at the G3(MP2)-RAD level of theory and the RMP2/6-311+G(3df,2p) method is used only to measure remote substituent effects. Based on a careful assessment study, the following general guidelines are suggested [10] for partitioning the core and full systems: in the addition–fragmentation reaction (2.6), a suitable core system should include all $\alpha$-substituents on the attacking radical R• but could replace the R′ group with a methyl substituent and, if necessary, also replace the Z group with a methyl substituent.

$$\bullet\text{R} + \text{S}{=}\text{C(Z)SR}' \rightarrow \text{RSC}\bullet\text{(Z)SR}' \tag{2.6}$$

Even when the Z group is itself a substituent capable of delocalizing the unpaired electron in the RAFT-adduct radical, its replacement with $CH_3$ in the core system does not introduce substantial error [10]. This is presumably because the radical is already highly delocalized by the thiyl groups (which are of course included in both the core and inner core systems), and the additional delocalization by the Z substituent is not as significant as in an ordinary carbon-centered radical [19]. This ONIOM-based procedure allows one to study quite large RAFT systems (ca. 30–40 nonhydrogen atoms) with kcal accuracy [10]. Similar performance has also been demonstrated for a range of other radical reactions including the bond dissociation energies of a range of organic compounds, propagation, hydrogen and halogen atom transfer reactions and radical ring opening [54].

### 2.2.2 Kinetics and Thermodynamics

Having obtained the geometries, energies and frequencies of the reactants, products and transition structures, it is possible to calculate the rates $k(T)$ and equilibrium constants $K(T)$ of chemical reactions, using the standard textbook formulae [55, 56]:

$$\begin{aligned} k(T) &= \kappa(T)\frac{k_B T}{h}(c^\circ)^{1-m}\, e^{(-\Delta G^\ddagger/RT)} \\ &= \kappa(T)\frac{k_B T}{h}(c^\circ)^{1-m}\frac{Q_\ddagger}{\prod_{\text{reactants}} Q_i}\, e^{(-\Delta E^\ddagger/RT)} \end{aligned} \tag{2.7}$$

$$K(T) = (c^\circ)^{\Delta n}\, e^{(-\Delta G/RT)} = (c^\circ)^{\Delta n}\left(\frac{\prod_{\text{products}} Q_j}{\prod_{\text{reactants}} Q_i}\right) e^{(-\Delta E/RT)} \tag{2.8}$$

In these formulae, $\kappa(T)$ is the tunneling correction factor, $T$ is the temperature (K), $k_B$ is Boltzmann's constant ($1.380658 \times 10^{-23}$ J·mol$^{-1}$·K$^{-1}$), $h$ is Planck's constant ($6.6260755 \times 10^{-34}$ J·s), $c^\circ$ is the standard unit of concentration (mol·L$^{-1}$), $R$ is the universal gas constant (8.3142 J·mol$^{-1}$·K$^{-1}$), $m$ is the molecularity of the reaction and $\Delta n$ the change in moles upon reaction, $Q_\ddagger$, $Q_i$ and $Q_j$ are the molecular partition functions of the transition structure, reactant $i$ and product $j$ respectively, $\Delta G^\ddagger$ is the Gibbs free energy of activation, $\Delta G$ is the Gibbs free energy of reaction, $\Delta E^\ddagger$ the 0-K, zero-point energy corrected energy barrier for the reaction and $\Delta E$ is the 0-K, zero-point energy corrected energy change for the reaction. The value of $c^\circ$ depends on the standard-state concentration assumed in calculating the thermodynamic quantities (and translational partition function). For example, if these quantities were calculated for 1 mol of an ideal gas at 333.15 K and 1 atm, then $c^\circ = P/RT = 0.0365971$ mol·L$^{-1}$. The tunneling coefficient

$\kappa(T)$ corrects for quantum effects in motion along the reaction path [57–60]. While tunneling is important in certain chemical reactions such as hydrogen abstraction, it is negligible (i.e. $\kappa \approx 1$) for the addition of carbon-centered radicals to thiocarbonyl compounds at typical polymerization temperatures (such as 333.15 K) because the masses of the rearranging atoms are large and the barriers for the reactions are relatively broad [9].

The molecular partition functions and their associated thermodynamic functions (i.e. enthalpy, $H$, and entropy, $S$) can be calculated using the standard textbook formulae [55, 56], based on the statistical thermodynamics of an ideal gas under the harmonic oscillator/rigid rotor approximation. These formulae require knowledge of the point group, multiplicity, geometry and vibrational frequencies of each species; the accuracy of the results depends upon both the accuracy of the calculated geometries and frequencies and the validity of the harmonic oscillator/rigid rotor approximation. As noted above, the geometries and frequencies are well described at relatively low levels of theory, such as B3-LYP/6-31G(d), provided that transition structures are corrected via IRCmax and frequencies are scaled by appropriate scale factors. However, the use of the harmonic oscillator/rigid rotor approximation can lead to errors of 1–2 orders of magnitude in both the kinetics and thermodynamics of the addition–fragmentation equilibrium [9, 18]. To address this problem, the partition functions for the low-frequency torsional modes ($<300$ cm$^{-1}$) should instead be treated as hindered internal rotations. Full details of these calculations are published elsewhere [9], but a short description is provided below.

For each low-frequency torsional mode, one first calculates the full rotational potential $V(\theta)$. This can be calculated at a relatively low level of theory, such as B3-LYP/6-31G(d), and is performed as a relaxed (rather than frozen) scan in steps of 10° through 360°. The potential is then fitted with a Fourier series of up to 18 terms, so that it can be interpolated to a finer numerical grid (typically 300 points instead of 36). The corresponding energy levels are found by numerically solving the one-dimensional Schrödinger equation 2.9 for a rigid rotor [31, 61, 62].

$$-\frac{h^2}{8\pi I_r}\frac{\partial^2\Psi}{\partial\theta^2} + V(\theta)\Psi = \varepsilon\Psi \tag{2.9}$$

The reduced moment of inertia ($I_r$) is calculated using the equation for $I^{2,3}$, as defined by East and Radom [63]. The resulting energy levels $\varepsilon_i$ are then summed to obtain the partition function at the specified temperature:

$$Q_{\text{int rot}} = \frac{1}{\sigma_{\text{int}}}\sum_i \exp\left(-\frac{\varepsilon_i}{k_B T}\right) \tag{2.10}$$

where $\sigma_{\text{int}}$ is the symmetry number associated with that rotation. It should be noted that in this method, the low-frequency torsional modes have been approximated as one-dimensional rigid rotors, while in practice these modes can be coupled with another. However, a recent study of coupled internal rotations in another radical addition reaction (ethyl benzyl radical addition to ethene) indicated that the errors incurred in using a one-dimensional treatment are relatively minor, particularly when compared with the errors incurred under the harmonic oscillator approximation [64].

In evaluating the rate coefficients, the method for identifying the transition structure is also important. Standard geometry optimization algorithms identify the transition structure as a first-order saddle point in the potential energy surface; that is, as the structure having the maximum internal energy ($E$) along the minimum energy path. However, ideally the rate calculations should be performed using the structure having the maximum Gibbs free energy ($G$). At nonzero temperatures, $E$ and $G$ are nonequal and thus the geometries corresponding to their maximum values are not necessarily equivalent. The corresponding methods for calculating the rate coefficients are known as standard transition state theory and variational transition state theory, respectively [65]. Variational transition state theory is more accurate but also more expensive, as it entails the calculation of the energies and partition functions at not just at the transition structure but at several points along the minimum energy path. In practice, for reactions with significant energy barriers, the differences between transition state theory and variational transition state theory are relatively minor, and the lower-cost standard method can be used. However, for barrierless reactions (and also for some low-barrier reactions), variational effects can become important.

Radical addition to the sulfur atom of a C=S double bond is typically a fast reaction, having a low or even negative barrier ($\Delta E^{\ddagger}$) in most cases. The positive Gibbs free energy barrier results from opposing enthalpic and entropic effects [9]. In other words, both $\Delta E$ and $\Delta H$ decrease along the reaction coordinate but $-T\Delta S$ increases and its opposing interaction leads to the maximum in $\Delta G$. Under these circumstances, one might have expected variational effects to be very important for these systems. However, in a recent assessment study for the prototypical RAFT systems R• + $S{=}C(Z)SCH_3$ (R = $CH_3$, $CH_2Ph$, $CH_2COOCH_3$, $C(CH_3)_2CN$; Z = $CH_3$, Ph, $CH_2Ph$), it was found that provided the transition structures were corrected via IRCmax, the use of variational transition state theory had little or no effect on the reaction barriers [9]. The effects on the entropies and hence reaction rates were somewhat larger (up to a factor of 16) in the case of the •$CH_3$ addition reactions, but for the reaction of the substituted radicals (which are more indicative of real polymerization systems), they remained relatively small (ca. a factor of 2). Thus, while variational transition state theory should always be used when possible for these systems, standard transition state theory may be adopted for large systems without incurring significant additional error.

### 2.2.3 Solvent Effects

The methodology described thus far is designed to reproduce chemically accurate values of the rate and equilibrium constants for gas-phase systems; however, the majority of RAFT polymerizations occur in solution. The development of cost-effective methods for treating the solvent in chemical reactions is an ongoing area of research [66, 67] and there have not yet been any benchmarking studies for the specific case of RAFT polymerization. However, it is worth making a few general comments on the types of solvent models that are currently available. The simplest and most computationally efficient methods are continuum models, in which each solute molecule is embedded in a cavity surrounded by a dielectric continuum of permittivity $\varepsilon$ [67]. Some of the more sophisticated continuum models, such as the ab initio conductor-like solvation model [68] and the polarizable continuum model (PCM) [69], also include terms for the nonelectrostatic contributions of the solvent, such as dispersion, repulsion and cavitation. Continuum models are designed to reproduce bulk or macroscopic behavior, and can fare extremely well in certain applications such as the calculation of the solvation free energies [70] and p$K_a$ values [71] of various organic molecules. However, the results obtained using continuum models are highly sensitive to the choice of cavities (which are typically parameterized to reproduce the free energies of solvation for a set of small organic molecules), and the choice of appropriate cavities for weakly bound species such as transition structures can be problematic [70]. Moreover, their description of important electronic effects is not generally adequate, particularly if there are explicit solute–solvent interactions such as complex formation and hydrogen bonding. Although this problem can be overcome by including a small number of explicit solvent molecules in the ab initio calculation, as in a cluster-continuum model [72], this adds significantly to the cost of the calculation. Such explicit solute–solvent interactions might be expected to be particularly important when studying aqueous-phase polymerizations and monomers capable of undergoing strong hydrogen-bonding interactions (such as acrylic acid).

Of greater concern for the study of the kinetics and thermodynamics of association and dissociation processes (as in RAFT) is the fact that continuum models completely ignore the changes to the (ideal gas) vibrational, translational and rotational partition functions upon solvation [73]. In the gas phase, the reactants and products have translational and rotational entropy, whereas in the solution phase this entropy is effectively 'lost' in collisions with the solvent. In place of the translational and rotational motion, the solution-phase molecules have additional internal degrees of freedom corresponding to their interaction with the solvent. However, these additional modes generally contribute less to the total entropy of a molecule than the corresponding external translational and rotational modes in the gas phase. The difference between the gas-phase and solution-phase entropy is significant in a bimolecular association reaction (i.e. $A + B \rightarrow C$), because three translational and three rotational modes are converted to internal modes on reaction. Since these lost modes contribute less to the total entropy in the solution phase, the solution-phase reaction is expected to be less exentropic than the corresponding gas-phase reaction, and thus it should have a larger equilibrium constant and a faster rate coefficient. Clear evidence for this 'entropic' solvent effect can be found in comparisons of corresponding *experimental* gas-phase and solution-phase rate coefficients for radical addition to alkenes. In these cases, the solution-phase rate coefficients generally exceed the gas-phase values by an order of magnitude [74].

Unfortunately there is no straightforward manner in which to quantify these entropic contributions to the solvent effect. Normally, one must include many explicit solvent molecules in the calculation and try to reproduce bulk behavior via molecular dynamics or Monte Carlo simulations, combined with the imposition of periodic boundary conditions [75]. Such calculations are hampered by problems such as the lack of potentials that can adequately describe both cluster and bulk behavior and the rapid increase in the conformational possibilities as number of individual components increases. Nonetheless, there is a growing body of work toward addressing these problems. In particular, the effective fragment potential method, in which the Coloumbic, induction and repulsive interactions are included as one-electron terms in the ab initio Hamiltonian, is showing promise as an accurate yet cost-effective method for developing potentials for any solute–solvent combination [76]. There is also some progress toward the ultimate goal of formulating a dynamical correction to gas-phase transition state theory rate coefficients [66, 77]. An examination of the accuracy and applicability of the various solvent models for radical reactions such as RAFT is currently underway, but until general guidelines for the accurate treatment of solvent effects are derived, gas-phase calculations are preferred. With the exception of systems displaying strong direct interactions, it is expected that the neglect of the solvent will not introduce substantial errors to the trends in the calculated rates and equilibrium constants, but may affect the absolute values. In particular, as noted above, the use of gas-phase calculations may lead to an underestimation of the solution-phase values of the rate and equilibrium constants for bimolecular association reactions by at least an order of magnitude.

### 2.2.4 Accuracy and Outstanding Challenges

Normally, one might establish the accuracy of ab initio calculations through comparison with reliable experimental data. However, in the case of the RAFT process this is difficult as there is no model-free manner in which to measure experimentally the rates and equilibrium constants of the individual addition–fragmentation processes. Even if one makes the assumption that addition and fragmentation are chain length independent beyond the dimer stage, a complete kinetic model would need to contain in excess of 100 adjustable parameters (see Scheme 2.1 [78]). As a result, various simplifications and approximations are made in obtaining experimental measurements, and these are a potential source of error. Alternative experimental measurements for ostensibly the same system can differ by several orders of magnitude, and choosing the 'correct' experimental value for comparison with the theoretical data is problematic. For example, although there is excellent agreement between calculated equilibrium constant of $4.2 \times 10^6$ L·mol$^{-1}$ (the originally reported value of $7.3 \times 10^6$ L·mol$^{-1}$ [3] corrected to W1 theory) for CDB-mediated polymerization of styrene at 303.15 K and experimental values ($1.06 \times 10^7$ L·mol$^{-1}$) obtained from model fitting to low-conversion kinetic data [3], both the experimental and theoretical values are in conflict with those estimated from

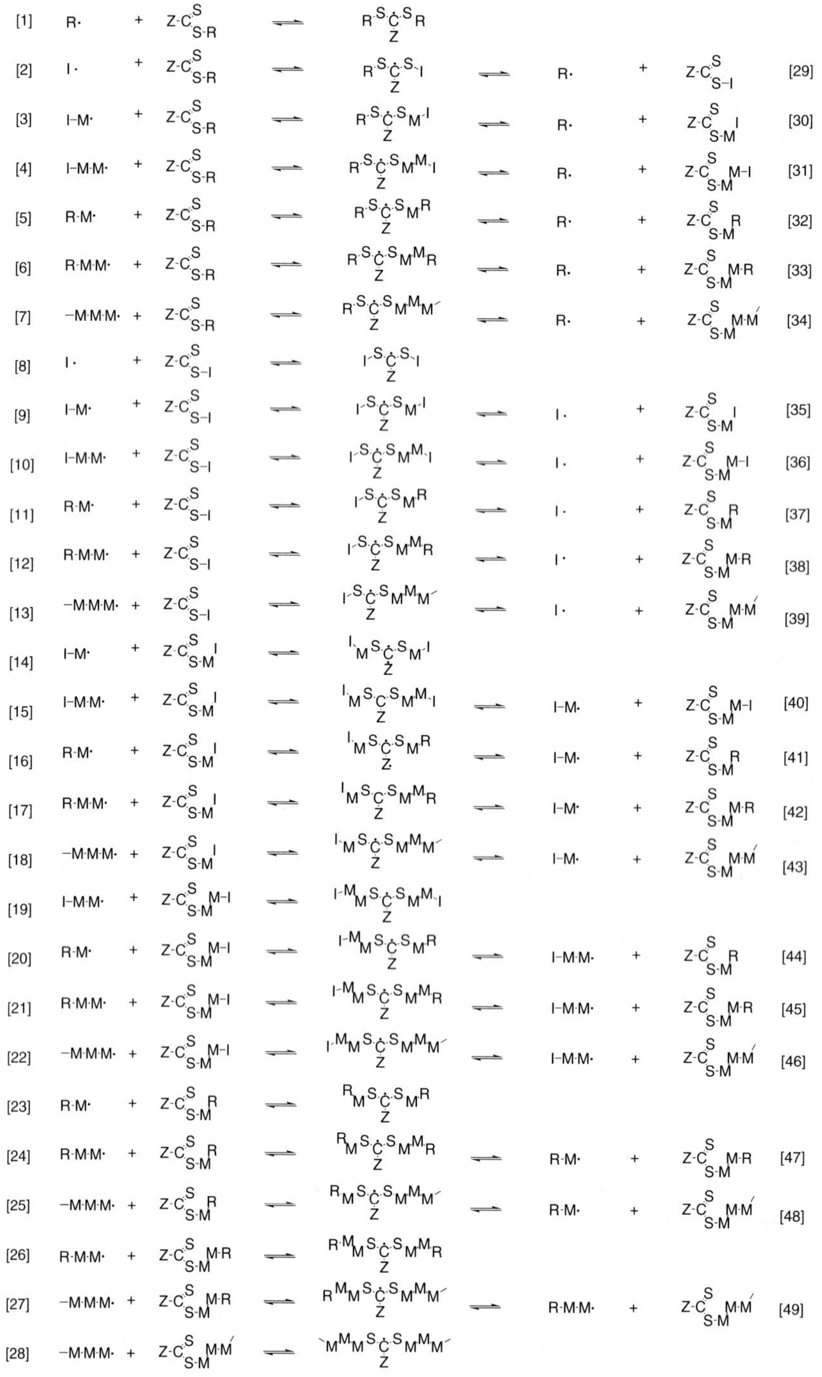

**Scheme 2.1** Set of addition–fragmentation reactions to be included in a complete RAFT reaction scheme if chain length effects extend to the trimer stage [78].

electron paramagnetic resonance spectroscopy data ($8 \times 10^1$ L·mol$^{-1}$) albeit under different polymerization conditions [4].

Nonetheless, there is greater consensus amongst the various experimental groups with regard to measurements of the rate constant for addition to the RAFT agent, and the values obtained thus far (ca. $10^6$ L·mol$^{-1}$·s$^{-1}$) also appear to be relatively insensitive to the nature of the dithioester and attacking radical (see, for example, Table 3.2 of Chapter 3). These experimental values are in excellent agreement with the theoretical calculations for related small radicals; for example, the calculated values for addition of R• to $S{=}C(CH_3)SCH_3$ are $1.2 \times 10^6$ and $3.8 \times 10^6$ L·mol$^{-1}$·s$^{-1}$ for the typical R groups $C(CH_3)_2CN$ and $CH(Ph)CH_3$, though slightly faster values are obtained (ca. $10^7$–$10^8$) for more reactive small radical species (such as $CH_2COOCH_3$ and $CH(CH_3)OCOCH_3$) [42].

Support for the accuracy of the computational predictions was also recently provided in a kinetic modeling study of the initialization period in cyanoisopropyl dithiobenzoate-mediated polymerization of styrene [79]. In this work, ab initio predictions of the equilibrium constants for the first eight addition–fragmentation reactions of the RAFT process were combined with reliable experimental values of the rate coefficient for radical addition to the RAFT agent and the rate coefficients for initiation, propagation and termination in styrene homopolymerization. These parameters were then used to predict, without any additional model fitting, the overall monomer conversion and individual concentration profiles of the low-molecular-weight thiocarbonyl compounds formed during the early stages of the process. These ab initio predictions showed excellent agreement with previously measured experimental data (obtained under the same reaction conditions used in the kinetic simulation), despite the fact that no adjustable fit parameters were used [79] (see Fig. 2.1). This work will be discussed in more detail in Section 2.3.4.

Other strategies have also been used for testing the accuracy of the theoretical calculations. To begin with, we recall that the levels of theory have been selected on the basis of assessment studies, and their accuracy established through comparison with high levels of theory, themselves benchmarked against gas-phase thermochemical data [8–10]. On the basis of these assessment studies, it is concluded that the selected procedures can model the absolute barriers and enthalpies for radical addition C=S double bonds to within approximately 4 kJ·mol$^{-1}$. It is also possible to benchmark the calculations against (relatively model-free) solution-phase experimental data for related small systems. For example, Scaiano and Ingold [81] have studied the addition of *tert*-butyl radicals to di-*tert*-butyl thioketone via laser flash photolysis, obtaining an equilibrium constant of $1.2 \times 10^6$ L·mol$^{-1}$ at 25 °C. This is in very good agreement with the corresponding theoretical value ($7.9 \times 10^5$ L·mol$^{-1}$) [82], the latter being slightly lower due to the neglect of the solvent in the calculations.

The computational methodology can also be benchmarked against experimental data for polymeric systems for the case of the propagation rate coefficients. Although not chemically identical to an addition–fragmentation reaction, a propagation reaction is still a radical addition reaction and has the advantage that reliable experimental data can be obtained using pulsed laser polymerization [83]. A recent

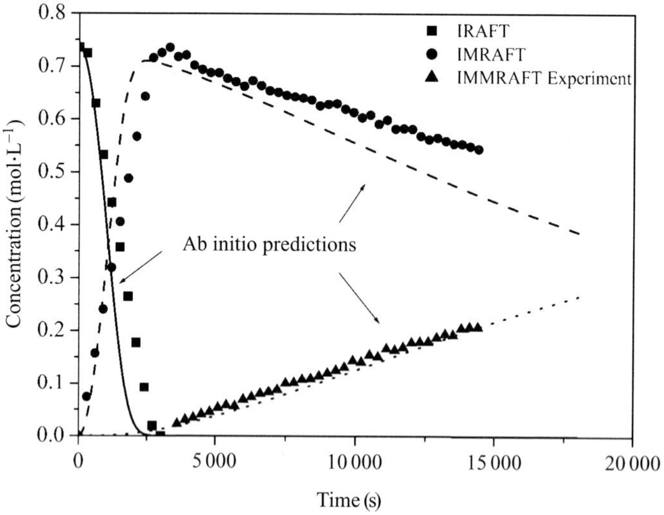


**Fig. 2.1** Experimental (McLeary et al. [80], symbols) and simulated (Coote et al. [79], lines) evolutions of the initial RAFT agent (cyanoisopropyl dithiobenzoate, IRAFT, full line), and the RAFT agents having one monomer unit (IMRAFT, dashed line) and two monomer units (IMMRAFT, dotted line) inserted. The concentrations of the reagents were $c^{\circ}_{\mathrm{IRAFT}} = 0.736$ mol·L$^{-1}$, $c^{\circ}_{\mathrm{AIBN}} =$ 0.10 mol·L$^{-1}$, $c^{\circ}_{\mathrm{Styrene}} = 3.65$ mol·L$^{-1}$ and $c^{\circ}_{\mathrm{C6D6}} = 5.4$ mol·L$^{-1}$, and the temperature was 70 °C.

computational study of the propagation of vinyl chloride and acrylonitrile yielded values within a factor of 2 of the experimental polymeric values, and demonstrated that the rate coefficient for the polymeric propagating species was adequately modeled using the corresponding rate coefficient for the dimer [7]. Further studies of a wider range of monomers are currently underway, but it does seem that 'chemical accuracy' is now attainable.

Despite this success, a number of challenges remain. In particular, as noted above, the treatment of solvent effects is not yet satisfactory, and larger errors might be expected for solvent-sensitive systems such as acrylic acids. It should also be stressed that the above methodology is not suitable for studying the diffusion-controlled termination processes. In principle, one could calculate the chemically controlled component of the termination reaction using computational chemistry, and this could be extremely helpful in some situations. For example, there is currently a debate about whether the RAFT-adduct radical is capable of undergoing termination [3, 4, 84–92]; establishing whether the chemically controlled component is fast or slow would be an important contribution to this debate. However, the above methodology, which makes use of methods with single-reference wavefunctions, is not likely to be suitable for studying the transition structure of a radical–radical process. Unfortunately, multireference methods are considerably more computationally intensive and further work is needed to identify reliable low-cost alternatives for such systems.

Even in the case of the addition–fragmentation reaction, the computational cost of the calculations remains an issue. Although the ONIOM-based approach facilitates the calculation of accurate energetics for relatively large systems, other computational bottlenecks are now emerging. Larger systems (such as the reactions of dimer radicals with RAFT agents) have a high degree of conformational complexity, and finding the minimum energy conformations of these species can be very computationally demanding. We currently identify these via a complete conformational search; that is, we optimize geometries for every (nonunique) conformation about every bond (or forming bond) in the molecule (or transition structure). For example, in a relatively simple species such as $(CH_3)_2C(Ph)—CH_2—CH(Ph)—S—C\bullet(Ph)—S—C(CH_3)_2Ph$, we might need to consider 324 possible conformations. Inserting just one extra styryl unit increases the number of possible conformers to 2916 and the calculations rapidly become impractical. To address this problem we have recently developed a more efficient algorithm for exploring conformational space called *energy-directed tree search* [93]. Preliminary indications are that it is possible to find the global minimum reliably at the cost of approximately 10% of a full search for typical RAFT systems, but further savings would be desirable. A related computational bottleneck stems from the need to treat the low-frequency torsional modes as hindered internal rotations. As explained above, this entails the calculation of full rotational potentials (typically 36 geometry optimizations) for each mode considered. This too can rapidly become computationally infeasible, and strategies for reducing this computational expense are currently being investigated. Nonetheless, despite these problems, it is now possible to perform accurate calculations on systems that are large enough (ca. 30–40 nonhydrogen atoms) to be of relevance to practical polymerization systems.

## 2.3
## Computational Modeling of RAFT Kinetics

One of the main advantages of a computational approach to studying RAFT polymerization kinetics is the ability to determine the rate and equilibrium constants of individual reactions directly, without having to fit a kinetic model to data. This is particularly useful for complex processes such as RAFT because the kinetic models for these processes potentially can contain more adjustable parameters than can practically be estimated from the available experimental data. For example, as noted above, even if we make the assumption that the addition–fragmentation equilibrium becomes chain length independent beyond the dimer stage, a 'complete' kinetic model for the process would need to contain 49 addition rate coefficients and 49 fragmentation rate coefficients [78] (see Scheme 2.1), in addition to various reinitiation rate coefficients, and the usual initiation, propagation, termination and transfer coefficients. Since this is impractical, experimental studies typically make simplifying assumptions—such as replacing all 98 rate coefficients by just 2 or 4—and these are a potential source of error. In particular, such simplifications are likely to affect the description of subtle yet important effects such as the concentration profiles of the various species during the early stages of the process [80, 94].

Computational chemistry can assist the kinetic modeling of RAFT in a number of ways. Firstly, it is possible to study the effects of chain length and other remote substituents on the addition–fragmentation equilibria. This information can be used to identify which simplifying assumptions are justifiable in a kinetic model, and thereby assist experimental studies. The same information can also be used to assist the computational studies by identifying which substituents can be replaced with smaller groups without affecting the calculated results. Secondly, it is possible to use computational chemistry to establish whether side reactions (such as alternative fragmentation pathways) are likely to be operative in particular systems, and thus whether they need to be taken into account in the kinetic modeling schemes. Finally, it is possible to calculate the rate and equilibrium constants for many of the early reactions in the RAFT process, and thereby help to minimize the number of adjustable parameters that need to be measured experimentally. Indeed, provided suitable small models for the polymeric reactions can be designed, it is possible in principle to calculate all chemically controlled rate coefficients for a RAFT process, so as to evaluate the performance of new RAFT agents prior to experimental studies. (One would need to use experimental data for the diffusion controlled termination rate coefficients but these could be obtained from a single homopolymerization of the specified monomer in the absence of RAFT agent.) These aspects are now examined in turn.

### 2.3.1
### Simplified Models for Theory and Experiment

Both theoretical and experimental studies of radical polymerization processes rely heavily on the fact that in a chemically controlled reaction, the effects of substituents diminish rapidly with their distance from the reaction center. From an experimental perspective, it allows one to treat the reactions of species differing only in their remote substituents as identical, thereby reducing the number of adjustable parameters in kinetic models. From a theoretical perspective, it allows one to replace these unimportant remote substituents with smaller chemical groups, thereby reducing the computational cost of the calculations. For example, in free radical copolymerization, the reactivity of the propagating radical is determined solely by the nature of its terminal and penultimate units, and is independent of the chemical composition of the remainder of the polymer chain [95]. As a result, kinetic models for copolymerization need only to consider the propagation reactions of four types of propagating radical instead of thousands, while computational studies can determine the radical and monomer reactivity ratios using dimers as chemical models of the propagating species. This has been confirmed in both experimental [96, 97] and theoretical [7] studies of the homopropagation rate coefficients, which indicate that the rate coefficients have largely converged to their long chain limit at the dimer radical stage and completely converged well before the decamer stage.

In the RAFT process, there are two key areas where simplifications to kinetic and theoretical studies are desirable. In the generic addition–fragmentation process R• + S=C(Z)SR′, it is important to discover whether the kinetics and thermodynamics are affected by the nature of the 'attacking group' (R•) and the 'nonparticipating group' (R′) and, if so, at what point the remote substituent effects become negligible. Recent experimental [1, 13, 14] and theoretical [9, 18–20, 78, 98] studies of the effects of substituents on the RAFT process indicate that the kinetics and thermodynamics are extremely sensitive to the nature of the attacking radical and the Z group of the RAFT agent (S=C(Z)SR′). For example, the equilibrium constant (at 333.15 K) for $•CH_3$ addition to $S{=}C(CH_3)SCH_3$ increases by approximately 6 orders of magnitude when the $CH_3$ Z group is replaced with a phenyl substituent, and decreases by 7 orders of magnitude when the methyl attacking group is replaced with the initiating radical derived from 2,2′-azobis(isobutyronitrile) (AIBN), $•C(CH_3)_2CN$ [18]. It is therefore essential that reactions differing in the nature of their Z groups and/or the primary substituents of their R groups are treated as being kinetically distinct. The influence of the R′ is less significant and some simplifications are possible. For example, the calculated equilibrium constants for $•CH_3$ addition to $S{=}C(CH_3)SR'$ for $R' = CH_2X$, $CH(CH_3)X$, $C(CH_3)_2X$ and $X = H$, $CH_3$, CN, Ph, $COOCH_3$, $OCOCH_3$ range over 3 orders of magnitude (see Table 2.1) [78]. However, this effect is mainly steric in origin, the bulky R′ groups destabilizing the thiocarbonyl compound to a greater extent than the more flexible RAFT-adduct radical. As a result, the $K$ values for individual classes of mono-, di- and trisubstituted R′ groups fall into much smaller ranges and, depending on the nature of the monomer and initial R′ group, it may be possible to ignore the effect of R′ in some kinetic models.

Given that the R, Z and, to some extent, the R′ groups all affect the addition–fragmentation equilibrium, it is not surprising that the addition–fragmentation equilibrium is also affected by the chain length of the R and R′ groups. Figure 2.2 shows the effects of chain length ($n$) on the equilibrium constants for addition of styryl radicals of varying chain length, $(NC)C(CH_3)_2{-}(CH_2CHPh)_n•$ ($n = 0, 1, 2, 3$), to $S{=}C(Ph)SCH_3$, and the addition of $•CH_3$ radicals to the corresponding poly-RAFT agents, $S{=}C(Ph)S{-}(CHPhCH_2)_n{-}C(CH_3)_2CN$ [78]. The $C(CH_3)_2CN$ end group is that imparted by the initiating species, when AIBN is used as the initiator. It should be noted that in the case of the trimer species, this end group is omitted entirely, owing to the large computational cost of the calculations. Not surprisingly, the equilibrium constants change by several orders of magnitude when the first monomer unit is inserted into the respective chains, as the primary substituents on the respective R and R′ groups are altered. In the case of the attacking radical (i.e. the R group), the equilibrium constants continue to change as successive styryl units are inserted into the chain. Thus from Fig. 2.2, it is seen that the equilibrium constant increases by a factor of 73 when the first styryl unit is inserted, increases by a further factor of 672 when the second unit is inserted, but increases by less than a factor of 4 when the third unit is inserted. Since the end group was omitted in calculating the latter species, it is possible that when the full trimer is investigated larger deviations will be observed, but the indications are that reasonable convergence occurs at the

**Table 2.1** Effect of R′ on the calculated enthalpies ($\Delta H$; kJ·mol$^{-1}$), entropies ($\Delta S$; J·mol$^{-1}$·K$^{-1}$) and equilibrium constants ($K$; L·mol$^{-1}$)[a] at 333.15 K for the model RAFT reaction, $\bullet CH_3 + S{=}C(CH_3)SR' \rightarrow CH_3SC\bullet(CH_3)SR'$

| | R′ = $CH_2X$ | | | R = $CH(CH_3)X$ | | | R′ = $C(CH_3)_2X$ | | |
|---|---|---|---|---|---|---|---|---|---|
| X | $\Delta H$ | $\Delta S$ | $K$ | $\Delta H$ | $\Delta S$ | $K$ | $\Delta H$ | $\Delta S$ | $K$ |
| H | 76.0 | 149.0 | $3.8 \times 10^7$ | | | | | | |
| $CH_3$ | 79.9 | 138.6 | $1.8 \times 10^7$ | 81.8 | 146.2 | $1.7 \times 10^7$ | 94.3 | 145.3 | $1.7 \times 10^9$ |
| CN | 92.9 | 161.5 | $1.8 \times 10^8$ | 93.2 | 163.2 | $1.5 \times 10^8$ | 102.4 | 146.4 | $2.9 \times 10^{10}$ |
| Ph | 87.8 | 166.4 | $1.5 \times 10^7$ | 89.9 | 151.0 | $1.8 \times 10^8$ | 104.5 | 153.3 | $2.9 \times 10^{10}$ |
| $COOCH_3$ | 90.0 | 168.1 | $3.3 \times 10^7$ | 97.1 | 175.4 | $1.8 \times 10^8$ | 97.5 | 158.6 | $1.4 \times 10^9$ |
| $OCOCH_3$ | 92.0 | 166.3 | $8.1 \times 10^7$ | 99.5 | 167.2 | $1.1 \times 10^9$ | 106.3 | 153.1 | $6.1 \times 10^{10}$ |

[a]The original data reported in Ref. [78] was recalculated at the W1-ONIOM//B3-LYP/6-31G(d) level of theory in conjunction with the harmonic oscillator approximation and taken from Ref. [42].

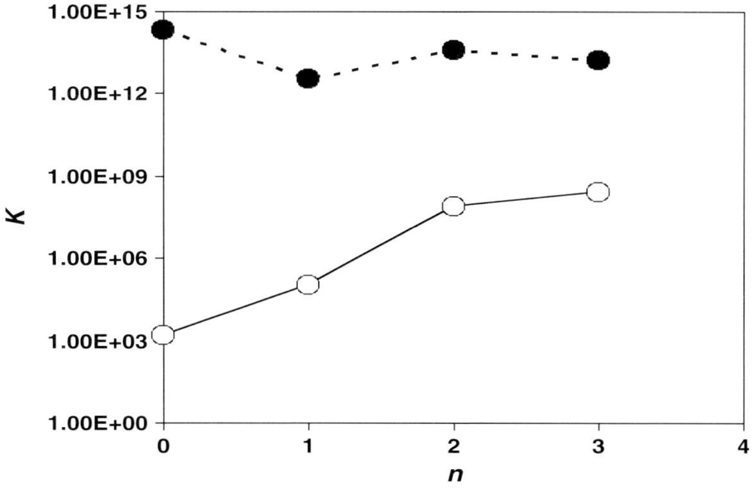


**Fig. 2.2** The effect of chain length (*n*) on the equilibrium constant ($K$; L·mol$^{-1}$) at 333.15 K for the addition of (NC)C($CH_3$)$_2$—($CH_2$CHPh)$_n$• to S=C(Ph)S$CH_3$ (o) and addition of $CH_3$• to S=C(Ph)S—(CH(Ph)$CH_2$)$_n$—C($CH_3$)$_2$CN (•) [42].

trimer stage. In the case of RAFT agent substituent (i.e. the R′ group), the chain length effects are much smaller but are nonetheless significant. Insertion of the first styryl unit causes the equilibrium constant to decrease by a factor of 67, while insertion of the second causes an increase by a factor of 10. Calculations of the full trimer species are again required in order to establish convergence; nonetheless, the H-terminated species has converged to within a factor of 3.

In summary, it is clear that in the early stages of the reaction chain length effects in the attacking radical, and to a lesser extent the RAFT agent substituent, are significant and their neglect leads a serious oversimplification of the initialization kinetics. Based on the currently available computational results, it seems reasonable to suppose that reactions of the various primary species (R• and I•), unimeric radicals (I—M• and R—M•), dimeric radicals (R—M—M• and I—M—M•) and remaining longer chain species do need to be treated as being kinetically distinct. It is also possible that chain length effects beyond the trimer position are also important, and it is also possible that for certain systems chain length effects are smaller than in the case of styrene polymerization. For highly accurate results, the RAFT agents derived from each of these species should also be treated as being kinetically distinct, though the chain length effects in these cases are much smaller. Nonetheless, a 'complete' kinetic model of the addition–fragmentation kinetics would thus resemble that provided in Scheme 2.1, and would contain far too many unknown parameters for fitting to experimental data alone. As will be seen below, computational chemistry offers the prospect of reducing the number of unknown parameters by providing direct calculations of the relevant values for the early small-molecule reactions of the RAFT process, leaving experiment to estimate a smaller number of parameters corresponding to the longer chain radicals.

### 2.3.2
### Side Reactions

In designing an appropriate kinetic model for RAFT polymerization, it is also necessary to determine whether additional side reactions need to be incorporated into the kinetic scheme. In principle, one could use computational chemistry to *find* all possible side reactions of a process from first principles. However, this would entail the calculation of a complete multidimensional potential energy surface for the chemical system, followed by molecular dynamics simulations for a wide variety of possible trajectories. Unfortunately, this is currently too computationally intensive for practical RAFT systems. However, it is possible to study specific postulated side reactions and test whether they are likely to occur. For example, if hydrogen abstraction reactions involving the RAFT-adduct radical were suspected to be a problem in certain polymerization systems, one could calculate rate coefficients for the various possible reactions (i.e. abstraction of the different hydrogen atoms on the monomer, polymer, RAFT agent and so forth). By comparing their rates with the normal β-scission reaction, one could thereby establish which, if any, of the possible abstraction reactions were likely to be competitive.

To date, only one type of side reaction has been identified computationally [17]. In this work, computational chemistry was used to study the competitive β-scission of the C—O bond of the alkoxy group in xanthate-mediated polymerization of vinyl acetate. Rate coefficients (333.15 K) for both the normal β-scission of the vinyl acetate radical and the side reaction were calculated for a series of model RAFT-adduct radicals, $CH_3SC\bullet(OZ')SCH_2OCOCH_3$ for Z′ = Me, Et, *i*-Pr and *t*-Bu (see Scheme 2.2) [42]. It was shown that for the Z′ = Me, Et and *i*-Pr systems, the normal $\beta$-scission reaction is favored and this is consistent with the experimental observation that these systems display normal RAFT behavior. In contrast, for the *t*-Bu case, the side reaction is preferred, and this provided a suitable explanation for the experimentally observed inhibition in this system. In general, it would be worth considering this side reaction when the Z′ substituent of the xanthate is a good radical leaving group (such as *t*-Bu) and/or the propagating radical is poor leaving group (as in vinyl acetate or ethylene polymerization).

Other side reactions might also be conceivable under certain circumstances, though their feasibility is yet to be confirmed computationally. As noted above, there is some debate as to whether the RAFT-adduct radical is capable of undergoing irreversible and/or reversible termination reactions (see Scheme 2.3) to any significant extent in normal polymerizing RAFT systems [3, 4, 84–92, 99]. Recently Buback and coworkers used DFT calculations to study the thermodynamics of several of these steps, showing that they are thermodynamically feasible [99]. However, calculations of the corresponding rate coefficients would be required to identify which, if any, of the reactions in Scheme 2.3 are capable of outcompeting

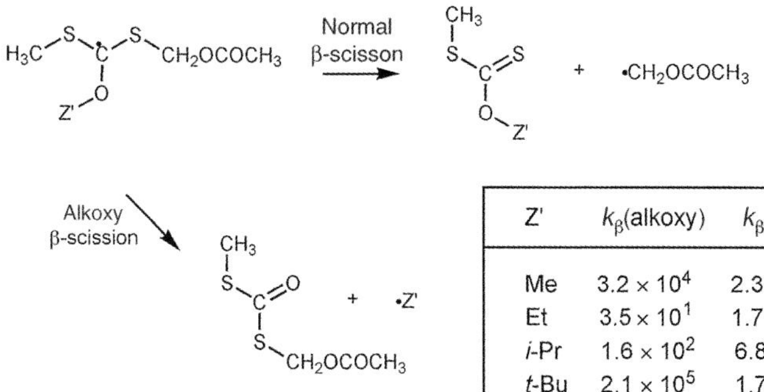


| Z' | $k_\beta$(alkoxy) | $k_\beta$(normal) | Ratio |
|---|---|---|---|
| Me | $3.2 \times 10^4$ | $2.3 \times 10^5$ | $1.4 \times 10^{-1}$ |
| Et | $3.5 \times 10^1$ | $1.7 \times 10^5$ | $1.2 \times 10^{-5}$ |
| *i*-Pr | $1.6 \times 10^2$ | $6.8 \times 10^2$ | $2.4 \times 10^{-1}$ |
| *t*-Bu | $2.1 \times 10^5$ | $1.7 \times 10^2$ | $1.2 \times 10^3$ |

**Scheme 2.2** Competitive $\beta$-scission processes in xanthate-mediated polymerization of vinyl acetate [42].

the β-scission reaction of the intermediate radical under normal polymerization conditions. Although computational quantum chemistry is not currently capable of predicting diffusion-controlled rate coefficients, computational calculations could nonetheless be used to calculate the chemically controlled component of the termination rate and establish a lower bound to the termination rate coefficient. To this

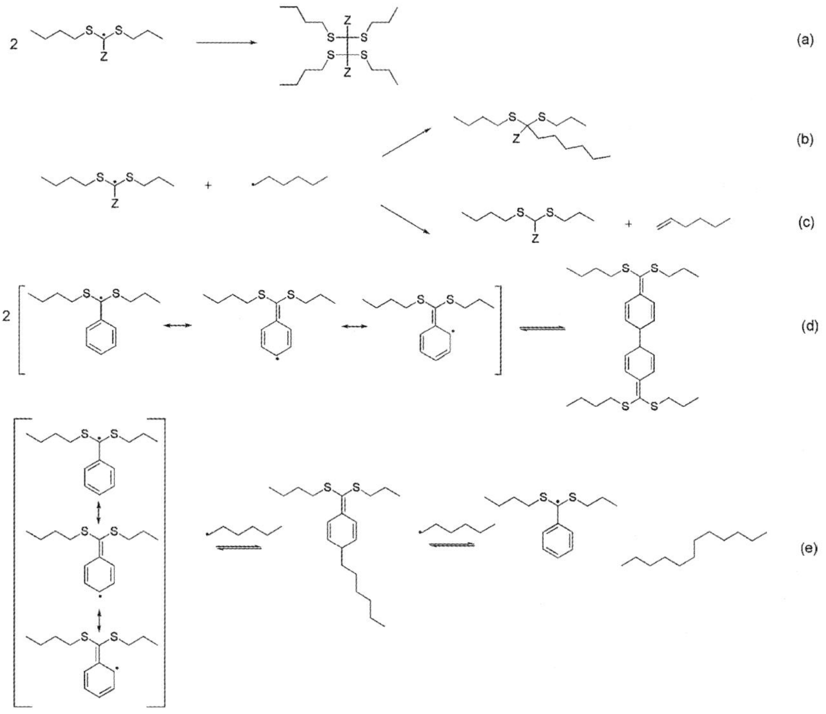


**Scheme 2.3** Possible reversible and irreversible termination reactions of the RAFT-adduct radical.

end, work is currently underway to identify cost-effective multireference methods for studying such systems.

Another side reaction that is worth considering is the attack of the propagating radical at the carbon center (rather than the sulfur center) of the C=S bond to produce a sulfur-centered radical. Studies of radical addition to simple thiocarbonyl compounds indicate that this side reaction is actually thermodynamically preferred, though it is kinetically less favorable by 1–2 orders of magnitude [8, 100]. This is due in part to the greater steric hindrance at the carbon center and in part to the stronger early bonding interaction when attack occurs at sulfur (see below) [100]. In RAFT systems, the steric hindrance at the carbon center is yet greater, and the kinetic preference for addition at carbon should normally be very small. Nonetheless, given that addition to the RAFT agent occurs many thousands of times during the lifetime of the propagating radical, even a low relative rate of addition at carbon might be sufficient to affect the polymerization kinetics in certain systems. Other side reactions that have been suggested include reaction of the RAFT-adduct radical with monomer (i.e. copolymerization), the *syn*-elimination of tertiary R groups (as the corresponding alkenes) at elevated temperatures and reactions with oxygen impurities [1, 2]. Computational calculations of these and other pathways could help to establish their feasibility and clarify their mechanism.

### 2.3.3 Computational Model Predictions

Provided calculations are performed at an appropriately high level of theory, computational chemistry can be used to determine the rates and equilibrium constants for the individual reactions in a RAFT polymerization. The resulting values can then be used to study structure–reactivity trends, reduce the number of adjustable parameters in a kinetic analysis of experimental data or, ultimately, predict the kinetic behavior of a RAFT polymerization system from first principles. Table 2.2 shows a compilation of the main equilibrium constants that have been reported to date [42]. Owing to their greater computational expense, the reaction rates for RAFT systems have been less widely studied. However, the rate coefficients have been reported for a number of small model systems and these are provided in Table 2.3 [42]. Owing to the significant errors that occur at low levels of theory, only high-level ab initio values are included in these tables. However, where possible, the original calculations were improved to the W1-ONIOM level of theory and recalculated at a consistent temperature (333.15 K) [42]. Values obtained under the harmonic oscillator approximation are reported for all systems and where possible the more accurate hindered rotor values are also reported.

The data confirm that the equilibrium constants (and thus the addition–fragmentation kinetics) are very dependent upon the nature of the attacking radical and the RAFT agent. In a practical RAFT system, it is not valid to treat all of the possible reactions as having the same rates. Thus, for example, when calculated at a uniform level of theory, the equilibrium constant for addition of a styryl radical to RAFT

**Table 2.2** Effect of R, Z and R′ on the calculated equilibrium constants $(K)^a$ at 333.15 K for the model RAFT reaction, •R + S=C(Z)SR′ → RSC•(Z)SR′

| R• | Z | R′ | K(HO) | K(HR) | R• | Z | R′ | K(HO) | K(HR) |
|---|---|---|---|---|---|---|---|---|---|
| $CH(CH_3)CN$ | $CH_3$ | $CH_3$ | $2.9 \times 10^{0}$ | | $CH_2COOCH_3$ | $CH_3$ | $CH_3$ | $6.0 \times 10^{3}$ | $6.7 \times 10^{4}$ |
| $C(CH_3)_2CN$ (AIBN) | $CH_3$ | $CH_3$ | $1.2 \times 10^{-1}$ | $9.6 \times 10^{-1}$ | | Ph | $CH_3$ | $4.1 \times 10^{7}$ | |
| | Ph | $CH_3$ | $1.5 \times 10^{3}$ | $2.6 \times 10^{4}$ | | $CH_2Ph$ | $CH_3$ | $1.1 \times 10^{4}$ | |
| | $CH_2Ph$ | $CH_3$ | $1.3 \times 10^{0}$ | | | $CH_3$ | $CH_2COOCH_3$ | $8.6 \times 10^{4}$ | |
| | Ph | $CH(CH_3)_2$ | $1.1 \times 10^{5}$ | $7.4 \times 10^{4}$ | | F | $CH_3$ | $2.1 \times 10^{1}$ | |
| | F | $CH_3$ | $9.4 \times 10^{-4}$ | | $CH(CH_3)COOCH_3$ (MA) | $CH_3$ | $CH_3$ | $3.5 \times 10^{2}$ | |
| | CN | $CH_3$ | $1.4 \times 10^{7}$ | | | $CH_3$ | $CH_2CN$ | $1.5 \times 10^{3}$ | |
| | $OCH_3$ | $CH_3$ | $3.8 \times 10^{-6}$ | | | Ph | $CH_3$ | $1.0 \times 10^{6}$ | $5.6 \times 10^{6}$ |
| | Ph | $C(CH_3)_2CN$ | $1.1 \times 10^{5}$ | | | Ph | $CH(CH_3)_2$ | $5.9 \times 10^{6}$ | $3.6 \times 10^{7}$ |
| $CH_2CH_3$ | $CH_3$ | $CH_3$ | $1.8 \times 10^{6}$ | | $C(CH_3)_2COOCH_3$ | $CH_3$ | $CH_3$ | $4.3 \times 10^{0}$ | |
| $CH(CH_3)_2$ | $CH_3$ | $CH_3$ | $4.7 \times 10^{5}$ | | $CH_2OCOCH_3$ | $CH_3$ | $CH_3$ | $1.3 \times 10^{6}$ | |
| $C(CH_3)_3$ | $CH_3$ | $CH_3$ | $1.5 \times 10^{5}$ | | | F | $CH_3$ | $2.3 \times 10^{3}$ | |

*(Continued)*

Table 2.2 (*Continued*)

| R• | Z | R′ | *K*(HO) | *K*(HR) | R• | Z | R′ | *K*(HO) | *K*(HR) |
|---|---|---|---|---|---|---|---|---|---|
| | F | $CH_3$ | $4.1 \times 10^{3}$ | | | CN | $CH_3$ | $4.5 \times 10^{14}$ | |
| $CH_2Ph$ | $CH_3$ | $CH_3$ | $2.7 \times 10^{0}$ | $2.8 \times 10^{2}$ | | $OCH_3$ | $CH_3$ | $5.6 \times 10^{0}$ | |
| | Ph | $CH_3$ | $2.8 \times 10^{5}$ | | | $OCH_2CH_3$ | $CH_3$ | $1.7 \times 10^{1}$ | |
| | $CH_2Ph$ | $CH_3$ | $4.1 \times 10^{1}$ | | | $OCH(CH_3)_2$ | $CH_3$ | $7.8 \times 10^{1}$ | |
| | F | $CH_3$ | $2.9 \times 10^{-3}$ | | | $OC(CH_3)_3$ | $CH_3$ | $3.7 \times 10^{2}$ | |
| | CN | $CH_3$ | $1.5 \times 10^{9}$ | | $CH(CH_3)OCOCH_3$ (VA) | $CH_3$ | $CH_3$ | $9.6 \times 10^{5}$ | $5.0 \times 10^{7}$ |
| | $OCH_3$ | $CH_3$ | $5.3 \times 10^{-5}$ | | | $OCH_3$ | $CH_3$ | $1.3 \times 10^{0}$ | $1.5 \times 10^{1}$ |
| $CH(CH_3)Ph$ (STY) | $CH_3$ | $CH_3$ | $1.1 \times 10^{0}$ | $5.1 \times 10^{1}$ | | F | $CH_3$ | $2.0 \times 10^{3}$ | $3.4 \times 10^{4}$ |
| | Ph | $CH_3$ | $1.5 \times 10^{5}$ | $3.4 \times 10^{6}$ | $C(CH_3)_2OCOCH_3$ | $CH_3$ | $CH_3$ | $6.6 \times 10^{4}$ | |
| | Ph | $CH(CH_3)_2$ | $2.9 \times 10^{5}$ | $6.6 \times 10^{5}$ | $CH(CH_3)COOH$ (AA) | $CH_3$ | $CH_3$ | $2.0 \times 10^{1}$ | |
| | Ph | $C(CH_3)_2CN$ | $6.5 \times 10^{7}$ | | | F | $CH_3$ | $2.1 \times 10^{-1}$ | $1.9 \times 10^{0}$ |
| | F | $CH_3$ | $1.0 \times 10^{-3}$ | $8.8 \times 10^{-3}$ | | *N*-pyrrole | $CH_3$ | $1.2 \times 10^{2}$ | |
| | $OCH_3$ | $CH_3$ | $1.1 \times 10^{-5}$ | $1.8 \times 10^{-4}$ | $CH(CH_3)CONH_2$ (AM) | $CH_3$ | $CH_3$ | $8.0 \times 10^{2}$ | |

| | | | | | | | | | |
|---|---|---|---|---|---|---|---|---|---|
| $C(CH_3)_2Ph$ (Cumyl) | $CH_3$ | $CH_3$ | $6.9 \times 10^{-2}$ | | | F | $CH_3$ | $1.2 \times 10^{0}$ | $7.9 \times 10^{0}$ |
| | Ph | $CH_3$ | $2.4 \times 10^{3}$ | $4.6 \times 10^{4}$ | | *N*-pyrrole | $CH_3$ | $4.1 \times 10^{3}$ | |
| | Ph | $CH(CH_3)_2$ | $1.3 \times 10^{5}$ | $2.5 \times 10^{5}$ | AIBN-STY | Ph | $CH_3$ | $1.1 \times 10^{5}$ | |
| Cumyl-STY | Ph | $CH_3$ | $1.3 \times 10^{7}$ | | | Ph | $CH(CH_3)_2$ | $1.2 \times 10^{6}$ | |
| | Ph | $CH(CH_3)_2$ | $8.7 \times 10^{7}$ | | | Ph | $C(CH_3)_2CN$ | $2.3 \times 10^{9}$ | |
| Cumyl-STY-STY | Ph | $CH_3$ | $2.8 \times 10^{8}$ | | AIBN-STY-STY | Ph | $CH_3$ | $7.4 \times 10^{7}$ | |
| STY-STY | Ph | $CH_3$ | $2.2 \times 10^{8}$ | | | Ph | $C(CH_3)_2CN$ | $1.8 \times 10^{11}$ | |
| | Ph | $CH(CH_3)_2$ | $3.6 \times 10^{9}$ | | STY-STY-STY | Ph | $CH_3$ | $2.5 \times 10^{8}$ | |

[a] Equilibrium constants *K* ($L{\cdot}mol^{-1}$) calculated at the W1-ONIOM//B3-LYP/6-31G(d) level of theory in conjunction with the harmonic oscillator approximation (HO) and, where data were available, also the hindered rotor model (HR). The original electronic structure data were taken from previous studies [3, 10, 17–19, 23, 24, 78, 98, 101] and corrected to the W1-ONIOM level in Ref. [42] using W1 calculations [8] for the core reaction. For most systems, a two-layer ONIOM approach was possible in which the inner core was calculated at W1 and the rest of the system at G3(MP2)-RAD. For a limited number of larger systems (such as reactions in which Z = Ph and reactions of dimer radicals with various substrates), a three-layer ONIOM approach was used in which the outer layer was calculated at RMP2/6-311+G(3df,2p).

**Table 2.3** Calculated addition ($k_{add}$), fragmentation ($k_{frag}$) and equilibrium ($K$) constants[a] at 333.15 K for the model RAFT reaction, •R + S=C(Z)SCH$_3$ → RSC•(Z)SCH$_3$

| R• | Z | $k_{add}$(HO) | $k_{add}$(HR) | $k_{frag}$(HO) | $k_{frag}$(HR) | $K$(HO) | $K$(HR) |
|---|---|---|---|---|---|---|---|
| $CH_3$ | $CH_3$ | $9.6 \times 10^{5}$ | $1.7 \times 10^{6}$ | $9.6 \times 10^{-2}$ | $3.4 \times 10^{-2}$ | $1.0 \times 10^{7}$ | $4.9 \times 10^{7}$ |
| | Ph | $4.2 \times 10^{7}$ | $7.7 \times 10^{7}$ | $2.5 \times 10^{-4}$ | $7.2 \times 10^{-6}$ | $1.7 \times 10^{11}$ | $1.1 \times 10^{13}$ |
| | $CH_2Ph$ | $2.6 \times 10^{6}$ | $1.2 \times 10^{7}$ | $1.4 \times 10^{-2}$ | $1.7 \times 10^{-3}$ | $1.9 \times 10^{8}$ | $7.3 \times 10^{9}$ |
| | $OCH_3$ | $1.2 \times 10^{4}$ | | $1.4 \times 10^{3}$ | | $8.2 \times 10^{0}$ | |
| | $OCH_2CH_3$ | $3.1 \times 10^{4}$ | | $8.2 \times 10^{2}$ | | $3.8 \times 10^{1}$ | |
| | $OCH(CH_3)_2$ | $3.4 \times 10^{4}$ | | $5.6 \times 10^{2}$ | | $6.1 \times 10^{1}$ | |
| | $OC(CH_3)_3$ | $3.2 \times 10^{4}$ | | $1.8 \times 10^{2}$ | | $1.8 \times 10^{2}$ | |
| $CH_2Ph$ | $CH_3$ | $6.6 \times 10^{5}$ | $3.9 \times 10^{6}$ | $1.2 \times 10^{5}$ | $1.4 \times 10^{4}$ | $5.6 \times 10^{0}$ | $2.8 \times 10^{2}$ |
| $CH(CH_3)Ph$ | $CH_3$ | $7.5 \times 10^{5}$ | $3.8 \times 10^{6}$ | $6.8 \times 10^{5}$ | $7.5 \times 10^{4}$ | $1.1 \times 10^{0}$ | $5.1 \times 10^{1}$ |
| | F | $2.3 \times 10^{5}$ | $1.8 \times 10^{6}$ | $2.3 \times 10^{8}$ | $2.1 \times 10^{8}$ | $1.0 \times 10^{-3}$ | $8.8 \times 10^{-3}$ |
| | $OCH_3$ | $6.2 \times 10^{3}$ | $4.4 \times 10^{4}$ | $5.8 \times 10^{8}$ | $2.5 \times 10^{8}$ | $1.1 \times 10^{-5}$ | $1.8 \times 10^{-4}$ |
| $CH_2COOCH_3$ | $CH_3$ | $5.7 \times 10^{7}$ | $1.7 \times 10^{8}$ | $1.1 \times 10^{4}$ | $2.5 \times 10^{3}$ | $5.2 \times 10^{3}$ | $6.8 \times 10^{4}$ |
| $C(CH_3)_2CN$ | $CH_3$ | $4.5 \times 10^{5}$ | $1.2 \times 10^{6}$ | $3.7 \times 10^{6}$ | $1.2 \times 10^{6}$ | $1.2 \times 10^{-1}$ | $9.6 \times 10^{-1}$ |
| $CH_2OCOCH_3$ | $OCH_3$ | $1.3 \times 10^{6}$ | | $2.3 \times 10^{5}$ | | $5.6 \times 10^{0}$ | |
| | $OCH_2CH_3$ | $2.8 \times 10^{6}$ | | $1.7 \times 10^{5}$ | | $1.7 \times 10^{1}$ | |
| | $OCH(CH_3)_2$ | $5.2 \times 10^{4}$ | | $6.8 \times 10^{2}$ | | $7.8 \times 10^{1}$ | |
| | $OC(CH_3)_3$ | $6.2 \times 10^{4}$ | | $1.7 \times 10^{2}$ | | $3.7 \times 10^{2}$ | |
| $CH(CH_3)OCOCH_3$ | $CH_3$ | $6.3 \times 10^{7}$ | $4.2 \times 10^{8}$ | $6.7 \times 10^{1}$ | $8.4 \times 10^{0}$ | $9.6 \times 10^{5}$ | $5.0 \times 10^{7}$ |
| | F | $1.5 \times 10^{8}$ | $5.6 \times 10^{8}$ | $7.5 \times 10^{4}$ | $1.7 \times 10^{4}$ | $2.0 \times 10^{3}$ | $3.4 \times 10^{4}$ |
| | $OCH_3$ | $3.9 \times 10^{6}$ | $5.0 \times 10^{7}$ | $3.0 \times 10^{6}$ | $3.3 \times 10^{6}$ | $1.3 \times 10^{0}$ | $1.5 \times 10^{1}$ |

[a] Rate coefficients for addition (L·mol$^{-1}$·s$^{-1}$) and fragmentation (s$^{-1}$) and equilibrium constants (L·mol$^{-1}$) calculated at the W1-ONIOM//B3-LYP/6-31G(d) level of theory in conjunction with the harmonic oscillator approximation (HO) and, where data were available, also the hindered rotor model (HR). The original electronic structure data were taken from previous studies [9, 16, 17, 101] and corrected to the W1-ONIOM level in Ref. [42] using W1 calculations [8] for the core reaction. Reactions of •$CH_3$ radical and •$CH_2OCOCH_3$ radical with the xanthates were studied via standard transition state theory; all other reactions were studied via variational transition state theory.

agents of the form $S{=}C(Ph)SCH(CH_3)_2$ vary from $2.9 \times 10^5$ to $3.6 \times 10^9$, depending on whether styryl is modeled as a unimer with an H end group, a $C(CH_3)_2CN$ end group, a $C(CH_3)_2Ph$ end group or as a dimer with an H end group. This explains why simplified models do not appear to be capable of describing the initialization period in RAFT systems; indeed, as will be shown below, once these effects are taken into account excellent agreement between the experimental data and model predictions is obtained [79, 102]. Chain length effects may also help to explain the large discrepancies in the experimental values for the rates and equilibrium constants in CDB-mediated polymerization that are obtained by fitting simplified models to experimental data, though it seems likely that additional factors (such as missing side reactions) may be required to resolve the problem fully [5].

Secondly, although the equilibrium constants vary considerably, this variability arises mainly in the fragmentation reaction. For most substituents, the addition rate coefficients fall into a relatively narrow range ($10^5$–$10^7$ L·mol$^{-1}$·s$^{-1}$). In these cases, it may be reasonable to treat all (or most) of the possible addition reactions in a specific process as having the same rate coefficients, provided the rates of their reverse fragmentation reactions are allowed to vary. However, there are a few important exceptions. In particular, the addition of propagating radicals to xanthates is typically much slower than to dithioesters (Z = alkyl, aryl) unless the propagating radical is highly reactive (as in vinyl acetate polymerization). As a result of their slow addition rate, monomers with stable propagating radicals (such as styrene) are not normally well controlled by xanthates. Moreover, within the xanthate reactions themselves, there is a large variation in the addition rate coefficient depending on whether the propagating radical is capable of undergoing hydrogen-bonding interactions with the alkoxy group of xanthate in the transition structure (see below). Indeed, this hydrogen-bonding interaction may be partially responsible for the success of xanthates in controlling vinyl acetate. In this regard, it is worth noting that the xanthates have been less successful in controlling ethylene polymerization, another system in which the propagating radical is relatively unstable but for which the hydrogen-bonding interaction is absent. From Table 2.3, it is seen that even for the methyl radical, which might reasonably be expected to be more reactive than the ethyl propagating species, the addition rate coefficient for the xanthate agent is relatively low (ca. $10^4$ L·mol$^{-1}$·s$^{-1}$).

Finally, the data indicate that slow fragmentation of the RAFT-adduct radical can help to explain the experimentally observed inhibition periods in CDB-mediated polymerizations [3]. Kinetic studies indicate that rate retardation is likely to be significant when $K$ is greater than or equal to approximately $10^6$ L·mol$^{-1}$ [103]. The equilibrium constants for addition of unimers of methyl acrylate ($3.6 \times 10^7$) and dimers ($3.6 \times 10^9$) of styrene satisfy this condition, while that for addition of cumyl ($2.5 \times 10^5$) and the styrene unimer ($6.6 \times 10^5$) lie within the level of possible error. In contrast, fragmentation of $\bullet C(CH_3)_2CN$ from the RAFT-adduct radical is not predicted to be retarded ($7.4 \times 10^4$), consistent with the experimental observation [104] that cyanoisopropyl dithiobenzoate does not display a significant inhibition period. On the basis of Table 2.2, other retarded processes would include the fragmentation reactions of the RAFT-adduct radicals bearing a cyano Z group, and the fragmentation of ethyl radicals and vinyl acetate radicals from the adduct radicals of dithioesters. Table 2.2 also correctly predicts that in the case of vinyl acetate this rate retardation is relieved when xanthates are used instead. They also indicate that similar success is likely with fluorodithioformates (Z = F), the new computationally designed class of RAFT agents [23]. In this way, such simple calculations can be used to evaluate quickly whether a polymerization is likely to be successful.

### 2.3.4 Ab Initio Kinetic Modeling

It is clear from both computational and experimental studies of RAFT kinetics that simplified kinetic models, in which side reactions and chain length effects are ignored, are not capable of providing an adequate description of all aspects of the RAFT process [5]. However, kinetic models in which these effects are properly taken into account would contain too many unknown adjustable parameters for fitting to experimental data. As noted above, computational chemistry offers a potential solution to this problem by allowing for the rates and equilibrium constants of the individual steps to be calculated independently. The calculated values could be used to reduce the number of fit parameters in the kinetic model to more manageable levels or, ultimately, replace the fit parameters altogether. In the latter case, one would have a truly ab initio kinetic modeling tool that could allow for predictions of the macroscopic properties of the polymerization process without recourse to experimental fit parameters. Such predictions could be compared directly with experimental data so as to provide a sensitive test of a kinetic model's validity. Moreover, having identified a suitable kinetic model, ab inito kinetic modeling could then be used to test new RAFT agents and optimize reaction conditions prior to experiment.

Proof of principle for ab initio kinetic modeling was recently provided for cyanoisopropyl dithiobenzoate-mediated polymerization of styrene at 70 °C in the presence of the initiator AIBN [79]. Based on computational studies of the chain length dependence of the addition–fragmentation reaction, a kinetic model that included a chain-length-dependent fragmentation reaction was constructed using modeling software PREDICI® [105]. In this model, chain length effects in the attacking radical were included up to the trimer stage and chain length effects in the poly-RAFT agent, which were found to be considerably smaller, were considered only up to the unimer stage (see Scheme 2.4). Equilibrium constants (at 70 °C) were then calculated for the eight kinetically distinct addition–fragmentation reactions and used in conjunction with the chain-length-independent experimental value for the addition rate coefficient so as to obtain the corresponding chain-length-dependent fragmentation rate coefficients. These were combined with reliable experimental data for the initiation, propagation and termination reactions, as taken from styrene homopolymerization experiments. A kinetic simulation of the process was then performed using identical reaction conditions to those in an

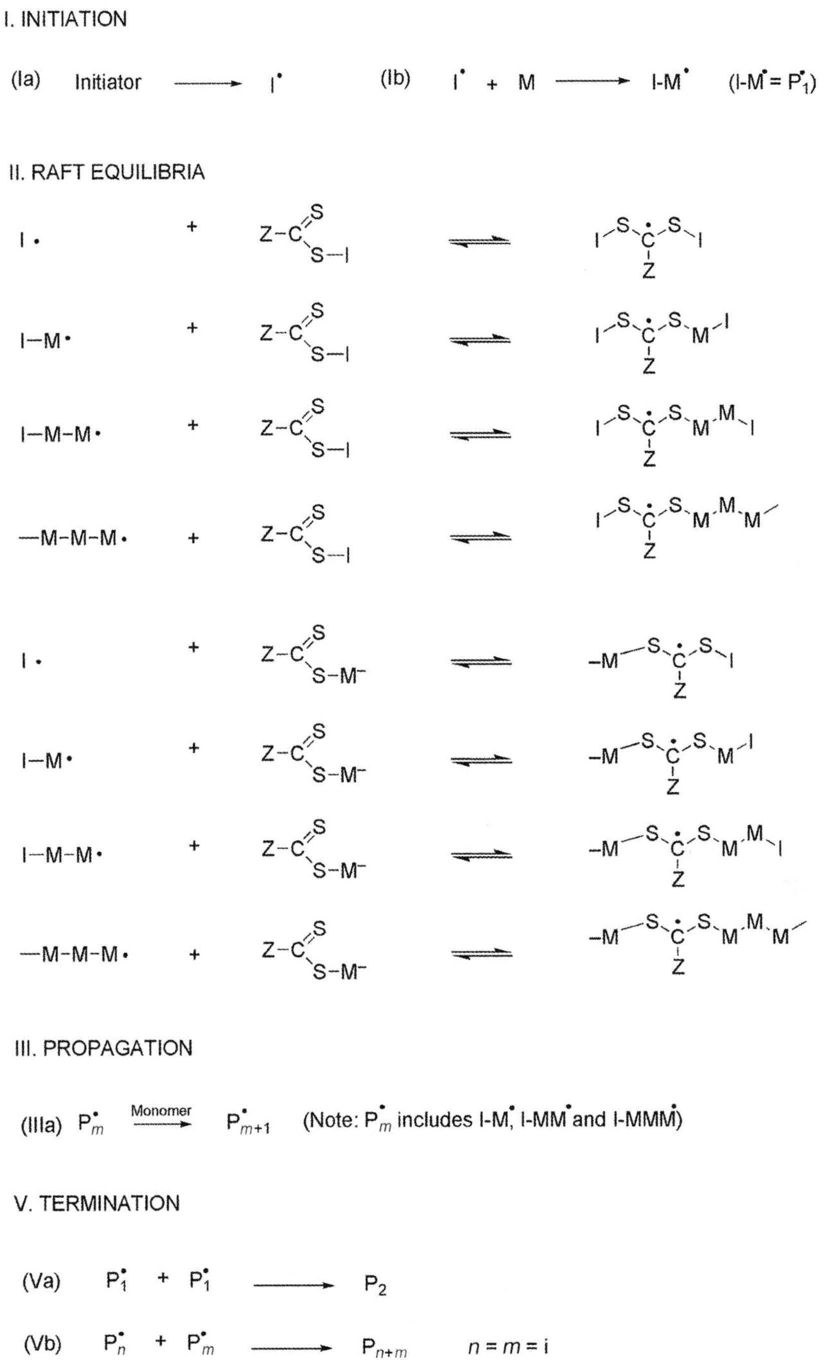


**Scheme 2.4** Reaction scheme for ab initio kinetic modeling of cyanoisopropyl dithiobenzoate-mediated polymerization of styrene in the presence of AIBN.

earlier experiment on the same system [80]. The concentration profiles obtained for the low-molecular-weight thiocarbonyl compounds formed during the early stages of the reaction, together with the corresponding experimental data, are shown in Fig. 2.1 [79]. The predicted and measured rates of overall monomer conversion were also compared [79]. As noted above, there is excellent agreement between the ab initio kinetic simulation and the experimental data, despite the fact that no model fitting of any kind was performed, and hence the model and calculations are sufficiently accurate for predictive purposes. Ab initio kinetic modeling is now a practical possibility for studying RAFT polymerization and the scope of its potential applications can be expected to expand further with increasing computer power.

## 2.4
## Structure–Reactivity Studies

Computational quantum chemistry also has an important role to play in providing an underlying understanding of the reaction mechanism and structure–reactivity trends. At one level, computational chemistry can be used as a tool for determining the rate or equilibrium constants of the individual addition–fragmentation processes for a variety of RAFT agent and monomer combinations. In this regard, it offers a complementary approach to experimental techniques, which are better suited to measuring more practical quantities, such as the extent of control or the apparent chain transfer constant. At a deeper level, computational chemistry can also offer convenient access to additional mechanistic information, such as the geometries, charges and spin densities, and to other relevant energetic quantities, such as radical stabilization energies of the adduct radical. This information can greatly assist in the interpretation of the structure–reactivity trends. Based on both ab initio calculations and experimental approaches, there is now a very good understanding of why C=S double bonds are so effective at addition–fragmentation reactions, and how the RAFT agent substituents affect this process. In what follows, these fundamental and practical aspects are summarized in turn, and we then show how this information can be used to design optimal RAFT agents.

### 2.4.1
### Fundamental Aspects

A key aspect to controlling free radical polymerization by RAFT is the establishment of an addition–fragmentation equilibrium in which there is rapid exchange between the propagating species and dormant poly-RAFT agent. This in turn entails that *both* the addition of the propagating radical to the RAFT agent and the fragmentation of the resulting RAFT-adduct radical (which is effectively the reverse reaction) are sufficiently fast. In radical addition to C=C bonds this is normally quite difficult to achieve. Typically, if addition is fast, the reaction is highly exothermic and fragmentation is too slow. If instead the addition reaction is not very exothermic, then fragmentation becomes relatively facile but addition is then too slow. Although some success has been achieved with macromonomers [106], alkenes are not generally suitable as practical RAFT agents. The question then arises: why are compounds containing C═S bonds so much more effective, and are there any other types of double bond that would serve the same purpose?

To understand the high reactivity of thiocarbonyl RAFT agents, it is helpful to examine a simple curve-crossing model analysis of prototypical radical additions to C═S and C═C bonds [100]. The curve-crossing model was developed by Pross and Shaik [107–109] as a theoretical framework for explaining barrier formation in chemical reactions. The basic idea is to think of a chemical reaction as comprising a rearrangement of electrons, accompanied by a rearrangement of nuclei (i.e. a geometric rearrangement). We can then imagine holding the arrangement of electrons constant in its initial configuration (which we call the reactant configuration) and examining how the energy changes as a function of the geometry. Likewise, we could hold the electronic configuration constant in its final form (the product configuration), and again examine the variation in energy as a function of the geometry. If these two curves (energy versus geometry) are plotted, we form a 'state correlation diagram'. The overall energy profile for the reaction, which is also plotted, is formed by the resonance interaction between the reactant and product configurations (and any other low-lying configurations). State correlation diagrams allow for a qualitative explanation for how the overall energy profile of the reaction arises, and can then be used to provide a graphical illustration of how variations in the relative energies of the alternative valence bond (VB) configurations affect the barrier height.

In radical addition to double bonds (X═Y), the principal VB configurations that may contribute to the ground-state wavefunction are the four lowest-doublet configurations of the three-electron–three-center system formed by the initially unpaired electron at the radical carbon (R) and the electron pair of the $\pi$ bond in X═Y (A) [74].

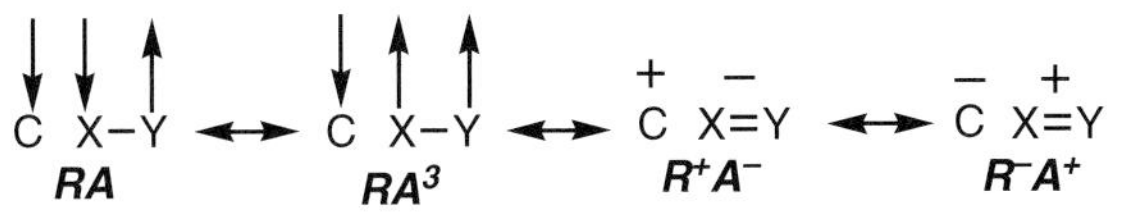


(2.11)

The first configuration ($RA$) corresponds to the arrangement of electrons in the reactants, the second to that of the products ($RA^3$) and the latter two ($R^+A^-$ and $R^-A^+$) to possible charge-transfer configurations. The state correlation diagram showing (qualitatively) how the energies of these configurations should vary as a function of the reaction coordinate is provided in Fig. 2.3 [42]. In plotting this figure we have arbitrarily designated the $R^+A^-$ configuration to be lower in energy than the $R^-A^+$ configuration.

In the early stages of the reaction, the reactant configuration ($RA$) is the lowest-energy configuration and dominates the reaction profile. This is due to the stabilizing influence of the bonding interaction in the $\pi$ bond of the $RA$ configuration, which is an antibonding interaction in the $RA^3$ configuration. However, as the

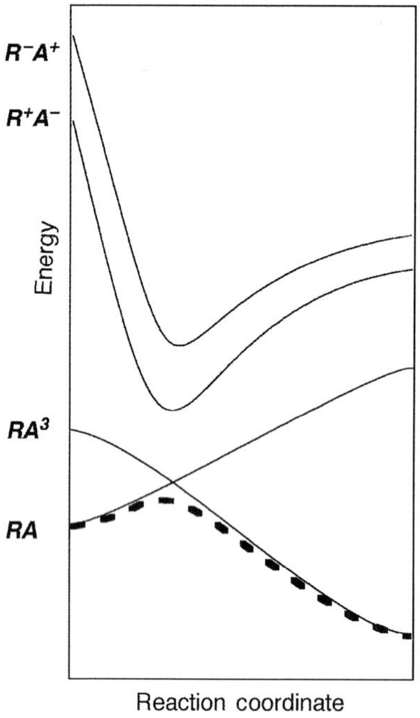


**Fig. 2.3** State correlation diagram for radical addition to double bonds [42].

reaction proceeds, the C···X=Y distance decreases and the unpaired electron on the radical is able to interact with X=Y species. This growing interaction destabilizes the *RA* configuration but stabilizes the $RA^3$ configuration due to the increasing bonding interaction in the forming C···X bond (which is an antibonding interaction in the *RA* configuration). As the relative energies of the *RA* and $RA^3$ configurations converge, the increasing interaction between the alternative configurations stabilizes the ground state wavefunction, with the strength of the stabilizing interaction increasing as the energy difference between the alternate configurations decreases. It is this mixing of the reactant and product configurations which leads to the avoided crossing, and accounts for barrier formation. Beyond the transition structure, the product configuration is lower in energy and dominates the wavefunction. The charge-transfer configurations of the isolated reactants and the isolated products are high in energy, but in the vicinity of the transition structure they are stabilized via favorable Coulombic interactions and can sometimes be sufficiently low in energy to interact with the ground state wavefunction. In those cases, the transition structure is further stabilized, and (if one of the charge transfer configurations is lower than the other) the mixing is reflected in a degree of partial charge-transfer between the reactants.

Using this state correlation diagram, in conjunction with simple VB arguments, the curve-crossing model can be used to predict the qualitative influence of various energy parameters on the reaction barrier [107–109]. In particular, the barrier is

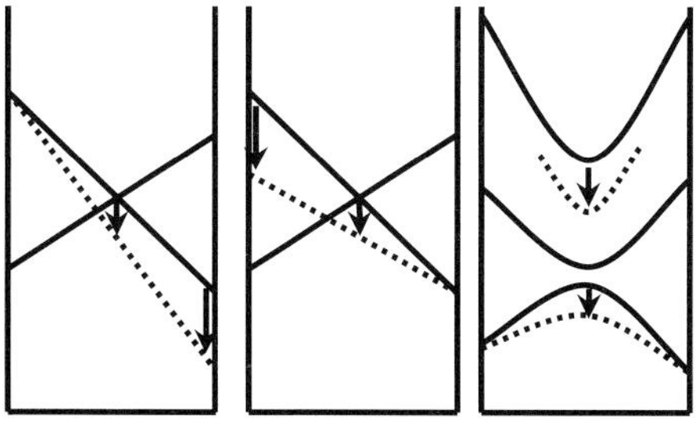


**Fig. 2.4** State correlation diagrams showing the qualitative effect on the barrier height of (a) increasing the exothermicity, (b) decreasing the singlet–triplet excitation gap and (c) decreasing the relative energy of one of the charge-transfer configurations [42].

lowered by an increase in the reaction exothermicity (see Fig. 2.4a) and/or a decrease in the $RA$–$RA^3$ separation in the reactants and/or products (see Fig. 2.4b) and/or a decrease in the relative energies of one or both of the charge-transfer configurations, provided that these are sufficiently low in energy to contribute to the ground-state wavefunction (see Fig. 2.4c). Of these parameters, the reaction exothermicity is of course directly accessible from ab initio molecular orbital calculations. For the radical addition reactions, the $RA$–$RA^3$ separation is related to the singlet–triplet excitation gap of the alkene or thiocarbonyl species. The relative importance of the charge transfer configurations can be assessed on the basis of the energy for charge transfer between the isolated reactants, and can be further assessed by examining the actual extent of charge transfer in the transition structures.

With these simple principles in hand, it then becomes obvious why thiocarbonyls make much more effective RAFT agents than alkenes. As is well known, the $\pi$ bond of a thiocarbonyl is much weaker than that of an alkene, due to the poorer overlap between the p–$\pi$ orbitals of the (third row) S atom and the (second row) C atom [110]. This reduced $\pi$ bond strength is reflected in a greatly reduced singlet–triplet gap for the thiocarbonyl species, and hence in lower barriers and earlier transition structures for radical addition [100]. Indeed, unless there is substantial stabilization of the C=S bond (via, for example, resonance with lone-pair donor substituents at carbon), radical addition to C=S bonds is typically barrierless (i.e. $\Delta H^{\ddagger} \approx 0$), the free energy barrier (i.e. $\Delta G^{\ddagger} > 0$) arising merely due to the opposing enthalpic and entropic effects [9]. Conversely, in the reverse direction, the transition state is very late and almost completely dominated by the reaction exothermicity (i.e. $\Delta H^{\ddagger} \approx \Delta H$).

The low singlet–triplet gap not only explains why dithioesters are generally very reactive to radical addition, it also explains how it is possible to manipulate the RAFT agent substituents so as to ensure that fragmentation is also sufficiently fast. The curve-crossing model predicts that the barrier is affected by the reaction enthalpy: the less exothermic the reaction, the greater the barrier. Indeed, in the absence of any additional factors (such as steric or polar effects), the well-known

Evans–Polanyi rule [111] ($\Delta H^{\ddagger} = \alpha \Delta H$) will hold. In radical addition to C=C bonds, this raises a problem. If one wishes to promote the fragmentation reaction, one needs to make the addition reaction less exothermic but, in doing so, this raises the barrier to radical addition. In radical addition to C=S bonds this is also the case, but there is an important difference. Due to the early transition structures, the influence of the exothermicity on the addition reaction is greatly reduced (i.e. the proportionality constant $\alpha$ is much smaller), and there is much greater scope for manipulating the reverse fragmentation reaction without compromising the addition rate.

The dominant influence of the triplet configuration of the $\pi$ bond also helps to explain the regioselectivity of the radical addition reaction. As explained above, radical addition to the carbon center of the C=S bond is actually the thermodynamically preferred reaction, though in practice addition occurs at S due to kinetic factors [8, 100]. The spin density is considerably greater on S than on C in the triplet configuration of the thiocarbonyl, and there is thus a much stronger early bonding interaction when addition occurs at this site. Although the preference for attack at sulfur is countered by the reduced exothermicity, the early transition structure ensures that influence of the triplet configuration dominates.

Finally, it is worth exploring whether compounds containing other types of double bonds would be suitable in RAFT processes. In particular, dithiophosphinate esters have been proposed as possible alternative RAFT agents [112]. In such systems, the propagating radical would add to the sulfur center of the P=S bond, generating a phosphoranyl radical (instead of a carbon-centered radical) as the intermediate.

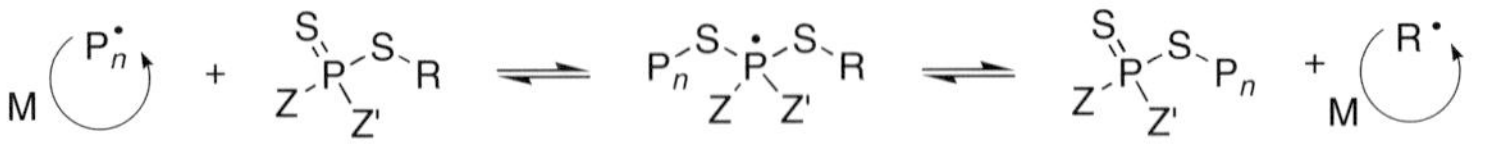

(2.12)

Although there is some experimental indication that styrene polymerization can be controlled using S=P(SCH($CH_3$)Ph))$_3$ and S=P(Ph)$_2$SCH($CH_3$)Ph [112], the observed control was by no means perfect and there was clear evidence of hybrid behavior. Computationally, it was shown that radical addition to the P=S bonds of dithiophosphinate esters was considerably less reactive and less exothermic than addition to corresponding C=S compounds [113]. This is due in part to the lower inherent stabilization energy of its phosphoranyl radical product and in part to its stronger double bond [113]. Indeed, unlike the C=S double bond of dithioesters, the P=S bond has a significant ionic character (i.e. a large resonance contribution from $P^+$—$S^-$), which generally renders it less susceptible to radical addition (unless the attacking radical is highly electrophilic). On this basis it was concluded that dithiophosphinate esters are likely to have only limited applicability as RAFT agents, but would be most suited to the control of electrophilic monomers [113]. The suitability of other types of compounds has yet to be explored computationally; however, studies of prototypical systems have revealed that C=C, C=O and C≡C bonds all have significantly larger singlet–triplet gaps than corresponding C=S bonds [100, 114]. As a result they are less reactive to radical addition than C=S bonds, and less likely to function as effective RAFT agents.

## 2.4.2
## Structure–Reactivity in Practical RAFT Systems

Notwithstanding the high reactivity of C=S bonds, the precise nature of the other substituents (R and Z) in a RAFT agent S=C(Z)SR is critical to a controlled polymerization. The degree of control that can be achieved is reliant on the propensity for addition of the propagating radical $P_n\bullet$ to the C=S bond, and the subsequent ability of $P_n\bullet$ to be released from the RAFT-adduct radical. Both of these depend on the steric and electronic properties of R and Z. As a result of numerous experimental studies [1], there now exists a broad understanding of how effective a given R or Z group is for a certain class of polymerization. For example, polymerizations of styrene, acrylates or methacrylates are well controlled by dithioester RAFT agents having simple alkyl or aryl Z groups, whereas the polymerization of vinyl acetate is not. For vinyl acetate, control can be achieved with xanthates (Z = OR) or dithiocarbamates ($Z = NR_2$). Equally, it is known [14] that a RAFT agent having R = benzyl would be suitable for polymerization of methyl acrylate, but not for polymerization of methyl methacrylate: the tertiary R group $C(CH_3)_2Ph$ is necessary in the latter case.

Building on the wealth of structure–reactivity data now available from experimental studies, computational chemistry adds a powerful tool that allows these trends to be explained, and this in turn has predictive power. Our computational investigations have been founded on the observation that, despite the mechanistic complexity of RAFT polymerization, the effects of R and Z can in large part be understood by considering simplified versions of the chain-transfer reaction (equation 2.13).

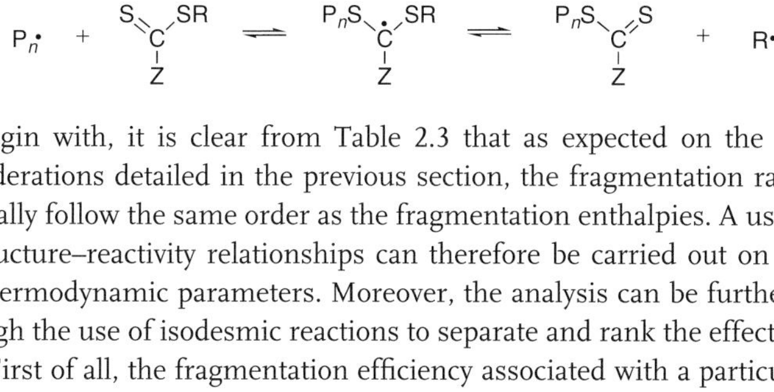

(2.13)

To begin with, it is clear from Table 2.3 that as expected on the basis of the considerations detailed in the previous section, the fragmentation rate constants generally follow the same order as the fragmentation enthalpies. A useful analysis of structure–reactivity relationships can therefore be carried out on the basis of the thermodynamic parameters. Moreover, the analysis can be further simplified through the use of isodesmic reactions to separate and rank the effects of R and Z [98]. First of all, the fragmentation efficiency associated with a particular Z group is measured by the enthalpy change ($\Delta H_{frag}$) for the reaction $CH_3SC\bullet(Z)SCH_3$ + $S{=}C(H)SCH_3 \rightarrow CH_3SC\bullet(H)SCH_3 + S{=}C(Z)SCH_3$. Next, the chain-transfer efficiency for a given R group is measured by the enthalpy change ($\Delta H_{CT}$) for the reaction $CH_3\bullet + S{=}C(CH_3)SR \rightarrow S{=}C(CH_3)SCH_3 + R\bullet$. The stabilities of RAFT agents with different Z groups are compared by considering the enthalpies of the reactions $S{=}C(Z)SCH_3 + CH_4 \rightarrow S{=}C(H)SCH_3 + CH_3Z$, while the stabilities of RAFT agents with different R groups are compared by the reactions $S{=}C(CH_3)SR +$ $CH_3SH \rightarrow S{=}C(CH_3)SCH_3 + RSH$. The well-known quantity, the radical stabilization energy (RSE) [19], is used to estimate the stabilities of the radical species. It should be noted that the use of radical stabilization energies to compare the stabilities of two radicals R• and R′• is only meaningful if the discrepancy between the R—H and R′—H bond strengths is negligible. In the case of carbon-centered radicals, this assumption is normally reasonable but minor discrepancies can arise due to noncanceling steric or polar interactions in the closed shell compounds. More generally, in other types of radical, such as phosphoranyl radicals, this assumption can break down altogether. For a more detailed discussion of this problem, see, for example, Ref. [115].

In the overall chain-transfer reaction (equation 2.13), the effect of Z involves both the RAFT agent (and the poly-RAFT agent) and the intermediate RAFT-adduct radical. One must therefore consider each addition–fragmentation step separately. In Fig. 2.5, a variety of Z groups are ranked in order of increasing fragmentation efficiency $-\Delta H_{frag}$. Shown on the same graph are the stabilities of the two relevant Z-containing species: the RAFT agent and the RAFT-adduct radical. The fragmentation efficiencies span a range of 100 kJ·mol$^{-1}$, making it clear that the choice of Z is critical. We have suggested [98] that in considering the effects of Z, RAFT agents may be divided into three broad classes based on their values of $\Delta H_{frag}$. In one class are the agents with Z groups favoring fragmentation (Z = OR or $NR_2$), in the second are those with low fragmentation efficiencies (Z = CN, Ph or $CF_3$) and in the third are those with moderate efficiencies (Z = SR, and modified OR or $NR_2$). As will be discussed below, the different classes are of varying utility depending on the monomer of interest.

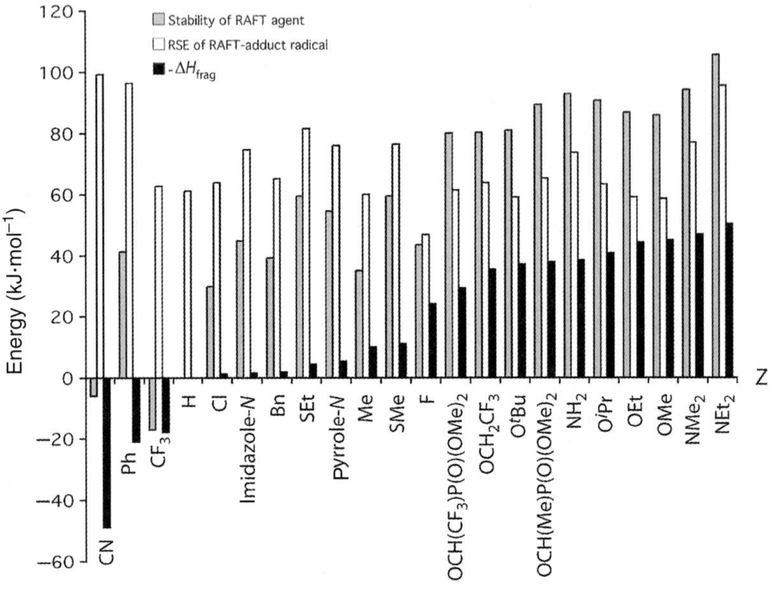


**Fig. 2.5** Effect of Z group on fragmentation efficiency, RAFT agent stability and RAFT-adduct radical stability [42].

For the RAFT agents (and poly-RAFT agents), there are two main ways in which Z affects stability. First, lone-pair donor Z groups such as OR, $NR_2$ and SR enhance stability, through delocalization of electron density into the C=S bond as shown in equation 2.14.

(2.14)

Second, RAFT agents are destabilized by $\sigma$-withdrawal. This property can be clearly seen with the RAFT agents that have Z = CN or $CF_3$. Within this stability scheme, the positions of several Z groups are worthy of note. For example, the groups Z = pyrrole and Z = imidazole confer much less stabilization on a RAFT agent than do aliphatic amino groups such as $NMe_2$. This is consistent with the notion that incorporating the nitrogen lone pair into an aromatic system reduces the capacity for electron delocalization according to equation 2.14 [116].

For RAFT-adduct radicals, the effects of Z are somewhat more complicated. Here, a major increase in stability is gained if Z is a $\pi$-acceptor group such as CN or Ph. Stabilization by $\pi$-acceptor groups is common for carbon-centered radicals, and in RAFT-adduct radicals the effect can be enhanced because the lone-pair donor SR groups can engage in captodative effects. In principle, the presence of a lone-pair donor Z group should be an alternative stabilizing feature (as in other carbon-centered radicals), but in RAFT-adduct radicals this is not always borne out. We have shown [19] that the delocalization of electron density from an SR group onto a carbon radical center places the unpaired electron into a higher-energy orbital, making further delocalization onto a second SR group much less favorable. A RAFT-adduct radical has two lone-pair donor SR groups even before Z is considered, and the capacity for a third interaction involving a lone-pair donor Z group is therefore small. Only when Z is a stronger lone-pair donor than SR will there be enhanced stabilization. Making the situation more complicated, RAFT-adduct radicals are strongly destabilized by $\sigma$-withdrawal (as demonstrated when Z = F), which means that only those Z groups for which lone-pair donation is stronger than $\sigma$-withdrawal (and stronger than the lone-pair donation by an SR group) will confer enhanced stabilization. In Fig. 2.5, the only Z groups that satisfy both criteria are the simple amino groups $NR_2$ (R = H, Me, Et). The alkoxy groups, although being strong lone-pair donors, are unable to enhance stability, because of their strong $\sigma$-withdrawing character.

Considering next the effects of R, we have found that it is usually sufficient to consider the overall chain-transfer equilibrium without giving separate attention to the individual steps. Several earlier computational studies have also adopted this approach. For example, semiempirical methods have been used to show that the chain-transfer efficiency of dithiobenzoate RAFT agents varied with R in the order $CH_2COOEt < CH_2COOH < CH(Me)COOEt$ [21]. A related study [20] of a wide range of R groups using DFT reached similar conclusions, suggesting that chain transfer is affected by both steric and polar effects. The overall chain-transfer

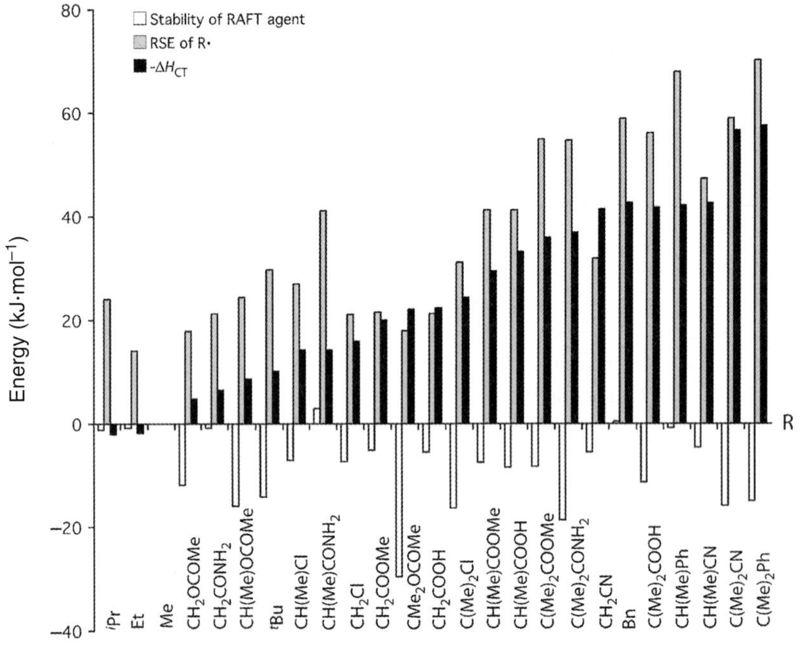


**Fig. 2.6** Effect of R on RAFT agent stability, R• stability and chain-transfer efficiency [42].

equilibrium is, of course, a competition between the radicals R• and $P_n$• for adding to the C=S bond. In most cases, the position of the equilibrium is determined mainly by the relative stabilities of the radicals R• and $P_n$•: the group whose release from the RAFT-adduct radical is favored is the one whose RSE is greatest. The values of $\Delta H_{CT}$ for a range of common R groups are shown in order of increasing chain-transfer propensity in Fig. 2.6, together with the stabilities of the relevant RAFT agents and radicals R•.

Considering first the R• radicals, the RSEs can be largely explained on the basis of factors normally associated with stability in carbon-centered radicals. Thus, the presence of $\pi$-acceptor groups (such as CN or Ph) as $\alpha$-substituents within R leads to enhanced stability, as does the capacity for hyperconjugative interactions provided by $\alpha$-$CH_3$ substituents. On the other hand, in the RAFT agents, the presence of $\alpha$-$CH_3$ groups or $\pi$-acceptor $\alpha$-substituents in R results in destabilization. Here, methylation primarily induces unfavorable steric interactions. The $\pi$-acceptor groups destabilize the RAFT agents by reducing the capacity for delocalization of the sulfur lone pair onto the double bond (equation 2.15).

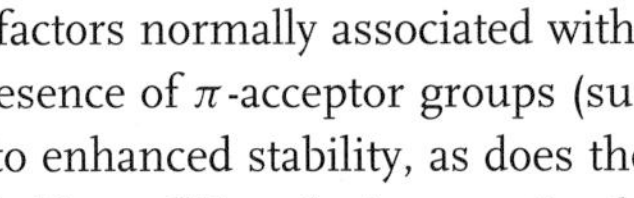


(2.15)

The steric and electronic properties of R therefore influence chain transfer in two reinforcing ways – one effect on the RAFT agent and the opposite effect on the R• radical. These combined effects render the commonly used R groups $CH(CH_3)Ph$, $C(CH_3)_2CN$ and $C(CH_3)_2Ph$ very good leaving group radicals. However, it is not correct to assume that a well-controlled polymerization will result if one simply chooses the best leaving group $R = C(CH_3)_2Ph$. If R• is *too* stable compared with the monomer, then the subsequent reinitiation step is disfavored, and polymerization will not proceed past the stage of the unimeric RAFT agent. When faced with a new polymerization, a sensible initial approach would therefore be first to take into account the stability of the propagating radical (which is roughly measured by the stability of the monomeric radicals shown in Fig. 2.6, and could be better measured using calculations on the dimers), and then choose an R group for which R• is only slightly more or less stable than M•.

The simplifications involved in deriving the structure–reactivity relationships just described are accompanied, of course, by a number of exceptional cases. One important case is where a system does not obey the usual correlation between kinetics and thermodynamics. For example, by using computational investigations, we have recently found [17] that in the xanthate-mediated polymerization of vinyl acetate, the rate constant for fragmentation of the RAFT-adduct radical $CH_3SC{\bullet}(Z)SCH_2OCOCH_3$ ($Z = OCH_3$) is 3 orders of magnitude larger than that for the radical where $Z = O^tBu$. This result cannot be due to solely to the steric effect of Z, however, because for the related RAFT-adduct radicals $CH_3SC{\bullet}(Z)SCH_3$, the rate constants for $Z = OCH_3$ and $Z = O^tBu$ are nearly identical. The source of the discrepancy lies instead in a hydrogen-bonding interaction in the transition state for fragmentation of $CH_3SC{\bullet}(OCH_3)SCH_2OCOCH_3$. A close contact takes place between the carbonyl oxygen and an $OCH_3$ hydrogen, leading to significant stabilization, whereas no analogous interactions take place for the other three species.

A breakdown of the structure–reactivity trends can also occur if there are synergistic interactions between the groups R, Z and $P_n$. For example, we have shown [19] that fragmentation of electron-withdrawing R groups is enhanced by a homoanomeric effect in which the withdrawal of electron density from sulfur toward the R group reduces the ability of sulfur to donate electron density to the radical center. If the R group contains a $\pi$-acceptor substituent, this induces an altered CS—R conformation in which the unpaired electron is delocalized into a CS—R antibonding orbital. This not only weakens the S—R bond but is accompanied by a strengthening of the S—$P_n$ bond in compensation. As a result, altered chain-transfer efficiencies arise that would not have been predicted on the basis of R• stabilities alone. Moreover, the homoanomeric interaction is further modulated by the nature of Z: for example, a fluorine substituent in the Z position can inhibit the effect of a $\pi$-acceptor-containing R group [24]. This example, and the vinyl acetate example given above, emphasizes the importance of conducting specific computational or experimental investigations once one has selected a range of candidate RAFT agents for a particular application.

### 2.4.3
### RAFT Agent Design

The ultimate goal of computational studies of RAFT polymerization is to design new and improved RAFT agents to tackle any specific control problem. One could envisage a two-stage approach in which one first utilizes the structure–reactivity relationships to select promising combinations of substituents, and then tests the candidate RAFT agents with direct calculations using model propagating radicals. Having established computationally that a certain RAFT agent was likely to be successful, one could then pursue experimental testing.

The kinetic requirements of a successful RAFT agent are now fairly well understood: it should have a reactive C=S bond ($k_{add}$ high) but the intermediate radical should undergo fragmentation at a reasonable rate (typically $K = k_{add}/k_{\beta} < 10^6$ L·mol$^{-1}$); the R group should fragment preferentially from the intermediate radical in the preequilibrium but also be capable of reinitiating polymerization; and there should be no side reactions [1, 103]. Computational data shown in Tables 2.2 and 2.3 and in Figs. 2.5 and 2.6 can assist in choosing RAFT agents that satisfy these criteria. For example, if one predicted that a propagating radical was relatively stable, then, in order to provide good control, its addition to the C=S bond would need to be fast enough to compete with addition to monomer. One could then select a Z group that destabilized the C=S bond, promoting addition (i.e. $\Delta H_{frag}$ in Fig. 2.5 is large). By contrast, if one predicted that the propagating radical was relatively unstable, then it would be necessary to ensure that the propagating radical did not become trapped for too long as the RAFT-adduct radical. In this case, one would choose a Z group that destabilized the RAFT-adduct radical relative to the C=S compound (i.e. $\Delta H_{frag}$ is small). Amongst substituents displaying acceptable $\Delta H_{frag}$ values, one could further optimize control by choosing the Z group that gave the lowest RAFT agent stability, and hence the fastest addition rates. Likewise, one could use $\Delta H_{CT}$ values (as in Fig. 2.6) to choose the R group – an optimal group usually having a $\Delta H_{CT}$ value marginally lower than that of the model propagating radical. Having selected candidate RAFT agents on the basis of the simplified isodesmic measures, they could then be tested computationally through direct calculation of the addition–fragmentation kinetics in more realistic model systems.

There is a growing interest in using computational calculations as a basis for RAFT agent design. Matyjaszewski and Poli [20] and Farmer and Patten [21] have advocated the use of calculated chain transfer energies (equation 2.13) as a basis for choosing appropriate R groups for specified monomers; Matyjaszewski and colleagues [20, 25, 26] and Tordo et al. [28] have applied similar strategies to atom transfer radical polymerization and nitroxide-mediated polymerization, respectively. Building on this work, the first computationally designed class of RAFT agents, the fluorodithioformates (S=C(F)SR or 'F-RAFT'), was recently proposed [23]. These were designed as multipurpose RAFT agents capable of controlling monomers with stable propagating radicals and also those with unstable propagating radicals. Other current RAFT agents for unstable monomers promote fragmentation of the propagating radicals by stabilizing the thiocarbonyl product of fragmentation (i.e. the RAFT agent). As a result, they are not sufficiently reactive for controlling monomers with stable propagating radicals. In contrast, F-RAFT promotes fragmentation by destabilizing the RAFT-adduct radical without deactivating the C=S bond of the RAFT agent. The fluorine Z group achieves this through its strong $\sigma$-withdrawing capacity, which contributes a destabilizing influence to both the radical and the C=S bond, and helps to counterbalance the stabilizing influence of its lone-pair donor capacity. It can be seen in Fig. 2.5 that the stability of the fluorinated RAFT agent is comparable to other dithioesters, and much lower than that of the xanthates. On this basis, it is predicted that the RAFT agent is sufficiently reactive for controlling stable propagating radicals, and calculated addition rate coefficients ($1.8 \times 10^6$ L·mol$^{-1}$·s$^{-1}$ for a styrene unimer adding to $CH_3SC(F)=S$ at 333.15 K) seem to confirm this [101]. Although the reduced stability renders fragmentation less facile, this is counterbalanced to some extent by the reduced stability of the RAFT-adduct radical. As a result, the equilibrium constant for fragmentation of a model vinyl acetate radical ($3.4 \times 10^4$ L·mol$^{-1}$ at 333.15 K) [101], though higher than that for xanthates, is still considerably lower than the threshold above which rate retardation occurs (ca. $10^6$ L·mol$^{-1}$ [103]). The computational calculations thus predict that F-RAFT should function as a multipurpose RAFT agent.

Experimental testing has already indicated that benzyl fluorodithioformate is capable of controlling styrene polymerization, though testing of vinyl acetate (and indeed other 'unstable' monomers such as ethylene) is still underway [24]. For these monomers, RAFT agents with alternative leaving groups are required; benzyl fluorodithioformate is not likely to be suitable for controlling vinyl acetate polymerization, due to the slow rate of reinitiation by the benzyl leaving group. Instead, based on further computational calculations [101], we predict that a RAFT agent with a $CH_2CN$ leaving group would be more suitable, and this is now being prepared and tested by experimental collaborators. With successful practical applications such as this, it is clear that computational quantum chemistry is showing promise as a kinetic tool for studying complicated radical polymerization processes such as RAFT.

## 2.5 Abbreviations

| | |
|---|---|
| AIBN | 2,2′-Azobis(isobutyronitrile) |
| B3-LYP | A DFT method with the Becke's B3 exchange and the correlation functional of Lee, Yang, and Parr |
| CDB | Cumyl dithiobenzoate |
| DFT | Density functional theory |
| G3 | Gaussian-3; a family of high-level composite methods, e.g. G3, G3B3, G3(MP2) |

| | |
|---|---|
| G3-RAD | A family of G3 methods designed for radical chemistry, e.g. G3(MP2)-RAD |
| HF | Hartree–Fock |
| MP2 | Second-order Møller–Plesset perturbation theory |
| ONIOM | Our own N-layered integrated molecular orbital + molecular mechanic |
| PCM | Polarizable continuum model; a method for incorporating solvent effects |
| RAFT | Reversible addition–fragmentation chain transfer |
| VB | Valence bond (theory) |
| W1 | Wiezmann-1; a high-level composite method |

## References


1. G. Moad, E. Rizzardo, S. H. Thang, *Aust. J. Chem.* **2005**, *58*, 379–410.
2. S. Z. Zard, *Aust. J. Chem.* **2006**, *59*, 663–668.
3. A. Feldermann, M. L. Coote, M. H. Stenzel, T. P. Davis, C. Barner-Kowollik, *J. Am. Chem. Soc.* **2004**, *126*, 15915–15923.
4. Y. Kwak, A. Goto, T. Fukuda, *Macromolecules* **2004**, *37*, 1219–1225.
5. C. Barner-Kowollik, M. Buback, B. Charleux, M. L. Coote, M. Drache, T. Fukuda, A. Goto, B. Klumperman, A. B. Lowe, J. B. Mcleary, G. Moad, M. J. Monteiro, R. D. Sanderson, M. P. Tonge, P. Vana, *J. Polym. Sci. A* **2006**, *44*, 5809–5831.
6. P. A. M. Dirac, *Proc. Roy. Soc. Lond.* **1929**, *123*, 714.
7. E. I. Izgorodina, M. L. Coote, *Chem. Phys.* **2006**, *324*, 96–110.
8. M. L. Coote, G. P. F. Wood, L. Radom, *J. Phys. Chem. A* **2002**, *106*, 12124–12138.
9. M. L. Coote, *J. Phys. Chem. A* **2005**, *109*, 1230–1239.
10. E. I. Izgorodina, M. L. Coote, *J. Phys. Chem. A* **2006**, *110*, 2486–2492.
11. R. Gómez-Balderas, M. L. Coote, D. J. Henry, L. Radom, *J. Phys. Chem. A* **2004**, *108*, 2874–2883.
12. M. L. Coote, *J. Phys. Chem. A* **2004**, *108*, 3865–3872.
13. J. Chiefari, R. T. A. Mayadunne, C. L. Moad, G. Moad, E. Rizzardo, A. Postma, M. A. Skidmore, S. H. Thang, *Macromolecules* **2003**, *36*, 2273–2283.
14. Y. K. Chong, J. Krstina, T. P. T. Le, G. Moad, A. Postma, E. Rizzardo, S. H. Thang, *Macromolecules* **2003**, *36*, 2256–2272.
15. C. L. Moad, G. Moad, E. Rizzardo, S. H. Thang, *Macromolecules* **1996**, *29*, 7717–7726.
16. M. L. Coote, L. Radom, *J. Am. Chem. Soc.* **2003**, *125*, 1490–1491.
17. M. L. Coote, L. Radom, *Macromolecules* **2004**, *37*, 590–596.
18. M. L. Coote, *Macromolecules* **2004**, *37*, 5023–5031.
19. M. L. Coote, D. J. Henry, *Macromolecules* **2005**, *38*, 1415–1433.
20. K. Matyjaszewski, R. Poli, *Macromolecules* **2005**, *38*, 8093–8100.
21. S. C. Farmer, T. E. Patten, *J Polym. Sci. A Polym. Chem.* **2002**, *A40*, 555–563.
22. A. Alberti, M. Benaglia, H. Fischer, M. Guerra, M. Laus, D. Macciantelli, A. Postma, K. Sparnacci, *Helv. Chim. Acta* **2006**, *89*, 2103–2118.
23. M. L. Coote, D. J. Henry, *Macromolecules* **2005**, *38*, 5774–5779.
24. A. Theis, M. H. Stenzel, T. P. Davis, M. L. Coote, C. Barner-Kowollik, *Aust. J. Chem.* **2005**, *58*, 437–441.
25. D. A. Singleton, D. T. Nowlan, III, N. Jahed, K. Matyjaszewski, *Macromolecules* **2003**, *26*, 8609–8616.
26. M. B. Gillies, K. Matyjaszewski, P.-O. Norrby, T. Pintauer, R. Poli, P. Richard, *Macromolecules* **2003**, *36*, 8551–8559.
27. C. Y. Lin, M. L. Coote, A. Petit, P. Richard, R. Poli, K. Matyjaszewski, *Macromolecules*, **2007**, *40*, 5985–5994.

28. P. Marsal, M. Roche, P. Tordo, P. de Sainte Claire, *J. Phys. Chem. A* **1999**, *103*, 2899–2905.
29. D. Siri, A. Gaudel-Siri, P. Tordo, *J. Mol. Struct. Theochem.* **2002**, *582*, 171–185.
30. J. P. A. Heuts, A. Pross, L. Radom, *J. Phys. Chem.* **1996**, *100*, 17087–17089.
31. J. P. A. Heuts, R. G. Gilbert, L. Radom, *Macromolecules* **1995**, *28*, 8771–8781.
32. J. P. A. Heuts, R. G. Gilbert, I. A. Maxwell, *Macromolecules* **1997**, *30*, 726–736.
33. D. M. Huang, M. J. Monteiro, R. G. Gilbert, *Macromolecules* **1998**, *31*, 5175–5187.
34. J. Filley, J. T. McKinnon, D. T. Wu, G. H. Ko, *Macromolecules* **2002**, *35*, 3731–3738.
35. J. S.-S. Toh, D. M. Huang, P. A. Lovell, R. G. Gilbert, *Polymer* **2001**, *42*, 1915–1920.
36. K. Van Cauter, K. Hemelsoet, V. Van Speybroeck, M. F. Reyniers, M. Waroquier, *Int. J. Quantum Chem.* **2004**, *102*, 454–460.
37. C.-G. Zhan, D. A. Dixon, *J. Phys. Chem. A* **2002**, *106*, 10311–10325.
38. J. Purmova, K. F. D. Pauwels, W. van Zoelen, E. J. Vorenkamp, A. J. Schouten, M. L. Coote, *Macromolecules* **2005**, *38*, 6352–6366.
39. M. L. Coote, T. P. Davis, L. Radom, *Macromolecules* **1999**, *32* (16), 5270–5276.
40. M. L. Coote, T. P. Davis, L. Radom, *Macromolecules* **1999**, *32* (9), 2935–2940.
41. M. L. Coote, T. P. Davis, L. Radom, *Theochemistry* **1999**, *461–462*, 91–96.
42. M. L. Coote, E. H. Krenske, E. I. Izgorodina, *Macromol. Rapid Commun.* **2006**, *27*, 473–497.
43. A. P. Scott, L. Radom, *J. Phys. Chem.* **1996**, *100*, 16502–16513.
44. D. K. Malick, G. A. Petersson, J. A. Montgomery, *J. Chem. Phys.* **1998**, *108*(14), 5704–5713.
45. M. Schwartz, P. Marshall, R. J. Berry, C. J. Ehlers, G. A. Petersson, *J. Phys. Chem. A* **1998**, *102*(49), 10074–10081.
46. J. M. L. Martin, S. Parthiban. In *Quantum Mechanical Prediction of Thermochemical Data*, J. Cioslowski, Ed., Kluwer-Academic, Dordrecht, The Netherlands, **2001**, pp. 31–65.
47. D. J. Henry, C. J. Parkinson, L. Radom, *J. Phys. Chem. A* **2002**, *106*, 7927–7936.
48. D. J. Henry, M. B. Sullivan, L. Radom, *J. Chem. Phys.* **2003**, *118*, 4849–4860.
49. E. I. Izgorodina, M. L. Coote, L. Radom, *J. Phys. Chem. A* **2005**, *109*, 7558–7566.
50. C. E. Check, T. M. Gilbert, *J. Org. Chem.* **2005**, *70*, 9828–9834.
51. S. Grimme, *Angew. Chem. Int. Ed.* **2006**, *45*, 4460–4464.
52. M. D. Wodrich, C. Corminboeuf, P. V. R. Schleyer, *Org. Lett.* **2006**, *8*, 3631–3634.
53. T. Vreven, K. Morokuma, *Theor. Chem. Acc.* **2003**, *109*, 125–132.
54. E. I. Izgorodina, D. R. B. Brittain, J. L. Hodgson, E. H. Krenske, C. Y. Lin, M. Namazian, M. L. Coote, *J. Phys. Chem. A* **2007**, ASAP Article; DOI: 10.1021/jp075837w.
55. P. W. Atkins, *Physical Chemistry*, Oxford University Press, Oxford, **1990**.
56. M. L. Coote. In *Encyclopedia of Polymer Science and Technology*, J. I. Kroschwitz, Ed.; Wiley, New York, **2004**, pp. 319–371.
57. R. P. Bell. *The Tunnel Effect in Chemistry*, Chapman and Hall, London and New York, **1980**.
58. D. G. Truhlar, A. D. Isaacson, R. T. Skodje, B. C. Garrett, *J. Phys. Chem.* **1982**, *86*(12), 2252–2261.
59. D. G. Truhlar, B. C. Garrett, *J. Am. Chem. Soc.* **1989**, *111*(4), 1232–1236.
60. D. G. Truhlar, B. C. Garrett, S. J. Klippenstein, *J. Phys. Chem.* **1996**, *100*, 12771–12800.
61. S. Nordholm, G. B. Bacskay, *Chem. Phys. Lett.* **1976**, *42*, 253–258.
62. J. P. A. Heuts, R. G. Gilbert, L. Radom, *J. Phys. Chem.* **1996**, *100*, 18997–19006.
63. A. L. L. East, L. Radom, *J. Chem. Phys.* **1997**, *106*, 6655.
64. V. Van Speybroeck, D. Van Neck, M. Waroquier, *J. Phys. Chem. A* **2002**, *106*, 8945–8950.
65. D. G. Truhlar, B. C. Garrett, S. J. Klippenstein, *J. Phys. Chem.* **1996**, *100*, 12771–12800.
66. G. A. Voth, R. M. Hochstrasser, *J. Phys. Chem.* **1996**, *100*, 13034–13049.

67. J. Tomasi, *Theor. Chem. Acc.* **2004**, *112*, 184–203.
68. V. Barone, M. Cossi, *J. Phys. Chem. A* **1998**, *102*, 1995–2001.
69. S. Miertus, E. Scrocco, J. Tomasi, *J. Chem. Phys.* **1981**, *55*, 117.
70. Y. Takano, K. N. Houk, *J. Chem. Theory Comput.* **2005**, *1*, 70–77.
71. A. M. Magill, K. J. Cavell, B. F. Yates, *J. Am. Chem. Soc.* **2004**, *126*, 8717–8724.
72. J. R. Pliego, Jr., J. M. Riveros, *J. Phys. Chem. A* **2001**, *105*, 7241–7247.
73. B. O. Leung, D. L. Reid, D. A. Armstrong, A. J. Rauk, *Phys. Chem. A* **2004**, *108*, 2720, and references cited therein.
74. H. Fischer, L. Radom, *Angew. Chem. Int. Ed.* **2001**, *40*, 1340–1371.
75. R. M. Levy, D. B. Kitchen, J. T. Blair, K. Krogh-Jespersen, *J. Phys. Chem.* **1990**, *94*, 4470–4476.
76. M. S. Gordon, M. A. Freitag, P. Bandyopadhyay, J. H. Jensen, V. Kairys, W. J. Stevens, *J. Phys. Chem. A* **2001**, *105*, 293–307.
77. M. Ben-Nun, R. D. Levine, *Int. Rev. Phys. Chem.* **1995**, *14*, 215–270.
78. E. I. Izgorodina, M. L. Coote, *Macromol. Theory Simul.* **2006**, *15*, 394–403.
79. M. L. Coote, E. I. Izgorodina, E. H. Krenske, M. Busch, C. Barner-Kowollik, *Macromol. Rapid Commun.* **2006**, *27*, 1015–1022.
80. J. B. McLeary, F. M. Calitz, J. M. McKenzie, M. P. Tonge, R. D. Sanderson, B. Klumperman, *Macromolecules* **2004**, *37*, 2382–2394.
81. J. C. Scaiano, K. U. Ingold, *J. Am. Chem. Soc.* **1976**, *98*, 4727–4732.
82. A. Ah Toy, H. Chaffey-Millar, T. P. Davis, M. H. Stenzel, E. I. Izgorodina, M. L. Coote, C. Barner-Kowollik, *Chem. Commun.* **2006**, 835–837.
83. O. F. Olaj, I. Bitai, F. Hinkelmann, *Makromol. Chem.* **1987**, *188*, 1689.
84. A. R. Wang, S. Zhu, Y. Kwak, A. Goto, T. Fukuda, M. S. Monteiro, *J. Polym. Sci. A* **2003**, *41*, 2833–2839.
85. C. Barner-Kowollik, M. L. Coote, T. P. Davis, L. Radom, P. Vana, *J. Polym. Sci. A Polym. Chem.* **2003**, *41*, 2828–2832.
86. F. M. Calitz, J. B. McLeary, J. M. McKenzie, M. P. Tonge, B. Klumperman, R. D. Sanderson, *Macromolecules* **2003**, *36*, 9687–9690.
87. J. A. M. de Brouwer, M. A. J. Schellekens, B. Klumperman, M. J. Monteiro, A. German, *J. Polym. Sci. A* **2000**, *19*, 3596.
88. Y. Kwak, A. Goto, K. Komatsu, Y. Sugiura, T. Fukuda, *Macromolecules* **2004**, *37*, 4434–4440.
89. A. Ah Toy, P. Vana, T. P. Davis, C. Barner-Kowollik, *Macromolecules* **2004**, *37*, 744–751.
90. S. W. Prescott, M. J. Ballard, E. Rizzardo, R. G. Gilbert, *Macromolecules* **2005**, *38*, 4901–4912.
91. A. Feldermann, A. Ah Toy, M. H. Stenzel, T. P. Davis, C. Barner-Kowollik, *Polymer* **2005**, *46*, 8448–8457.
92. R. Venkatesh, B. B. P. Staal, B. Klumperman, M. J. Monteiro, *Macromolecules* **2004**, *37*, 7906–7917.
93. E. I. Izgorodina, C. Y. Lin, M. L. Coote, *Phys. Chem. Chem. Phys.* **2007**, *9*, 2507–2516.
94. J. B. McLeary, F. M. Calitz, J. M. McKenzie, M. P. Tonge, R. D. Sanderson, B. Klumperman, *Macromolecules* **2005**, *38*, 3151–3161.
95. M. L. Coote, T. P. Davis, *Prog. Polym. Sci.* **1999**, *24*(9), 1217–1251.
96. P. Zetterlund, W. Busfield, I. Jenkins, *Macromolecules* **1999**, *32*, 8041–8045.
97. M. Deady, A. W. H. Mau, G. Moad, T. H. Spurling, *Makromol. Chem.* **1993**, *194*, 1691.
98. E. H. Krenske, E. I. Izgorodina, M. L. Coote. In *Controlled/Living Radical Polymerization: From Synthesis to Materials*, K. Matyjaszewski, Ed., American Chemical Society, Washington, DC, **2006**, pp. 406–420.
99. M. Buback, O. Janssen, R. Oswald, S. Schmatz, P. Vana, *Macromol. Symp.* **2007**, *248*, 158–167.
100. D. J. Henry, M. L. Coote, R. Gómez-Balderas, L. Radom, *J. Am. Chem. Soc.* **2004**, *126*, 1732–1740.
101. M. L. Coote, E. I. Izgorodina, G. E. Cavigliasso, M. Roth, M. Busch, C. Barner-Kowollik, *Macromolecules* **2006**, *39*, 4585–4591.

**102.** J. B. McLeary, M. P. Tonge, B. Klumperman, *Macromol. Rapid Commun.* **2006**, *27*, 1233–1240.

**103.** P. Vana, T. P. Davis, C. Barner-Kowollik, *Macromol. Theory Simul.* **2002**, *11*, 823–835.

**104.** S. Perrier, C. Barner-Kowollik, J. F. Quinn, P. Vana, T. P. Davis. *Macromolecules* **2002**, *35*, 8300–8306.

**105.** M. Wulkow. *Macromol. Theory Simul.* **1996**, *5*, 393–416.

**106.** J. Krstina, G. Moad, E. Rizzardo, C. L. Winzor, C. T. Berge, M. Fryd. *Macromolecules* **1995**, *28*, 5381–5385.

**107.** A. Pross, S. S. Shaik, *Acc. Chem. Res.* **1983**, *16*, 363.

**108.** A. Pross, *Adv. Phys. Org. Chem.* **1985**, *21*, 99–196.

**109.** S. S. Shaik, *Prog. Phys. Org. Chem.* **1985**, *15*, 197–337.

**110.** M. W. Schmidt, P. H. Truong, M. S. Gordon, *J. Am. Chem. Soc.* **1987**, *109*, 5217–5227.

**111.** M. G. Evans, *Discuss. Faraday. Soc.* **1947**, *2*, 271–279.

**112.** D. Gigmes, D. Bertin, S. Marque, O. Guerret, P. Tordo, *Tetrahedron Lett.* **2003**, *44*, 1227–1229.

**113.** J. L. Hodgson, K. A. Green, M. L. Coote, *Org. Lett.* **2005**, *7*, 4581–4584.

**114.** R. Gómez-Balderas, M. L. Coote, D. J. Henry, H. Fischer, L. Radom, *J. Phys. Chem. A* **2003**, *107*, 6082–6090.

**115.** J. L. Hodgson, M. L. Coote, *J. Phys. Chem. A* **2005**, *109*, 10013–10021, and references cited therein.

**116.** R. T. A. Mayadunne, E. Rizzardo, J. Chiefari, Y. K. Chong, G. Moad, S. H. Thang, *Macromolecules* **1999**, *32*, 6977–6980.

# 3
# The Mechanism and Kinetics of the RAFT Process: Overview, Rates, Stabilities, Side Reactions, Product Spectrum and Outstanding Challenges

*Graeme Moad and Christopher Barner-Kowollik*

## 3.1
## History

RAFT is an acronym for reversible addition–fragmentation chain transfer. Radical addition–fragmentation processes have been exploited in synthetic organic chemistry since the early 1970s [1–3]. Allyl transfer reactions with allyl stannanes and the Barton–McCombie deoxygenation process with xanthates are just two examples. Chain transfer to monomer in vinyl chloride polymerization may be considered as the first reported example of addition–fragmentation chain transfer in a polymerization context. However, the first reports of the use of deliberately added addition–fragmentation transfer agents in controlling polymerization appeared in the late 1980s [4–10]. Polymerizations with reversible addition–fragmentation chain transfer which showed some of the characteristics of living polymerization were first reported in 1995 [11]. The macromonomer RAFT agents used in that work are described in section 3.1.2. The term RAFT polymerization was coined in 1998 when the use of thiocarbonylthio RAFT agents was first reported in the open literature [12].

Unsaturated compounds with general structure 1 may act as transfer agents by a two-step addition–fragmentation mechanism, as shown in Scheme 3.1. In structure **1**, C=X is a reactive double bond (X is most often methylene or sulfur). Z is a group chosen to give the transfer agent **1** (and **3**) an appropriate reactivity with respect to the monomer(s) and stability for the intermediate (**2**). In this chapter we are primarily concerned with agents that provide living characteristics to the polymerization. That is, the polymers formed by the RAFT process retain the propensity for chain growth given the presence of monomer and a source of free radicals.

Thus, an essential feature of RAFT is that the product of chain transfer (**3**) is also a RAFT agent and, for living characteristics to be associated with the polymerization,

*Handbook of RAFT Polymerization.* Edited by Christopher Barner-Kowollik

ISBN: 978-3-527-31924-4

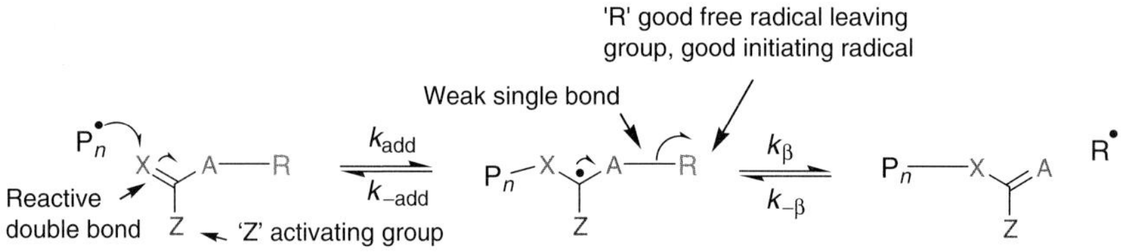


1 2 3

**Scheme 3.1** Reversible addition–fragmentation chain transfer (RAFT) mechanism.

this transfer agent must have activity that is similar to or greater than that of the precursor transfer agent (**1**). For this reason, A will also usually be methylene or sulfur and identical to X. The RAFT process has also been termed a degenerate or degenerative chain transfer since the polymeric starting materials and products have equivalent properties and differ only in molecular weight.

In Scheme 3.1, R is a radical leaving group and must be effective in reinitiating polymerization for chain transfer to be effective and R of the initial RAFT agent should be rapidly converted to a propagating species.

In proposing the RAFT mechanism, several alternative schemes whereby compounds of generic structure **1** might provide living characteristics were considered. These mechanisms include (a) reversible addition–fragmentation, (b) reversible homolytic substitution chain transfer and (c) reversible coupling.

The reversible addition–fragmentation mechanism is shown in Scheme 3.2. There is some historical precedent for this process in that certain 1,1-disubstituted ethylenes that are slow to propagate can act as deactivation agents [13]. Examples include diphenylethylene derivatives (**1**, X = $CH_2$; Z, A-R = aryl) [14–17] and certain captodative monomers (**1**, X = $CH_2$; A-R = *S*-butyl; Z = CN) [18]. In these examples the substituents Z and A-R activate the double bond to radical addition and provide stability for the adduct radical (**4**) and R is a poor homolytic leaving group. The kinetics of polymerization in the presence of reagents that provide deactivation by reversible addition–fragmentation should show similarities to those of nitroxide-mediated polymerization (NMP) and other processes that provide deactivation by reversible coupling. Note that polymeric product will be different from that derived from the RAFT mechanism shown in Scheme 3.1 since $P_n$ are always attached to atom X and never to atom A. To obtain living characteristics by this mechanism, it

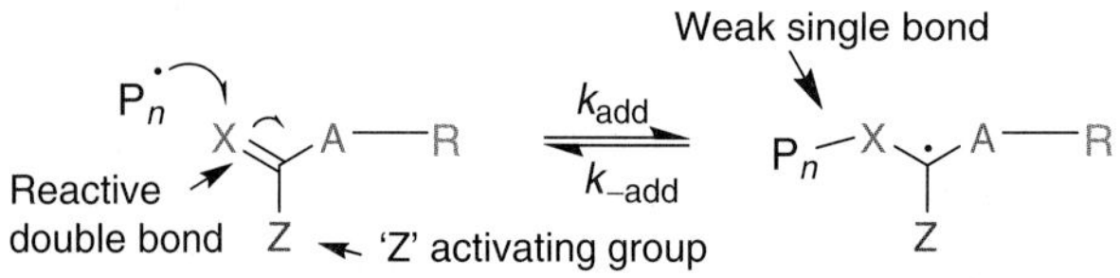


1 4

**Scheme 3.2** Reversible addition–fragmentation mechanism.

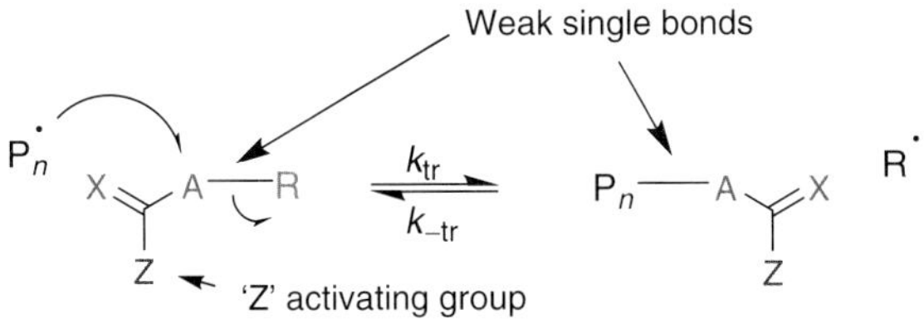


**1** **5**

**Scheme 3.3** Reversible homolytic substitution mechanism.

is a requirement that the dormant species are maintained in a high concentration. For the examples mentioned above, it is also observed that the 'stable radical' (4) undergoes reversible coupling. Recently, it has been suggested that the control exerted by certain thioketones (X = S; A-R, Z = alkyl or aryl) over free radical polymerizations may proceed via such a mechanism [19–21].

Scheme 3.3 shows a process that involves reversible homolytic substitution chain transfer. One feature distinguishing the products of this process from that shown in Schemes 3.1 and 3.2 is that R and $P_n$ are always attached to A, never to X. However, since A and X are normally identical, this is not easily established but might be detected by isotopic or other labeling of A and/or X. Iodine-transfer polymerization [22] and controlled polymerizations mediated by organoselenides, tellurides [23] and stibines [24] are believed to provide reversible homolytic substitution. Chain transfer to disulfides may proceed by this mechanism and the process was thought at one stage to be involved in the control of polymerization by dithiocarbamate photoiniferters. If it occurs during these polymerizations, it is only a minor pathway. It was also proposed that such a process might be involved in controlling polymerization via NMP [25].

A fourth process to be considered is reversible coupling–dissociation (Scheme 3.4). This is the main mechanism thought to be involved in the control of polymerization by dithiocarbamate and xanthate photoiniferters. Most species considered as RAFT agents are thermally stable at temperatures usually encountered in polymerization processes (<150 °C) and thus this mechanism can be discounted. While there are reports that tertiary dithiobenzoate [26, 27] and cyanoisopropyl 1*H*-pyrrole-1-carbodithioate [28] undergo slow dissociation at lower temperatures, the process is still likely to be, at best, a minor pathway.

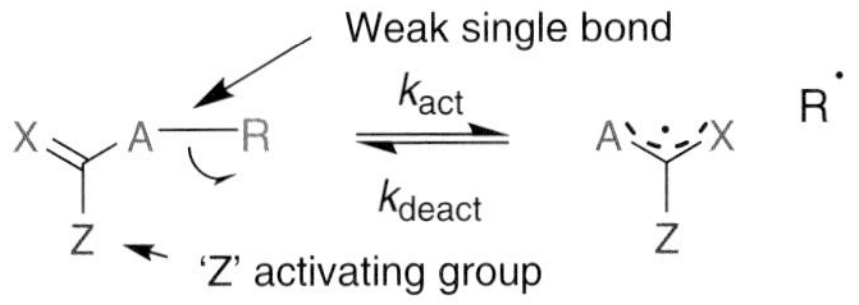


**1** **6**

**Scheme 3.4** Reversible coupling–dissociation mechanism.

While it should not be concluded that the processes described in Schemes 3.2–3.4 do not operate, a variety of experimental and theoretical evidence attests to addition–fragmentation chain transfer being dominant mechanism in controlling polymerization in the RAFT process.

### 3.1.1
### Macromonomer RAFT Agents

Certain macromonomers (where X is $CH_2$) have been known as transfer agents in radical polymerization since the mid-1980s [4]. However, radical polymerizations which involved a reversible (degenerate) chain transfer step for chain equilibration and which displayed some of the characteristics associated with living polymerization were not reported until 1995 [11, 29]. One of the best-known examples of macromonomer RAFT agents is the methyl methacrylate (MMA) macromonomer **7**.

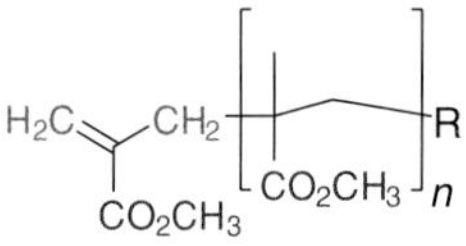


**7**

The term macromonomer is a misnomer since, in acting as a control agent, it does not function as a monomer. Rates of addition to transfer agents (where X is $CH_2$) are similar to and are determined by the same factors that determine the rates of addition to monomers. The double bonds of the transfer agents often have a reactivity toward propagating radicals that is comparable with that of the common monomers they resemble, and with efficient fragmentation, transfer constants can be close to unity. The radicals formed by addition should have a low reactivity toward further propagation and other intermolecular reactions because of steric crowding about the radical center. For this reason, the macromonomer RAFT agents are most used, and are most effective, in polymerization of methacrylic monomers. Copolymerization of the macromonomer is a significant complication in polymerizations of monosubstituted monomers.

Efficient transfer requires that the radicals formed by addition undergo facile $\beta$-scission to form a new radical that can reinitiate polymerization. The driving force for fragmentation of the intermediate radical is provided by cleavage of a weak $CH_2$—R bond and/or formation of a strong $C{=}CH_2$ bond. If fragmentation partitions in favor of the starting materials then the transfer constant will be low, as they are for the case of the MMA dimer (**7**, $n = 1$, R = H) [30] and other methacrylate dimers in MMA polymerization [31]. If the overall rate of $\beta$-scission is slow relative to propagation, then retardation is the likely result. The adducts then have greater potential to undergo side reactions by addition (e.g. copolymerization of the transfer agent) or radical–radical termination. Retardation is an issue, particularly for high-$k_p$ monomers such as vinyl acetate (VAc) and methyl acrylate (MA). In designing transfer agents and choosing an R group, a balance must be achieved between the leaving group ability of R and reinitiation efficiency by R•.

## 3.1.2
## Thiocarbonylthio RAFT Agents

The use of thiocarbonylthio RAFT agents (where X and A are both sulfur) to control polymerization was introduced in 1998 when patents covering the process were published by CSIRO [32] and shortly afterward by Rhodia [33], and the first paper on RAFT polymerization appeared in the open literature [12]. Since that time the literature on RAFT polymerization with thiocarbonylthio compounds has greatly expanded as evidenced by recent reviews, which include those by Moad and Solomon [34], Moad and coworkers [35–37], Mayadunne and Rizzardo [38], Chiefari and Rizzardo [39], Barner-Kowollik and coworkers [40], Perrier and Takolpuckdee [41] as well as Favier and Charreyre [42]. Many other reviews deal with specific applications of RAFT polymerization.

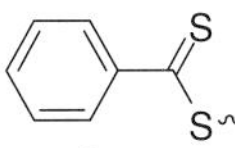

**8**

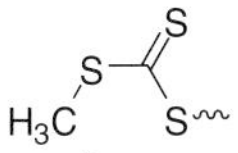

**9**

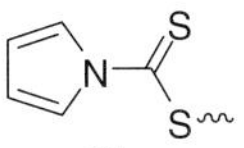

**10**

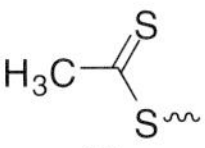

**11**

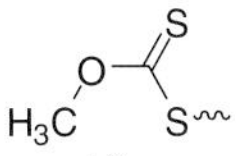

**12**

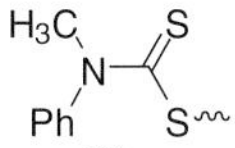

**13**

Rate coefficients for addition to thiocarbonyl groups of these RAFT agents depend strongly on the substituent Z and are typically several orders of magnitude higher than those for analogous carbon–carbon double bonds of the macromonomer RAFT agents mentioned above. As a consequence, transfer constants and the potential for precisely controlling radical polymerization are also higher. Effective control can be achieved with most monomers polymerizable by radical polymerization. Thiocarbonylthio compounds used in this context include aromatic (**8**) and aliphatic dithioesters (**11**), trithiocarbonates (**9**), xanthates (**12**) and dithiocarbamates (**10**, **13**). Rate constants for addition and transfer constants of the RAFT agents increase in the series **8** > **9** > **10** > **11** > **12** > **13** and the potential for retardation tends to decrease in the same series (see later discussion) [28, 43]. Factors governing the choice of R are similar to those already mentioned for macromonomer RAFT agents requiring that a balance between the leaving group ability of R and reinitiation efficiency by R• be achieved [44].

## 3.1.3
## Kinetic Features of RAFT Polymerizations: Phenomenological Observations

Radicals are neither formed nor destroyed as a consequence of the RAFT equilibria. Once a steady state is established, the RAFT agent can behave as an ideal transfer agent such that its presence in a polymerization medium does not affect the rate of polymerization. The RAFT process does not rely on a persistent radical effect to induce living characteristics [45]. In this respect, the control mechanism in

RAFT is fundamentally different from that associated with NMP or ATRP (atom transfer radical polymerization), in which retardation is inherent and the dormant species is also the polymerization initiator. Thus, the RAFT mechanism does not by itself always induce any inherent rate retardation, provided that the fragmentation of the RAFT adduct radical and subsequent reinitiation are rapid and not rate determining.

The RAFT process generates short propagating chains that progressively increase in size as the conversion of monomer increases. Therefore the chain length dependence of the kinetic rate coefficients (in particular, of termination and propagation) can affect the rate of polymerization and cause the polymerization kinetics to be different from that of an analogous conventional radical polymerization. For example, it is established that the (diffusion-controlled) termination rate coefficient is chain length dependent and increases with decreasing length of the terminating macroradicals [46]. For low-monomer-conversion RAFT polymerizations, the average chain length of the propagating macroradicals will be short. As a consequence, termination rate coefficients should be higher than in conventional radical polymerization for the same monomer conversion and the rate of polymerization will also therefore be lower. Indeed, Barner-Kowollik and coworkers [47–51] have made use of this and employed accurate rate measurements during RAFT polymerizations to deduce chain-length-dependent termination rate coefficient (see Chapter 4 for a detailed account on employing the RAFT process as a kinetic tool).

Notwithstanding the above discussion, there are situations where the addition of a RAFT agent to a polymerizing mixture has a profound effect on the reaction kinetics outside of that induced by the above-mentioned phenomena caused by chain length dependencies of rate coefficients. Typically, these rate effects are manifest either as

- an inhibition period or a period of slow polymerization where no or extremely little polymerization activity is observed over a defined period of time at the beginning of the polymerization; or
- rate retardation, where the rate of polymerization is reduced compared to the polymerization that does not contain the RAFT agent possibly for the duration of the experiment [52].

In many cases the inhibition phenomena are associated with the preequilibrium, otherwise known as initialization, where the initial RAFT agent is converted to a polymeric RAFT agent (further details appear in section 3.2.4.2). In this context it is important to differentiate between inhibition periods that are caused by the RAFT agent itself through its interactions with (propagating or initiator-derived) radicals and those that are caused by impurities that either arise in low quantities as by-products of RAFT agent synthesis or are formed subsequently as decomposition products of the RAFT agent. One of the most studied RAFT agents is cumyl dithiobenzoate (CDB), which carries a stabilizing phenyl Z group and a cumyl R leaving group. This RAFT agent is known to cause inhibition and its use gives rate retardation phenomena with a range of monomers. The effect of addition of CDB to

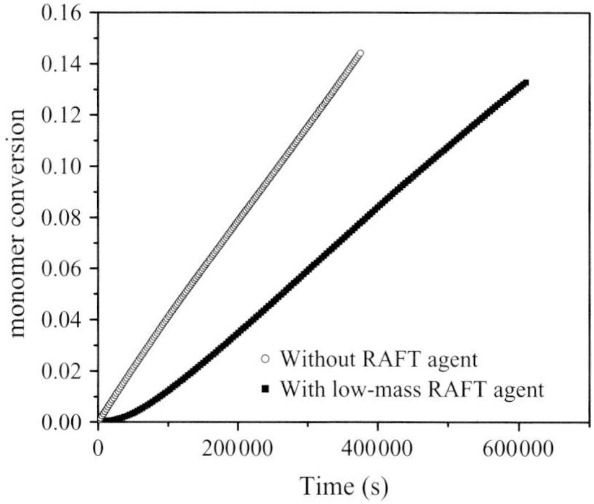


**Fig. 3.1** Evolution of the monomer to polymer conversion with reaction time in the γ-radiation-initiated radical polymerization of styrene in bulk at 30 °C for initial CDB concentrations of $[CDB]_0 = 0$ mol·L$^{-1}$ (open circles) and $[CDB]_0 = 5.5 \times 10^{-3}$ mol·L$^{-1}$ (full squares). Reproduced with kind permission from Ref. [53]. Copyright 2004 American Chemical Society.

a low-temperature polymerization of styrene is illustrated in Fig. 3.1 with reference to the corresponding non-RAFT system. As expected, the rate of polymerization in the absence of the RAFT agent is constant with time as a steady state is rapidly established. However, in the RAFT polymerization the rate of polymerization is initially very small with reference to the non-RAFT polymerization and steadily increases. A period of non-steady-state polymerization is observed [53]. Depending on the reaction temperature and other reaction conditions, the rate of polymerization of the RAFT and non-RAFT systems can appear identical for longer reaction times. Thus, for a higher reaction temperature (>60 °C) and similar RAFT agent concentration, it has been shown that CDB induces no discernable retardation following consumption of the initial RAFT agent [44]. The exact cause of the initially slower rate of polymerization is still subject of conjecture and there are a number of possible contributing factors. These include the slow fragmentation of the RAFT adduct radicals formed during the preequilibrium, termination of the adduct radicals with themselves or propagating radicals and slow reinitiation. These possibilities will be discussed in Section 3.2 in detail. In extreme cases, the initial rate of polymerization may be so slow as to cause complete inhibition.

One way of completely avoiding issues associated with the preequilibrium step and the process of initialization is to employ a polymeric RAFT or macro-RAFT agent, e.g. with the aim of generating block copolymers. Rate retardation may still be observed but often without the pronounced initial period of non-steady-state polymerization. Such main-equilibrium rate retardation phenomena have sometimes been rationalized via the concept of fast fragmenting intermediate radicals and their rapid termination with propagating macroradicals. This should lead to a steady decrease in the RAFT end group concentration and to the formation of significant amounts of terminated (three- and four-arm star) by-products. In both cases, the situation can be improved by employing a RAFT agent with a more reactive C=S double bond; e.g., by the use a trithiocarbonate or dithiobenzoate.

It is important to note that RAFT agents most prone to causing rate retardation and inhibition effects are those carrying Z groups that most effectively stabilize the adduct radicals. Notable amongst these are dithiobenzoate (Z = phenyl) and other aromatics (e.g. Z = naphthyl). In addition to the Z group, the R group plays an important role in determining the early (preequilibrium) polymerization kinetics, as the R group co-determines the stability of the preequilibrium adduct radical and must efficiently initiate macromolecular growth (see also section 3.2.4.6). A RAFT agent that carries a very poor leaving group (e.g. ethyl) or one that inefficiently reacts with monomer will either not control the polymerization process or induce strong inhibition phenomena.

A further key experimental observation from RAFT-mediated polymerization is sometimes called hybrid behavior. This behavior is manifest as a high initial molecular weight which then approaches the molecular weight calculated based on complete conversion of the RAFT agent as conversion increases. In addition, an initially broad polydispersity is observed which narrows with conversion. The effect is caused by the transfer constant of the initial RAFT agent being relatively low. A low transfer constant can be a consequence of the rate of addition ($k_{add}$) being slow or the intermediate partitioning in favor of starting materials. The behavior is a characteristic of macromonomer RAFT polymerization and is also observed with less active thiocarbonylthio compounds [44, 54]. The departure of the found molecular weight from that calculated assuming complete consumption of RAFT agent has been used as a method of estimating transfer constants (see section 3.2.1) (Fig. 3.2).

Examples include the use of a xanthate to mediate styrene polymerization [28] or the use of a dithioacetate to mediate the polymerization of MMA where linear increase in molecular weight with conversion is not observed and the polydispersities are relatively large [57]. The situation can be improved by employing a more reactive C=S double bond containing RAFT agent for styrene as well as for MMA, e.g. by increasing the stability of the result adduct radical via the use of dithiobenzoates. The underpinning kinetic details of the hybrid effect will be explored in greater depth in section 3.2.1.

Generally, the following experimental observations and approximate trends are valid:

(i) In situations (which are in the majority) where the RAFT adduct radicals are fast to fragment, the RAFT process should not to induce any rate retardation/inhibition effects.

(ii) The more reactive a propagating radical (i.e. the less reactive the monomer), the higher the equilibrium constants in both the pre- and main equilibrium. While dithiobenzoate RAFT agents are able to effectively mediate polymerization styrene or MMA at 60 °C under conditions of thermal initiation – albeit with some small rate retardation depending on the RAFT agent concentration – they almost completely inhibit the polymerization of acrylate esters (e.g. MA)

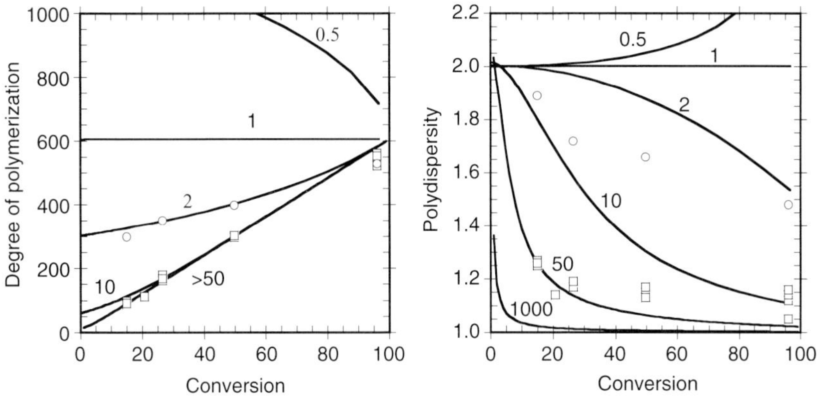


**Fig. 3.2** Predicted dependence of (a) polydispersity and (b) degree of polymerization on conversion in polymerizations involving reversible chain transfer as a function of the chain transfer constant. Predictions are based on equations proposed by Müller and coworkers [55, 56] with $\alpha = 10^{-7}$ (the concentration of active species), $\beta$ (the transfer constant) as indicated and $\gamma = 605$ (the ratio monomer to transfer agent). Experimental data points are for MMA polymerization in the presence of dithiobenzoate esters (0.0116 M) where R = $C(Me)_2CO_2Et$ (o) and $C(Me)_2Ph$ (□). Reproduced with kind permission from Ref. [44]. Copyright 2003 American Chemical Society.

at ambient temperatures [44]. Dithiobenzoate RAFT agents also control the polymerization of acrylates or acrylamides (e.g. MA); however, they also dramatically slow the rate of polymerization [44]. They inhibit the polymerization of vinyl esters (e.g. VAc).

(iii) For a given monomer, the potential for the occurrence of inhibition/rate retardation effects increases with an increasing stability of the RAFT adduct radical and inability of the leaving group to reinitiate the polymerization.

(iv) Less reactive propagating radicals (derived from more reactive monomers, e.g. styrene or MMA) require a RAFT agent with an increased C=S double bond reactivity to ensure that hybrid behavior is avoided (e.g. dithioester, trithiocarbonate).

In the following sections the above experimental observations will be rationalized in terms of a detailed analysis of the pre- and main equilibrium.

## 3.2 Preequilibrium Kinetics and Mechanism

The now well-known mechanism of RAFT polymerization as proposed in the first publication on this topic [12] is shown in Scheme 3.5. The step labeled reversible chain transfer in which the initial RAFT agent (**14**) is converted to an analogous polymeric species (**16**) is variously known as the preequilibrium or as initialization. The step labeled chain equilibration is also known as the main equilibrium.

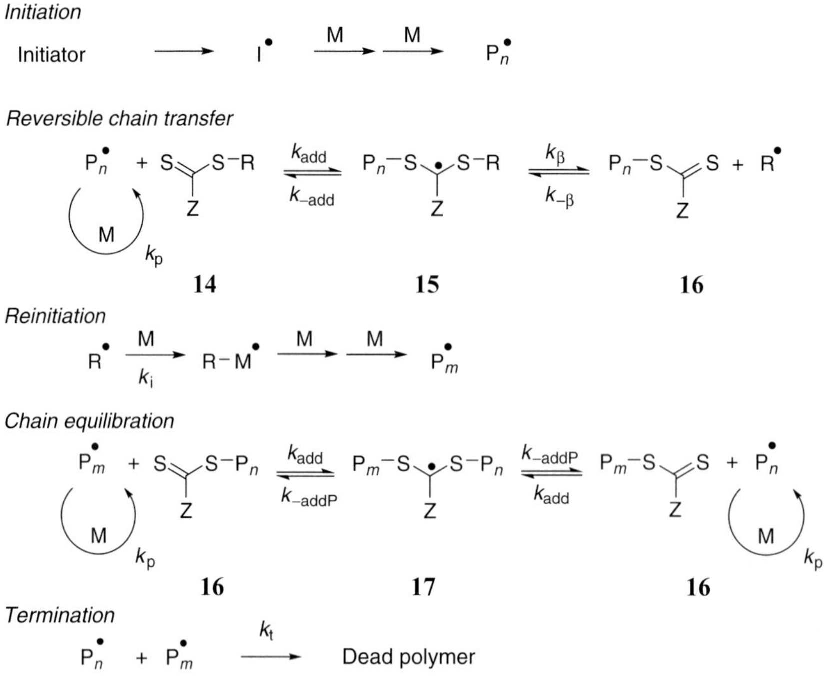


**Scheme 3.5** Mechanism of RAFT polymerization showing the steps of initiation, propagation, reversible chain transfer (also known as preequilibrium, initialization), reinitiation, chain equilibration (also known as main equilibrium) and termination. Note that the indicated rate coefficients are (potentially) chain length dependent.

## 3.2.1 Equilibrium Constants/Transfer Constants/Rate Coefficients

### 3.2.1.1 Equilibrium Constants

In a given RAFT polymerization, there are at least four equilibrium constants that need to be considered:

- $K$ (=$k_{add}/k_{-add}$) and $K_\beta = (k_{-\beta}/k_\beta)$ associated with the preequilibrium (Scheme 3.5);
- $K_P$ (=$k_{addP}/k_{-addP}$) associated with the main equilibrium (Scheme 3.5);
- $K_R$ (=$k_{addR}/k_{-addR}$) associated with the reaction of the expelled radical (R•) with the initial RAFT agent (Scheme 3.6).

The last reaction (Scheme 3.6) is degenerate and is usually ignored on the presumption that the lifetime of the intermediate **18** is negligible. If fragmentation is slow or if there are side reactions which involve the intermediate, this reaction should not be neglected.

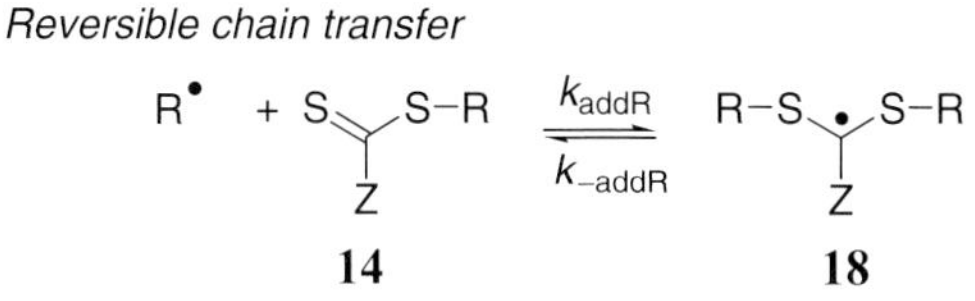

Scheme 3.6 Reversible chain transfer.

There may be other equilibrium constants to consider when penultimate group effects are significant. There is both theoretical data and some experimental evidence to indicate that penultimate group effects can be substantial for short chain lengths (<5) [58, 59]. Examples of such scenarios will be discussed in Section 3.2.3.3. There is also a further series of reactions that needs to be considered which involve initiator-derived RAFT agents. In principle, RAFT agents of differing reactivity might be derived from each radical species present.

#### 3.2.1.2 Transfer Constants

It can be shown that the rate of consumption of the transfer agent depends on two transfer coefficients: $C_{tr}$ ($=k_{tr}/k_p$) and $C_{-tr}$ ($=k_{-tr}/k_i$), which describe the reactivity of the propagating radical ($P_n\bullet$), and the expelled radical ($R\bullet$) respectively ($C_{tr} = k_{tr}/k_p$).

The rate constant for chain transfer can be defined in terms of the rate constant for addition to the RAFT agent ($k_{add}$) and a partition coefficient ($\phi$) which defines how the adduct is partitioned between products and starting materials (equation 3.1).

$$k_{tr} = k_{add} \frac{k_\beta}{k_{-add} + k_\beta} = k_{add}\phi \tag{3.1}$$

We can similarly define $k_{-tr}$ for the reverse process (equation 3.2).

$$k_{-tr} = k_{-\beta} \frac{k_{-add}}{k_{-add} + k_\beta} = k_{-\beta}(1 - \phi) \tag{3.2}$$

Where the expelled radical is a propagating chain of sufficiently long chain length, these will be identical. Thus, $C_{tr} = C_{-tr}$, $k_{tr} = k_{-tr}$, $k_{-add} = k_\beta$ and the partition coefficient ($\phi$) will be 0.5.

Many have assumed when estimating transfer constants that $C_{-tr}$, $k_{-tr}$, $k_{-\beta}$ are zero and that the expelled radical does not react with RAFT agent (polymeric or initial). These reactions are known to be important particularly with low-$k_p$ monomers (e.g. styrene, MMA), with more reactive RAFT agents (e.g. dithiobenzoates), with high RAFT agent concentrations, and with more nucleophilic or less reactive expelled radicals (e.g. benzyl, cumyl radicals). These constants (based on the assumption that $C_{-tr}$, $k_{-tr}$, $k_{-\beta}$ are zero) should be termed apparent constants (i.e. $C_{tr}^{app}$, $k_{tr}^{app}$, $k_{add}^{app}$) [43, 44, 60]. One proposed test for the importance of the reaction of expelled radicals with RAFT agent is the (in)dependence of $C_{tr}$, $k_{tr}$ or $k_{add}$ [44] on RAFT agent concentration.

#### 3.2.1.3 Rate Coefficients/Constants

The absolute values of the individual rate coefficients associated with the RAFT equilibria have proved more difficult to access. Reported values derived from experimental data generally require assumptions regarding the values of partition coefficients ($\phi$) and the absence of significant penultimate unit effects both in the RAFT equilibria and on propagation. These points should be carefully considered before accepting published rate coefficients data including those reported herein. An advantage of values derived by theoretical methods (section 3.2.3.2) is that they do not suffer from these problems.

### 3.2.2
### Molecular Weights and Molecular Weight Distributions

As with living polymerization, the degree of polymerization and the molecular weight can be estimated from the concentration of monomer and reagents as shown in equations 3.3 and 3.4 respectively [35, 60]:

$$\overline{X}_n = \frac{[\mathrm{M}]_0 - [\mathrm{M}]_t}{[\mathrm{T}]_0 + df([\mathrm{I}_2]_0 - [\mathrm{I}_2]_t)} \tag{3.3}$$

$$\overline{M}_n = \frac{[\mathrm{M}]_0 - [\mathrm{M}]_t}{[\mathrm{T}]_0 + df([\mathrm{I}_2]_0 - [\mathrm{I}_2]_t)}(m_\mathrm{M} + m_\mathrm{T}) \tag{3.4}$$

where $m_\mathrm{M}$ and $m_\mathrm{T}$ are the molecular weights of the monomer (M) and the RAFT agent (T) respectively, $d$ is the number of chains produced in a radical–radical termination event ($d \sim 1.67$ for MMA polymerization and $\sim 1.0$ for styrene polymerization) and $f$ is the initiator efficiency. The form of the term in the denominator is suitable for initiators that produce radicals in pairs but will change for other types of initiators [44].

The fraction of living chains ($L$) (assuming no side reactions other than those associated with initiation and termination) is given by equation 3.5 [35]:

$$L = \frac{[\mathrm{T}]_0}{[\mathrm{T}]_0 + df([\mathrm{I}_2]_0 - [\mathrm{I}_2]_t)} \tag{3.5}$$

which can be compared with the actual fraction of living chains estimated from the degree of polymerization obtained and the concentration of RAFT agent used (equation 3.6).

$$L = \frac{(\overline{M}_n - m_\mathrm{RAFT})[\mathrm{T}]_0}{([\mathrm{M}]_0 - [\mathrm{M}]_t)m_\mathrm{M}} \tag{3.6}$$

Reaction conditions should usually be chosen such that the fraction of initiator-derived chains (should be greater than or equal to the number of chains formed by radical–radical termination) is negligible (i.e. $L \sim 1$). In a well-designed RAFT polymerization, $L$ will be >0.95 [35]. Under these conditions, expressions for number-average degree of polymerization and molecular weight (equations 3.3 and 3.4) then simplify to equations 3.7 and 3.8:

$$\overline{X}_n = \frac{[M]_0 - [M]_t}{[T]_0} \tag{3.7}$$

$$\overline{M}_n = \frac{[M]_0 - [M]_t}{[T]_0}(m_M + m_T) \tag{3.8}$$

These equations suggest that a plot of $\overline{M}_n$ vs conversion should be linear. A positive deviation from the line predicted by equation 3.8 indicates incomplete usage of transfer agent (T) (the phenomenon sometimes called hybrid behavior). Under these conditions, the $[T]_0$ in the denominator of equations 3.3–3.8 should be replaced by $[T]_0 - [T]_t$ (the amount of RAFT agent consumed). A negative deviation from the line predicted by equation 3.8 indicates that other sources of polymer chains are significant (e.g. the initiator, RAFT agent homolysis, chain transfer to solvent, monomer).

Analytical expressions have been derived for calculating dispersities of polymers formed by polymerization with reversible chain transfer. Expression 3.9 applies in circumstances where the contributions to the molecular weight distribution by termination between propagating radicals, external initiation and differential activity of the initial transfer agent are negligible [61].

$$\frac{\overline{X}_w}{\overline{X}_n} = 1 + \frac{1}{\overline{X}_n} + \left(\frac{2-c}{c}\right)\frac{1}{C_{tr}} \tag{3.9}$$

where $c$ is the fractional conversion of monomer.

The transfer constant governs the number of propagation steps per activation cycle and should be small for a narrow molecular weight distribution. Rearrangement of equations 3.9 to 3.10 suggests a method of estimating transfer constants on the basis of measurements of the conversion, molecular weight and dispersity [61].

$$\left(\frac{\overline{X}_w}{\overline{X}_n} - 1 - \frac{1}{\overline{X}_n}\right)^{-1} = C_{tr}\left(\frac{c}{2-c}\right) \tag{3.10}$$

In more complex cases, kinetic simulation has been used to predict the time/conversion dependence of the polydispersity. Moad et al. [62] first published on kinetic simulation of the RAFT process in 1998. Many papers have now been published on this subject. Zhang and Ray [63] and also Wang and Zhu [64, 65] applied a method of moments to estimate molecular weights and dispersities. Peklak et al. [66] used a coarse-graining approach while Shipp and Matyjaszewski [67] and the CAMD group [54, 68–70] used a commercial software package (PREDICI®)

[71] to evaluate complete molecular weight distributions. Moad et al. [28, 44, 62] applied a hybrid scheme in which the differential equations are solved directly to give the complete molecular weight distribution to a finite limit ($\overline{X}_n < 500$) and a method of moments is then used to provide closure to the equations and accurate molecular weights and polydispersities. Much of the research in this area has been carried out with a view to understand the factors that influence retardation. The main difficulty in kinetic modeling of RAFT polymerization lies in choosing values for the various rate constants.

## 3.2.3 Kinetic Data

### 3.2.3.1 Measurement

A number of methods have been reported for determining values of the rate and equilibrium constants associated with the RAFT process.

The usual methods for measuring transfer constants, for example, the Mayo method and the log molecular weight distribution method can be applied but only when transfer constants of RAFT agents are low (<1). These methods are not appropriate in the case of more active transfer agents.

The reversibility of chain transfer means that apparent chain transfer coefficients are dependent on the transfer agent concentration and on conversion of the monomer and the transfer agent. One of the ways of estimating the transfer coefficient is to determine the rates of consumption of RAFT agent and monomer. In the case of RAFT polymerization, the rate of consumption of the transfer agent depends on two transfer coefficients: $C_{tr}$ ($=k_{tr}/k_p$) and $C_{-tr}$ ($=k_{-tr}/k_i$), which describe the reactivity of the propagating radical ($P_n\bullet$) and the expelled radical ($R\bullet$) respectively [28, 44, 60].

$$\frac{d[T]}{d[M]} \approx C_{tr}\frac{[T]}{[M] + C_{tr}[T] + C_{-tr}[P]} \tag{3.11}$$

where $[P]$ ($= [T]_0 - [T]$) is the concentration of macro-RAFT agent formed from initial RAFT agent (T). This expression simplifies for the case of macro-RAFT agents when $C_{tr} \sim C_{-tr}$; if reaction with the initial RAFT agent is irreversible, then $C_{-tr} = 0$, and for less active RAFT agents at low monomer conversion, $[P] \sim 0$.

$$\frac{d[T]}{d[M]} \approx C_{tr}\frac{[T]}{[M]} \tag{3.12}$$

$$C_{tr} = \frac{k_{tr}}{k_p} \approx \frac{[M]d[T]}{[T]d[M]} = \frac{d\ln[T]}{d\ln[M]} \tag{3.13}$$

In this case, the slope of a plot of ln [M] vs ln [T] yields the transfer coefficient. Transfer constants based on an assumption that $C_{-tr}$ is zero should be considered apparent constants (i.e. $C_{tr}^{app}$) [43, 44, 60] until proven otherwise (see section 3.2.1.2).

The transfer constant can also be calculated from the discrepancy between the found molecular weight and that estimated based on complete consumption of the RAFT agent (T) [28, 44].

$$\text{Fraction T consumed} = \frac{[\mathrm{T}]_0 - [\mathrm{T}]_t}{[\mathrm{T}]_0} = \left\{\frac{[\mathrm{M}]_0 - [\mathrm{M}]_t}{[\mathrm{T}]_0}\right\} \Big/ \left\{\frac{[\mathrm{M}]_0 - [\mathrm{M}]_t}{[\mathrm{T}]_0 - [\mathrm{T}]_t}\right\}$$

$$\approx \frac{\overline{X}_n(\text{calc})}{\overline{X}_n(\text{found})}$$

where $\overline{X}_n$(calc) is the expected number-average degree of polymerization assuming complete consumption of transfer agent and $\overline{X}_n$(found) is the measured number-average degree of polymerization. It is important to correct $\overline{X}_n$(found) (using equation 3.3) to allow for the formation of initiator-derived chains when this fraction is significant.

Rate constants for chain transfer to RAFT agents can also be calculated if $k_p$ is known. Experimental methods for evaluation of constants for the addition of radicals to low-molecular-weight RAFT agents ($k_{add}$) are strongly model dependent in that they are implicitly or explicitly derived from transfer rate coefficients ($k_{tr}$) by assuming a value of the partition coefficient ($\phi$) and often involve other assumptions.

Barner-Kowollik and coworkers [51] proposed a method for estimating the addition rate coefficient in the preequilibrium from molecular weight measurements. The method might be considered equivalent to the method mentioned above that was introduced by Moad and coworkers [28, 44] to determine transfer constants but was derived differently. The calculation involves an implicit assumption that penultimate unit effects on addition (e.g. $k_p$, $k_{add}$, $k_{add\,P}$) and fragmentation (e.g. $k_{-add}$, $k_{-add\,P}$ and $k_\beta$) rate constants are either negligible or cancel. If the R leaving group of the initial RAFT agent behaves identically to the propagating radical, it is possible to derive a set of equations that can be employed to deduce the addition rate coefficient $k_{add}$ in systems that display hybrid behavior. The initial rate of propagation, $R_p$, is given by equation 3.14:

$$R_p = k_p[P_n][\mathrm{M}]_0 \tag{3.14}$$

where $[P_n]$ is the propagating radical and $[\mathrm{M}]_o$ is the monomer concentration. The rate of radical addition to the carbon–sulfur double bond of the initial (or polymeric) RAFT agent, $R_{add}$, can be quantified in a similar fashion via equation 3.15:

$$R_{add} = k_{add}[P_n][\mathrm{T}]_0 \tag{3.15}$$

$[\mathrm{T}]_0$ is the initial (or polymeric) RAFT agent concentration. In RAFT systems that display hybrid behavior, both rates are correlated by the degree of polymerization of the polymer formed instantaneously, $X_n^{inst}$. $X_n^{inst}$ can be obtained via plotting the number-average molecular weight $\overline{M}_n$ vs the monomer conversion and extrapolating to zero monomer conversion.

$$X_n^{inst} = \frac{R_p}{R_{add}\phi} + 1 \tag{3.16}$$

The partition coefficient $\phi$ in the denominator of equation 3.16 is associated with the fact that the reaction product of the addition reaction, i.e. the intermediate macro-RAFT radicals, can also fragment to yield the starting materials. The probability of the macro-RAFT radical undergoing a radical transfer (in contrast to fragmenting back to the starting materials) is 50%, since both fragmentation pathways are assumed to have identical rate coefficients. Thus, in order for equation 3.16 to yield the correct value for $X_n^{inst}$, the addition rate $R_{add}$ has to be multiplied by a factor of 0.5. When some trithiocarbonates are employed (e.g. dimethoxy carbonylethyl trithiocarbonate), there exist three different fragmentation pathways with the probability of a radical transfer being 67%. This can be compensated by multiplying $R_{add}$ by a factor of 0.67. Following a similar rationale, a hypothetical *ideal* RAFT agent that would only allow fragmentation of the intermediate radical to release the R group (and never fragment back to the starting materials) would not require multiplication of $R_{add}$ by any factor other than unity. In contrast, a RAFT agent that leads to an intermediate radical that fragments preferentially to the starting material would have a factor of smaller than 0.5 on $R_{add}$. Furthermore, it has to be noted that the first propagation step already leads to a chain length of 2. Therefore, one needs to be added to the right-hand side of equation 3.16. Equation 3.16 can subsequently be rearranged to equation 3.17, which allows for the deduction of $k_{add}$, when the propagation rate coefficient $k_p$, the monomer concentration $[M]_o$, the instantaneous degree of polymerization $X_n^{inst}$ and the initial RAFT agent concentration $[T]_0$ are known.

$$k_{add} = \frac{k_p[M]_0}{(X_n^{inst} - 1)[T]_0\phi} \tag{3.17}$$

Since it is very difficult – at least with the data presently at hand in the literature – to quantify the individual reaction channels of the intermediate radical, the only reliable avenue to $k_{add}$ determination is the use of a RAFT agent that features an R group radical identical to the propagating radical. While the above approach does not give access to a (potential) chain length dependency of $k_{add}$ with regard to the macro-RAFT agent, the length of the adding macroradical varies with $X_n^{inst}$. Of course, a similar rationale as outlined above can also be employed to a macro-RAFT agent that is chain extended with a monomer of its own kind. However, this approach will invariably give a value for the main equilibrium addition rate coefficient $k_{add}$ at the corresponding chain length. A typical set of results obtained when applying the above procedure in MA polymerizations is given in Table 3.1. Inspection of Table 3.1 indicates that – as required – the addition rate coefficient is independent of the initial RAFT agent concentration and also (within experimental

**Table 3.1** Summary of the initial MCEPDA [1-(methoxycarbonyl)ethyl phenyldithioacetate (**23**)] and DMCETC [di-(1-(methoxycarbonyl)ethyl) trithiocarbonate] concentrations in MA bulk-free radical polymerizations as well as the associated degrees of polymerization of the instantaneously generated polymer, $X_n^{inst}$, at 60 °C [51]

| RAFT reagent | $[RAFT]_0$ ($mol \cdot L^{-1}$) | $X_n^{inst}$ | $k_{add}$ ($L \cdot mol^{-1} \cdot s^{-1}$) |
|---|---|---|---|
| MCEPDA | $4.37 \times 10^{-3}$ | 103 | $1.5 \times 10^6$ |
| MCEPDA | $1.27 \times 10^{-2}$ | 46 | $1.2 \times 10^6$ |
| MCEPDA | $3.68 \times 10^{-2}$ | (8) | ($2.6 \times 10^6$) |
| DMCETC | $4.31 \times 10^{-3}$ | 39 | $3.1 \times 10^6$ |
| DMCETC | $1.05 \times 10^{-2}$ | 34 | $1.5 \times 10^6$ |
| DMCETC | $3.04 \times 10^{-2}$ | 11 | $1.7 \times 10^6$ |
| DMCETC | $1.02 \times 10^{-1}$ | (7) | ($0.8 \times 10^6$) |

The initiator (AIBN) concentration was close to $3.0 \times 10^{-3}$ $mol \cdot L^{-1}$. The $k_{add}$ values have been deduced via equation 3.17 and the method described in the text.

accuracy) independent of the type of RAFT agent employed. This suggests that the implicit assumption that the expelled radical does not react with polymeric RAFT agent to re-form the original RAFT agent is valid (i.e. $C_{-tr}$ is zero) for the case of acrylate polymerization with these RAFT agents. It must be noted that the accuracy of the method is determined by the accuracy by which the initial molecular weights ($X_n^{inst}$) can be determined by extrapolating the number-average molecular weight $M_n$ vs conversion evolution as well as the accuracy of the method employed (usually size exclusion chromatography) to measure the molecular weights.

Transfer constants may also be estimated via direct kinetic modeling approaches, where given kinetic scheme is fitted to a set of experimental parameters such as the full molecular weight distribution evolution and the conversion vs time evolutions [54, 57]. The kinetic models employed in such fitting procedure are sometimes simplified (see also section 3.2.3.3) due to difficulty of having to assess a large number of variable parameters in the more complete models. Thus, the transfer constants obtained are often average, model-dependent quantities that, however, agree surprisingly well with the transfer constants assessed by other methods.

#### 3.2.3.2 **Calculation**

The first application of molecular orbital methods in RAFT polymerization was reported by Moad and coworkers who applied both ab initio and semiempirical methods in a qualitative way to gain understanding of the relative effectiveness of macromonomer [30] and thiocarbonylthio RAFT agents [28, 44]. Matyjaszewski and Poli [72] have reported the results of density functional theory (DFT) calculations used to estimate R—S bond dissociation energies in thiocarbonylthio RAFT agents (R—S(C=S)Z) and suggest these may be useful as a guide for the design of RAFT agents. Even though some parallels might be expected, it should be pointed out that it is the bond dissociation energy of R—S bond in the radical intermediate

(R—S(C(·)—SR′)Z) rather than the RAFT agent (R—S(C=S)Z) that is important in determining fragmentation rates. DFT calculations have also been applied in understanding the properties of radical adducts to phosphoryldithioacetate RAFT agents [73, 74]. The suitability of DFT methods for calculations relating to RAFT polymerization has, however, been questioned [58].

Recently, Coote and colleagues have applied ab initio methods to gain quantitative information about the equilibrium and rate constants associated with the RAFT process. Quantitative information about the kinetics and thermodynamics of chemical reactions can be obtained via ab initio molecular orbital theory [75]. In addition to the rates and equilibrium constants, the charge distributions, spin densities, geometries and vibrational frequencies of all species including transition structures may be estimated by such calculations. Ab initio approaches consequently represent useful partners to experiment. The principal disadvantage of quantum chemistry is that the accuracy of the predictions is strongly dependent on the simplifications and approximations introduced in solving the Schrödinger equation. Given an infinite basis of orbitals, a wavefunction that includes contributions from all possible configurations of those orbitals and the computer power necessary to process the data, one could predict the chemical behavior of any chemical system with absolute accuracy. In practice, of course, this is not feasible and it is necessary to introduce simplifications, which are a potentially large source of error. It is extremely important to note that there is a range of possible methods to use and that these vary considerably in their accuracy and their computational expense. Errors can typically range from the sub-kJ level (for methods such as W1 or W2) to more than 50 $kJ \cdot mol^{-1}$ (in some applications of DFT or semiempirical methods), and are dependent on the type of chemical reaction under investigation. Moreover, the most accurate methods are extremely time expensive, with their computational cost scaling exponentially with the size of the chemical system. Hence, applying computational chemistry to polymeric reactions necessarily involves a compromise in which reliable low-cost procedures are applied to small chemical models of the polymeric reaction under investigation. Both the method and the model reaction, however, must be chosen carefully on the basis of assessment studies.

Such assessment studies have been performed in great detail by Coote and coworkers, both for the RAFT-relevant case of radical addition to S=C bonds [76–78] and for other reactions of significance in radical polymerization, such as propagation and chain transfer [78–80]. In general, it appears that geometries and frequencies can be calculated at relatively low levels of theory such as B3-LYP/6-31G(d), provided transition structures are corrected via IRCmax. However, very high level composite procedures such as W1 are essential for accurate absolute values of the barriers and enthalpies. Low-cost approximations to this level of theory (to within kcal-accuracy) can be obtained via an ONIOM-based procedure in which the inner core is studied at W1, the core at G3(MP2)-RAD and the full system at ROMP2/6-311+G(3df,2p) [78]. It should be stressed that popular DFT methods (such as B3-LYP and BB1K) fail comprehensively to model the energetics of these polymerization-relevant reactions, with errors in excess of 50 $kJ \cdot mol^{-1}$ in some cases [81]. As in the case of propagation [82], the harmonic oscillator approximation is not adequate and it is therefore important to treat low-frequency torsional modes as hindered internal rotations [76, 83]. It is also advisable to use variational transition state theory rather than standard transition state theory, though the latter can be adopted for order of magnitude estimates [76]. Coote and colleagues demonstrated that when following these recommended procedures, chemical accuracy is achievable for small radical reactions in RAFT systems. For example, the experimental equilibrium constant for *tert*-butyl radical addition to di-*tert*-butyl thioketone at 25 °C was recently reproduced to within a factor of 2 [19, 21].

With respect to the RAFT process, the $k_{add}$, $k_\beta$ and $K$ values for a variety of R and Z groups of small compounds, as occurring in the preequilibrium, have been calculated by Coote and collaborators using high-level ab initio methods [53, 76, 83–85]. (A complete listing of the data from these various studies, updated to a consistent level of theory, is provided in Ref. [78] and is in detail explained in Chapter 1 of this book.) The results demonstrate that $k_\beta$ and $K$ depend strongly on the nature of R and Z, and can range over 10 orders of magnitude at 60 °C. For Z = Ph, the $K$ for the RAFT reaction R• + S=C(Ph)SCH$_3$ ⇌ RS-C(•)(Ph)SCH$_3$ was, e.g., $7.3 \times 10^6$ L·mol$^{-1}$ (30 °C) for R = cumyl [53], $8.8 \times 10^5$ L·mol$^{-1}$ (60 °C) for R = benzyl [83] and $1.5 \times 10^6$ L·mol$^{-1}$ (60 °C) for R = CH$_2$(COOCH$_3$) [83]. The $K$ values calculated for these pre-equilibrium-type reactions involving small species generally agree with those expected for slow fragmentation (typically $10^6$–$10^7$ L·mol$^{-1}$). Most recently, calculations performed for R groups of different oligomeric chain length indicated a pronounced impact of the leaving group size on the obtained $K$ values, which were found going from $2.6 \times 10^4$ L·mol$^{-1}$ when R = C(CH$_3$)$_2$CN (chain length = 0) to $7.4 \times 10^7$ L·mol$^{-1}$ when R = CH(Ph)CH$_2$CH(Ph)CH$_2$C(CH$_3$)$_2$CN (chain length = 2) in a simplified model reaction in which the nonparticipating R′ group is CH$_3$. To date, it is not fully clear which physical effects may be the cause of such a pronounced effect, making generalization to all RAFT systems difficult. The finding of a pronounced chain length dependency of $K$, however, highlights the importance of considering the preequilibrium and the main equilibrium independently. Observations such as the chain length dependence of the equilibrium constants – whether caused by a variation in $k_{-add}$, $k_{add}$ or both – make direct comparison of the quantum chemically obtained rate coefficients sometimes difficult. The experimental determination of such quantities lacks behind the coefficients that are obtainable quantum chemically. An example of how rate coefficients determined via quantum chemical methods can be employed in exploring the RAFT process is provided in Section 3.2.3.3; several more examples can be found in Chapter 1 of this book.

#### 3.2.3.3 **Kinetic Modeling**

Kinetic modeling is an important tool for probing how polymerization reactions behave under variable kinetic assumptions; in addition, kinetic modeling can be employed to deduce kinetic rate coefficients by assuming a given model and fitting it to a set of experimentally observed data. All model-fitting to experimental data is to some extent model dependent. Section 3.2.1 gave a brief overview about the groups that have used kinetic modeling to arrive at a deeper understanding of the RAFT process; in this section, we will consider the approaches of some of these modeling procedures and the associated advantages and disadvantages in the context of the preequilibrium. It is clear that the more detailed the kinetic data available, the higher the precision with which a model can be fitted, i.e. the more accurate the derived rate parameters will be. For example, one can attempt to derive kinetic information solely from the overall rate of polymerization in conjunction with the molecular weight evolution (including the number-average molecular weight as well as the polydispersity). In such cases, it will only be possible to extract (model-dependent) kinetic information from the periods of nonconstant rates of polymerization [53, 57, 86]. One of the most critical issues in modeling kinetic rate and molecular weight data is the selection of which reactions should be included in the model and what simplifications are allowable. High-level quantum chemical calculations of Coote and coworkers suggest that the equilibrium constant is strongly dependent on chain length as well as the chemical nature of the groups adjacent to the adduct radical center [87]. It thus seems that simplification of the RAFT mechanism into a single pre- and main equilibrium should be avoided in modeling studies. For example, Moad and coworkers [44, 60, 88] demonstrated that it was not possible to fit the rate of consumption of the initial RAFT agent as a function of RAFT agent concentration or the evolution of the molecular weight distribution with time for styrene and MMA polymerization using a simplified model.

As a consequence, one should employ a kinetic scheme which incorporates a highly complex preequilibrium as illustrated in Scheme 3.7. Here, the chain length

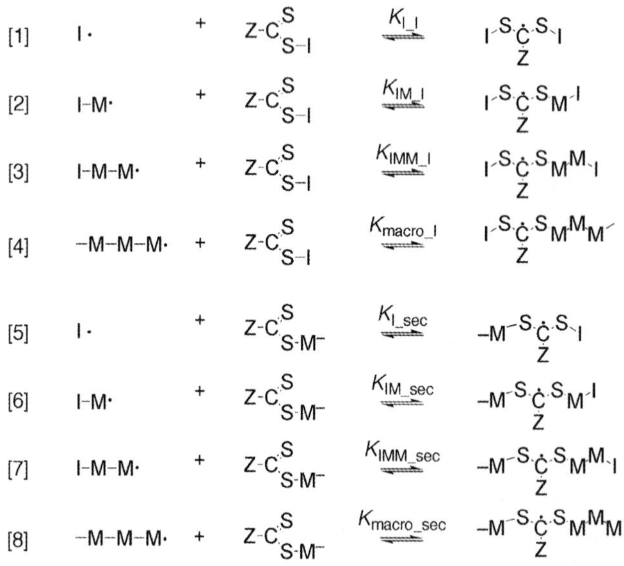


**Scheme 3.7** Enhanced preequilibrium reaction scheme for the case that the RAFT agent leaving group (R in S=C(Z)SR) and initiating species (I) are identical. The scheme incorporates the standard kinetic model for RAFT, but is augmented to include eight addition–fragmentation equilibria (instead of one or two) in order to model substituent and chain length effects up to the trimer in the attacking radical and up to the unimer in the RAFT agent.

dependency of the equilibrium constant is taken into account, but only for cases where the initiating species and the leaving group are identical. Clearly, the scheme will be more complex if these two entities are nonidentical.

Given the above detailed complexity of the situation in the preequilibrium, it seems that the fitting of simplified models can only give approximate and average values for the rate coefficients. Yet, complete modeling of the preequilibrium with the aim of deducing individual rate coefficients from overall monomer to polymer conversion time data is difficult. The situation can be improved if individual conversion versus time trace of each species are available (see Section 3.2.4.2). Indeed, Klumperman and coworkers have shown that model-dependent kinetic rate parameters may be derived by fitting a given model to the individual species evolutions [123], while Coote and Barner-Kowollik have – in view of the model-dependent nature of analysis – suggested that rather than a fitting exercise be carried out, all rate coefficients be calculated by high-level ab initio quantum mechanical methods and subsequently employed in arriving at an ab initio kinetic prediction of the experimentally observed species concentration evolutions [58, 89].

The situation in modeling main-equilibrium kinetics for chain lengths exceeding a certain size should be a much simpler task (see section 3.2.6) as for longer polymer chains the rate coefficients should be chain length independent and the adduct radical is essentially symmetrical. At present, it is not possible to provide a defined number at which chain lengths the chain–length-dependent effects on the equilibrium constant subside or becomes less significant.

#### 3.2.3.4 **Transfer Constants/Equilibrium Constants**

Most reported transfer constants for initial RAFT agents should be considered as apparent transfer constants (Table 3.2).

We have refrained from listing experimentally obtained equilibrium constants explicitly in Table 3.2, although there is now a considerable body of experimental and ab initio molecular orbital calculation work carried out on the value of the equilibrium constants. The reason for not tabulating equilibrium constants lies in the large discrepancy in their value reported using different experimental approaches, with up to 6 orders of magnitude difference [97]. The currently published data on the size of the equilibrium constants also indicates that tabulating single values for this parameter is highly problematic (and this is shown in greater detail in Section 3.2.3.3), because for every RAFT agent/initiator/monomer system the preequilibrium is constituted by a multitude of reactions involving the initiator fragments, the R group as well as variable chain lengths of the propagating radical chains. Thus, the equilibrium constant is highly likely not a true constant but variable with chain lengths as well as highly influenced by the chemical nature of the substituents on the intermediate radical. As will be explored further below in this chapter, the seemingly disparate values for equilibrium constants obtained via variable approaches may in fact be a reflection of the reaction conditions and chain length regimes under which these values were obtained [97]. It should also be pointed out that even in situations where the equilibrium constants have been assessed under main-equilibrium conditions [98, 99] (i.e. employing macro-RAFT

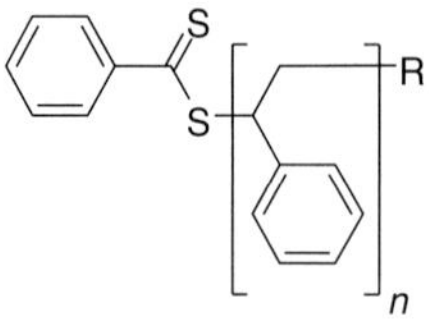

**19**

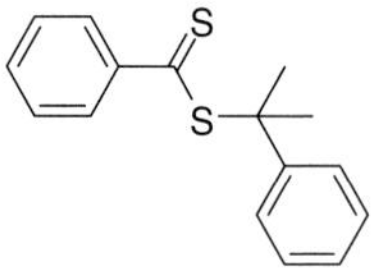

**20**

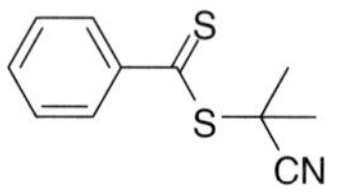

**21**

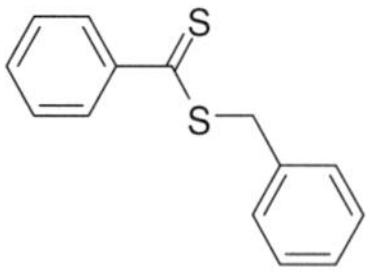

**22**

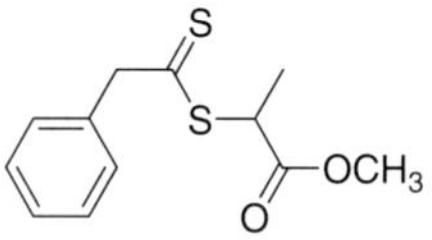

**23**

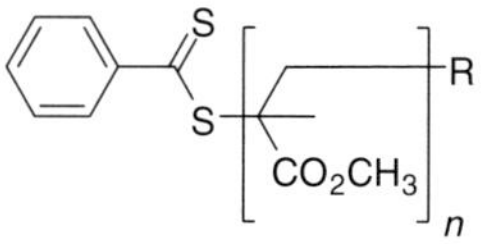

**24**

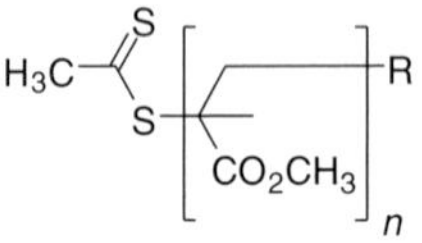

**25**

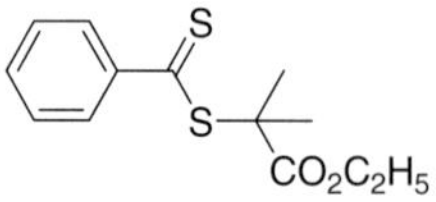

**26**

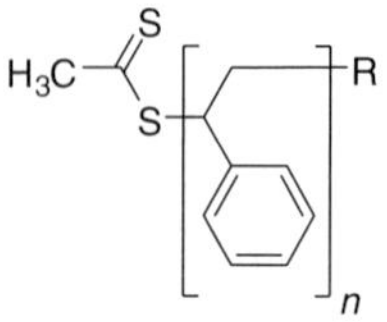

**27**

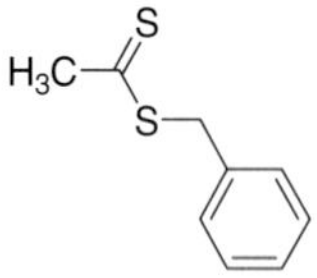

**28**

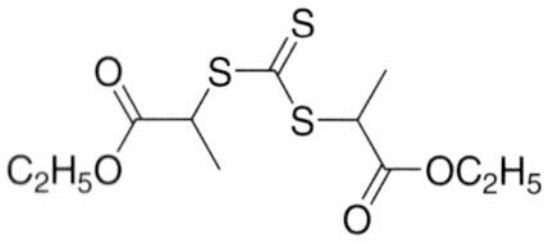

**29**

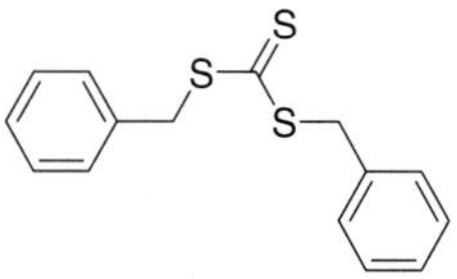

**30**

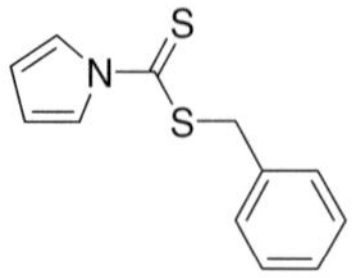

**31**

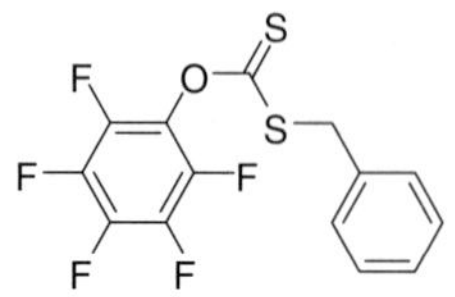

**32**

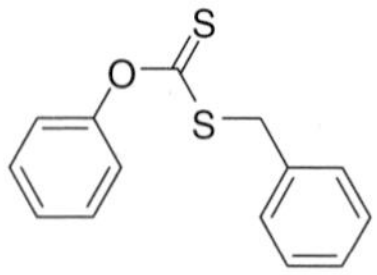

**33**

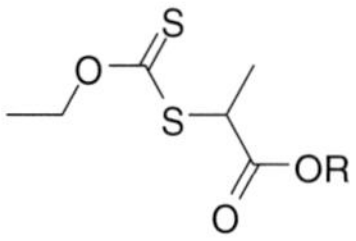

**34a** R=$CH_3$
**b** R=$C_2H_5$

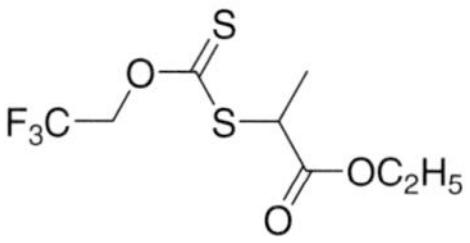

**35**

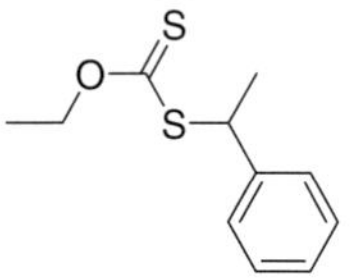

**36**

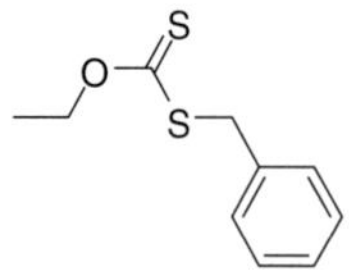

**37**

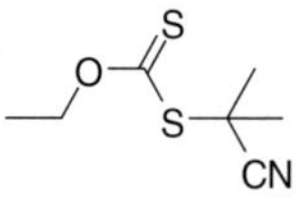

**38**

**Table 3.2** Transfer constants, transfer rate coefficients and addition rate coefficients of select RAFT agents with various monomers (i.e. the associated propagating radicals)[a]

| Agent | M | $T$ (°C) | $C_{tr}^{app}$ | $C_{tr}$ | $C_{-tr}$ | $k_{add}$[a] | $\phi$ | Reference[b] |
|---|---|---|---|---|---|---|---|---|
| **24** | MMA | 60[d] | 140 | 140 | 140 | — | 0.5 | [90] |
| **25** | MMA | 40 | 40 | 40 | 40 | — | 0.5 | [91] |
| **27** | MMA | 40 | 0.83 | 0.83 | — | — | — | [91] |
| **20** | MMA | 60 | 5.9 | 56 | 2500 | — | — | [44] |
| **21** | MMA | 60 | 6.8 | 25 | 450 | — | — | [44] |
| **26** | MMA | 60 | 1.7 | — | — | — | — | [44] |
| **22** | MMA | 60 | <0.03 | — | — | — | — | [44] |
| **38** | MMA | 60 | 0.04 | — | — | — | — | [92] |
| **23** | MA | 60 | 70[c] | — | — | 1.3 | 0.5 | [51] |
| **29** | MA | 60 | 112[c] | — | — | 2.1 | 0.67 | [51] |
| **22** | MA | 60 | 105[c] | — | — | — | — | [44] |
| **29** | BA | 60 | 72[c] | — | — | 1.7 | 0.67 | [93] |
| **29** | BA | −30 | 98[c] | — | — | 0.2 | 0.67 | [94] |
| **36** | BA | 60 | 2 | — | — | — | — | [92] |
| **38** | tBA | 60 | 7.2 | — | — | — | — | [92] |
| **32** | tBA | 60 | 2.7 | — | — | — | — | [92] |
| **29** | Dodecyl acrylate | 60 | 98[c] | — | — | 2.4 | 0.67 | [50] |
| **19** | Styrene | 60 | ~6000 | ~6000 | ~6000 | ~2 | 0.5 | [90] |
| **27** | Styrene | 60[d] | 160 | 160 | 160 | 0.05 | 0.5 | [90] |
| **27** | Styrene | 40 | 220 | 220 | 220 | — | 0.5 | [91] |
| **25** | Styrene | 40 | 420 | 420 | — | — | — | [91] |
| **20** | Styrene | 60 | 50 | 2000 | 10 000 | — | — | [44] |
| **20** | Styrene | 60 | 25000[c] | — | — | 4.0 | (1.0) | [57] |
| **36** | Styrene | 60[d] | 3.92 | — | — | — | — | [95] |
| **34b** | Styrene | 60[d] | 0.67 | — | — | — | — | [95] |
| **35** | Styrene | 60[d] | 0.65 | — | — | — | — | [95] |
| **20** | Styrene | 110 | 29 | — | — | — | — | [28] |
| **30** | Styrene | 110 | 18 | — | — | — | — | [28] |
| **28** | Styrene | 110 | 10 | — | — | — | — | [28] |
| **31** | Styrene | 110 | 11 | — | — | — | — | [28] |
| **32** | Styrene | 110 | 2.3 | — | — | — | — | [28] |
| **33** | Styrene | 110 | 0.72 | — | — | — | — | [28] |
| **37** | Styrene | 110 | 0.11 | — | — | — | — | [28] |
| **34a** | VAc | 80 | 14 | — | — | 0.5 | (0.75) | [49] |

[a] Estimated assuming the value of $\phi$ shown. The units of $k_{add}$ are $M^{-1}{\cdot}s^{-1} \times 10^6$.
[b] Note that a range of experimental protocols have been applied to arrive at these data (see the indicated references for details).
[c] $C_{tr}$ corresponding to value of $k_{add}$ shown ($C_{tr} = k_{add}/\phi/k_p$) with $k_p(\text{MA}, 60°\text{C}) = 2.78 \times 10^4\ M^{-1}s^{-1}$ [98], $k_p(\text{BA}, 60°\text{C}) = 3.5 \times 10^4\ M^{-1}s^{-1}$, $k_p(\text{BA}, -30°\text{C}) = 3.0 \times 10^3\ M^{-1}s^{-1}$, $k_p(\text{DA}, 60°\text{C}) = 3.8 \times 10^4\ M^{-1}s^{-1}$, $k_p(\text{S}, 60°\text{C})$ 160 $M^{-1}s^{-1}$ $k_p(\text{VAc}, 80°\text{C}) = 47400\ M^{-1}s^{-1}$ [174].
[d] Arrhenius parameters are provided in the reference indicated.

agents as opposed to initial low-molecular-weight RAFT agent), strictly speaking no single equilibrium constant can be measured, as the employed systems also make use of low-molecular-weight initiator and its corresponding derived radicals (i.e. cyanoisopropyl) that react to give intermediate radicals of variable chemical compositions. For an in-depth overview on the specific problems of reporting equilibrium constants, the reader is referred to Ref. [97].

### 3.2.4
### Methods for Product Analysis

The above sections have made clear that the RAFT process can be extremely complex. Both from a synthetic point of view and to provide greater understanding of the kinetics and mechanism, it has become a matter of priority to reliably determine the product distributions generated during RAFT polymerization. Techniques that have been applied include high-resolution mass spectrometry, especially soft ionization mass spectrometry techniques such as electrospray ionization–mass spectrometry (ESI–MS) and matrix-assisted laser desorption and ionization–time of flight–mass spectrometry (MALDI–TOF–MS). Mass spectrometry is particularly useful in determining whether side products (such as three- or four-arm star polymer) have been formed during the RAFT process. Similarly, nuclear magnetic resonance (NMR) spectroscopy in both its $^{1}H$ and $^{13}C$ variants has been extensively used, especially in a dynamic fashion to probe the initial phases of the RAFT process. Concomitantly, electron spin resonance (ESR) spectroscopy has been employed to gain information about the concentration of the intermediate RAFT radicals that occur during the polymerization process. In the following, we will highlight in greater detail how each of these methods have helped in constructing a more comprehensive picture of the RAFT process.

#### 3.2.4.1 **Soft Ionization Mass Spectrometry**

Powerful tools to achieve this are soft ionization mass spectrometry techniques, which have found ample use in the past in elucidating mechanistic issues in a range of free radical polymerization mechanisms. The most prominent of the soft ionization mass spectrometry techniques are ESI–MS and MALDI–TOF–MS, which have both been applied by many research groups [100–110] to assess the polymer species formed in RAFT polymerization under conditions that yield good living characteristics (i.e. typical RAFT agent concentrations in the range of $10^{-3}$–$10^{-2}$ mol·L$^{-1}$ and initiator concentrations an order of magnitude lower). Barner-Kowollik and colleagues [111] have recently reviewed the use of soft ionization mass spectrometric techniques for the elucidation of radical polymerization mechanisms and discussed the specific strengths of both ESI–MS and MALDI–TOF–MS for this purpose. MALDI–TOF–MS has frequently been found to lead to relatively large degrees of fragmentation of polymer generated in RAFT systems [102] which only recently could be minimized by careful tuning of the matrix system as well as laser power and wavelength [110, 112]. However, MALDI–TOF–MS potentially offers

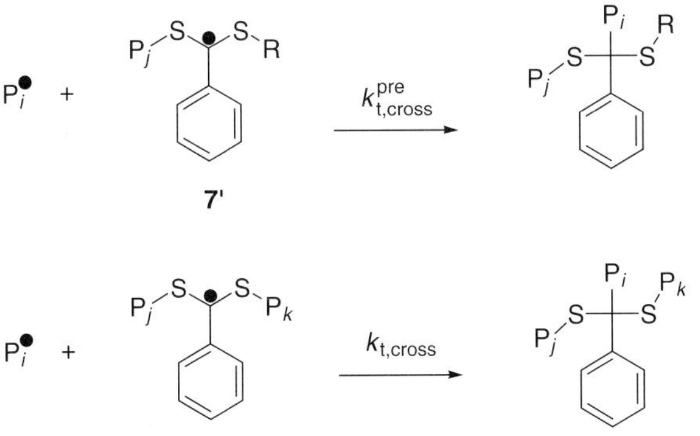


**Scheme 3.8** Bimolecular termination between propagating radicals and intermediate RAFT radicals in the preequilibrium (upper part) and main equilibrium (lower part). Note that in principle each intermediate RAFT radical listed in Scheme 3.7 in section 3.2.3.3 can terminate via the above reaction pathway.

the possibility to analyze relatively high-molecular-weight polymers. ESI–MS is a considerably softer ionization procedure, leading to no fragmentation of the polymer chain or end group. In its usual combination with an ion-trap mass analyzer, however, ESI–MS is limited to relatively small chain lengths of generally below 4000 $m/z$. While chemical mass bias due to varying ionization potentials can always cause problems in any mass spectrometric technique (and thus potentially suppress or exaggerate certain peaks), this problem was considered to be small in the case of acrylic polymers from RAFT polymerization, which are assumed to ionize via the polyacrylate backbone, without being affected by the potential end group. Mass spectrometric techniques should thus be highly useful in establishing an accurate image of the products generated during the RAFT process and thus also allow mechanistic deductions.

One aspect of interest during RAFT polymerization is the question whether three-arm cross-terminated and four-arm self-terminated star polymers, respectively, are formed during the polymerization process (for details see section 3.3.2) via the bimolecular termination reactions depicted in Scheme 3.8.

In monomer-free RAFT model reactions such star polymers have been produced and it has been demonstrated that they can readily be ionized in soft ionization mass spectrometry [100, 103]. Monomer-free experiments are considered favorable for observing termination products, as the spreading of the polymeric termination products over a wide range of molecular weight, which occurs due to propagation, is eliminated. The kinetic similarity of the polymerization and the monomer-free reaction was suggested by similar values for $K(k_{t,cross}/k_t)$ obtained in both systems [100]. Cross- and self-termination products, however, cannot be identified in polymer samples drawn from actual dithiobenzoate-mediated RAFT polymerizations of acrylates [103, 108, 109] and only trace quantities could be identified in styrene systems [110]. Such findings are particularly interesting as the ESR-measured intermediate radical concentrations should give rise to very large excesses of three- and four-arm star polymer over conventionally terminated linear polymer (up to 30 times in excess at any given chain length) in case cross- and self-termination occurs with a termination rate coefficient similar in magnitude to that of the conventional termination process [86, 109]. For example, if the RAFT-specific rate coefficients for the MA/CDB system [86], i.e., $k_{\text{add}} = 1 \times 10^7$ L·mol$^{-1}$·s$^{-1}$ and $k_{-\text{add P}} = 1 \times 10^3$ s$^{-1}$, which were derived via modeling rate data, molecular weight evolution, and ESR-measured intermediate radical concentration, are employed to calculate the overall concentration ratio of three-arm star polymer to conventionally terminated product, a pronounced cumulative excess of star polymer over linear termination products is predicted [109]. While the conventionally terminated polymer can be clearly observed by ESI–MS, the forecast excess of star polymer is not observed [108, 109]. A recent potential explanation of why such three- and four-arm star polymers cannot be observed – apart from the fact that they may simply not form in sufficient quantities or form reversibly only – will be discussed in section 3.3.2.6.

#### 3.2.4.2 **NMR Spectroscopy**

Real-time $^1$H NMR has previously been used to study initiation of polymerization of styrene with 2,2′-azobis(isobutyronitrile) (AIBN) initiator and with cumyl [113] or cyanoisopropyl dithiobenzoate [114] as the RAFT agent and more recently polymerization of MA [115], VAc [116] NVP [116] and styrene–maleic anhydride [117, 118] and styrene–acrylonitrile copolymerization [118] with various RAFT agents. Real-time $^{13}$C NMR was also used to study initiation of styrene polymerization with AIBN initiator and CDB-$\alpha$-$^{13}$C as the RAFT agent [119]. Most recently, RAFT polymerization of styrene and MMA has been studied with AIBN-$\alpha$-$^{13}$C as initiator and with benzyl dithiobenzoate-$\alpha$-$^{13}$C, CDB, cyanoisopropyl dithiobenzoate or dodecyl cyanoisopropyl trithiocarbonate as the RAFT agent [71]. The in situ NMR method enables the concentration–time profiles of the various species formed during the polymerization process to be recorded. Most of the early experiments were carried out using relatively high RAFT agent concentrations to facilitate the monitoring of the various reagents and products. High initial RAFT agent concentrations result in significant retardation with dithiobenzoate RAFT agents that are used. Differentiation between the individual oligomeric species is limited to short chains.

For styrene polymerization at 70 or 84 °C with high concentrations of cumyl or cyanoisopropyl dithiobenzoate, complete conversion of the initial RAFT agent to a single unit 'chain' was observed prior to any significant formation of two unit and higher chains [113, 114, 119]. This was called 'efficient initialization'. The process has successfully been modeled by kinetic simulation based on assumption of slow fragmentation and rate constants estimated by ab initio calculations [58] or with faster fragmentation (so as not to cause retardation directly) and intermediate radical termination [120]. Moad et al. have shown by kinetic modeling that the observation of 'efficient initialization' is not dependent on slow fragmentation or the occurrence of intermediate radical fragmentation. It is observed when initiation by 'R' is rapid with respect to further propagation (as is usually the case; see Table 3.3 [44, 121]) and RAFT agent concentrations such that less than one monomer unit is added per activation cycle. Very rapid reaction of the expelled radical 'R' with RAFT agent is another likely cause of retardation during this conversion regime. Coote and Barner-Kowollik [58] proposed that the 'efficient initialization' and retardation might be a consequence of slow fragmentation and significant chain length dependence of the equilibrium constant for the RAFT equilibria suggested by ab initio molecular orbital calculations.

The dynamic $^1$H NMR technique has also been applied to acrylate systems, where very strong inhibition periods are often observed [122]. The CDB-mediated polymerization of MA as a retarding and cumyl phenyldithioacetate (CPDA) as a nonretarding RAFT agent indicated a limited influence of the stabilizing group on the initial rate of polymerization in this particular case. The low addition rate coefficient of the leaving group [123] compared to the propagation rate coefficient of MA macroradicals was hence concluded to be responsible for the induction period with both types of RAFT agents [124]. The rate of consumption of the initial RAFT agent for both the retarding and nonretarding RAFT agents was almost identical, suggesting that the rate-determining step for both polymerizations was the same during the initialization period, which supported the conclusions of the styrene polymerization studies. After the initialization period, the rates of polymerization were significantly different with the nonretarding RAFT agent, reaching high monomer conversions within a short period of time, while the rate of monomer consumption with the retarding RAFT agent showed only a small change from the rate during initialization [124].

The dynamic NMR experiments clearly indicate that the kinetic behavior observed during the preequilibrium period is a strong function of the leaving groups and of the monomer used.

#### 3.2.4.3 Electron Spin Resonance Spectroscopy

The first direct evidence of the presence of the dithioketal radicals as discrete intermediates and the addition–fragmentation mechanism was obtained by ESR spectrometry [125] when intermediates were observed during butyl acrylate (BA) and styrene polymerization with CDB as RAFT agent. The intermediates have now been observed directly by ESR spectrometry in RAFT polymerizations of styrene and acrylate esters and in model reactions with dithiobenzoate, trithiocarbonate and dithiophosphonate RAFT agents [74, 94, 99, 119, 125–132]. Such intermediates have not been detectable during MMA polymerizations or in polymerizations with aliphatic dithioester, xanthate or dithiocarbamate RAFT agents. This is attributed to the greater rate of fragmentation and the correspondingly shorter lifetime of the adduct radicals in those polymerizations. In many cases, the initiator concentration (e.g. AIBN) had to be relatively large (i.e. $> 1 \times 10^{-1}$ mol·L$^{-1}$) in order to enable the observation of a sufficiently strong signal (see, for example, Refs. [99, 125, 126]); the observed spectrum changes with polymerization time/conversion.

Quantification of the ESR signals when combined with the rate of polymerization data and analyzed via kinetic modeling provides an access to the addition rate and intermediate fragmentation rate coefficients. Intermediate radical concentrations measured by ESR spectroscopy in both the pre- and main-equilibrium regimes disagree with those expected on the basis of the equilibrium constants predicted by quantum mechanical ab initio calculations (see section 3.2.3.2) [97]. This discrepancy is one of the outstanding open questions in the RAFT mechanism (see also section 3.4).

#### 3.2.4.4 **Chromatography**

Interaction chromatography (liquid chromatography at the critical point of adsorption, LC-CC) and two-dimensional chromatography (2D GPC/LC-CC) have also been applied to establish the presence of RAFT end groups and the quality of RAFT polymerization. Pasch et al. [133] used 2D GPC/LC-CC to examine a poly(methyl methacrylate)-*block*-polystyrene (PMMA-*b*-PS) prepared by RAFT with CDB as the RAFT agent and were able to 'confirm the remarkable molar mass and compositional homogeneity' of the sample. Jiang et al. have determined the end group composition (fraction of RAFT agent and initiator-derived chain ends) of PMMA [134, 135] and PBA [136] prepared by RAFT with functional RAFT agents by LC-CC coupled with mass spectrometry.

Size exclusion chromatography in combination with RAFT agents containing UV chromophores has been used to provide information on the mechanism of RAFT polymerization of styrene [43] and BA [137]. Ah Toy et al. and Feldermann et al. have employed coupled SEC–UV–ESI–MS to map end groups generated during RAFT polymerization [109, 138].

#### 3.2.4.5 **Rate of Polymerization**

Rate measurements (of the overall rate of polymerization) during the preequilibrium sequence provide one of the main avenues for gathering information about the underpinning mechanistic processes. However, rate measurement can be very challenging to interpret (see section 3.2.3.3), since the knowledge of several rate coefficients and parameters is required for deducing the desired information, e.g. on the size of the RAFT equilibrium constant. However, it can be very instructive to record time-dependent monomer conversion data in RAFT-mediated polymerizations under varying experimental conditions and to analyze these data under different model assumptions. It has been shown in section 3.1.3 that the addition of low-molecular-weight (e.g. initial) RAFT agents can lead to periods of non-steady-state polymerization; the observation of such periods alone suggests that the propagating radical concentration is not constant. The cause of such nonconstancy can be multiple: (i) either an apparent non-steady-state behavior induced by rate coefficients changing with chain length, (ii) a true non-steady-state polymerization induced by slowly fragmenting RAFT adduct radicals in the preequilibrium or (iii) a hypothesized multifaceted situation in which the radical addition rates of various radicals to both RAFT agent and monomer differ largely, which basically relates to a composite situation containing the concept of a different stability of the first intermediate in comparison to that of the main equilibrium, of slow

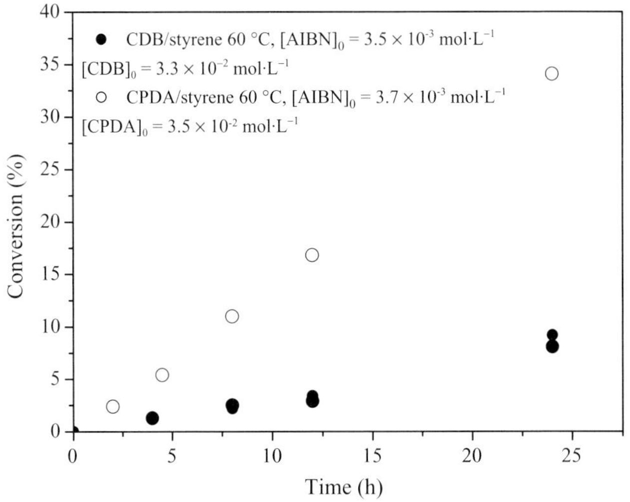


**Fig. 3.3** Conversion vs time evolution in a CPDA (open circles) and a CDB (closed circles) mediated styrene polymerization at 60 °C initiated with AIBN as thermally decaying initiator. The reaction conditions are identical for both polymerizations. Data are taken from Refs. [54, 57], respectively.

reinitiation and of potentially different cross- and self-termination rates for small intermediate radicals. It is clear that rate of polymerization measurements with one RAFT agent and one monomer yields only limited information about the possibly complex underpinning kinetics. However, rate measurements become particularly useful when rates of polymerization during the preequilibrium of different RAFT agents under otherwise identical reaction conditions are compared. For example, while the CDB-mediated polymerization of styrene is severely rate retarded with a pronounced period featuring a nonconstant rate of polymerization [57], the corresponding polymerization under identical reaction conditions carried out with CPDA as the RAFT agent displays a constant rate of polymerization (see Fig. 3.3) [54].

Since both RAFT agents are identical accept for the different Z groups (phenyl vs benzyl) and all other experimental conditions are identical, too, one can conclude that the strongly increased stability of the RAFT adduct radical when going from CPDA to CDB leads to a lowered and nonconstant radical concentration. The cause of this nonconstancy in the radical concentration has been interpreted differently by different groups and has largely been correlated with either high equilibrium constants in the preequilibrium (as foreshadowed by the quantum mechanical calculations described in Chapter 1) or/and selective reactivities of the variable species occurring during the preequilibrium as mapped by detailed $^1$H NMR spectroscopy (see section 3.2.4.2), possibly in conjunction with side reactions of the RAFT adduct radicals. As has been highlighted in section 3.2.3.3, such rate data from RAFT polymerizations can be employed to deduce kinetic rate coefficients via preassumed kinetic models; however, such approaches are model dependent. Yet, any model that is proposed to describe the RAFT mechanism has to describe the overall rate of polymerization.

#### 3.2.4.6 Radical Storage

It has been established that under certain conditions the addition of a RAFT agent to the polymerization can not only lead to strong rate retardation effects but also allow for radicals to be stored – without further propagation activity – in the reaction mixture. Thus, radical storage experiments are a method to probe for species in RAFT systems which can store radicals and release them at a later stage. In a typical radical storage experiment, a mixture of monomer and initial RAFT agent is exposed to a source of radicals (e.g. $\gamma$-radiation) at ambient temperature $T_1$ for a time period $t_1$. Subsequently, the reaction mixture is left without initiation for a time period $t_2$ and then maintained at a temperature $T_2 > T_1$ for a time period $t_3$. After each step of the sequence, i.e. radical feed at ambient temperature, waiting period and heating interval, the reaction mixture is analyzed with respect to monomer conversion and molecular weight distribution of any formed polymer. This experimental sequence has been applied both to CDB/styrene and CDB/MA systems [139, 140], as well as to non-rate-retarding RAFT agents [141]. The findings of these studies, which were accompanied by reference experiments containing no RAFT agent, indicate that systems containing low-mass dithiobenzoates as the RAFT agent are capable of storing free radicals for variable lengths of time, signifying that transient but relatively stable species are formed, which are capable of inducing monomer to polymer conversion at a later stage without any initiation source present. While radical storage experiments show that there are long-lived species present in certain RAFT systems, they were not designed to identify the nature of the radical sink, i.e. whether it be radical or nonradical in nature. It should be noted that all reported radical storage experiments relate to the preequilibrium situation in which only small radicals are involved. To what extent the effect is occurring in the main equilibrium has yet to be examined.

In a related context, it is noteworthy that low-molecular-weight thioketone compounds (which have been demonstrated to be effective radical spin traps) can form relatively stable adduct radicals upon reaction with other radicals (giving rise to equilibrium constants, $K = k_{add}/k_{-add}$, between $10^5$ and $10^8$ L·mol$^{-1}$), with reported apparent lifetimes are similar to those conceptualized in the slow fragmentation model [142]. As all RAFT agents are a subset of the thioketone compound class, the experimental observation [19] that certain thioketones are capable of imparting living characteristics on a free radical polymerization in styrene [19] and $n$-BA systems [21] – possibly by simply acting as effective spin traps – may also be significant in the context of dithiobenzoate-mediated RAFT polymerizations.

### 3.2.5 Reinitiation and RAFT Agent Selection – Choice of R

Many thiocarbonylthio RAFT agents have now been described [35, 36]. Transfer constants are known to be strongly dependent on the Z and R substituents. For an efficient RAFT polymerization:

- both the initial and polymeric RAFT agents should have a reactive C=S double bond (high $k_{add}$);
- the intermediate radicals should fragment rapidly (high $k_{\beta}$, weak S—R bond) and give no side reactions;
- the intermediate should partition in favor of products ($k_{\beta} \geq k_{-add}$); and
- the expelled radicals (R•) should efficiently reinitiate polymerization ($k_i \geq k_p$).

Details of the dependence of RAFT agent activity on R and Z as a function of monomer are provided in Chapter 6.

The R group of the initial RAFT agent should be chosen such that R• is a similar or better homolytic leaving group than the propagating radical derived from the monomer being polymerized. One ideal choice for R would be a propagating radical. In this case $k_{\beta} = k_{-add}$, $k_{add}$ is the same for both initial and polymeric RAFT agents and $k_i = k_p$.

For the case of acrylates it has been proposed that a suitable R group is a unimer of the propagating radical. Thus, MA polymerization is effectively controlled by RAFT agents with R = 1-(methoxycarbonyl)ethyl (e.g. 1-(methoxycarbonyl)ethyl phenyldithioacetate, **23**). The suggestion that the R group might be chosen to be a unimer of the propagating radical is unfortunately not general since $k_{\beta}$ and $k_{-add}$ can be subject to substantial penultimate unit effects. This is particularly true for the case of tertiary propagating radicals. RAFT agents with R = 2-(ethoxycarbonyl)prop-2-yl (e.g. 2-(ethoxycarbonyl)prop-2-yl dithiobenzoate, **26**) provide only poor control over polymerization of MMA.

It is also important that the expelled radicals (R•) should efficiently reinitiate polymerization of the particular monomer. The leaving ability of the substituent R must be balanced with the ability of the radical R• to reinitiate polymerization. The triphenylmethyl radical, for instance, would be an excellent leaving group but would be a poor reinitiator of chains and its use would result in retardation of polymerization [143]. For similar reasons, benzylic species (cumyl, 1-phenylethyl, benzyl) should not be chosen for vinyl monomer polymerization (e.g. VAc, NVP) since rates of reinitiation are very low with respect to $k_p$. Retardation in MMA polymerization when R is —CHPh(CN) has also been attributed to slow reinitiation by R• [144]. Rate constants for the reaction of primary radicals with monomers are known for many of the species (R•) likely to be encountered. Good sources of such information are the tabulation by Moad and Solomon [143] and Fischer and Radom's [121] review. Values of rate constants for reaction of common R• with MMA, MA and styrene are provided in Table 3.3. From the point of view of reinitiation efficiency, *tert*-butyl (•$C(CH_3)_3$) is an excellent choice for R with MMA, MA and styrene (*tert*-butyl RAFT agents are not suitable for use in MMA polymerization as *tert*-butyl radicals are a very poor leaving group with respect to the PMMA propagating radical).

As with conventional chain transfer, for no retardation the following relationship must hold:

$$k_i \geq \frac{[M]_0}{[T]_0} k_p \tag{3.18}$$

**Table 3.3** Absolute rate constants ($k_i$) for primary radical addition to various monomers*

| | Rate constant for R• ($M^{-1}\cdot s^{-1}$) | | | | |
|---|---|---|---|---|---|
| Monomer | •$C(CH_3)_3$ (300 K) [145, 146] | •$C(CH_3)_2CO_2Me$ (294 K) [147] | •$C(CH_3)_2CN$ (315 K) [148] | •$C(CH_3)_2Ph$ (296 K) [145] | •$CH_2Ph$ (296 K) [145] |
| MMA | | | | | |
| a | 660 000 | 3 170 | 1590 | 2700 | 2100 |
| b | — | — | — | 10 225 | — |
| c | 952 970 | 9 722 | 3 291 | 7 829 | 8 348 |
| MA | | | | | |
| a | 110 000 | 1 150 | 367 | 800 | 430 |
| b | — | 3 129 | — | — | 2 134 |
| c | 1 523 800 | 3 412 | 865 | 2 650 | 2 042 |
| Styrene | | | | | |
| a | 132 000 | 5 500 | 2 410 | 1 100 | 1 200 |
| b | 226 622 | 12 508 | 4 041 | 3 803 | 4 557 |
| c | 224 850 | 13 455 | 4 896 | 3 802 | 4 685 |

(a) Value provided in original paper for temperature indicated. (b) Value at 60 °C calculated using Arrhenius parameters provided in the reference indicated (— indicates no Arrhenius parameters given). (c) Value at 60 °C estimated using log *A* 7.5 (tertiary radicals) or log *A* 8.5 (benzyl radical) and activation energy as reported [123].
*Data are reproduced from Ref. [44].

Further details of the exact dependence of RAFT agent activity on R and Z as a function of monomer are provided in Chapter 6.

### 3.2.6
### Structure–Property Correlation – Choice of Z

The chemical nature of the R and Z groups determines the magnitude of the individual rate coefficients that govern the preequilibrium for a given monomer. Most of the variation in the preequilibrium seems to stem from variations in the fragmentation rates. The addition rate coefficient appears largely insensitive to variations in the RAFT agent structure as evident from the published experimental values for polymeric systems, which show close agreement with one another, despite significant changes in the attacking radical and in some cases also the RAFT agent and temperature. For example, addition rate coefficients ($L\cdot mol^{-1}\cdot s^{-1}$) for styrene propagating radicals adding to a dithiobenzoate at 60 °C ($4 \times 10^6$) [57] VAc propagating radicals adding to a xanthate at 80 °C ($5 \times 10^5$) [49], MA propagating radicals adding to a dithioacetate ($1.5 \times 10^6$) at 80 °C [51] BA ($1.7 \times 10^6$) [93] and dodecyl acrylate propagating radicals ($2.4 \times 10^6$) [50] adding to a trithiocarbonate at 80 °C, are within an order of magnitude with one another, and are similar to the ab initio values [149] for the addition of various small radicals to RAFT agents. Temperature seems to have a relatively small influence on the addition rate coefficients with BA propagating radicals adding to *S,S'*-bis(methyl-2-propionate)trithiocarbonate at $-30^{\circ}C$ ($2 \times 10^5$) showing only a limited reduction over a wide temperature range [150]. Such findings are not unexpected as the reaction could potentially come under diffusion control for longer propagating radicals and macro-RAFT agents; however, such situations will be primarily found during the main equilibrium (see section 3.2.3.4).

The fragmentation rate coefficient of the preequilibrium adduct radical is, on the contrary, highly dependent on the nature of both the Z and R groups as is also indicated by the quantum chemical calculations mentioned in section 3.2.3.2 and Chapter 1 of this book. Indeed, it is a simplification to assume a stability for the preequilibrium adduct radical when the R group, free radical initiator fragment and propagating radical are of different electronic and/or steric nature. Even in cases when the initiator-derived radical and the R group are identical, a strong chain length dependence caused by penultimate effects on the preequilibrium adduct radicals is likely operational (see section 3.2.3.3 for a detailed example). Thus, it is preferable to compare the effects of the Z group on the stability of the adduct radical for given side chains or one specific R group. For a given R group (e.g. cyanoisopropyl), the intermediate radical stability increases strongly when going from methyl to phenyl and the polymerization becomes increasingly prone to rate retardation and inhibition effects. However, the degree by which rate retardation/inhibition is to be expected is also a function of the monomer. Acrylates for a given end group will induce a greater degree than styrene, due to the higher reactivity of the propagating macroradicals. The correlation between propagating radical reactivity and the stabilizing ability of the Z group is depicted in Fig. 3.4. For a given Z group and monomer combination (e.g. Z = phenyl and M = MA), the ability of the R group to fragment off the intermediate radical will co-determine its stability; i.e., a benzyl R group would lead to a more stabilized intermediate radical than a *tert*-butyl R group. Note that the ability of a particular R group to fragment easily from the intermediate radical (in the preequilibrium) does not necessarily ensure that the polymerization will not be rate retarded or inhibited, as the co-condition for an efficient RAFT process is also an effective reinitiation ability of the leaving group radical (see also section 3.2.5, especially Table 3.3).

## 3.3 Main Equilibrium Kinetics and Mechanism

From a kinetic perspective, the main-equilibrium kinetics for RAFT polymerizations, in which control is exerted by a polymeric RAFT agent of similar composition, are simpler to study than those of the preequilibrium [97]. This is true since, as the above discussion illustrates, many of the complexities relate to (poor) selection of 'R'. It should be noted that since RAFT polymerization requires a source of (typically low-molecular-weight) free radicals to maintain the process, preequilib-

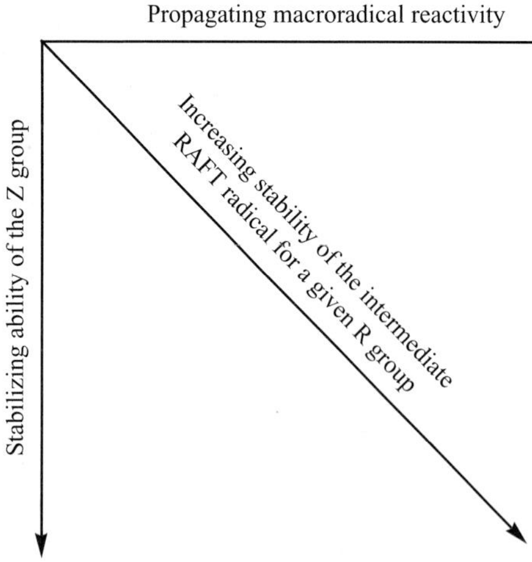


**Fig. 3.4** Effect of the stabilizing ability of the Z group and the reactivity of the propagating macroradical on the stability of the intermediate macro-RAFT radical for a given R group. An increase in stability in the intermediate radical is often correlated with the magnitude of rate retardation and inhibition effects observed during the polymerization.

rium phenomena cannot be completely avoided. In this context, selection of the initiator for RAFT polymerization is extremely important especially when making kinetic measurements. In what follows, we will investigate the RAFT main equilibrium in great detail, especially with a view to the potential side reactions that can occur. We will also comment on transfer constants for polymeric RAFT agents and initiator selection for chain extension experiments (i.e. block copolymer formation).

### 3.3.1
### Initiator Selection

Radicals are neither formed nor destroyed in the RAFT process. Thus, RAFT polymerization, like conventional radical polymerization, requires a source of radicals. In a well-designed RAFT experiment, the initiator is a minor component of the reaction mixture (typically the ratio of RAFT agent (or RAFT end group) to initiator concentration is larger than 10:1). Furthermore, with less active RAFT agents and during the synthesis of higher-molecular-weight polymers, the initiator-derived radicals will most likely react preferentially with monomer. For these reasons, the consequences of poor initiator selection may not be detectable and will often cause only a minor perturbation of 'ideal' kinetics [35].

However, side reactions involving the initiator and the initiator-derived radicals known to complicate conventional radical polymerization will also occur during

RAFT polymerization. These reactions include transfer to initiator and primary radical termination. Since these reactions lead to dead chains, for optimal control it is important that the initiator should be chosen such that any side reactions of the initiator or the initiator-derived radicals are minimized. Selectivity in the reaction of the initiator-derived radical with monomer can also be an important consideration. The specificity of the reactions of initiator-derived radicals with monomers has been extensively studied [151].

It is also important to avoid adverse reactions between the RAFT agent and the initiator and the initiator-derived radicals. For example, some peroxide initiators (dibenzoyl peroxide, potassium peroxydisulfate) and derived radicals may oxidize the RAFT agent [152]. Other radicals may react with the RAFT agent to form a stable product. It is important that the initiator-derived radical is a good leaving group with respect to the propagating radical. Thus, 2,2′-azobis(methylisobutyrate) is not a suitable choice for RAFT polymerization of MMA. During RAFT polymerization of MMA with 2,2′-azobis(methylisobutyrate), formation of the initiating radical-derived RAFT agent as a relatively stable by-product is observed [60].

The mechanism of AIBN and other azonitriles decomposition is complicated by the formation of ketenimines as unstable intermediates. In the presence of high concentrations of RAFT agents, the ketenimine is intercepted and converted to by-products which reduces the initiator efficiency and the polymerization rate [71].

## 3.3.2 Potential Side Reactions

A variety of actual and potential side reactions have been proposed to complicate the basic RAFT mechanism.

### 3.3.2.1 Reaction with Oxygen

RAFT polymerizations, like conventional radical polymerizations, are air sensitive. Both propagating radicals and intermediate radicals should be anticipated to react with oxygen at diffusion-controlled rates. These side reactions can be avoided by taking the usual precautions to exclude air. The effect of air on the course of RAFT polymerization will depend on the system. In the case of solution polymerization of MMA, substantial retardation and poorer control over molecular weight and polydispersity are observed [35]. Some RAFT emulsion polymerizations are reported to be unaffected by air (see also Chapter 8 for detailed account of (mini)emulsion polymerization and RAFT).

### 3.3.2.2 Intermediate Radical Propagation

Intermediate radical propagation (or copolymerization of the RAFT agent) is a major issue with the use of macromonomer RAFT agents when used with monosubstituted monomers such as acrylates and styrene [153]. Intermediate radical propagation is unknown with thiocarbonylthio RAFT agents. This is underpinned by the findings in recent thioketone-mediated polymerization, where copolymerization of the thioketone is not observed [21].

#### 3.3.2.3 Irreversible Intermediate Radical Termination

All radical polymerization including those that go under the heading living radical polymerization will be complicated by irreversible termination by radical–radical reaction between propagating radicals. To achieve good control it is important to choose reaction conditions such that the incidence of termination is negligible.

The intermediate is a radical species and therefore it may also be involved in radical–radical reaction either by self-reaction or by reaction with propagating radicals or other radical species. Monteiro [154, 155] proposed that what he termed 'intermediate radical termination' might explain retardation seen in styrene polymerization with CDB. Support for this hypothesis has come from the work of Fukuda and coworkers [61, 98–100]. Irreversible termination reactions will lead to a depletion of the initial RAFT agent and can thus affect the efficiency of the mediation process and – in extreme cases – lead to the loss of molecular weight control. To what extent truly irreversible cross-termination reactions are operative in the RAFT main equilibrium remains to be established. Theoretical small molecule studies (up to the trimer stage) by Coote and coworkers indicate high stabilities for dithiobenzoate-derived RAFT adduct radicals that would lead to a large rate of intermediate radical termination, while theoretical suggestions by Buback and coworkers indicate that the potentially generated three-arm star polymers are intermediates themselves (see section 3.3.2.6).

#### 3.3.2.4 Reversible Intermediate Radical Termination

The possibility of the bimolecular termination reactions discussed in section 3.3.2.3 has been proposed in numerous studies, largely driven by attempts to harmonize the discrepancy between the observed (via ESR spectroscopy) relatively low intermediate radical concentrations and the predicted large equilibrium constants obtained from the analysis of low conversion RAFT polymerization [53, 86, 111] and high-level ab initio quantum chemical calculations [156].

Reversible termination of the intermediate possibly proceeds via self-termination pathways (as well as cross-termination) as depicted in Scheme 3.9.

A termination reaction of intermediate RAFT radicals with their radical centers being delocalized into the aromatic ring may suffer less steric hindrance and the stability of the resonance-stabilized radical may allow for reversibility [54, 157]. When using a dithiobenzoate with a substituent on the para position of the Z group, such as cumyl-*para*-methyl dithiobenzoate, a significant reduction of the rate retardation effect has been observed [158]. As the *para*-methyl group is not expected to change the stability of the intermediate significantly, but makes the para position less prone to radical attack due to steric congestion, an involvement of delocalized radical sites in rate-retarding reactions can be concluded. Direct evidence for such reactions, however, has yet to be found. The radical storage

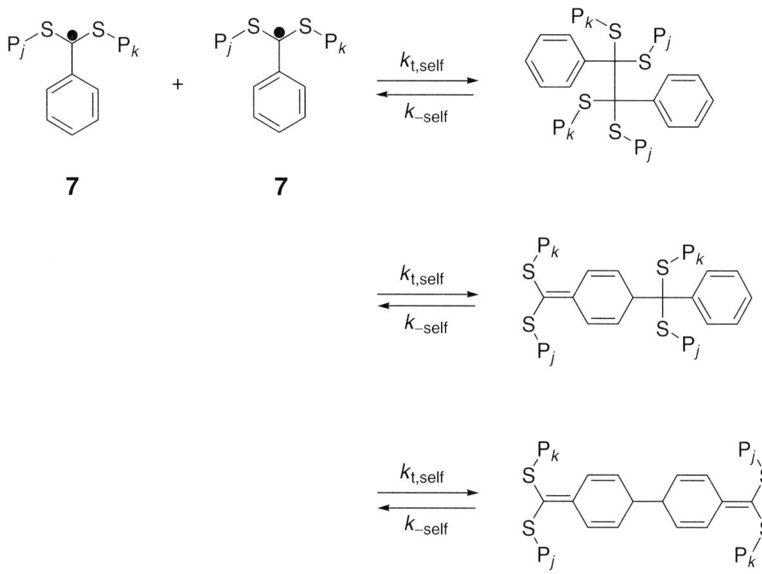


**Scheme 3.9** Reversible self-termination reactions between intermediate radicals during the main equilibrium proceeding via variable resonance structures.

experiments described above designed to probe the preequilibrium sequence (see section 3.2.4.6) have been rationalized by invoking the presence of self-reversibly terminated intermediate radicals, yet they provide no information of the possibility of such reactions during the main-equilibrium stage. In addition, a satisfactory embedment of such reaction channels into kinetic schemes has not been achieved so far.

#### 3.3.2.5 **Undesired Fragmentation**

For some RAFT agents there are multiple pathways for β-scission involving cleavage of a bond within the 'Z' group. This is true in the case of xanthates where the substituent on O is a good homolytic leaving group where two possible fragmentation pathways are possible. Examples are *O-tert*-butyl xanthates [159]. The same situation may pertain in the case of unsymmetrical trithiocarbonates [137].

#### 3.3.2.6 **Other Intermediate Radical Termination**

Buback and Vana [160] have proposed another mechanism for disappearance of the product of intermediate radical termination during RAFT polymerization with dithiobenzoates. In that it does not lead to by-products, the proposal has the potential of resolving the current impasse regarding the mechanism of retardation observed in these polymerizations [97]. The proposed mechanism involves the intermediate radical **39** (or an isomer) reacting with a propagating radical to regenerate the intermediate **40** and the product **41** as the only by-product, which is

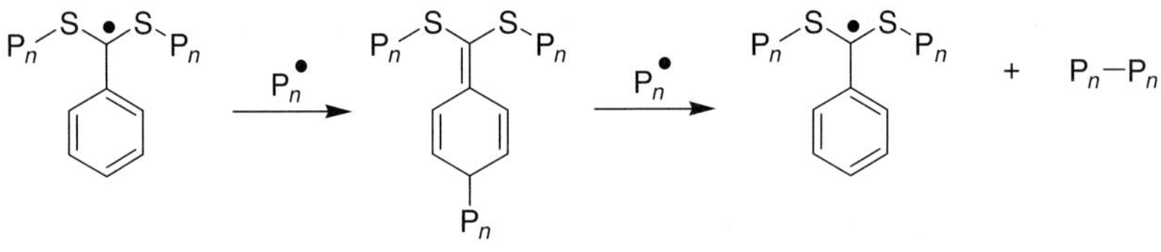

**39** **40** **39** **41**

**Scheme 3.10** Intermediate radical termination and subsequent reactivation of star polymer.

identical to the normal product of radical–radical termination. While the process is thermodynamically favorable being driven by aromatization and formation of a relatively stable radical, there appears no literature precedent for the reaction of quinonoid species such as **40** with radicals to proceed as shown in Scheme 3.10.

Another possibility is that the quinonoid species (**40**), if formed, reacts with radicals by hydrogen atom transfer. This process would result in formation of a new intermediate (**42**) to fragment to a new RAFT agent (**44**) and a dead polymer chain (**43**) (Scheme 3.11). Quinonoid species are well known as excellent hydrogen atom donors. The effect on kinetics would be similar to the Buback–Vana proposal [160] but would lead to the formation of a new RAFT agent (**44**) which might be detectable under appropriate conditions but, for the case of acrylate polymerization, may not be readily distinguished from the by-product formed by other processes such as that described in section 3.3.3.1.

The above processes (Schemes 3.10 and 3.11) are only relevant to RAFT polymerizations with dithiobenzoates and closely related RAFT agents. They are not relevant to RAFT polymerizations carried out with trithiocarbonates, aliphatic dithioesters and most other RAFT agents, many of which also give retarded polymerization.

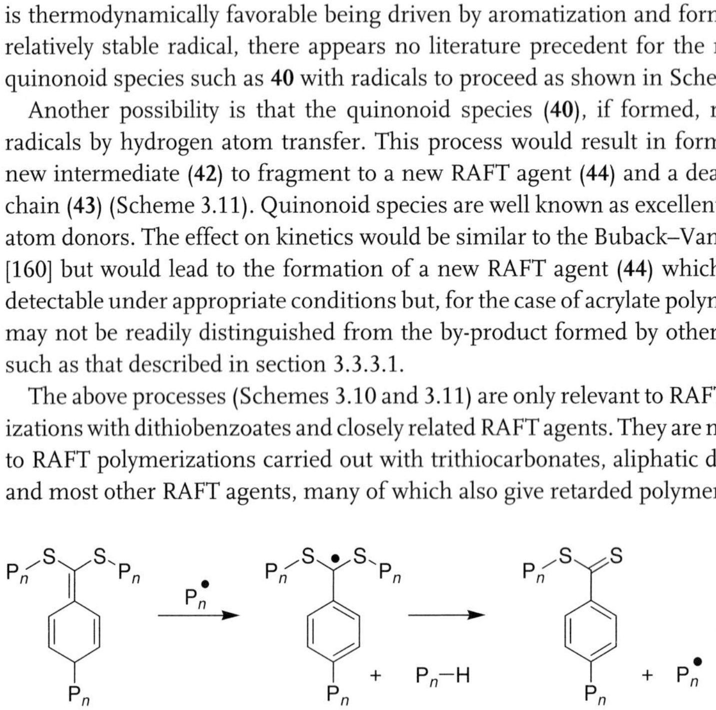

**40** **42** **43** **44**

**Scheme 3.11** Formation of polymeric RAFT agent from quinonoid structures.

### 3.3.3 Non-RAFT-Dependent Side Reaction Which Nonetheless Influence the RAFT Process

Side reactions which occur in conventional radical polymerization can be expected to also complicate RAFT polymerization. These side reactions include transfer to monomer and both inter- and intramolecular transfer to polymer. It is not appropriate to ignore these when trying to understand the kinetics and mechanism of RAFT polymerization.

#### 3.3.3.1 **Backbiting**

It is now well established that backbiting occurs during radical polymerization of acrylates. Backbiting may be followed by propagation or chain scission.

Bimodal molecular weight distributions are often observed in RAFT polymerization of acrylate monomers when polydispersities are less than 1.2 [88]. They are manifest as a higher-molecular-weight shoulder with peak molecular weight approximately twice that of the parent peak and are most pronounced for higher conversion and higher-molecular-weight polymers. Note that if the polydispersity is greater than 1.2, such bimodal distributions will be masked. They have been observed for dithiobenzoate, dithioacetate, trithiocarbonate and dithiocarbamate RAFT agents and are most noticeable for higher monomer conversions (>50%) and for higher molecular weights (>50 000). Postma and coworkers [137, 161] have provided evidence that the mechanism of formation is copolymerization of the macromonomer formed by the propagating radical undergoing backbiting β-scission (Scheme 3.12). The incidence of long-chain branching by this mechanism was compared to that of short-chain branching (by backbiting followed by propagation). Since β-scission competes with propagation the number of branches per molecule is greater for high monomer conversions and for higher molecular weights. Note that the proposed process does not depend on the RAFT mechanism and should complicate all radical polymerizations of acrylate monomers. The process may also complicate the polymerizations of other monosubstituted

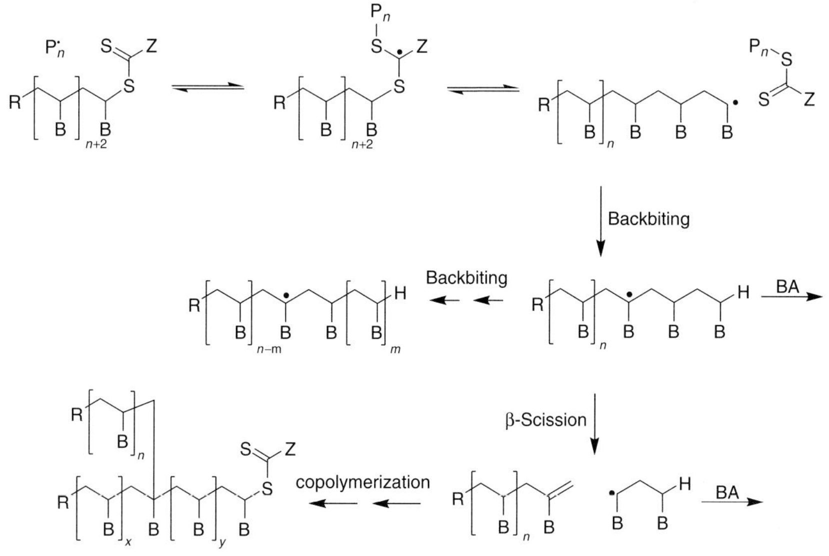


**Scheme 3.12** B = $CO_2C_4H_9$.

monomers (e.g. VAc) but is likely to be less prevalent because other monomers generally display a reduced propensity for backbiting β-scission.

## 3.4
## Mechanisms for Rate Retardation/Inhibition – Outstanding Questions

From the above analysis of the mechanism and kinetics of the RAFT process in both the pre- and main equilibrium, it has become clear that facets of the kinetics and mechanism – particularly relating to rate retardation and inhibition effects in dithiobenzoate-mediated RAFT processes – are not fully established and many important questions remain unresolved. There is experimental and computational (see Chapter 1) evidence to indicate that some of the issues are related to the differences in reactivity between the initial 'low mass' and the polymeric RAFT agent. The staggering complexity of the experimental observations and theoretical calculations suggests that the RAFT process is probably not explicable via a simple 'unified' model. The reactions observed are dependent on the specific RAFT agent. There are significant differences between the pre- and main equilibrium. The initiator and initiator-derived RAFT agents are important. The intermediate species may undergo reactions other than fragmentation. It is likely that other hitherto unknown mechanistic features of the RAFT process will need to be incorporated. Indeed, the situation in 2006 has been found to be sufficiently complex that an IUPAC task group was established to define the dilemma at hand [97]. Since then some progress has been made in suggesting possible solutions to the outstanding inconsistencies (including the idea that potentially generated three- and four-arm star polymers are only intermediates themselves capable of reactivation by propagating radicals; see section 3.3.2.6) as well as the emergence of novel ways to probe the intermediate radical stability via pulsed laser experiments coupled with ESR spectrometry [150] as well as the approach to reduce the complexity of the RAFT system by a structural change leading to thiocarbonyl compounds whose ability to control polymerizations has been tested [19–21]. However, the summary conclusions drawn by the IUPAC task force for dithiobenzoate systems still hold valid:

(a) A model featuring short-lived intermediate RAFT radicals in conjunction with their irreversible intermediate termination is in agreement with the experimentally determined equilibrium constant and with the observed steady state in the main equilibrium; it provides a consistent kinetic picture for the main equilibrium. However, this model predicts concentrations of intermediate termination products, which cannot be identified in correct quantities.
(b) The quantum-chemical calculations referred to in section 3.2.3.2 and Chapter 1 of this book, which are performed for small model species, as occurring in the preequilibrium, predict a very high stability of the intermediate RAFT radicals resulting in their slow fragmentation. However, the predicted extremely high intermediate radical concentrations cannot be observed experimentally.

(c) Mass spectrometric studies reveal the occurrence of intermediate termination in model systems, indicating the possibility of cross- and self-termination in dithiobenzoate-mediated polymerizations. In polymerizing systems, however, the amount of intermediate termination products predicted from the model of fast fragmentation in conjunction with cross-termination cannot be found (or can only be found in insufficient quantities) by mass spectrometry.

The unresolved complexity of the RAFT processes, in particular dithiobenzoate-mediated polymerizations, may seem a substantial disadvantage of the process over others such as ATRP and NMP. However, the consequences of the variable mechanistic underpinnings are of relatively low importance for the synthetically working chemist, as will be highlighted in section 3.7. The resolutions of the outstanding mechanistic questions, however, are of critical importance on a fundamental level as they touch upon more general questions of radical stability ('how stable can thiocarbonylthio derived radicals be?') as well as to the accuracy of high-level quantum chemical calculations ('are quantum chemical calculations a true replacement for experiment?').

## 3.5
## RAFT Copolymerization: Block and Statistical Copolymers

There are now numerous reports to show that the RAFT process is applicable to statistical copolymerizations. In most copolymerizations, the monomers are consumed at different rates dictated by the steric and electronic properties of the reactants. Consequently, both the monomer feed and copolymer composition will drift with conversion. Thus conventional copolymers are generally not homogeneous in composition at the molecular level. In RAFT polymerization processes, where all chains grow throughout the polymerization, the composition drift is captured within the chain structure. All chains have similar composition and are called gradient or tapered copolymers. Thus one difference between copolymers synthesized by RAFT copolymerization and those synthesized by conventional radical copolymerization is that copolymers formed by RAFT copolymerization are homogeneous gradient copolymers whereas those formed by the conventional process are blends.

High conversion copolymers when formed in the presence of a RAFT agent have the same overall composition and monomer sequence distribution as those formed in the absence of a RAFT agent. Reactivity ratios appear unaffected by the RAFT process. However, for very low conversions when the molecular weight of all chains is low, copolymer composition may be slightly, yet significantly, different for that seen in conventional copolymerization because of the specificity shown in initiation by R and in capture of the propagating radicals by the RAFT agent. The same phenomenon is observed in radical polymerization with conventional chain transfer when molecular weights are low [162–164]. This means that the usual methods for evaluation of reactivity ratios which require low conversions and do not take monomer usage during initiation into account are not directly applicable to RAFT copolymerization. Modified equations which are suitable in these circumstances have been reported [164].

In the CDB-mediated copolymerizations of MMA–styrene, MA–styrene and MMA–BA, the polymer mole fraction of the monomer with the larger reactivity ratio is increased compared to the conventional copolymerization [165]. Simulations demonstrate that the RAFT process itself may alter the macroradical populations and the copolymer composition.

Note that while RAFT and other chain transfer processes may affect the relative concentration of the propagating species, this should not have a direct influence on the copolymer composition [164].

A detailed study of the kinetics of RAFT copolymerization of styrene and MMA with dithioacetate RAFT agents has been reported by Fukuda and coworkers [91]. Transfer constants at 40 °C were reported for the PMMA and polystyrene macro-RAFT agents in polymerizations of MMA, styrene and the azeotropic copolymerization of MMA and styrene. The data indicate that in styrene polymerization, 50% conversion of the PMMA macro-RAFT agent is achieved at very low monomer conversion (~0.16%), while in MMA polymerization, the polystyrene macro-RAFT agent is half consumed at a much higher conversion (~57%). The results provide further quantitative support to the observation that when preparing block copolymers of methacrylates with styrene (and other monosubstituted monomers) it is best to prepare the methacrylate block first.

The results also further understanding of the observation that copolymerizations of monomers may be controlled even though homopolymerization of the same monomer may be uncontrolled or poorly controlled by particular RAFT agents [35]. Other examples of these phenomena are copolymerizations of the captodative monomer ethyl $\alpha$-acetoxyacrylate with acrylic monomers (BA, acrylic acid (AA), DMA, DMAEA) and a xanthate RAFT agent [166] (ethyl $\alpha$-acetoxyacrylate homopolymerization is not controlled with such RAFT agents), and the above-mentioned copolymerizations of octadecyl acrylate and other acrylic monomers with maleic anhydride or NVP with a trithiocarbonate RAFT agent (maleic anhydride does not readily homopolymerize and NVP homopolymerization is not controlled with trithiocarbonates) [167].

While statistical copolymerizations are important examples for the application of the RAFT process, its true strength – just as the strength of ATRP and NMP – lies in the formation of block copolymer structures. In this chapter we will not go into great detail into the various examples of block copolymers that have been generated (please refer to Chapter 9), but we will briefly remark on the mechanistic aspects underlying the correct monomer sequence that needs to be employed for a successful block copolymer formation. The formation of macro-RAFT agents – polymers with thiocarbonylthio end groups – as a result of a RAFT process is the key to the successful formation of block copolymers. Two possible mechanisms can be considered leading to block copolymers of varying chemical compositions: (i) chain extension of a macro-RAFT agent using a monomer disparate from the original monomer and (ii) attachment of a RAFT agent via covalent linkage to a polymer with functional end groups preferably to combine blocks prepared via different polymerization techniques. In here, we will focus on the first approach only, as the second approach will be discussed in Chapter 9.

The final product of the RAFT process is ideally a polymer chain carrying thiocarbonylthio end groups, the so-called macro-RAFT agent, which resembles the initial RAFT agent with a polymeric leaving group. Employing such a macro-RAFT agent in the polymerization of a new monomer results in chain extension, and hence the formation of block copolymers. A crucial step for the successful formation of block copolymers via this method is the evaluation of the stability of the corresponding propagating macroradical associated with each block: Fragmentation of the RAFT adduct radical in the block copolymer formation preequilibrium must be able to proceed toward both sides forming either a macroradical based on the monomer of the original macro-RAFT agent $P(M_1)$ or the macroradical based on the new monomer $P(M_2)$. A preferred fragmentation toward the $P(M_1)$ macroradical is crucial. Fragmentation of the radical intermediate generating $P(M_2)$ macroradicals will only result in homopolymers. In general, methacrylate-type monomers should always be prepared as the first block followed by chain extension with styrene or acrylate-type monomers. Once the initial transfer from one macroradical to the second macroradical has succeeded, the formation of well-defined block copolymers structures will only be prevented by competing reaction such as unwanted terminations and/or chain transfer reactions. The kinetic parameters that lead to favorable conditions for well-defined block copolymer formation have recently been explored in a simulation study [168]. For example, when the first dormant block displays a (idealized) PDI of unity and the attacking propagating radical has a low reactivity to the initial RAFT end group, the overall polydispersity index (PDI) will be significantly greater than 1 and vary strongly with the weight fraction of each block (Scheme 3.13).

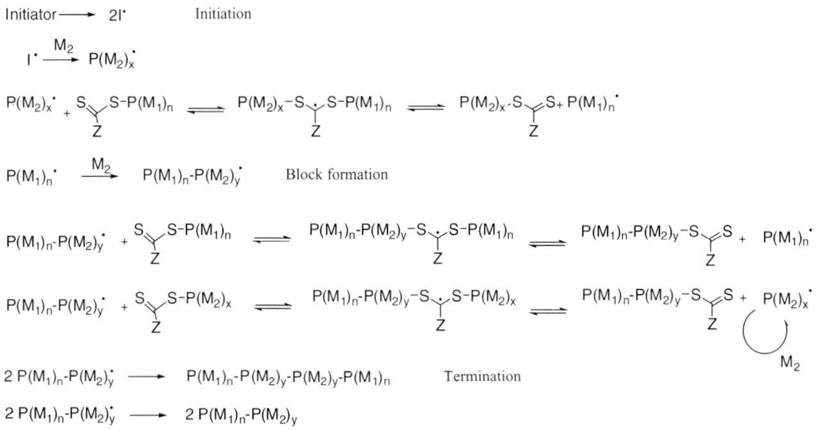


**Scheme 3.13** Formation of block copolymers via the chain extension of macro-RAFT reagents.

## 3.6
## The Kinetics and Mechanism of Star and Graft Polymer Formation Processes

The possibility of forming complex macromolecular architectures via the RAFT process is addressed in great detail in Chapter 9 and thus these findings will not be reiterated here in great detail. Yet, it seems apt in a kinetically oriented chapter to briefly outline the key factors leading to well-defined star polymers. In principle, there exist two design options in the construction of complex macromolecules via the RAFT process in which part of the thiocarbonylthio compound is to be linked to the scaffold that will serve as the core of the architecture. The question of which linking strategy should be followed is not trivial and must be considered carefully.

Two key approaches exist: In the Z-group approach [169], the RAFT agent stays permanently tethered to the scaffold structure (unless hydrolysis reactions interfere) and consequently the core will not carry any propagating radical functions. Propagation occurs exclusively in the solution surrounding the core. In contrast, an R-group approach implies that the dithioester moiety leaves the core structure and mediates the polymerization detached from the core, while the core itself becomes a radical. As a consequence, the core can undergo coupling reactions and this may in turn prevent the formation of well-defined macromolecular material. At a first inspection, it may thus seem clear that a Z-group architecture is always to be preferred over an R-group approach as it avoids the formation of higher-order coupling products. To some extent, such an observation is correct as the Z-group approach offers the in-principle opportunity of generating extremely pure star polymer products up to high monomer to polymer conversions without the added complexity of cross-coupling reactions. A deeper theoretical analysis of the difference between R- and Z-group approach architectures, however, reveals that under certain circumstances the R-group approach is excellently suited for the generation of very well defined star polymer material that contains virtually no side products in the form of higher-order couples and/or linear chain contaminants. It is paramount for the synthetic polymer chemist to have sound guiding principles that allow for rational decisions regarding which design approach (i.e. Z vs R) is preferred for a given monomer and set of reaction conditions. Recently, an approach has been developed for a modeling procedure making use of a combination of the PREDICI® software package and a probability method to construct star polymer molecular weight distributions from the chain length distributions of individual star polymer arms [170]. Subsequently, this methodology was employed in a combined experimental and modeling study to arrive at a set of design criteria that should be followed to obtain well-defined star polymers when using R-group approach polymerizations [171] (Scheme 3.14).

Four key factors can be identified that determine the success of the polymerization process in terms of arriving at a large fraction of well-defined star polymer:

*Minimal linear chain contaminants*: Linear contaminants can be minimized by employing a monomer with a high propagation rate coefficient while at the same

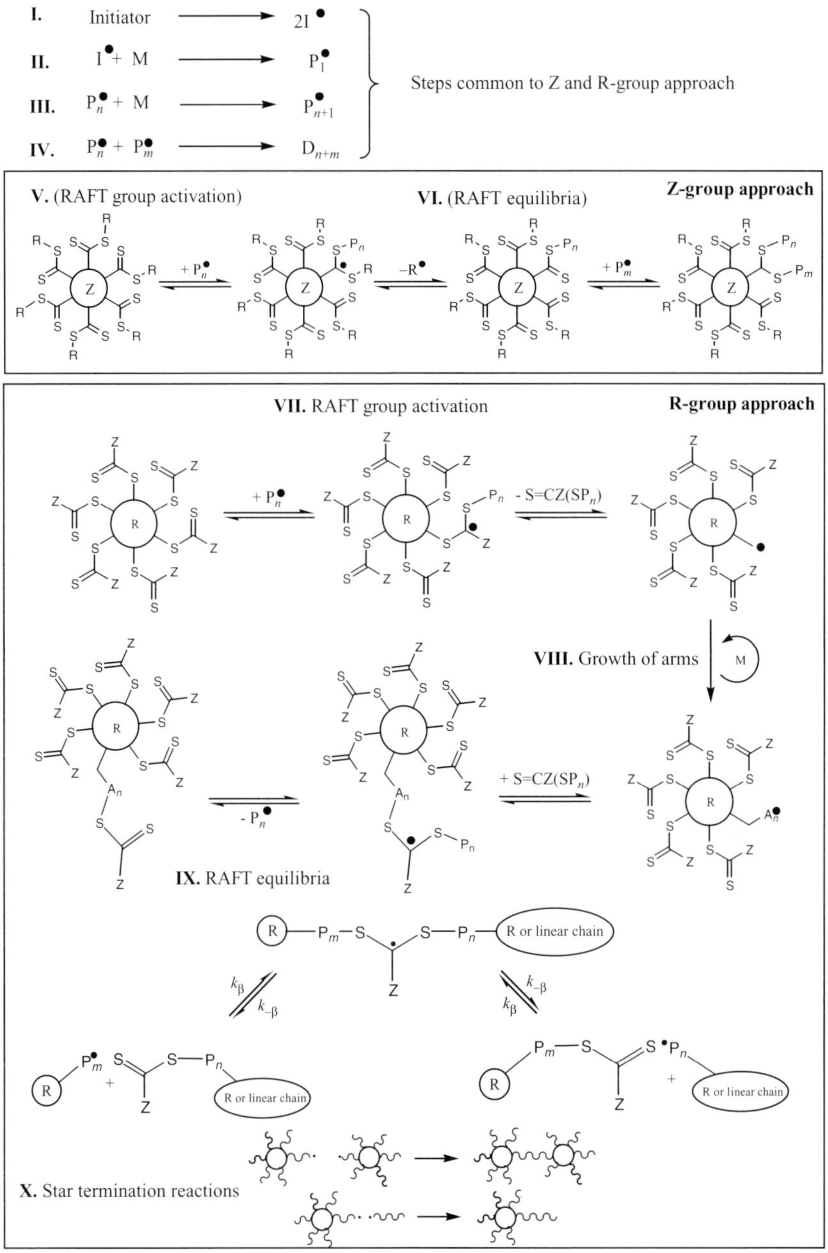


**Scheme 3.14** Summarized RAFT star polymerization mechanisms: Z-group vs R-group approach. In the Z-group approach, the arms of the star grow as linear chains and are only arms in the true sense when they have undergone additive reaction with the RAFT agent (VI). The R-group approach features arm growth on the star itself (VIII) with RAFT equilibria occurring between stars and stars, stars and linear chains, or linear chains and linear chains. Reprinted with kind permission from Ref. [171]. Copyright 2006 American Chemical Society.

time having a low rate of primary radical delivery into the system. However, having excessively high propagation rate coefficients may lead to an irregular activation of the RAFT groups contained on the initial multifunctional RAFT agent.

*Minimization of star–star coupling reactions*: Star–star coupling is a bimolecular radical termination event and its minimization can be achieved by (a) minimizing the number of radicals introduced into the polymerization system to achieve a given conversion after a given time (as determined by the primary radical delivery rate and the magnitude of the propagation rate coefficient). (b) Further, star–star coupling reactions may be minimized by a reduction of the propagating radical concentration in the RAFT polymerization via the formation of relatively stable intermediate RAFT radicals. The conversion of propagating radicals into intermediate (or adduct) RAFT radicals which may serve as a radical storage reservoir and undergo bimolecular radical terminations to a lesser extent (possibly due to steric shielding of the radical center) can offer protection to the propagating radicals from termination events, thus minimizing the amount of higher-order couples. (c) In addition, an effective reduction in the amount of star–star coupling can be achieved by a reduction of the number of arms that a multifunctional R-group approach RAFT agent carries. Such a reduction in the number of star–star couples via a reduction in the number of arms can be understood in terms of there being a fixed probability of one (propagating radical) arm forming a star–star couple. Thus, a reduction in the number of arms proportionally lowers the amount of star–star coupling events.

The above R-group approach guidelines have generally been found to hold true for a wide range of architectural designs and monomer systems. For selected examples demonstrating the abilities of the above kinetic guidelines to arrive at well-defined star polymer structures, the reader is referred to Chapter 9.

The situation in the case of the Z-group approach polymerization should – in principle – be simpler: Z-group approach polymerization should (provided monomer-induced site reactions such as branching do not interfere) exclusively yield well-defined and unimodal molecular weight distributions. However, in the case of the Z-group approach, the RAFT group is permanently bonded directly to the core (or polymer backbone), while the growing macroradicals are detached. To undergo transfer to the RAFT agent, the radical has to reach the RAFT functionality close to the core. With increasing conversion and therefore increasing length of the (star) arms an effective chain equilibration is increasingly more hindered due to the shielding effects of the polymer arms. Under certain conditions the propagating macroradical will rather terminate with another radical – instead of reacting with the RAFT agent – to generate a dead polymer. As a consequence, the molecular weight may deviate from the theoretical values as calculated from monomer conversion. Although the molecular weight development does not necessarily follow the theoretical values in the system with the core/backbone in the Z group, the RAFT polymerization often leads to unimodal molecular weight distributions and therefore – at least in theory – to a better-defined polymer. Termination reactions in such systems become less pronounced in core structures that feature a lower arm density. Here, less steric hindrance occurs and the RAFT process is less influenced by the growing arm size resulting in products following the molecular weight evolution predicted from the conversion and monomer–RAFT agent concentration ratio.

## 3.7 Mechanism and Kinetics as a Guide for the Synthetic Polymer Chemist

The synthetic polymer chemist is confronted with a seemingly complex mechanistic scenario that governs the RAFT process and the correct choice of RAFT agent for the monomer to be polymerized seems often daunting. However, it is very important to note that the RAFT process is arguably the most versatile of the living/controlled radical polymerization processes with regard to the monomer functionalities that are tolerated. Despite the fact that much attention has been drawn to the observation that a few RAFT-mediated polymerizations are strongly rate retarded, can display variable inhibition periods and may be subject to other kinetic anomalies, it is important to reiterate the comment that was made at the beginning of this chapter. Conceptually, the RAFT mechanism possesses no inherent rate retardation or inhibition phenomena. Thus, the RAFT process has the critical advantage, that it proceeds at a rate identical to that of a conventional radical polymerization, with the only difference being a possible mitigation of the gel or Tromsdorf effect.

As a 'fast guide' it seems apt to summarize a few key guidelines to enable the synthetic polymer chemist to carry out a straightforward polymerization leading to a desired polymeric product with the minimum of fuss (i.e. by avoiding complex reaction kinetics).

(a) The R group of the initial RAFT agent should be chosen such that R is a similar or better homolytic leaving group than the propagating radical. Thus, a tertiary propagating radical should be polymerized with a RAFT agent carrying a tertiary leaving group. In the case of acrylates the R group might be chosen to be a unimer of the propagating radical e.g., MA (i.e. methyl acrylate mediated by methoxycarbonylethyl phenyldithioacetate).

(b) The Z group should be chosen such that it does not 'more than necessary' stabilize the intermediate radical. 'Necessary' in this context means that the intermediate radical must be stable enough to allow for a high rate of addition of the propagating radical to the initial RAFT agent, so that the ratio between the addition rate coefficient and the propagation rate coefficient is high in order to avoid hybrid behavior [54]. For example, MMA polymerization cannot effectively be mediated with CPDA. If the stabilizing effect on the Z group is too great (e.g. when attempting to mediate VAc polymerization with CDB), large inhibition periods ensue due to highly stable adduct radicals that are prone – already in the preequilibrium – to undergo cross-termination reaction with propagating radicals.

(c) The ratio of initiator to initial RAFT agent should be kept as low as possible while still allowing for an adequate rate of polymerization. The yield of termination products approximates the amount of initiator decomposed.

While a wide variety of RAFT agents have been made featuring variable Z and R groups, it is important to point out that in principle it should be possible to control polymerization of all monomers using just two RAFT agent structures (e.g. a cyanoisopropyl dithiobenzoate for styrenics, acrylates, methacrylates, acrylamides) and a xanthate or dithiocarbamate (e.g. *O*-ethyl-*S*-cyanomethyl xanthate) for VAc, NVP and similar monomers [35]. Moreover, if recent theoretical predictions prove correct and synthetic challenges are overcome, there is a prospect of controlling polymerization of all monomers with a single RAFT agent with Z = F (e.g. cyanoisopropyl fluorodithioformate) [172, 173].

**Acknowledgements**

Some sections of this review where indicated are adapted with permission from the reviews of Moad, Rizzardo and Thang [35, 36] first published in the *Australian Journal of Chemistry* © 2005, 2006 CSIRO or Moad and Solomon *The Chemistry of Radical Polymerization* [34] © 2006 Elsevier.

**Abbreviations**

| | |
|---|---|
| AIBN | 2,2′-Azobis(isobutyronitrile) |
| BA | Butyl acrylate |
| B3-LYP | Becke 3-parameter (exchange), Lee, Yang and Parr |
| CDB | Cumyl dithiobenzoate |
| CPDA | Cumyl phenyl dithioacetate |
| DFT | Density functional theory |
| DMA | Dimethyl acrylamide |
| DMAEA | Dimethylaminoethyl acrylate |
| ESI | Electrospray ionization |
| ESR | Electron spin resonance |
| GPC | Gel permeation chromatography |
| IRC | Intrinsic reaction coordinate |
| LC-CC | Liquid chromatography at the critical point of adsorption |
| M | Monomer |
| MA | Methyl acrylate |
| MALDI | Matrix-assisted laser desorption and ionization |
| MMA | Methyl methacrylate |
| MS | Mass spectrometry |
| NMP | Nitroxide-mediated polymerization |
| NMR | Nuclear magnetic resonance |

| | |
|---|---|
| NVP | *N*-Vinylpyrrolidone |
| Ph | Phenyl |
| PREDICI | Polyreaction Distributions by Countable System Integration |
| SEC | Size exclusion chromatography |
| T | RAFT agent |
| tBA | *tert*-Butyl acrylate |
| UV | Ultraviolet |
| VAc | Vinyl acetate |

## References


1. S. N. Lewis, J. J. Miller, S. Winstein, *J. Org. Chem.* **1972**, *37*, 1478.
2. B. Giese, *Radicals in Organic Synthesis: Formation of Carbon–Carbon Bonds*, Pergamon Press, Oxford, **1986**.
3. W. B. Motherwell, D. Crich, *Free Radical Chain Reactions in Organic Synthesis*, Academic Press, London, **1992**.
4. P. Cacioli, D. G. Hawthorne, R. L. Laslett, E. Rizzardo, D. H. Solomon, *J. Macromol. Sci. Chem. A* **1986**, *23*, 839–852.
5. G. F. Meijs, E. Rizzardo, S. H. Thang, *Macromolecules* **1988**, *21*, 3122–3124.
6. G. F. Meijs, E. Rizzardo, *Die Makromol. Chem. Rapid Commun.* **1988**, *9*, 547–551.
7. D. Colombani, *Prog. Polym. Sci.* **1999**, *24*, 425–480.
8. D. Colombani, *Prog. Polym. Sci.* **1997**, *22*, 1649–1720.
9. D. Colombani, P. Chaumont, *Prog. Polym. Sci.* **1996**, *21*, 439–503.
10. B. Yamada, S. Kobatake, *Prog. Polym. Sci.* **1994**, *19*, 1089–1152.
11. J. Krstina, G. Moad, E. Rizzardo, C. L. Winzor, C. T. Berge, M. Fryd, *Macromolecules* **1995**, *28*, 5381–5385.
12. J. Chiefari, Y. K. Chong, F. Ercole, J. Krstina, J. Jeffery, T. P. T. Le, R. T. A. Mayadunne, G. F. Meijs, C. L. Moad, G. Moad, E. Rizzardo, S. H. Thang, *Macromolecules* **1998**, *31*, 5559–5562.
13. G. Moad, D. H. Solomon, *The Chemistry of Radical Polymerization*, 2nd edn, Elsevier, Oxford, **2006**, pp. 470–471.
14. P. C. Wieland, O. Nuyken, Y. Heischkel, B. Raether, C. Strissel, *ACS Symp. Ser.* **2003**, *854*, 619–630.
15. P. C. Wieland, B. Raether, O. Nuyken, *Macromol. Rapid Commun.* **2001**, *22*, 700–703.
16. S. Viala, M. Antonietti, K. Tauer, W. Bremser, *Polymer* **2003**, *44*, 1339–1351.
17. S. Viala, K. Tauer, M. Antonietti, R. P. Kruger, W. Bremser, *Polymer* **2002**, *43*, 7231–7241.
18. H. Tanaka, *Prog. Polym. Sci.* **2003**, *28*, 1171–1203.
19. A. Ah Toy, H. Chaffey-Millar, T. P. Davis, M. H. Stenzel, E. I. Izgorodina, M. L. Coote, C. Barner-Kowollik, *Chem. Commun. (Camb. U K)* **2006**, 835–837.
20. H. Chaffey-Millar, E. I. Izgorodina, C. Barner-Kowollik, M. L. Coote, *J. Chem. Theor. Comput.* **2006**, *2*, 1632–1645.
21. T. Junkers, M. H. Stenzel, T. P. Davis, C. Barner-Kowollik, *Macromol. Rapid Commun.* **2007**, *28*, 746–753.
22. G. Moad, D. H. Solomon, *The Chemistry of Radical Polymerization*, 2nd edn, Elsevier, Oxford, **2006**, pp. 521–522.
23. G. Moad, D. H. Solomon, *The Chemistry of Radical Polymerization*, 2nd edn, Elsevier, Oxford, **2006**, pp. 522–524.
24. G. Moad, D. H. Solomon, *The Chemistry of Radical Polymerization*, 2nd edn, Elsevier, Oxford, **2006**, pp. 524–525.
25. D. Greszta, D. Mardare, K. Matyjaszewski, *Macromolecules* **1994**, *27*, 638–644.
26. Y. Liu, J. P. He, J. T. Xu, D. Q. Fan, W. Tang, Y. L. Yang, *Macromolecules* **2005**, *38*, 10332–10335.
27. J. Xu, J. He, D. Fan, W. Tang, Y. Yang, *Macromolecules* **2006**, *39*, 3753–3759.

**28.** J. Chiefari, R. T. A. Mayadunne, C. L. Moad, G. Moad, E. Rizzardo, A. Postma, M. A. Skidmore, S. H. Thang, *Macromolecules* **2003**, *36*, 2273–2283.
**29.** K. Matyjaszewski, S. Gaynor, J.-S. Wang, *Macromolecules* **1995**, *28*, 2093–2095.
**30.** C. L. Moad, G. Moad, E. Rizzardo, S. H. Thang, *Macromolecules* **1996**, *29*, 7717–7726.
**31.** L. Hutson, J. Krstina, C. L. Moad, G. Moad, G. R. Morrow, A. Postma, E. Rizzardo, S. H. Thang, *Macromolecules* **2004**, *37*, 4441–4452.
**32.** T. P. Le, G. Moad, E. Rizzardo, S. H. Thang, WO 9801478, *1998*, E. I. Du Pont De Nemours and Co., USA [*Chem. Abstr. 128*, 115390f.]
**33.** P. Corpart, D. Charmot, S. Z. Zard, O. Biadatti, D. Michelet, US6153705, **2000** [*Chem. Abstr. 130*, 082018b.]
**34.** G. Moad, D. H. Solomon, *The Chemistry of Radical Polymerization*, 2nd edn, Elsevier, Oxford, **2006**.
**35.** G. Moad, E. Rizzardo, S. H. Thang, *Aust. J. Chem.* **2005**, *58*, 379–410.
**36.** G. Moad, E. Rizzardo, S. H. Thang, *Aust. J. Chem.* **2006**, *59*, 669–692.
**37.** G. Moad, *Aust J. Chem.* **2006**, *59*, 661–662.
**38.** R. T. A. Mayadunne, E. Rizzardo, In *Living and Controlled Polymerization: Synthesis, Characterization and Properties of the Respective Polymers and Copolymers*, J. Jagur-Grodzinski, Ed., Nova Science Publishers, Hauppauge, NY, **2006**, pp. 65–107.
**39.** J. Chiefari, E. Rizzardo, In *Handbook of Radical Polymerization*, John Wiley & Sons, Hoboken, NY, **2002**, pp. 263–300.
**40.** C. Barner-Kowollik, T. P. Davis, J. P. A. Heuts, M. H. Stenzel, P. Vana, M. Whittaker, *J. Polym. Sci., Part A: Polym. Chem.* **2003**, *41*, 365–375.
**41.** S. Perrier, P. Takolpuckdee, *J. Polym. Sci., Part A: Polym. Chem.* **2005**, *43*, 5347–5393.
**42.** A. Favier, M.-T. Charreyre, *Macromol. Rapid Commun.* **2006**, *27*, 653–692.
**43.** G. Moad, J. Chiefari, J. Krstina, A. Postma, R. T. A. Mayadunne, E. Rizzardo, S. H. Thang, *Polym. Int.* **2000**, *49*, 993–1001.
**44.** Y. K. Chong, J. Krstina, T. P. T. Le, G. Moad, A. Postma, E. Rizzardo, S. H. Thang, *Macromolecules* **2003**, *36*, 2256–2272.
**45.** H. Fischer, *Chem. Rev.* **2001**, *101*, 1885-3581-3610.
**46.** C. Barner-Kowollik, M. Buback, M. Egorov, T. Fukuda, A. Goto, O. F. Olaj, G. T. Russell, P. Vana, B. Yamada, P. B. Zetterlund, *Prog. Polym. Sci.* **2005**, *30*, 605–643.
**47.** P. Vana, T. P. Davis, C. Barner-Kowollik, *Macromol. Rapid Commun.* **2002**, *23*, 952–956.
**48.** A. Theis, T. P. Davis, M. H. Stenzel, C. Barner-Kowollik, *Macromolecules* **2005**, *38*, 10323–10327.
**49.** A. Theis, T. P. Davis, M. H. Stenzel, C. Barner-Kowollik, *Polymer* **2006**, *47*, 999–1010.
**50.** A. Theis, A. Feldermann, N. Charton, T. P. Davis, M. H. Stenzel, C. Barner-Kowollik, *Polymer* **2005**, *46*, 6797–6809.
**51.** A. Theis, A. Feldermann, N. Charton, M. H. Stenzel, T. P. Davis, C. Barner-Kowollik, *Macromolecules* **2005**, *38*, 2595–2605.
**52.** P. Vana, T. P. Davis, C. Barner-Kowollik, *Macromol. Theory Simul.* **2002**, *11*, 823–835.
**53.** A. Feldermann, M. L. Coote, M. H. Stenzel, T. P. Davis, C. Barner-Kowollik, *J. Am. Chem. Soc.* **2004**, *126*, 15915–15923.
**54.** C. Barner-Kowollik, J. F. Quinn, T. L. U. Nguyen, J. P. A. Heuts, T. P. Davis, *Macromolecules* **2001**, *34*, 7849–7857.
**55.** A. H. E. Müller, G. Litvenko, *Macromolecules* **1997**, *30*, 1253–1266.
**56.** A. H. E. Müller, R. Zhuang, D. Yan, G. Litvenko, *Macromolecules* **1995**, *28*, 4326–4333.
**57.** C. Barner-Kowollik, J. F. Quinn, D. R. Morsley, T. P. Davis, *J. Polym. Sci., Part A: Polym. Chem.* **2001**, *39*, 1353–1365.
**58.** M. L. Coote, E. I. Izgorodina, E. H. Krenske, M. Busch, C. Barner-Kowollik, *Macromol. Rapid Commun.* **2006**, *27*, 1015–1022.
**59.** M. L. Coote, E. H. Krenske, E. I. Izgorodina, *Macromol. Rapid Commun.* **2006**, *27*, 473–497.

**60.** G. Moad, J. Chiefari, C. L. Moad, A. Postma, R. T. A. Mayadunne, E. Rizzardo, S. H. Thang, *Macromol. Symp.* **2002**, *182*, 65–80.
**61.** A. Goto, T. Fukuda, *Prog. Polym. Sci.* **2004**, *29*, 329–385.
**62.** G. Moad, F. Ercole, C. H. Johnson, J. Krstina, C. L. Moad, E. Rizzardo, T. H. Spurling, S. H. Thang, A. G. Anderson, *ACS Symp. Ser.* **1998**, *685*, 332–360.
**63.** M. Zhang, W. H. Ray, *Ind. Eng. Chem. Res.* **2001**, *40*, 4336–4352.
**64.** A. R. Wang, S. P. Zhu, *J. Polym. Sci. Part A: Polym. Chem.* **2003**, *41*, 1553–1566.
**65.** A. R. Wang, S. Zhu, *Macromol. Theory Simul.* **2003**, *12*, 196–208.
**66.** A. D. Peklak, A. Butte, G. Storti, M. Morbidelli, *Macromol. Symp.* **2004**, *206*, 481–494.
**67.** D. A. Shipp, K. Matyjaszewski, *Macromolecules* **1999**, *32*, 2948–2955.
**68.** C. Barner-Kowollik, *Aust. J. Chem.* **2001**, *54*, 343.
**69.** P. Vana, T. P. Davis, C. Barner-Kowollik, *Macromol. Theory Simul.* **2002**, *11*, 823–835.
**70.** M. Wulkow, M. Busch, T. P. Davis, C. Barner-Kowollik, *J. Polym. Sci. Part A: Polym. Chem.* **2004**, *42*, 1441–1448.
**71.** Y. K. Chong, G. Moad, R. Mulder, E. Rizzardo, S. H. Thang, *Macromolecules*, **2007**, submitted.
**72.** K. Matyjaszewski, R. Poli, *Macromolecules* **2005**, *38*, 8093–8100.
**73.** A. Alberti, M. Guerra, P. Hapiot, T. Lequeux, D. Macciantelli, S. Masson, *Phys. Chem. Chem. Phys.* **2005**, *7*, 250–257.
**74.** A. Alberti, M. Benaglia, H. Fischer, M. Guerra, M. Laus, D. Macciantelli, A. Postma, K. Sparnacci, *Helv. Chim. Acta* **2006**, *89*, 2103–2118.
**75.** M. L. Coote, In *Encyclopedia of Polymer Science and Technology*, H. F. Mark, Ed., Wiley-Interscience, New Jersey, **2004**, pp. 319–371.
**76.** M. L. Coote, *J. Phys. Chem. A* **2005**, *109*, 1230–1239.
**77.** M. L. Coote, G. P. F. Wood, L. Radom, *J. Phys. Chem. A* **2002**, *106*, 12124–12138.
**78.** E. I. Izgorodina, M. L. Coote, *J. Phys. Chem. A* **2006**, *110*, 2486–2492.
**79.** M. L. Coote, *J. Phys. Chem. A* **2004**, *108*, 3865–3872.
**80.** R. Gomez-Balderas, M. L. Coote, D. J. Henry, L. Radom, *J. Phys. Chem. A* **2004**, *108*, 2874–2883.
**81.** E. I. Izgorodina, M. L. Coote, L. Radom, *J. Phys. Chem. A* **2005**, *109*, 7558–7566.
**82.** J. P. A. Heuts, R. G. Gilbert, L. Radom, *J. Phys. Chem. A* **1996**, *100*, 18997–19006.
**83.** M. L. Coote, *Macromolecules* **2004**, *37*, 5023–5031.
**84.** M. L. Coote, L. Radom, *J. Am. Chem. Soc.* **2003**, *125*, 1490–1491.
**85.** M. L. Coote, D. J. Henry, *Macromolecules* **2005**, *38*, 1415–1433.
**86.** M. Drache, G. Schmidt-Naake, M. Buback, P. Vana, *Polymer* **2005**, *46*, 8483–8493.
**87.** E. I. Izgorodina, M. L. Coote, *Macromol. Theor. Simul.* **2006**, *15*, 394–403.
**88.** G. Moad, R. T. A. Mayadunne, E. Rizzardo, M. Skidmore, S. Thang, *ACS Symp. Ser.* **2003**, *854*, 520–535.
**89.** M. L. Coote, C. Barner-Kowollik, *Aust. J. Chem.* **2006**, *59*, 712–718.
**90.** A. Goto, K. Sato, Y. Tsujii, T. Fukuda, G. Moad, E. Rizzardo, S. H. Thang, *Macromolecules* **2001**, *34*, 402–408.
**91.** K. Kubo, A. Goto, K. Sato, Y. Kwak, T. Fukuda, *Polymer* **2005**, *46*, 9762–9768.
**92.** J. Chiefari, R. T. Mayadunne, G. Moad, E. Rizzardo, S. H. Thang, US6747111, **2004**, E.I. Du Pont De Nemours and Co, USA; Commonwealth Scientific and Industrial Research Organization [*Chem. Abstr. 131*, 45250w.]
**93.** T. Junkers, A. Theis, M. Buback, T. P. Davis, M. H. Stenzel, P. Vana, C. Barner-Kowollik, *Macromolecules* **2005**, *38*, 9497–9508.
**94.** M. Buback, P. Hesse, T. Junkers, P. Vana, *Macromol. Rapid Commun.* **2006**, *27*, 182–187.
**95.** M. Adamy, A. M. van Herk, M. Destarac, M. J. Monteiro, *Macromolecules* **2003**, *36*, 2293–2301.
**96.** M. Buback, C. H. Kurz, C. Schmaltz, *Macromol. Chem. Phys.* **1998**, *199*, 1721–1727.
**97.** C. Barner-Kowollik, M. Buback, B. Charleux, M. L. Coote, M. Drache,

T. Fukuda, A. Goto, B. Klumperman, A. B. Lowe, J. B. Mcleary, G. Moad, M. J. Monteiro, R. D. Sanderson, M. P. Tonge, P. Vana, *J. Polym. Sci., Part A: Polym. Chem.* **2006**, *44*, 5809–5831.
98. Y. Kwak, A. Goto, T. Fukuda, *Macromolecules* **2004**, *37*, 1219–1225.
99. Y. Kwak, A. Goto, Y. Tsujii, Y. Murata, K. Komatsu, T. Fukuda, *Macromolecules* **2002**, *38*, 3026–3029.
100. Y. Kwak, A. Goto, K. Komatsu, Y. Sugiura, T. Fukuda, *Macromolecules* **2004**, *37*, 4434–4440.
101. A. Favier, C. Ladaviere, M.-T. Charreyre, C. Pichot, *Macromolecules* **2004**, *37*, 2026–2034.
102. X. Jiang, P. J. Schoenmakers, J. L. J. van Dongen, X. Lou, V. Lima, J. Brokken-Zijp, *Anal. Chem.* **2003**, *75*, 5517–5524.
103. R. Venkatesh, B. B. P. Staal, B. Klumperman, M. J. Monteiro, *Macromolecules* **2004**, *37*, 7906–7917.
104. C. Schilli, M. G. Lanzendoerfer, A. H. E. Mueller, *Macromolecules* **2002**, *35*, 6819–6827.
105. W. Mellon, D. Rinaldi, E. Bourgeat-Lami, F. D'Agosto, *Macromolecules* **2005**, *38*, 1591–1598.
106. P. Vana, L. Albertin, L. Barner, T. P. Davis, C. Barner-Kowollik, *J. Polym. Sci, Part A: Polym. Chem.* **2002**, *40*, 4032–4037.
107. D. Boschmann, P. Vana, *Polym. Bull.* **2005**, *53*, 231–242.
108. A. Ah Toy, P. Vana, T. P. Davis, C. Barner-Kowollik, *Macromolecules* **2004**, *37*, 744–751.
109. A. Feldermann, A. Ah Toy, T. P. Davis, M. H. Stenzel, C. Bamer-Kowollik, *Polymer* **2005**, *46*, 8448–8457.
110. G. Zhou, I. I. Harruna, *Anal. Chem.* **2007**, *79*, 2722–2727.
111. C. Barner-Kowollik, T. P. Davis, M. H. Stenzel, *Polymer* **2004**, *45*, 7791–7805.
112. M. Bathfield, F. D'Agosto, R. Spitz, C. Ladaviere, M.-T. Charreyre, T. Delair, *Macromol. Rapid Commun.* **2007**, *28*, 856–862.
113. J. B. McLeary, F. M. Calitz, J. M. McKenzie, M. P. Tonge, R. D. Sanderson, B. Klumperman, *Macromolecules* **2005**, *38*, 3151–3161.
114. J. B. McLeary, F. M. Calitz, J. M. McKenzie, M. P. Tonge, R. D. Sanderson, B. Klumperman, *Macromolecules* **2004**, *37*, 2383–2394.
115. J. B. McLeary, J. M. McKenzie, M. P. Tonge, R. D. Sanderson, B. Klumperman, *Chem. Commun.* **2004**, 1950–1951.
116. G. Pound, J. B. McLeary, J. M. McKenzie, R. F. M. Lange, B. Klumperman, *Macromolecules*, **2006**, *39*, 7796–7797.
117. E. T. A. Van Den Dungen, J. Rinquest, N. O. Pretorius, J. M. McKenzie, J. B. McLeary, R. D. Sanderson, B. Klumperman, *Aust. J. Chem.*, **2006**, *59*, 742–748.
118. B. Klumperman, J. B. McLeary, E. T. A. Van Den Dungen, W. J. Soer, J. Bozovic, *ACS Symp. Ser.* **2005**, *944*, 501–513.
119. F. M. Calitz, J. B. McLeary, J. M. McKenzie, M. P. Tonge, B. Klumperman, R. D. Sanderson, *Macromolecules* **2003**, *36*, 9687–9690.
120. J. B. McLeary, M. P. Tonge, B. Klumperman, *Macromol. Rapid Commun.* **2006**, *27*, 1233–1240.
121. H. Fischer, L. Radom, *Angew. Chem. Int. Ed. Engl.* **2001**, *40*, 1340–1371.
122. S. Perrier, C. Barner-Kowollik, J. F. Quinn, P. Vana, T. P. Davis, *Macromolecules* **2002**, *35*, 8300–8306.
123. H. Fischer, L. Radom, *Angew. Chem. Int. Ed.* **2001**, *40*, 1340–1371.
124. J. B. McLeary, J. M. McKenzie, M. P. Tonge, R. D. Sanderson, B. Klumperman, *Chem. Commun.* **2004**, 1950–1951.
125. D. G. Hawthorne, G. Moad, E. Rizzardo, S. H. Thang, *Macromolecules* **1999**, *32*, 5457–5459.
126. F. M. Calitz, M. P. Tonge, R. D. Sanderson, *Macromolecules* **2003**, *36*, 5–8.
127. F. S. Du, M. Q. Zhu, H. Q. Guo, Z. C. Li, F. M. Li, M. Kamachi, A. Kajiwara, *Macromolecules* **2002**, *35*, 6739–6741.
128. A. Alberti, M. Benaglia, M. Laus, D. Macciantelli, K. Sparnacci, *Macromolecules* **2003**, *36*, 736–740.
129. E. Chernikova, A. Morozov, E. Leonova, E. Garina, V. Golubev, C. O. Bui, B.

Charleux, *Macromolecules* **2004**, *37*, 6329–6339.
**130.** E. V. Chernikova, P. S. Terpugova, E. S. Garina, V. B. Golubev, *Polym. Sci. Ser. A* **2007**, *49*, 108–119.
**131.** M. P. Tonge, F. M. Calitz, R. D. Sanderson, *Macromol. Chem. Phys.* **2006**, *207*, 1852–1860.
**132.** V. B. Golubev, E. V. Chernikova, E. A. Leonova, A. V. Morozov, *Polym. Sci. Ser. A* **2005**, *47*, 678–685.
**133.** H. Pasch, K. Mequanint, J. Adrian, *e-Polymers* **2002**, 005.
**134.** X. L. Jiang, P. J. Schoenmakers, J. L. J. van Dongen, X. W. Lou, V. Lima, J. Brokken-Zijp, *Anal. Chem.* **2003**, *75*, 5517–5524.
**135.** X. L. Jiang, A. Van Der Horst, V. Lima, P. J. Schoenmakers, *J. Chromatogr. A* **2005**, *1076*, 51–61.
**136.** X. L. Jiang, P. J. Schoenmakers, X. W. Lou, V. Lima, J. L. J. van Dongen, J. Brokken-Zijp, *J. Chromatogr. A* **2004**, *1055*, 123–133.
**137.** A. Postma, T. P. Davis, G. Li, G. Moad, M. O'Shea, *Macromolecules* **2006**, *39*, 5307–5318.
**138.** A. Ah Toy, P. Vana, T. P. Davis, C. Barner-Kowollik, *Macromolecules* **2004**, *37*, 744–751.
**139.** C. Barner-Kowollik, P. Vana, J. F. Quinn, T. P. Davis, *J. Polym. Sci. Polym. Chem.* **2002**, *40*, 1058–1063.
**140.** P. Vana, J. F. Quinn, T. P. Davis, C. Barner-Kowollik, *Aust. J. Chem.* **2002**, *55*, 425–431.
**141.** S. W. Prescott, M. J. Ballard, E. Rizzardo, R. G. Gilbert, *Macromolecules* **2005**, *38*, 4901–4912.
**142.** J. C. Scaiano, K. U. Ingold, *J. Am. Chem. Soc.* **1976**, *98*, 4727–4732.
**143.** G. Moad, D. H. Solomon, *The Chemistry of Radical Polymerization*, 2nd edn, Elsevier, Oxford, **2006**, pp. 114–115.
**144.** E. Rizzardo, M. Chen, B. Chong, G. Moad, M. Skidmore, S. H. Thang, *Macromol. Symp.* **2007**, *248*, 104–116.
**145.** M. Walbiner, J. Q. Wu, H. Fischer, *Helv. Chim. Acta* **1995**, *78*, 910–924.
**146.** K. Münger, H. Fischer, *Int. J. Chem. Kinet.* **1985**, *17*, 809–829.
**147.** T. Zytowski, B. Knuehl, H. Fischer, *Helv. Chim. Acta* **2000**, *83*, 658–675.
**148.** K. Heberger, H. Fischer, *Int. J. Chem. Kinet.* **1993**, *25*, 249–263.
**149.** M. L. Coote, *J. Phys. Chem. A* **2005**, *109*, 1230–1239.
**150.** M. Buback, P. Hesse, T. Junkers, P. Vana, *Macromol. Rapid Commun.* **2006**, *27*, 182–187.
**151.** G. Moad, D. H. Solomon, *The Chemistry of Radical Polymerization*, 2nd edn., Elsevier, Oxford, **2006**, pp. 111–133.
**152.** P. Vana, L. Albertin, L. Barner, T. P. Davis, C. Barner-Kowollik, *J. Polym. Sci., Part A: Polym. Chem.* **2002**, *40*, 4032–4037.
**153.** J. Krstina, C. L. Moad, G. Moad, E. Rizzardo, C. T. Berge, M. Fryd, *Macromol. Symp.* **1996**, *111*, 13–23.
**154.** M. J. Monteiro, H. de Brouwer, *Macromolecules* **2001**, *34*, 349–352.
**155.** M. J. Monteiro, *J. Polym. Sci., Part A: Polym. Chem.* **2005**, *43*, 3189–3204.
**156.** M. L. Coote, E. H. Krenske, E. I. Izgorodina, *Macromol. Rapid Commun.* **2006**, *27*, 473–497.
**157.** M. J. Monteiro, R. Bussels, S. Beuermann, M. Buback, *Aust. J. Chem.* **2002**, *55*, 433–437.
**158.** T. P. Davis, C. Barner-Kowollik, P. Vana, M. Stenzel, J. F. Quinn, T. L. Uyen Nygen, *ACS Symposium Series*, **2003**, *854*, 557–569.
**159.** M. L. Coote, L. Radom, *Macromolecules* **2004**, *37*, 590–596.
**160.** M. Buback, P. Vana, *Macromol. Rapid Commun.* **2006**, *27*, 1299–1305.
**161.** G. Moad, G. Li, R. Pfaendner, A. Postma, E. Rizzardo, S. Thang, H. Wermter, *ACS Symp. Ser.* **2006**, *944*, 514–532.
**162.** M. N. Galbraith, G. Moad, D. H. Solomon, T. H. Spurling, *Macromolecules* **1987**, *20*, 675–679.
**163.** T. H. Spurling, M. Deady, J. Krstina, G. Moad, *Makromol. Chem., Macromol. Symp.* **1991**, *51*, 127–146.
**164.** J. Chiefari, J. Jeffery, J. Krstina, C. L. Moad, G. Moad, A. Postma, E. Rizzardo, S. H. Thang, *Macromolecules* **2005**, *38*, 9037–9054.
**165.** A. Feldermann, A. Ah Toy, H. Phan, M. H. Stenzel, T. P. Davis, C. Barner-Kowollik, *Polymer* **2004**, *45*, 3997–4007.

**166.** D. Batt-Coutrot, J.-J. Robin, W. Bzducha, M. Destarac, *Macromol. Chem. Phys.* **2005**, *206*, 1709–1717.

**167.** G. Moad, K. Dean, L. Edmond, N. Kukaleva, G. Li, R. T. A. Mayadunne, R. Pfaendner, A. Schneider, G. Simon, H. Wermter, *Macromol. Symp.* **2006**, *233*, 170–179.

**168.** M. J. Monteiro, *J. Polym. Sci., Part A: Polym. Chem.* **2006**, *43*, 5643–5651.

**169.** C. Barner-Kowollik, T. P. Davis, M. H. Stenzel, *Aust. J. Chem.*, **2006**, *59*, 719–727.

**170.** H. Chaffey-Millar, M. Busch, T. P. Davis, M. H. Stenzel, C. Barner-Kowollik, *Macromol. Theor. Simul.* **2005**, *14*, 143–157.

**171.** H. Chaffey-Millar, M. H. Stenzel, T. P. Davis, M. L. Coote, C. Barner-Kowollik, *Macromolecules* **2006**, *39*, 6406–6419.

**172.** M. L. Coote, E. I. Izgorodina, G. E. Cavigliasso, M. Roth, M. Busch, C. Barner-Kowollik, *Macromolecules* **2006**, *39*, 4585–4591.

**173.** A. Theis, M. H. Stenzel, T. P. Davis, M. L. Coote, C. Barner-Kowollik, *Aust. J. Chem.* **2005**, *58*, 437–441.

**174.** Beuermann S. Buback M. *Progress in Polymer Science* **2002**, 27, 191–254

# 4
# The RAFT Process as a Kinetic Tool: Accessing Fundamental Parameters of Free Radical Polymerization

*Thomas Junkers, Tara M. Lovestead, and Christopher Barner-Kowollik*

## 4.1
## Introduction

Living/controlled free radical polymerization (LFRP) techniques such as nitroxide-mediated polymerization, atom transfer polymerization and reversible addition–fragmentation chain transfer (RAFT) polymerization gave rise to new designs and syntheses of well-defined materials with respect to size, shape, polydispersity and functionality. For example, macromolecular architectures such as stars, combs, block copolymers, core–shell nanoparticles and branched structures have become feasible [1–15], rendering these materials ideal for applications in biomaterials (e.g. drug delivery devices and diagnostic nanoparticles) and molecular electronics [1, 8, 16–19].

The reason why LFRPs give access to such a vast number of highly specific microstructure is that they yield materials with high-functionality and low-polydispersity indices, PDI, as all radical chains nearly grow uniformly with increasing monomer consumption, i.e. polymerization time (see equation 4.1). Ideally, the chain length $i$ depends solely on the fraction of monomer consumed, $c_{\mathrm{M}}^{0}\,X$ (where $X$ denotes monomer conversion and $c_{\mathrm{M}}^{0}$ the initial monomer concentration) and $c_{\mathrm{LFRP}}^{0}$, which is the concentration of the mediating agent before the reaction commenced, that is at $t = 0$ [1, 5–7, 20–24].

$$i(t) = \frac{c_{\mathrm{M}}^{0}\,X(t)}{c_{\mathrm{LFRP}}^{0}} + 1 \qquad (4.1)$$

This feature is utilized in many ways for synthetic purposes because the chain length of the polymer product can be conveniently tuned – widely independent from the outer reaction conditions – simply by changing the monomer and mediating agent concentrations. Less realized is, however, that the correlation of equation 4.1 can also be a powerful tool in the hands of a physical chemist for the very same reason: it gives easy access to reactions and kinetic information of radicals with known, and moreover, adjustable size.

*Handbook of RAFT Polymerization.* Edited by Christopher Barner-Kowollik

ISBN: 978-3-527-31924-4

All LFRP techniques can be categorized with respect to the underlying mechanism involved to achieve close-to living conditions in a radical process; most techniques are of the kind in which termination of two radicals is avoided by largely reducing the free radical concentration. RAFT, in contrast, is different in that the RAFT-specific equilibria (see Chapter 3) governing the process do not change the radical concentration nor have, at least under ideal conditions, any influence on the rate of polymerization [25–31]. Thus, RAFT chemistry offers the advantage (provided that the initial RAFT agent is chosen judiciously) of nominal impact on the radical concentration and, concomitantly, the propagation and termination rates [25, 32]. This attribute is unique to RAFT, rendering it an ideal method for aiding current experimental techniques and developing novel techniques for ascertaining the chain length dependence of free radical polymerization (FRP) rate coefficients, especially for the chain-length-dependent termination (CLD-T) reaction.

Scheme 4.1 presents a typical RAFT-mediated FRP mechanism. Reaction I entails a two-step initiation mechanism, where the generation of initiating radicals, I• (Ia), depends on the rate coefficient $k_d$ and on the initiator efficiency $f$. Additionally, the first addition of a monomer unit to form $P_1$• (Ib) depends on the initiation rate coefficient $k_i$. Reactions II and IV are unique to RAFT polymerizations and represent the preequilibrium, where the initial RAFT agent is transformed into macro-RAFT agent, and the core equilibrium, where the poly-RAFT species grows uniformly with monomer to polymer conversion. The preequilibrium also entails monomer addition to the RAFT agent-derived radicals, R•, which proceeds with a rate coefficient $k_{p,rein}$. The preequilibrium and the core equilibrium are governed by the magnitude of the individual addition and fragmentation rate coefficients $k_{add}$ and $k_{frag}$. Reactions III and V represent the conventional FRP propagation and termination reactions, respectively; however, only termination by combination is assumed in the present scheme. Depending on the type of monomer, the reaction scheme would need to be extended for termination by disproportionation. Both reactions are usually considered to be chain length dependent, while the degree of this dependence is smaller for the propagation reaction. Because termination is a bimolecular reaction, its rate coefficient depends on the length of both radicals; however, in an effectively functioning RAFT process both radicals are of almost the same size ($i \approx j$). Additional reactions such as termination of the intermediate radicals (2, 4) could potentially occur. For a more in-depth analysis of the kinetics and mechanism of the RAFT process, the reader is referred to Chapter 3, where the underpinning complexities of the RAFT processes are discussed in detail. It is important for the purpose of employing the RAFT process as a kinetic tool that it functions ideally, i.e. that the often discussed (yet relatively rare) inhibition and/or rate retardation phenomena are absent [25]. In particular, it must be ensured that the intermediate RAFT radicals in the pre- and main equilibrium are fragmenting rapidly and do not undergo bimolecular termination reactions with themselves or propagating macroradicals as such reactions would provide an additional radical loss pathway in competition to conventional bimolecular termination. It can also not be stated enough that in an ideal RAFT process, the rate of termination is

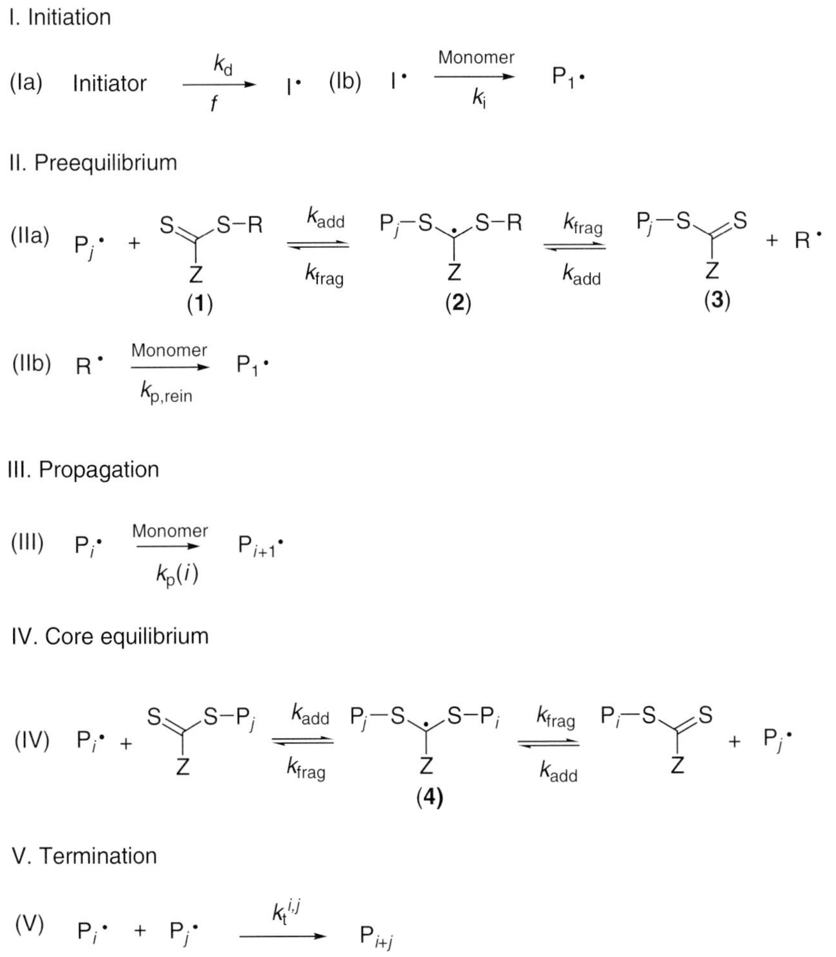


**Scheme 4.1** Reaction scheme for the RAFT process. Note that in an effectively working RAFT process, $i \approx j$.

*not* reduced compared to conventional FRP, although this notion is sometimes propagated in the literature.

## 4.2
## Chain-Length-Dependent Termination: A Brief Overview

Several decades of research have been aimed at improving the understanding of the termination processes – arguably the single most complex reaction in chemistry – in FRPs [33–40]. While, initially, the termination reaction may appear to be a simple bimolecular reaction between two highly reactive radical species, the scientific

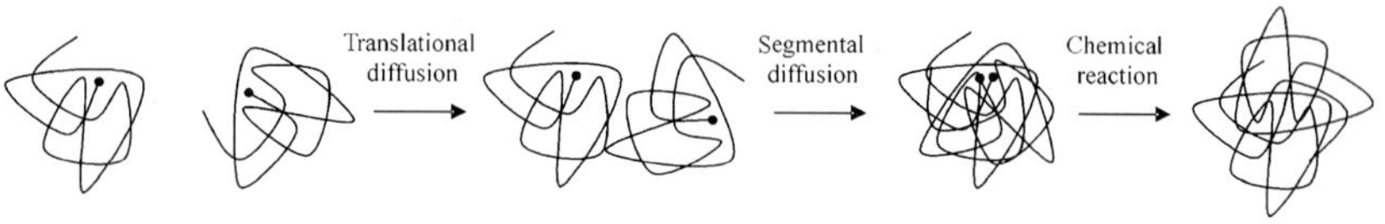


**Scheme 4.2** The complex multistep diffusion-controlled termination process during FRPs.

community has produced many reviews that focus solely on FRP termination processes and the variety of methods that allow for its quantification [37]. The termination process is inherently complex to characterize because it is a multistep diffusion-controlled event that requires translational diffusion of the center of masses of large macroradicals and subsequent segmental rearrangement of their chain ends before chemical reaction can occur (see Scheme 4.2) [41–55]. Additionally, so-called reaction diffusion, which means the approach of two radical sites via chain growth toward each other, was shown to contribute to the termination rate [56, 57]. Thus, the termination rate coefficient depends on all factors that influence the mobility of the radical chain, e.g. viscosity, monomer to polymer conversion, temperature, pressure, solvent concentration and degree of polymerization to name only a few. One of the most difficult, because only indirectly accessible, and hence one of the least understood – at least until recently – aspects of the diffusion-controlled nature of the termination reaction is the termination rate coefficient's, $k_t$, dependence on the terminating radicals' chain length.

Frequently, bimolecular termination, pseudo-steady-state radical concentration and termination kinetics that are independent of kinetic chain length (or initiation rate) are assumed in the kinetic equations describing radical polymerizations [54, 58–60]. When utilized together, these assumptions predict that the polymerization rate increases with the square root of the initiation rate (equation 4.2),

$$R_p = \frac{k_p}{\sqrt{2k_t}} c_M R_i^{1/2} \tag{4.2}$$

where $R_i$ is the initiation rate, $k_p$ is the propagation rate coefficient and $k_t$ is the termination rate coefficient. This relationship represents the classical kinetic relationship; however, equation 4.2 incorrectly predicts most polymerization behavior.

Often, equation 4.2 fails to predict experimental behavior because, in its current form, CLD-T is not accounted for. By including CLD-T, radicals diffuse and terminate according to their length, i.e. shorter chain radicals more readily diffuse and terminate than longer, bulkier chains. Thus, the termination rate, i.e. the termination rate coefficient, is averaged over the entire distribution of chains. Monomers that exhibit CLD-T kinetics have an average termination rate coefficient, $k_t^{avg}$, that depends on the molecular weight distribution (MWD), and thus any cure condition that impacts the MWD [34, 41, 42, 50, 58, 60–72]. For example, increasing the initiation rate will shift the MWD toward shorter average kinetic chain lengths, and thus when CLD-T is important, the average termination rate will also increase, resulting in a less than classically expected increase in the polymerization rate, i.e. $R_p \propto R_i^{\alpha}$, where $\alpha < 1/2$. In principle, a discrete termination rate coefficient exists for each radical chain length combination. However, a continuous function can be assumed to express the dependency, and a power law dependency of $k_t$ on radical chain length has shown to yield a reasonable description on the basis of an exponent $\alpha$:

$$k_t^{i,j} = k_t^{1,1} i, j^{-\alpha/2} = k_t^{1,1} i^{-\alpha} \quad \text{when} \quad i = j \tag{4.3}$$

It should be noted that equation 4.3 is based on the so-called geometric mean model to account for reactions between different chain lengths and other forms of the equation may be used when other weightings are assumed, such as the diffusion or the harmonic mean model [40, 72, 73]. The drawback of the above equation is that the power law is – while in principle having a physical background – somehow arbitrary in that the diffusion mechanism may change during chain growth and thus may involve a change in the exponent $\alpha$. To account for such variation, a so-called composite model was introduced by Russell and Smith [55] on the basis of earlier experimental data compilations [74], in which $\alpha$ adopts different values for the short- and the long-chain regime. Therefore, the chain length axis is divided by the so-called critical chain length $i_c$. When the chain length $i$ (and $j$, respectively) is less than $i_c$, equation 4.3 is applied where $\alpha$ is a short-chain-regime-specific constant. For larger chain length, $k_t$ is given by

$$k_t^{i,i} = k_t^{1,1} i_c^{-\alpha_s + \alpha_l} i^{-\alpha_l} \tag{4.4}$$

whereby $\alpha_s$ denotes the short-chain exponent and $\alpha_l$ the long-chain exponent. Within the composite model, $\alpha_s = 0.5$ and $\alpha_l = 0.16$ have been proposed (for dilute solution). The value of 0.5 is based on theoretical predictions [75, 76] and on measurements of (center-of-mass) diffusion coefficients of smaller molecules of various sizes [77]. The value of 0.16 originates from theoretical predictions carried out by Khokhlov [78] and by O'Shaughnessy and coworkers [79, 80] based on the dynamics of macromolecular coils. Although the $\alpha_s$ value was recently verified by experiments [81–83], the value is still the topic of discussion and may be different for different monomers. The $\alpha_l$ value has been confirmed in several experiments at least for styrene and methyl methacrylate [60, 71, 81, 84–87] and can be regarded as a good approximation for the long-chain regime in the case of macromolecular coils in which the radical center is located on the terminal chain unit.

## 4.3
## RAFT Chemistry as a Tool for Elucidating the Chain Length Dependence of $k_t$

Numerous techniques have been developed to quantify the extent that the termination rate coefficient depends on the radical chain length. Characterization of the termination rate coefficient's dependence on chain length is complex as is evident from the development of almost as many techniques to measure the conversion and chain length dependence of $k_t$ as there are research groups working in the field [33, 37]. However, most of these techniques make use of pulsed-laser polymerization, or in its somewhat older form of the rotating sector method. Both methods provide instationary reaction conditions and allow for a situation where the chain length is to some extent controlled by the frequency of laser pulsing (or dark time period in the rotating sector, respectively) giving access to chain-length-dependent rate coefficients.

Among the various techniques, two promising methods have been developed that exploit RAFT FRP as a tool for determining $k_t$ as a function of polymerization time (i.e. as a function of radical chain length, $k_t^{i,i}$, where $i$ is the radical chain length, as given in equation 4.1): the (stationary) RAFT chain-length-dependent termination (RAFT-CLD-T) technique [88–93] introduced by Barner-Kowollik and coworkers, and the (instationary) single pulse–pulsed-laser polymerization–RAFT (SP–PLP–RAFT) procedure developed later by Buback et al. [94, 95]. Once again, while all LFRP techniques can yield polymer materials with narrow polydispersity, the RAFT FRP technique adds the advantage that the radical concentration and, concomitantly, the propagation and termination rates are impacted nominally [25, 32], hence allowing for the determination of $k_t^{i,i}$ by determining $k_t(X)$, which is much simpler to do.

In here, the RAFT-CLD-T method is described in detail, followed by a thorough discussion of the SP–PLP–RAFT technique and the implications on the RAFT equilibrium caused by laser pulsing, as well as a section that compares and contrasts the advantages and disadvantages of each technique.

### 4.3.1 The RAFT-CLD-T Technique

The RAFT-CLD-T technique is based on the direct and experimentally nondemanding measurement of the polymerization rate as a function of time, $R_p(t)$ (equation 4.5). Thus, the RAFT-CLD-T characterizes the chain length dependence of $k_t$ via a stationary technique that is experimentally very simple.

$$R_p(t) = -\frac{d[M]}{dt} = k_p[M][P\bullet] \tag{4.5}$$

The polymerization rate is related to the termination rate coefficient via equation 4.6, which indicates that the radical concentration at any point in time depends on the initiation rate $R_i$ minus the termination rate $R_t$ [48].

$$\frac{d[P\bullet]}{dt} = 2fk_d[I] - 2k_t[P\bullet]^2 \tag{4.6}$$

The factor of 2 is necessary because initiator decomposition results in two initiating fragments. Solving equation 4.6 for $k_t$ provides the termination rate coefficient averaged as a function of time (equation 4.7; note that $\int_0^t R_P(t)\,dt = c_M$), when the initiator decomposition rate coefficient, the propagation rate coefficient as well as the initiator efficiency are available [92, 96].

$$\langle k_t \rangle (t) = \frac{2 f k_d [I]_0\, e^{-k_d t} - \dfrac{d\left( R_p(t) \Big/ k_p \left( [M]_0 - \int_0^t R_p(t)\,dt \right) \right)}{dt}}{2 \left( \dfrac{R_p(t)}{k_p \left( [M]_0 - \int_0^t R_p(t)\,dt \right)} \right)^2} \quad (4.7)$$

Additionally, highly reactive radicals may undergo chain transfer to the polymer backbone forming midchain radicals. Midchain radicals display a lower reactivity toward monomer addition, leading to an apparent propagation rate coefficient $k_p^*$ (equation 4.8) that can be expressed on the basis of a virtual monomer reaction order, $\omega$, that is empirically found to be greater than unity [90, 95, 97, 98].

$$k_p^* = k_p (1 - X)^{\omega - 1} \quad (4.8)$$

Equation 4.8 leaves a virtually conversion-dependent propagation rate coefficient. Such conversion dependency is taken into account as the probability to form midchain radicals is greatly enhanced with increasing polymer content due to the increase in intermolecular chain transfer reactions. Indeed, an increase in midchain radicals with conversion in acrylate polymerization has been observed by electron spin resonance (ESR) spectroscopy [99–101].

Equation 4.9 depicts the important components from the classical polymerization rate equation that are necessary to evaluate the relationship between the polymerization rate, the initiator and monomer concentrations and the termination rate coefficient when the assumption of a steady-state radical concentration is made. The virtual monomer reaction order can be determined by measuring the rate of polymerization, $R_p(t)$, for varying monomer concentrations up to high conversions and plotting $\log(R_p(t)) - \log([I]^{0.5})$ vs $\log([M])$ [90]. The slope of the resulting graph is the virtual monomer reaction order, which then translates directly to $k_p^*$ using equation 4.8 [90, 102].

$$R_p \approx [M]^{\omega} \left( \frac{[I]}{k_t^{i,i}} \right)^{0.5} \quad (4.9)$$

$$\log(R_p) - \log([I]^{0.5}) - \log(i^{\alpha/2}) = \omega \log([M]) \quad (4.10)$$

Equation 4.6 makes no assumption of a steady-state radical concentration. The RAFT-CLD-T method is versatile not only because it is experimentally easy to access the polymerization rate, but also because it affords the ability to control the chain length distribution as a function of time and monomer conversion, even though reliable values for $\langle k_t \rangle (t)$ are accessible only when *all* of the parameters in equation

4.7 are known. The initial monomer, initiator and RAFT agent concentrations are accessible to a high degree of accuracy via measuring the density of the polymerization mixture at the reaction temperature. Additionally, the initiator decomposition rate, $k_d$, for the initiator 2,2′-azobisbutyronitrile (ABIN) is measured with relative ease and good accuracy using UV/Vis spectrometry [90, 103]. More difficult to obtain is an accurate initiator efficiency $f$, a crucial parameter that can affect the absolute value of $\langle k_t \rangle(t)$ as well as the shape of the $\langle k_t \rangle(t)$ curve since $f$ is a function of viscosity, and thus often of monomer conversion, $f(X)$; however, the problem of knowing $f(X)$ can be avoided by restricting the analysis to low conversions where a constant value might be assumed to good accuracy. A reliable propagation rate coefficient, $k_p$, can be obtained with good accuracy (for true end-chain propagation) using the IUPAC-recommended PLP–SEC (pulsed-laser polymerization–size exclusion chromatography) method, and $\omega$ and, subsequently, $k_p^*$ can be ascertained using the procedure outlined in the previous paragraph [90, 102]. While the impact of chain-length-dependent propagation (CLD-P) on the polymerization kinetics is currently under investigation, the impact of CLD-P on the polymerization kinetics is, for the most part, assumed to be nominal, at least in nonliving systems. It will, however, be shown later that the RAFT process may also be used to assess detailed information on CLD-P.

The unique attributes of RAFT-mediated FRP, i.e. a linear increase in the average radical length ($i$), a nearly monodisperse radical chain length distribution ($i \approx j$), and nominal impact on the radical concentration and, concomitantly, the propagation and termination rates [25, 32], allow for relating the chain-length-averaged $k_t$ directly to the microscopic $k_t^{i,i}$ at any point in time. Equation 4.11 describes how the average termination rate coefficient (equation 4.7) correlates with the microscopic CLD-T kinetic coefficient $k_t^{i,j}$ [59, 68].

$$\langle k_t \rangle = \frac{\sum_{i=1}^{n} \sum_{j=1}^{n} k_t^{i,j} [P_i\cdot][P_j\cdot]}{\left(\sum_{i=1}^{n} [P_i\cdot]\right)^2} \tag{4.11}$$

Here, the numerator is the total termination rate that takes into account a different $k_t$ possible for each combination of radical chains, $n$ is the longest radical chain length that undergoes termination and $\sum_{i=1}^{n} [P_i\cdot]$ is the total radical concentration. Even though a truly monodisperse distribution is impossible to achieve, RAFT chemistry allows for a *nearly* monodisperse chain length distribution (i.e. $i \approx j$). It should further be noted that for a successful correlation of $k_t(X)$ with $k_t^{i,i}$, no narrow radical distribution is required as long as the average $M_n$ increases with $X$ as $i$ does not necessarily needs to equal $j$ but only needs to be equal *in average*.

Equation 4.1 provides the theoretical kinetic chain length for a RAFT-mediated FRP where $c_{\mathrm{LFRP}}^0$ is the RAFT agent concentration at $t = 0$, i.e. $c_{\mathrm{RAFT}}^0$. However, the kinetic chain length depends on the RAFT agent's kinetic parameters (i.e. the magnitude of the addition and fragmentation rate coefficients in the pre- and core equilibrium), and thus equation 4.12 describes the chain length as a function of time more accurately in cases where the experimental evolution of molecular weight does not follow the theoretical change with $X$.

$$s(t) = \frac{M_n^{ex}}{M_{Mon}} \tag{4.12}$$

where $M_n^{ex}$ is the experimentally determined number-average molecular weight and $M_{Mon}$ is the molecular weight of the monomer species.

To summarize, the following procedure is used to determine the evolution of the CLD-T rate coefficient $k_t^{i,i}$, i.e. $\alpha$:

1. Monitor the polymerization rate as a function of conversion/time for the reaction mixture containing the RAFT agent, initiator and monomer.
2. Determine the MWDs as a function of conversion/time via gel permeation chromatography (based on size exclusion principles).
3. Construct a double-log plot of equation 4.3 and obtain a best fit to the slope.

It has been mentioned above that the RAFT-CLD-T method will not be applicable if the lifetime of the intermediate RAFT radical is very long and/or this radical undergoes significant side termination reactions. The effects of these two scenarios on the RAFT-CLD-T method have been investigated by Theis et al. [88], conclusively demonstrating that if slow fragmentation or intermediate radical termination occur, the RAFT-CLD-T method does not yield consistent results. The solid line in Fig. 4.1 represents the resulting log $k_t^{i,i}$ vs log $i$ plots for a fragmentation rate coefficient $k_{frag} = 1 \times 10^5\ s^{-1}$ (i.e. fast fragmentation, short intermediate radical lifetime), whereby no rate retardation is observed.

Inspection of Fig. 4.1 clearly indicates that – irrespective of the initial RAFT agent concentration – the given CLD-T rate coefficient $k_t^{i,i}$ is returned. The dotted line results for a value of $1 \times 10^2\ s^{-1}$ for $k_{frag}$ (slow fragmentation, longer intermediate radical lifetime, causing considerable rate retardation), which preferentially manifests itself at high initial RAFT agent concentrations and in the initial period of each polymerization. It is evident that the overestimation of the termination level results from an additional – albeit reversible – radical loss pathway inducing non-steady-state conditions. Rate retardation as an effect of cross-termination (see dashed lines in Fig. 4.1) should appear preferentially at high initial RAFT agent concentrations, but, in contrast to slow fragmentation, in the later stages of the polymerization. Both effects – slow fragmentation and cross-termination – result in a severe overestimation of the chain-length-dependent $k_t$ and give inconsistent (i.e. nonmatching) data for the different RAFT agent concentrations.

Originally, the RAFT-CLD-T technique was exemplified on styrene [93] and was later applied successfully to map the chain length dependence of $k_t$ for methyl acrylate (MA) [91], butyl acrylate (BA) [95], dodecyl acrylate (DA) [90], methyl methacrylate (MMA) [83] and vinyl acetate (VAc) [89]. Figure 4.2 shows an overview of the obtained chain length dependencies via the RAFT-CLD-T method for various monomers at 80°C along with a structural image of the RAFT agent that was

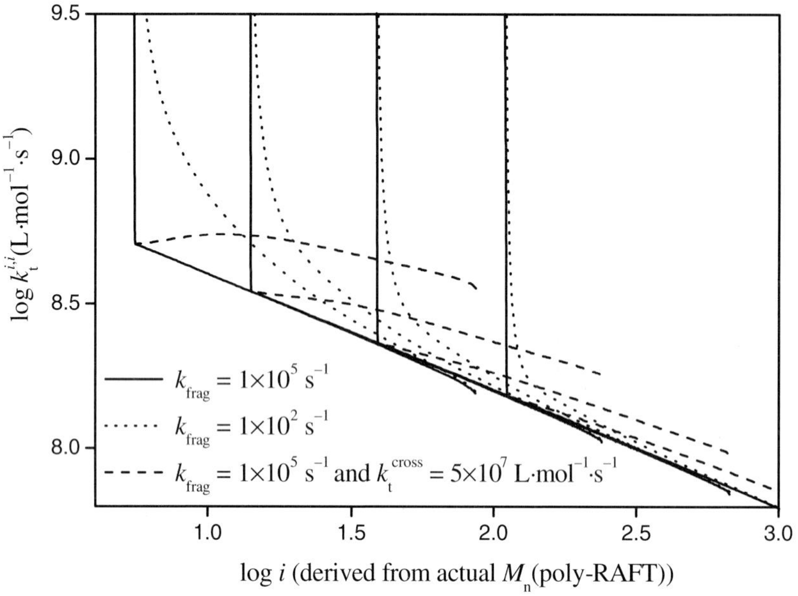


**Fig. 4.1** Simulated log $k_t^{i,i}$ vs chain length $i$ plots obtained by analyzing the time-dependent rate of polymerization data by equation 4.1 for different RAFT reagent concentrations at 60 °C. The graph depicts the expected log $k_t^{i,i}$ vs log $i$ plots under the assumption of slow fragmentation (pre- and main equilibrium) as well as cross-termination. Further, a CLD-T rate coefficient ($k_t^{i,i} = 10^9$ L·mol$^{-1}$ s$^{-1}$·$i^{-0.4}$) and the average experimental determined $k_{add}$ of 1.4 × 10$^6$ L·mol$^{-1}$·s$^{-1}$ for the system MA/MCEPDA was used. For a list of the specific simulation parameters, the reader is referred to Ref. [91]. Reproduced from Ref. [91], with kind permission from the American Chemical Society.

employed in the measurement and the (independently determined) value of $\omega$ that was used in the evaluation procedure. Figure 4.2 reveals that acrylates display significantly higher $\alpha$ values in the long-chain region (starting from approximately $i > 10$) than MMA and, interestingly, VAc. Typically, $\alpha$ values for the acrylates range from 0.36 for MA to 0.22 for BA and 0.28 for DA. In the small-chain regime, $\alpha$ is increasing with increasing chain length of the side chain from 0.36 in the case of MA to 1.2 in DA, which may be attributed to an increased shielding of the radical toward other polymeric radical chains during the first few propagation steps. At longer chain lengths, the $\alpha$ value of MA is significantly higher than the ones observed for BA and DA, indicating a significantly different flexibility and coil structure of MA compared to BA and DA. The relatively high average $\alpha$ values in acrylate systems are in good agreement with earlier findings via SP–PLP ($\alpha$(MA) = 0.32 [86] and $\alpha$(DA) = 0.4 [87]). In addition, the results are in excellent agreement with previous data obtained using pulsed-laser polymerization by de Kock [104, 105] for MA and BA, where a short-chain $\alpha$ values ($i$ = 6–10) of 0.41 for MA and 0.80–0.85 for BA as well as long-chain values ($i$ = 50–100) of 0.35–0.36 for MA and 0.17–0.19 for BA

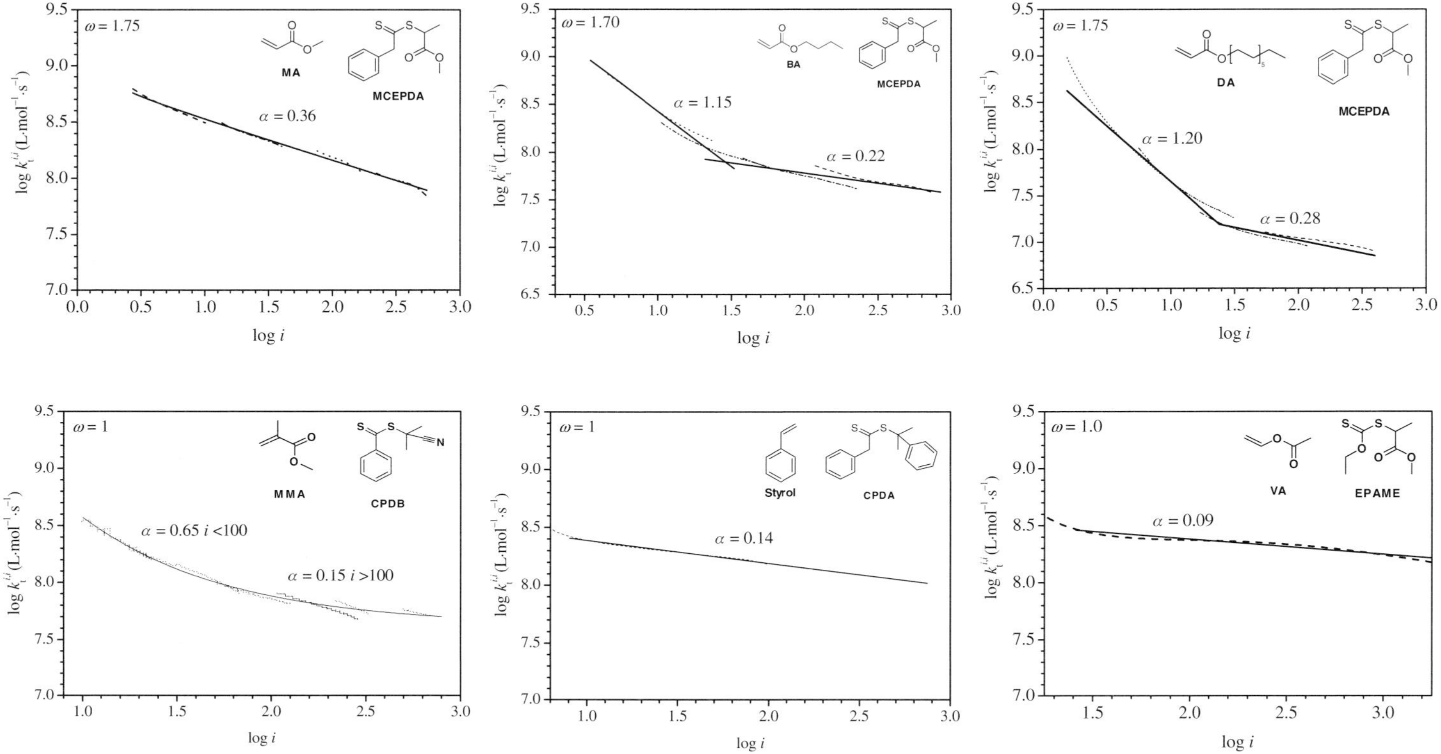


**Fig. 4.2** Chain length dependence of the termination rate coefficient of two macroradicals of nearly similar size, $k_t^{i,i}$, for MA, BA, DA, styrene, MMA and VAc at 80 °C. The employed RAFT/MADX agent is depicted within each subfigure alongside the monomer reaction order used in the data evaluation procedure according to equation 4.1. Reproduced from Ref. [92], with kind permission from the American Chemical Society.

were found. However, these former results display increased $\alpha$ values above unity for medium chain lengths of $i = 15$–30, which may be a result of transfer reactions being an aspect that was not considered by de Kock.

The overall higher level of $\alpha$ in acrylate systems may be attributed tentatively to an increased presence of midchain radicals in acrylate polymerizations generated via inter- and intramolecular chain transfer reactions [106]. Work of Friedman and O'Shaughnessy suggests that such a concept may indeed be compatible with larger $\alpha$ values [79, 80]. Thus, it is not surprising that the long-chain values obtained for monomers with less reactive propagating radicals are significantly lower than those observed for acrylates, with $\alpha$ for MMA reading 0.15, in excellent agreement with the theoretical predictions made for systems where no midchain radicals are present. Nevertheless, the short-chain regime also displays for these monomers larger $\alpha$ values (in the vicinity of those observed for acrylates), indicating center-of-mass diffusion to be the dominating process.

It is worthwhile to consider VAc separately from the other monomers. To our knowledge, the RAFT-CLD-T method was the first method employed to deduce CLD-T rate coefficients for this monomer. The VAc propagating radicals are highly reactive and thus it would reasonable to assume a priori that backbiting reactions (similar to those observed in acrylate systems) occur to a significant extent. Somewhat surprisingly, this is not the case for VAc, which shows a far decreased tendency to undergo both inter- as well as intramolecular chain transfer reactions [107]. This notion is underpinned by measurements of the monomer reaction order for VAc, which lead to an $\omega$ of 1.17 (at 80°C) [89], far lower than the $\omega$ values observed for the homologous series of acrylates ($\omega \approx 1.70$ at 80°C). Under such circumstances, it can thus be expected that the chain length dependence of $k_t$ is more in line with those determined for MMA and styrene. Inspection of the comparative Fig. 4.2 largely confirms this notion: analysis of $k_t$ in the long-chain regime (the short-chain regime is inaccessible due to hybrid behavior in the RAFT process and overlapping rate retardation effects) returns $\alpha$ values close to 0.09.

#### 4.3.1.1 **The Termination Rate Coefficient as a Function of Chain Length and Conversion**

The RAFT-CLD-T methodology has also been extended to quantify the simultaneous conversion and chain length dependence of $k_t$, i.e. $k_t(X, i)$, via constructing a three-dimensional surface of $k_t$ as a function of both these variables for MA [88, 92], MMA [108] and VAc [89]. Figure 4.3 shows a typical ($X$, $i$, $k_t$) surface for an MA polymerization at 80 °C using four RAFT agent concentrations [88, 92]. Specifically, the $k_t$ values are plotted vs monomer conversion and chain length $P$, and are fitted to a surface plot using arbitrary functions.

To examine Fig. 4.3 in more detail, slash graphs can be constructed for $k_t$ vs either conversion or chain length for various (fixed) chain lengths or conversions, respectively. For example, Fig. 4.4 shows the $k_t$ vs $i$ data from Fig. 4.3 for four different conversions. Figure 4.4 reveals that $k_t$ decreases more with chain length at higher conversions.

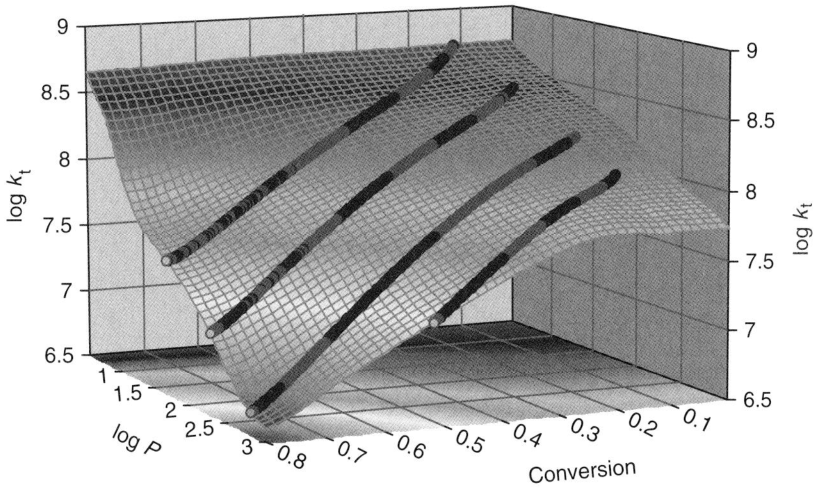


**Fig. 4.3** Three-dimensional surface of the termination rate coefficient for MA as a function of both conversion and chain length $P$ at 80 °C. Four initial RAFT agent concentrations ($4.37 \times 10^{-3}$, $1.27 \times 10^{-2}$, $3.68 \times 10^{-2}$ and $1.07 \times 10^{-1}$ mol·L$^{-1}$) were employed using $3.0 \times 10^{-3}$ mol·L$^{-1}$ AIBN. The $k_t$ (L · mol$^{-1}$·s$^{-1}$) values are plotted vs monomer conversion and chain length $P$, and are fitted to a surface plot using the TableCurve® 3D software using a correlation fit of $r^2 > 0.999\,9992$ with the lighter and darker parts of the experimental data indicating very minor deviations. Reproduced from Ref. [88], with kind permission from the American Chemical Society.

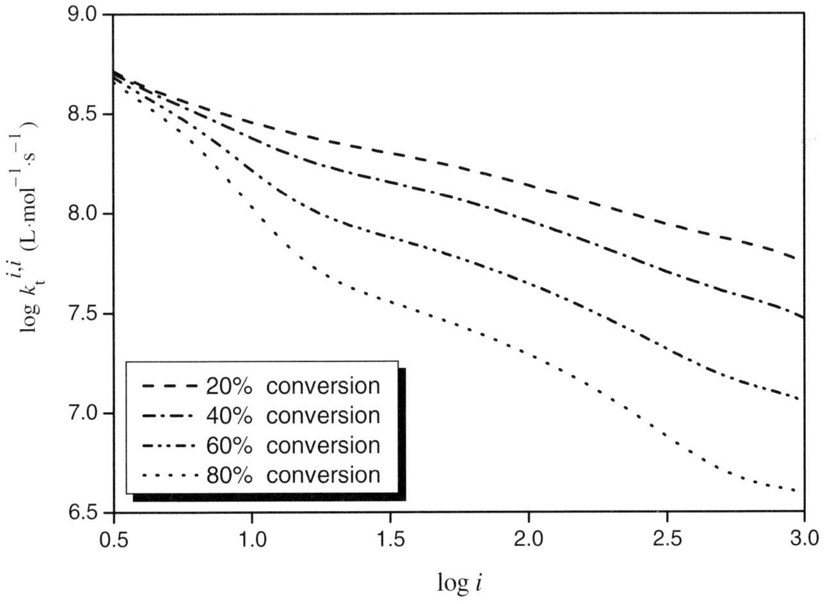


**Fig. 4.4** Extracted $k_t$ vs chain length data from Fig. 4.3 for four different conversions. Reproduced from Ref. [88], with kind permission from the American Chemical Society.

#### 4.3.1.2 RAFT Chemistry as a Tool for $k_t^{i,j}$ Elucidation

The success of the RAFT-CLD-T method for elucidation of the chain length dependence of $k_t$ led to investigating the RAFT-mediated FRP as a tool for obtaining the termination rate coefficient for disparate length radicals, where *s* and *l* are arbitrary chain lengths that differ in magnitude. Modeling is a useful tool for developing new techniques, and thus was used to demonstrate that reliable information about $k_t^{i,j}$ can be obtained from direct measurement of the polymerization rate [35, 96]. To the best of our knowledge, this method and analytical approach is the first thorough analysis of a method and procedure that quantifies $k_t^{s,l}$. de Kock et al. proposed a similar method that uses PLP rather than RAFT – the 'TR-echo-PLP' method; however, they offered no theoretical or experimental justification [66].

The method is based on the original RAFT-CLD-T method, which was modified for the parallel polymerization of two RAFT species of disparate average chain length, *s* and *l*. To quantify $k_t^{s,l}$, two distributions of disparate average chain lengths are generated via accounting for two complete RAFT FRPs (see Scheme 4.1). Within a given RAFT distribution, the individual chain lengths are denoted *i* and *j* and are assumed to be of approximately equal length; however, chains belonging to different distributions (denoted *s* and *l* for the average of the 'short-' and 'long-'chain distributions) can have different average chain lengths. To distinguish between termination events taking place between chains of approximately identical size (*ss* or *ll*) and those of disparate size (*sl* and *ls*), we denote the extent of similar size termination (within macroradicals associated with the same polyRAFT distribution) $\alpha^{s,s}$ (or $\alpha^{l,l}$ or simply $\alpha$) and the extent of disparate size termination (within macroradicals associated with different polyRAFT distributions) $\varphi^{s,l}$ (or $\varphi^{l,s}$ or simply $\varphi$). Initiation, propagation, macro-RAFT agent and poly-RAFT species generation, as well as termination occur for each distribution. In addition, the 'long' (*l*) or 'short' (*s*) chain poly-RAFT species and reactive radical, respectively, can enter the core equilibrium and/or terminate according to the reactions presented in Scheme 4.3.

This methodology was first tested using the model monomer system, styrene, because it has been modeled extensively and its rate coefficients and material properties are readily available in the literature. The material properties, kinetic parameters and rate coefficients for the RAFT-mediated FRP of styrene using AIBN as the initiator at 80°C are utilized. The rate coefficients for initiator decompositions, $k_d$ [109], and propagation, $k_p$ [110], are available from the literature. An estimation of the initiator efficiency $f$ [111] for AIBN has also been detailed. Values for the addition and fragmentation rate coefficients $k_{add}$ and $k_{frag}$ [32] are more difficult to assess; however, when the RAFT agent does not induce significant rate retardation and inhibition phenomena (such as cumyl phenyl dithioacetate in styrene polymerization), it can be assumed safely that fragmentation is a fast process. For this work, a fragmentation rate coefficient of $10^5\ s^{-1}$ was employed as done before in the context of the RAFT-CLD-T method. Additionally, the rate coefficients for initiation and for monomer addition to the RAFT agent-derived radical, $R\bullet$, $k_{p,rein}$, do not impact the model output as long as they are selected to be greater than $k_p$. All parameters used in the kinetic model are listed in Table 4.1.

To develop a simulation that provides insight into termination kinetics between disparate length radical chains, the program package PREDICI® can be utilized

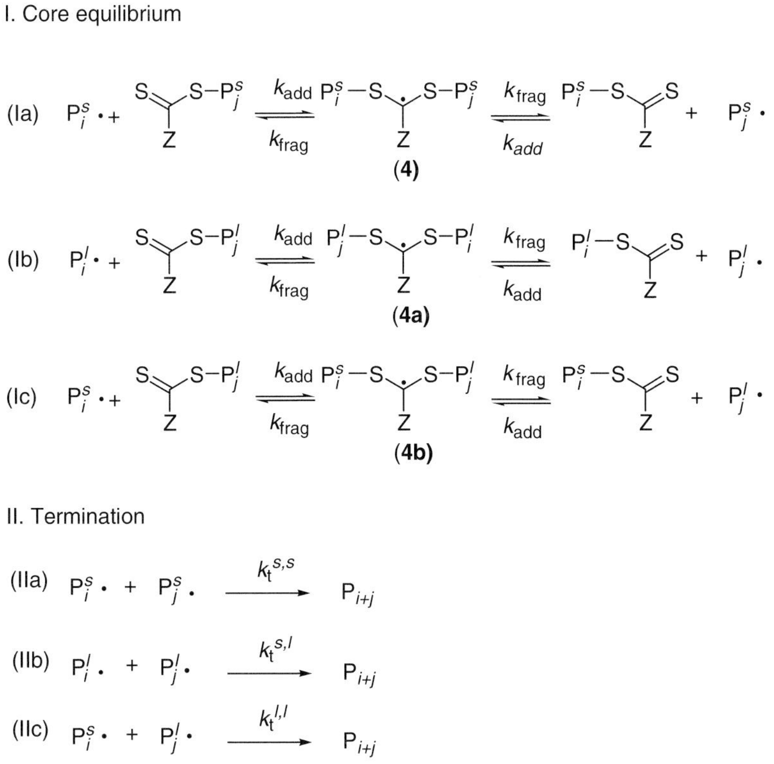

I. Core equilibrium

(Ia) $P_i^s \cdot + \underset{Z}{\overset{S}{\underset{\|}{\diagdown}}\mkern-6mu\overset{}{\diagup}}S-P_j^s \quad \underset{k_{frag}}{\overset{k_{add}}{\rightleftharpoons}} \quad P_i^s-S\underset{Z}{\overset{}{\diagdown}}\mkern-6mu\overset{\bullet}{\diagup}S-P_j^s \quad \underset{k_{add}}{\overset{k_{frag}}{\rightleftharpoons}} \quad P_i^s-S\underset{Z}{\overset{}{\diagdown}}\mkern-6mu\overset{}{\diagup}S \quad + \quad P_j^s \cdot$

(4)

(Ib) $P_i^l \cdot + \underset{Z}{\overset{S}{\underset{\|}{\diagdown}}\mkern-6mu\overset{}{\diagup}}S-P_j^l \quad \underset{k_{frag}}{\overset{k_{add}}{\rightleftharpoons}} \quad P_i^l-S\underset{Z}{\overset{}{\diagdown}}\mkern-6mu\overset{\bullet}{\diagup}S-P_j^l \quad \underset{k_{add}}{\overset{k_{frag}}{\rightleftharpoons}} \quad P_i^l-S\underset{Z}{\overset{}{\diagdown}}\mkern-6mu\overset{}{\diagup}S \quad + \quad P_j^l \cdot$

(4a)

(Ic) $P_i^s \cdot + \underset{Z}{\overset{S}{\underset{\|}{\diagdown}}\mkern-6mu\overset{}{\diagup}}S-P_j^l \quad \underset{k_{frag}}{\overset{k_{add}}{\rightleftharpoons}} \quad P_i^s-S\underset{Z}{\overset{}{\diagdown}}\mkern-6mu\overset{\bullet}{\diagup}S-P_j^l \quad \underset{k_{add}}{\overset{k_{frag}}{\rightleftharpoons}} \quad P_i^s-S\underset{Z}{\overset{}{\diagdown}}\mkern-6mu\overset{}{\diagup}S \quad + \quad P_j^l \cdot$

(4b)

II. Termination

(IIa) $P_i^s \cdot + P_j^s \cdot \quad \xrightarrow{k_t^{s,s}} \quad P_{i+j}$

(IIb) $P_i^l \cdot + P_j^l \cdot \quad \xrightarrow{k_t^{s,l}} \quad P_{i+j}$

(IIc) $P_i^s \cdot + P_j^l \cdot \quad \xrightarrow{k_t^{l,l}} \quad P_{i+j}$

**Scheme 4.3** The core equilibrium and termination reactions for two disparate distributions of radical chains generated via RAFT-mediated FRP.

[112]. A RAFT-mediated FRP is simulated until its poly-RAFT species has achieved a certain preset chain length. Subsequently, a function is used to add initiator and RAFT agent to generate the second, 'short', poly-RAFT species mid-simulation. Thus, $t = 0$ in Fig. 4.5 effectively corresponds to $t = 620$ s simulation time, i.e. the time necessary to generate a poly-RAFT species having an average length $l$ equal to 43. The latter times presented in Fig. 4.5 are 620 s less than the total simulation time. After the 'long' poly-RAFT species is generated, a second RAFT agent is administered and two distributions of poly-RAFT species coexist. Figure 4.5 illustrates the simultaneous growth of two poly-RAFT species of disparate average chain and that the long- and short-chain length distributions continue to shift toward higher degrees of polymerization with increasing polymerization time.

The average chain length of the poly-RAFT species, and concomitantly the average chain length of the propagating radical distribution, $s$ or $l$, is assumed to be equal to the ratio of the first moment to the zeroth moment of the MWD of the poly-RAFT species. Figure 4.6 illustrates the average chain length of each distribution ($s$ and $l$) from Fig. 4.5 as a function of double bond conversion. Initially, only the long distribution is present and thus its average chain length increases. At 20% conversion, the second RAFT agent is added and the average of both distributions

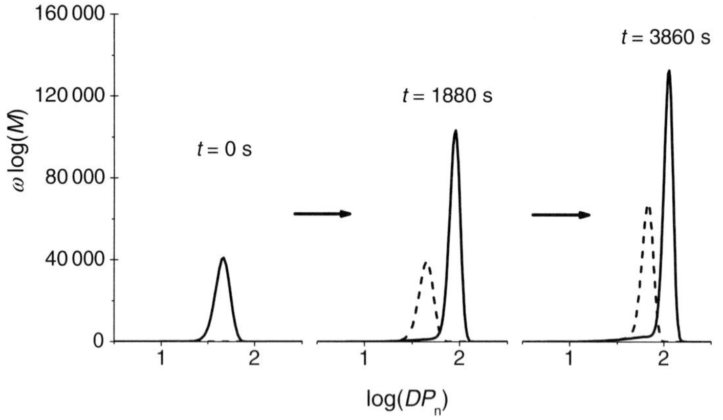


**Fig. 4.5** Simulated (radical) MWD evolution of the two RAFT species having an initial average chain length *l* equal to 43 (—) and 0 (. . .) is presented at $t = 0$, 1880 and 3860 s. The rate coefficients and other parameters relevant for the simulation are collated in Table 4.1. The simulation has been parameterized on the example of a styrene bulk polymerization at 80 °C. All $\varphi$ input values, $\varphi^{s,s}$ (and $\varphi^{l,l}$) or $\varphi^{s,l}$, are set to 0.16 and the geometric mean is used (see equation 4.3 for a definition of the geometric mean). Note that the concentrations of the individual radical distributions associated with each poly-RAFT distribution are identical. Reproduced from Ref. [35], with kind permission from the American Chemical Society.

increases albeit the long distribution's average chain length increases more slowly than prior to the addition of the second RAFT agent, as is evident from the change in slope (see Fig. 4.5).

RAFT agents are administered at preselected time points solely to generate disparate length chains as SEC differentiates chains based only on their size, i.e. molecular weight. However, the model keeps track of each corresponding poly-RAFT species separately and the core equilibrium reaction (Ic) mixes the chains generated at early time points with the shorter, younger chains. Even though

**Table 4.1** Input parameters used for the kinetic modeling of the RAFT-mediated free radical bulk polymerization of styrene at 80 °C[a]

| $k_d$ ($s^{-1}$) [109] | $k_p$ [110] | $f$ [111] | $k_{add}$ [32] | $k_{frag}$ ($s^{-1}$) [32] | $k_{t0}$ |
|---|---|---|---|---|---|
| $1.36 \times 10^{-4}$ | 663 | 0.713 | $5.0 \times 10^5$ | $1.0 \times 10^5$ | $1.0 \times 10^7$ |
| $k_i$ | $k_{p,rein}$ | $[I]_0$ | $[RAFT]_0$ | $[Sty]_0$ [32] | $T$ (°C) |
| $k_i \geq k_p$ | $k_{p,rein} \geq k_p$ | $2.0 \times 10^{-2}$ | $4.0 \times 10^{-2}$ | 8.73 | 80 |

[a] All rate coefficients are given in $L{\cdot}mol^{-1}{\cdot}s^{-1}$ and all concentrations are given in $mol{\cdot}L^{-1}$ unless otherwise indicated.
Sty = styrene.

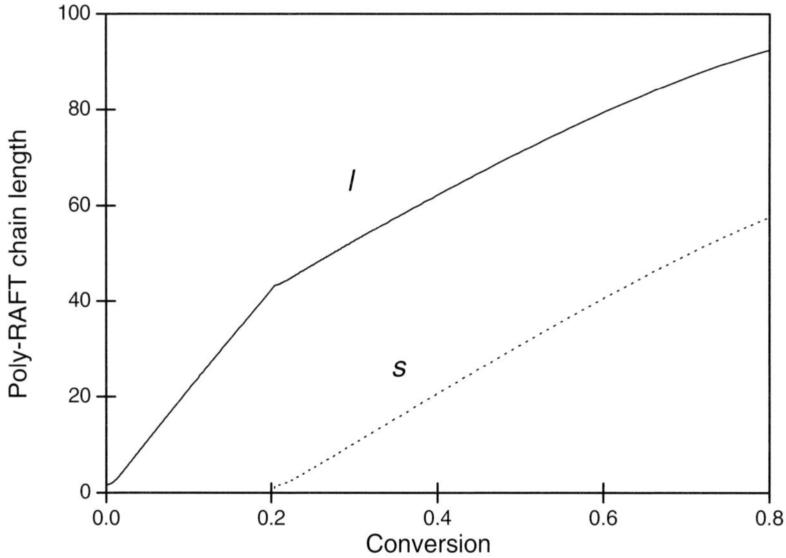


**Fig. 4.6** The average chain length of the prepolymerized poly-RAFT species (*l*, —) and of the poly-RAFT species that is initiated at $t = 620$ s (*s*, . . .) as a function of conversion is presented as predicted via simulation at 80 °C. All $\varphi$ and $\alpha$ input values are set to 0.16 and the geometric mean is used (see equation 4.3 for a definition of the geometric mean). The rate coefficients and other parameters relevant for the simulation are collated in Table 4.1. Reproduced from Ref. [35], with kind permission from the American Chemical Society.

disparate distributions of chain lengths are generated, for ease of data presentation and due to the PREDICI internal data output structure, all simulations are conducted without accounting for the core equilibrium reaction (Ic) in Scheme 4.3. This simplification does not impact the predicted reaction kinetics and/or kinetic chain lengths and allows for MWD predictions analogous to that provided by an SEC experiment.

Accounting for the termination event between propagating radicals from distributions of disparate average chain length requires modification of equation 4.11. Assuming that the two distributions are represented adequately by their respective average chain length, equation 4.13 provides the relationship for the average termination kinetic coefficient and the termination kinetic coefficients for two 'short' chains, $k_t^{s,s}$, two 'long' chains, $k_t^{l,l}$, and one 'short' and one 'long' chain, $k_t^{s,l}$.

$$\langle k_t \rangle = \frac{k_t^{s,s}[P_s\bullet]^2 + 2k_t^{s,l}[P_s\bullet][P_l\bullet] + k_t^{l,l}[P_l\bullet]^2}{([P_s\bullet] + [P_l\bullet])^2} \tag{4.13}$$

Here $[P_s\bullet]$ and $[P_l\bullet]$ are the concentration of short and long radical chains, respectively. When equal concentrations of reacting species exist (i.e. $[P_s\bullet] = [P_l\bullet]$), equation 4.13 simplifies to the following:

$$\langle k_t \rangle = \frac{1}{4}k_t^{s,s} + \frac{1}{2}k_t^{s,l} + \frac{1}{4}k_t^{l,l} \tag{4.14}$$

Equal concentrations of reacting species are achieved by employing equal concentrations of RAFT agent (for the poly-RAFT and initial RAFT species), resulting in a simple relationship for the dependence of $k_t^{s,l}$ on the average termination rate coefficient and the termination kinetic coefficients for equal length radical chains (equation 4.15). While this assumption may seem counterintuitive because CLD-T involves small radicals terminating faster than long radicals, which should lead to concentration differences between poly-RAFT species with disparate average chain lengths, the RAFT polymerization is a steady-state process and terminating radicals are replaced by newly generated radicals. Thus, as long as there is initiator present in the reacting system, each population of poly-RAFT species remains constant.

$$k_t^{s,l} = 2\langle k_t \rangle - \frac{1}{2}k_t^{s,s} - \frac{1}{2}k_t^{l,l} \tag{4.15}$$

Figure 4.7 illustrates equation 4.15: that is, how the average termination rate coefficient $\langle k_t \rangle$ (deduced from equation 4.5 from the rate of polymerization data; Fig. 4.7a) and $k_t^{s,s}$ (deduced from prior knowledge of the extent to which similar size radical chains terminate, $\alpha^{s,s}$, and thus $\alpha^{l,l}$; Fig. 4.7b) combine to provide the termination rate coefficient for disparate length radicals, $k_t^{s,l}$ (Fig. 4.7c). The simulation conditions presented here are identical to those employed in Fig. 4.5.

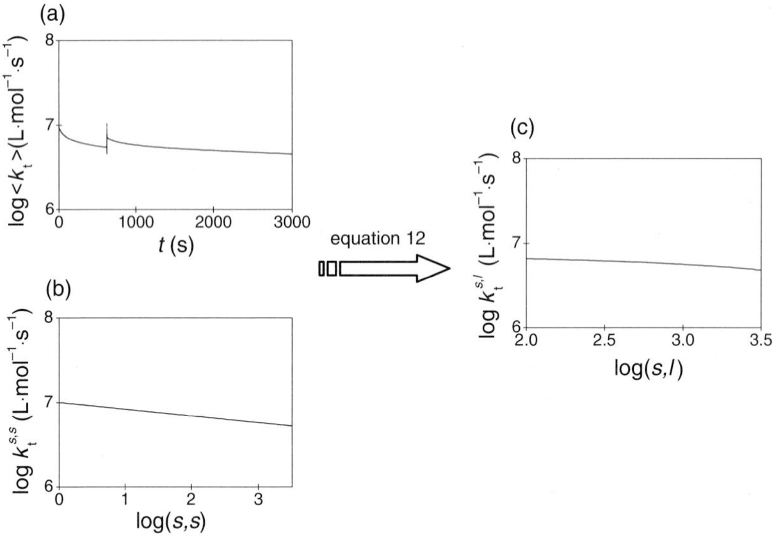


**Fig. 4.7** A log plot of the average termination kinetic coefficient $\langle k_t \rangle$ vs polymerization time (a) and double-log plots of the short–short, $k_t^{s,s}$, and short–long, $k_t^{s,s}$, termination rate coefficients vs the products of the radical chain lengths terminating (b and c, respectively) are presented for the simulated RAFT-mediated styrene polymerization at 80 °C with a poly-RAFT species that has been prepolymerized to an average chain length, $l$, equal to 43. All $\alpha$ and $\varphi$ input values are set to 0.16, the geometric mean is used and data up to 65% conversion are presented. Note that when $\alpha^{s,s}$ is equal to $\alpha^{l,l}$, the model predicts the same $k_t^{s,s}$ and $k_t^{l,l}$ for equivalent chain lengths. Reproduced from Ref. [35], with kind permission from the American Chemical Society.

Figure 4.7a also reveals that the average termination rate coefficient increases upon addition of the second RAFT agent, which generates a younger, shorter-chain poly-RAFT species that terminates more readily than the older, longer-chain poly-RAFT species (born at $t = 0$). The addition of the second RAFT agent and initiator is instantaneous and the termination rate, and concomitantly the initiation and polymerization rates, achieves steady state in 1 s of simulated polymerization time, after which $k_t$ is predicted to decrease with increasing polymerization time, i.e. increasing chain length (Fig. 4.7).

To test the robustness of the method, the harmonic mean model (equation 4.16) is also used instead of the geometric mean model in equation 4.3. In both equations the extent or magnitude of the chain length dependence of the short–long termination is expressed via the exponent $\varphi$ (identical to the role that $\alpha$ has in equation 4.3).

$$k_t^{s,l} = k_{t0}\left(\frac{2sl}{s+l}\right)^{-\varphi} \tag{4.16}$$

The extent that the termination rate coefficient depends on disparate length radicals, $\varphi$, is obtained via construction of a double-log plot of either equation 4.3 or equation 4.16, where the $k_t^{s,l}$ values are obtained from equation 4.15. Note that when equation 4.3 is used the $x$-axis is the log of the product of the chain lengths and when equation 4.16 is employed then the double-log plot will present $k_t^{s,l}$ vs $\frac{2sl}{s+l}$. The data are fit typically from approximately 3% conversion after the second RAFT agent is administered to approximately 85% monomer conversion or until the data no longer exhibit a linear relationship for the combination of the short- and long-chain lengths relevant to either equation 4.3 or equation 4.16 vs polymerization time.

Figure 4.8 presents an example of a double-log plot of equation 4.3. Both the model predictions and the slope of the best linear fit are presented, indicating that the method returns $\varphi_{out}$ (i.e. the method deduced input parameter for the extent of the chain length dependence) equal to 0.152 and 0.466. These values agree well with the model input values of $\varphi$ equal to 0.16 and 0.5, respectively. Interestingly, when $\varphi$ is 0.5, the plot is nonlinear. This nonlinearity is due to the high initiation and concomitant termination rate, which results in more polydisperse populations. Reducing the initiation rate decreases the termination rate, generating more monodisperse populations that yield linear plots, i.e. improve the method's accuracy. However, the higher initiation condition is presented here because it represents more accurately the necessary condition for obtaining experimentally reliable rate information, e.g. when using differential scanning calorimetry. In both cases, all parameters except $\varphi$ are equal and $\alpha^{s,s}$ (and $\alpha^{l,l}$) is 0.16.

The method's ability to predict $\varphi_{out}$ within reasonable accuracy was shown to be independent of the extent of prepolymerization of the poly-RAFT species, the $\varphi$ and/or $\alpha$ input values, and/or whether or not the geometric or the harmonic mean is employed [35]. Interestingly, when the harmonic mean is used and the termination rate coefficient has a strong chain length dependence (i.e. the scaling exponent $\varphi$ is 0.5) the method's predictive capability decreases as is evident for a $\varphi_{out}$ value

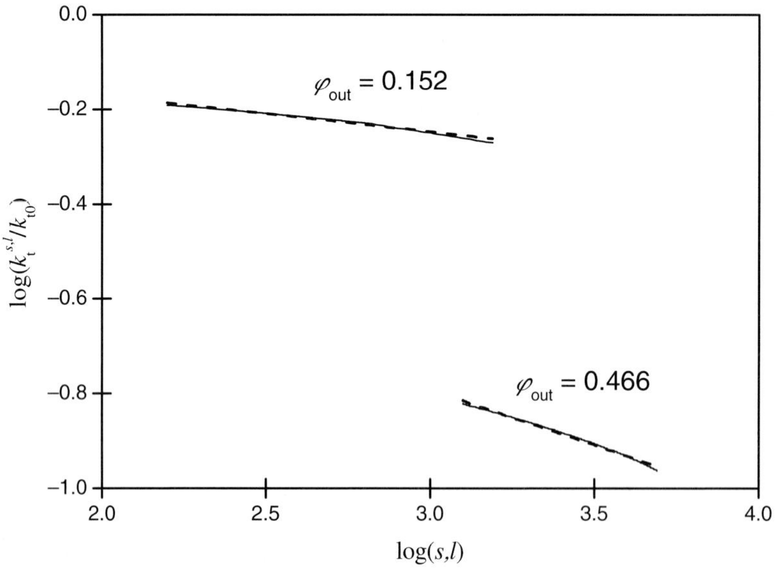


**Fig. 4.8** Double-log plot of $k_t^{s,l}$ normalized by $k_{t0}$ vs the product of the average chain length of the two poly-RAFT species (of disparate chain length s and l) is presented as predicted by the simulated polymerization of styrene, RAFT agent and an average poly-RAFT species of initial average chain length equal to 43. Two cases are presented: one simulation where $\varphi = 0.16$ and one where $\varphi = 0.50$. The dashed line is the best linear fit to the data (continuous line). The slope of the linear fit is equal to $-(1/2)\varphi_{out}$ (see equation 4.3) and gives the extent of $k_t^{s,l}$'s chain length dependence. In both cases all parameters except $\varphi$ are equal and $\alpha^{s,s}$ (and $\alpha^{l,l}$) is 0.16. Reproduced from Ref. [35], with kind permission from the American Chemical Society.

equal to 0.358. When CLD-T is important, one would expect that the concentration of the shorter length species will deplete more rapidly than the concentration of the longer length species. Additionally, $k_t$ decreases less with increasing chain length (or combination of chain lengths) when the harmonic mean vs the geometric mean is used. Since the analytical technique is based on the assumption of equal concentration of reacting species (an assumption necessary for the simplification of equation 4.13), the model's predictive capability may weaken for very strong chain length dependence and/or increasing difference in each distribution's average chain length coupled with a very high termination rate coefficient. However, when the difference in the reacting species chain length is increased the method's performance improves [35].

#### 4.3.1.3 **Fast-Propagating Monomers That Undergo Chain Transfer to the Polymer Backbone**

Testing the method – at least theoretically – to elucidate $k_t^{s,l}$ for fast-propagating monomers such as acrylates is an interesting problem. For one, acrylates undergo side reactions such as inter- and intramolecular chain transfer (i.e. chain transfer to polymer and backbiting, respectively) [113–118], and thus whether or not these side reactions impact the method's ability to elucidate the termination rate coefficient from only the polymerization rate data warrants examination. Additionally, the RAFT-CLD-T method relies on accurate online determination of the polymerization rate, and given their rapid polymerization, acrylates are an attractive option for experimental validation of this procedure. Also, acrylates are used extensively in industry and complete characterization of their FRP kinetics would be advantageous as is evident from the numerous investigations of acrylate kinetics found in the literature [33, 38, 87, 88, 90, 91, 95, 97, 119–126]. Thus, the impact of fast propagation, backbiting and midchain radical reactions on the method's ability to obtain accurately the chain length dependence of $k_t$ for both similar and disparate size radicals is investigated with the goal of aiding the experimentalist in choosing the optimum polymerization system for validating the recently introduced $k_t^{s,l}$ methodology.

The material properties and kinetic parameters for MA, the initiator AIBN and the RAFT agent methoxy carbonylethyl phenyldithioacetate (MCEPDA) [88] are incorporated into the model (Table 4.2), including the addition, fragmentation, initiation and the initial termination rate coefficients ($k_{add}$, $k_{frag}$, $k_i$ and $k_t^0$) along with the initiator decomposition, propagation and reinitiation rate coefficients ($k_d$, $k_p$ and $k_{p,rein}$) and the monomer, RAFT agent and initiator concentrations at time zero. For simplicity, the gel effect is not taken into account.

Backbiting is accounted for via inclusion of the reaction steps for tertiary radical, $P_{i,t}\bullet$, formation (Ia), propagation (Ib) and termination (Ic and Id) into the PREDICI simulation (see Scheme 4.4). These reactions depend on the rate coefficients for backbiting, $k_{bb}$, tertiary radical propagation, $k_{p,t}$, and tertiary radical termination,

**Table 4.2** Input parameters used for the kinetic modeling of the RAFT-mediated acrylate FRP initiated with AIBN[a].

| $k_{add}$ [90] | $k_{frag}$ ($s^{-1}$) [90] | $k_i$ [90] | $k_t^0$ [91] |
|---|---|---|---|
| $1.4 \times 10^6$ | $1.0 \times 10^5$ | $1.57 \times 10^3$ | $1.0 \times 10^9$ |
| **$k_d$ ($s^{-1}$) [92, 104]** | **$k_p$ [128][b]** | **$k_{p,rein}$ [92]** | **$F$ [89]** |
| $8.4 \times 10^{-6}$ | $3.3 \times 10^4$ | $3.3 \times 10^4$ | 0.7 |
| **$[MA]_0$ [128]** | **$[MCEPDA]_0$ [91]** | **$[AIBN]_0$ [89]** | **$T$ (°C)** |
| 10.2 | $3.7 \times 10^{-2}$ | $3.0 \times 10^{-3}$ | 60 |

[a] All rate coefficients are given in $L \cdot mol^{-1} \cdot s^{-1}$ and all concentrations are given in $mol \cdot L^{-1}$ unless otherwise indicated.
[b] Propagation rate coefficient here is for end-chain propagation only.

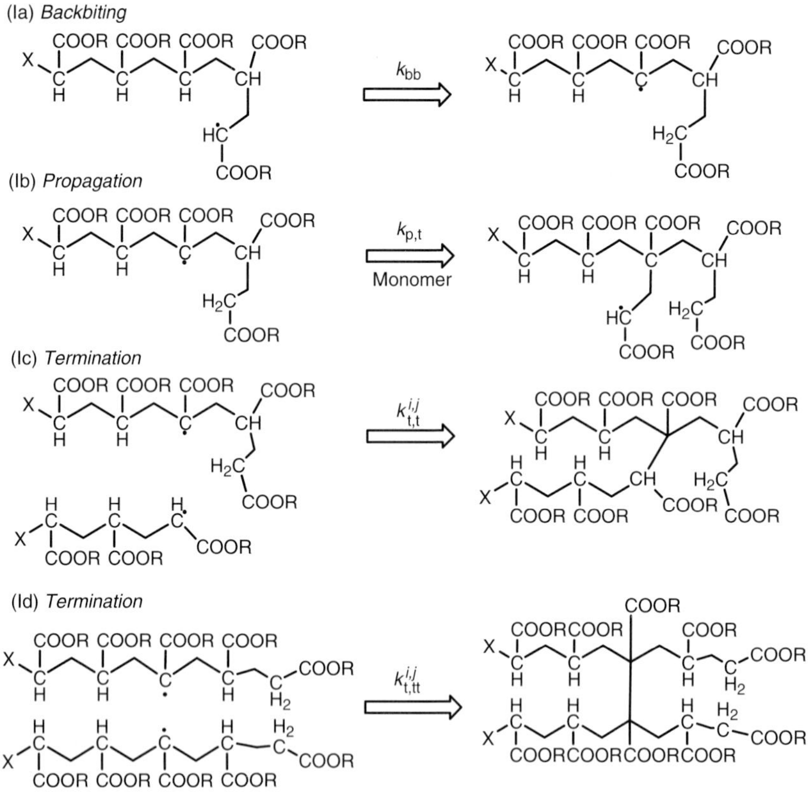


**Scheme 4.4** Tertiary radical formation (backbiting) and midchain radical propagation and termination.

which occurs either between two tertiary radicals, $k_{t,tt}^{i,j}$, or between a midchain and an end-chain radical, $k_{t,t}^{i,j}$, where the moiety X represents the continuing chain. Additionally, the model was expanded to account for the reactions for the RAFT equilibria of tertiary radicals (see Scheme 4.5). The preequilibrium of a tertiary radical with the initial RAFT agent (Ia) yields a macro-RAFT species that is attached midchain. Also, the core equilibrium, where the macro-RAFT species is formed from an end chain (Ib) or a midchain (Ic) radical, is taken into account. The preequilibrium and the core equilibrium are governed by $k_{add,t}$ and $k_{frag,t}$, respectively. All necessary parameters for the kinetic modeling of intramolecular chain transfer are given in Table 4.3. There are no kinetic parameters to date for how backbiting occurs during the RAFT-mediated AIBN-initiated MA polymerization; thus, the kinetic parameters for the RAFT-mediated AIBN-initiated DA FRP were used and addition and fragmentation rate coefficients were assumed independent of location on the chain.

Accessing the chain length dependence of the termination rate coefficient for *disparate* length radicals is a process that is significantly more complex when

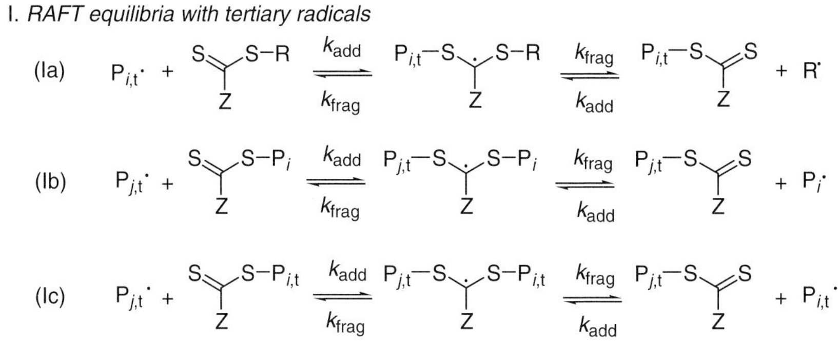


**Scheme 4.5** Basic reactions for tertiary radical formation (backbiting) and midchain radical propagation, termination and RAFT equilibria.

intramolecular chain transfer occurs. As mentioned during the discussion of equation 4.5, accounting for midchain radical formation has been shown to lead to virtual monomer reaction orders (i.e. $\omega$) greater than 1 [97]. Additionally, two new termination rate coefficients ($k_{t,t}^{i,j}$ and $k_{t,tt}^{i,j}$) are accounted for that may differ in value and chain length dependence from conventional $k_t^{i,j}$. To investigate the impact of backbiting and tertiary radicals on the method's ability to predict the chain length dependence of $k_t$ for disparate length radicals, i.e. $\varphi$, where $k_t^{s,l} \propto (sl)^{-\varphi/2}$, the method was expanded to include the necessary reactions for accounting for the core equilibrium and termination events when two RAFT distributions polymerize simultaneously (see Scheme 4.6). Elucidation of $\varphi_{out}$ when backbiting is important first requires determination of the virtual monomer reaction order for the specific polymerization condition.

Simulation was used previously to show that $\omega$ increases with increasing CLD-T, i.e. $\omega$ is equal to 1.2, 1.5 and 2.5 when $\alpha_{in}$ and $\varphi_{in}$ are equal to 0.16, 0.4 and 0.8, respectively [96]. Additionally, when backbiting and virtual monomer reaction orders greater than 1 are accounted for, the method predicts $\varphi_{out}$ better than the method that considers only fast propagation (see Figs. 4.9 and 4.10, respectively).

Figures 4.9 and 4.10 illustrate that the method's predictive capability is sensitive to the polymerization rate's dependence on monomer concentration, i.e. the apparent propagation rate coefficient (virtual monomer reaction order). Simulation

**Table 4.3** Backbiting and tertiary radical formation parameters necessary for the kinetic modeling of the RAFT-mediated acrylate FRP initiated with AIBN[a]

| $k_{bb}$ ($s^{-1}$) [90] | $k_{p,t}$ [90] | $k_{t,t}^{i,j}$ [90] | $k_{t,tt}^{i,j}$ [90] | $k_{add,t}$ [90] | $k_{frag,t}$ ($s^{-1}$) [90] |
|---|---|---|---|---|---|
| $1.623 \times 10^3$ | 55 | $1.0 \times 10^8$ | $1.0 \times 10^7$ | $1.4 \times 10^6$ | $1.0 \times 10^5$ |

[a] All rate coefficients are given in $L \cdot mol^{-1} \cdot s^{-1}$ and all concentrations are given in $mol \cdot L^{-1}$ unless otherwise indicated.

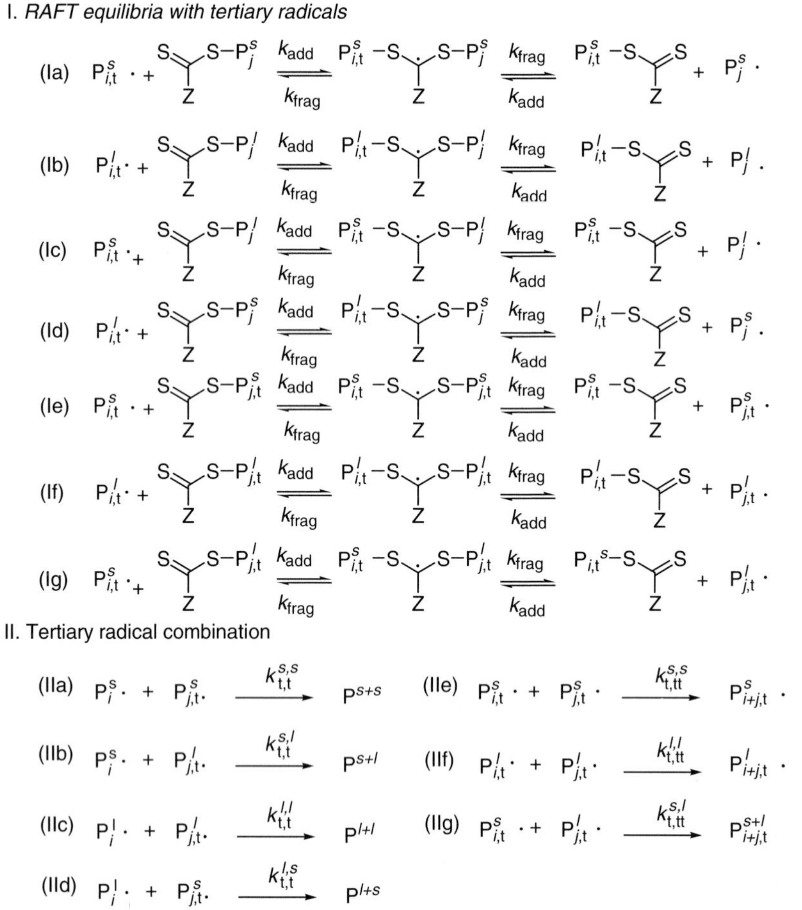


**Scheme 4.6** The core equilibrium and termination reactions for two simultaneous RAFT FRPs when midchain radical formation and subsequent reactions are taken into account.

is used to illustrate that knowledge of the virtual monomer reaction order may indeed allow for accurate determination of the extent that $k_t$ depends on radical size (both $\alpha$ and $\varphi$) for the acrylate FRP (see Figs. 4.9 and 4.10). Additionally, increasing the extent that $k_t$ depends on the radical's chain length increases the virtual monomer reaction order, i.e. decreases the apparent propagation rate coefficient. Since the assumption that equal concentrations of reacting species are guaranteed via employing equal concentrations of RAFT agents (and neglecting a potential CLD of the RAFT equilibrium reactions), the only other assumption that could cause the model to inaccurately predict $\varphi_{out}$ is the assumption that each RAFT distribution is represented adequately by its average chain length. In fact, when backbiting is neglected and fast propagation is accounted for, the model predicts a more polydisperse 'short' macro-RAFT distribution that increases in polydispersity

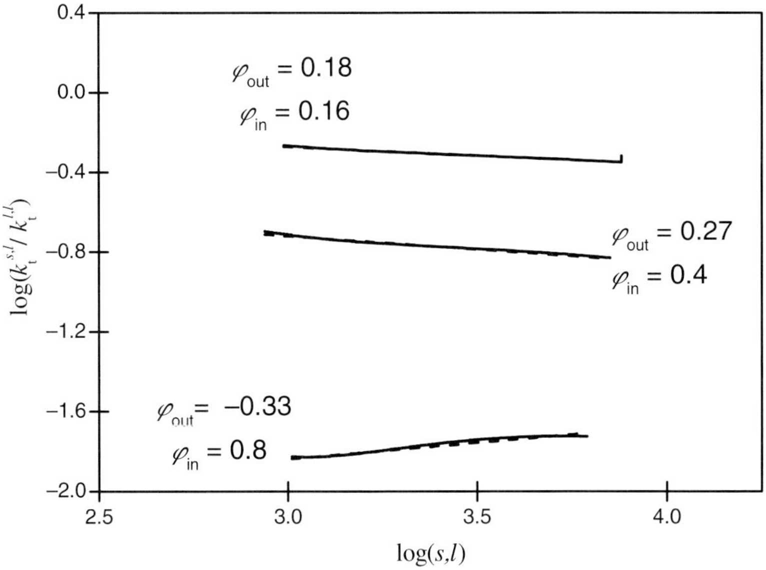


**Fig. 4.9** Double-log plot of $k_t^{s,l}$ normalized by $k_{t0}$ vs the product of the average chain length of the two poly-RAFT species (of disparate chain length $s$ and $l$) is presented as predicted by the simulated polymerization of a fast-propagating monomer *neglecting backbiting*, RAFT agent and an average poly-RAFT species of initial average chain length equal to 82. Three cases are presented: where $\varphi$ is varied from 0.16, 0.4 and 0.8. The dashed line is the best linear fit to the data (continuous line). The slope of the linear fit is equal to $-(1/2)\varphi_{out}$ (see equation 4.3) and gives the extent of $k_t^{s,l}$'s chain length dependence. In all cases, all parameters except $\varphi$ are equal and $\alpha^{s,s}$ is equal to $\alpha^{l,l}$, which is equal to $\varphi$. All termination events involving midchain radicals are assumed chain length independent (i.e. $\alpha^{t}$ and $\alpha^{tt} = 0$).

when the termination rate decreases more rapidly (i.e. with increasing chain length dependence, the geometric mean and greater radical size disparity $(s - l)$). When backbiting is accounted for, a more monodisperse 'short' macro-RAFT distribution is predicted and consequently the method predicts more accurately $\varphi_{out}$. In this context, it is important to note that the macro-RAFT distributions' polydispersity can be controlled via changing the initiation conditions. Thus, when the data are analyzed with extreme care and the reaction thoroughly characterized (i.e. intramolecular chain transfer is accounted for and $\omega$ is determined), determining the extent that the termination rate coefficient depends on disparate size radicals for the acrylate polymerization may be possible using this methodology.

### 4.3.2
### The SP–PLP–RAFT Technique

While the RAFT-CLD-T technique provides an easy and well-reproducible pathway to probe the chain length dependency of the termination rate coefficient, it

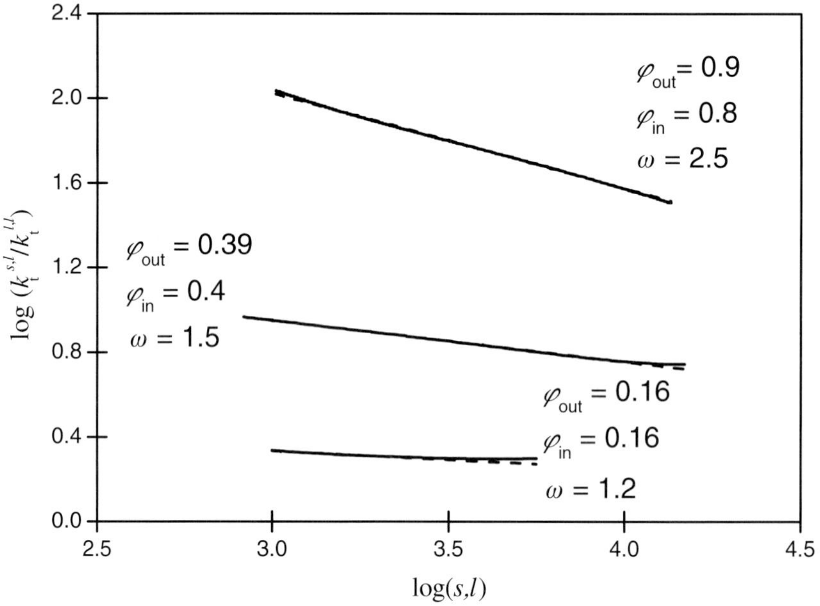


**Fig. 4.10** Double-log plot of $k_t^{s,l}$ normalized by $k_{t0}$ vs the product of the average chain length of the two poly-RAFT species (of disparate chain length $s$ and $l$) is presented as predicted by the simulated polymerization of a fast-propagating monomer *accounting backbiting*, RAFT agent and an average poly-RAFT species of initial average chain length equal to 82. Three cases are presented: where $\varphi$ is varied from 0.16, 0.4 and 0.8. The dashed line is the best linear fit to the data (continuous line). The slope of the linear fit is equal to $-(1/2)\varphi_{out}$ (see equation 4.3) and gives the extent of $k_t^{s,l}$ 's chain length dependence. In all cases, all parameters except $\varphi$ are equal and $\alpha^{s,s}$ is equal to $\alpha^{l,l}$, which is equal to $\varphi$. All termination events involving midchain radicals are assumed chain length independent (i.e. $\alpha^t$ and $\alpha^{tt} = 0$).

employs stationary reaction conditions to evaluate the individual rate coefficients, and such methods generally suffer from the lack of precise knowledge of the product of initiator decomposition rate and of initiator efficiency $fk_d$, which can be overcome by employing single laser pulse (SP) techniques. With the so-called single pulse–pulsed-laser polymerization (SP–PLP) technique equipped with near-infrared (NIR) detection, the decrease in monomer concentration after illumination of a sample with a laser pulse is followed with microsecond time resolution [87, 128]. The termination rate coefficient is then deduced by fitting the so-obtained concentration trace to the integrated rate law given in equation 4.3, which reads

$$\frac{c_M(t)}{c_M^0} = (1 + 2k_t c_R^0 t)^{-k_p/2k_t} \tag{4.17}$$

From the fitted parameters, the termination rate coefficient is obtained from the coupled parameter $k_p/k_t$; thus, only knowledge on the propagation rate coefficient is required for which data are available to high accuracy from the PLP–SEC method.

Because no reaction, in principle, will occur in the dark time between two consecutive laser pulses (after all radicals have undergone termination), the reaction can be stopped at any time allowing for a pointwise probing of the rate coefficient as a function of monomer conversion.

If a photo-initiator is chosen that decomposes upon laser illumination into two fragments of equal initiating reactivity, such as $\alpha$-methyl-4-(methylmercapto)-$\alpha$-morpholino propiophenone, a monodisperse growth of radicals may be assumed to hold up to the point where chain transfer reaction distort the narrow distribution. Hence, at short delay times after a laser pulse (usually around 0.1 s), termination occurs only between radicals of the same size. As the chain length of the radicals can easily be deduced from the simple relation of $i = (k_p c_M)^{-1}$, $k_t(i)$ can directly be yielded by fitting more complex equation to the data [87]. While this approach yields reliable data, it is limited in that the chain length dependence is analyzed only with respect to the relatively simple monoexponential power law given in equation 4.3. Additionally, the method works significantly better for the relatively slowly propagating methacrylates than for acrylate monomers, which also suffer from the advent of extensive chain transfer to polymer reactions that prohibits a true monodisperse growth of the radicals from almost chain length unity on. The limitation with respect to the applied CLD-T model in the fit functions can be overcome by observing the decay in radical concentration via switching to (ESR) detection [81, 119, 129]. Knowledge on $c_R(t)$ gives more direct access to $k_t(i)$, and a differentiation between short-chain and long-chain behavior as proposed by Smith and Russell [55] becomes feasible. However, also this technique does not allow for comprehensive analysis of acrylate polymerization kinetics. Thus, correlating the average chain length of radicals with monomer conversion in a RAFT-controlled polymerization system poses an advantage over the conventional SP–PLP methods as the pointwise probing of the conversion dependence is changed to a pointwise probing of the chain length dependence of the termination rate coefficient. Hence, no assumptions on the CLD-T model needs to be made and, moreover, chain transfer reactions do not interfere.

In the attempt to combine the RAFT-CLD-T method with SP–PLP–NIR, conditions suitable for RAFT need to be identified so as to proceed effectively under laser pulsing. For successful RAFT–SP–PLP experiments, (a) a RAFT agent needs to be identified that is stable under laser pulse irradiation and (b) it needs to be verified whether the reversible transfer mechanism is still able to control the polymerization under non-steady-state reactions conditions. It goes without saying that these requirements come on top of the requirements for a RAFT agent to be used in RAFT-CLD-T, i.e. more or less ideal RAFT conditions in that no retardation of the overall polymerization rate is, in principle, allowed to occur. If a retarding RAFT agent is used, the measured apparent $k_t$ may not reflect the true termination rate of the propagating radicals.

#### 4.3.2.1 **Ultraviolet Radiation Stability of RAFT Agents**

Most of the common RAFT agents such as the dithiobenzoates are only of limited UV stability and show pronounced absorption in the UV/Vis wavelength region.

Hence, those compounds are not overly suitable for use in photo-initiated polymerizations. However, photo-induced free radical polymerization is of interest for various applications as it allows for production of polymers at mild reaction conditions under easily controllable conditions. Because of the relative UV instability of most RAFT agents, $\gamma$-ray-initiated RAFT polymerization has recently gained a lot of attention, as equal advantages are provided [130–134]. However, not all RAFT agents decompose when irradiated with UV light and successful RAFT polymerizations were carried out for various monomers and the most suitable mediating agents seem to be of the substance class of trithiocarbonates [135–139].

It should be noted that even a RAFT agent absorbing UV light in the relevant wavelength regime does not necessarily decompose. As a certain amount of input power may be redistributed via relaxation processes, low UV doses may be tolerable if the absorption is not too high.

Figure 4.11 depicts the UV spectra of two RAFT agents, namely cumyl dithiobenzoate (CDB) and *S,S'*-bis(methyl-2-propionate) trithiocarbonate (BMPT). Both compounds show a maximum in absorption around 300 nm. A relatively large difference is however seen toward higher wavelength. While CDB absorbs significantly up to around 385 nm, only very low absorptivity is observed for BMPT from about 350 nm on. This difference is of high importance for most photo-initiated polymerization systems as wavelengths in the region of 350–370 nm are often used; for example, in conventional SP–PLP polymerization, the reaction is started by Excimer laser pulses at 351 nm (XeF line). As is expected from the higher absorptivity of

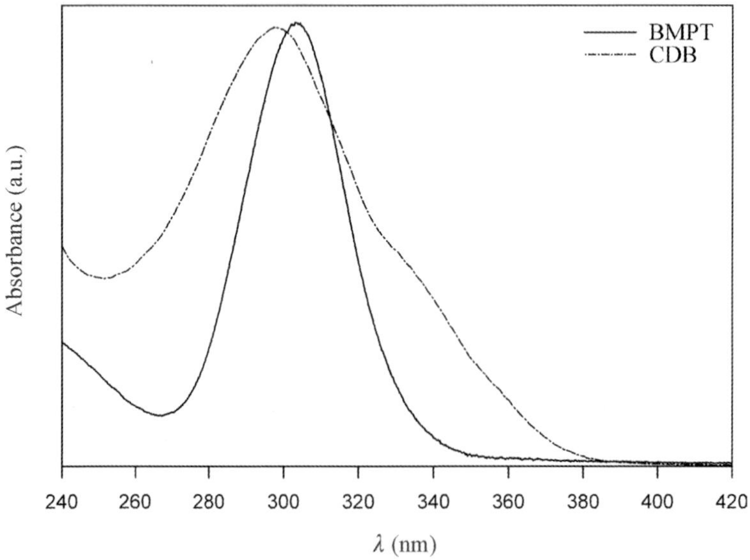


**Fig. 4.11** UV spectra of the RAFT agents CDB and BMPT. Partially reproduced from Ref. [140], with kind permission from the American Chemical Society.

CDB at 351 nm, the dithiobenzoate undergoes relatively fast decomposition with laser irradiation. While the BMPT concentration decreases only by a few percent after application of 2500 laser pulses, almost complete CDB decay is seen after an equal amount of pulses. About 10% CDB is lost after application of about 200 laser pulses, which appears to be the maximum tolerable decay for any reasonably RAFT-controlled reaction [98]. Thus, the trithiocarbonate might safely be used in a laser-initiated RAFT polymerization under the assumption of a constant RAFT agent concentration, while CDB is less suitable for such experiments due to RAFT agent decomposition. The good compatibility of BMPT with laser pulsing at 351 nm opens a path for various kinetic experiments as instationary reaction conditions allow for a different viewpoint on the polymerization kinetics by providing a way to virtually separate specific reactions. Consequently, the kinetics governing the RAFT process itself has been examined by SP–PLP experiments with ESR detection. Under certain conditions, the intermediate radical concentration as a function of time is accessible from such experiments, allowing for unambiguous determination of the addition and the fragmentation rate coefficient [141].

#### 4.3.2.2 **The RAFT Process under Instationary Reaction Conditions: Control over Molecular Weight**

That BMPT is indeed capable of effectively controlling molecular weight in laser-induced BA polymerization is shown in Figs. 4.12a and 4.12b. Figure 4.12a depicts the MWD evolution of poly(BA) samples taken at various degrees of

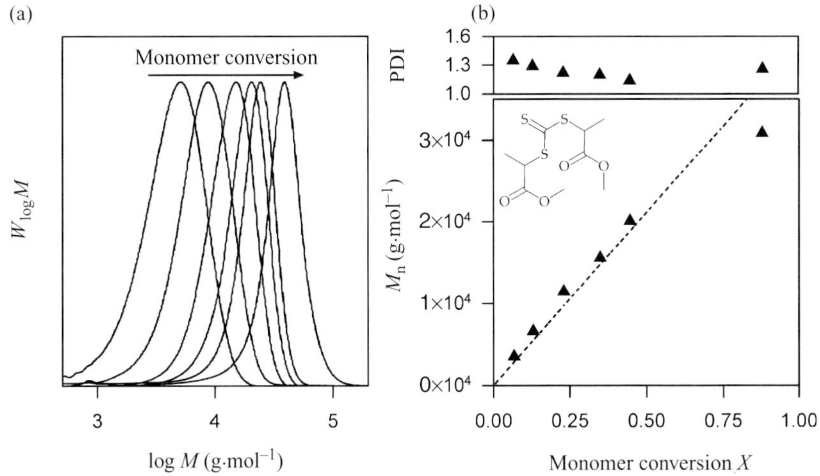


**Fig. 4.12** (a) Evolution of MWDs with monomer conversion from SEC of poly(BA) samples obtained via pulsed-laser-induced RAFT polymerization at 60 °C, 1000 bar and $c_{BMPT} = 2.1 \times 10^{-2}$ mol·L$^{-1}$ with BMPT (see structure) as the mediating agent. Part (b) depicts the associated $M_n$ and PDI values as a function of conversion. Reproduced from Ref. [140], with kind permission from the American Chemical Society.

monomer conversion (in a range from 7 to 88%) from BMPT-mediated bulk high-pressure polymerization at 1000 bar, 60 °C and $c_{\mathrm{BMPT}} = 2.1 \times 10^{-2}$ mol·L$^{-1}$. The MWDs shift to higher molecular weights with increasing monomer consumption. The associated $M_{\mathrm{n}}$ values as well as the PDIs of the SEC curves are given in Fig. 4.12b.

The PDI of the 1000 bar data, as is typical for RAFT polymerization, decreases upon increasing monomer conversion, resulting in a value of about 1.15 at about $X = 0.45$. At high degrees of monomer conversion, at about 90%, a slightly broadened distribution is observed. However, even in this late stage of polymerization, a PDI that is well below 1.5 is observed, which demonstrates effective RAFT control. It should, however, be noted that the RAFT control is aided by the usage of high pressure (which is favorable for SP–PLP–NIR experiments as well). That the RAFT control is somewhat enhanced under high-pressure conditions was demonstrated for CDB-mediated styrene polymerization before [142]. As is expected from well-controlled RAFT polymerization, $M_{\mathrm{n}}$ increases linearly with monomer conversion. Moreover, up to 50% monomer conversion, the data are in excellent agreement with the theoretically expected increase of average molecular weight, as is indicated by the dashed line. Only for the sample at the highest experimentally accessed monomer conversion, the measured $M_{\mathrm{n}}$ is slightly below the theoretical prediction, given by equation 4.18:

$$i = \frac{c_{\mathrm{M}}^{0} X}{c_{\mathrm{RAFT}}^{0} \sigma} + 1 \tag{4.18}$$

where $\sigma$ is introduced in addition to equation 4.1 to account for the number of leaving groups per RAFT molecule, which is $\sigma = 2$ for any trithiocarbonate. However, the assumption of equation 4.18 should hold for any RAFT polymerization process as long as no dithioester compound is lost by termination processes during polymerization and no hybrid behavior [30] is observed. As neither hybrid behavior nor significant deviation of the experimental $M_{\mathrm{n}}$ data from the theoretical line by loss of BMPT compounds is observed up to 50% monomer conversion, equation 4.18 may hence be used for the calculation of the size of terminating propagating macroradicals at any stage of polymerization. When the BA polymerization is controlled by the addition of CDB which has been identified to be not overly suitable due to its decomposition upon laser pulsing, reasonable control over molecular weight is retained for lower conversions (Figure 4.13).

In comparison to the trithiocarbonate data, two differences may be identified. One is that $M_{\mathrm{n}}$ generally deviates toward slightly higher values (as is expected when significant amounts are consumed by decomposition), and secondly, higher PDIs are observed which might also be seen as a consequence from loss of active RAFT agent. It should, however, be noted that, with high probability, almost no control would be observed in the laser-induced CDB-mediated polymerization of slower-propagating monomers as a largely increased number of laser pulses would be required to reach similar conversions, hence being accompanied by much further decomposition of CDB.

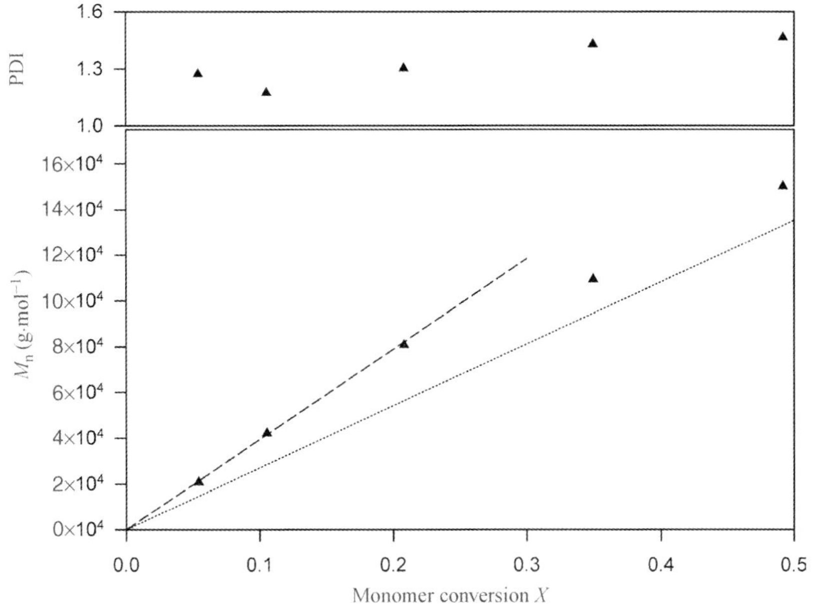


**Fig. 4.13** Evolution of MWDs with monomer conversion in preceding BA polymerization obtained via pulsed-laser-induced RAFT polymerization at 60 °C, 1000 bar and $c_{CDB} = 3.3 \times 10^{-3}$ mol·L$^{-1}$ with CDB as the mediating agent. (Data taken from Ref. [98].) The dashed line gives the best fit to the data up to 30% monomer conversion where the dotted line represents the theoretical $M_n$ evolution.

#### 4.3.2.3 **The RAFT Process under Instationary Reaction Conditions: Transformation of Radical Size Distributions**

Figure 4.14 demonstrates the change in the individual concentration vs time traces obtained by SP–PLP–NIR when BMPT is added to the reaction solution [94, 98]. While systematic deviations between the experimental data and their best fit to equation 4.17 are identified, almost no deviation is seen for the trace obtained from RAFT-controlled polymerization. Moreover, the deviation of the fit in Fig. 4.14a is in good agreement with the expected deviation when a chain-length-dependent $k_t$ is assumed [87, 94]. The close resemblance of the data of the RAFT-mediated experiment with the ideal (chain-length-independent) termination kinetics equation is hence strongly indicative for a radical distribution that does not significantly change with the time after application of the laser pulse.

However, with each laser pulse, a population of initiator-derived radicals is produced that subsequently start a chain growth reaction. In other words, with each laser pulse, small radicals are generated that need to be transformed via the RAFT process. In order to allow for conditions as observed in Fig. 4.14b, such interchange of the radicals must be a particularly fast process.

Figure 4.15 depicts the change in the free radical distribution after application of a laser pulse as simulated by the program package PREDICI® on the basis of a simple

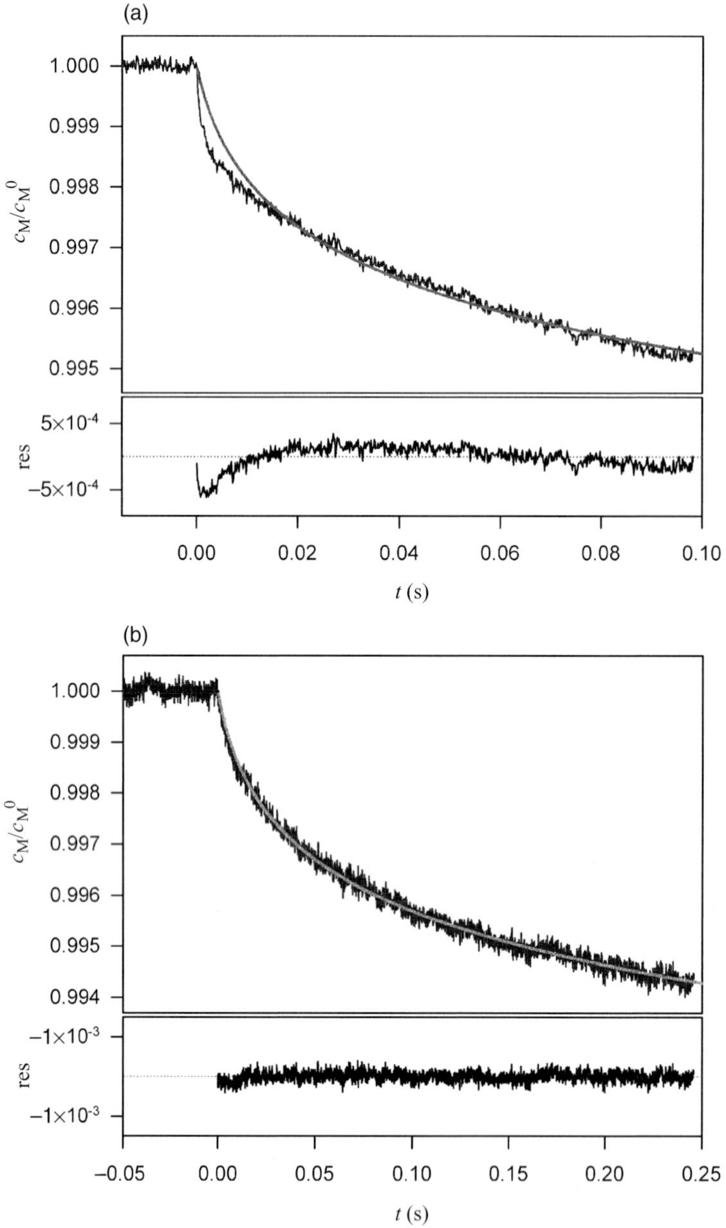


**Fig. 4.14** Monomer concentration vs time traces and their deviation to the ideal polymerization kinetics equation (equation 4.17) as deduced via the SP–PLP–NIR technique from conventional (a) and BMPT-controlled (b) BA polymerization at 1000 bar. Part (a) has been deduced at 40 °C and an average monomer conversion of $X = 33\%$. The underlying experiment in part (b) was performed at 60 °C, $X = 31\%$ and $c_{BMPT} = 2.1 \times 10^{-2}$ mol·L$^{-1}$.

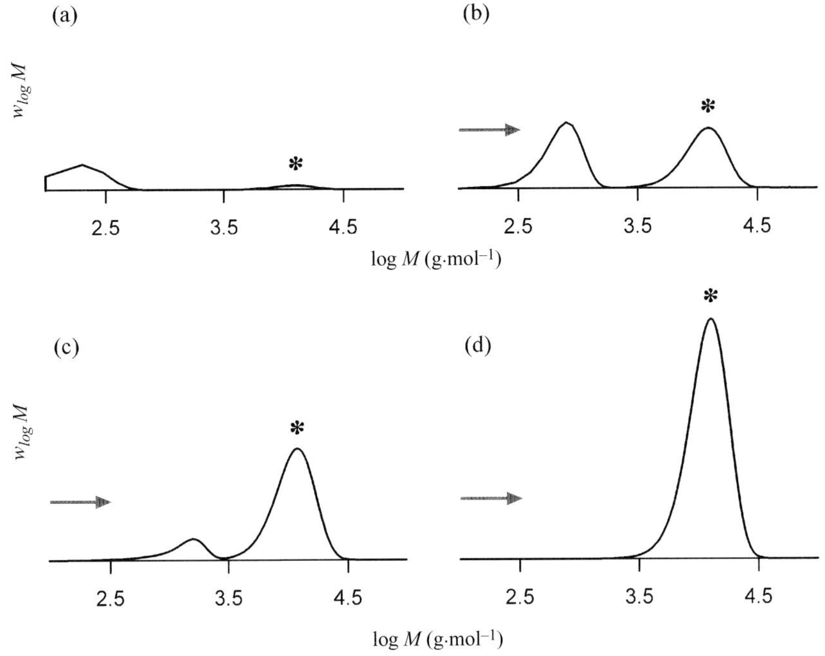


**Fig. 4.15** Simulated evolution with time of the MWD of propagating radicals at very short reaction times below 1 ms after application of a single laser pulse. The asterisk marks radicals that emerge from the poly-RAFT species. Reproduced from Ref. [95], with kind permission from the American Chemical Society.

RAFT model [95, 98]. Almost instantly after the laser pulse, radicals that emerged from initiator decomposition start growing. While these radicals grow in size, they are available to terminate, but, provided the addition rate of the propagating rate toward a poly-RAFT agent is large enough, they are more likely to be involved into the RAFT main equilibrium as the number of available poly-RAFT molecules largely exceeds the number of radicals ready for termination. If fragmentation of the RAFT intermediate radical is also fast (with a $K_{RAFT}$ of 1000 L·mol$^{-1}$ being assumed for the simulation), almost only radicals of the RAFT concentration and conversion-specific chain length are released from the intermediates, also on a short time scale. Such a situation is depicted in Figs. 4.15b and 4.15c where a strong decrease in the growing radical concentration is seen while the narrow RAFT-specific radical distribution (marked by the asterisk) increases until all radicals have been transformed at the time corresponding to Figure 4.14d. When $k_{add}$ is in the order of $10^6$ L·mol$^{-1}$·s$^{-1}$ as was assumed for the simulation shown above, the full change of the radical distribution has taken place within a few milliseconds. As for a successful control of the RAFT mechanism over molecular weight, the addition to the RAFT agent needs to be favored over chain propagation, and it may be assumed that the depicted transformation is fast in any RAFT polymerization. However, if fragmentation is a slow process, the radicals would be consumed and the RAFT-specific distribution would only be released over an extended time range. Nevertheless, the PLP-derived chain length distribution would also be removed almost instantly.

In conclusion, from the viewpoint of an SP–PLP–NIR experiment, the transformation can be regarded to take place almost instantly and the situation of Fig. 4.15d might be seen to be the starting point of the actual experiment, which explains the ideality with respect to CLD-T seen in Fig. 4.14b. While the RAFT-specific average chain length increases over time, or conversion respectively, it may be regarded to be constant on the time scale of one individual SP–PLP probing as the conversion per pulse is, even for acrylate monomers, very low.

#### 4.3.2.4 **The RAFT Process under Instationary Reaction Conditions: Impact of the Intermediate Radical Lifetime**

RAFT polymerization may lead to rate retardation effects (see also Chapter 3), which, however, occur to different extends. Such retardation is caused by a loss of radicals available for propagation either by cross-termination and self-termination of the intermediate radical or by radical storage caused by slow fragmentation of the same species [31, 141, 143]. The outcome of both kinetic scenarios have been tested by simulations with PREDICI® with respect to the $k_t$ determination via SP–PLP–NIR–RAFT experiments. It should be noted that both mechanisms potentially have an influence on the apparent $k_t$ or $k_t(1,1)$ and thus on the outcome for $\alpha$, respectively, without apparently inducing rate retardation of the overall polymerization rate in a steady-state polymerization.

The influence of nonideal RAFT kinetics on the experiment is estimated by simulating a series of individual SP–PLP experiments where a chain-length-dependent $k_t$ is implemented in the model in conjunction with different assumptions on the fragmentation rate as well as on the cross-termination rate coefficient for the reaction of a RAFT intermediate with a propagating radical. The obtained series of SP–PLP experiments can then be analyzed with equation 4.17, yielding an apparent set of $k_t^0$ and $\alpha$ values which can then be compared to the input values. As the RAFT intermediate in trithiocarbonate polymerization consists of three arms, self-termination of this species can safely be neglected because the formation of a six-arm star is more than unlikely due to the largely decreased probability of the active centers coming close enough to each other [144]. In principle, the addition rate coefficient may also have an influence on the resulting chain length dependence of $k_t$. However, a significant impact is expected only when $k_{add}$ is low and thus under conditions where no low-molecular-weight material can be expected.

From the simulation of RAFT polymerization under stationary reaction conditions, one can expect that both slow fragmentation as well as cross-termination result in an apparently increased absolute termination rate coefficient as both the mechanisms lower the amount of radicals in the system, hence implicating an increased termination rate. However, under instationary reaction conditions, slow fragmentation of the intermediate radical results in an apparently lower termination rate coefficient. The radicals that are produced at $t = 0$ are partially stored in the intermediate species and subsequently released over time in the dark time period. In consequence, radicals that were not available for termination at short delay times reappear at later stages, which causes an overall increased consumption of monomer. A higher monomer consumption per pulse is again correlated with a smaller apparent termination rate coefficient. While a $K_{RAFT} = 10$ L·mol$^{-1}$ has no significant effect on the determination of $k_t^0$, this value is already decreased by 1 order of magnitude when $K_{RAFT}$ is in the order of 1000 L·mol$^{-1}$ [95, 98, 140]. At lower fragmentation rates, no typical SP–PLP signal shape can be obtained due to the nonceasing reaction, hence invalidating the method and leaving no result at all. While the analysis of $k_t^0$ is clearly dependent on the fragmentation rate coefficient governing the RAFT process, no significant influence on the chain length dependency itself and thus on the parameter $\alpha$ is observed. So regardless of the size of the equilibrium constant, $\alpha$ is either determined in its true size or may not be determined at all due to failure of the experiment, i.e. losing the characteristic signal shape.

When cross-termination is taken into account, an apparent increase in $k_t^0$ is observed as is expected when a second pathway is opened that irreversibly removes radicals from the system. While only a slight increase in the apparent $k_t^0$ is observed when the cross-termination rate coefficient is assumed to be a factor of 100 smaller than the conventional termination, a more than 10-fold increase is identified for the situation where the conventional and intermediate termination rate coefficients are equally high, which may be regarded as the theoretical limit such reaction could reach (it should, at the same place, be noted that the described change is observed only when $K_{RAFT} = 1000$ L·mol$^{-1}$ is assumed. When the equilibrium constant is small, no significant effect of the cross-termination level is expected to occur). On the other hand, an increase in $\alpha$ is noted when the cross-termination rate coefficient is close to the conventional rate coefficient. The increase, however, is not as high (a factor of 2 in maximum) and may be regarded as a minor effect considering the uncertainty usually associated with this parameter and considering that the limiting case of an equally high cross- and conventional termination rate coefficient is more than unlikely.

In conclusion from the kinetic simulations of the RAFT process under laser pulse conditions, three major statements may be drawn:

- The RAFT process can be initiated by single laser pulsing without losing the characteristic features of molecular weight control (and narrowness of the MWDs).
- Conversion vs time traces from SP–PLP–NIR–RAFT experiments which pass the ideality requirement (good description of the experimental data by the ideal termination rate equation) are governed by an equilibrium constant of around 1000 L·mol$^{-1}$ or below.
- RAFT agents that do not exhibit strong rate retardation in chemical-induced polymerization may not interfere with the determination of chain-length-dependent $k_t$ as long as $K_{RAFT}$ is below 1000 L·mol$^{-1}$ whereas the absolute value $k_t^0$ may be shifted toward lower values with increasing equilibrium constant and toward higher values in the case of a rising cross-termination rate.

#### 4.3.2.5 The CLD-T Rate Coefficient for Butyl Acrylate Determined by SP–PLP–NIR–RAFT

As BMPT has been shown to be suitable for an unambiguous determination of $k_t(i)$ in BA polymerization due to the excellent control over molecular weight as well as for the finding of a relatively high fragmentation rate coefficient being in place [141], SP–PLP–RAFT experiments have been carried out for this system. Figure 4.16 depicts the change in the conversion dependency of $k_t$ when the RAFT agent is added to reaction mixture. The termination rate coefficient in the uncontrolled BA polymerization follows the general $k_t(X)$ identified for many monomers in that a constant average $k_t$ is observed in the initial conversion regime up to about 25% conversion followed by a slight decrease toward higher $X$, indicating a change in the type of diffusion being rate determining [56]. The $k_t$ values obtained from the RAFT-mediated polymerization experiments, however, display a completely different behavior in that there is a steep decrease in the termination rate coefficients observed at very low monomer conversions followed by a less pronounced, almost linear decrease up to high conversions. Interestingly, the low-conversion $k_t$ in the RAFT-mediated experiment is much higher than the average termination rate coefficient in the uncontrolled polymerization, while at high $X$ the average $k_t$ falls below the one derived from the RAFT. This finding is generally in good agreement

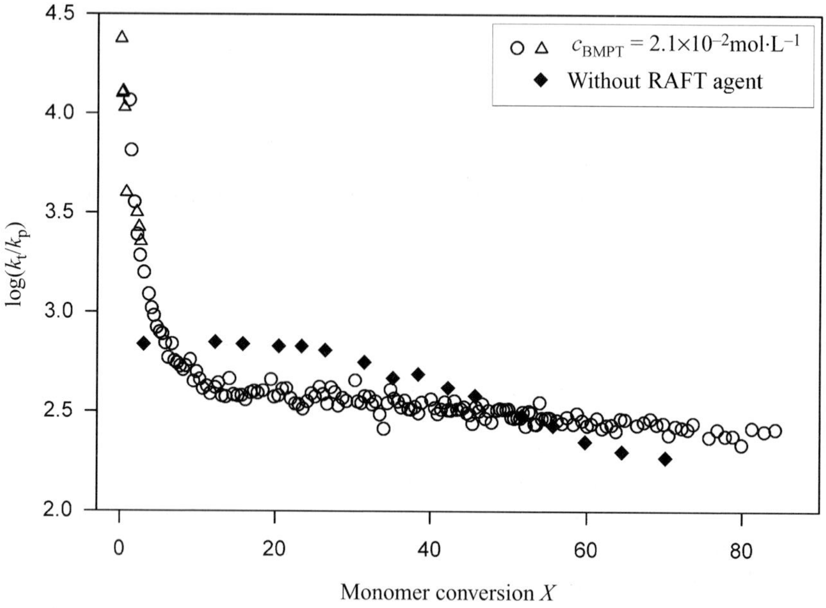


**Fig. 4.16** $\log(k_t/k_p)$ deduced via the SP–PLP–NIR–RAFT technique at 1000 bar and 60 °C from BA bulk polymerization using BMPT as the mediating agent at $c_{BMPT} = 2.1 \times 10^{-2}$ mol·L$^{-1}$ and from uncontrolled BA bulk polymerization at 1000 bar and 40 °C, extrapolated to 60 °C via the literature activation energy [145]. Different markers display independent series of experiments.

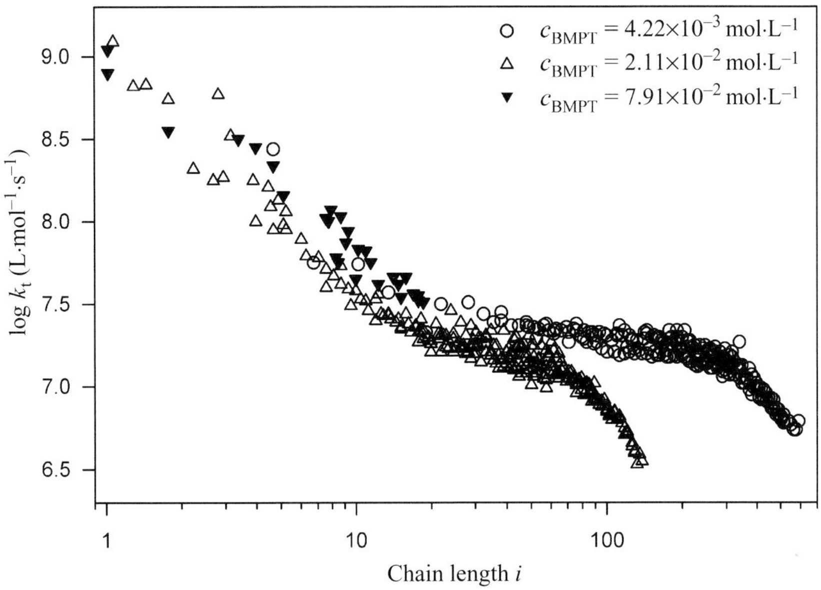


**Fig. 4.17** $k_t$ as a function of chain length in BA bulk polymerization derived via the SP–PLP–NIR–RAFT technique at 60 °C and 1000 bar using various $c_{BMPT}$. Reproduced from Ref. [95], with kind permission from the American Chemical Society.

with the idea of CLD-T following a composite model as was proposed before [55] and equally identified from the RAFT–CLD-T method [90].

According to the RAFT-CLD-T method, the $k_t(i)$ data are best analyzed when given in a double-log plot after converting the conversion axis into a chain length axis via equation 4.18. Also, as the primary result from fitting SP–PLP data is $k_t/k_p$, the 'reaction-order-corrected' $k_p^*$ value is used to calculate actual $k_t$ values. Again, as the conversion dependence of $k_t$ overlaps the chain length dependence, data for several RAFT agent concentrations are required to yield a complete picture and to identify the regions where $k_t$ is mainly governed by the conversion dependency (indicated by a stronger decrease at higher $X$, or $i$, respectively). As shown in Fig. 4.17, the $k_t$ values evaluated from experiments at three different RAFT agent concentrations are in good agreement with each other, where at the two lower RAFT agent concentrations a more pronounced decrease in $k_t$ is observed at larger $i$, which is not backed up by the other concentration data. That now such conversion-dependent regime is identified for the experiments at the highest RAFT agent concentration can be explained by the drastic change in viscosity with the overall degree of polymerization. As much shorter chains are produced at high RAFT agent concentrations, viscosity of the bulk solution remains low in an extended conversion regime and hence no change in diffusion control takes place. Generally, a large dependence of $k_t$ in the short-chain regime (up to about chain length 15) is observed, while a much weaker dependency is then seen for $i$, as was already expected from analysis of the data in Fig. 4.16. Thus, data fully consistent with the composite model are obtained. Rather similar behavior has also been found for MA and for DA via the same procedure [146(a) and (b)].

### 4.3.3
### Comparing Results from RAFT-CLD-T and SP–PLP–NIR–RAFT

Both the RAFT-CLD-T and SP–PLP–RAFT methods have been shown to yield reliable $k_t(i)$ data. It is therefore straightforward to compare the results from both techniques. Figure 4.18 shows an overlap of the $k_t(i)$ data determined for BA polymerization with the SP–PLP–RAFT data in Fig. 4.17, where the single pulse data have been adjusted to match the reaction conditions of the steady-state experiment via literature activation parameters [145]. In the case of SP–PLP–NIR–RAFT, data are presented for a conversion interval of $0 < X < 0.3$ for the intermediate and the low RAFT agent concentration, and for $0 < X < 0.4$ for the highest RAFT agent concentration. The data from RAFT-CLD-T are given for the interval $0.05 < X < 0.3$ for four different RAFT agent concentrations. According to the PREDICI®

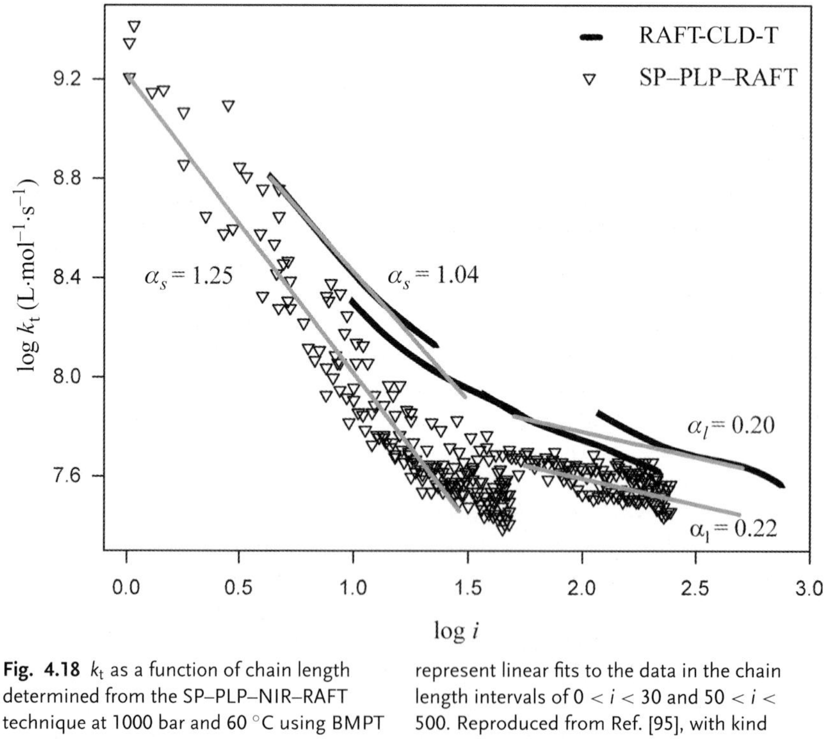


**Fig. 4.18** $k_t$ as a function of chain length determined from the SP–PLP–NIR–RAFT technique at 1000 bar and 60 °C using BMPT extrapolated to ambient pressure and 80 °C and from RAFT-CLD-T at ambient pressure and 80 °C using MCEPDA. The gray lines represent linear fits to the data in the chain length intervals of $0 < i < 30$ and $50 < i < 500$. Reproduced from Ref. [95], with kind permission from the American Chemical Society.

simulations carried out with respect to the influence of intermediate radical stability, it seems justified to compare data which were derived by different RAFT agents, as the slightly increased stability of the BMPT intermediate radical compared to intermediates being involved in MCEPDA-mediated BA polymerization should have no significant effect on the determination of the chain length dependency of $k_t$ via SP–PLP–RAFT as the single pulse method appears to be slightly more robust with respect to the size of the equilibrium constant.

Inspection of Fig. 4.18 demonstrates close agreement between the two data sets. The initial $k_t^{1,1}$ values of about $2.0 \times 10^9$ L·mol$^{-1}$·s$^{-1}$, determined from SP–PLP–RAFT, and $2.5 \times 10^9$ L·mol$^{-1}$·s$^{-1}$, extrapolated from medium chain-length RAFT-CLD-T data, are in good agreement with predictions via the Smoluchowski equation [147] and the monomer self-diffusion coefficient. The difference in absolute $k_t$ of about 0.2 logarithmic units between both methods may be due to uncertainties in $k_p$ (as the steady-state method yields $k_t/k_p^2$ in contrast to $k_t/k_p$ deduced from SP–PLP) and the initiator efficiency $f$. However, the degree of agreement between the two data sets is remarkable regarding the very different reaction conditions, method applied as well as RAFT agent in use. Thus, both techniques support the result of each other, allowing for the conclusion that the 'true' chain length dependency is captured by either method without any significant impact from the RAFT kinetics themselves.

### 4.3.4
### Specific Advantages of the Individual Techniques

Both RAFT-based methods for the determination of $k_t(i)$ are associated with specific advantages and disadvantages. As it is based on simple rate measurements via isothermal differential scanning calorimetry, the RAFT-CLD-T method is clearly easily to carry out than the experimentally more sophisticated SP–PLP technique, which requires not only an expensive laser setup, but also carefully adjusted optics and a highly sensitive electronic setup. The SP–PLP–RAFT method is, however, better suited for investigations into the small-chain regime with rapidly polymerizing monomers, such as acrylates, because $k_t$ data are deduced from $X = 0$ on, whereas the initial polymerization phase must be disregarded in RAFT-CLD-T experiments because of the heating time of the samples. Also, the RAFT-CLD-T method displays a slightly higher sensitivity toward large equilibrium constants. The SP–PLP–RAFT method on the other hand requires RAFT agents, which are transparent in the wavelength region of the pulsing laser but is more robust with respect to slow fragmentation of the intermediate and to cross-termination of this species. The largest advantage of the SP–PLP–RAFT methodology, as is also true for conventional SP–PLP–NIR, originates from the fact that $f\,k_d$ knowledge is not necessary and that uncertainties in the effective propagation rate coefficient are only linearly transferred onto $k_t$, whereas they are squared in experiments carried out under stationary reaction conditions. Regardless of these disturbing influences, it may be stated that both methodologies yield either complementary data on the chain length dependencies of the termination rate coefficient or data that are in very good agreement with each other when the reaction conditions are carefully chosen [146(b)]. Hence, the employment of RAFT polymerization as a kinetic instead of a synthetic tool has proven to be very effective and has the potential to yield benchmark values for $k_t(i)$ and $k_t(i,j)$ for a broad range of monomers and reaction conditions.

## 4.4 Chain-Length-Dependent Propagation Rate Coefficients

Recently, there has been a lively discussion on whether the propagation rate coefficient $k_p$ is equally beset by a chain length dependence [61, 148–152]. A potential chain length dependence of $k_p$ may alter the outcome of $k_t$ in the short-chain regime. Due to the rise in $k_p$, the estimated $k_t$ values would increase with decreasing chain length and thus also short-chain $\alpha$ would increase. Since the effect of a chain-length-dependent $k_p$ on $k_t$ is linear in the SP–PLP–NIR–RAFT method and squared in the case of the RAFT-CLD-T method, the utilization of a chain-length-dependent $k_p$ could affect the $\alpha$ values obtained for both methods. While $k_p$ is in all likelihood chain length dependent at small $i$, there is significant disagreement to what extent. Smith et al. [82] included a chain length dependence of $k_p$ to reasonably fit experimental $k_t/k_p^2$ data from MMA polymerization for varying chain length regions with also varying chain-length-dependent $k_t$ values. Most studies on the propagation rate taking chain length into account have been carried out for styrene and MMA, with no report on acrylates. Of course, a chain-length-dependent $k_p$ has a distinctive influence on the resulting $k_t(i)$ function from SP–PLP–RAFT as well as from RAFT-CLD-T. On the other hand, if the exact $k_t(i)$ functionality is known, the experimental results can be used to fit $k_p(i)$. To circumvent the problem of needing to know $k_t$ as an input parameter, the combination of SP–PLP–RAFT with RAFT-CLD-T data as shown in Fig. 4.18 may, in principle, be used to directly yield $k_p(i)$. Such a pathway is given by the primary result of $k_t/k_p^2$ in one case and $k_t/k_p$ in the other. Thus, when being plotted on a logarithmic scale, the difference between both experimental results directly yields $k_p$ as a function of chain length. However, it goes without saying that such a procedure not only requires accurate data from CLD-T at low chain length, that is down to chain length unity (which might not easily be accessed due to the heating-up phase in the experiments allowing for steady-state conditions only after a certain chain length is reached), but also affords for highly precise measurements.

## References

1. C. Barner-Kowollik, T. P. Davis, J. P. A. Heuts, M. H. Stenzel, P. Vana, M. Whittaker, *J. Polym. Sci., Part A: Polym. Chem.* **2003**, *41*, 365–375.
2. G. Moad, E. Rizzardo, S. H. Thang, *Aust. J. Chem.* **2005**, *58*, 379–410.
3. H. Gao, S. Ohno, K. Matyjaszewski, *J. Am. Chem. Soc.* **2007**, *40*, 399–401.

4. E. Harth, B. Van Horn, V. Y. Lee, D. S. Germack, C. P. Gonzales, R. D. Miller, C. J. Hawker, *J. Am. Chem. Soc.* **2002**, *124*, 8653–8660.
5. S. Perrier, P. Takolpuckdee, *J. Polym. Sci., Part A: Polym. Chem.* **2005**, *43*, 5347–5393.
6. K. Matyjaszewski, J. Xia, *Chem. Rev.* **2001**, *101*, 2921–2990.
7. C. J. Hawker, A. W. Bosman, E. Harth, *Chem. Rev.* **2001**, *101*, 3661–3688.
8. G. Chen, D. Huynh, P. L. Felgner, Z. Guan, *J. Am. Chem. Soc.* **2006**, *128*, 4298–4302.
9. H. Gao, K. Matyjaszewski, *Macromolecules* **2006**, *39*, 4960–4965.
10. R. T. A. Mayadunne, E. Rizzardo, J. Chiefari, J. Kristina, G. Moad, A. Postma, S. H. Thang, *Macromolecules* **2000**, *33*, 243–245.
11. Y. K. Chong, T. P. T. Le, G. Moad, E. Rizzardo, S. H. Thang, *Macromolecules* **1999**, *32*, 2071–2074.
12. J. Chiefari, Y. K. Chong, F. Ercole, J. Krstina, J. Jeffery, T. P. T. Le, R. T. A. Mayadunne, G. F. Meijs, C. L. Moad, G. Moad, E. Rizzardo, S. H. Thang, *Macromolecules* **1998**, *31*, 5559–5562.
13. D. Quemener, T. P. Davis, C. Barner-Kowollik, M. H. Stenzel, *Chem. Commun.* **2006**, *48*, 5051–5053.
14. C. Barner-Kowollik, T. P. Davis, M. H. Stenzel, *Aust. J. Chem.* **2006**, *59*, 719–727.
15. Y.-K. Goh, M. J. Monteiro, *Macromolecules* **2006**, *39*, 4966–4974.
16. C.-A. Fustin, C. Colard, M. Filali, P. Guillet, A.-S. Duwez, M. A. R. Meier, U. S. Schubert, J.-F. Gohy, *Langmuir* **2006**, *22*, 6690–6695.
17. M. L. Becker, J. Liu, K. L. Wooley, *Biomacromolecules* **2005**, *6*, 220–228.
18. M. J. Yanjarappa, K. V. Gujraty, A. Joshi, A. Saraph, R. S. Kave, *Biomacromolecules* **2006**, *7*, 1665–1670.
19. H. Fischer, *Chem. Rev.* **2001**, *101*, 3581–3610.
20. S. W. Prescott, M. J. Ballard, E. Rizzardo, R. G. Gilbert, *Aust. J. Chem.* **2002**, *55*, 415–424.
21. X. Hao, J. P. A. Heuts, C. Barner-Kowollik, T. P. Davis, E. Evans, *J. Polym. Sci., Part A: Polym. Chem.* **2003**, *41*, 2949–2963.
22. R. T. A. Mayadunne, J. Jeffery, G. Moad, R. Ezio, *Macromolecules* **2003**, *36*, 1505–1513.
23. S. G. Boyes, A. M. Granville, M. Baum, B. Akgun, B. K. Mirous, W. J. Brittain, *Surf. Sci.* **2004**, *570*, 1–12.
24. D. Hua, X. Ge, R. Bai, W. Lu, C. Pan, *Polymer* **2005**, *46*, 12696–12702.
25. C. Barner-Kowollik, M. Buback, B. Charleux, M. L. Coote, M. Drache, T. Fukuda, A. Goto, B. Klumperman, A. B. Lowe, J. McLeary, G. Moad, M. J. Monteiro, R. D. Sanderson, M. P. Tonge, P. Vana, *J. Polym. Sci., Part A: Polym. Chem.* **2006**, *44*, 5809–5831.
26. M. Buback, P. Vana, *Macromol. Rapid Commun.* **2006**, *27*, 1299–1305.
27. M. Buback, O. Janssen, R. Oswald, S. Schmatz, P. Vana, *Macromol. Symp.* **2007**, *248*, 158–167.
28. P. Vana, *Macromol. Symp.* **2007**, *248*, 71–81.
29. G. Moad, E. Rizzardo, S. H. Thang, *Aust. J. Chem.* **2006**, *59*, 669–692.
30. C. Barner-Kowollik, J. F. Quinn, T. L. U. Nguyen, J. P. A. Heuts, T. P. Davis, *Macromolecules* **2001**, *34*, 7849–7857.
31. A. Feldermann, M. L. Coote, M. H. Stenzel, T. P. Davis, C. Barner-Kowollik, *J. Am. Chem. Soc.* **2004**, *126*, 15915–15923.
32. A. Feldermann, M. H. Stenzel, T. P. Davis, P. Vana, C. Barner-Kowollik, *Macromolecules* **2004**, *37*, 2404–2410.
33. M. Buback, R. G. Gilbert, G. T. Russell, D. J. T. Hill, G. Moad, K. F. O'Driscoll, J. Shen, M. A. Winnik, *J. Polym. Sci., Part A: Polym. Chem.* **1992**, *30*, 851–863.
34. O. F. Olaj, A. Kornherr, P. Vana, M. Zoder, G. Zifferer, *Macromol. Symp.* **2002**, *182*, 15–30.
35. T. M. Lovestead, A. Theis, T. P. Davis, M. H. Stenzel, C. Barner-Kowollik, *Macromolecules* **2006**, *39*, 4975–4982.
36. K. A. Berchtold, T. M. Lovestead, C. N. Bowman, *Macromolecules* **2002**, *35*, 7968–7975.

**37.** C. Barner-Kowollik, M. Buback, M. Egorov, T. Fukuda, A. Goto, O. F. Olaj, G. T. Russell, P. Vana, B. Yamado, P. B. Zetterlund, *Prog. Polym. Sci.* **2005**, *30*, 605–643.

**38.** M. Buback, M. Egorov, R. G. Gilbert, V. A. Kaminsky, O. F. Olaj, G. T. Russell, *Macromol. Chem. Phys.* **2002**, *203*, 2570–2582.

**39.** G. B. Smith, J. Heuts, G. T. Russell, *Macromol. Symp.* **2005**, *226*, 133–146.

**40.** G. T. Russell, *Aust. J. Chem.* **2002**, *55*, 399–414.

**41.** S. Zhu, A. E. Hamielec, *Macromolecules* **1989**, *22*, 3093–3098.

**42.** T. J. Tulig, M. Tirrell, *Macromolecules* **1981**, *14*, 1501–1511.

**43.** H. Tobita, *Macromolecules* **1996**, *29*, 3073–3080.

**44.** S. K. Soh, D. C. Sundberg, *J. Polym. Sci., Part A: Polym. Chem.* **1982**, *20*, 1299–1313.

**45.** S. K. Soh, D. C. Sundberg, *J. Polym. Sci., Part A: Polym. Chem.* **1982**, *20*, 1315–1329.

**46.** S. K. Soh, D. C. Sundberg, *J. Polym. Sci., Part A: Polym. Chem.* **1982**, *20*, 1331–1344.

**47.** S. K. Soh, D. C. Sundberg, *J. Polym. Sci., Part A: Polym. Chem.* **1982**, *20*, 1345–1371.

**48.** G. Odian, *Principles of Polymerization*, 3rd edn, John Wiley & Sons, Inc., New York, **1991**, p. 768.

**49.** A. M. North, G. A. Reed, *Trans. Faraday Soc.* **1961**, *57*, 859–870.

**50.** A. N. Nikitin, A. V. Evseev, *Macromol. Theory Simul.* **1999**, *8*, 296–308.

**51.** G. I. Litvinenko, V. A. Kaminsky, *Prog. React. Kinet.* **1994**, *19*, 139–193.

**52.** P. J. Flory, *Principles of Polymer Chemistry*, 1st edn, Cornell University Press, Ithaca, **1953**.

**53.** S. W. Benson, A. M. North, *J. Am. Chem. Soc.* **1962**, *84*, 935–940.

**54.** P. E. M. Allen, C. R. Patrick, *Macromol. Chem. Phys.* **1961**, *47*, 154–167.

**55.** G. B. Smith, G. T. Russell, *Macromol. Theory Simul.* **2003**, *12*, 299–314.

**56.** M. Buback, *Makromol. Chem.* **1990**, *191*, 1575–1587.

**57.** M. Buback, B. Huckestein, G. T. Russell, *Macromol. Chem. Phys.* **1994**, *195*, 539–554.

**58.** G. T. Russell, R. G. Gilbert, D. H. Napper, *Macromolecules* **1992**, *25*, 2459–2469.

**59.** G. T. Russell, *Macromol. Theory Simul.* **1995**, *4*, 519–548.

**60.** H. Mahabadi, *Macromolecules* **1991**, *24*, 606–609.

**61.** O. F. Olaj, A. Kornherr, G. Zifferer, *Macromol. Theory Simul.* **2001**, *10*, 881–890.

**62.** O. F. Olaj, A. Kornherr, G. Zifferer, *Macromol. Theory Simul.* **1999**, *8*, 561–570.

**63.** O. F. Olaj, P. Kremminger, I. Schnöll-Bitai, *Macromol. Rapid Commun.* **1988**, *9*, 771–779.

**64.** G. A. O'Neil, J. M. Torkelson, *Macromolecules* **1999**, *32*, 411–422.

**65.** M. Buback, C. Kowollik, *Macromol. Chem. Phys.* **1999**, *200*, 1764–1770.

**66.** J. B. L. de Kock, B. Klumperman, A. M. van Herk, A. L. German, *Macromolecules* **1997**, *30*, 6743–6753.

**67.** P. A. G. M. Scheren, G. T. Russell, D. F. Sangster, R. G. Gilbert, A. L. German, *Macromolecules* **1995**, *28*, 3637–3649.

**68.** G. T. Russell, *Macromol. Theory Simul.* **1994**, *3*, 439–468.

**69.** G. T. Russell, R. G. Gilbert, D. H. Napper, *Macromolecules* **1993**, *26*, 3538–3552.

**70.** C. H. Bamford, *Eur. Polym. J.* **1989**, *25*, 683–689.

**71.** H. Mahabadi, *Macromolecules* **1985**, *18*, 1319–1324.

**72.** O. F. Olaj, G. Zifferer, *Macromolecules* **1987**, *20*, 850.

**73.** O. F. Olaj, A. Kornherr, G. Zifferer, *Macromol. Theory Simul.* **1998**, *7*, 501.

**74.** C. Barner-Kowollik, P. Vana, T. P. Davis, In *The Handbook of Radical Polymerization*, K. Matyjaszewski, T. P. Davis, Eds., **2002**, Wiley and Sons, New York, NY, p. 187.

**75.** H. K. Mahabadi, K. F. O'Dricsoll, *Macromolecules* **1977**, *10*, 55.

**76.** H. K. Mahabadi, K. F. O'Dricsoll, *J. Polym. Sci., Chem. Ed.* **1977**, *15*, 283.

**77.** M. C. Pinton, R. G. Gilbert, B. E. Chapman, P. W. Kuchel, *Macromolecules* **1993**, *26*, 4472.

**78.** A. R. Khokhlov, *Macromol. Chem., Rapid Commun.* **1981**, *2*, 633.

**79.** B. Friedman, B. O'Shaughnessy, *Macromolecules* **1993**, *26*, 5726.

**80.** E. Karatekin, B. O'Shaughnessy, *Macromol. Symp.* **2002**, *182*, 81.

**81.** M. Buback, M. Egorov, T. Junkers, E. Panchenko, *Macromol. Rapid Commun.* **2004**, *25*, 1004–1009.

**82.** G. B. Smith, G. T. Russell, M. Yin, J. P. A. Heuts, *Eur. Polym. J.* **2005**, *41*, 225.

**83.** G. Johnston-Hall, A. Theis, M. J. Monteiro, T. P. Davis, M. H. Stenzel, C. Barner-Kowollik, *Macromol. Chem. Phys.* **2005**, *206*, 2047–2053.

**84.** O. F. Olaj, P. Vana, *Macromol. Rapid Commun.* **1998**, *19*, 433.

**85.** O. F. Olaj, P. Vana, *Macromol. Rapid Commun.* **1998**, *19*, 533.

**86.** M. Buback, M. Busch, C. Kowollik, *Macromol. Theory Simul.* **2000**, *9*, 442–452.

**87.** M. Buback, M. Egorov, A. Feldermann, *Macromolecules* **2004**, *37*, 1768–1776.

**88.** A. Theis, T. P. Davis, M. H. Stenzel, C. Barner-Kowollik, *Macromolecules* **2005**, *38*, 10323–10327.

**89.** A. Theis, T. P. Davis, M. H. Stenzel, C. Barner-Kowollik, *Polymer* **2006**, *47*, 999–1010.

**90.** A. Theis, A. Feldermann, N. Charton, T. P. Davis, M. H. Stenzel, C. Barner-Kowollik, *Polymer* **2005**, *46*, 6797–6809.

**91.** A. Theis, A. Feldermann, N. Charton, M. H. Stenzel, T. P. Davis, C. Barner-Kowollik, *Macromolecules* **2005**, *38*, 2595–2605.

**92.** A. Theis, M. H. Stenzel, T. P. Davis, C. Barner-Kowollik, In *ACS Symposium Series on Living/Controlled Free Radical Polymerization*, Vol. *944*, K. Matyjaszewski, Ed., ACS Press, Washington, DC, **2006**, pp. 486–500.

**93.** P. Vana, T. P. Davis, C. Barner-Kowollik, *Macromol. Rapid Commun.* **2002**, *23*, 952–956, and references within.

**94.** M. Buback, T. Junkers, P. Vana, *Macromol. Rapid Commun.* **2005**, *26*, 796–802.

**95.** T. Junkers, A. Theis, M. Buback, T. P. Davis, M. H. Stenzel, P. Vana, C. Barner-Kowollik, *Macromolecules* **2005**, *38*, 9497–9508.

**96.** T. M. Lovestead, T. P. Davis, M. H. Stenzel, C. Barner-Kowollik, *Macromol. Symp.* **2007**, *248*, 82–93.

**97.** A. N. Nikitin, R. A. Hutchinson, *Macromolecules* **2005**, *38*, 1581–1590.

**98.** T. Junkers, PhD Thesis; Georg-August-Universität Göttingen, **2006**.

**99.** M. Azukizawa, B. Yamada, D. J. T. Hill, P. J. Pommery, *Macromol. Chem. Phys.* **2000**, *201*, 774.

**100.** B. Yamada, M. Azukizawa, H. Yamazoe, D. J. T. Hill, P. J. Pommery, *Polymer* **2000**, *41*, 5611.

**101.** E. Sato, T. Emoto, P. B. Zetterlund, B. Yamada, *Macromol. Chem. Phys.* **2004**, *205*, 1829.

**102.** S. Beuermann, M. Buback, T. P. Davis, R. G. Gilbert, R. A. Hutchinson, O. F. Olaj, G. T. Russell, J. P. A. Heuts, A. M. van Herk, *Macromol. Chem. Phys.* **1997**, *198*, 1545–1560.

**103.** N. Charton, A. Feldermann, A. Theis, M. H. Stenzel, T. P. Davis, C. Barner-Kowollik, *J. Polym. Sci., Part A: Polym. Chem.* **2004**, *42*, 5170–5179.

**104.** J. B. L. de Kock, A. M. van Herk, A. L. German, *J. Macromolecular Sci., Polym. Rev.* **2001**, *C41*, 199–252.

**105.** J. B. L. de Kock, Ph.D. Thesis Technische Universiteit, Eindhoven, **1999**.

**106.** R. X. E. Willemse, A. M. van Herk, E. Panchenko, T. Junkers, M. Buback, *Macromolecules* **2005**, *38*, 5098–5103.

**107.** D. Britton, F. Heatley, P. A. Lovell, *Macromolecules* **1998**, *31*, 2828–2837.

**108.** G. Johnston-Hall, M. H. Stenzel, T. P. Davis, C. Barner-Kowollik, M. J. Monteiro, *Macromolecules* **2007**, *40*, 2730–2736.

**109.** J. P. Van Hook, A. V. Tobolsky, *J. Am. Chem. Soc.* **1958**, *80*, 779.

**110.** M. Buback, R. G. Gilbert, R. A. Hutchinson, B. Klumperman, F.-D. Kuchta, B. G. Manders, K. F. O'Driscoll, G. T. Russell, J. Schweer, *Macromol. Chem. Phys.* **1995**, *196*, 3267–3280.

**111.** M. Buback, B. Huckestein, F.-D. Kuchta, G. T. Russell, E. Schmidt, *Macromol. Chem. Phys.* **1994**, *195*, 2117–2140.

**112.** M. Wulkow, M. Busch, T. P. Davis, C. Barner-Kowollik, *J. Polym. Sci., Part A: Polym. Chem.* **2004**, *42*, 1441–1448.

**113.** A. N. F. Peck, R. A. Hutchinson, *Macromolecules* **2004**, *37*, 5944–5951.

**114.** E. Chernikova, A. Morozov, E. Leonova, E. Garina, V. Golubev, C. Bui, B. Charleux, *Macromolecules* **2004**, *37*, 6329–6339.

**115.** C. Quan, M. Soroush, M. C. Grady, J. E. Hansen, W. J. Simonsick, Jr, *Macromolecules* **2005**, *38*, 7619–7628.

**116.** A. Postma, T. P. Davis, G. Li, G. Moad, M. S. O'Shea, *Macromolecules* **2006**, *39*, 5307–5318.

**117.** E. Sato, T. Emoto, P. B. Zetterlund, B. Yamada, *Macromol. Chem. Phys.* **2004**, *205*, 1829–1839.

**118.** A. N. Nikitin, P. Castignolles, B. Charleux, J.-P. Vairon, *Macromol. Rapid Commun.* **2003**, *24*, 778–782.

**119.** M. Buback, E. Müller, G. T. Russell, *J. Phys. Chem. A* **2006**, *110*, 3222–3230.

**120.** M. Buback, A. Kuelpmann, C. Kurz, *Macromol. Chem. Phys.* **2002**, *203*, 1065–1070.

**121.** S. Perrier, C. Barner-Kowollik, J. F. Quinn, P. Vana, T. P. Davis, *Macromolecules* **2002**, *35*, 8300–9306.

**122.** J. B. McLeary, J. M. McKenzie, M. P. Tonge, R. D. Sanderson, B. Klumperman, *Chem. Commun.* **2004**, 1950–1951.

**123.** S. Beuermann, D. A. Paquet, J. H. McMinn, R. A. Hutchinson, *Macromolecules* **1996**, *29*, 4206–4215.

**124.** K. Tanaka, B. Yamada, C. M. Fellows, R. G. Gilbert, T. P. Davis, L. H. Yee, G. B. Smith, M. T. L. Rees, G. T. Russell, *J. Polym. Sci., Part A: Polym. Chem.* **2001**, *39*, 3902–3915.

**125.** J. M. Asua, S. Beuermann, M. Buback, P. Castignolles, B. Charleux, R. G. Gilbert, R. A. Hutchinson, J. R. Leiza, A. N. Nikitin, J.-P. Vairon, A. M. van Herk, *Macromol. Chem. Phys.* **2004**, *205*, 2151–2160.

**126.** M. Busch, M. Müller, *Macromol. Symp.* **2004**, *206*, 399–418.

**127.** M. Buback, C. H. Kurz, C. Schmaltz, *Macromol. Chem. Phys.* **1998**, *199*, 1721–1727.

**128.** M. Buback, H. Hippler, J. Schweer, H. P. Vögele, *Makromol. Chem. Rapid Commun.* **1986**, *7*, 261–265.

**129.** M. Buback, M. Egorov, T. Junkers, E. Panchenko, *Macromol. Chem. Phys.* **2005**, *206*, 333–341.

**130.** J. F. Quinn, L. Barner, T. P. Davis, S. H. Thang, E. Rizzardo, *Macromol. Rapid Commun.* **2002**, *23*, 717–721.

**131.** J. F. Quinn, L. Barner, E. Rizzardo, T. P. Davis, *J. Polym. Sci., Part A: Polym. Chem.* **2002**, *40*, 19–25.

**132.** L. Barner, J. F. Quinn, C. Barner-Kowollik, P. Vana, T. P. Davis, *Eur. Polym. J.* **2003**, *39*, 449–459.

**133.** P. E. Millard, L. Barner, M. H. Stenzel, T. P. Davis, C. Barner-Kowollik, A. H. E. Muller, *Macromol. Rapid Commun.* **2006**, *27*, 821–828.

**134.** T. M. Lovestead, G. Hart-Smith, T. P. Davis, M. H. Stenzel, C. Barner-Kowollik, *Macromolecules* **2007**, *40*, 4142–4153.

**135.** S. Muthukrishnan, E. H. Pan, M. H. Stenzel, C. Barner-Kowollik, T. P. Davis, D. Lewis, L. Barner, *Macromolecules* **2007**, *40*, 2978–2980.

**136.** J. F. Quinn, L. Barner, C. Barner-Kowollik, E. Rizzardo, T. P. Davis, *Macromolecules* **2002**, *35*, 7620–7627.

**137.** L. Lu, H. J. Zhang, N. F. Yang, Y. L. Cai, *Macromolecules* **2006**, *39*, 3770–3776.

**138.** L. C. Lu, N. F. Yang, Y. L. Cai, *Chem. Commun.* **2005**, 5287–5288.

**139.** Y. Z. You, C. Y. Hong, R. K. Bai, C. Y. Pan, J. Wang, *Macromol. Chem. Phys.* **2002**, *203*, 477–483.

**140.** M. Buback, T. Junkers, P. Vana, In *ACS Symposium Series on Living/*

*Controlled Free Radical Polymerization*, Vol. *944*, K. Matyjaszewski Ed., ACS Press, Washington, D.C., **2006**, pp. 455–472.

**141.** M. Buback, P. Hesse, T. Junkers, P. Vana, *Macromol. Rapid Commun.* **2006**, *27*, 182–187.

**142.** T. Arita, M. Buback, O. Janssen, P. Vana, *Macromol. Rapid Commun.* **2004**, *25*, 1376.

**143.** Y. Kwak, A. Goto, Y. Tsuji, Y. Murata, K. Komatsu, T. Fukuda, *Macromolecules* **2002**, *35*, 3026.

**144.** G. Zifferer, *Macromol. Theory Simul.* **2000**, *9*, 479.

**145.** S. Beuermann, M. Buback, *Prog. Polym. Sci.* **2002**, *27*, 191–254.

**146a.** M. Buback, T. Junkers, T. Theis, P. Vana, In *4th IUPAC Sponsored International Symposium on Radical Polymerization: Kinetics and Mechanism, SML '06*, **2006**, Poster 27.

**146b.** M. Buback, P. Hesse, T. Junkers, T. Theis, P. Vana, *Aust. J. Chem.* **2007**, *60*, in press.

**147.** M. Smoluchowski, *Z. Phys. Chem.* **1917**, *92*, 129.

**148.** J. P. A. Heuts, G. T. Russell, *Eur. Polym. J.* **2006**, *42*, 3.

**149.** R. X. E. Willemse, B. B. P. Staal, A. M. van Herk, J. Pierik, B. Klumpermann, *Macromolecules* **2003**, *36*, 9797.

**150.** P. B. Zetterlund, W. K. Busfield, I. D. Jenkins, *Macromolecules* **2002**, *35*, 7232.

**151.** O. F. Olaj, P. Vana, M. Zoder, A. Kornherr, G. Zifferer, *Macromol. Rapid Commun.* **2000**, *21*, 913.

**152.** O. F. Olaj, P. Vana, M. Zoder, *Macromolecules* **2002**, *35*, 1208.

# 5
# The Radical Chemistry of Thiocarbonylthio Compounds: An Overview

*Samir Z. Zard*

## 5.1
## Historical Overview and Early Chemistry

Thiocarbonylthio compounds of general formula [Z—(C=S)S—R] emerged quite early in the history of chemistry because of the ready accessibility of carbon disulfide, a substance made by merely heating native sulfur with powdered charcoal. The Danish pharmacist William C. Zeise thus reported the synthesis of the ethyl xanthates (the more systematic nomenclature is *O*-ethyl dithiocarbonates) of potassium, barium, lead and copper in 1815 [EtOC(=S)SM, M = K, Ba, Pb, Cu], nearly two centuries ago, by adding carbon disulfide to an ethanolic solution of potassium or barium hydroxide [1]. The potassium or barium can then be easily exchanged with other metal cations. The synthesis of related derivatives of carbon disulfide, such as dithiocarbamates [R′R″N—(C=S)S—R], trithiocarbonates [R′S—(C=S)S—R] and dithiocarboxylates [R—(C=S)S—R] followed later. Various industrial applications for these compounds were also ultimately found. Xanthate salts are extensively used as flotation agents for separating ores (or the native metal) of thiophilic metals such as gold, copper, lead and zinc from nonvaluable minerals and gangue. Xanthates are intermediates in the manufacture of regenerated cellulose products such as viscose and cellophane. Other applications of xanthates and dithiocarbamates concern the vulcanization of rubber and in the formulation of insecticides and fungicides for use in plant protection (the widely employed maneb and zineb are ethylene (bis)dithiocarbamates of manganese and zinc, respectively) [2].

Most of the early studies centered on the ionic chemistry of these carbon disulfide derivatives. Indeed, many of the ionic reactions parallel those of the corresponding and more common oxygen analogues. Thus, xanthates (or dithiocarbonates), dithiocarbamates, trithiocarbonates and dithioesters undergo saponification, alcoholysis or aminolysis much in the same way as carbonates, carbamates and carboxylic esters. There are, however, many differences due to the presence of the two sulfur atoms, and especially the thiocarbonyl group. The field is much too vast to be

*Handbook of RAFT Polymerization.* Edited by Christopher Barner-Kowollik

ISBN: 978-3-527-31924-4

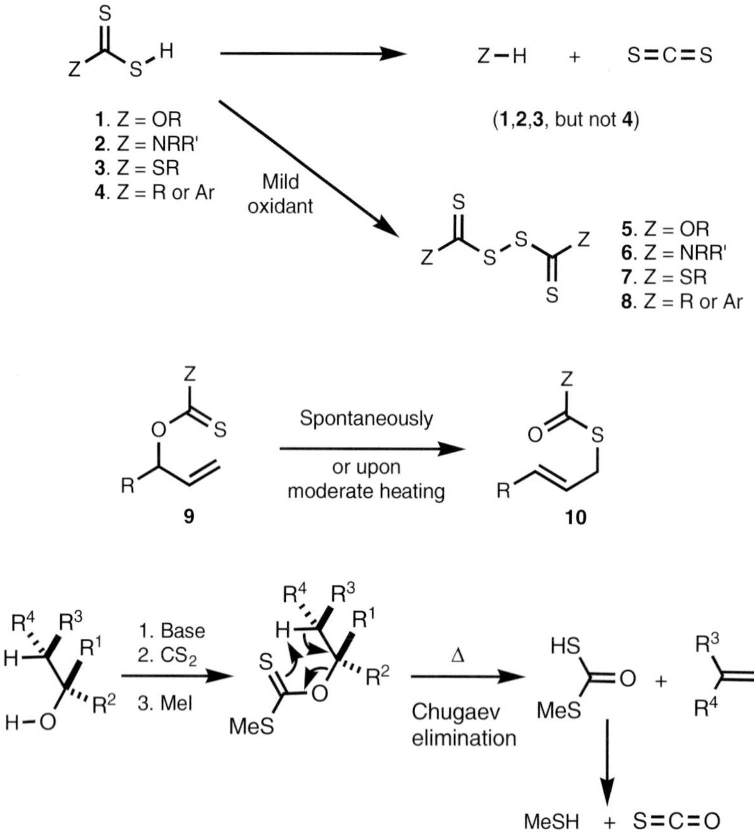


**Scheme 5.1** Some general reactions of thiocarbonylthio compounds.

covered adequately in such a short overview, but the transformations in Scheme 5.1 perhaps give a general idea [3].

The parent acids **1**, **2** and **3** of xanthates, dithiocarbamates and trithiocarbonates (but not dithiocarboxylic acids **4**) are generally unstable toward loss of carbon disulfide, in the same way as their corresponding oxygen analogues are unstable toward loss of carbon dioxide. In stark contrast, however, they are quite sensitive to oxidation (even by air), which leads to the formation of disulfides **5**, **6**, **7** and **8**, as with thiols in general. Derivatives of allylic alcohols such as **9** often undergo a spontaneous or, upon gentle warming, a [3, 3] sigmatropic rearrangement to give the transposed isomer **10** [4]. The driving force is the great difference in bond strength between the carbonyl and thiocarbonyl groups, as well as the enhanced nucleophilicity of the sulfur terminus of the thiocarbonyl. The analogous rearrangement of allylic carbonates vastly requires vastly more drastic conditions to occur.

Another reaction that nicely illustrates the special reactivity of the thiocarbonyl group is the Chugaev elimination, depicted at the bottom of Scheme 5.1 [5]. In this reaction, an alcohol, **11**, is converted into the xanthate and the latter heated to give the corresponding olefin **12** by a *syn*-elimination (in fact a retro-ene reaction). Xanthates derived from secondary alcohols ($R^1$ or $R^2 = H$) can be decomposed at temperatures around 150 °C, but xanthates of tertiary alcohols ($R^1$, $R^2 \neq H$) frequently decompose at room temperature or below. The corresponding *syn*-elimination of carbonates or esters requires temperatures at least 100 °C higher.

## 5.2 The Barton–McCombie Deoxygenation

Perhaps the most remarkable difference between thiocarbonylthio derivatives and their oxygen analogues concerns the tremendous radicophilicity of the thiocarbonyl group as compared with that of a carbonyl. This property was ingeniously exploited by Barton and McCombie to effect the deoxygenation of secondary alcohols, a little over 30 years ago [6]. Alcohols are ubiquitous in nature, and in synthetic intermediates, yet, in many substances, it is quite difficult, if not impossible, to replace the C—O bond by a C—H bond under conditions that are tolerated by the other functional groups present in the structure. One case in point is carbohydrates and cyclitols, where hydroxy groups are densely packed around the carbon backbone, and the reductive removal of one of the groups is frequently a tedious process requiring many separate steps.

In the original Barton–McCombie deoxygenation, the alcohol is converted into a xanthate in the same way as for the Chugaev reaction and then treated with tributyltin hydride and a small amount of a radical initiator, usually 2,2′-azobis(isobutyronitrile) (AIBN). The spectacular transformation of the aminoglycoside antibiotic derivative pictured at the top of Scheme 5.2 [7] would be almost impossible to achieve by any other route and testifies to the power of this method. The deoxygenation procedure is most suitable for secondary alcohols. Xanthates of primary alcohol react sluggishly (but in some special case can indeed be reduced), and the xanthates of tertiary alcohols are generally difficult to handle because of their susceptibility to the Chugaev fragmentation, as mentioned above. The xanthate group can also be replaced by thiocarbonyl imidazolides, as illustrated by the second example in Scheme 5.2, but simple thionocarbamates are not normally effective precursors.

Numerous alcohols have been deoxygenated in this manner, and the original paper by Barton and McCombie [6a] has been cited more than 1200 times [8]. Indeed, the discovery of this superb deoxygenation method represented a watershed in radical chemistry, since it has opened the eyes of the synthetic community to the unique potential of radical reactions for the synthesis and modification of small molecules. It may seem surprising, but in the early 1970s radical reactions were seldom employed in classical organic synthesis, in contrast to the polymer field, where radical processes were routinely applied on a very large scale.

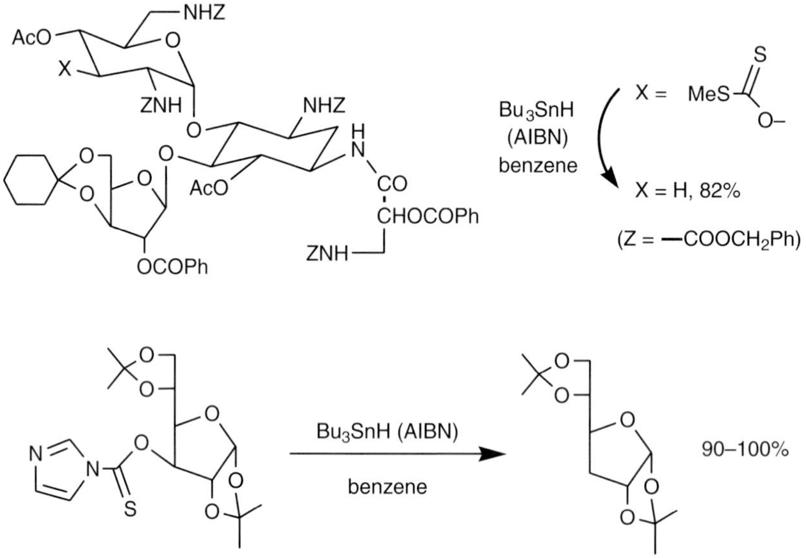


Scheme 5.2 Examples of Barton–McCombie deoxygenations.

## 5.3
## A Minor Mechanistic Controversy

Originally, Barton and McCombie proposed the mechanism depicted in the upper part of Scheme 5.3, whereby the stannyl radicals attack the sulfur of the thiocarbonyl portion of the xanthate to form intermediate radical **14**, which undergoes β-scission of the carbon–oxygen bond [6a]. The ensuing carbon radical R• is reduced into the corresponding alkane, R—H, by hydrogen abstraction from the stannane. The coproduct **15** is unstable and decomposes into tributyltin methylsulfide (**16**) and carbon oxysulfide. This, seemingly reasonable, mechanism was challenged a decade later by Beckwith and Barker who proposed an alternative mechanism, based on the observation of alkoxythiocarbonyl radicals **17** by electron spin resonance (ESR) when a solution of a xanthate and hexamethyl ditin is irradiated in the cavity of the ESR spectrometer [9]. Beckwith and Barker postulated that the attack of the stannyl radicals occurs on the sulfide carbon to give directly tributyltin methylsulfide (**16**) and alkoxythiocarbonyl radical **17**, which then extrudes carbon oxysulfide to give radical R•.

There were various arguments that could be put forward against this alternative mechanism. One is that a thiocarbonyl group is much more radicophilic than a sulfide sulfur, as had been shown by various kinetic studies. The second is that the sulfide sulfur is not in fact necessary for the deoxygenation, as indicated by the second example in Scheme 5.2. However, in order to distinguish unambiguously between these two mechanisms, various experiments were designed and tested [10]. One competition experiment, in particular, had unintended and far-reaching

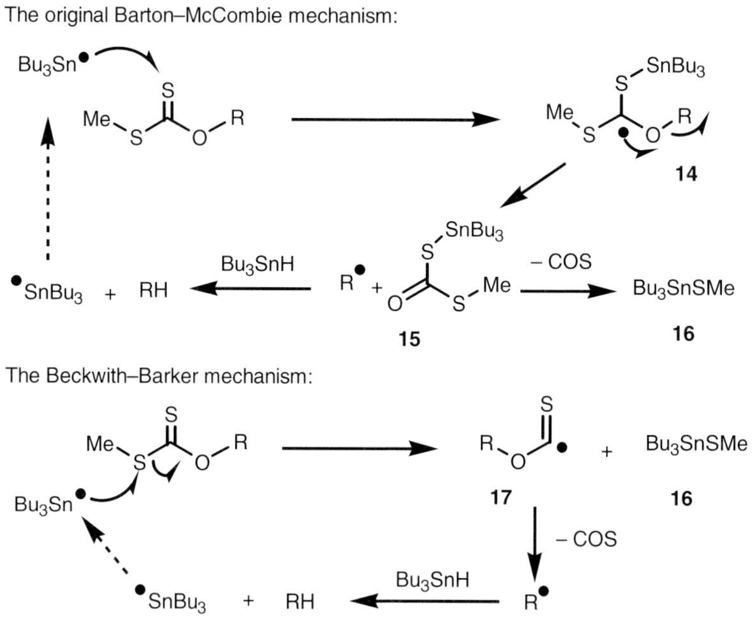


**Scheme 5.3** Two different mechanisms for the Barton–McCombie reaction.

consequences. It consisted in treating the two cholestane-derived xanthates **18** and **19** with only one equivalent of tributylstannane [10a]. The reasoning was that if the original mechanism were the correct one, then both xanthates should react at approximately the same rate since the thiocarbonyl sulfur is too far to be hindered by the isopropyl group, whereas the bulkier *S*-isopropyl xanthate (**19**) should react significantly more slowly, if the initial attack occurred on the neighboring sulfide sulfur as proposed by Beckwith and Barker (Scheme 5.4).

In the event, neither of these two expected outcomes was observed. *S*-Isopropropyl xanthate (**19**) reacted *much faster* than the less hindered methyl analogue **18**, and the products of the reaction were stannyl xanthate (**22**) and propane, instead of the expected cholestane. It was immediately clear that the addition of the stannyl radical had to take place on the thiocarbonyl sulfur and that this step had to be *fast and reversible* leading to intermediates **20** and **21**. The subsequent fragmentation is the rate-limiting step at the usual reaction temperatures. The collapse of radical **21** is significantly faster than that of **20** and proceeds by β-scission of the C—S bond instead of the C—O bond to give the isopropyl radical. In the case of adduct **20**, derived from *S*-methyl xanthate (**18**), which would be the normal substrate for a deoxygenation reaction, the (slower) fragmentation occurs preferentially by cleavage of the C—O bond, because rupture of the C—S bond would give a high-energy methyl radical. The original mechanism postulated by Barton and McCombie therefore appeared to be broadly correct, except for the reversibility of the addition of the stannyl radical on the thiocarbonyl group, which was not

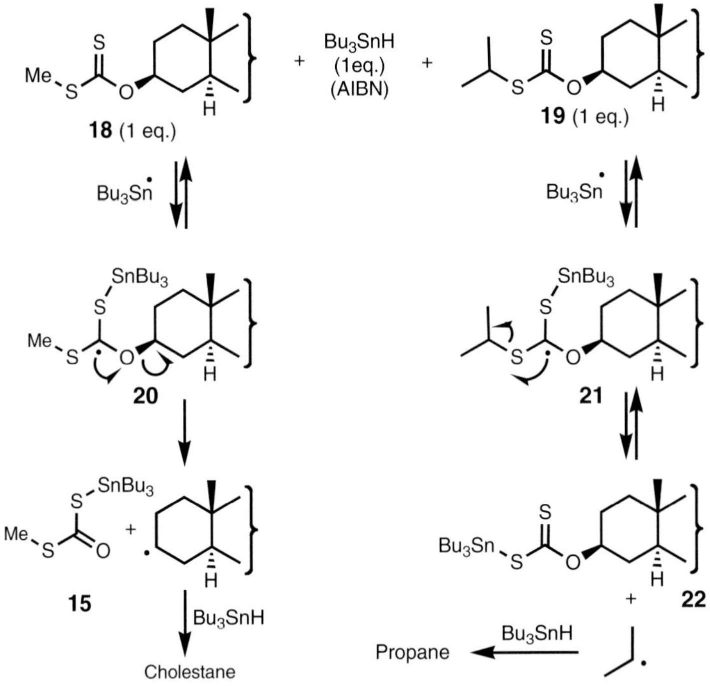


**Scheme 5.4** A key competition experiment.

appreciated at the time and was not invoked in mechanistic discussions or kinetic studies.

## 5.4 A New Degenerative Radical Process

Perhaps the most important revelation of the failed competition experiment described above is that when the groups present in the starting xanthate lead to radicals of comparable stability (e.g. isopropyl and cholestanyl in **19**), and then collapse of the intermediate (e.g. **21**) proceeds preferentially by scission of the C—S rather than the C—O bond. This raises various mechanistic questions and it is interesting to explore some of the consequences of this serendipitous observation.

One obvious application is to extend the Barton–McCombie deoxygenation to the reductive desulfurization of thiols. This is illustrated by the efficient reduction of compound **23** into sulfur-free derivative **24** shown in Scheme 5.5 [11]. In this case, the collapse of the intermediate adduct radical **25** is clearly biased toward fragmentation of the C—S bond as indicated by the arrows. Scission of the C—O bond would lead to a higher-energy ethyl radical and is therefore not favored. From a synthetic standpoint, generating alkanes by reductive cleavage of the C—S bond in xanthates is much less useful than reductive cleavage of the C—O bond, as in the

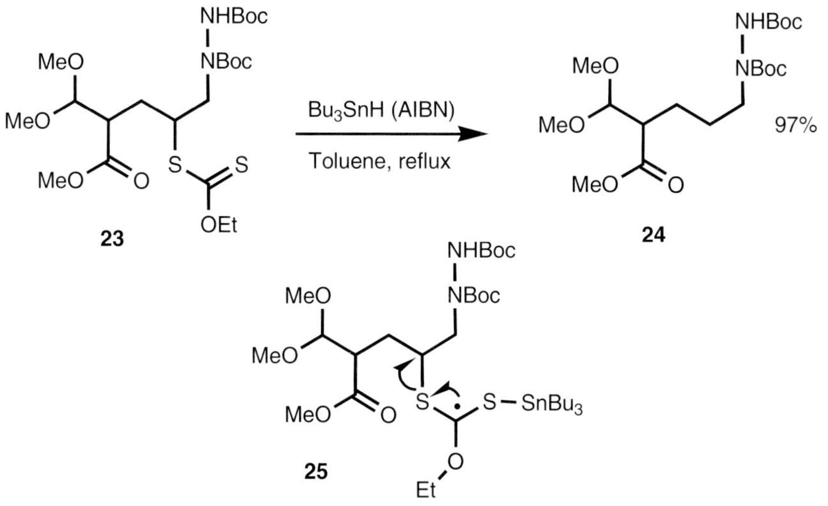

**Scheme 5.5** The sulfur variant of the Barton–McCombie reaction.

original Barton–McCombie deoxygenation. Alcohols are ubiquitous whereas thiols are less common in nature or as synthetic intermediates.

A more interesting situation arises when the attacking stannyl radical is replaced by a carbon-centered radical. This gives rise to the mechanistic manifold displayed in Scheme 5.6. Addition of R′• to the thiocarbonyl group of xanthate **26** leads to intermediate **27**, which can fragment back to **26** (path (a)) or forward to xanthate **28** and radical R• (path (b)). Both of these steps are reversible and the position of the equilibrium will depend, to a first approximation, on the relative stabilities of R• and R′•. Scission of the C—O bond (path c) is irreversible and leads to an *S*,*S*′-dithiocarbonate **29**. This last step can be made to prevail by a proper selection of the various substituents and the experimental conditions. This possibility can be exploited to convert xanthates of the Barton–McCombie-type [R—OC(=S)SMe] into *S*,*S*′-dithiocarbonates [R—S(C—O)SMe] by a tin-free radical chain process [12]. Overall, this represents a convenient procedure for converting secondary alcohols

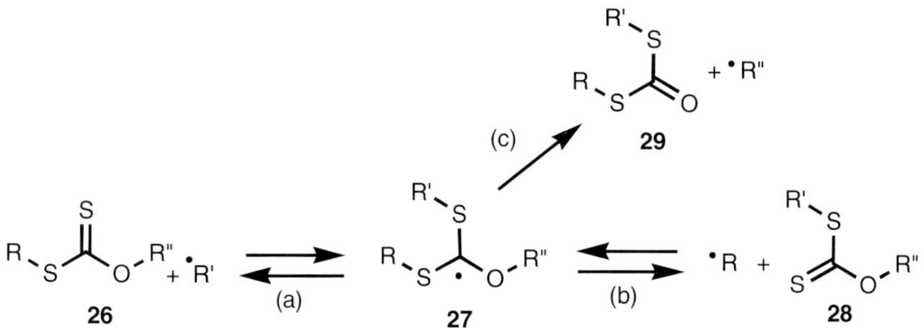

**Scheme 5.6** Possible fragmentations of radical adducts to a xanthate.

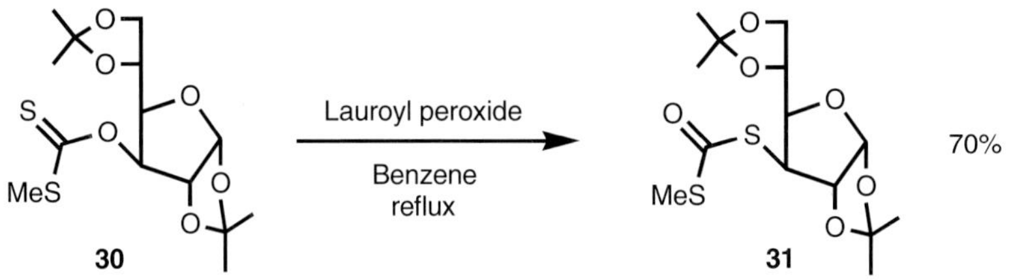


**Scheme 5.7** Oxygen to sulfur rearrangement.

into the corresponding thiols in cases where a traditional ionic approach would be difficult (Scheme 5.7).

In contradistinction, if the R″ group on the oxygen corresponds to a radical of relatively high energy (R″ = Me or Et, for example) and if the incoming radical R′• is less stable than radical R•, then attack of R′• on xanthate **26** would result in the expulsion of R• by a reversible addition–fragmentation on the thiocarbonyl group, *without untoward complications due to fragmentation from the oxygen side*. In other words, only paths (a) and (b) will operate but not path (c). On this basis, a chain process can, in principle, be constructed by generating incoming radical R′• from radical R• through any typical radical sequence (addition, fragmentation or any combination thereof). This possibility is detailed in Scheme 5.8 for the synthetically most interesting case of radical addition to an olefin [13].

Thus, radicals R•, produced in the initiation step, will rapidly add to the highly radicophilic thiocarbonyl group of the starting xanthate **32** to give intermediate

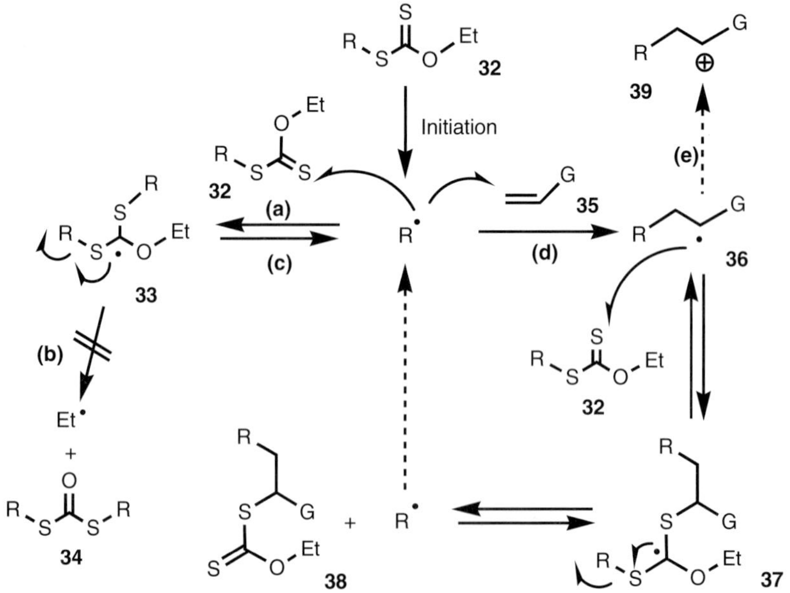


**Scheme 5.8** The degenerative xanthate transfer addition to an alkene.

**33** (path (a)). Rupture of the C—O to give *S,S′*-dithiocarbonate **34** (path (b)) is difficult because it leads to a high-energy ethyl radical and requires breaking a particularly strong bond. Scission of the C—S bond (path (c)) is normally easier for most R groups, but simply gives back the starting xanthate **32** and the same radical R•. Thus, *the reaction of the initial radical R• with its xanthate precursor is fast, but reversible and degenerate.* As a consequence, the effective lifetime of R• in the medium increases considerably, since it is continuously being regenerated. Now, addition even to simple, nonactivated alkenes becomes possible (path (d)). More generally, the radical is able to undergo comparatively slow inter- or intramolecular processes not easily achievable with other methods. In the case of addition to alkene **35** (path (d)), a new radical **36** is created, which in turn reacts *rapidly and reversibly* with the starting xanthate **32** to produce intermediate radical **37**. Reversible collapse of this species finally furnishes adduct **38**, as well as the initial radical R• to propagate the chain.

This process embodies numerous advantages that will not be discussed in detail here [13]. Suffice it to note that it does not involve tin or other undesirable heavy metals; it does not require high dilution conditions and can be easily scaled up; and the reagents are cheap and readily available. Furthermore, the product **38** is itself a xanthate; it is therefore possible to implement another radical process or to exploit the exceedingly rich chemistry of sulfur for subsequent modifications of the initial adduct. The fact that the exchange of the xanthate group occurs by two reversible addition–fragmentation steps, it is necessary to choose the starting xanthate **32** and the olefinic trap **35** in such a way that the adduct radical **36** is *less* stable than radical R•, so as to ensure that the fragmentation of intermediate **37** will favor the formation of the latter, and thus push the chain sequence in the desired direction. This is a key consideration that allows control of the process and must be constantly kept in mind, especially when dealing with intermolecular additions. In some cases, if the adduct radical is sufficiently electron rich, it is possible to oxidize adduct radical **35** into corresponding cation **39**, for example by a peroxide, and hence enter into an ionic manifold (path (e) in Scheme 5.8). This possibility expands considerably the range of useful olefinic traps and proves especially valuable in additions to aromatic and heteroaromatic rings (see below for some examples). The peroxide can therefore act both as an initiator and as a stoichiometric oxidant.

An *O*-ethyl xanthate is depicted in Scheme 5.8 for simplicity, and because most of the examples in small molecule chemistry involve xanthates. These substrates show good reactivity and most can be readily prepared starting from potassium *O*-ethyl xanthate, a commercially available and very cheap reagent (as mentioned above, it is used mostly as a flotation agent in mining). The ethoxy group in **32**, however, remains essentially a spectator, whose influence is only indirect, albeit important. It can be replaced by various substituents, and exactly the same reversible addition–fragmentation transfer (RAFT) mechanism can be applied to other thiocarbonylthio compounds [14]. This is summarized in Scheme 5.9 in a condensed version of the reaction manifold. Ideally, substituent Z must at the same time favor the first radical addition to the thiocarbonyl group and not hinder the subsequent fragmentation step. In the case of ordinary dithiocarbamates (**40**, Z = NR′R″),

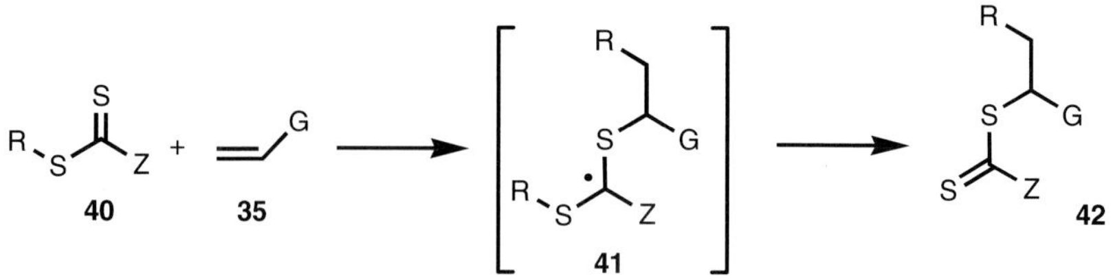


**Scheme 5.9** Condensed mechanism for the addition of thiocarbonylthio derivatives.

for example, the thiocarbonyl group is not sufficiently reactive because of strong donation by the nitrogen lone pair (cf. amides). Placing an electron-withdrawing group on the nitrogen atom restores the reactivity. *S*-acyl dithiocarbamates (**40**, R = R‴CO-; Z = NR′R″), where the carbonyl group on the sulfur counteracts somewhat the donating effect of the nitrogen, can also be interesting substrates (see second example in Scheme 5.27 below). Trithiocarbonates (**40**, Z = SR′) appear to be a little more reactive than xanthates [14], but this slightly enhanced reactivity is neutralized by a diminished accessibility and the inconvenience of having to handle obnoxious thiols. Aliphatic dithioesters (**40**, Z = R′) are also quite reactive but again suffer from lack of availability. In the case of aromatic dithioesters (**40**, Z = Ar), the first addition becomes a little too efficient, since the corresponding adduct **41** is now stabilized by conjugation with the aromatic ring but the subsequent fragmentation step is slowed down. This class of thiocarbonylthio derivatives is useful when dealing with particularly stabilized radicals, where the reluctance to fragment of adduct **41**, when Z = Ar, is compensated by the special stability of the departing radical (see below).

## 5.5
## Synthetic Routes to Thiocarbonylthio Derivatives

### 5.5.1
### Ionic Reactions

Before presenting examples of radical additions, it is perhaps useful to survey the main synthetic routes to thiocarbonylthio derivatives. Most of the examples will involve xanthates, simply because of their more widespread use in synthesis as compared with dithiocarbamates, trithiocarbonates, dithioesters and so forth. Nevertheless, the methods developed for xanthates can often be extended to the other congeners, even if the efficiency or ease of transposition will necessarily vary with the nature of the substituents.

By far the most important route to thiocarbonylthio compounds is by nucleophilic substitution of a leaving group by reaction with the corresponding salt of the thiocarbonylthio derivatives [3]. The salts are generally made from carbon disulfide with the appropriate ‘$Z^-$’, as pictured in Scheme 5.10. Thus, reaction of alcoholates, amines, thiolates or carbanions (including organometallics) with carbon disulfide

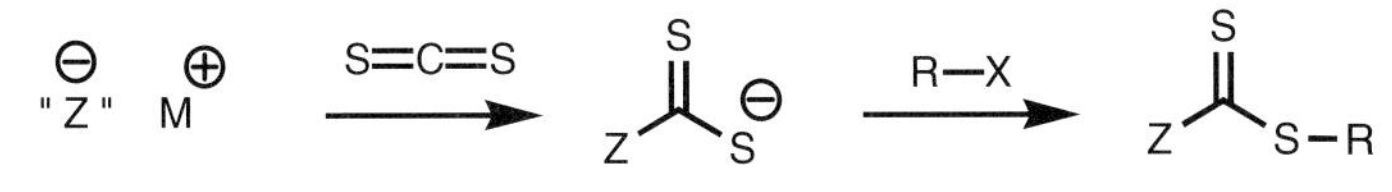


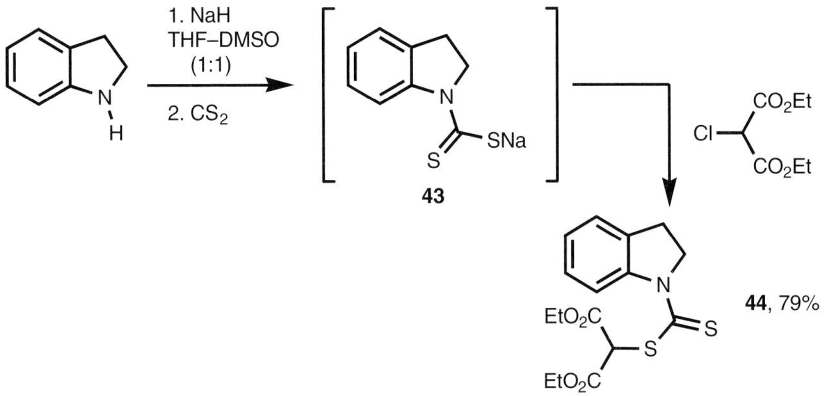


**Scheme 5.10** Example of a synthesis of a dithiocarbamate.

would give the corresponding xanthates, dithiocarbamates, trithiocarbonates and dithioesters, respectively. These can then be made to interact with the alkylating agent, often in situ, without need for prior isolation of the thiocarbonylthiolate salt. In the case of xanthates and dithiocarbamates, several of the simpler alkaline salts are cheap commodity chemicals that need not be prepared. This makes the use of xanthates in particular very convenient. In any case, it is wise to store the thiocarbonylthio salts in dark, tightly closed recipients to avoid slow oxidation with air, which leads, among other compounds, to the corresponding disulfides, and which degrades the quality of the material. An example of the synthesis of a dithiocarbamate, **44**, derived from indoline is displayed in Scheme 5.10 (G. Bouhadir, X. Franck, S. Z. Zard, unpublished observations). The intermediate sodium salt **43** was not isolated.

Thiolates are powerful nucleophiles but are nonetheless subject to the limitations of $S_N2$ reactions: they react, in general, readily with primary or secondary but not tertiary alkylating agents. Access to tertiary thiocarbonylthio derivatives thus usually requires alternative and, often, indirect approaches. In some limited cases, it is possible to obtain tertiary derivatives through an $S_N1$ mechanism, by capture of a stabilized cation. Two examples illustrating the synthesis of cumyl dithioesters are depicted in Scheme 5.11. Prolonged heating of $\alpha$-methyl styrene in neat thioacetic acid leads to a moderate yield of cumyl dithioacetate (**45**) [15]. Addition of sulfuric acid shortens considerably the reaction time and allows the use of lower temperatures as shown for the synthesis of the corresponding dithiobenzoate **46** (G. Bouhadir, X. Franck, S. Z. Zard, unpublished observations).

Another route is to thioacylate a thiol with a chlorothiocarbonyl derivative of general formula [Z—(C=S)—Cl] in the presence of a base. An example is presented

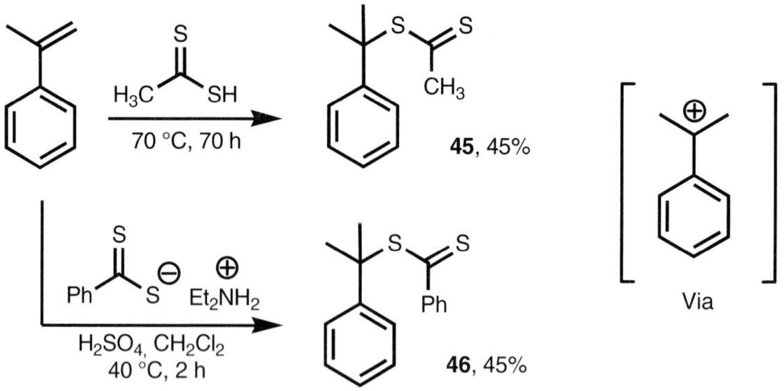


**Scheme 5.11** $S_N1$-type addition of dithiocarboxylic acids to cumene.

in the upper part of Scheme 5.12 [16]. This strategy is quite general for preparing primary, secondary and tertiary derivatives but is constrained by the unpleasantness of handling the reagents and their very limited availability. Rannard, Perrier and their collaborators have recently reported a more convenient alternative using as the key reagent 1,1′-thiocarbonyl diimidazole (**48**) a commercially available crystalline solid that is easy to handle [17]. As shown in Scheme 5.12, 1,1′-thiocarbonyl

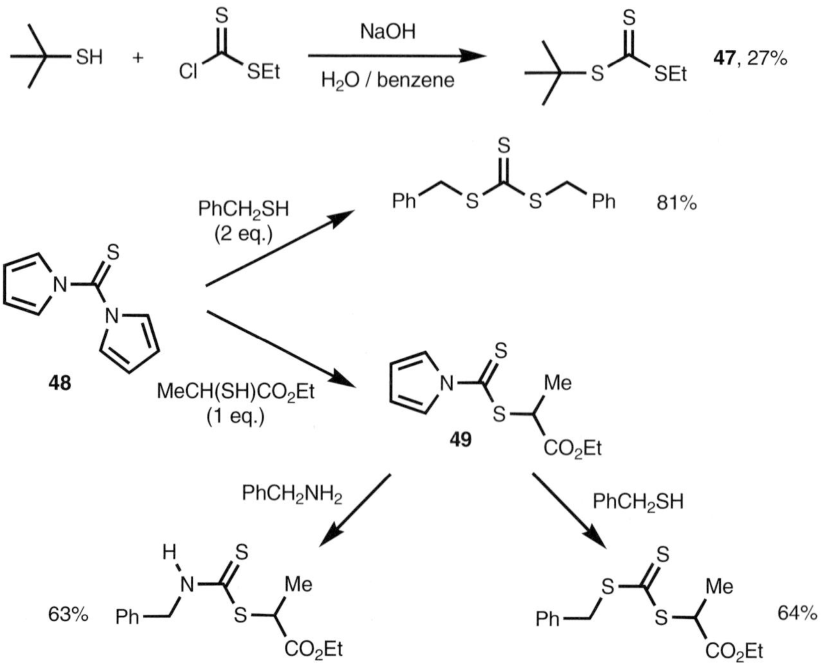


**Scheme 5.12** Syntheses starting with thiols.

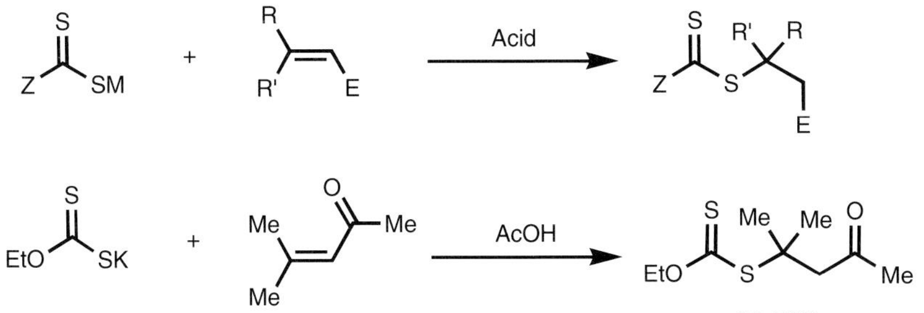


**50**, 76%

**Scheme 5.13** Example of a Michael addition of xanthic acid.

diimidazole reacts readily with a thiol such as benzyl mercaptan to give directly the expected symmetrical trithiocarbonate; alternatively, the reaction can be stopped at the monosubstituted product if only 1 eq. of thiol is used, as in the case of compound **49**, derived from ethyl thiolactate. Without prior isolation, this compound can then be reacted with another thiol or an amine to give the unsymmetrical trithiocarbonate or the dithiocarbamate in a one-pot procedure.

Conjugate additions to activated olefins are much less sensitive to steric hindrance than $S_N2$ substitutions and advantage can be taken of the presence of an electron-withdrawing group to mediate a Michael addition of the thiocarbonylthiolate salt, as outlined in Scheme 5.13 [18]. The reaction has to be conducted in an acidic medium to avoid the reversal of the addition. Thiocarbonylthiols [Z—(C=S)—SH] are more acidic than simple carboxylic acids and therefore undergo β-eliminations quite readily under basic conditions. Interesting derivatives can nevertheless be prepared as shown by the efficient addition of a xanthate to mesityl oxide to give **50** in 76% yield. It is thus possible to access tertiary derivatives by going through a conjugate addition rather than by an $S_N2$ substitution.

An indirect manner to secure tertiary thiocarbonylthio compounds relies on exploiting the acidity of secondary derivatives containing two electron-withdrawing groups to generate and alkylate the parent anion. An example of this approach is displayed in Scheme 5.14. Even though the need for electron-withdrawing groups is in itself a serious limitation, the method provides a series of otherwise inaccessible structures [19]. The synthesis of tertiary xanthate **52** by methylation of readily available xanthate **51** is a representative example. It must be pointed out that it is important to have the alkylating agent, preferably in excess amounts, present before addition of the base; otherwise, the reaction follows a different course to give a ketene monothioketal such as **53** in the present case. The latter can be obtained as the sole compound by inverting the mode of addition [19].

Instead of introducing an alkyl group by alkylating a carbanion already containing the thiocarbonylthio group, as above, it is possible to append the thiocarbonylthio by reacting the anion with a bis(thiocarbonylthio) derivative **54**, as illustrated in Scheme 5.15 [20]. The synthesis of malonyl xanthate (**55**) thus contrasts with that presented in the preceding scheme for xanthate **52** [20a]. If only one ester group

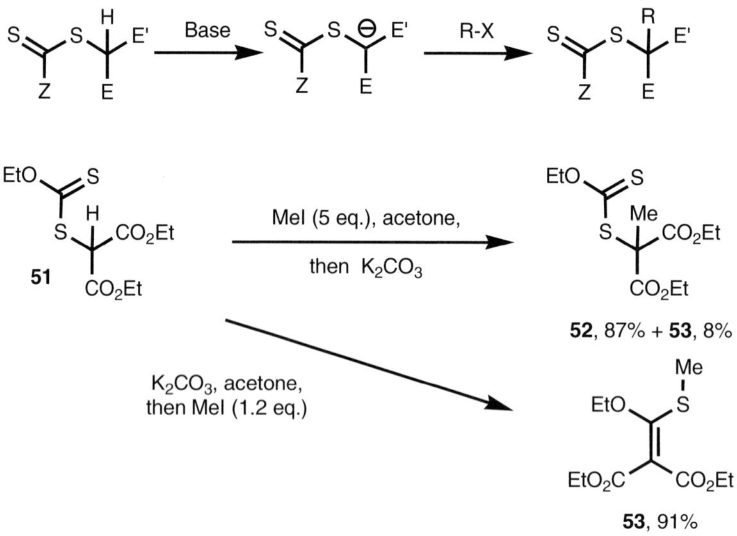


**Scheme 5.14** Alkylation of xanthate-substituted carbon acids.

is present, it is necessary to use a much stronger base, as shown by the second example [20a]. This approach has not been often applied because of its capricious nature and, so far, ill-defined scope. Nevertheless, it is a useful addition to the methodological toolbox that can prove handy for the synthesis of certain specific derivatives.

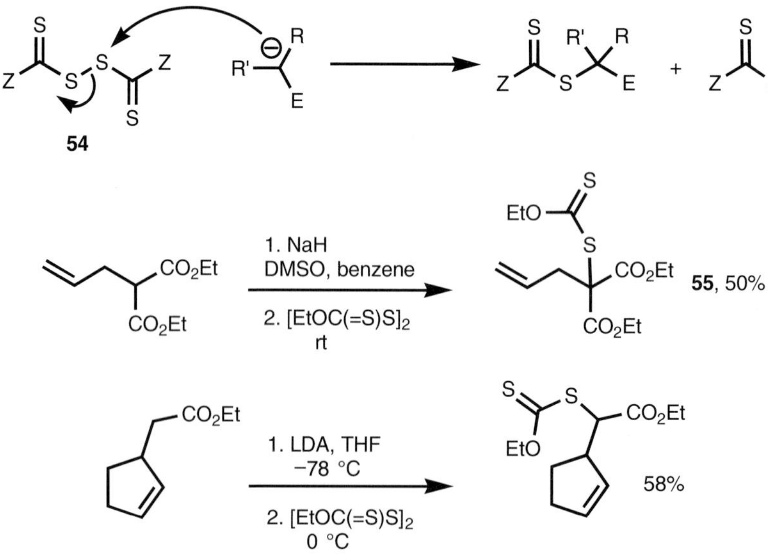


**Scheme 5.15** Xanthylation of carbanions with a bisxanthate.

## 5.5.2 Radical-Based Approaches

A recently developed radical variant of the above route proved more efficient and general, as far as the disulfide component is concerned [21, 22]. The conception, outlined in Scheme 5.16, involves the thermal decomposition of a diazo derivative in the presence of a bis(thiocarbonylthio) derivative. Radical **57**, derived from the diazo compound, reacts by an addition–fragmentation with the bis(thiocarbonylthio) reagent to give the desired product **58** and a sulfur-centered radical **59** that dimerizes back to the initial disulfide. It can also couple with radical **57** to give the same product **58**. Overall, only nitrogen in principle is lost in the process. Some examples of application are shown in the same scheme, illustrating the synthesis of xanthates, dithiocarbamates and dithioesters. A typical experimental procedure is as follows: a mixture of the diazo derivative (1.2 mmol) and disulfide (1 mmol) in cyclohexane or dioxane (2 mL) is heated to reflux under an inert atmosphere for

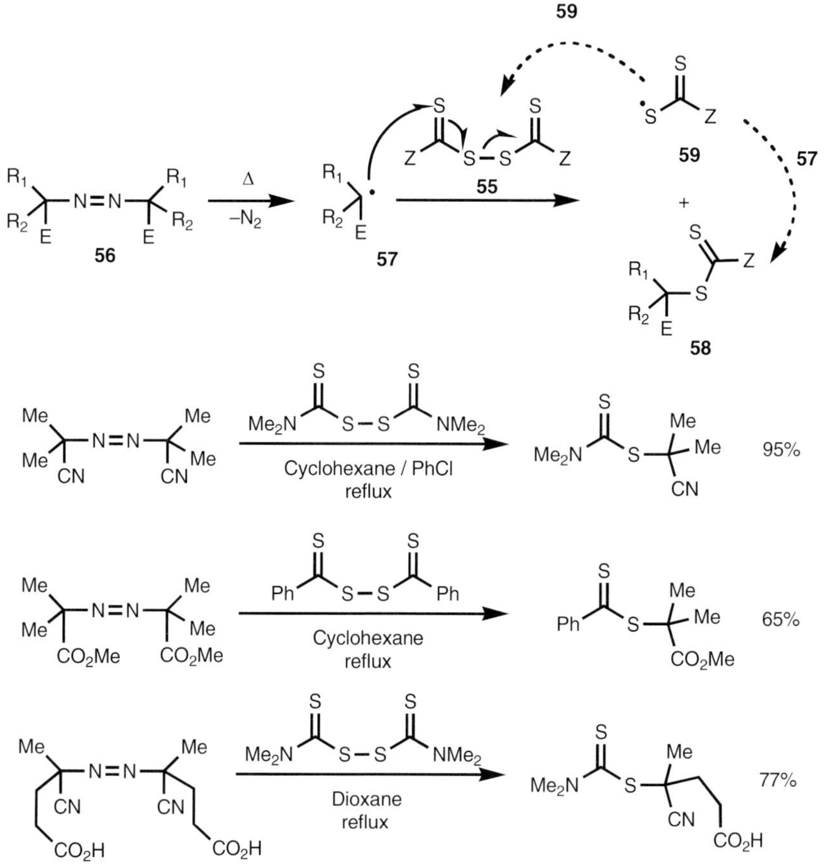


**Scheme 5.16** Syntheses starting with diazo compounds.

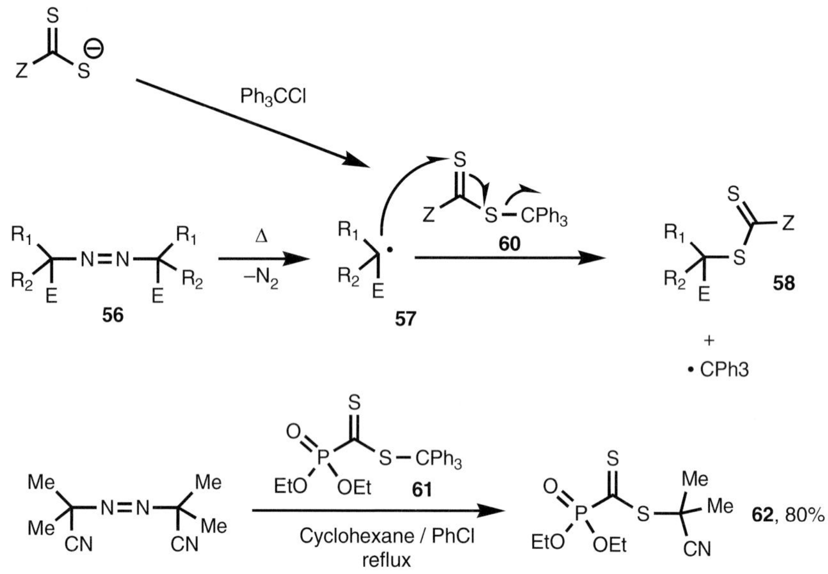


**Scheme 5.17** Using the trityl radical as a leaving group.

3–4 h. If necessary, additional diazo derivative is added in portions (0.2 mmol) with further heating until the reaction is complete. The solvent is then evaporated and the residue purified by chromatography or recrystallization.

For some substituents, the corresponding bis(thiocarbonylthio) reagent is not readily accessible and a variant of the above approach was devised by Benaglia and coworkers to overcome this bottleneck [23]. The thiocarbonylthio salt is first made to react with trityl chloride to give the expected triphenylmethyl derivative **60**, which is heated, as above, with the requisite diazo component **56** (Scheme 5.17). In this case, the addition of radical **57** derived from the thermal decomposition of the diazo substrate adds to the thiocarbonyl group and causes the departure of the highly stabilized trityl radical. Even if this route is not very 'atom efficient', since the large trityl portion is ultimately wasted, it can be the only synthetic entry for certain individual compounds. Indeed, phosphonodithioester **61** (the synthesis of which is shown in the same scheme) could not be prepared by any other route.

In yet another variation, the powerful Barton decarboxylation reaction can be combined with the radicophilicity of bis(thiocarbonylthio) derivatives to prepare thiocarbonylthio structures, starting with carboxylic acids [24]. The principle of this approach is outlined in Scheme 5.18 for the generic Barton ester **63**. The process can be triggered by light, heat or by the use of various initiators. Furthermore, the mixed disulfide side product **64** can itself act as a thiocarbonylthio transfer agent, which should increase the overall atom efficiency. The example given concerns the synthesis of an *S*-trifluoromethyl xanthate (**66**) via the corresponding Barton ester **65**, which is not isolated. It is worth pointing out that owing to the presence of the highly electronegative fluorine atoms, *S*-perfluoroalkyl xanthates (and other

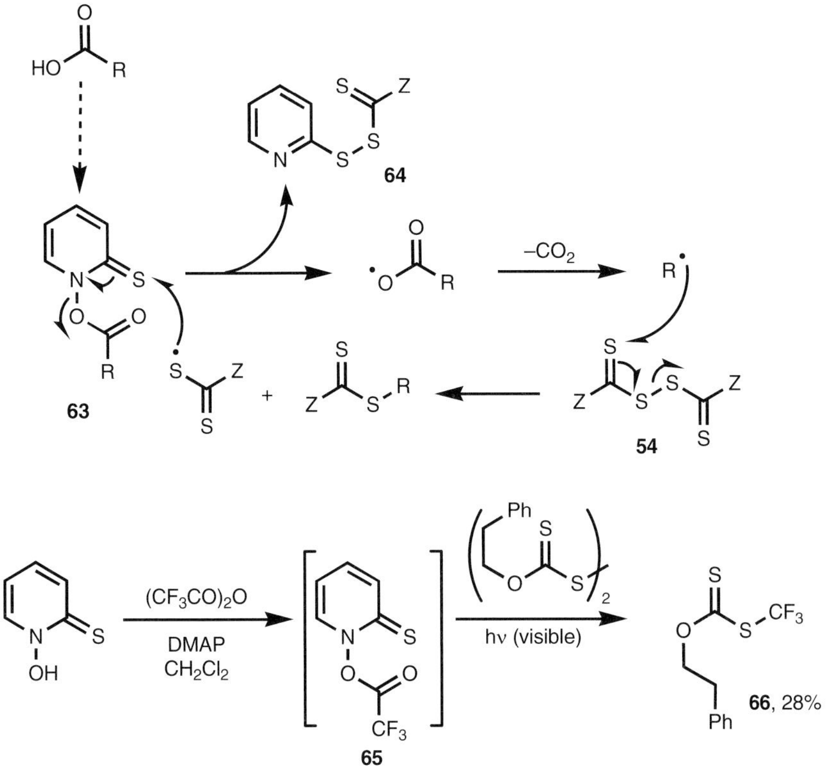


**Scheme 5.18** Decarboxylative xanthylation using Barton esters.

thiocarbonylthio derivatives) cannot be made by the usual nucleophilic displacement from the corresponding perfluoroalkyl halides. Indirect routes, such as the present, must therefore be considered to access these compounds. The scope and flexibility of the decarboxylation method still need to be defined, as well as the optimum experimental conditions; nevertheless, judging from the broad applicability of the Barton decarboxylation to aliphatic and alicyclic carboxylic acids, as well as its remarkable functional group tolerance, this approach appears to be quite promising.

An alternative route also starting from carboxylic acids exploits the decarbonylation of acyl radicals. The decabonylation is a much slower process than the decarboxylation, but can become synthetically useful if it leads to a stabilized entity, such as a tertiary or a benzylic radical. For example, irradiation of *S*-pivaloyl xanthate (**67**) with visible light, or heating in the presence of a suitable peroxide, causes its conversion into *S*-*tert*-butyl xanthate (**69**) [25]. This transformation proceeds by way of the radical chain process presented in Scheme 5.19 and involves the intermediacy of tertiary acyl radical **68**. It is worth mentioning that trifluoromethyl xanthate (**66**) can be prepared more conveniently and in a better yield (53%) by using this variant [25].

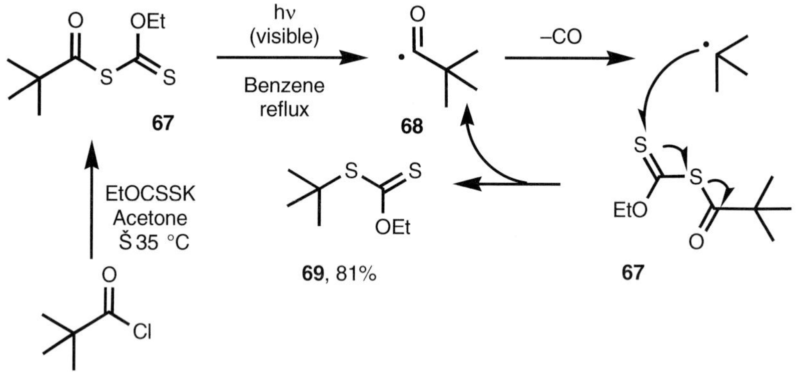


**Scheme 5.19** Decarbonylative rearrangement of *S*-acyl xanthates.

Finally, a conceptually related reaction involves decarboxylation of an alkoxycarbonyl radical derived from the corresponding *S*-alkoxycarbonyl xanthate **70**, as shown by the reaction sequence in Scheme 5.20 [26]. Irradiation of **70** with visible light triggers an efficient chain reaction proceeding by way of alkoxycarbonyl radical **71**. *S*-alkoxycarbonyl xanthates are readily obtained from the chloroformate of the requisite alcohol. In the example displayed in Scheme 5.20, tertiary xanthate **73** is efficiently prepared from precursor **72**, itself made from β-caryophyllene alcohol. It is interesting to note that the loss of carbon dioxide from radical **71** is sufficiently slow for most substrates to allow their inter- or intramolecular capture by an olefin to furnish esters or lactones respectively. Even though only xanthates have been made by the last three approaches, extension to other thiocarbonylthio derivatives should be possible.

## 5.6 Some Synthetic Applications of the Degenerative Radical Transfer to Small Molecules

Having surveyed briefly the synthetic methods for the preparation of thiocarbonylthio compounds, the next task is to overview some of the unique radical chemistry that accrues from the degenerate transfer process outlined in Scheme 5.8 and summarized in Scheme 5.9.

The absence of a major, nonproductive, competing pathway and the degeneracy of the reaction of the initial radical with its dithiocarbonylthio precursor allows even comparatively slow radical processes to take place. In particular, *intermolecular* additions to *nonactivated* alkenes become feasible, thus solving a long-standing problem in organic synthesis [13]. By far the greatest number of reactions involves xanthates, on account of their cheapness, ready availability and very respectable reactivity. Indeed, for small molecule synthesis there is very rarely a need to replace xanthates by other thiocarbonylthio derivatives, in contrast to the controlled radical polymerization, where the nature of the monomer may dictate the type of group

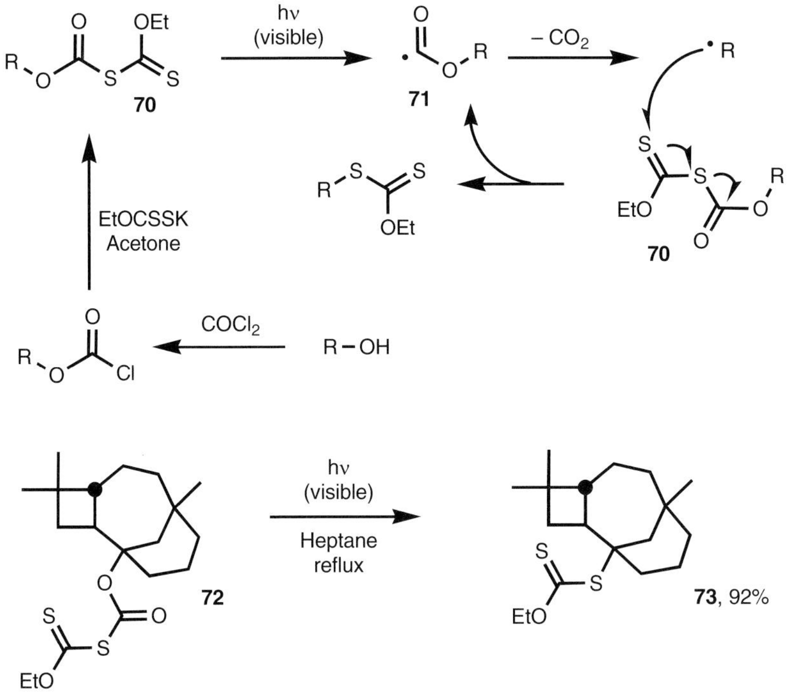


**Scheme 5.20** Decarboxylative rearrangement of S-alkoxycarbonyl xanthates.

that must be employed. In the case of small molecule chemistry, the requirement is to stop after one addition to the olefin, and this is achieved by choosing the reacting partners in such a way that the adduct radical **36** is less stable than the initial radical R•, so that in the fragmentation of intermediate **37**, the latter is favored. In fact, one of the key considerations is the behavior of intermediate **37**, how fast it is formed, how fast and in what preferred direction does it break up, and does it interact with other radicals in the medium.

Perhaps some of these aspects can be better appreciated by examining a simple case where the process actually fails. Let us consider the possibility of preparing compound **76** by generating a methyl radical from xanthate **74** and adding it to an alkene **75**, as depicted in Scheme 5.21. If the methyl radical can be generated, say by a peroxide-mediated initiation, then its addition to the alkene should proceed to give adduct radical **77**. In turn, the latter will add swiftly but reversibly to the starting xanthate **74** to give intermediate **78**. The last propagation step, however, fails because the *methyl* radical is normally much less stable than a *secondary* radical such as **77**, whatever the nature of substituent G. Fragmentation forward to produce the propagating methyl radical is strongly disfavored in comparison with the reverse process leading back to adduct radical **77**. The desired, addition product **76** cannot therefore be reached efficiently by this route.

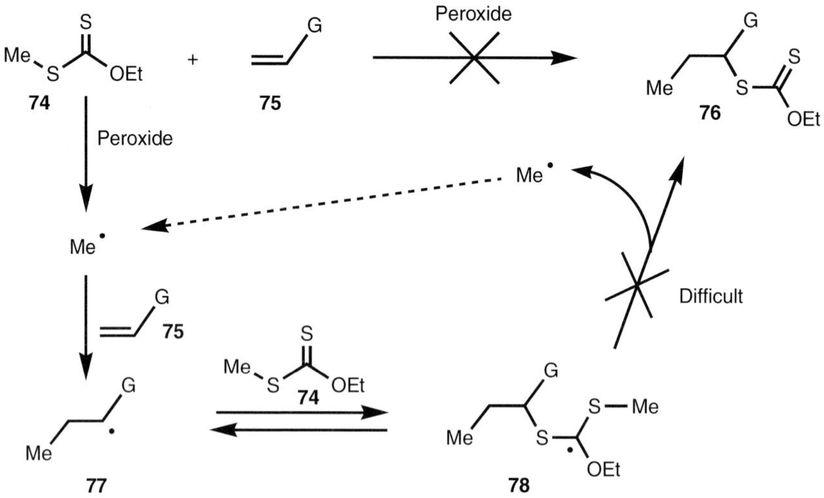


**Scheme 5.21** Chain rupture in the case of methyl radicals.

In order to circumvent this difficulty, temporary substituents can be placed on the methyl group to increase the stability of the ensuing radical, and thus encourage fragmentation of the intermediate in the correct direction. This is vindicated by the behavior of xanthate **79** in Scheme 5.22, which undergoes a smooth addition to olefin **80** to give the expected adduct **81** in good yield [27]. In contrast to the situation in the preceding scheme, the initial radical **82** is now much more stable than adduct radical **83**, and so collapse of intermediate **84** will occur preferentially in the desired direction of product **81**.

The fact that intermediate radical **84** will fragment more easily to give radical **82** than to give adduct radical **83** means that the product xanthate **81** will not react further to any significant extent, as long as the starting xanthate **79** is still present. However, if the product xanthate is separately exposed to the action of the peroxide, then there is no alternative but to generate radical **83**, even if it is not particularly stabilized. In the present case, this radical has the possibility of adding to the aromatic ring to give the stabilized cyclohexadienyl radical **85**, as shown in Scheme 5.23 [27]. This cyclization is not especially rapid, and is very probably reversible, but there are no other seriously competing pathways. Because of its high stability, cyclized radical **85** is incapable of propagating the chain, for the same reason invoked to explain the case of the methyl radical in Scheme 5.21 (i.e. the product radical is more stable than the initial radical). However, radical **85** is easily oxidized to the equally stabilized corresponding cation **86** by electron transfer to the peroxide (this corresponds to step (e) in Scheme 5.8). Aromatization by loss of a proton quickly follows to give indoline **87** in good yield. The peroxide is therefore acting as an initiator and as an oxidant; it must therefore be used in stoichiometric amounts.

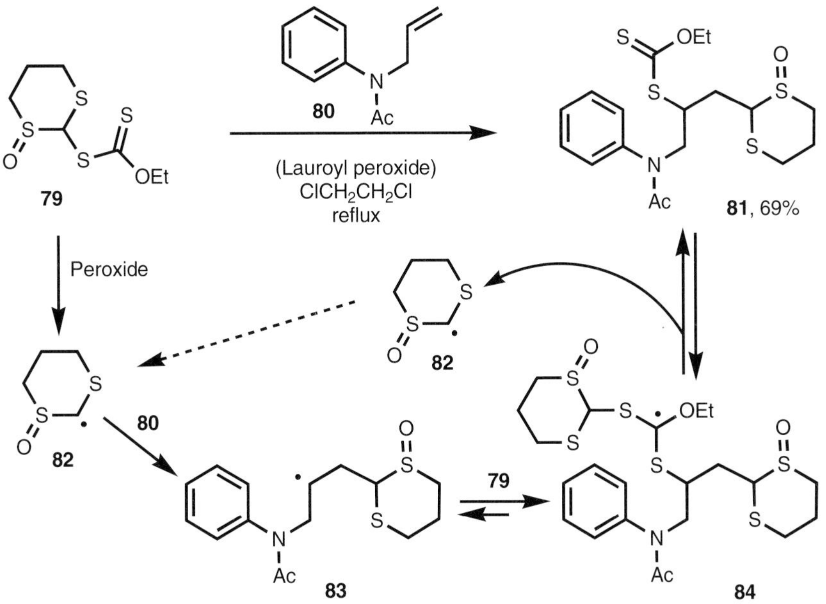


**Scheme 5.22** Dithiane monoxide as a one-carbon radical synthetic equivalent.

Treatment of indoline **87** with Raney nickel removes the sulfur atoms and gives indoline **88**, which formally derives by addition of a methyl radical to olefin **80**, followed by cyclization to the aromatic ring. Thus, xanthate **79** is a synthetic surrogate for unsuitable xanthate **74**. Moreover, dithianes possess an extremely rich chemistry and compound **87** can be converted into a plethora of other derivatives. One example is its efficient transformation into aldehyde **89**. In fact, radical **82** is the synthetic equivalent to a number of high-energy or inaccessible radicals: methyl, formyl (•CH=O), carboxyl (•COOH) and so forth.

The xanthate group has served to mediate two difficult radical processes: an intermolecular radical addition to a nonactivated alkene and a ring closure to an aromatic ring. Both of these transformations are not easy to perform with other methods. Indeed, this xanthate transfer process is of quite broad scope and represents possibly the most general solution to the long-standing problem in organic synthesis of how to create a carbon–carbon bond on nonactivated olefins in an *intermolecular* fashion. The synthetic potential of the xanthate transfer can be appreciated by glancing at the examples of Scheme 5.24, concerning only additions to allyltrimethylsilane and chosen from several hundred different additions to a variety of olefins that have been performed over the years [13].

Experimentally, the procedure is particularly simple, involving merely heating a concentrated solution (1–2 M) of the xanthate and olefin (2–3 eq.) in refluxing 1,2-dichloroethane (DCE; but other solvents can also be used) under an inert atmosphere and adding solid lauroyl peroxide (5–15 mol%) portionwise from the top of the condenser. The reaction is monitored for completion by thin-layer

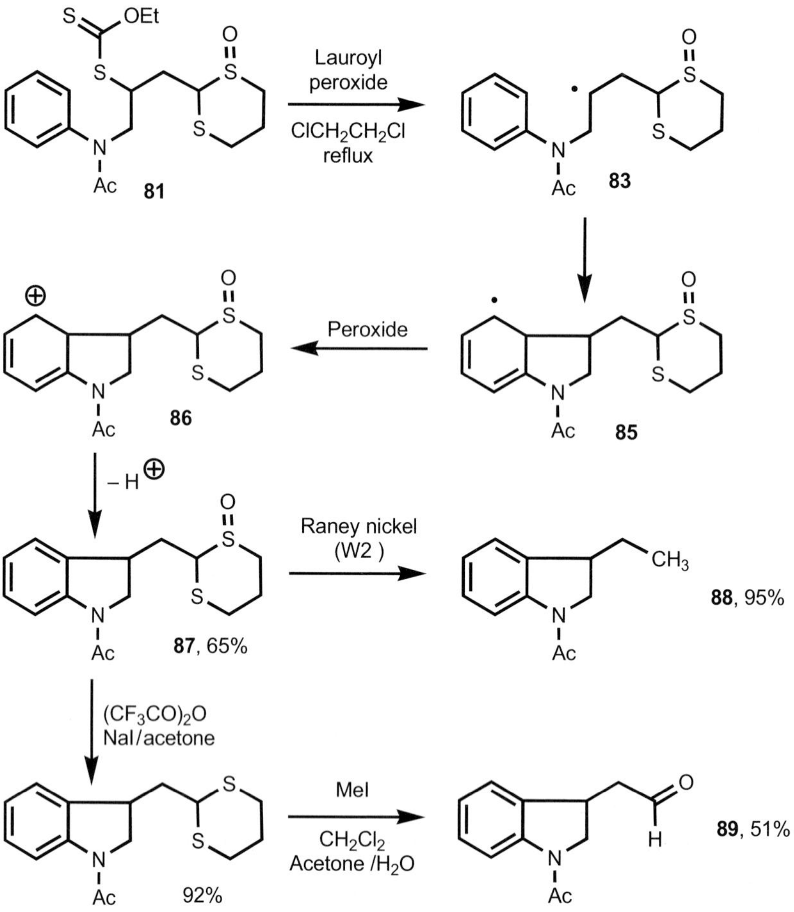


**Scheme 5.23** Synthesis of an indoline.

chromatography, the solvent evaporated and the residue purified by chromatography or recrystallization. It is perhaps worth pointing out the ease with which fluorinated groups can be introduced. Thus, xanthate **66**, the preparation of which is described in Scheme 5.18, adds smoothly to allyl trimethylsilane to give the corresponding adduct **90** [25d]. In the same manner, xanthates **91**, **92** (*neo*Pn = neopentyl) and **93** furnish fluorinated adducts **94**, **95** and **96** respectively in generally high yield [27–29]. The efficient synthesis of bisphosphonate **98** from precursor **97** is also worthy of mention [20b].

Using the same simple experimental procedure, it is possible to perform the radical addition on much more complex olefins, as demonstrated by the three transformations in Scheme 5.25 involving trifluoromethyl ketone xanthate (**92**), a very convenient source of trifluoroacetonyl radicals [28]. Trifluoromethyl ketones attached to structures as complex as pleuromutilin, a terpene antibiotic, can be

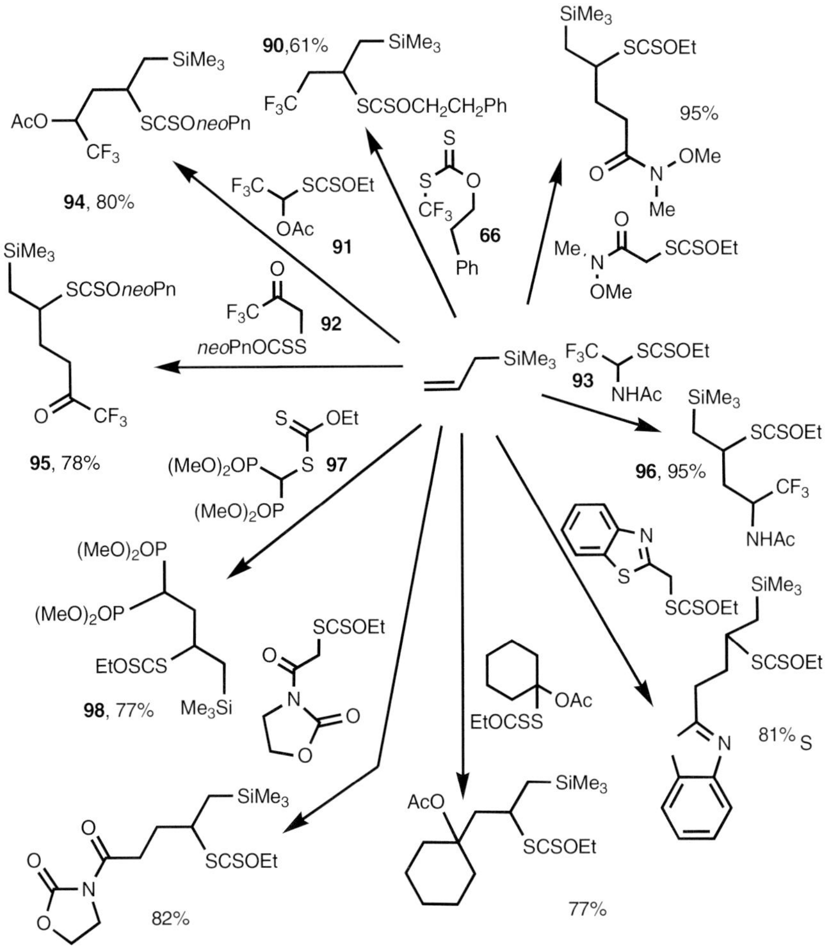


**Scheme 5.24** Radical additions to allyl trimethylsilane.

readily assembled without the need for protection of the other functional groups present [30]. Such derivatives would be extremely tedious to obtain by more conventional radical or nonradical routes.

It is not the purpose of this short overview to comprehensively discuss the synthetic potential and applications of the xanthate transfer process. The interested reader is directed to recent reviews on the subject [13]. However, a few further examples will help delineate some of the more interesting aspects. The degenerative xanthate addition can be perceived as a means to bring together functional groups that do not interact with each other under the mild, neutral conditions of the radical addition, but which can be made to react by a change in pH or by the addition of an appropriate external reagent. The two examples in Scheme 5.26 illustrate the possibility of combining the radical process with the powerful

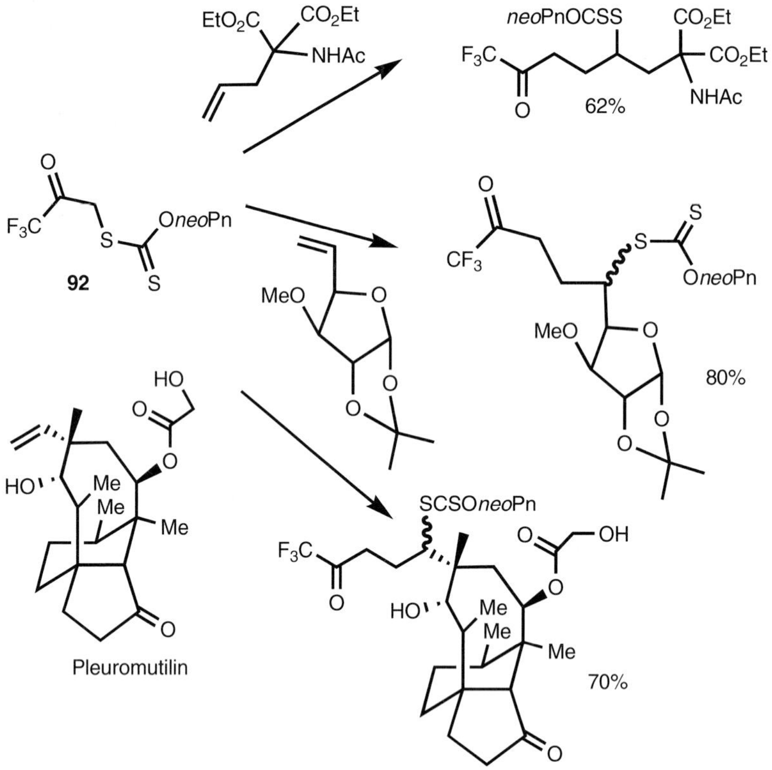


**Scheme 5.25** Trifluoromethyl ketones from complex alkenes.

Horner–Wadsworth–Emmons reaction to construct various carbocycles [31]. Thus, depending on the distance between the phosphonate and the olefin, it is possible to access cyclohexenes or cycloheptenes and possibly other ring-sized cycloalkenes.

By far the majority of the additions have been performed using xanthates, but, as stated above, similar reactions can be accomplished more or less efficiently with other thiocarbonylthio derivatives. This possibility is illustrated by the transformations displayed in Scheme 5.27. Dithiocarbamates with ordinary aliphatic substituents on the nitrogen are normally not sufficiently reactive to undergo smooth intermolecular addition to nonactivated alkenes. However, diminishing the electron density on the nitrogen atom by placing an electron-withdrawing group dramatically increases the reactivity and allows a highly efficient addition. This is demonstrated by the addition of dithiocarbamate **99** to vinyl acetate to give the expected adduct **100** in high yield (G. Bouhadir, X. Franck, S. Z. Zard, unpublished observations). For the generation and intramolecular capture of carbamoyl radicals, simple dithiocarbamates proved, unexpectedly, better than xanthates. The remarkably efficacious formation of β-lactam **103** from acyl dithiocarbamate **101** via acyl radical **102** has recently been reported by Grainger and Innocenti [32]. Lactams of various sizes were prepared by the same approach. Finally, an instance of

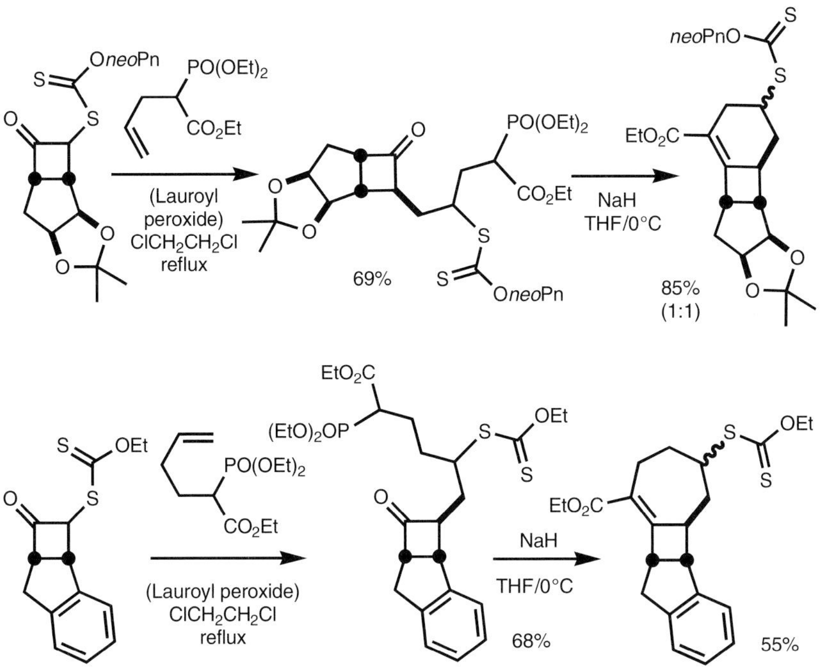


**Scheme 5.26** Syntheses of cycloalkenes.

addition of a dithioester, **103**, to a coumarin-substituted styrene is also provided in Scheme 5.27. The resulting adduct **104** was used in the synthesis of light-harvesting macromolecules [33].

The presence of the thiocarbonylthio group in the radical addition product opens access to the tremendously varied chemistry of sulfur. The thiocarbonylthio group is easily converted into a thiol, a sulfide, a sulfoxide, a sulfone, a sulfonyl chloride, a sulfonamide, a sulfonium salt, a sulfur ylide and so forth [3]. It is also possible to exploit the rich radical chemistry of the thiocarbonylthio group to expand considerably the range of useful modifications. The fact that it is possible to go back from the thiocarbonylthio product to the precursor radical (for example from product xanthate **38** to adduct radical **36** in Scheme 5.8) allows the implementation of a second radical process leading to a functional group interchange or to the formation of a new carbon–carbon bond.

A xanthate, for example, can be reductively removed with various reagents, including tributylstannane. The nearly quantitative dexanthylation of compound **23** in Scheme 5.5 is one such instance. Incidentally, xanthate **23**, itself, was made by way of the radical addition discussed at length above [11]. Other reducing systems are salts of hypophosphorus acid [34, 35] or a combination of lauroyl peroxide and 2-propanol [36]. In the latter case, 2-propanol acts as both the solvent and the source of hydrogen atoms. A xanthate can also be exchanged for a bromine atom [37], or, by using the leaving group ability of sulfones, replaced with an azide [38], an oxime

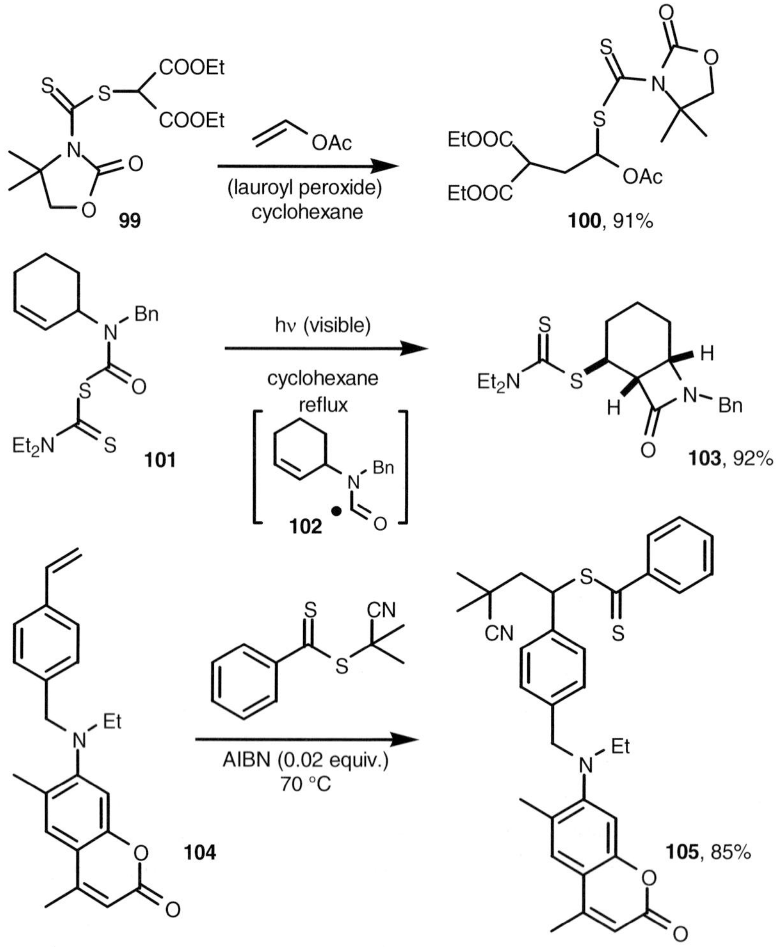


**Scheme 5.27** Radical additions of dithiocarbamates and dithioesters.

derivative [39], an allyl [40] or a vinyl group [41]. These various transformations are summarized in Scheme 5.28.

In Scheme 5.23, radical **83** was regenerated from the first addition product **81** and made to cyclize onto the aromatic ring to give indoline **89**. Indeed, this approach can be generalized to the synthesis of numerous aromatic and heteroaromatic structures fused to five-, six- and even seven-membered rings and containing a wide variety of substituents and functional groups [13]. One enormous advantage of the radical cyclization is that it takes place with relatively little regard as far as the nature of the substituents on the aromatic ring is concerned. It is thus more broadly applicable, in its intramolecular version, than traditional ionic processes such as the Friedel–Crafts reaction.

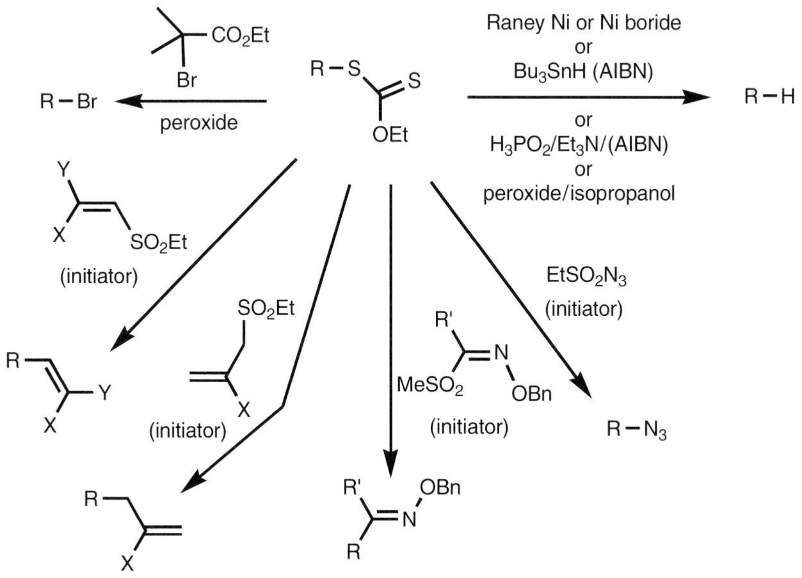


**Scheme 5.28** Radical-based transformations of the xanthate group.

It is also possible for the product xanthate to participate in another 'normal' addition to a second olefin, as long as the rule of thumb discussed above in connection with Schemes 5.8 and 5.21 concerning the requirements in the relative stabilities of the initial and adduct radicals is respected. Two examples are laid out in Scheme 5.29. In the first, xanthate **93** is added to vinylidene carbonate and the product, **106**, thus obtained is reacted with allyl trimethylsilane to give the highly functionalized compound **107** in good overall yield [29]. In the second transformation, xanthate **50** adds to vinyl pivalate to furnish **108**, which is then reacted with allyl cyanide. The expected adduct **109** is not isolated but exposed to the action of a base, 1,8-diazabicyclo [5.4.0] undec-7-ene DBU, which triggers two successive β-eliminations leading to diene **110** [42].

## 5.7
## Applications to Controlled Radical Polymerizations

### 5.7.1
### Synthesis of Block Polymers

The sequential addition to two different alkenes allows the swift assembly of complex, densely functionalized structures. This is feasible because of the difference in the stability of the intermediate radicals. If the same olefin is employed for the second addition (for example reacting xanthate **106** with another molecule of vinylidene carbonate instead of allyl trimethylsilane), then it will be difficult to stop at one

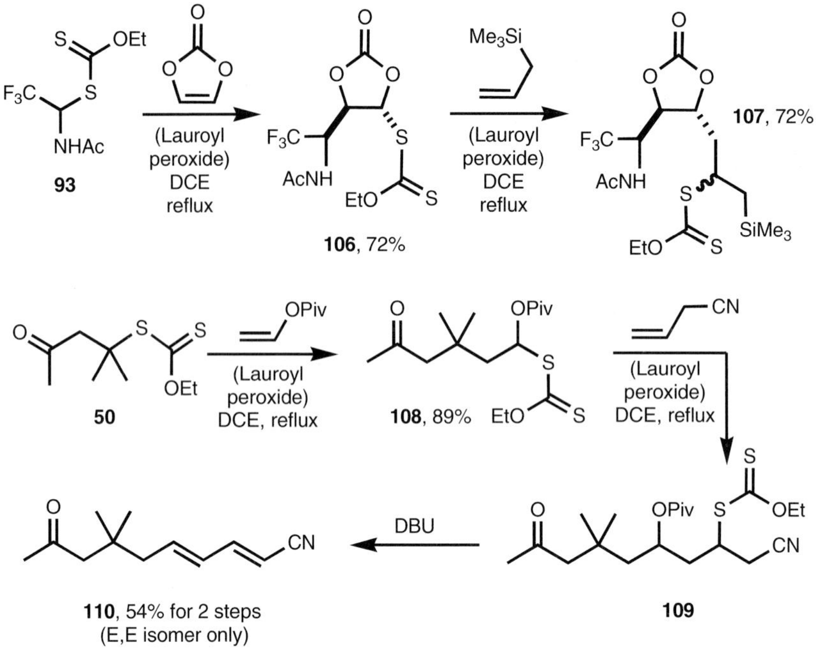


**Scheme 5.29** Examples of sequential additions.

addition because the ensuing adduct radical will be of essentially identical stability as the starting radical. The process now cannot be easily controlled in the case of ordinary olefins and mixtures of oligomers are invariably obtained. However, if the olefinic trap is a readily polymerizable monomer, then a very interesting situation arises. As shown in Scheme 5.30, the first addition of xanthate **40a** to give **111a** is followed by successive additions leading ultimately to polymer **112a** that is now capped by the xanthate group. This 'macro-xanthate' can in turn be used as the starting point for another polymerization involving a different monomer and resulting in the formation of diblock polymer **113a**. The procedure can of course be repeated to provide a triblock and so on and so forth. In contrast to classical radical polymerizations, the chains in this new approach can keep growing until the monomer is consumed, resulting in considerable narrowing of the molecular weight distribution. In favorable cases, polydispersity indexes close to 1 can be observed.

A new radical polymerization technique is now at hand, which can deliver block polymers and which can benefit from all the advantages of proceeding via radical intermediates: neutral conditions, compatibility with numerous functional and especially polar groups; tolerance for the presence of water, no need for a strict purification of the monomers as for anionic polymerization, low cost of initiators and controlling agent and so forth [43]. A fruitful collaboration with Rhodia (initially part of Rhône-Polenc Rorer) over several years led to the MADIX process (*ma*cromolecular *d*esign by *i*nterchange of *x*anthates) [44]. Concomitantly, polymer

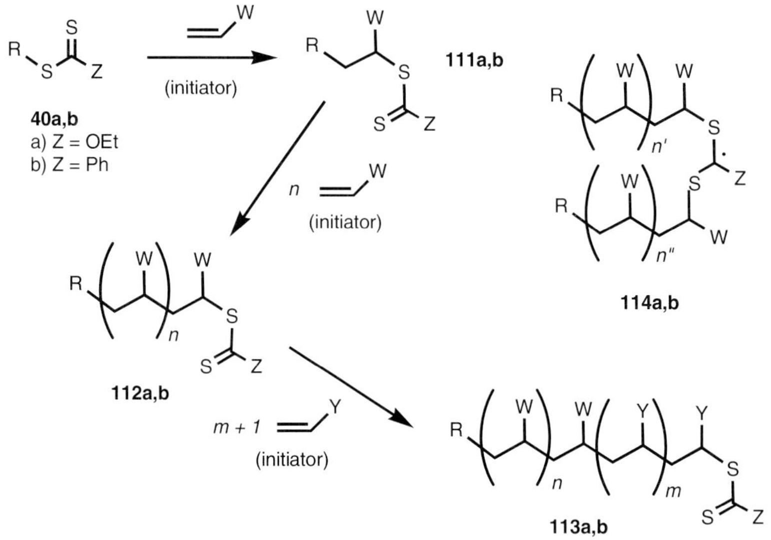


**Scheme 5.30** MADIX/RAFT synthesis of block polymers.

chemists at CSIRO developed an analogous technique which they called RAFT [45]. Whereas the study with Rhodia centered chiefly on xanthates (**40a**), that of CSIRO employed mostly dithioesters (**40b**), even though the respective patents covered more broadly the various possibilities for group Z. Both are based on the same addition–fragmentation on the thiocarbonylthio system, with the Z group playing only an indirect, albeit important, role. The nature of the Z group has a strong influence on the rates of the addition and fragmentation steps and on the stability of radicals such as **114**, which correspond to the intermediate in the exchange sequence of the thiocarbonylthio group between the growing chains.

In order to control the molecular weight distribution, this exchange process has to be faster than addition to a monomer molecule. This influence can be seen in the polymerization of methyl methacrylate, which can be controlled with dithioesters but not (so far) with xanthates. In contrast, the polymerization of vinyl acetate is more easily controlled with xanthates than with dithioesters. Xanthates have the additional advantage of being very cheap and the polymers obtained are essentially colorless. Xanthates are also more amenable to emulsion polymerization, a feature that simplifies large-scale operations. Indeed, Rhodia has industrialized the MADIX process and now produces block polymers on the tens of metric tonnes scale.

The RAFT/MADIX polymerization has attracted a great deal of interest within the polymer community, with articles on the topic now numbering in the hundreds [43]. Polymers of various architectures and compositions have been made, some with evocative names: stars, brushes, combs and so on. The properties of the material can be further modified and tuned by exploiting the presence of both the R and the thiocarbonylthio groups at the extremities of the block polymer. The sulfur group can be reductively removed or partially or totally oxidized to forestall any possible

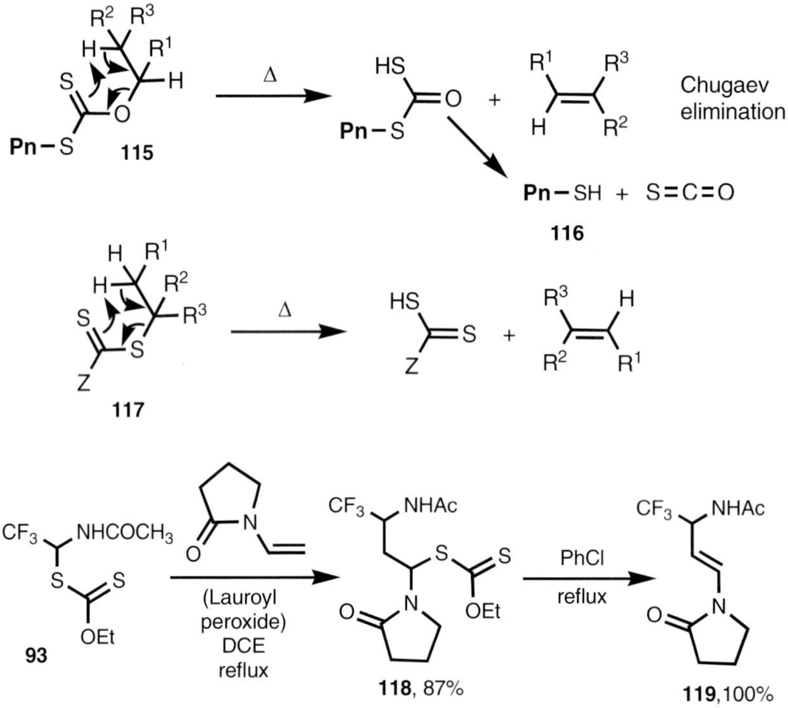

Scheme 5.31 Thermal elimination of the thiocarbonylthio function.

liberation of unpleasant volatile sulfur compounds from the finished material [46]. In this respect, it is possible to exploit the fact that xanthates can undergo upon heating the Chugaev elimination described briefly in Scheme 5.1 at the beginning of this chapter. Thus, by using a xanthate derived from a suitable secondary alcohol to mediate the polymerization, it is possible to eliminate the xanthate group from the resulting polymer **115** (**Pn** = polymer chain; Scheme 5.31) by merely heating at around 150 °C [47]. The process leads ultimately to thiol **116**, which has no realistic possibility to undergo loss of the remaining sulfur. The analogous elimination *from the sulfur side* occurs much less readily. It does nevertheless take place with tertiary derivatives (**117**, $R^2$, $R^3 \neq$ H) if the temperature exceeds 150 °C (M. Bingham, S. Z. Zard, unpublished observations), and at even lower temperatures if one of the substituents is an aryl (**117**, $R^2$ or $R^3$ = Ar) or a similar stabilizing group [48]. Fragmentation of a secondary derivative can occur when a heteroatom such as nitrogen is present *geminal* to the sulfur, causing a weakening of the C—S bond by an anomeric-type effect. An example of the latter case is provided by the quantitative conversion of **118** into enamide **119** upon refluxing in chlorobenzene. This Chugaev-like elimination can be a complicating factor, and the results of polymerizations performed at high temperature with cumyl dithiobenzoate, for example, must be interpreted with caution [49].

## 5.7.2
## Some Mechanistic Considerations

The emergence of RAFT/MADIX as an important polymerization technique has elicited several detailed mechanistic studies. A number of theoretical calculations and kinetic studies have appeared, attempting to paint a coherent picture of intimate workings of the polymerization process. Despite these efforts, there is still disagreement concerning the role of the so-called intermediate macro-RAFT radicals (such as **114**) in the observed slowing down of the polymerization when using dithiobenzoates and related structures (Z = aryl) and its possible involvement in termination reactions. Some groups have proposed that such intermediates possess a relatively long lifetime, sometimes of the order of seconds, and it is the fragmentation step that can become sluggish [50]. Others have defended the idea that these intermediates engage in termination processes (radical recombinations) causing a break in the propagating chain [51]. The former vision is more coherent with the various experimental observations that have been made in the case of small molecule chemistry, in particular with a key series of experiments summarized in Scheme 5.32.

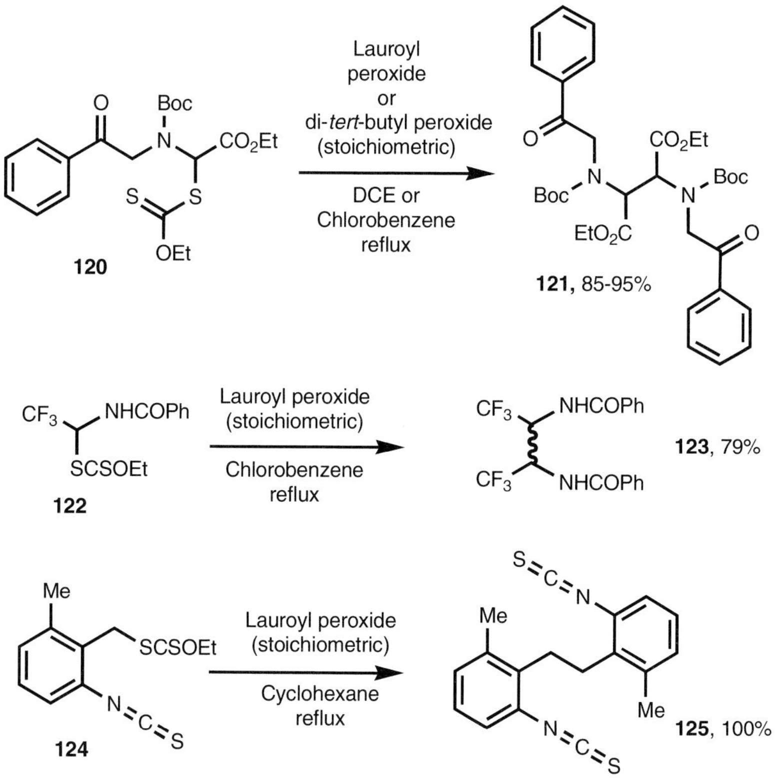


**Scheme 5.32** Efficient formation of homodimers.

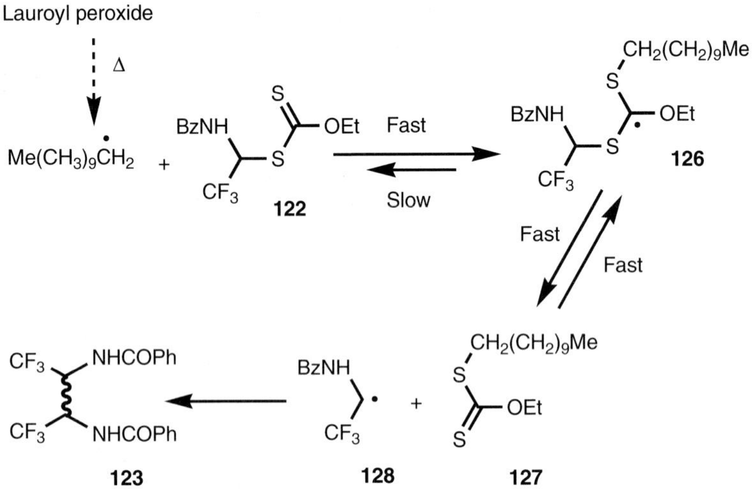


**Scheme 5.33** Selective control of the relative concentration of radicals.

When, in an attempt to bring about ring closure onto the benzene ring, xanthate **120** was treated with a stoichiometric amount of lauroyl peroxide in refluxing DCE or chlorobenzene, or with di-*tert*-butyl peroxide in refluxing chlorobenzene, no cyclization took place. Instead, the corresponding dimer **121** was formed in high yield (M. Lampilas, B. Quiclet-Sire, S. Z. Zard, unpublished observations). Later, a similar clean conversion of xanthate **122** into dimer **123** was observed [29]. Finally, Vidal and coworkers reported that benzylic xanthate (**124**) produced a quantitative yield of dibenzyl bisisothiocyanate **125** upon exposure to stoichiometric quantities of lauroyl peroxide in refluxing cyclohexane [52]. The formation of dimers, when working with radicals, is not abnormal; what is astonishing is the highly selective and high yielding formation of *a homodimer* in a medium supposedly containing a number of different radical species. It must be emphasized that these experiments were performed in a fairly concentrated medium under a high flux of initiating radicals from the peroxide (the half-life of lauroyl peroxide in boiling chlorobenzene is only a few minutes), yet no significant cross-coupling of radicals is observed.

This, initially puzzling, observation can be understood by looking at the mechanistic manifold displayed in Scheme 5.33 and detailing the case of xanthate **122**. The thermal decomposition of lauroyl peroxide leads to primary undecyl radicals, which are rapidly captured by the xanthate to give intermediate **126**. This intermediate will preferentially fragment to the more stabilized radical **128** as well as *S*-undecyl xanthate (**127**). At any given time, the concentration of radicals **128** is therefore much higher than that of the undecyl radicals derived from lauroyl peroxide and, in the absence of a suitable trap, dimer **123** is selectively formed. Thus, the xanthate regulates the relative concentration of the various radical species in the medium by scavenging the reactive and less stabilized primary undecyl radicals and releasing the more stabilized radicals **128**.

*The high yield observed for homodimers also means that radical adduct 126 does not participate significantly in any radical–radical interactions.* This is a key observation and several factors or combination of factors can be invoked to rationalize this behavior. Intermediate **126** is a tertiary radical stabilized by three heteroatoms that cannot undergo disproportionation reactions. It is hindered and appears to be unable to couple with other such tertiary radicals (or does so reversibly). In fact, it may be sufficiently hindered to react (relatively) sluggishly even with 'ordinary' radicals such as **128**. The formation of homodimer **123** is hence not complicated by possible cross-coupling between radicals **126** and **128**. The ability of the even more hindered 'macro'-version **114a** (or **114b**) to capture macroradicals during the polymerization must be even much less pronounced and very unlikely.

One could argue that radicals such as **126** or **114a,b** fragment so rapidly that their concentration becomes too low to allow them to participate to any significant extent in radical–radical interactions. This may be true for cases where the radical generated in the fragmentation is highly stabilized, but even for benzyl radicals, ab initio calculations have shown that the rate constant for the addition to methyl dithioacetate [Me—C(=S)SMe] is about four times greater than for the fragmentation step ($2.76 \times 10^6\ M^{-1} \cdot s^{-1}$ vs $7.15 \times 10^5\ s^{-1}$ at 60 °C) [53]. This means that the concentration of the adduct radical at 60 °C is expected to be greater than that of benzyl radicals. The *quantitative* formation of dibenzyl derivative **125** would be very difficult to explain if the corresponding adduct (of type **126**) had a normal propensity to cross-couple. It is also worth pointing out that Scaiano and Ingold could easily detect by ESR the tertiary radical formed by addition of a *tert*-butyl radical to di-*tert*-butyl thioketone, despite its lesser stability (only one geminal heteroatom stabilising the radical) as compared with, say, **126** and its increased tendency to fragment because of its incredibly congested structure [54]. It is also interesting to note that, very recently, thioketones were used as radical sinks to control polymerizations [55]. These various observations do not support an important role for macroradicals **114a,b** as terminating species. It seems more likely that retardation in the case of polymerizations mediated by dithioesters **40b** is due to the increased stability of the corresponding intermediate **114b** because of conjugation with the aromatic ring and the consequently slower fragmentation step.

### 5.7.3
### Further Reflections on the Mechanism of the Barton–McCombie Deoxygenation

These considerations, incidentally, also provide a simple rationalization to the observations of Beckwith and Barker, which, indirectly, elicited the discovery of the degenerative addition–fragmentation process of thiocarbonylthio derivatives. The persistency of intermediate **130**, produced under Beckwith and Barker's conditions by addition of a trimethylstannyl radical to xanthate **129**, gives it a long enough lifetime to pursue an ionic pathway. Thus, it can collapse into alkoxythiocarbonyl radical **17**, observed by ESR spectroscopy by Beckwith and Barker, and methyl trimethyltin sulfide (**131**), as pictured in Scheme 5.34.

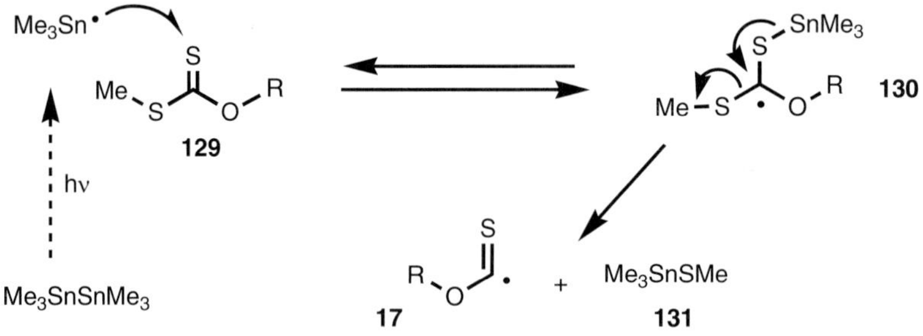

**Scheme 5.34** Formation of an alkoxythiocarbonyl radical.

Furthermore, it is also now possible to propose a hybrid mechanism for the Barton–McCombie deoxygenation that reconciles the observations of both the Barton's group and Beckwith and Barker. As shown in Scheme 5.35, the addition of tributyltin radicals occurs *reversibly* on the thiocarbonyl group of the xanthate to give a persistent radical intermediate **14**, which can evolve through two pathways. It either undergoes a *radical* fragmentation (path (a)) or collapses by an *ionic* mechanism (path (b)). The former corresponds to the original Barton–McCombie mechanism; the latter leads to alkoxythiocarbonyl radical **17**, an intermediate proposed by Beckwith and Barker (and observed under specific conditions). The propensity for alkoxythiocarbonyl radicals **17** to extrude carbon oxysulfide was postulated by Beckwith and Barker, without the support of direct experimental evidence. No method existed at the time for unambiguously generating these species in order to study their behavior. This problem was later solved by finding that xanthic

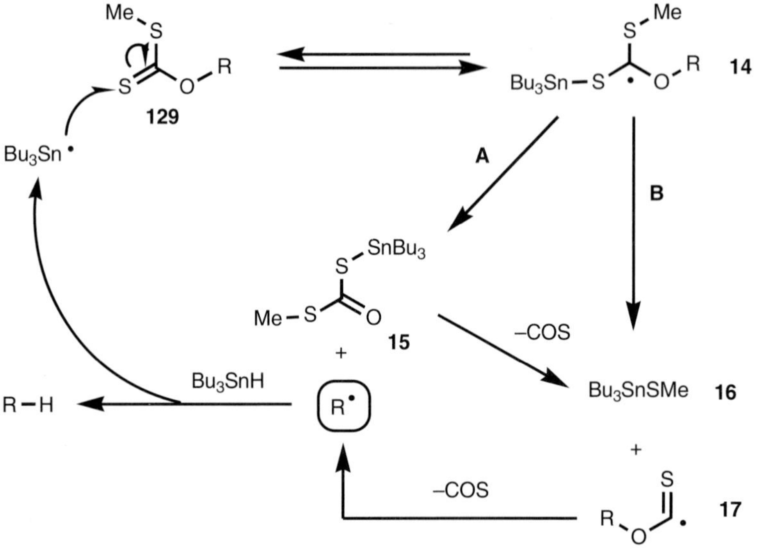

**Scheme 5.35** A hybrid mechanism for the Barton–McCombie deoxygenation.

anhydrides [ROC(=S)—S—C(=S)OR], because of their symmetrical structure, provided alkoxythiocarbonyl radical **17** cleanly and unambiguously upon irradiation with visible light [56]. It could indeed be shown that alkoxythiocarbonyl radicals readily expel carbon oxysulfide and that this fragmentation was at least 1 order of magnitude faster than the loss of carbon dioxide from the corresponding alkoxycarbonyl species (ROC•=O). These, somewhat counterintuitive, experimental observations have recently been buttressed by theoretical calculations [57].

Pathways (a) and (b) are not mutually exclusive. Which one prevails or whether both operate simultaneously in a given situation will depend chiefly on the nature of the R group in the starting xanthate **129** and the experimental conditions, especially the reaction temperature. These factors will strongly influence the rates of the homolytic (path (a)) or ionic (path (b)) rupture of intermediate **14**, both of which are unimolecular process with a high, positive entropy term. The hybrid mechanism displayed in Scheme 5.35 represents a rare instance where a purely ionic process (in contrast to mechanisms involving radical ions) is in competition with a homolytic process *within* the same mechanistic manifold. Elementary steps involving radicals are normally so much faster than the common elementary ionic reactions that the respective reactive species are often made to evolve in two *separate* reaction manifolds with a crossover step between the two (for example step (e) in Scheme 5.8).

## 5.8 Concluding Remarks

In summary, the unexpected behavior of *S*-isopropyl xanthate (**19**) has had far-reaching consequences, completely unimagined at the time. It has resulted in the uncovering of a powerful, versatile radical process, and provided a solution to the long-standing problem of creating C—C bonds in an intermolecular fashion, starting with ordinary, nonactivated alkenes. It has also led to the emergence of an efficient controlled radical polymerization technology that has started to find its way into the market. It has perhaps furnished, ultimately, a better insight into the Barton–McCombie deoxygenation itself. Finally, and as a result of this work, a number of new ionic reactions of xanthates were also found, some proceeding by way of totally novel intermediates [13f].

## References


1. (a)W. C. Zeise, *Rec. Mem. Acad. R. Sci. Copenhagen* **1815**, *1*, 1; (b) W. C. Zeise, *J. Chem. Phys.* **1822**, *35*, 173.
2. S. Ramachandra Rao, *Xanthates and Related Compounds*, Marcel Dekker Inc., New York, **1971**.
3. F. Duus, In *Comprehensive Organic Chemistry, Vol. 3*, D. H. R Barton, W. D. Ollis, Eds., Pergamon Press, Oxford, **1979**, pp. 373–487.
4. K. Harano, M. Eto, K. Ono, K. Misaka, T. Hisano, *J. Chem. Soc. Perkin Trans. 1* **1993**, 299.
5. L. Chugaev, *Chem. Ber.* **1899**, *32*, 3332; (b) H. R. Nace, *Org. React.* **1962**, *12*, 57.

**6.** (a) D. H. R Barton, S. W. McCombie, *J. Chem. Soc. Perkin Trans. 1* **1975**, 1574; (b) D. H. R Barton, *Half a Century of Free Radical Chemistry*, Cambridge University Press, Cambridge, UK, **1993**.

**7.** T. Hayashi, T. Iwaoka, N. Takeda, E. Ohki, *Chem. Pharm. Bull.* **1978**, *26*, 1786.

**8.** (a) W. Hartwig, *Tetrahedron* **1983**, *39*, 2609; (b) D. Crich, L. Quintero, *Chem. Rev.* **1989**, *89*, 1413.

**9.** A. L. J Beckwith, P. J. Barker, *J. Chem. Soc., Chem. Commun.* **1984**, 683.

**10.** (a) D. H. R Barton, D. Crich, A. Löbberding, S. Z. Zard, *J. Chem. Soc., Chem. Commun.* **1985**, 646; *Tetrahedron* **1986**, *42*, 2329; (b) M. D. Bachi, E. Bosch, *J. Chem. Soc., Perkin Trans. 1* **1988**, 1517; M. D. Bachi, E. Bosch, D. Denenmark, D. Girsh, *J. Org. Chem.* **1992**, *57*, 6803.

**11.** B. Quiclet-Sire, B. Sortais, S. Z. Zard, *Synlet* **2002**, 903.

**12.** B. Quiclet-Sire, S. Z. Zard, *Tetrahedron Lett.* **1998**, *39*, 9435.

**13.** (a) B. Quiclet-Sire, S. Z. Zard, *Chem. Eur. J.* **2006**, *12*, 6002; (b) B. Quiclet-Sire, S. Z. Zard, *Top. Curr. Chem.* **2006**, *264*, 201–236; (c) S. Z. Zard, In *Radicals in Organic Synthesis, Vol. 1*, P. Renaud, M. P. Sibi, Eds., Wiley-VCH, Weinheim, **2001**, pp. 90–108; (d) B. Quiclet-Sire, S. Z. Zard, *Phosphorus Sulfur Silicon* **1999**, *153–154*, 137; (e) B. Quiclet-Sire, S. Z. Zard, *J. Chin. Chem. Soc.* **1999**, *46*, 139; (f) S. Z. Zard, *Angew. Chem., Int. Ed. Engl.* **1997**, *36*, 672; *Angew. Chem.* **1997**, *109*, 724.

**14.** (a) H. Henry, DEA Report, Université Paris XI, Orsay, France, **1994**; (b) M. Destarac, X. Franck, D. Charmot, S. Z. Zard, *Macromol. Rapid Commun.* **2000**, *21*, 1035.

**15.** J. Chiefari, R. T. A Mayadunne, C. L. Moad, G. Moad, E. Rizzardo, A. Postma, M. A. Skidmore, S. H. Thang, *Macromolecules* **2003**, *36*, 2273.

**16.** H. C. Godt, R. E. Wann, *J. Org. Chem.* **1961**, *26*, 4047.

**17.** M. R. Wood, D. J. Duncalf, S. P. Rannard, S. Perrier, *Org. Lett.* **2006**, *8*, 553.

**18.** (a) N. Kreutzkamp, H. Peschel, *Pharmazie*, **1970**, *25*, 322; (b) G. Binot, B. Quiclet-Sire, T. Saleh, S. Z. Zard, *Synlett* **2003**, 382.

**19.** S. Fabre, X. Vila, S. Z. Zard, *Chem. Commun.* **2006**, 4964.

**20.** (a) V. Maslak, Z. Cekovic, R. N. Saicic, *Synlett* **1998**, 1435; (b) F. Gagosz, S. Z. Zard, *Synlett* **2003**, 387.

**21.** G. Bouhadir, N. Legrand, B. Quiclet-Sire, S. Z. Zard, *Tetrahedron Lett.* **1999**, *40*, 277.

**22.** S. H. Thang, Y. K. Chong, R. T. A Mayadunne, G. Moad, E. Rizzardo, *Tetrahedron Lett.* **1999**, *40*, 2435.

**23.** A. Alberti, M. Benaglia, M. Laus, K. Sparnacci, *J. Org. Chem.* **2002**, *67*, 7911.

**24.** F. Bertrand, PhD Thesis, Université Paris-Sud, Orsay, France, **2000**.

**25.** (a) D. H. R Barton, M. V. George, M. Tomoeda, *J. Chem. Soc.* **1962**, 1967; (b) P. Delduc, C. Tailhan, S. Z. Zard, *J. Chem. Soc., Chem. Commun.* **1988**, 308; (c) B. Quiclet-Sire, S. Z. Zard, *Tetrahedron Lett.* **1998**, *39*, 1073; (d) F. Bertrand, V. Pevere, B. Quiclet-Sire, S. Z. Zard, *Org. Lett.* **2001**, *3*, 1069.

**26.** (a) J. E. Forbes; S. Z. Zard, *J. Am. Chem. Soc.* **1990**, *112*, 2034; (b) J. E. Forbes, R. N. Saicic, S. Z. Zard, *Tetrahedron* **1999**, *55*, 3791.

**27.** L. Tournier, S. Z. Zard, *Tetrahedron Lett.* **2005**, *46*, 455.

**28.** M.-P. Denieul, B. Quiclet-Sire, S. Z. Zard, *Chem. Commun.* **1996**, 2511.

**29.** F. Gagosz, S. Z. Zard, *Org. Lett.* **2003**, *5*, 2655.

**30.** E. Bacqué, F. Pautrat, S. Z. Zard, *Chem. Commun.* **2002**, 2312.

**31.** G. Binot, S. Z. Zard, *Tetrahedron Lett.* **2005**, *46*, 7503.

**32.** R. S. Grainger, P. Innocenti, *Angew. Chem. Int. Ed.* **2004**, *43*, 3445.

**33.** M. Chen, K. P. Ghiggino, A. W. H Mau, E. Rizzardo, W. H. F Sasse, S. H. Thang, G. F. Wilson, *Macromolecules* **2004**, *37*, 5479.

**34.** D. H. R. Barton, D. O. Jang, J. Cs. Jaszberenyi, *Tetrahedron Lett.*, **1992**, *33*, 5709.

**35.** J. Boivin, R. Jrad, S. Jugé, V. T. Nguyen, *Org. Lett.* **2003**, *5*, 1645.

**36.** A. Liard, B. Quiclet-Sire, S. Z. Zard, *Tetrahedron Lett.* **1996**, *37*, 5877.

**37.** F. Barbier, F. Pautrat, B. Quiclet-Sire, S. Z. Zard, *Synlet* **2002**, 811.

**38.** L. Chabaud, Y. Landais, P. Renaud, *Org. Lett.* **2005**, *7*, 2587; L. Chabaud, Y.

Landais, P. Renaud, *Org. Lett.* **2002**, *4*, 4257; P. Panchaud, P. Renaud, *J. Org. Chem.* **2004**, *69*, 3205; P. Panchaud, C. Ollivier, P. Renaud, S. Zigmantas, *J. Org. Chem.* **2004**, *69*, 2755; C. Ollivier, P. Renaud, *J. Am. Chem. Soc.* **2001**, *123*, 4717; C. Ollivier, P. Renaud, *J. Am. Chem. Soc.* **2000**, *122*, 6496.

**39.** S. Kim, H.-J. Song, T.-L. Choi, J.-Y. Yoon, *Angew. Chem., Int. Ed. Engl.* **2001**, *40*, 2524; S. Kim, *Adv. Synth. Catal.* **2004**, *346*, 19.

**40.** B. Quiclet-Sire, S. Seguin, S. Z. Zard, *Angew. Chem., Int. Ed. Engl.* **1998**, *37*, 2864.

**41.** (a) F. Bertrand, B. Quiclet-Sire, S. Z. Zard, *Angew. Chem., Int. Ed. Engl.* **1999**, *38*, 1943; (b) F. Bertrand, F. Le Guyader, L. Liguori, G. Ouvry, B. Quiclet-Sire, S. Seguin, S. Z. Zard, *C. R. Acad. Sci. Paris* **2001**, *II4*, 547.

**42.** S. Bagal, L. Tournier, S. Z. Zard, *Synlett* **2006**, 1485.

**43.** (a) G. Moad, E. Rizzardo, S. H. Thang, *Aust. J. Chem.* **2005**, *58*, 379; (b) S. Perrier, T. Takolpuckdee, *J. Polym. Sci., Part A: Polym. Chem.* **2005**, *43*, 5347.

**44.** P. Corpart, D. Charmot, T. Biadatti, S. Z. Zard, D. Michelet, W.O 9858974, **1998**.

**45.** T. P. Le, G. Moad, E. Rizzardo, S. H. Thang, Int. Pat. 9801478, **1998**.

**46.** (a) M. Destarac, C. Kalai, L. Petit, A. Z. Wilczewska, G. Mignani, S. Z. Zard, *Abst. Pap. Am. Chem. Soc.* **2005**, *230*, 533-POLY; (b) S. Z. Zard, B. Sire, P. Jost, WO 2005040233 **2005**.

**47.** M. Destarac, C. Kalai, A. Z. Wilczewska, G. Mignani, S. Z. Zard, *Abst. Pap. Am. Chem. Soc.* **2005**, *230*, 535-POLY.

**48.** (a) J. Xu, J. He, D. Fan, W. Tang, Y. Yang, *Macromolecules* **2006**, *39*, 3573; Y. Liu, J. He, J. Xu, D. Fan, W. Tang, Y. Yang, *Macromolecules* **2005**, *38*, 10332; G. Moad, Y. K. Chong, A. Postma, E. Rizzardo, S. H. Thang, *Polymer* **2005**, *46*, 8458; A. Postma, T. P. Davis, G. Moad, M. S. O'Shea, *Macromolecules* **2005**, *38*, 5371.

**49.** T. Arita, M. Buback, P. Vana, *Macromolecules* **2005**, *38*, 7935.

**50.** C. Barner-Kowollik, J. F. Quinn, T. L. U Nguyen, J. P. A Heuts, T. P. Davis, *Macromolecules* **2001**, *34*, 7849; S. Perrier, C. Barner-Kowollik, J. F. Quinn, P. Vana, T. P. Davis, *Macromolecules* **2002**, *35*, 8300; P. Vana, T. P. Davis, C. Barner-Kowollik, *Macromol. Theory Simul.* **2002**, *11*, 823; C. Barner-Kowollik, J. F. Quinn, D. R. Morsley, T. P. Davis, *J. Polym. Sci., Part A: Polym. Chem.* **2001**, *39*, 1353; C. Barner-Kowollik, T. P. Davis, J. P. A Heuts, M. H. Stenzel, P. Vana, M. Whittaker, *J. Polym. Sci., Part A: Polym. Chem.* **2003**, *41*, 365; P. Vana, J. F. Quinn, T. P. Davis, C. Barner-Kowollik, *Aust. J. Chem.* **2002**, *55*, 425.

**51.** Y. Kwak, A. Goto, T. Fukuda, *Macromolecules* **2004**, *37*, 1219; Y. Kwak, A. Goto, Y. Tsujii, Y. Murata, K. Komatsu, T. Fukuda, *Macromolecules* **2002**, *35*, 3026; M. J. Monteiro, H. de Brouwer, *Macromolecules* **2001**, *34*, 349.

**52.** M. Alajarin, A. Vidal, M.-M. Ortin, *Org. Biomol. Chem.* **2003**, *3*, 4282.

**53.** (a) M. L. Coote, *J. Phys. Chem. A.* **2005**, *109*, 1230; (b) E. I. Izgorodina, M. L. Coote, *J. Phys. Chem. A* **2006**, *110*, 2486; M. L. Coote, D. J. Henry, *Macromolecules* **2005**, *38*, 1415; M. L. Coote, L. Radom, *Macromolecules* **2004**, *37*, 590.

**54.** J. C. Scaiano, K. U. Ingold, *J. Am. Chem. Soc.* **1976**, *98*, 4727.

**55.** A. A. Toy, H. Chaffey-Millar, T. P. Davis, M. H. Stenzel, E. I. Izgorodina, M. L. Coote, C. Barner-Kowollik, *Chem. Commun.* **2006**, 835; T. Junkers, M. H. Stenzel, T. P. Davis, C. Barner-Kowollik, *Macromol. Rapid Commun.* **2007**, *28*, 746.

**56.** J. E. Forbes, S. Z. Zard, *Tetrahedron Lett.* **1989**, *30*, 4367.

**57.** M. L. Coote, C. J. Easton, S. Z. Zard, *J. Org. Chem.* **2006**, *71*, 4996.

# 6
# RAFT Polymerization in Bulk Monomer or in (Organic) Solution

*Ezio Rizzardo, Graeme Moad, and San H. Thang*

## 6.1
## Introduction

Radical polymerization is one of the most widely used processes for the commercial production of high-molecular-weight polymers [1]. The main factors responsible for the preeminent position of radical polymerization are as follows (a) it can be used with a large variety of monomers; (b) it is tolerant of a wide range of functional groups and reaction conditions; (c) it is simple to implement and inexpensive in relation to competitive technologies. However, the conventional process has some notable limitations with respect to the degree of control over the molecular weight distribution, polymer composition and architecture.

The recent emergence of techniques for implementing living radical polymerization has provided a new set of tools that allow very precise control over the polymerization process while retaining much of the versatility of conventional radical polymerization [2–5]. New materials that have the potential of revolutionizing a large part of the polymer industry are beginning to appear. Possible applications range from novel surfactants, dispersants, coatings and adhesives, through to biomaterials, membranes, drug delivery media and materials for microelectronics.

The living radical polymerization techniques that have received greatest attention are nitroxide-mediated polymerization (NMP), atom transfer radical or metal-mediated polymerization (ATRP) and reversible addition–fragmentation chain transfer (RAFT). The NMP technique was devised at CSIRO in the early 1980s [6, 7] and in recent years has been exploited extensively for the synthesis of narrow-molecular-weight-distribution homopolymers and block copolymers of styrene and acrylates [8–10]. Recent developments have made NMP applicable to a wider, though still restricted, range of monomers [10]. ATRP is substantially more versatile [11, 12]; however, it requires unconventional initiating systems with poor compatibility with some polymerization media. Again, substantial advances have been made to redress this and other issues [5]. RAFT polymerization, also devised at CSIRO, is one of the

*Handbook of RAFT Polymerization.* Edited by Christopher Barner-Kowollik

ISBN: 978-3-527-31924-4

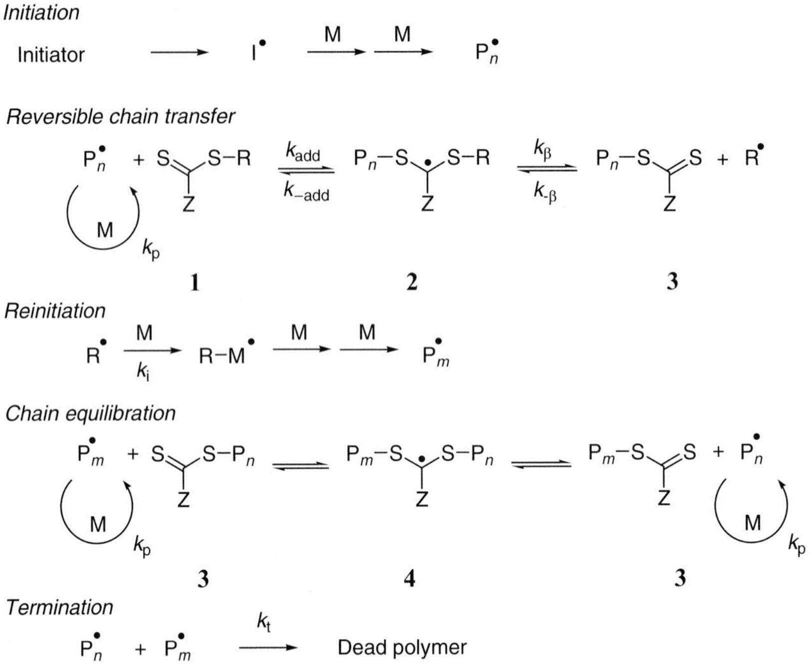


**Scheme 6.1** Mechanism of RAFT polymerization.

more recent entrants in this field and arguably the most convenient and versatile [13, 14].

Macromonomer RAFT agents were first reported in the open literature in 1986 [15] and their use in polymer synthesis to produce block copolymers and polymers with narrow molecular weight distribution was published in 1995 [16, 17]. The use of xanthates to give RAFT organic synthesis was reported in 1988 [18, 19] while the use of thiocarbonylthio RAFT agents in the context of conferring living characteristics on radical polymerization was first communicated in 1998 [20, 21]. Thiocarbonylthio RAFT agents with general structure **1** may act as transfer agents by a two-step addition–fragmentation mechanism as shown in Scheme 6.1.

RAFT polymerization with thiocarbonylthio compounds has been found compatible with the vast majority of monomers polymerizable by free radical processes. This includes (meth)acrylates, (meth)acrylamides, acrylonitrile, styrene and derivatives, butadiene, vinyl acetate and *N*-vinylpyrrolidone. It is tolerant of functionality in monomer, solvent and initiator (e.g. OH, $NR_2$, $CO_2H$, $SO_3H$, $CONR_2$). In 2000 we provided a review [22] of RAFT polymerization in which we surveyed RAFT polymerizations of various monomers. This chapter can be considered as an update on that review. We focus on RAFT polymerization in bulk and in (organic) solution and deal with aspects of the kinetics and mechanism of RAFT polymerization and its relationship with the selection of RAFT agent and reaction conditions for particular polymerizations and copolymerizations. The kinetics of RAFT polymerization are considered in greater detail in Chapter 3. RAFT polymerization in homogeneous aqueous solution will be mentioned but is given more comprehensive coverage in Chapter 7.

Recent reviews which relate specifically to RAFT polymerization include general reviews by Moad and coworkers [13, 14, 23], Mayadunne and Rizzardo [24], Chiefari and Rizzardo [25], Perrier and Takolpuckdee [26], Favier and Charreyre [27] and Barner-Kowollik, Davis, Stenzel and coworkers [28]. Many other reviews deal with specific applications of RAFT polymerization such as computational studies related to RAFT agents and RAFT polymerization [29, 30], the kinetics and mechanism of RAFT polymerization [31], the use of RAFT in organic synthesis [19, 32, 33], the control of molecular weight distributions produced by RAFT polymerization [34], the use of RAFT polymerization in aqueous media [35] and in heterogeneous media [36–39], the synthesis of end functional polymers by RAFT polymerization [40, 41], star polymer synthesis [42], the synthesis and properties of stimuli responsive block and other polymers [41, 43] and the preparation of honeycomb structures [44].

RAFT polymerization is also reviewed within works that deal more generically with radical polymerization. The process is comprehensively reviewed within the chapter 'Living radical polymerization' in *The Chemistry of Radical Polymerization* [1] and is given substantial coverage in many recent works that relate more generically to polymer synthesis, living polymerization or novel architectures [5, 45–55]. The literature is expanding very rapidly; an update review [14] covering the period mid-2005 to mid-2006 revealed more than 200 papers dealing directly with the use and application of RAFT polymerization.

The essential features of the ideal RAFT polymerization can be summarized as follows [13]:

- RAFT polymerization (Scheme 6.2) can be performed by simply adding a chosen quantity of an appropriate RAFT agent to a conventional free radical polymerization. Usually the same monomers, initiators, solvents and temperatures can be used.
- RAFT polymerization possesses the characteristics usually associated with living polymerization. Essentially chains begin growth at the commencement of polymerization and continue to grow until the monomer is consumed. Molecular weights increase linearly with conversion. Active chain ends are retained.
- Molecular weights in RAFT polymerization can be estimated using equation 6.1:

$$\overline{M}_n(\text{calc}) = \frac{\text{Monomer consumed}}{\text{Chains initiated}} \approx \frac{[\text{M}]_o - [\text{M}]_t}{[\mathbf{1}]_o} m_M \tag{6.1}$$

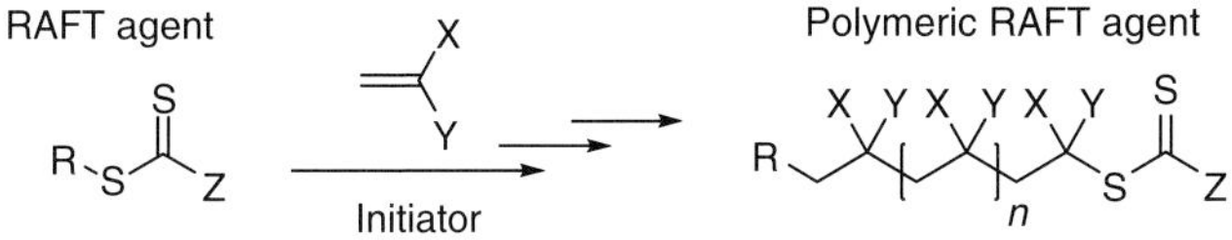


**1** **5**

**Scheme 6.2** Overall reaction in RAFT polymerization.

where $[M]_o - [M]_t$ is the monomer consumed and $m_M$ is the monomer molecular weight.

- Narrow molecular weight distributions are achievable.
- Blocks, stars and complex molecular architectures are accessible.

## 6.2 RAFT Agents

With appropriate selection of the RAFT agent for the monomers used and the reaction conditions, all or most of the aforementioned features can be routinely achieved. An understanding of the kinetics and mechanism of the RAFT process provides insight and allows the formulation of some simple guidelines for successfully implementing RAFT polymerization.

### 6.2.1 RAFT Agent Design

A wide variety of thiocarbonylthio RAFT agents (ZC(=S)SR, **1**) have now been reported. Our initial communication on this form of RAFT polymerization [20] focused on the utility of dithiobenzoate (Z = Ph) and other dithioesters. However, our patents [21, 56, 57] and many subsequent papers demonstrate that a wide range of thiocarbonylthio compounds can be used. These include certain trithiocarbonates, xanthates, dithiocarbamates and other compounds. The effectiveness of the RAFT agent depends on the monomer being polymerized and depends strongly on the properties of the free radical leaving group R and the group Z which can be chosen to activate or deactivate the thiocarbonyl double bond and to modify the stability of the intermediate radicals (Fig. 6.1) [22, 58, 59]. For an efficient RAFT polymerization (Scheme 6.1) [13],

- the RAFT agents **1** and **3** should have a reactive C=S double bond (high $k_{add}$);
- the intermediate radicals **2** and **4** should fragment rapidly (high $k_\beta$, weak S—R bond) and give no side reactions;
- the intermediate **2** should partition in favor of products ($k_\beta \geq k_{-add}$); and
- the expelled radicals R• should efficiently reinitiate polymerization.

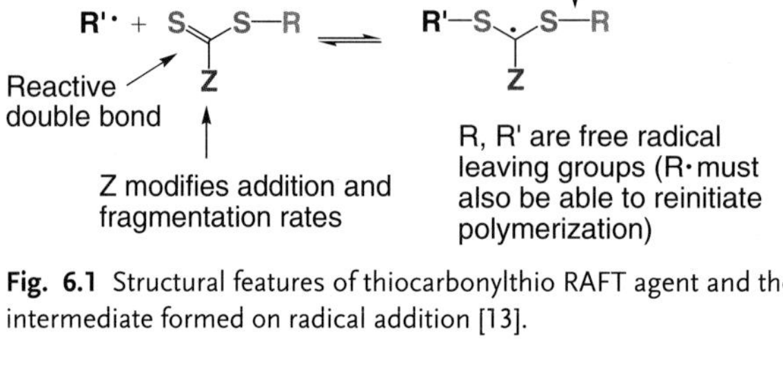


**Fig. 6.1** Structural features of thiocarbonylthio RAFT agent and the intermediate formed on radical addition [13].

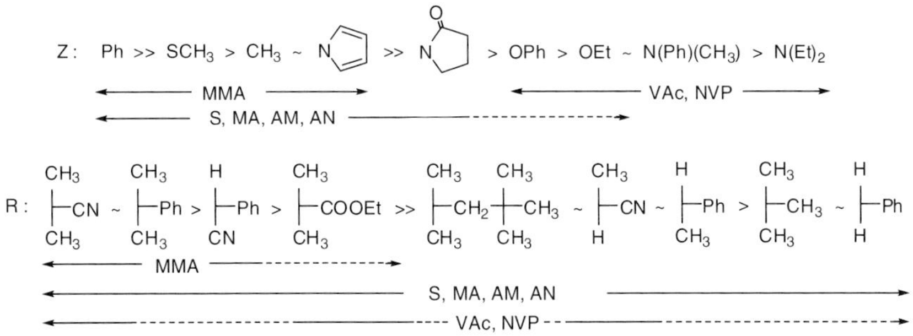


**Fig. 6.2** Guidelines for selection of RAFT agents for various polymerizations [13, 14, 60]. For Z, addition rates decrease and fragmentation rates increase from left to right. For R, fragmentation rates decrease from left to right. Dashed line indicates partial control (i.e. control of molecular weight but poor polydispersity or substantial retardation in the case of VAc).

Figure 6.2 provides general guidance on how to select the appropriate RAFT agent for a particular monomer. It should be clear that with just two RAFT agents it should be possible to exert effective control over the vast majority of polymerizations. For example, a tertiary cyanoalkyl trithiocarbonate (e.g. **77**, **78**) provides excellent control and no or little retardation in RAFT polymerizations of (meth)acrylates, (meth)acrylamides and styrene. A cyanoalkyl dithiocarbamate (e.g. **136**) or xanthate (e.g. **116**) enables similar control in RAFT polymerizations of vinyl acetate, vinyl pyrrolidone and similar monomers. Other RAFT agents may be required for solubility or compatibility with particular polymerization media or to provide specific end group functionality.

A tabulation of RAFT agents already described in the literature and that are mentioned in the succeeding sections is provided in Tables 6.1–6.5. RAFT agents are organized by Z (dithioesters, trithiocarbonates, xanthates, dithiocarbamates, others) and by R (tertiary, secondary, primary) within each table. Multifunctional RAFT agents with more than two thiocarbonylthio groups are not included in the summary. Where available, a reference in which the synthesis of the RAFT agent is described is provided in Tables 6.1–6.5. References to the use of the RAFT agents are provided in the tables that follow in Section 6.3. The list of RAFT agents is certainly not comprehensive as it continues to increase, but it, nonetheless, demonstrates the wide range of functionality that may be introduced into the RAFT agent structure. Inclusion of a compound in Tables 6.1–6.5 should not be taken as an indication that the compound is an effective RAFT agent. Discussion of the dependence of effectiveness on RAFT agent structure is provided in Chapter 3 of this Handbook and in our other recent reviews [13, 14].

Dithiobenzoates and similar dithioesters with Z = aryl are amongst the most active RAFT agents and with appropriate choice of R have general applicability in the polymerization of (meth)acrylic and styrenic monomers [13, 14]. However, their use can give retardation, particularly when used in high concentrations (to provide lower-molecular-weight polymers) and with high-$k_p$ monomers (acrylates, acrylamides). They are also sensitive to hydrolysis and decomposition by Lewis acids

**Table 6.1** Dithioesters (Z = aryl) used as RAFT agents[a]

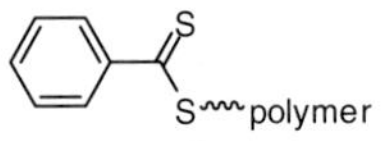


**6**

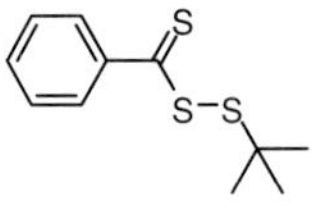


**7**[59]

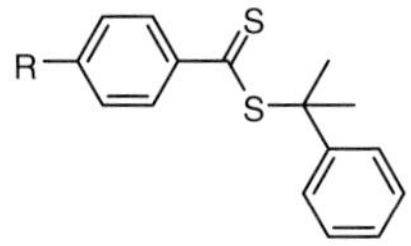


**8a** R=H[61]
**b** R=F

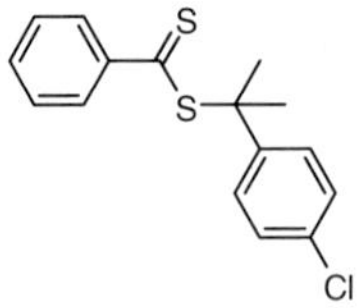


**9**[59]

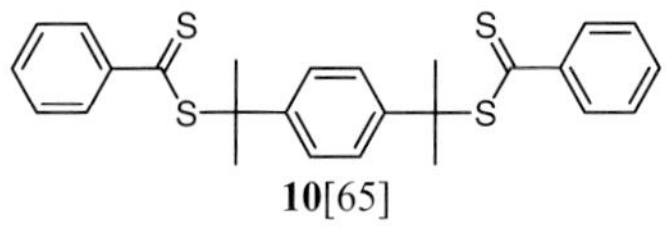


**10**[65]

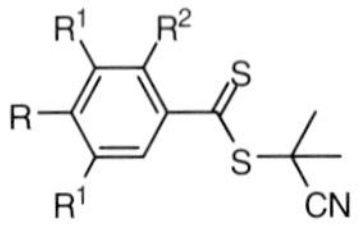


**11a** R,R[1],R[2]=H[59,61,66]
**b** R=F,R[1]=H,R[2]=H[64]
**c** R=CN,R[1]=H,R[2]=H[64]
**d** R= $OCH_3$,R[1]=H,R[2]=H[64]
**e** R= Ph,R[1]=H,R[2]=H[64]
**f** R= F,R[1]=H,R[2]=F[64]
**g** R= $CH_3$,R[1]=H,R[2]=$CH_3$[64]
**h** R= $OCH_3$,R[1]=H,R[2]=$OCH_3$[64]
**i** R= H,R[1]=$CF_3$,R[2]=H[64]
**j** R= H,R[1]=CN,R[2]=H[64]

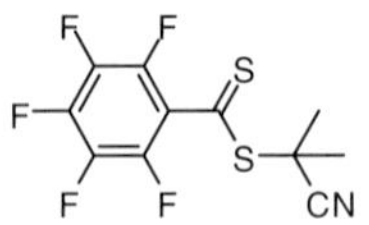


**12**[64]

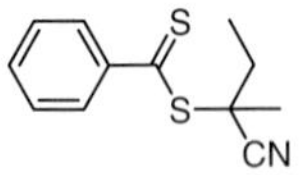


**13**[57,68]

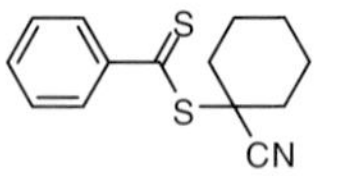


**14**[57]

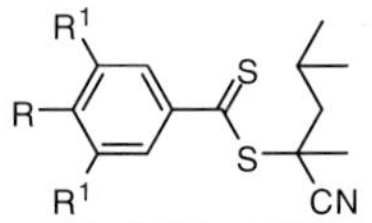


**15a** R,R[1]=H[64]
**b** R=CN,R[1]=H[64]
**c** R=H,R[1]=$CF_3$[64]
**d** R=$OCH_3$,R[1]=H [64]

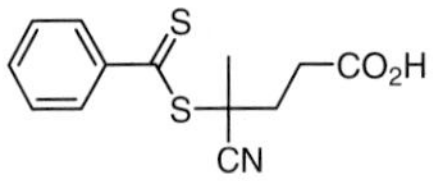


**16**[66,57]

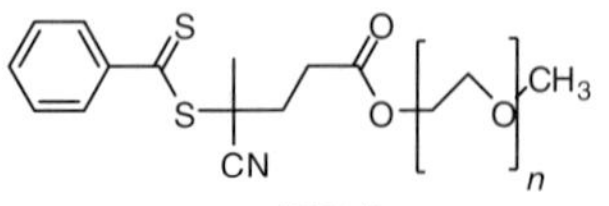


**17**[24]

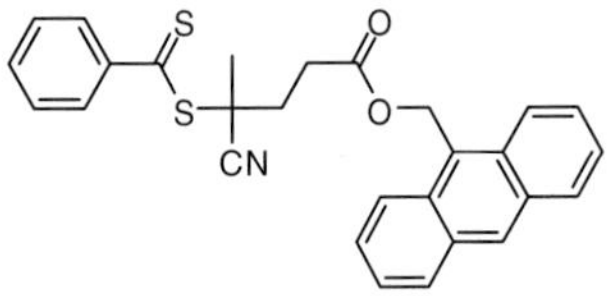


**18**[68]

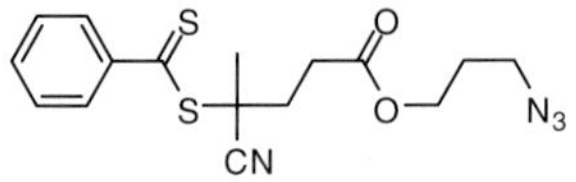


**19**[69]

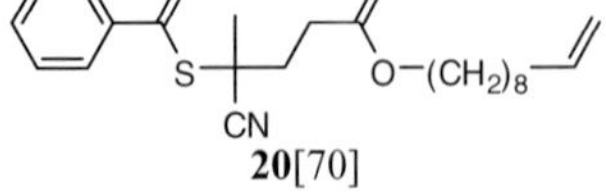


**20**[70]

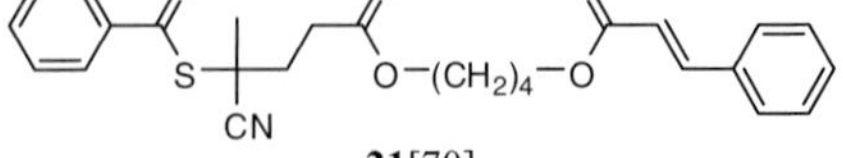


**21**[70]

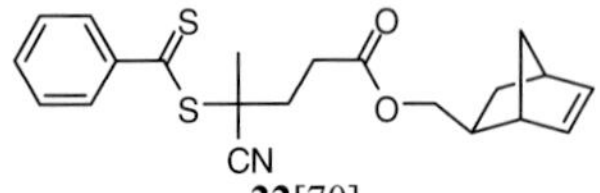


**22**[70]

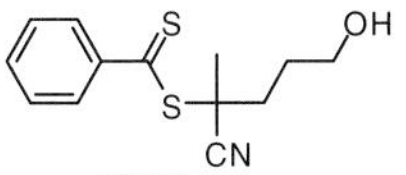

**23**[59,66]

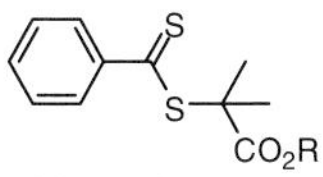

**24a** R=$CH_3$
**b** R=$C_2H_5$[59]
**c** R=$CH_2C_6F_{13}$[72]

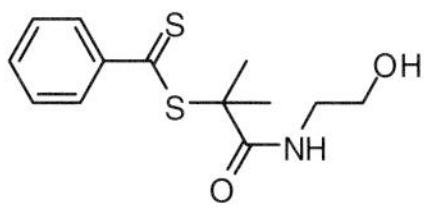

**25**[59]

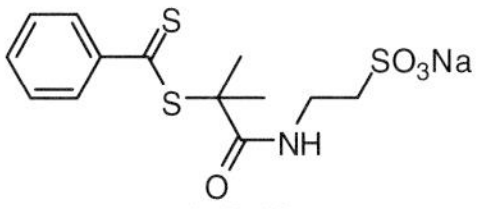

**26**[73]

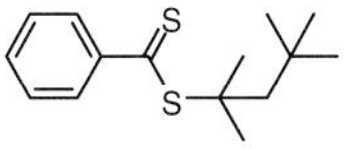

**27**[59]

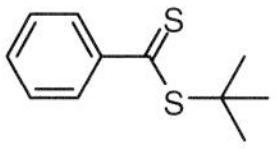

**28**[59]

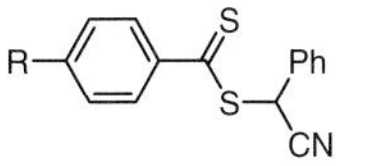

**29a** R=H[74]
**b** R=$OCH_3$[74]
**c** R=F[74]
**d** R=Ph[74]

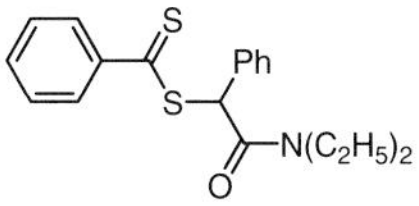

**30**[75,76]

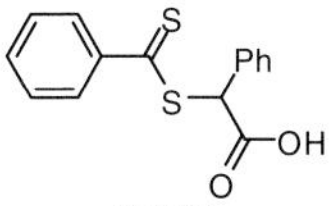

**31**[77]

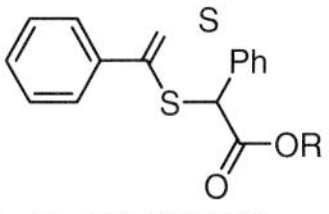

**32a** R=$CH_3$[76,78]
**b** R=PEGM[76,79]
**c** R=PLA[76]
**d** R=$CH_2CH_2C_6F_{13}$[72]

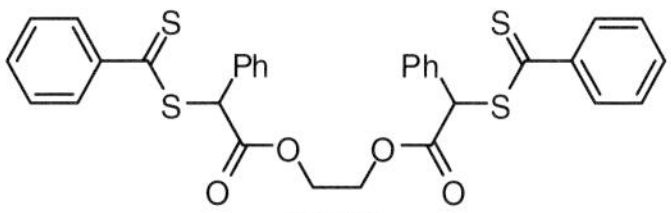

**33**[77]

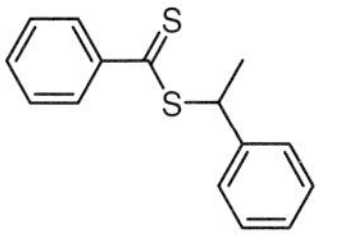

**34**[59,80]

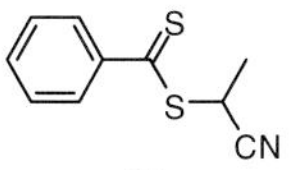

**35**

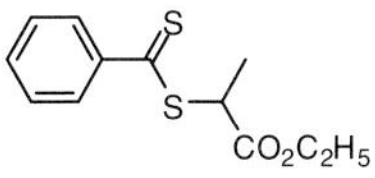

**36**[81]

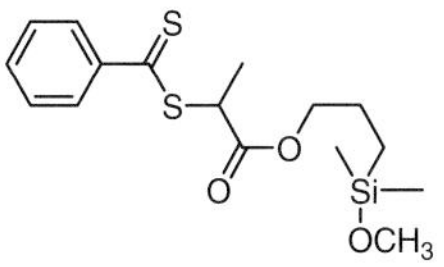

**37**[82]

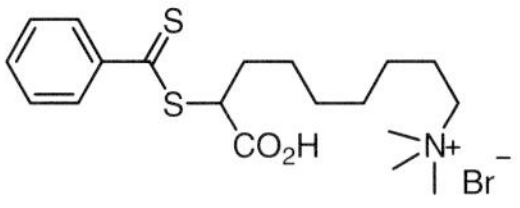

**38**[82]

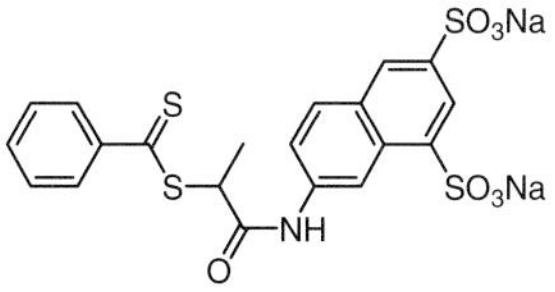

**39**[73]

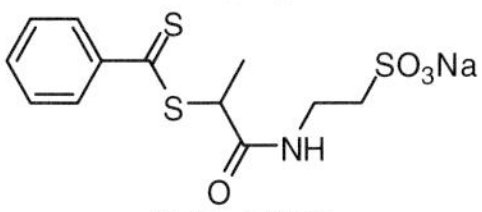

**40** R=H[73]

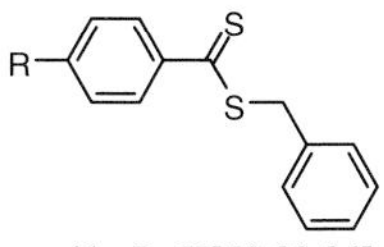

**41a** R=H[59,83,84]
**b** R=CN

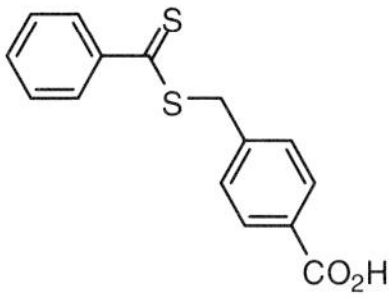

**42**[85]

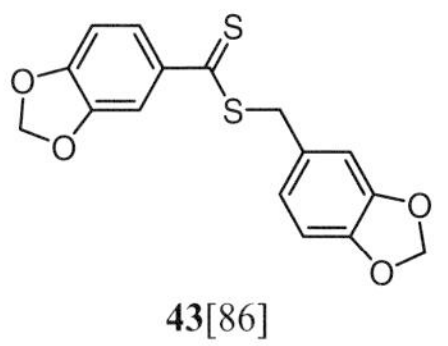

**43**[86]

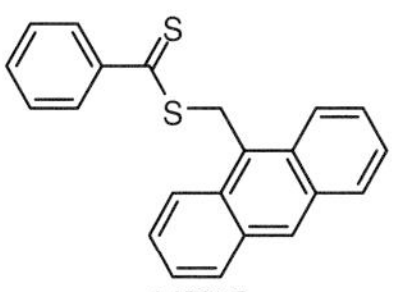

**44**[87]

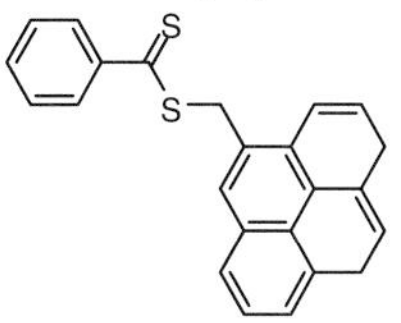

**45**[87]

(*Continued*)

**Table 6.1** *(Continued)*

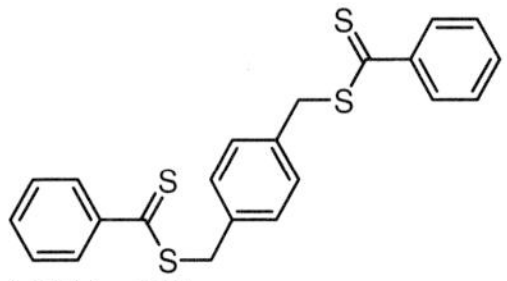

**46**[21, 65]

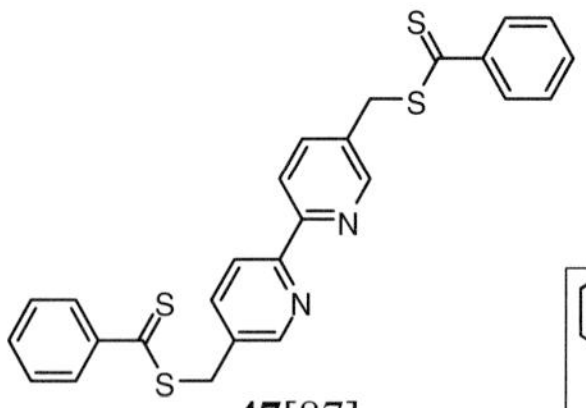

**47**[87]

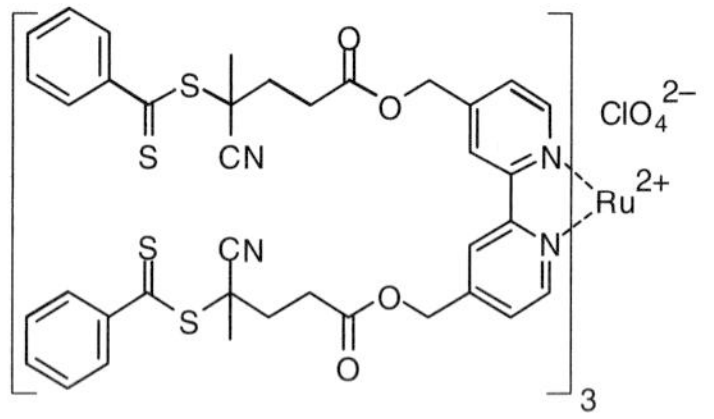

**50**[69]

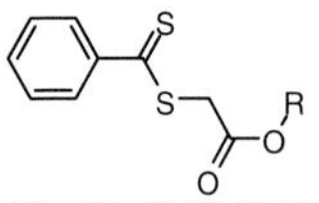

**48a** R=$C_2H_5$[80]
**b** R=$CH_2CH_2C_6F_{13}$[81]

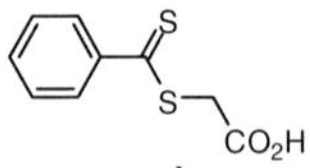

**49**[b]

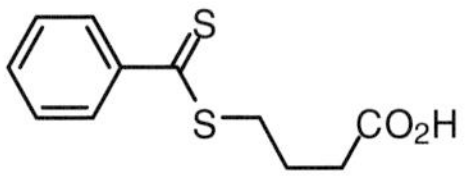

**51**[88]

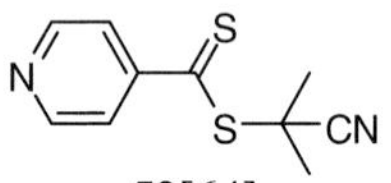

**52**[64]

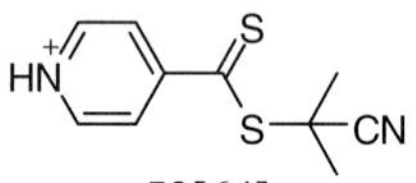

**53**[64]

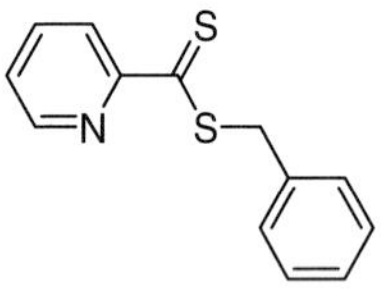

**54**[89]

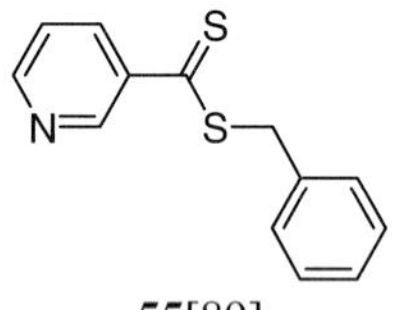

**55**[89]

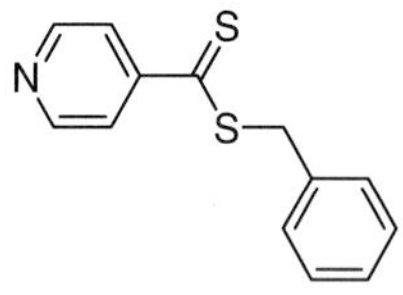

**56**[89]

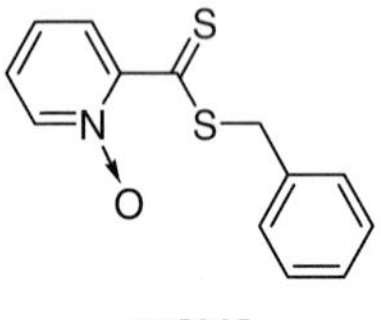

**57**[89]

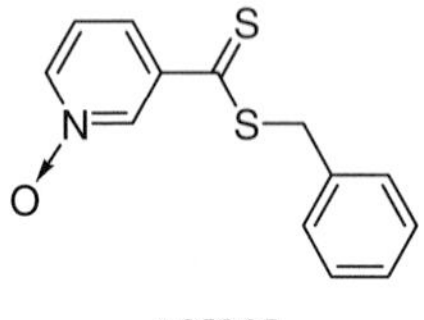

**58**[89]

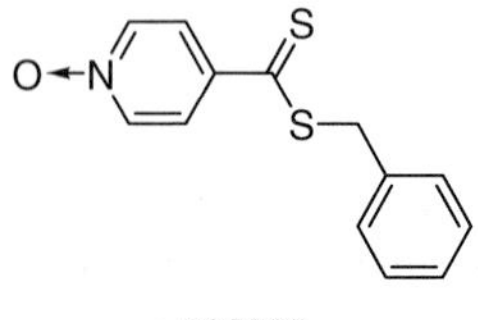

**59**[89]

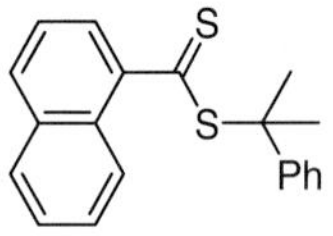

**60**

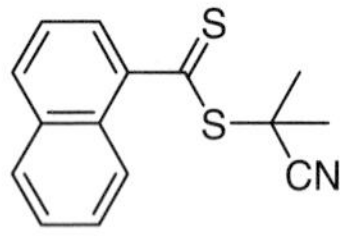

**61**[64]

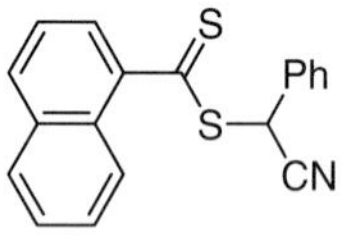

**62**[73]

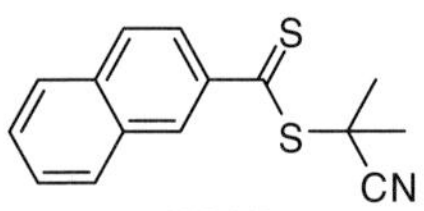

**63**[64]

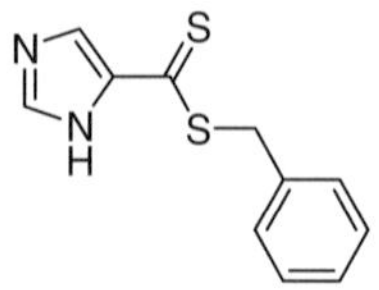

**64**[90]

[a] References provided in this table relate to the synthesis of RAFT agent. For refernce to use of RAFT agent, see also the tables that follow.
[b] RAFT agent is commercially available.

**Table 6.2** Dithioesters (Z = alkyl or araalkyl) used as RAFT agents[a]

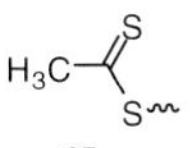
**65**

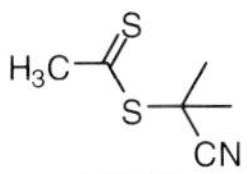
**66**[58]

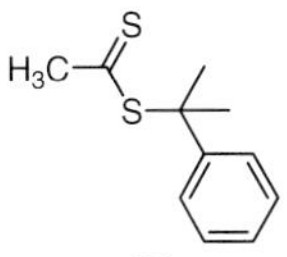
**67**

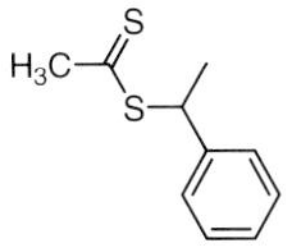
**68**

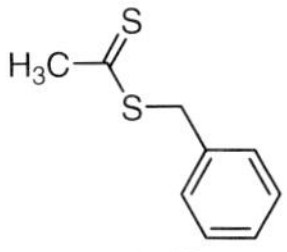
**69**[58]

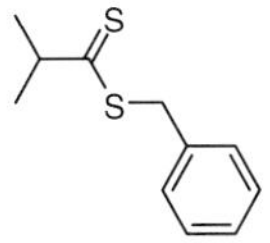
**70**[91]

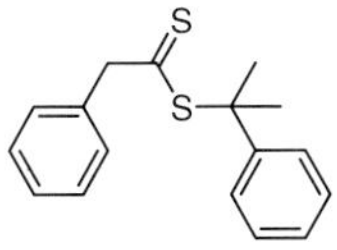
**71**

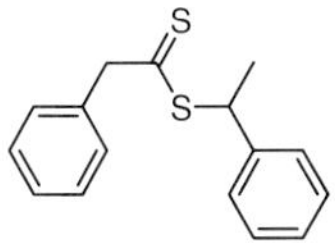
**72**

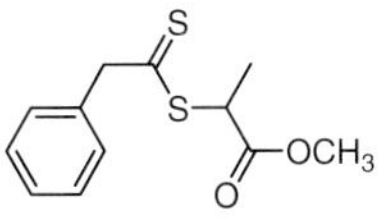
**73**

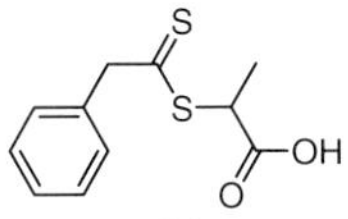
**74**

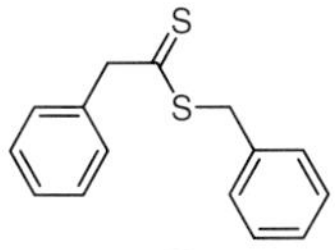
**75**

[a] References provided in this table relate to the synthesis of RAFT agent. For refernce to use of RAFT agent, see also the tables that follow.

[60]. The utility of trithiocarbonate RAFT agents was disclosed in the first RAFT patent [21] and many papers now describe their application (vide infra). Trithiocarbonates are less active yet still provide good control over the polymerization of (meth)acrylic and styrenic monomers, give substantially less retardation and are less prone to hydrolytic degradation. Ideally, to avoid odor issues with the RAFT agent and polymer, the 'Z', and preferably the 'R(S)', groups should be based on thiols with low volatility (e.g. dodecanethiol) [40, 60].

*O*-Alkyl xanthates and *N,N*-dialkyl dithiocarbamates are most suited for polymerization of VAc, NVP and related vinyl monomers where the propagating radical is a poor homolytic leaving group. The relatively low activity of *O*-alkyl xanthates and simple *N,N*-dialkyl dithiocarbamate derivatives in polymerization of styrenic and (meth)acrylic monomers can be qualitatively understood in terms of the importance of the zwitterionic canonical forms (Fig. 6.3) which arise through interaction between the O or N lone pairs and the C=S double bond [58, 61]. Electron-withdrawing

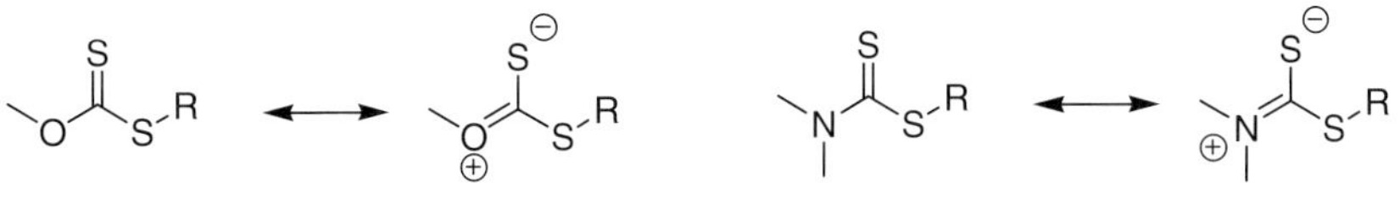

**Fig. 6.3** Canonical forms of xanthates and dithiocarbamates.

**Table 6.3** Trithiocarbonates (Z = thioalkyl) used as RAFT agents[a]

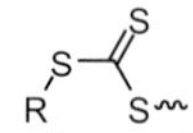

**76** R=primary alkyl

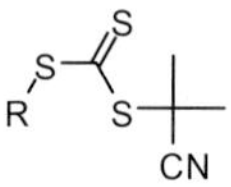

**77a** R=$CH_3$[58,66]
**b** R=$C_{12}H_{25}$

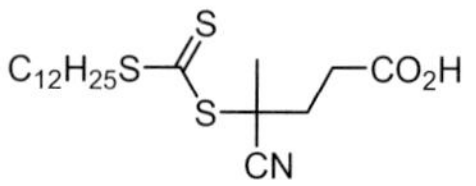

**78**[40]

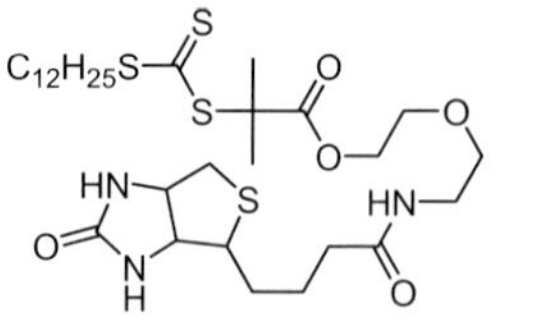

**79**[92]

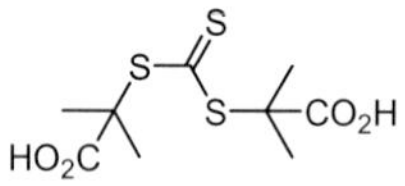

**80**[93]

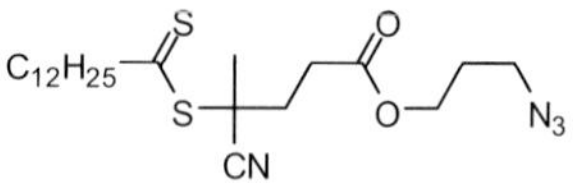

**81**[69]

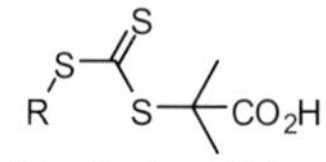

**82a** R=$C_2H_5$[94]
**b** R=$C_{12}H_{25}$[95]
**c** R=$CH_2CH_2CO_2H$[95]

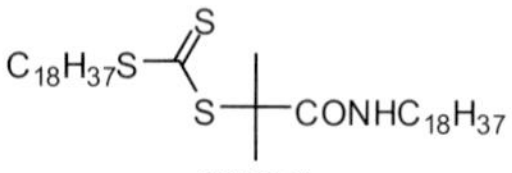

**83**[96]

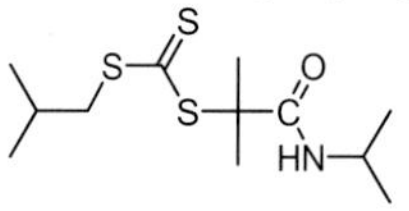

**84**[96]

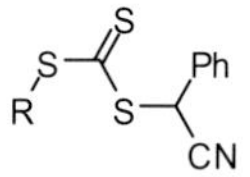

**85a** R=$CH_3$[14]
**b** R=$C_{12}H_{25}$[60]

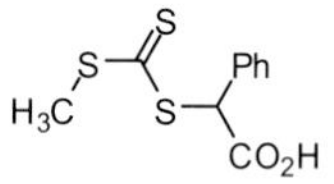

**86**

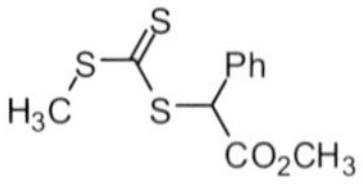

**87**[75,77]

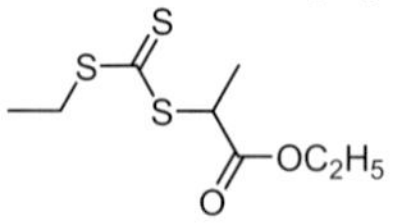

**88**[97]

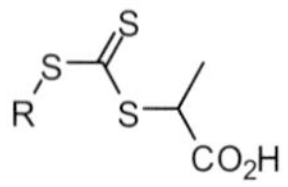

**89a** R=$C_4H_9$[98]
**b** R=$C_{12}H_{25}$[98]
**c** R=$CH_2CH_2CO_2H$[95]

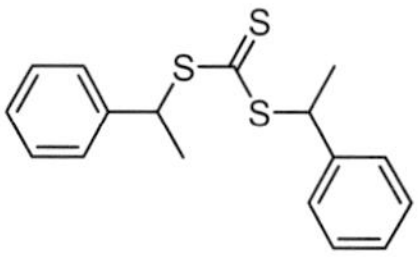

**90**

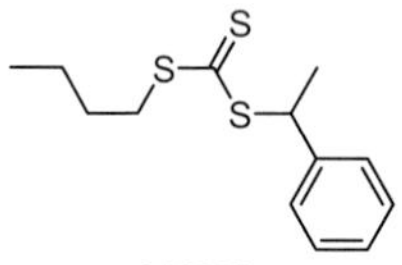

**91**[99]

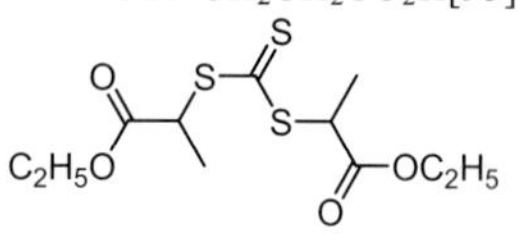

**92**

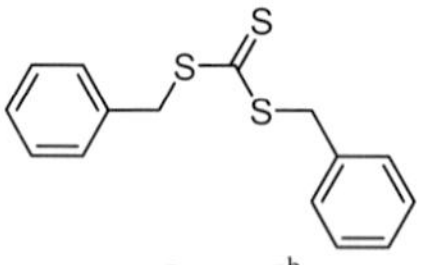

**93**[58,97][b]

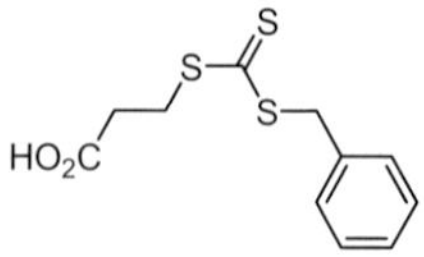

**94**[95]

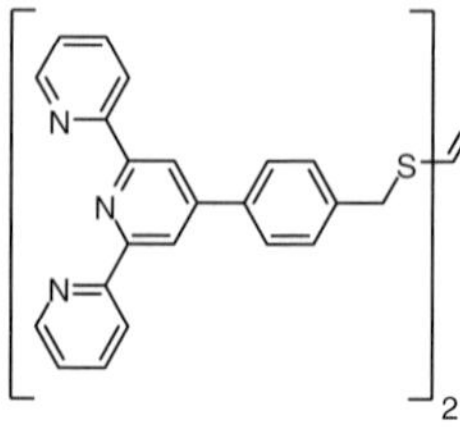

**95**[100]

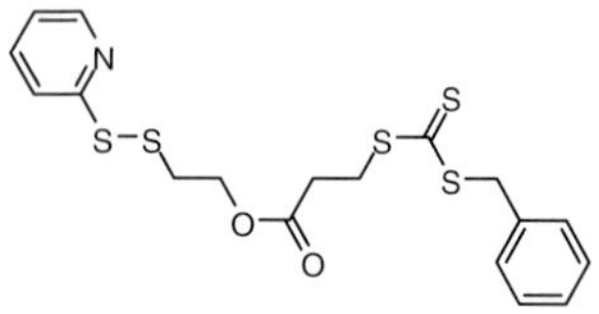

**96** [101]

**Table 6.3** (*Continued*)

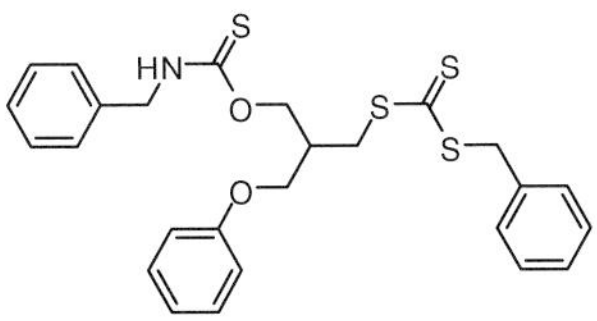

**97**[102]

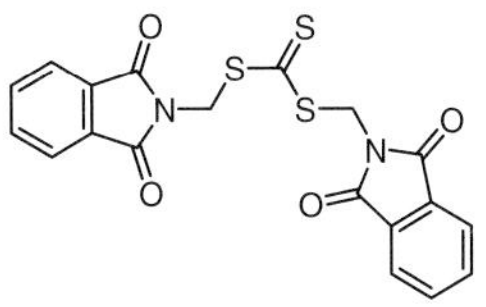

**98**[99]

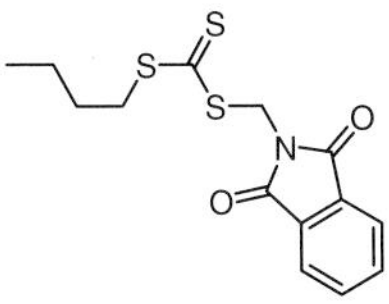

**99**[99]

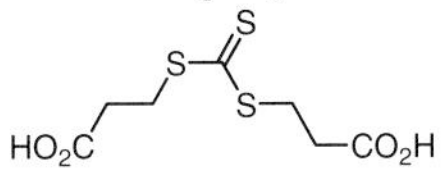

**100**[95]

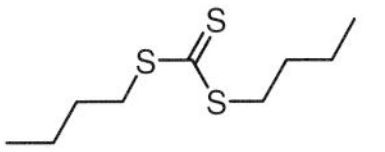

**101**[99]

[a] References provided in this table relate to the synthesis of RAFT agent. For refernce to use of RAFT agent, see also the tables that follow.
[b] RAFT agent is commercially available.

substituents on Z can enhance the activity of RAFT agents to modify the above order [58, 62, 63]. Thus, xanthate and dithiocarbamate RAFT agents where the oxygen or nitrogen lone pair is less available for delocalization with the C=S by virtue of being part of an aromatic ring or by possessing an adjacent electron-withdrawing substituent can be very effective in polymerization of styrenic and (meth)acrylic monomers.

Electron-withdrawing groups can also enhance the activity of dithiobenzoate RAFT agents. For ring-substituted cyanoisopropyl dithiobenzoate RAFT agents in MMA polymerization, electron-withdrawing groups, which render the thiocarbonyl sulfur more electrophilic, enhance the rate of addition to the C=S double bond and provide narrower polydispersities from the early stages of polymerization [13, 64].

For fragmentation to occur efficiently in the desired direction, the substituent R must be a good homolytic leaving group, relative to the attacking radical $P_n^{\bullet}$. For example, the RAFT agent with R = $CH_2Ph$ (e.g. benzyl dithiobenzoate (**41a**)) functions as a suitable chain transfer agent in polymerization with styryl and acrylyl propagating radicals but not in those with methacrylyl propagating radical. The benzyl radical is a reasonable leaving group with respect to the styryl and acrylyl propagating radicals but is a very poor leaving group with respect to the methacrylyl propagating radical. In MMA polymerization, RAFT agents such as benzyl dithiobenzoate (**41a**) can appear almost inert because R is a poor leaving group with respect to the PMMA propagating radical [59].

The rate of fragmentation of intermediate **2** is enhanced by increasing steric hindrance, by the presence of electron-withdrawing groups and by radical stabilizing groups on R. Examples illustrating the importance of these factors can be found in the subsequent sections (see in particular Section 6.3.1). More detail is also provided in Chapter 3 and in our other recent reviews [13, 14].

During chain extension, the attacking and leaving propagating radicals ($P_n^{\bullet}$ and $P_m^{\bullet}$, $n$, $m > 2$) are, in essence, identical in all respects other than chain length and

**Table 6.4** Xanthates (dithiocarbonates) (Z = alkoxy, aryloxy) used as RAFT agents[a]

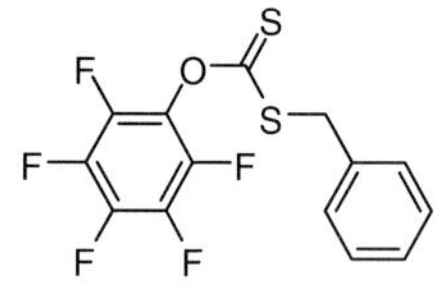

**102**[58,103]

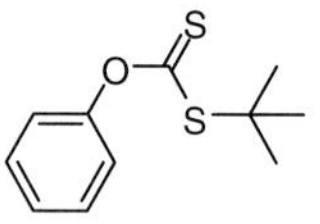

**103**[79]

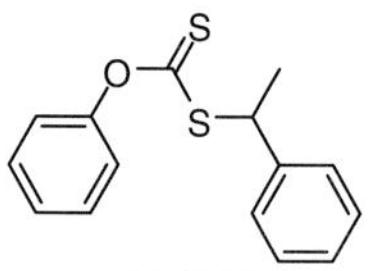

**104**[79]

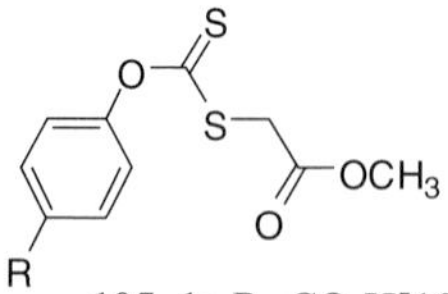

**105 1a** R=$CO_2H$[105]
**b** R=$CO_2CH_3$
**c** R=F
**d** R=$OCH_3$

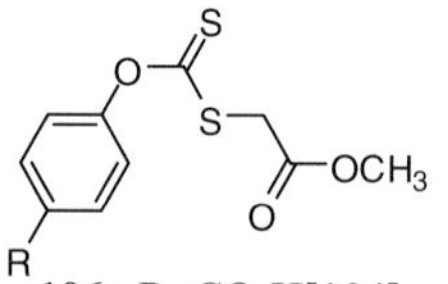

**106a** R=$CO_2H$[104]
**b** R=$CO_2CH_3$
**c** R=F
**d** R=$OCH_3$

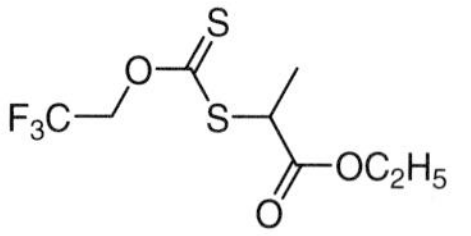

**107**

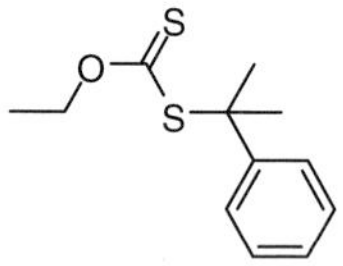

**108**[79]

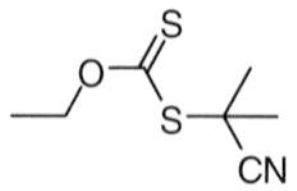

**109**[66,103]

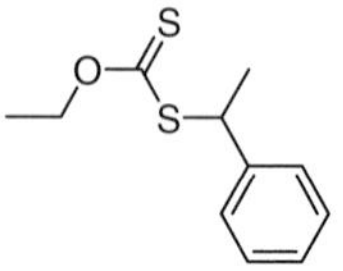

**110**[103,105,106]

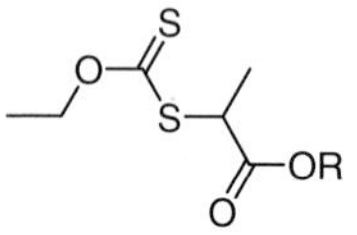

**111a** R=$C_2H_5$[97]
**b** R=$CH_3$[261]

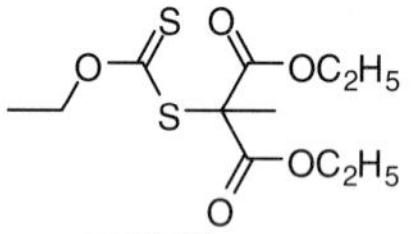

**112**[79]

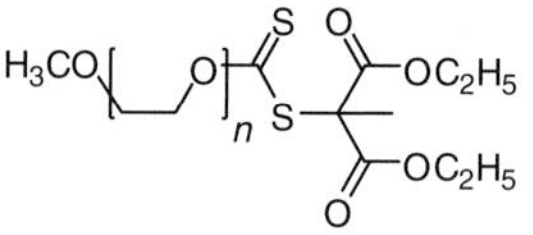

**113a** *n*=1[79]
**b** *n*=3[79]

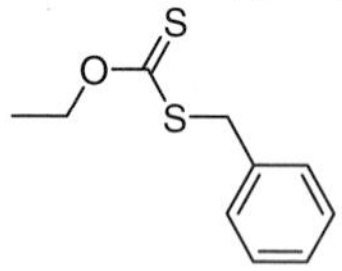

**114**[58,79,105,106]

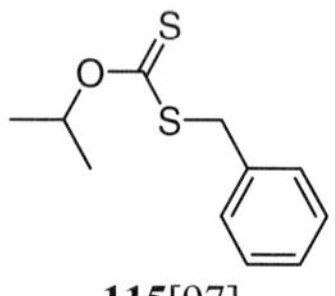

**115**[97]

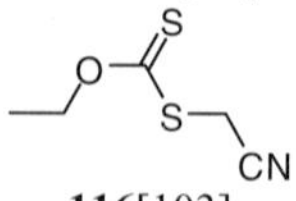

**116**[103]

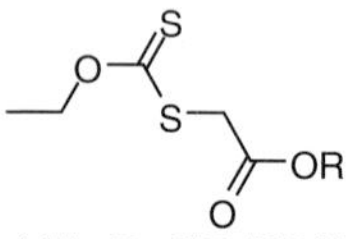

**117a** R=$CH_3$[104]
**b** R=$C_2H_5$

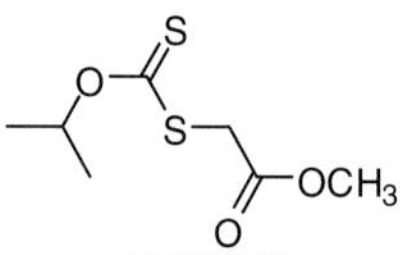

**118**[104]

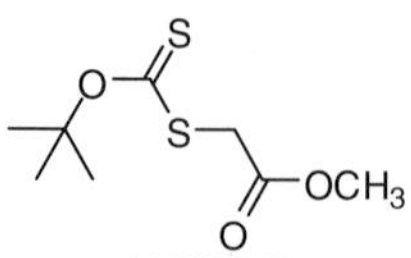

**119**[104]

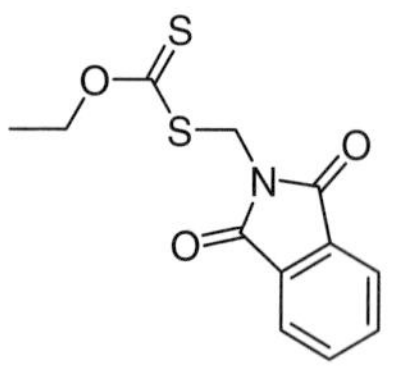

**120**[107]

[a] References provided in this table relate to the synthesis of RAFT agent. For refernce to use of RAFT agent, see also the tables that follow.

**Table 6.5** Dithiocarbamates (Z = N<) used as RAFT agents[a]

| | | |
|---|---|---|
| 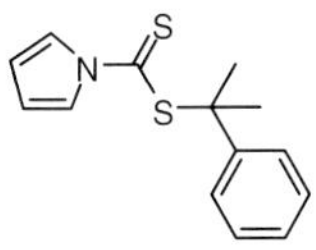 **121**[108] | 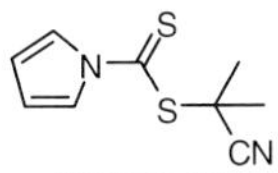 **122**[58,66] | 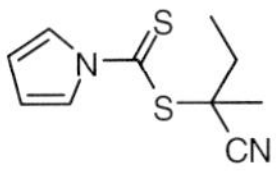 **123**[66] |
| 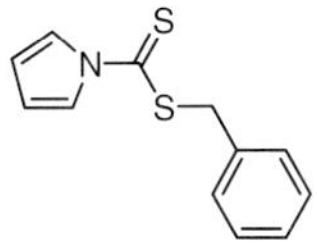 **124**[58] | 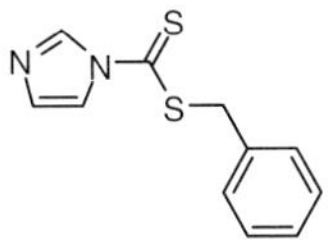 **125**[103] | 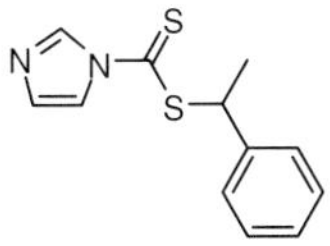 **126**[79] |
| 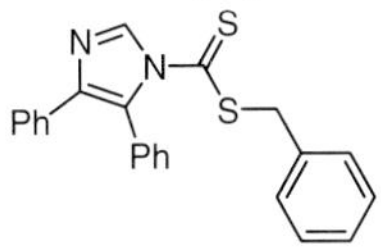 **127**[109] | 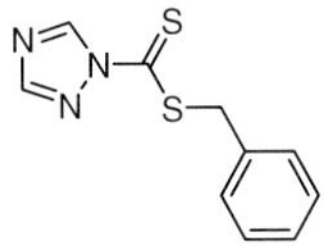 **128**[109] | 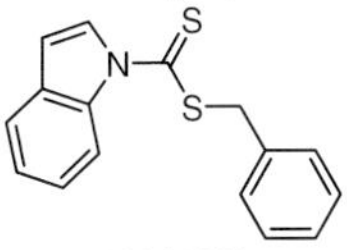 **129**[109] |
| 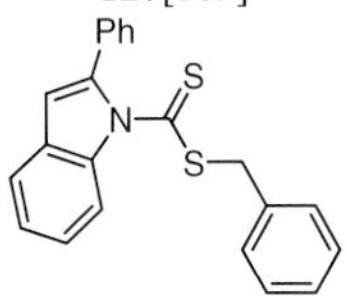 **130**[109] | 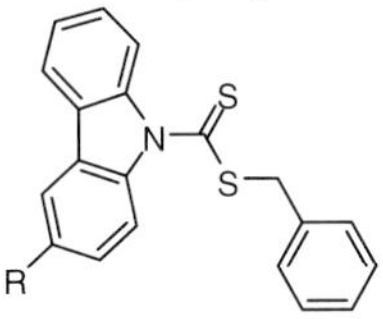 **131a** R=H [109]<br>**b** R=PhN$_2$[110] | 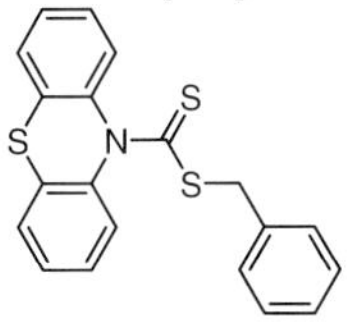 **132**[109] |
| 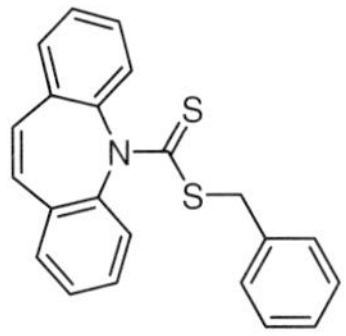 **133**[109] | 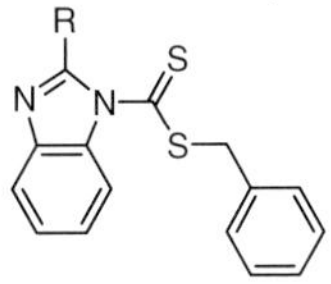 **134a** R=CH$_3$<br>**b** R=Ph<br>**c** R=C$_9$H$_{19}$[111] | 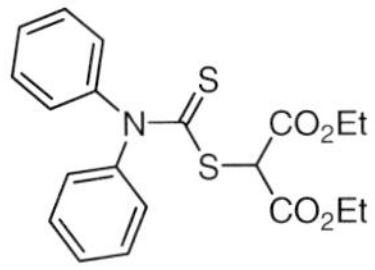 **135** |
| 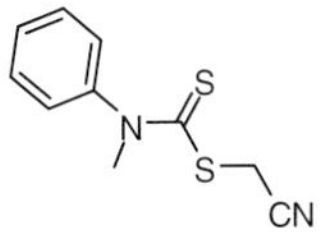 **136** | 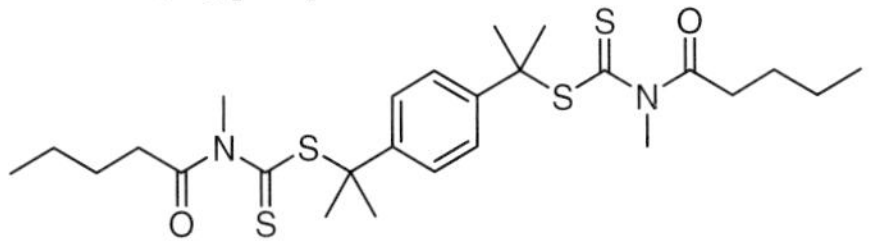 **137**[112] | |
| 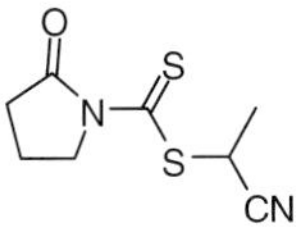 **138** | 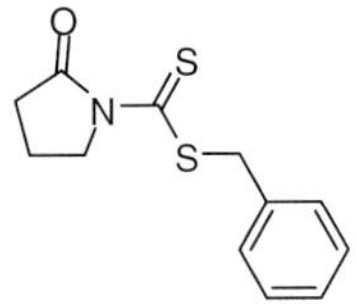 **139**[58,103] | 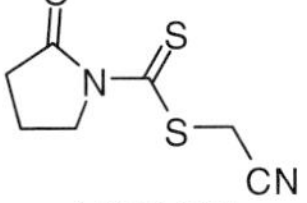 **140**[103] |

(*Continued*)

Table 6.5 (*Continued*)

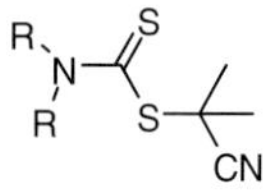


**141a** R=$CH_3$[66,103]
**b** R=$C_2H_5$

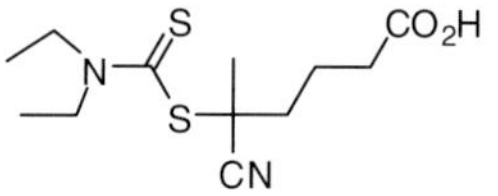


**142**[113]

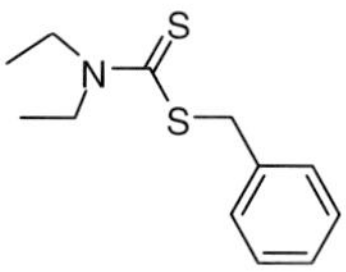


**143**[58]

[a] References provided in this table relate to the synthesis of RAFT agent. For refernce to use of RAFT agent, see also the tables that follow.

therefore should have equivalent rates in addition to RAFT agent, fragmentation and propagation.

The leaving ability of the substituent R must also be balanced with the ability of the radical R• to reinitiate polymerization. The triphenylmethyl radical, for instance, would be an excellent leaving group but would be a poor reinitiator of chains and its use would result in retardation of polymerization. Inhibition periods observed when RAFT agents with R = benzylic species (1-phenylethyl, benzyl) for vinyl monomer polymerization (e.g. VAc, NVP) can also be rationalized in terms of the low rate of reinitiation (see Section 6.3.7).

## 6.2.2 RAFT Agent Synthesis

This section is taken largely from our other recent reviews [13, 14]. Currently, few RAFT agents are available commercially. Arkema have indicated they have commenced producing dibenzyl trithiocarbonate (**93**) on a pilot scale and have reported it as commercially available [114]. However, RAFT agents are available in moderate to excellent yields by a variety of methods, and syntheses are generally straightforward. References to syntheses of specific RAFT agents are provided in Tables 6.1–6.5.

The methods most commonly exploited include the following:

- Reaction of a carbodithioate salt with an alkylating agent (Scheme 6.3) [21, 58, 59, 80, 115, 116]. Often this will involve sequential treatment of an anionic species with carbon disulfide and an alkylating agent in a one-pot reaction. For example, this process was used to prepare benzyl dithiobenzoate (**41a**) [58],

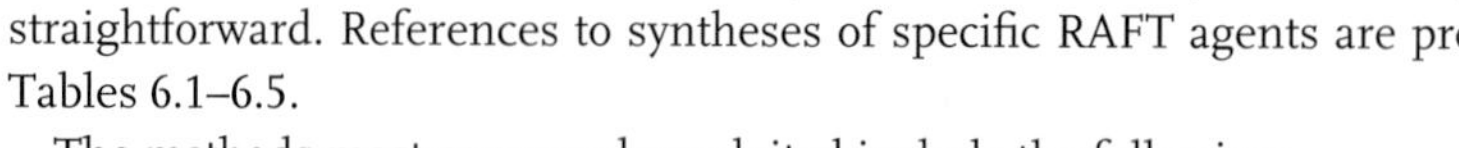


**41a**

**Scheme 6.3** Synthesis of benzyl dithiobenzoate (**41a**) by reaction of a carbodithioate salt with an alkylating agent.

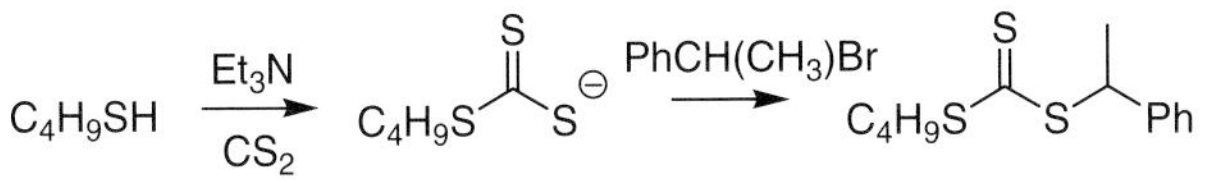

91

**Scheme 6.4** Synthesis of butyl 1-phenylethyl trithiocarbonate (**91**) by reaction of a carbodithioate salt with an alkylating agent.

2-(ethoxycarbonyl)prop-2-yl dithiobenzoate [58] and 2-cyanoprop-2-yl dithiobenzoate [22].

Similar chemistry is used in the synthesis of unsymmetrical trithiocarbonates (Scheme 6.4) [99, 117, 118]. Yields are generally high (>70%) for substitution of primary and secondary alkyl halides but can be low for tertiary halides (5–40%).

Thiocarbonylbisimidazole may be used as alternative to carbon disulfide in the synthesis of some RAFT agents (e.g. Scheme 6.5) [97, 103].

- Addition of a dithioacid across an olefinic double bond [59, 119–121]. This procedure has been used to prepare cumyl dithiobenzoate (**8a**) (Scheme 6.6) [21, 22]. Electron-rich olefins (styrene, AMS, isooctene, VAc) give the desired Markownikov addition (sulfur at substituted position). However, similar reactions with electron-deficient olefins (MMA, MA, AN) unfortunately give Michael-like addition (sulfur at unsubstituted position) and therefore do not give useful RAFT agents.
- Radical-induced decomposition of a bis(thioacyl) disulfide [40, 57, 58, 66, 122]. This is probably the most used method for the synthesis of RAFT agents requiring tertiary R groups. An example is the synthesis of the tertiary trithiocarbonate (**78**) (Scheme 6.7) [40]. It is also possible to use this chemistry to generate a RAFT agent in situ during polymerization [57]. A new synthesis of bis(thioacyl) disulfides has appeared [123].
- Sulfuration of a thioloester (Scheme 6.8) [21, 59] or a mixture of a carboxylic acid with an alcohol, halide or olefin [124–126] with $P_4S_{10}$, Davey or Lawesson's reagent [127].
- Radical-induced ester exchange [22, 58, 59, 128]. For this method to be effective the R group of the precursor RAFT agent should be a good free radical leaving group with respect to that of the product RAFT agent. For example, the cyanoisopropyl radical generated from 2,2′-azobis(isobutyronitrile) (AIBN) can replace the cumyl group of cumyl dithiobenzoate (**8a**) (Scheme 6.9) [22].

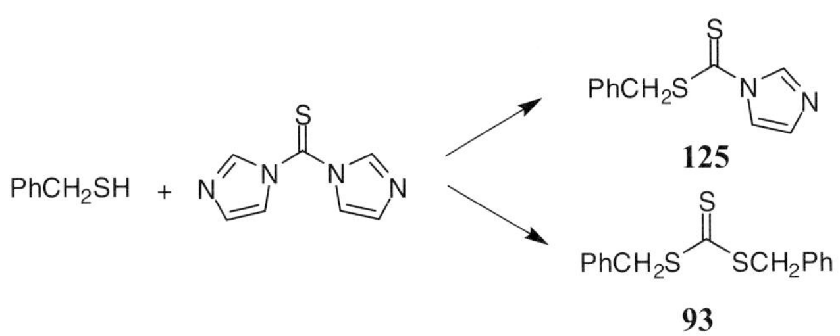

**Scheme 6.5** Synthesis of benzyl 1*H*-imidazole-1-carbodithioate (**125**) and dibenzyl trithiocarbonate (**93**) from 1,1′-thiocarbonylbisimidazole.

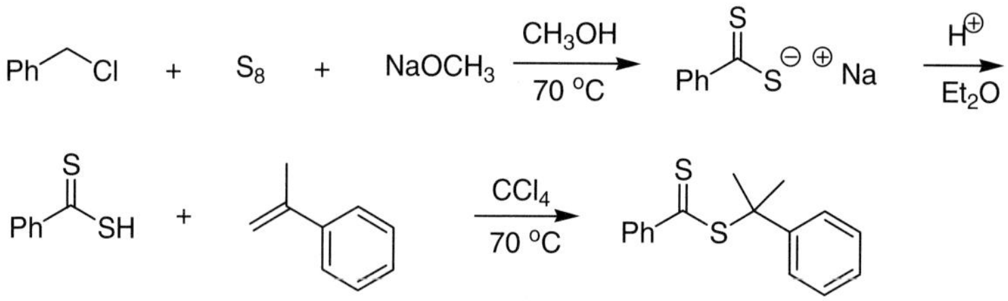


**8a**

**Scheme 6.6** Synthesis of cuyml dithiobenzoate (**8a**) by addition of a dithioacid across a double bond.

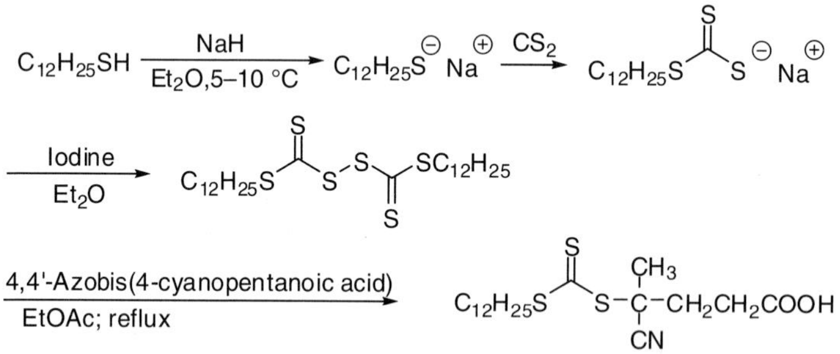


**78**

**Scheme 6.7** Synthesis of the unsymmetrical trithiocarbonate, 4-cyano-4-(dodecylthiocarbonothioylthio)pentanoic acid (**78**) by radical-induced decomposition of a bis(thioacyl) disulfide.

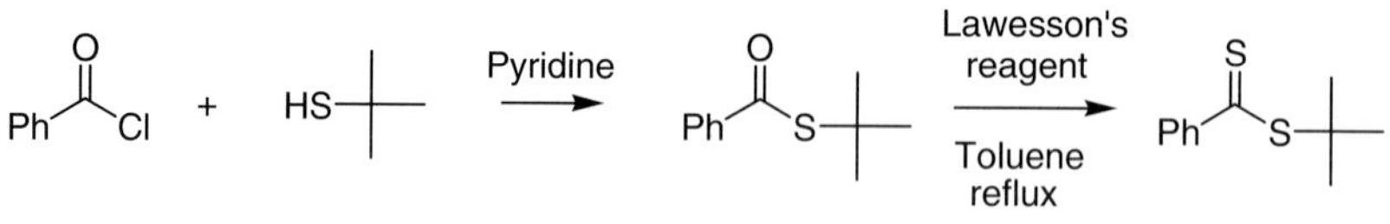


**28**

**Scheme 6.8** Synthesis of *tert*-butyl dithiobenzoate (**28**) by sulfuration of a thioloester.

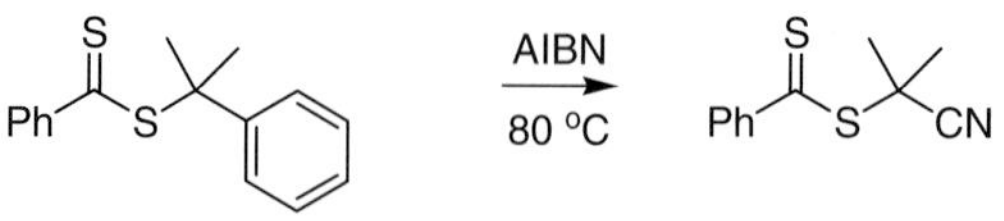


**8a** **11a**

**Scheme 6.9** Synthesis of cyanoisopropyl dithiobenzoate (**11a**) by radical-induced ester exchange.

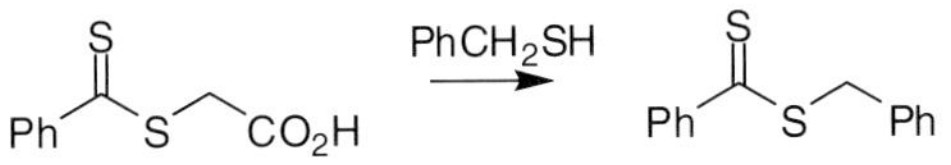


**49** **41a**

**Scheme 6.10** Synthesis of benzyl dithiobenzoate (**41a**) by thiol exchange.

- Transesterification [79, 83] (thiol exchange by reaction of a dithioester with a thiol). Thioglycolic acid-based dithioesters are poor RAFT agents [80]. However, they can serve as precursors to other RAFT agents as they undergo facile reaction with other thiols to provide a new dithioesters. For example, reaction of **49** with benzyl mercaptan provides benzyl dithiobenzoate (**41a**) in high yield (Scheme 6.10) [83].

## 6.3 RAFT Polymerization

A summary of RAFT polymerization is provided in the sections that follow. For selected monomers MMA, MA, BA, AA and styrene, we have categorized the results according to the level of control observed. The type of control observed will depend on the reaction conditions used, for example, the polymerization temperature, the concentration of the RAFT agent and the reaction medium. However, even taking these differences into account, the experiments are not always consistent with, for example, some reporting retardation under conditions where others observe no or little retardation. This most likely reflects unstated differences in reaction conditions such as the level of degassing or the presence of impurities in RAFT agent or other components of the polymerization medium. Refer to the original papers for further details.

### 6.3.1 Methacrylates

The choice of the substituents R and Z is crucial to success in RAFT polymerization of MMA polymerization. The RAFT agents first shown to be suitable for polymerization of MMA and other methacrylate esters include cumyl dithiobenzoate (**8a**) and tertiary cyanoalkyl dithiobenzoates, e.g. cyanoisopropyl dithiobenzoate (**11a**) [20]. It has been shown that with appropriate choice of R, aromatic dithioester (Z = aryl), trithiocarbonate (Z = thioalkyl) and certain activated dithiocarbamate RAFT agents (e.g. Z = pyrrole) provide good control. Alkyl or araalkyl dithioesters and derivatives (Table 6.2), unactivated dithiocarbamates (e.g. last line of Table 6.3) and xanthates (Table 6.4) provide poor or no control (refer Fig. 6.2) [13, 14].

The suggestion that R might be selected to be a monomeric analog of the propagating radical is flawed with reference to MMA polymerization since penultimate unit effects are substantial. Thus, RAFT agent 2-ethoxycarbonyl-2-propyl dithiobenzoate (**24b**), which may be considered as a monomeric propagating radical, provides only poor control over the polymerization of MMA and other methacrylates because R (2-ethoxycarbonyl-2-propyl radical) is a poor leaving group with respect to the

PMMA propagating radical [59] (interestingly, the fluorinated dithiobenzoate (**24c**) shows signs of offering slightly better control [71]). Similarly, *tert*-octyl dithiobenzoate (**27**) has a substantially higher transfer constant in MMA polymerization than *tert*-butyl dithiobenzoate (**28**); both offer only poor control [59]. These differences in RAFT agent activity indicate a strong penultimate unit effect and are attributed mainly to steric factors. During chain extension, the attacking and leaving propagating radicals ($P_n^{\bullet}$ and $P_m^{\bullet}$, $n$, $m > 2$) are, in essence, identical in all respects other than chain length and therefore should have equivalent rates in addition to RAFT agent, fragmentation and propagation. When R is secondary, penultimate unit effects on addition and fragmentation reactions are likely to be smaller but still should not be ignored.

Secondary aromatic dithioesters with R = —CHPh(CN) (**29**, **62**) [73] and —CHPh($CO_2$R) (e.g. **32a–c**) [26, 77, 78] and analogous trithiocarbonates with R = —CHPh(CN) (**85a** [14] or **85b** [60]) and —CHPh($CO_2$R) (**86**) [77] have also been shown to have utility in controlling polymerization of methacrylates. A slow initialization is seen for RAFT agents with R = —CHPh(CN), which has been attributed to a slow rate of reinitiation by the expelled radical R• [14, 60, 73].

Marked retardation of MMA polymerization is observed for high RAFT agent concentrations and is most pronounced for the case of cumyl dithiobenzoate (**8a**) [129]. The retardation appears to be associated with usage of the initial RAFT agent. Substantially less retardation is seen with tertiary cyanoalkyl dithiobenzoates (**11–14**) or with cumyl dithioacetate (**67**) [129]. Minimal retardation is observed for lower concentrations of the RAFT agent. Thus, RAFT polymerization of MMA with **8a** (<0.01 M) or **11a** shows the usual half-order dependence on initiator concentration seen with conventional radical polymerization (Fig. 6.4) [59]. RAFT polymerization of MMA with **11a** has been used to determine chain-length-dependent

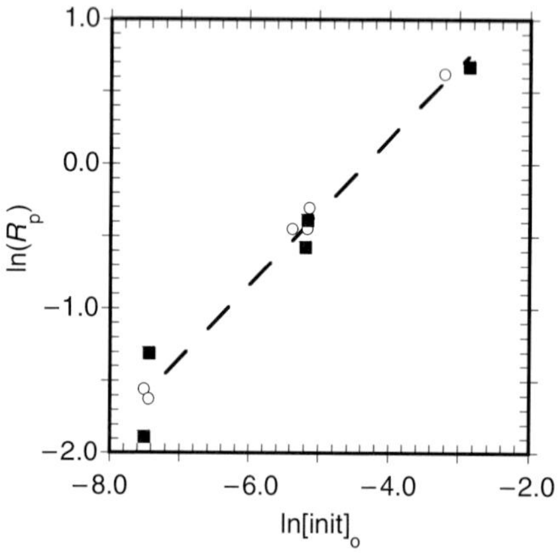


**Fig. 6.4** Plot of log(initial $R_p$) vs log(initial initiator concentration). Data are for bulk MMA polymerization at 60 °C with AIBN initiator (0.0005–0.045 M) and either cumyl dithiobenzoate (**8a**) (■) or 2-cyanoprop-2-yl dithiobenzoate (**11a**) (○) as RAFT agent (0.006–0.03 M). A least squares fit provides slope 0.507, $R = 0.98655$. Reproduced from Ref. [59]. © 2003 American Chemical Society.

termination rate constants on the basis that the kinetics of radical polymerization are not influenced by the RAFT process [130].

Improved control in MMA polymerization (reduced polydispersity) is observed at higher polymerization temperatures (up to 110 °C) [64]. Note, however, that some RAFT agents and the end groups of macro-RAFT agents appear unstable at higher temperatures (>140 °C depending on the RAFT agent) [131].

RAFT polymerization of MMA has also been shown to be air sensitive [13]. While controlled polymerization is observed in the presence of air, significant retardation, slightly broader molecular weight distributions and some departure from expected molecular weights are observed for polymerization under air or for polymerization performed with inadequate removal of air.

RAFT polymerization of MMA can be carried out in the presence of Lewis acids to obtain simultaneous control over molecular weight and tacticity and an enhanced rate of polymerization (Figure 6.5) [60, 132, 133]. Cumyl dithiobenzoate and cyanoisopropyl dithiobenzoate are, however, unstable in the presence of some Lewis acids [60, 132]. The trithiocarbonate RAFT agent (**77a**) appears stable and provides good control [60, 132].

A summary of RAFT polymerizations of methacrylate esters is provided in Table 6.6. For MMA the results have been categorized according to the level of control observed. Tolerated monomer functionality includes tertiary amino (in DMAEMA), quaternary amino (in TMAEMA), carboxylic acid (in MAA), hydroxyl (in HEMA, **146**, **149–153** [72]), epoxy (in GMA), thiirane (in **147** [134]) (see Table 6.6 for references). RAFT polymerization of functional methacrylates has been used as a route to materials as diverse as glycopolymers (e.g. **148–153**) and possible hole- or electron-transport materials (e.g. **154**, **155**). High-throughput syntheses of methacrylate-based polymers by RAFT polymerization have been developed by Schubert and coworkers [135, 136].

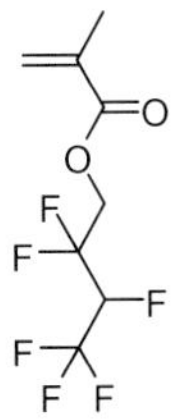

**144**[137]

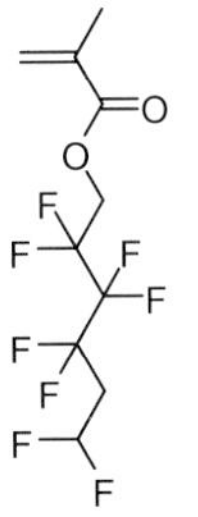

**145**[137]

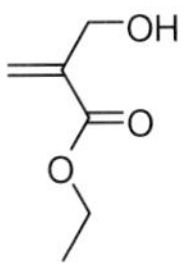

**146**[138,139]

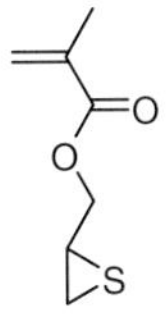

**147**[134]

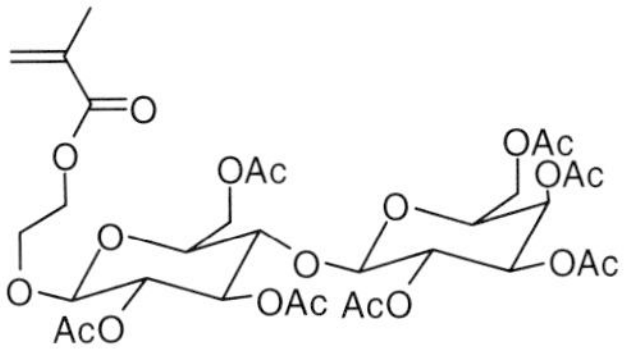

**148**[140]

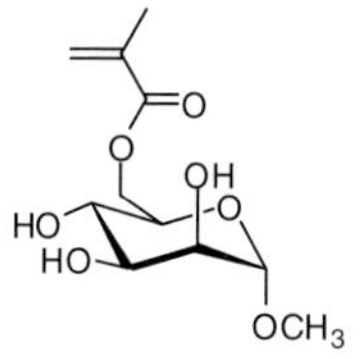

**149**[72,141]

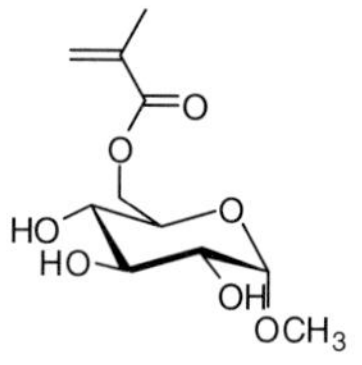

**150**[72,141]

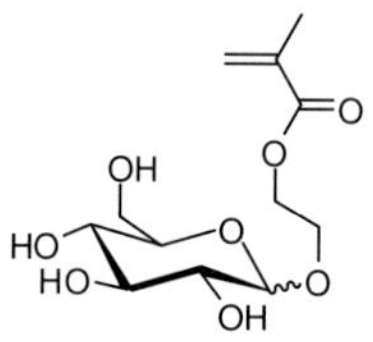

**151**[72,141]

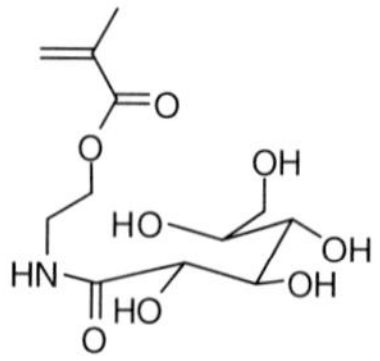

**152**[142]

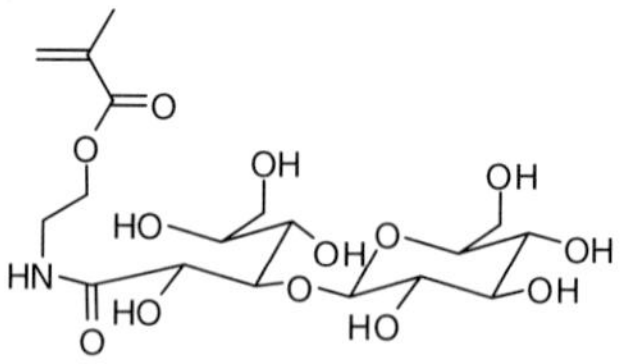

**153**[142]

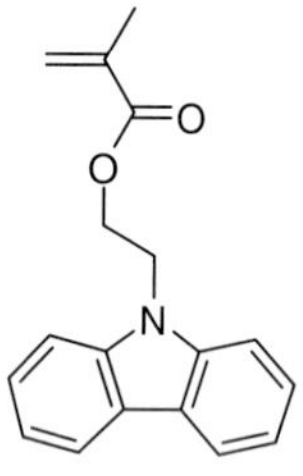

**154**[143]

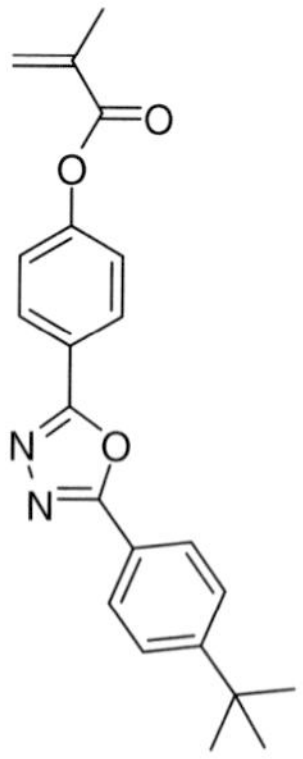

**155**[143]

## 6.3.2
## Acrylates

A wide variety of RAFT agents have been successfully used to control polymerization of acrylates. There is some discrepancy in reports on the level of control achieved in acrylate polymerization with dithiobenzoate RAFT agents. While dithiobenzoates can provide good control, they also give substantial retardation; and inhibition may be observed for low reaction temperatures or very high concentrations of the RAFT agent. Xanthates generally provide only limited control. The best balance is obtained with the use of trithiocarbonate, aliphatic dithioester or activated dithiocarbamates as RAFT agents.

Even though very narrow molecular weight distributions can be produced, bimodal molecular weight distributions are frequently observed and become more pronounced for higher molecular weights and higher monomer conversions (Figure 6.6).

**Table 6.6** RAFT polymerization of methacrylate esters and methacrylic acid in bulk or solution

| Monomer | RAFT agent[a] | Comments[b] |
|---|---|---|
| MMA | **8a** [20, 59, 129, 144, 145], **29** [73], **10** [144], **85a** [14], **85b** [60] | A, B |
| | **11a** [59, 129, 130, 146], **13** [136], **17** [144], **20** [70], **21** [70], **22** [70], **23** [20, 59], **24c** [71], **32a** [75, 77], **32b,c** [75], **33** [76], **77a** [146, 147], **77b** [13, 40], **122** [63], PMMA (**6**) [148], PMMA (**65**) [149], PBMA (**10**) [144] | B |
| | **24b** [59], **71** [150], **67** [129, 149], **86** [77], **87** [77] | C |
| | **25** [59], **27** [59], **28** [59], **34** [59], **41a** [59], PS (**6**) [59], PS (**65**) [149], **80** [93] | D |
| BMA | **8a** [151], **10** [144], **13** [136], PMMA (**33**) [76] | — |
| BzMA | **8a** [144], **13** [136] | — |
| GMA | **8a** [145], **61** [152] | — |
| PFMA | **16** [153] | — |
| HEMA | **8a** [145, 154], PBMA (**10**) [144], P**149** (**16**) [141] | — |
| PEGMA | **16** [155], **13** [135] | — |
| **144, 145** | **11a** [137] | |
| **146** | **8a** [138, 139], **13** [139] | — |
| **147** | **8a** [134], **11** [134] | — |
| **148** | **8a** [140] | — |
| **149, 150, 151** | **16** [73, 141] | — |
| DMAEMA | **8a** [145, 156], **11** [156], **13** [135, 136], PbzMA (**6**) [144] | — |
| TMAEMA | **26** [157] | — |
| **152, 153** | [142] | — |
| **154, 155** | **8a** [143] | — |
| MAA | PMMA (**6**) [144], PbzMA (**6**) [144] | — |

[a] Polymeric or macro-RAFT agents (used in block copolymer synthesis) take the form 'P'–'monomer'–'(RAFT end group)', where the latter end group is derived from the RAFT agent used in the synthesis of the polymeric RAFT agent.
[b] A – Marked retardation observed for higher RAFT agent concentrations. Some retardation for lower RAFT agent concentrations. B – Predicted molecular weights. Narrow molecular weight distributions ($\overline{M}_w/\overline{M}_n$ typically <1.2 at high conversion). Some retardation for higher RAFT agent concentrations. C – Molecular weight control observed only at high monomer conversion. Broader molecular weight distributions (but $\overline{M}_w/\overline{M}_n$ typically <1.4 at high conversion). Bimodal molecular weight distributions may be observed during initialization. D – Some molecular weight reduction. Little evidence of living characteristics.

There is significant retardation in the polymerization of acrylate esters in the presence of dithiobenzoate esters [59, 61, 129, 158–162]. Rates of polymerization of MA were identical with those of benzyl dithiobenzoate (**41a**) or cyanoisopropyl dithiobenzoate (**11a**) as RAFT agent at 60 °C, with substantial retardation being observed from the beginning of the experiment (Fig. 6.7). The retardation did appear to be directly related to consumption of the initial RAFT agent which was rapid, with the dithioester being completely consumed at the first time/conversion point. An aliphatic dithioester, benzyl dithioacetate (**69**), was found to give

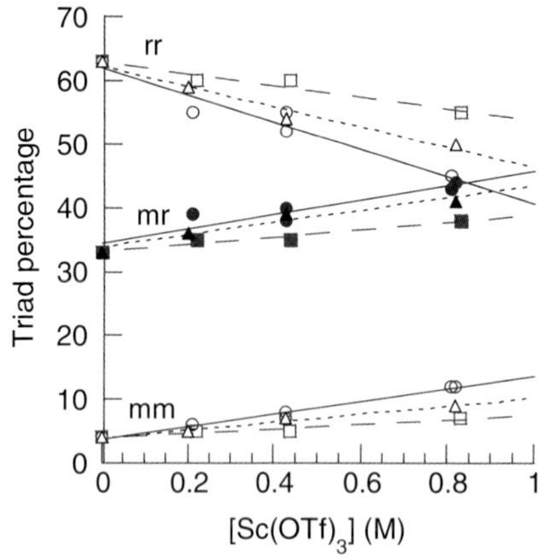


**Fig. 6.5** Percentage of syndiotactic (rr), heterotactic (*mr*) and isotactic (*mm*) triads as a function of the concentration of scandium triflate during RAFT polymerization of MMA (7.01 M in benzene) at 60 °C with cyanoisopropyl methyl trithiocarbonate (**77a**) for 20–30% (— — —), 40–50% (- - - -) and 85–95% monomer conversion (———). Reproduced from Ref. [132]. © 2007 American Chemical Society.

substantially less retardation under the same reaction conditions. The observation of less retardation in RAFT polymerization of acrylate esters with aliphatic and trithiocarbonate RAFT agents that is seen with dithiobenzoate RAFT agents has also been reported under other circumstances [61, 129, 158, 160, 162]. Quinn et al. [163] observed that 1-phenylethyl dithiophenylacetate (**72**) enabled RAFT polymerization of MA at ambient temperature whereas 1-phenylethyl dithiobenzoate (**34**)

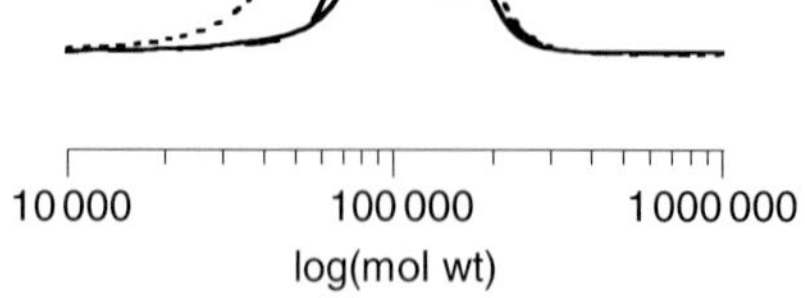


**Fig. 6.6** GPC traces of high conversion PMA prepared in the presence of various RAFT agents. Molecular weight distributions shifted so as to correct for differences in conversion to facilitate comparison (*X*-axis correct for sample made with **12**). Samples prepared with **139** (0.0019 M) $M_n$ 87 000, $M_w/M_n$ 1.19, 72% conversion (·······); **138** (0.0036 M) $M_n$ 110 100, $M_w/M_n$ 1.08, 89% conversion (– – – –); **77a** (0.0037 M) $M_n$ 123 700, $M_w/M_n$ 1.08, 92% conversion (———). Molecular weights are in polystyrene equivalents. Initiator is AIBN (0.00033 M). Reproduced from Ref. [158]. © 2003 American Chemical Society.

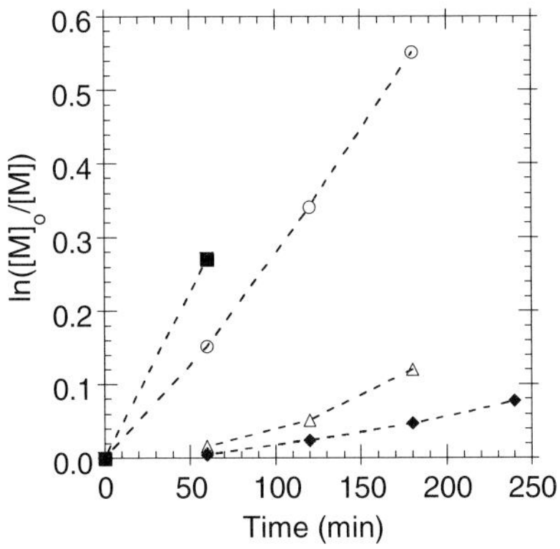


**Fig. 6.7** Pseudo-first-order rate plot for bulk polymerization of MA (4.45 M in benzene) at 60 °C with ~3.3 × $10^{-4}$ M AIBN in the absence (■) or presence of MeC(=S)$CH_2$Ph (**69**, 0.00306 M) (○); MeC(=S)$CH_2$Ph (**69**, 0.0306 M) (Δ); PhC(=S)SC$(Me)_2$CN (**11a**, 0.003 66 M) and PhC(=S)S$CH_2$Ph (**41a**) (◇). The data points for RAFT polymerization with **11a** and **41a** are superimposed. Reproduced from Ref. [59]. © 2003 American Chemical Society.

inhibited polymerization under those conditions. McLeary et al. [162] observed that RAFT polymerization of MA with cumyl dithiophenylacetate (**71**) was subject to an inhibition period corresponding to the consumption of the initial RAFT agent. This was attributed to slow reinitiation by the cumyl radical during what was called the initialization period. Available data indicate that the rate constant for addition of cumyl radicals to MA is slow with respect to that for propagation [59, 164]. However, the reported rate constants for benzyl and cyanoisopropyl radicals adding to MA is similarly slow or slower with respect to propagation [59, 164], yet no similar substantial inhibition period is seen with these RAFT agents. Moad and coworkers [59] proposed that the inhibition period relates not to slow reinitiation by cumyl radical in itself but rather to the importance of the back reaction of cumyl radicals with the RAFT agent. Cumyl radicals add to RAFT agents at close to diffusion-controlled rates, which has the effect of magnifying any problems associated with slow reinitiation and also leads to concentration dependence of the apparent transfer constants (see also Chapter 3).

A wide range of functional acrylate esters have been polymerized. Functionality includes tertiary amino (in DMAEA), quaternary amino (in TMAEA), carboxylic acid (in AA), hydroxyl (in HEA) and epoxy (in GA).

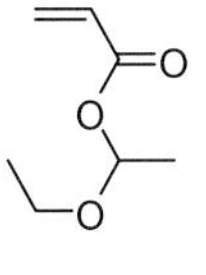


**156**[165]

### 6.3.3
### Acrylonitrile

There are only a few reports of AN polymerization in the literature [166, 167, 175, 176, 178]. Examples of AN polymerization are included as the final entries in Table 6.7. Marked retardation was observed with cumyl dithiobenzoate as RAFT agent. Better control was obtained with either 1-cyanoethyl dithiobenzoate (**35**) or cyanoisopropyl dithiobenzoate (**11a**) [175, 176] and with the trithiocarbonate **82b** [177]. Polyacrylonitrile block copolymers have been prepared from poly(acrylic acid) macro-RAFT agent PAA (**76**) [177].

One difficulty in polymerization of acrylonitrile is the poor solubility of PAN in most polymerization media. It has been suggested that transfer to solvent is an issue

**Table 6.7** RAFT polymerization of acrylic monomers in bulk or solution

| Monomer | RAFT agent[a] | Comments[b] |
|---|---|---|
| MA | **11a** [59, 158], **13** [136], **18** [68], **32a** [77], **41a** [59, 158] | A |
| | **77a** [158], **69** [158], **87** [77], **139** [158], **138** [13, 158], PAN(**6**) [166, 167] | B |
| EA | **80** [93], **82b** [93] | — |
| BA | **7** [59], **8a** [61], **34** [59], **36** [80], **41a** [59], **48a** [80], **49** [80], **69** [20], **137** [112] | A |
| | P(PEGA)(**76**) [101], **80** [93], **82b** [93] | B |
| | **111a** [168] | C |
| TBA | **13** [136] | |
| BzA | **13** [136] | |
| EHA | **13** [136] | |
| HEA | **80** [93] | — |
| DEGEA | **111a** [168] | |
| PEGA | **96** [101] | — |
| **155** | **13** [165] | |
| AA | **34** [20], **41a** [79, 169] | A |
| | **80** [93], **82b** [93], **89a** [170, 171], **93** [79] **117b** [172], **138** [13, 173, 174] | |
| | PBA(**111a**) [168], PDEGEA(**111a**) [168], **105** [79] | C |
| | **103** [79] | D |
| AN | **11** [175, 176], **35** [167], PAA(**76**) [177], **82b** [177], **93** [178] | B |
| | **7a** [167] | C |
| | PBA(**6**) [167] | D |

[a] Polymeric or macro-RAFT agents (used in block copolymer synthesis) take the form 'P'–'monomer'–'(RAFT end group)', where the latter end group is derived from the RAFT agent used in the synthesis of the polymeric RAFT agent.
[b] A – Predicted molecular weights. Narrow molecular weight distributions ($\overline{M}_w/\overline{M}_n$ typically <1.2). Marked retardation or inhibition observed for higher RAFT agent concentrations. Significant retardation for lower RAFT agent concentrations. B – Predicted molecular weights. Narrow molecular weight distributions ($\overline{M}_w/\overline{M}_n$ typically <1.2). Some retardation for higher RAFT agent concentrations may be observed. C – Molecular weight control observed only at high monomer conversion. Broader molecular weight distributions (but $\overline{M}_w/\overline{M}_n$ typically >1.4). D – Some molecular weight reduction. Little evidence of living characteristics.

with either *N,N*-dimethylformamide or dimethyl sulfoxide as solvent and the use of ethylene carbonate has been recommended particularly for higher polymerization temperatures (90 °C) [176].

### 6.3.4 Acrylamides and Methacrylamides

Many papers have appeared on RAFT polymerization of acrylamides with most work focused on NIPAM and DMAM. RAFT agents used include dithioesters [107, 179–181], dithiocarbamates [108] and trithiocarbonates [94, 96, 182] as summarized in Table 6.8. Many acrylamides are water soluble and thus are often polymerized in aqueous solution. This topic is the subject of another chapter of this book and the reader is referred to Chapter 7 for further examples. Large numbers of acrylamide block copolymers have also been reported.

Some features of RAFT polymerization of acrylamides are similar to those already discussed for the case of acrylate polymerization. RAFT polymerization of acrylamides can be carried out in the presence of Lewis acids to obtain simultaneous control over molecular weight and tacticity and an enhanced rate of polymerization [11, 180, 183, 184].

For methacrylamides, similar considerations to those discussed for methacrylate esters apply. Choice of R is crucial to controlling polymerization since R needs to be a good homolytic leaving group with respect to the propagating radical. Thus, R should be tertiary cyanoalkyl or similar. In the synthesis of block copolymers with blocks based on monosubstituted monomers, the methacrylamide block should be prepared first [185].

**Table 6.8** RAFT polymerization of methacrylamides and acrylamides in bulk or solution

| Monomer | RAFT agent[a] | Comments |
|---|---|---|
| MAM | **16** [185] | |
| HPMAM | **16** [186] | |
| AM | **16** [185], PAM (**16**) [185], **80** [93] | |
| DMAM | **19** [69], **32a** [77], **81** [69], **87** [77], PS (**76**) [69] | |
| NIPAM | **121** [108] | Inhibition period, good control |
| | **47** [87], **64** [90], **74** [181], **82a** [94], **82b** [93], **99** [107], **124** [108] | Good control |
| TBAM | **28** [187], **82b** [93] | |
| **157** | **28** [187] | |
| **158, 159** | **82** [188], **P158** (**83**) [188] | |

[a] Polymeric or macro-RAFT agents (used in block copolymer synthesis) take the form 'P'–'monomer'–'(RAFT end group)' where the latter end group is derived from the RAFT agent used in synthesis of the polymeric RAFT agent.

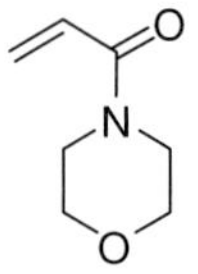

157 (NAM)

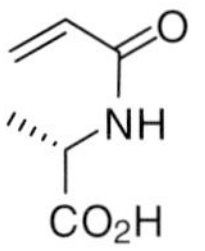

158

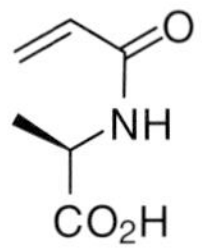

159

6.3.5
### Styrene and Related Monomers

Styrene is one of the most polymerized monomers by the RAFT process. There have been many studies on the kinetics of styrene polymerization in the presence of various RAFT agents. These studies are mentioned in greater detail in Chapter 3. Retardation may be observed when high concentrations of the RAFT agent are used [61].

Styrene is conveniently polymerized in a purely thermal process in which monomer and RAFT agent are heated to >100 °C preferably in the absence of air. The results for a series of thermal styrene polymerizations performed with a range of concentrations of cumyl dithiobenzoate (**8a**) (0.001–0.0029 M) are shown in Fig. 6.8 [61]. The polystyrene formed with the highest concentration of the RAFT agent appears to be of very narrow polydispersity and has a unimodal molecular weight distribution (see Fig. 6.8). The styrene conversion (~55%) is reduced with respect to the control experiment (~72%). This can be largely attributed to a reduced gel or Trommsdorf effect. As the concentration of the RAFT agent used in the experiments is decreased, polydispersities increase and a high-molecular-weight shoulder in the molecular weight distribution becomes more evident (Fig. 6.8). The peak molecular weight at the shoulder is approximately twofold the molecular weight at the main peak. This is consistent with this shoulder arising from termination by coupling of polystyryl propagating radicals. For lower dithioester concentration, the molecular weight distribution is also seen to tail to lower molecular weights. The shoulder corresponds in amount to that anticipated to arise from termination by combination and analysis by UV demonstrates that eluting polymer responsible for the shoulder contains no thiocarbonylthio chain ends [61]. This behavior observed for styrene polymerization appears in marked contrast to what is observed during polymerization of high-$k_p$ monomer such as acrylate esters (see above).

The polymerization behavior is strongly dependent on the RAFT agent concentration, the polymerization temperature and the specific RAFT agent. With high concentrations of cumyl dithiobenzoate (**8a**) retardation is manifest, as an 'inhibition period' is observed during which the RAFT agent is only slowly consumed.

For an experiment with 0.02 M **8a** (50% v/v styrene in toluene at 110 °C), the molecular weight is significantly greater than that expected based on complete consumption of the RAFT agent during this period [61]. For longer reaction times, after

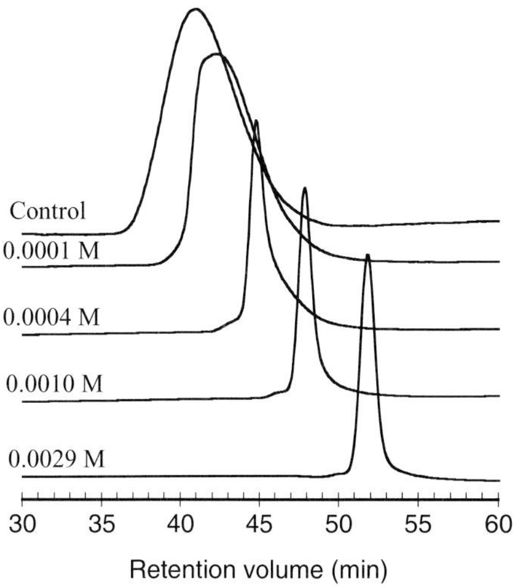


**Fig. 6.8** GPC elution profiles for polystyrenes prepared by thermal polymerization of styrene in the presence of various concentrations of cumyl dithiobenzoate (**8a**) at 110 °C for 16 h. From top to bottom are the control ($\overline{M}_n$ 323 700, $\overline{M}_w/\overline{M}_n$ 1.74, 72% conversion), 0.0001 M **8a** ($\overline{M}_n$ 189 300, $\overline{M}_w/\overline{M}_n$ 1.59, 59% conversion), 0.0004 M **8a** ($\overline{M}_n$ 106 600, $\overline{M}_w/\overline{M}_n$ 1.21, 60% conversion), 0.0010 M **8a** ($\overline{M}_n$ 48 065, $\overline{M}_w/\overline{M}_n$ 1.07, 55% conversion) and 0.0029 M **8a** ($\overline{M}_n$ 14 400, $\overline{M}_w/\overline{M}_n$ 1.04, 55% conversion). Reproduced from Ref. [61]. © 2000 Society of Chemical Industry first published by John Wiley & Sons, Inc.

the initial RAFT agent has been converted to a polymeric species, the polymerization rate increased. This phenomenon has been termed 'hybrid behavior' and is described in greater detail in Chapter 3. The use of the RAFT agent **11a**, which contain a more effective leaving group/initiating species R (the cyanoisopropyl radical), alleviated retardation [61].

With even higher concentrations of **8a** at lower temperature (0.1 M, 50% v/v styrene in benzene-$d_6$ at 70 °C), no polymerization was observed until all of the initial RAFT agent had been converted to 'unimer' RAFT agent when formation of 'dimer' RAFT agent was observed. The behavior has been called efficient initialization [189, 190] and is also observed with the RAFT agent **11a**. Again further discussion can be found in Chapter 3.

For thermal polymerization of styrene at 110 °C with RAFT agent concentrations <0.02 M, inhibition periods are short (<10 min) and longer-term monomer conversions appear to be largely independent of the RAFT agent and its effectiveness in controlling polymerization [58].

A variety of styrene derivatives, vinyl aromatics and related monomers have been subjected to RAFT polymerization. These include the carbazole derivative **160** [191] and acenaphthalene (**161**). A wide range of RAFT agents can be used, including aromatic dithioesters (dithiobenzoates), trithiocarbonates, aliphatic dithioesters or

**Table 6.9** RAFT polymerization of styrene and derivatives in bulk or solution

| Monomer | RAFT agent | Comments[a] |
|---|---|---|
| Styrene | **8a** [20, 58, 59, 61, 195], **11a** [13, 58, 59, 61], **19** [69], **20** [70], **21** [70], **22** [70], **24b** [59], **27** [59], **32a** [75, 77], **34** [59], **36** [80], **41a** [59], **42** [88], **47** [196], **61** [197], **66** [58], **69** [58], **77a** [58], **80** [93], **81** [69], **87** [77], **93** [58], **95** [100], **122** [58], **124** [58], **127–130** [109], **131a** [109], **131b** [110], **134a,b** [198], **134c** [111], **139** [58] | A |
| | **48a** [80], **49** [80, 88], **102** [58], **105** [58], **107** [199], **110** [199], **111a** [199], **114** [58], **132** [109], **133** [109] | B |
| | **51** [88], **143** [58] | C |
| SSO$_3$Na | **16** [20] | |
| **160** | **8a** [191] | |
| 2VP, 4VP | **8a** [200] | |
| **161** | **11a** [192, 193], **18** [68], PMA(**18**) [68] | |
| **162** | **50** [68] | |
| **163** | **124** [194] | |

[a] A – Predicted molecular weights. Narrow molecular weight distributions observed at high conversion ($\overline{M}_w/\overline{M}_n$ typically <1.2). Some retardation for higher RAFT agent concentrations may be observed. B – Molecular weight control observed only at high monomer conversion (so-called hybrid behavior). Broader molecular weight distributions ($\overline{M}_w/\overline{M}_n$ typically >1.4 even at high conversion). Bimodal molecular weight distributions may be observed during initialization. C – Some molecular weight reduction. Little evidence of living characteristics.

activated dithiocarbamates as RAFT agents (Table 6.9). Xanthates may also be used but show substantially poorer control (broader molecular weight distributions).

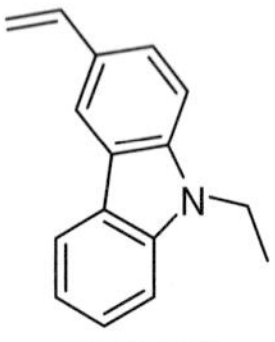


**160**[191]

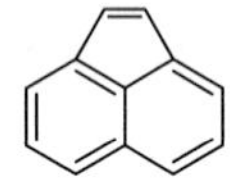

**161**[68,192,193]

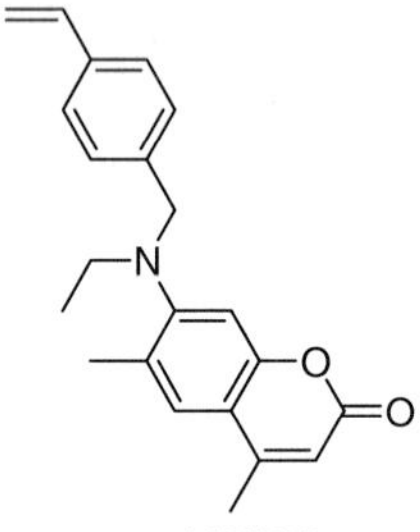


**162**[68]

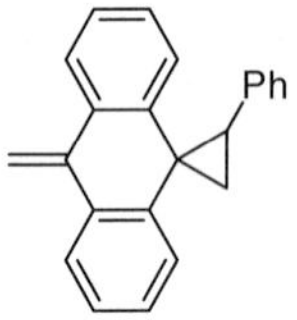


**163**[194]

## 6.3.6
## Diene Monomers

Little has been reported on RAFT polymerization of diene monomers (butadiene, isoprene).

Lebreton et al. [71] reported on RAFT polymerization of butadiene using various fluorinated dithiobenzoates **24c** or **32d** as RAFT agents.

Jitchum and Perrier [201] reported on RAFT polymerization of isoprene. They observed marked retardation. Reasonable control was obtained with the use of the trithiocarbonate **88** at 115 °C. Use of lower temperatures (60 and 90 °C) gave very slow polymerization. Use of higher temperatures (130 °C) gave more rapid polymerization and broader molecular weight distributions. Poorer results were obtained with cyanoisopropyl dithiobenzoate (**11a**), which was attributed to instability of the RAFT agent under the polymerization conditions (120 °C).

The difficulty lies with the high polymerization temperatures that are required to obtain acceptable rates of fragmentation and perhaps to the stability of RAFT agents.

### 6.3.7
### Vinyl Acetate, *N*-Vinylpyrrolidone and Related Monomers

The RAFT agents most suited for use with VAc, NVP and related monomers are xanthates and unactivated (or less activated) dithiocarbamates (refer Fig. 6.2). Use of dithioesters and trithiocarbonates gives inhibition, which is attributed to the relative stability of the intermediate radical which in turn is a consequence of the propagating radicals derived from VAc, NVP and related monomers that are comparatively poor homolytic leaving groups.

Related monomers that have been subjected to RAFT polymerization include other vinyl esters, such as vinyl benzoate, vinyl propionate and the glycomonomer **164**, and VCBz. These examples are included in Table 6.10.

**Table 6.10** RAFT polymerization of vinyl acetate and related monomers in bulk or solution

| Monomer | RAFT agent | Comments[a] |
|---|---|---|
| VAc | **119** [104] | A |
| | **106** [104], **117** [104], **118** [104], **120** [107] | B |
| | **111a** [22, 207], **111b** [213], **116** [22] **117** [209, 210] | C |
| | **141** [22] | D |
| VBz | **11a** [20] | D |
| **164** | **142** [113], **147** [113] | B |
| NVP | **68** [105], **99** [107] | A |
| | **110** [106, 107], **111a** [214], **116** [13], **135** [215] | B |
| | **120** [107] | C |
| | **114** [106, 107] | D |
| VCBz | **110** [216] | C |

[a] A – Little or no polymerization. B – Predicted molecular weights. Narrow molecular weight distributions (typically <1.2). Inhibition period observed. C – Predicted molecular weights. Narrow molecular weight distributions (typically <1.2). No inhibition period. D – Poorer control, broader molecular weight distribution (typically <1.4).

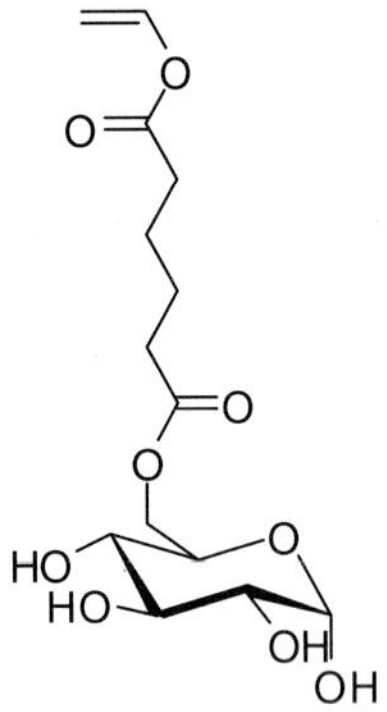


**164** [114]

#### 6.3.7.1 Vinyl Acetate

Poly(vinyl acetate) (PVAc) and its derivative, poly(vinyl alcohol) (PVA), are extremely important industrially for the production of adhesives and paints and have recently been investigated for various biomedical applications [202, 203]. Living radical polymerization of VAc is often problematic and techniques such as ATRP and NMP are generally not effective. Recently, cobalt-mediated living radical polymerization of VAc has been reported [204]. Polymerization with reversible chain transfer with certain iodo compounds [205] or organostibine derivatives [206] has also been shown to provide some control over VAc polymerization, allowing preparation of PVAc with narrow molecular weight distribution.

Early work showed that RAFT polymerization with certain xanthate [22, 207] and certain (nonactivated) dithiocarbamate RAFT agents [22, 62] can be successful and a significant number of papers on the use of these reagents have now appeared [104, 208–211]. RAFT polymerization with xanthates is sometimes called MADIX (macromolecular design by interchange of xanthate) [13, 207]. RAFT polymerization with dithioester or trithiocarbonate RAFT agents is strongly retarded [13, 22].

For polymerization of VAc with xanthate RAFT agents, the choice of the *O*-alkyl substituent is important [104, 212]. For example, control (predicted $\overline{M}_n$, low $\overline{M}_w/\overline{M}_n$) can be obtained in RAFT polymerization with *O*-methyl, *O*-ethyl (**117**), *O*-isopropyl (**118**) and *O*-aryl xanthates, but not with the *O*-*tert*-butyl xanthates (**119**) [104]. It is important that the alkyl on oxygen is a very poor homolytic leaving group with respect to the alkyl group on sulfur for cleavage of the 'S—R' bond to be favored over cleavage of the 'O—alkyl' bond [212]. Electron-withdrawing substituents on oxygen enhance the transfer constant of xanthates [58]. The choice of the R group is also extremely important. It is necessary to choose R such that the radical R• is able to efficiently reinitiate VAc polymerization. For example, benzyl radical is slow to add to VAc and is a poor R group. RAFT polymerization of VAc with *S*-phthalimidomethyl xanthate (**120**) gave good control over both molecular weight and polydispersity [107].

#### 6.3.7.2 *N*-Vinylpyrrolidone

Until recently little has been published on living radical polymerization of NVP. In their review, Moad et al. [13] mentioned RAFT polymerization of NVP in methanol with xanthate **116** at 60 °C to provide $\overline{M}_n$ of 17 000 and $\overline{M}_w/\overline{M}_n$ of 1.35 for $[NVP]_o/[RAFT]_o = 50$ and 53% conversion. Devasia et al. have recently described RAFT polymerization of NVP in dioxane with xanthate **111a** [214] or dithiocarbamate **135** [215] at 80 °C. For **111a** they reported $\overline{M}_n$ of 8000 and $\overline{M}_w/\overline{M}_n$ of 1.3 for $[NVP]_o/[RAFT]_o = 100$ and 80% conversion. Poor control ($\overline{M}_w/\overline{M}_n > 1.5$) was obtained using higher $[NVP]_o/[RAFT]_o$. Wan et al. [105] reported that the dithioacetate **68** inhibited NVP polymerization. They [105] and Nguyen et al. [106] explored the use of benzyl xanthate (**114**) and 1-phenylethyl xanthate (**110**) for controlling NVP polymerization. Wan et al. reported inhibition periods of 6 h and 1 h respectively (AIBN initiator, 60 °C). Nguyen et al. [106] observed shorter inhibition periods. The inhibition periods seen with **114** and **110** are most likely attributed to benzylic radicals having poor reinitiating ability in NVP polymerization. It was also clear that the xanthate RAFT agent with R = benzyl (**114**) had a significantly lower transfer constant than that with R = 1-phenylethyl (**110**) (high initial molecular weight, broader molecular weight distribution).

Postma et al. [107] found that polymerization of NVP in the presence of the xanthate RAFT agent **120** with R = phthalimidomethyl provided good control with no discernable inhibition period. An analogous trithiocarbonate (**99**) gave inhibition.

### 6.3.8 Gradient Copolymers

In most copolymerizations, the monomers are consumed at different rates dictated by the steric and electronic properties of the reactants. Consequently, both the monomer feed and copolymer composition will drift with conversion. Thus, conventional copolymers are generally not homogeneous in composition at the molecular level. In RAFT polymerization processes, where all chains grow throughout the polymerization, any composition drift is captured within the chain structure. All chains have similar composition and are called gradient or tapered copolymers.

The overall composition of RAFT-synthesized copolymers will generally be the same as that of copolymers formed by a similar conventional radical copolymerization. Reactivity ratios are not affected. The exception is at very low conversion when all chains are short and when specificity shown in the transfer and reinitiation steps will affect composition. This corresponds to the conditions used for reactivity ratio determination.

It is possible to synthesize gradient block polymers in a batch polymerization by taking advantage of disparate reactivity ratios between particular monomer pairs (BA–MMA [25, 217], MA–VAc [13], tBA–VAc [218], styrene–MAH [82, 83, 219], styrene–NPMI [13]). The composition can be further tailored by the use of suitable monomer feed protocols.

A detailed study of the kinetics of RAFT copolymerization of styrene and MMA with dithioacetate RAFT agents has been reported [149]. Transfer constants ($C_{tr} = k_{tr}/k_p$, where $k_{tr} = k_{add}[k_\beta/(k_{-add} + k_{-\beta})]$ at 40 °C were reported for the PMMA and polystyrene macro-RAFT agents in polymerizations of MMA, styrene and the azeotropic copolymerization of MMA and styrene. The data indicate that in styrene polymerization, 50% conversion of the PMMA macro-RAFT agent is achieved at very low monomer conversion (~0.16%), while in MMA polymerization, the polystyrene macro-RAFT agent is half consumed at a much higher conversion (~57%). The results provide further quantitative support for the observation that when preparing block copolymers of methacrylates with styrene (and other monosubstituted monomers) it is best to prepare the methacrylate block first (see below). The result may be related to the observation that RAFT polymerization of MMA with benzyl dithiobenzoate (**41a**) provides very poor control, yet for copolymerization of styrene with MMA control is retained while the medium contains some (at least 5%) styrene.

RAFT copolymerization can be successful (provide molecular weight control and narrow molecular weight distributions) even when one of the monomers is not amenable to direct homopolymerization using a particular RAFT agent. Examples include MAH, VAc and NVP with trithiocarbonate RAFT agents and EAA (see Table 6.11 for leading references).

**Table 6.11** Syntheses of gradient copolymers by RAFT polymerization

| Monomers | RAFT agent | Comments |
|---|---|---|
| MMA/HEMA | **8a** [13] | |
| MMA/BA | **8a** [25, 217] | Gradient block synthesis |
| MMA/styrene | | |
| MA/VAc | **116** [13] | Gradient block synthesis |
| AA/EAA | **117b** [262] | |
| BA/EAA | **117b** [262] | |
| DMA/EAA | **117b** [262] | |
| DMAEA/EAA | **117b** [262] | |
| BA/styrene | **70** [91] | Controlled feed addition |
| BA/VDC | **77a** [13] | |
| ODA/NVP | **37** [221] | |
| ODA/MAH | **37** [221] | |
| HPMA/NMS | **11a** [224] | |
| NAM/NAS | **28** [225], TBAM(**28**) [225] | |
| AN/IP | **77a** [13] | |
| Styrene/AN | **8a** [13, 20], PMMA(**8a**) [145] | |
| Styrene/MAH | **41a** [82, 83], **8a** [227], **11a** [219, 227] | Gradient block synthesis |
| Styrene/NPMI | **8a** [13] | Gradient block synthesis |
| Styrene/MAH/NVP | **93** [222] | |
| SCl/MAH | **41a** [82] | |
| SCl/NPMI | **24b** [228] | |
| SMe/MAH | **41a** [82] | |
| SOMe/MAH | **41a** [82] | |

Other examples of gradient copolymers include hydrophilic–hydrophobic copolymers based on long-chain acrylates (e.g. ODA, DA) [220, 221]. The copolymers P(ODA-*grad*-NVP) and P(ODA-*grad*-MAH) prepared with trithiocarbonate **91** [221] and which find use as dispersants in polypropylene–clay nanocomposites. Other examples are the terpolymer of styrene, MAH and NVP [222] and various copolymers based on phosphonated monomers [223].

Copolymers of *N*-(meth)acryloyl succinimide or pentafluorophenyl methacrylate have been produced by RAFT polymerization and served as substrates for biofunctionalization or other grafting reactions using *grafting-to* processes (NAS, NMS and PFMA are referred to as active esters of (meth)acrylates) [153, 224–226].

### 6.3.9 Block Copolymers

RAFT polymerization is recognized as one of the most versatile methods for block copolymer synthesis and numerous examples of block synthesis have now appeared in the literature. RAFT polymerization proceeds with retention of the thiocarbonylthio group. This allows an easy entry to the synthesis of AB diblock copolymers by the simple addition of a second monomer [144]. Higher-order (ABA, ABC, etc.) blocks are possible by sequential addition of further monomer(s). Examples of block copolymers formed in this way are included in the tables above.

In RAFT polymerization, the order of constructing the blocks of a block copolymer can be very important [59, 144]. The propagating radical for the first-formed block must be a good homolytic leaving group with respect to that of the second block. For example, in the synthesis of a methacrylate–acrylate or methacrylate–styrene diblock, the methacrylate block should be prepared first [144, 148]. The styrene or acrylate propagating radicals are very poor leaving groups with respect to methacrylate propagating radicals. In some cases it is possible to prepare block copolymers in the reverse direction (e.g. growing an MMA block from a polystyrene macro-RAFT agent) by making use of a feed addition protocol which minimizes the concentration of the monomer with respect to the RAFT agent [61].

Block copolymers based on polymers formed by other mechanisms can be prepared by forming a prepolymer containing thiocarbonylthio groups and using this as a macro-RAFT agent [144, 218]. The first example of applying this methodology involved preparation of PEO-*block*-PS from commercially available hydroxy end-functional PEO [144, 218]. Many other examples of the preparation of block copolymers using related strategies have now been reported. They include poly(ethylene-*co*-butylene)-*block*-poly(styrene-*co*-MAH) [219], PEO-*block*-PMMA [77], PEO-*block*-poly(*N*-vinyl formamide) [229], PEO-*block*-PNIPAM [230], PEO-*block*-poly(1,1,2,2-tetrahydroperfluorodecyl acrylate) [231], PEO-*block*-PMMA-*block*-PS [78], PLA-*block*-PMMA [77], PLA-*block*-PNIPAM [232, 233] and PLA-*block*-PDMA-*block*-PS [234].

Star and graft copolymers and more complex architectures can be prepared by applying a similar strategy using a multifunctional precursor RAFT agent.

## 6.4 RAFT Polymerization Conditions

Aspects of the polymerization conditions that are monomer specific have been mentioned in the sections above. In this section we discuss generic aspects of the polymerization conditions in bulk or (organic) solution.

### 6.4.1 Temperature

Temperatures reported for RAFT polymerization range from ambient to 140 °C. There is evidence with dithiobenzoates that retardation, when observed, is less at higher temperatures and also some data that show narrower molecular weight distributions can be achieved at higher temperatures [25]. This is consistent with rate constants for fragmentation of the RAFT intermediates and over transfer constants of RAFT agents increasing with reaction temperature. However, for MMA polymerization with trithiocarbonate **78** at 60 and at 90 °C there appears to be no significant effect of temperature on the molecular weight or molecular weight distribution observed [13]. Higher temperature does allow higher rates of polymerization, allowing a given conversion to be achieved in a shorter reaction time.

RAFT polymerization of 'polar' monomers (MMA, MA) was reported to be substantially accelerated by microwave heating [235]. No similar acceleration is observed for styrene polymerization [235, 236]. It is expected that monomers with a higher dielectric constant will be more effectively heated by microwave irradiation. However, the effect with MMA and MA was more than that expected for an effect of temperature alone [235]. An explanation for the microwave effect was not provided [235].

Cumyl dithiobenzoate (**8a**) appears substantially less stable than benzyl or 1-phenylethyl dithiobenzoate and degrades rapidly at temperatures $>100$ °C [237]. The instability was attributed to reversible dissociation to form AMS and dithiobenzoic acid. The success of high-temperature polymerization (of, for example, styrene) was attributed to the fact that the RAFT agent **8a** was rapidly consumed and converted to more stable polymeric RAFT agents (**6**). It was also suggested that the poor control observed in synthesis of PMMA with dithiobenzoate RAFT agents at higher temperatures ($\geq 120$ °C) might be attributed to the lability of the dithiobenzoate end group [131].

### 6.4.2 Pressure

RAFT polymerization of styrene with cumyl dithiobenzoate (**8a**) under very high pressure (5 kbar) has been reported [238–241]. At very high pressure, radical–radical termination is slowed down and this allows the formation of higher-molecular-weight polymers and higher rates of polymerization than are achievable at ambient pressure.

### 6.4.3
### Solvent Selection

Generally, the polymerization conditions for solution or bulk RAFT polymerization are the same as those for conventional radical polymerization. The RAFT process is compatible with a wide range of reaction media including all common organic solvents, protic solvents such as alcohols and water [20, 242, 243] (see also Chapter 7), and less conventional solvents such as ionic liquids [244] and supercritical carbon dioxide [245, 246]. It is important that the RAFT agent should be selected for solubility in the reaction medium. In polar media and in the presence of Lewis acids, RAFT agents can show hydrolytic sensitivity [72, 157]. We have found that this order roughly correlates with the RAFT agent activity (dithiobenzoates > trithiocarbonates ~ aliphatic dithioesters).

### 6.4.4
### Initiator Selection

For optimal control of the RAFT process, it is important to pay attention to such factors as initiator concentration and selection [13]. RAFT polymerization is usually carried out with conventional radical initiators. In principle, any source of free radicals can be used [21] but most often thermal initiators (e.g. AIBN, ACP, $K_2S_2O_8$) are used. Styrene polymerization may be initiated thermally between 100 and 120 °C. Polymerizations initiated with UV irradiation [195, 247, 248], a gamma source [249–256] or a plasma field [257] have also been reported. In the latter polymerizations, radicals may be generated directly from the RAFT agent and these may be responsible for initiation. It was initially suggested by Pan and coworkers that the mechanism for molecular weight control in UV- [247] and gamma-initiated [251, 252] processes might involve only reversible coupling and be similar to that proposed by Otsu [258] to describe the chemistry of dithiocarbamate photoiniferters (see Chapter 3). However, Quinn et al. [195, 253] demonstrated that the living behavior observed in these polymerizations can be attributed to the standard RAFT mechanism.

The initiator concentration and rate of radical generation in RAFT polymerization should be chosen to provide a balance between an acceptable rate of polymerization and an acceptable level of dead chains (radical–radical termination). One useful guideline is to choose conditions such that the target molecular weight is ~10% of that which would have been obtained in the absence of the RAFT agent. A common misconception is that it is necessary to use very low rates of polymerization in order to achieve narrow molecular weight distributions. Sometimes, using a high rate of polymerization and a correspondingly short reaction time can provide excellent results (a narrow molecular weight distribution; see, for example, [58]). However, it is very important not to use prolonged reaction times when retention of the RAFT functionality is important. Once the monomer is fully converted, continued radical generation may not change the molecular weight distribution, but it can lead to loss of the RAFT end group and formation of dead chains.

Side reactions of the initiator or initiator-derived radicals with the RAFT agent are possible. However, these are not always readily discernable or of significance because of the high RAFT agent/initiator ratios used in well-designed experiments. It follows from the mechanism of the RAFT process that there should be a fraction of dead chains formed which relates directly to the number of initiator-derived radicals. Ideally, this fraction should be taken into account when calculating the molecular weights of polymers formed by the RAFT process [58]. The molecular weight of the polymer formed can usually be estimated knowing the concentration of the monomer consumed and the initial RAFT agent concentration using the relationship 6.1. Positive deviations from equation 6.1 indicate incomplete usage of the RAFT agent. Negative deviations indicate that other sources of polymer chains are significant. These include the initiator-derived chains.

If initiator-derived chains are significant, equation 6.2 should be used to calculate molecular weights [13]:

$$\overline{M}_n(\text{calc}) = \frac{[\text{M}]_o - [\text{M}]_t}{[\mathbf{1}]_o + df([\text{I}]_o - [\text{I}]_t)} m_\text{M} + m_\text{RAFT} \tag{6.2}$$

where $m_\text{M}$ and $m_\text{RAFT}$ are the molecular weight of the monomer and the RAFT agent respectively, $d$ is the number of chains produced from radical–radical termination ($d \sim 1.67$ in MMA and $d \sim 1.0$ in styrene polymerization), $[\text{I}]_o - [\text{I}]_t$ is the concentration of the initiator consumed and $f$ is the initiator efficiency.

If the initiator decomposition rate constant is known, the initiator consumption can be estimated using equation 6.3:

$$[\text{I}]_o - [\text{I}]_t = [\text{I}]_o(1 - e^{-k_d t}) \tag{6.3}$$

The fraction of living chains ($L$) in RAFT polymerization (assuming no other side reactions) is given by equation 6.4.

$$L = \frac{[\mathbf{1}]_o}{[\mathbf{1}]_o + df([\text{I}]_o - [\text{I}]_t)} \tag{6.4}$$

Some initiators (e.g. dibenzoyl peroxide, potassium peroxydisulfate) and the derived radicals may oxidize RAFT agents to the sulfine or other products [259]. Other initiator radicals may react with the RAFT agent to form a stable thiocarbonylthio compound. It is important that the initiator-derived radical is a good leaving group with respect to the propagating radical. For example, use of an aliphatic diacyl peroxides (e.g. dilauroyl peroxide) will provide a relatively stable 'RAFT agent' with R = primary alkyl. Similarly, azobis(methyl isobutyrate) (AIBMe) is not a suitable choice for RAFT polymerization of MMA. During RAFT polymerization of MMA with AIBMe and **8a** or **11a**, formation of the initiating radical-derived RAFT agent **24a** as a relatively stable by-product is observed [129].

The mechanism of AIBN and other azonitriles decomposition is complicated by the formation of ketenimines as unstable intermediates [260]. In the presence of high concentrations of RAFT agents, the ketenimine is intercepted and converted to by-products with consequences for the initiator efficiency and the polymerization kinetics.


## Acknowledgements

Sections of this review where indicated are adapted with permission from the reviews of Moad, Thang and Rizzardo [13, 14] first published in the *Australian Journal of Chemistry* © 2005, 2006 CSIRO or Moad and Solomon *The Chemistry of Radical Polymerization* [1] © 2006 Elsevier.


## Abbreviations

| | |
|---|---|
| AA | Acrylic acid |
| ACP | Azobis(2-cyanopentanoic acid) |
| AIBN | 2,2′-azobis(isobutyronitrile) |
| AIBMe | Azobis(methyl isobutyrate) |
| AM | Acrylamide |
| AMS | $\alpha$-Methylstyrene |
| AN | Acrylonitrile |
| ATRP | Atom transfer radical or metal-mediated polymerization |
| BA | Butyl acrylate |
| BMA | Butyl methacrylate |
| BzMA | Benzyl methacrylate |
| DA | Dodecyl acrylate |
| DMAEMA | *N,N*-(Dimethylamino)ethyl methacrylate |
| DMAM | *N,N*-Dimethylacrylamide |
| EA | Ethyl acrylate |
| EAA | Ethyl-$\alpha$-acetoxyacetate |
| EHA | 2-Ethylhexyl acrylate |
| GMA | Glycidyl methacrylate |
| GPC | Gel permeation chromatography |
| HEA | 2-Hydroxyethyl acrylate |
| HEMA | 2-Hydroxyethyl methacrylate |
| I | Initiator |
| M | Monomer |
| MA | Methyl acrylate |
| MAA | Methacrylic acid |

| | |
|---|---|
| MAH | Maleic anhydride |
| MAM | Methacrylamide |
| MMA | Methyl methacrylate |
| NAM | *N*-Acryloylmorpholine |
| NAS | *N*-Acryloylsuccinimide |
| NIPAM | *N*-Isopropyl acrylamide |
| NMP | Nitroxide-mediated polymerization |
| NMS | *N*-Methacryloylsuccinimide |
| NPMI | *N*-Phenylmaleimide |
| NVP | *N*-Vinylpyrrolidone |
| ODA | Octadecyl acrylate |
| RAFT | Reversible addition–fragmentation chain transfer |
| PEO | Poly(ethylene oxide), poly(ethylene glycol) |
| PEGM | Poly(ethylene glycol) monomethyl ether |
| PFMA | Pentafluorophenyl methacrylate |
| PLA | Poly(lactic acid) |
| S | Styrene |
| SCl | 4-Chlorostyrene |
| SMe | 4-Methylstyrene |
| SOMe | 4-Methoxystyrene |
| $SSO_3Na$ | Sodium styrene-4-sulfonate |
| tBA | *tert*-Butyl acrylate |
| TBAM | *N*-*tert*-Butyl acrylamide |
| TMAEA | 2-(Trimethylamino)ethyl acrylate |
| UV | Ultraviolet |
| VAc | Vinyl acetate |
| VBz | Vinyl benzoate |
| VCBz | *N*-Vinylcarbazole |
| VDC | Vinylidene chloride |
| 2VP | 2-Vinylpyridine |
| 4VP | 4-Vinylpyridine |

Abbreviations of polymers are formed by suffixing the abbreviation for the corresponding monomer with 'P'. For example, PMMA – poly(methyl methacrylate), PS – polystyrene.

## References

1. G. Moad, D. H. Solomon, *The Chemistry of Radical Polymerization*, 2nd edn, Elsevier, Oxford, **2006**.
2. T. P. Davis, K. Matyjaszewski, Eds., *Handbook of Radical Polymerization*, John Wiley & Sons, Hoboken, **2002**.
3. G. Moad, D. H. Solomon, *The Chemistry of Radical Polymerization*, 2nd edn, Elsevier, Oxford, **2006**, pp. 451–585.
4. K. Matyjaszewski, Ed., *Controlled/Living Radical Polymerization: From Synthesis to Materials*, American Chemical Society, Washington, DC, **2006**.

**5.** W. A. Braunecker, K. Matyjaszewski, *Prog. Polym. Sci.* **2007**, *32*, 93–146.
**6.** D. H. Solomon, E. Rizzardo, P. Cacioli, US 4581429, **1986**.
**7.** D. H. Solomon, *J. Polym. Sci., Part A: Polym. Chem.* **2005**, *43*, 5748–5764.
**8.** Y. K. Chong, F. Ercole, G. Moad, E. Rizzardo, S. H. Thang, A. G. Anderson, *Macromolecules* **1999**, *32*, 6895–6903.
**9.** G. Moad, E. Rizzardo, *Macromolecules* **1995**, *28*, 8722–8728.
**10.** C. J. Hawker, A. W. Bosman, E. Harth, *Chem. Rev.* **2001**, *101*, 3661–3688.
**11.** M. Kamigaito, T. Ando, M. Sawamoto, *Chem. Rev.* **2001**, *101*, 3689–3745.
**12.** K. Matyjaszewski, J. Xia, *Chem. Rev.* **2001**, *101*, 2921–2990.
**13.** G. Moad, E. Rizzardo, S. H. Thang, *Aust. J. Chem.* **2005**, *58*, 379–410.
**14.** G. Moad, E. Rizzardo, S. H. Thang, *Aust. J. Chem.* **2006**, *59*, 669–692.
**15.** P. Cacioli, D. G. Hawthorne, R. L. Laslett, E. Rizzardo, D. H. Solomon, *J. Macromol. Sci., A: Chem.* **1986**, *23*, 839–852.
**16.** J. Krstina, G. Moad, E. Rizzardo, C. L. Winzor, C. T. Berge, M. Fryd, *Macromolecules* **1995**, *28*, 5381–5385.
**17.** G. Moad, C. L. Moad, J. Krstina, E. Rizzardo, C. T. Berge, T. R. Darling, WO9615157, 1996, E. I. Du Pont De Nemours and Company, USA; Commonwealth Scientific and Industrial Research Organisation.
**18.** P. Delduc, C. Tailhan, S. Z. Zard, *Chem. Commun.* **1988**, 308–310.
**19.** S. Z. Zard, *Angew. Chem., Int. Ed. Engl.* **1997**, *36*, 672–685.
**20.** J. Chiefari, Y. K. Chong, F. Ercole, J. Krstina, J. Jeffery, T. P. T. Le, R. T. A. Mayadunne, G. F. Meijs, C. L. Moad, G. Moad, E. Rizzardo, S. H. Thang, *Macromolecules* **1998**, *31*, 5559–5562.
**21.** T. P. Le, G. Moad, E. Rizzardo, S. H. Thang, WO9801478, **1998**, E. I. Du Pont De Nemours and Co., USA.
**22.** E. Rizzardo, J. Chiefari, R. T. A. Mayadunne, G. Moad, S. H. Thang, *ACS Symp. Ser.* **2000**, *768*, 278–296.
**23.** G. Moad, *Aust J. Chem.* **2006**, *59*, 661–662.
**24.** R. T. A. Mayadunne, E. Rizzardo, In *Living and Controlled Polymerization: Synthesis, Characterization and Properties of the Respective Polymers and Copolymers*, J. Jagur-Grodzinski, Ed., Nova Science Publishers, Hauppauge, NY, **2006**, pp. 65–107.
**25.** J. Chiefari, E. Rizzardo, In *Handbook of Radical Polymerization*, T. P. Davis, K. Matyjaszewski, Eds., John Wiley & Sons, Hoboken, NY, **2002**, pp. 263–300.
**26.** S. Perrier, P. Takolpuckdee, *J. Polym. Sci., Part A: Polym. Chem.* **2005**, *43*, 5347–5393.
**27.** A. Favier, M.-T. Charreyre, *Macromol. Rapid Commun.* **2006**, *27*, 653–692.
**28.** C. Barner-Kowollik, T. P. Davis, J. P. A. Heuts, M. H. Stenzel, P. Vana, M. Whittaker, *J. Polym. Sci., Part A: Polym. Chem.* **2003**, *41*, 365–375.
**29.** M. L. Coote, E. H. Krenske, E. I. Izgorodina, *Macromol. Rapid Commun.* **2006**, *27*, 473–497.
**30.** M. L. Coote, C. Barner-Kowollik, *Aust. J. Chem.* **2006**, *59*, 712–718.
**31.** C. Barner-Kowollik, M. Buback, B. Charleux, M. L. Coote, M. Drache, T. Fukuda, A. Goto, B. Klumperman, A. B. Lowe, J. B. Mcleary, G. Moad, M. J. Monteiro, R. D. Sanderson, M. P. Tonge, P. Vana, *J. Polym. Sci., Part A: Polym. Chem.* **2006**, *44*, 5809–5831.
**32.** S. Z. Zard, *Aust. J. Chem.* **2006**, *59*, 663–668.
**33.** B. Quiclet-Sire, S. Z. Zard, *Top. Curr. Chem.* **2006**, *264*, 201–236.
**34.** M. J. Monteiro, *J. Polym. Sci., Part A: Polym. Chem.* **2005**, *43*, 3189–3204.
**35.** A. B. Lowe, C. L. McCormick, *Prog. Polym. Sci.* **2007**, *32*, 283–351.
**36.** F. J. Schork, Y. W. Luo, W. Smulders, J. P. Russum, A. Butte, K. Fontenot, *Adv. Polym. Sci.* **2005**, *175*, 129–255.
**37.** J. B. McLeary, B. Klumperman, *Soft Matter* **2006**, *2*, 45–53.
**38.** M. Save, Y. Guillaneuf, R. G. Gilbert, *Aust. J. Chem.* **2006**, *59*, 693–711.
**39.** B. S. Lokitz, A. B. Lowe, C. L. McCormick, *ACS Symp. Ser.* **2006**, *937*, 95–115.
**40.** G. Moad, Y. K. Chong, E. Rizzardo, A. Postma, S. H. Thang, *Polymer* **2005**, *46*, 8458–8468.
**41.** C. W. Scales, A. J. Convertine, B. S. Sumerlin, A. B. Lowe, C. L. McCormick, *ACS Symp. Ser.* **2005**, *912*, 43–54.

**42.** C. Barner-Kowollik, T. P. Davis, M. H. Stenzel, *Aust. J. Chem.* **2006**, *59*, 719–727.

**43.** C. L. McCormick, S. E. Kirkland, A. W. York, *Polym. Rev.* **2006**, *46*, 421–443.

**44.** M. H. Stenzel, C. Barner-Kowollik, T. P. Davis, *J. Polym. Sci., Part A: Polym. Chem.* **2006**, *44*, 2363–2375.

**45.** A. V. Yakimansky, *Polym. Sci. Ser. C* **2005**, *47*, 1–49.

**46.** D. A. Shipp, *J. Macromol. Sci., Polym. Rev.* **2005**, *C45*, 171–194.

**47.** N. Hadjichristidis, H. Iatrou, M. Pitsikalis, S. Pispas, A. Avgeropoulos, *Prog. Polym. Sci.* **2005**, *30*, 725–782.

**48.** B. Ochiai, T. Endo, *Prog. Polym. Sci.* **2005**, *30*, 183–215.

**49.** H. Frauenrath, *Prog. Polym. Sci.* **2005**, *30*, 325–384.

**50.** H. Mori, A. H. E. Mueller, In *Living and Controlled Polymerization: Synthesis, Characterization and Properties of the Respective Polymers and Copolymers*, J. Jagur-Grodzinski, Ed., Nova Science Publishers, Hauppauge, NY, **2006**, pp. 257–288.

**51.** Y. Tsujii, K. Ohno, S. Yamamoto, A. Goto, T. Fukuda, *Adv. Polym. Sci.* **2006**, *197*, 1–45.

**52.** S. Garnier, A. Laschewsky, J. Storsberg, *Tenside Surfact. Deterg.* **2006**, *43*, 88–102.

**53.** M. Ali, S. Brocchini, *Adv. Drug Deliv. Rev.* **2006**, *58*, 1671–1687.

**54.** N. Gaillard, J. Claverie, A. Guyot, *Prog. Org. Coat.* **2006**, *57*, 98–109.

**55.** J. F. Lutz, *Polym. Int.* **2006**, *55*, 979–993.

**56.** J. Chiefari, R. T. Mayadunne, G. Moad, E. Rizzardo, S. H. Thang, WO9931144, **1999**, E. I. Du Pont De Nemours and Company, USA; Commonwealth Scientific and Industrial Research Organisation.

**57.** E. Rizzardo, S. H. Thang, G. Moad, WO9905099, **1999**, Commonwealth Scientific and Industrial Research Organisation, Australia; E. I. Du Pont deNemours & Co.

**58.** J. Chiefari, R. T. A. Mayadunne, C. L. Moad, G. Moad, E. Rizzardo, A. Postma, M. A. Skidmore, S. H. Thang, *Macromolecules* **2003**, *36*, 2273–2283.

**59.** Y. K. Chong, J. Krstina, T. P. T. Le, G. Moad, A. Postma, E. Rizzardo, S. H. Thang, *Macromolecules* **2003**, *36*, 2256–2272.

**60.** E. Rizzardo, M. Chen, B. Chong, G. Moad, M. Skidmore, S. H. Thang, *Macromol. Symp.* **2007**, *248*, 104–116.

**61.** G. Moad, J. Chiefari, J. Krstina, A. Postma, R. T. A. Mayadunne, E. Rizzardo, S. H. Thang, *Polym. Int.* **2000**, *49*, 993–1001.

**62.** M. Destarac, D. Charmot, X. Franck, S. Z. Zard, *Macromol. Rapid Commun.* **2000**, *21*, 1035–1039.

**63.** R. T. A. Mayadunne, E. Rizzardo, J. Chiefari, Y. K. Chong, G. Moad, S. H. Thang, *Macromolecules* **1999**, *32*, 6977–6980.

**64.** M. Benaglia, E. Rizzardo, A. Alberti, M. Guerra, *Macromolecules* **2005**, *38*, 3129–3140.

**65.** D. L. Patton, M. Mullings, T. Fulghum, R. C. Advincula, *Macromolecules* **2005**, *38*, 8597–8602.

**66.** S. H. Thang, Y. K. Chong, R. T. A. Mayadunne, G. Moad, E. Rizzardo, *Tetrahedron Lett.* **1999**, *40*, 2435–2438.

**67.** J. S. Song, M. A. Winnik, *Macromolecules* **2006**, *39*, 8318–8325.

**68.** M. Chen, K. P. Ghiggino, A. W. H. Mau, E. Rizzardo, S. H. Thang, G. J. Wilson, *Chem. Commun.* **2002**, 2276–2277.

**69.** S. R. Gondi, A. P. Vogt, B. S. Sumerlin, *Macromolecules* **2007**, *40*, 474–481.

**70.** D. L. Patton, R. C. Advincula, *Macromolecules* **2006**, *39*, 8674–8683.

**71.** P. Lebreton, B. Ameduri, B. Boutevin, J. M. Corpart, *Macromol. Chem. Phys.* **2002**, *203*, 522–537.

**72.** L. Albertin, M. H. Stenzel, C. Barner-Kowollik, T. P. Davis, *Polymer* **2006**, *47*, 1011–1019.

**73.** C. Z. Li, B. C. Benicewicz, *J. Polym. Sci., Part A: Polym. Chem.* **2005**, *43*, 1535–1543.

**74.** P. Takolpuckdee, C. A. Mars, S. Perrier, S. J. Archibald, *Macromolecules* **2005**, *38*, 1057–1060.

**75.** P. Takopuckdee, J. Westwood, D. M. Lewis, S. Perrier, *Macromol. Symp.* **2004**, *216*, 23–35.

**76.** T. M. Legge, A. T. Slark, S. Perrier, *Macromolecules* **2007**, *40*, 2318–2326.

**77.** S. Perrier, P. Takolpuckdee, J. Westwood, D. M. Lewis, *Macromolecules* **2004**, *37*, 2709–2717.

**78.** J. Bang, S. H. Kim, E. Drockenmuller, M. J. Misner, T. P. Russell, C. J. Hawker, *J. Am. Chem. Soc.* **2006**, *128*, 7622–7629.

**79.** C. Ladaviere, N. Dorr, J. P. Claverie, *Macromolecules* **2001**, *34*, 5370–5372.

**80.** S. C. Farmer, T. E. Patten, *J. Polym. Sci., Part A: Polym. Chem.* **2002**, *40*, 555–563.

**81.** C. Z. Li, B. C. Benicewicz, *Macromolecules* **2005**, *38*, 5929–5936.

**82.** M. C. Davies, J. V. Dawkins, D. J. Hourston, *Polymer* **2005**, *46*, 1739–1753.

**83.** E. Chernikova, P. Terpugova, C. O. Bui, B. Charleux, *Polymer* **2003**, *44*, 4101–4107.

**84.** S. E. Shim, Y. Shin, J. W. Jun, K. Lee, H. Jung, S. Choe, *Macromolecules* **2003**, *36*, 7994–8000.

**85.** N. C. Zhou, L. D. Lu, X. L. Zhu, X. J. Yang, X. Wang, J. Zhu, Z. P. Cheng, *J. Appl. Polym. Sci.* **2006**, *99*, 3535–3539.

**86.** N. C. Zhou, L. D. Lu, X. L. Zhu, X. J. Yang, X. Wang, J. Zhu, D. Zhou, *Polym. Bull.* **2006**, *57*, 491–498.

**87.** G. C. Zhou, I. I. Harruna, C. W. Ingram, *Polymer* **2005**, *46*, 10672–10677.

**88.** J. M. Lee, O. H. Kim, S. E. Shim, B. H. Lee, S. Choe, *Macromolecular Res.* **2005**, *13*, 236–242.

**89.** A. Alberti, M. Benaglia, M. Guerra, M. Gulea, P. Hapiot, M. Laus, D. Macciantelli, S. Masson, A. Postma, K. Sparnacci, *Macromolecules* **2005**, *38*, 7610–7618.

**90.** S. Carter, B. Hunt, S. Rimmer, *Macromolecules* **2005**, *38*, 4595–4603.

**91.** X. Y. Sun, Y. W. Luo, R. Wang, B. G. Li, B. Liu, S. P. Zhu, *Macromolecules* **2007**, *40*, 849–859.

**92.** C. Y. Hong, C. Y. Pan, *Macromolecules* **2006**, *39*, 3517–3524.

**93.** J. T. Lai, D. Filla, R. Shea, *Macromolecules* **2002**, *35*, 6754–6756.

**94.** A. J. Convertine, B. S. Lokitz, Y. Vasileva, L. J. Myrick, C. W. Scales, A. B. Lowe, C. L. McCormick, *Macromolecules* **2006**, *39*, 1724–1730.

**95.** R. Wang, C. L. McCormick, A. B. Lowe, *Macromolecules* **2005**, *38*, 9518–9525.

**96.** P. Kujawa, F. Segui, S. Shaban, C. Diab, Y. Okada, F. Tanaka, F. M. Winnik, *Macromolecules* **2006**, *39*, 341–348.

**97.** M. R. Wood, D. J. Duncalf, S. P. Rannard, S. Perrier, *Org. Lett.* **2006**, *8*, 553–556.

**98.** C. J. Ferguson, R. J. Hughes, D. Nguyen, B. T. T. Pham, R. G. Gilbert, A. K. Serelis, C. H. Such, B. S. Hawkett, *Macromolecules* **2005**, *38*, 2191–2204.

**99.** A. Postma, T. P. Davis, R. A. Evans, G. Li, G. Moad, M. O'Shea, *Macromolecules* **2006**, *39*, 5293–5306.

**100.** L. W. Zhang, Y. H. Zhang, Y. M. Chen, *Eur. Polym. J.* **2006**, *42*, 2398–2406.

**101.** J. Liu, V. Bulmus, C. Barner-Kowollik, M. H. Stenzel, T. P. Davis, *Macromol. Rapid Commun.* **2007**, *28*, 305–314.

**102.** A. Suzuki, D. Nagai, B. Ochiai, T. Endo, *J. Polym. Sci., Part A: Polym. Chem.* **2005**, *43*, 5498–5505.

**103.** J. Chiefari, R. T. Mayadunne, G. Moad, E. Rizzardo, S. H. Thang, US6747111, **2004**, E. I. Du Pont De Nemours and Company, USA; Commonwealth Scientific and Industrial Research Organisation.

**104.** M. H. Stenzel, L. Cummins, G. E. Roberts, T. R. Davis, P. Vana, C. Barner-Kowollik, *Macromol. Chem. Phys.* **2003**, *204*, 1160–1168.

**105.** D. C. Wan, K. Satoh, M. Kamigaito, Y. Okamoto, *Macromolecules* **2005**, *38*, 10397–10405.

**106.** T. L. U. Nguyen, K. Eagles, T. P. Davis, C. Barner-Kowollik, M. H. Stenzel, *J. Polym. Sci., Part A: Polym. Chem.* **2006**, *44*, 4372–4383.

**107.** A. Postma, T. P. Davis, G. Li, G. Moad, M. O'Shea, *Macromolecules* **2006**, *39*, 5307–5318.

**108.** C. Schilli, M. G. Lanzendoerfer, A. H. E. Mueller, *Macromolecules* **2002**, *35*, 6819–6827.

**109.** D. Zhou, X. L. Zhu, J. Zhu, H. S. Yin, *J. Polym. Sci., Part A: Polym. Chem.* **2005**, *43*, 4849–4856.

**110.** D. Zhou, X. L. Zhu, J. Zhu, L. H. Hu, Z. P. Cheng, *e-Polymers* **2006**, *59*.

**111.** H. S. Yin, Z. P. Cheng, J. Zhu, X. L. Zhu, *J. Macromol. Sci., Part A: Pure Appl. Chem.* **2007**, *44*, 315–320.

**112.** R. Bussels, C. E. Koning, *Tetrahedron* **2005**, *61*, 1167–1174.

**113.** L. Albertin, C. Kohlert, M. Stenzel, L. J. R. Foster, T. P. Davis, *Biomacromolecules* **2004**, *5*, 255–260.

**114.** L. Couvreur, O. Guerret, J.-A. Laffitte, S. Magnet, *Polym. Prepr. (Am. Chem. Soc., Div. Polym. Chem.)* **2005**, *46*, 219–220.

**115.** J. Meijer, P. Vermeer, L. Brandsma, *Recueil* **1973**, *92*, 601–604.

**116.** R. T. A. Mayadunne, unpublished results.

**117.** R. A. Mayadunne, G. Moad, E. Rizzardo, *Tetrahedron Lett.* **2002**, *43*, 6811–6814.

**118.** G. Moad, G. P. Simon, K. M. Dean, G. Li, R. T. A. Mayadunne, E. Rizzardo, R. A. Evans, H. Wermter, R. Pfaendner, 04113436, **2004**, Polymers Australia Pty. Limited, Australia.

**119.** S. Kanagasabapathy, A. Sudalai, B. C. Benicewicz, *Tetrahedron Lett.* **2001**, *42*, 3791–3794.

**120.** S. Oae, T. Yagihara, T. Okabe, *Tetrahedron* **1972**, *28*, 3203–3216.

**121.** Y. K. Goh, M. R. Whittaker, M. J. Monteiro, *J. Polym. Sci., Part A: Polym. Chem.* **2005**, *43*, 5232–5245.

**122.** G. Bouhadir, N. Legrand, B. Quiclet-Sire, S. Z. Zard, *Tetrahedron Lett.* **1999**, *40*, 277–280.

**123.** W. G. Weber, J. B. McLeary, R. D. Sanderson, *Tetrahedron Lett.* **2006**, *47*, 4771–4774.

**124.** A. Sudalai, S. Kanagasabapathy, B. C. Benicewicz, *Org. Lett.* **2000**, *2*, 3213–3216.

**125.** A. Dureault, D. Taton, M. Destarac, F. Leising, Y. Gnanou, *Macromolecules* **2004**, *37*, 5513–5519.

**126.** A. Dureault, Y. Gnanou, D. Taton, M. Destarac, F. Leising, *Angew. Chem., Int. Ed. Engl.* **2003**, *42*, 2869–2872.

**127.** H. Davy, *J. Chem. Soc., Chem. Commun.* **1982**, 457–458.

**128.** A. Alberti, M. Benaglia, M. Laus, K. Sparnacci, *J. Org. Chem.* **2002**, *67*, 7911–7914.

**129.** G. Moad, J. Chiefari, C. L. Moad, A. Postma, R. T. A. Mayadunne, E. Rizzardo, S. H. Thang, *Macromol. Symp.* **2002**, *182*, 65–80.

**130.** G. Johnston-Hall, A. Theis, M. J. Monteiro, T. P. Davis, M. H. Stenzel, C. Barner-Kowollik, *Macromol. Chem. Phys.* **2005**, *206*, 2047–2053.

**131.** J. Xu, J. He, D. Fan, W. Tang, Y. Yang, *Macromolecules* **2006**, *39*, 3753–3759.

**132.** B. Chong, G. Moad, E. Rizzardo, M. Skidmore, S. H. Thang, *Macromolecules*, **2007**, *40*, in press.

**133.** J. F. Lutz, W. Jakubowski, K. Matyjaszewski, *Macromol. Rapid Commun.* **2004**, *25*, 486–492.

**134.** M. L. T. de Sordi, A. Ceschi, C. L. Petzhold, A. H. E. Muller, *Macromol. Rapid Commun.* **2007**, *28*, 63–71.

**135.** D. Fournier, R. Hoogenboom, H. M. L. Thijs, R. M. Paulus, U. S. Schubert, *Macromolecules* **2007**, *40*, 915–920.

**136.** M. W. M. Fijten, R. M. Paulus, U. S. Schubert, *J. Polym. Sci., Part A: Polym. Chem.* **2005**, *43*, 3831–3839.

**137.** X. Zhou, P. Ni, Z. Yu, F. Zhang, *J. Polym. Sci., Part A: Polym. Chem.* **2007**, *45*, 471–484.

**138.** T. Y. J. Chiu, M. H. Stenzel, T. P. Davis, C. Barner-Kowollik, *J. Polym. Sci., Part A: Polym. Chem.* **2005**, *43*, 5699–5703.

**139.** R. Cuervo-Rodriguez, V. Bordege, M. Sanchez-Chaves, M. Fernandez-Garcia, *J. Polym. Sci., Part A: Polym. Chem.* **2006**, *44*, 5618–5629.

**140.** T. Y. Guo, P. Liu, J. W. Zhu, M. D. Song, B. H. Zhang, *Biomacromolecules* **2006**, *7*, 1196–1202.

**141.** L. Albertin, M. H. Stenzel, C. Barner-Kowollik, L. J. R. Foster, T. P. Davis, *Macromolecules* **2005**, *38*, 9075–9084.

**142.** A. Housni, H. Cai, S. Liu, S. H. Pun, R. Narain, *Langmuir* **2007**, *23*, 5056–5061.

**143.** P. Zhao, Q. D. Ling, W. Z. Wang, J. Ru, S. B. Li, W. Huang, *J. Polym. Sci., Part A: Polym. Chem.* **2007**, *45*, 242–252.

**144.** B. Y. K. Chong, T. P. T. Le, G. Moad, E. Rizzardo, S. H. Thang, *Macromolecules* **1999**, *32*, 2071–2074.

**145.** D. Q. Fan, J. P. He, J. T. Xu, W. Tang, Y. Liu, Y. L. Yang, *J. Polym. Sci., Part A: Polym. Chem.* **2006**, *44*, 2260–2269.

**146.** B. Chong, G. Moad, E. Rizzardo, M. Skidmore, S. H. Thang, *Aust. J. Chem.* **2006**, *59*, 755–762.

147. R. T. A. Mayadunne, E. Rizzardo, J. Chiefari, J. Krstina, G. Moad, A. Postma, S. H. Thang, *Macromolecules* **2000**, *33*, 243–245.
148. A. Goto, K. Sato, Y. Tsujii, T. Fukuda, G. Moad, E. Rizzardo, S. H. Thang, *Macromolecules* **2001**, *34*, 402–408.
149. K. Kubo, A. Goto, K. Sato, Y. Kwak, T. Fukuda, *Polymer* **2005**, *46*, 9762–9768.
150. C. Barner-Kowollik, J. F. Quinn, T. L. U. Nguyen, J. P. A. Heuts, T. P. Davis, *Macromolecules* **2001**, *34*, 7849–7857.
151. D. G. Hawthorne, G. Moad, E. Rizzardo, S. H. Thang, *Macromolecules* **1999**, *32*, 5457–5459.
152. J. Zhu, D. Zhou, X. L. Zhu, G. J. Chen, *J. Polym. Sci., Part A: Polym. Chem.* **2004**, *42*, 2558–2565.
153. M. Eberhardt, P. Théato, *Macromol. Rapid Commun.* **2005**, *26*, 1488–1493.
154. R. Plummer, Y. K. Goh, A. K. Whittaker, M. J. Monteiro, *Macromolecules* **2005**, *38*, 5352–5355.
155. M. Mertoglu, S. Garnier, A. Laschewsky, K. Skrabania, J. Storsberg, *Polymer* **2005**, *46*, 7726–7740.
156. M. Sahnoun, M. T. Charreyre, L. Veron, T. Delair, F. D'Agosto, *J. Polym. Sci., Part A: Polym. Chem.* **2005**, *43*, 3551–3565.
157. M. Mertoglu, A. Laschewsky, K. Skrabania, C. Wieland, *Macromolecules* **2005**, *38*, 3601–3614.
158. G. Moad, R. T. A. Mayadunne, E. Rizzardo, M. Skidmore, S. Thang, *ACS Symp. Ser.* **2003**, *854*, 520–535.
159. P. Vana, T. P. Davis, C. Barner-Kowollik, *Macromol. Theory Simul.* **2002**, *11*, 823–835.
160. S. Perrier, C. Barner-Kowollik, J. F. Quinn, P. Vana, T. P. Davis, *Macromolecules* **2002**, *35*, 8300–8306.
161. E. Chernikova, A. Morozov, E. Leonova, E. Garina, V. Golubev, C. O. Bui, B. Charleux, *Macromolecules* **2004**, *37*, 6329–6339.
162. J. B. McLeary, J. M. McKenzie, M. P. Tonge, R. D. Sanderson, B. Klumperman, *Chem. Commun.* **2004**, 1950–1951.
163. J. F. Quinn, E. Rizzardo, T. P. Davis, *Chem. Commun.* **2001**, 1044–1045.
164. H. Fischer, L. Radom, *Angew. Chem., Int. Ed. Engl.* **2001**, *40*, 1340–1371.
165. R. Hoogenboom, U. S. Schubert, W. Van Camp, F. E. Du Prez, *Macromolecules* **2005**, *38*, 7653–7659.
166. Q. F. An, J. W. Qian, C. J. Gao, *Chin. Chem. Lett.* **2006**, *17*, 365–368.
167. C. Tang, T. Kowalewski, K. Matyjaszewski, *Macromolecules* **2003**, *36*, 8587–8589.
168. M. Jacquin, P. Muller, G. Lizarraga, C. Bauer, H. Cottet, O. Theodoly, *Macromolecules* **2007**, *40*, 2672–2682.
169. M. F. Llauro, J. Loiseau, F. Boisson, F. Delolme, C. Ladaviere, J. Claverie, *J. Polym. Sci., Part A: Polym. Chem.* **2004**, *42*, 5439–5462.
170. C. J. Ferguson, R. J. Hughes, B. T. T. Pham, B. S. Hawkett, R. G. Gilbert, A. K. Serelis, C. H. Such, *Macromolecules* **2002**, *35*, 9243–9245.
171. B. T. T. Pham, D. Nguyen, C. J. Ferguson, B. S. Hawkett, A. K. Serelis, C. H. Such, *Macromolecules* **2003**, *36*, 8907–8909.
172. D. Taton, A. Z. Wilczewska, M. Destarac, *Macromol. Rapid Commun.* **2001**, *22*, 1497–1503.
173. C. M. Schilli, A. H. E. Muller, E. Rizzardo, S. H. Thang, Y. K. Chong, *ACS Symp. Ser.* **2003**, *854*, 603–618.
174. C. M. Schilli, M. F. Zhang, E. Rizzardo, S. H. Thang, Y. K. Chong, K. Edwards, G. Karlsson, A. H. E. Muller, *Macromolecules* **2004**, *37*, 7861–7866.
175. Q. F. An, J. W. Qian, L. Y. Yu, Y. W. Luo, X. Z. Liu, *J. Polym. Sci., Part A: Polym. Chem.* **2005**, *43*, 1973–1977.
176. X. H. Liu, Y. G. Li, Y. Lin, Y. S. Li, *J. Polym. Sci., Part A: Polym. Chem.* **2007**, *45*, 1272–1281.
177. A. Aqil, C. Detrembleur, B. Gilbert, R. Jerome, C. Jerome, *Chem. Mater.* **2007**, *19*, 2150–2154.
178. X. H. Liu, G. B. Zhang, X. F. Lu, J. Y. Liu, D. Pan, Y. S. Li, *J. Polym. Sci., Part A: Polym. Chem.* **2006**, *44*, 490–498.
179. F. Ganachaud, M. J. Monteiro, R. G. Gilbert, M. A. Dourges, S. H. Thang, E. Rizzardo, *Macromolecules* **2000**, *33*, 6738–6745.
180. B. Ray, Y. Isobe, K. Matsumoto, S. Habaue, Y. Okamoto, M. Kamigaito,

M. Sawamoto, *Macromolecules* **2004**, *37*, 1702–1710.

**181.** H. S. Bisht, D. S. Manickam, Y. Z. You, D. Oupicky, *Biomacromolecules* **2006**, *7*, 1169–1178.

**182.** P. Kujawa, H. Watanabe, F. Tanaka, F. M. Winnik, *Eur. Phys. J.* **2005**, *17*, 129–137.

**183.** B. Ray, Y. Isobe, K. Morioka, S. Habaue, Y. Okamoto, M. Kamigaito, M. Sawamoto, *Macromolecules* **2003**, *36*, 543–545.

**184.** B. Ray, Y. Okamoto, N. Kamigaito, M. Sawamoto, K. Seno, S. Kanaoka, S. Aoshima, *Polym. J.* **2005**, *37*, 234–237.

**185.** Y. A. Vasilieva, C. W. Scales, D. B. Thomas, R. G. Ezell, A. B. Lowe, N. Ayres, C. L. McCormick, *J. Polym. Sci., Part A: Polym. Chem.* **2005**, *43*, 3141–3152.

**186.** C. W. Scales, Y. A. Vasilieva, A. J. Convertine, A. B. Lowe, C. L. McCormick, *Biomacromolecules* **2005**, *6*, 1846–1850.

**187.** M. Nuopponen, J. Ojala, H. Tenhu, *Polymer* **2004**, *45*, 3643–3650.

**188.** B. S. Lokitz, J. E. Stempka, A. W. York, Yuting Li, H. K. Goel, G. R. Bishop, C. L. McCormick, *Aust. J. Chem.*, **2006**, *59*, 749–754.

**189.** J. B. McLeary, F. M. Calitz, J. M. McKenzie, M. P. Tonge, R. D. Sanderson, B. Klumperman, *Macromolecules* **2004**, *37*, 2383–2394.

**190.** J. B. McLeary, F. M. Calitz, J. M. McKenzie, M. P. Tonge, R. D. Sanderson, B. Klumperman, *Macromolecules* **2005**, *38*, 3151–3161.

**191.** H. Mori, S. Nakano, T. Endo, *Macromolecules* **2005**, *38*, 8192–8201.

**192.** M. Chen, K. P. Ghiggino, T. A. Smith, S. H. Thang, G. J. Wilson, *Aust. J. Chem.* **2004**, *57*, 1175–1177.

**193.** M. Chen, K. P. Ghiggino, S. H. Thang, J. White, G. J. Wilson, *J. Org. Chem.* **2005**, *70*, 1844–1852.

**194.** H. Mori, S. Masuda, T. Endo, *Macromolecules* **2006**, *39*, 5976–5978.

**195.** J. F. Quinn, L. Barner, C. Barner-Kowollik, E. Rizzardo, T. P. Davis, *Macromolecules* **2002**, *35*, 7620–7627.

**196.** G. Zhou, I. I. Harruna, *Macromolecules* **2005**, *38*, 4114–4123.

**197.** J. Zhu, X. L. Zhu, D. Zhou, J. Y. Chen, *e-Polymers* **2003**, 043.

**198.** H. S. Yin, X. L. Zhu, D. Zhou, J. Zhu, *J. Appl. Polym. Sci.* **2006**, *100*, 560–564.

**199.** M. Adamy, A. M. van Herk, M. Destarac, M. J. Monteiro, *Macromolecules* **2003**, *36*, 2293–2301.

**200.** A. J. Convertine, B. S. Sumerlin, D. B. Thomas, A. B. Lowe, C. L. McCormick, *Macromolecules* **2003**, *36*, 4679–4681.

**201.** V. Jitchum, S. Perrier, *Macromolecules* **2007**, *40*, 1408–1412.

**202.** J. Bernard, A. Favier, T. P. Davis, C. Barner-Kowollik, M. H. Stenzel, *Polymer* **2006**, *47*, 1073–1080.

**203.** M. H. Stenzel, T. P. Davis, C. Barner-Kowollik, *Chem. Commun.* **2004**, 1546–1547.

**204.** A. Debuigne, J. R. Caille, R. Jerome, *Macromolecules* **2005**, *38*, 5452–5458.

**205.** M. C. Iovu, K. Matyjaszewski, *Macromolecules* **2003**, *36*, 9346–9354.

**206.** S. Yamago, B. Ray, K. Iida, J. Yoshida, T. Tada, K. Yoshizawa, Y. Kwak, A. Goto, T. Fukuda, *J. Am. Chem. Soc.* **2004**, *126*, 13908–13909.

**207.** D. Charmot, P. Corpart, H. Adam, S. Z. Zard, T. Biadatti, G. Bouhadir, *Macromol. Symp.* **2000**, *150*, 23–32.

**208.** A. Favier, C. Barner-Kowollik, T. P. Davis, M. H. Stenzel, *Macromol. Chem. Phys.* **2004**, *205*, 925–936.

**209.** R. W. Simms, T. P. Davis, M. F. Cunningham, *Macromol. Rapid Commun.* **2005**, *26*, 592–596.

**210.** J. P. Russum, N. D. Barbre, C. W. Jones, F. J. Schork, *J. Polym. Sci., Part A: Polym. Chem.* **2005**, *43*, 2188–2193.

**211.** D. Boschmann, P. Vana, *Polym. Bull.* **2005**, *53*, 231–242.

**212.** M. L. Coote, L. Radom, *Macromolecules* **2004**, *37*, 590–596.

**213.** A. Theis, T. P. Davis, M. H. Stenzel, C. Barner-Kowollik, *Polymer* **2006**, *47*, 999–1010.

**214.** R. Devasia, R. L. Bindu, N. Mougin, Y. Gnanou, *Polym. Prepr. (Am. Chem. Soc., Div. Polym. Chem.)* **2005**, *46*, 195–196.

**215.** R. Devasia, R. L. Bindu, R. Borsali, N. Mougin, Y. Gnanou, *Macromol. Symp.* **2005**, *229*, 8–17.

**216.** H. Mori, H. Ookuma, S. Nakano, T. Endo, *Macromol. Chem. Phys.* **2006**, *207*, 1005–1017.

**217.** E. Rizzardo, J. Chiefari, Y. K. Chong, F. Ercole, J. Krstina, J. Jeffery, T. P. T. Le, R. T. A. Mayadunne, G. F. Meijs, C. L. Moad, G. Moad, S. H. Thang, *Macromol. Symp.* **1999**, *143*, 291–307.

**218.** G. Moad, R. T. A. Mayadunne, E. Rizzardo, M. Skidmore, S. Thang, *Macromol. Symp.* **2003**, *192*, 1–12.

**219.** H. de Brouwer, M. A. J. Schellekens, B. Klumperman, M. J. Monteiro, A. L. German, *J. Polym. Sci., Part A: Polym. Chem.* **2000**, *38*, 3596–3603.

**220.** G. Moad, G. Li, R. Pfaendner, A. Postma, E. Rizzardo, S. Thang, H. Wermter, *ACS Symp. Ser.* **2006**, *944*, 514–532.

**221.** G. Moad, K. Dean, L. Edmond, N. Kukaleva, G. Li, R. T. A. Mayadunne, R. Pfaendner, A. Schneider, G. Simon, H. Wermter, *Macromol. Symp.* **2006**, *233*, 170–179.

**222.** Z. Q. Hu, Z. C. Zhang, *Macromolecules* **2006**, *39*, 1384–1390.

**223.** B. Rixens, R. Severac, B. Boutevin, P. Lacroix-Desmazes, *J. Polym. Sci., Part A: Polym. Chem.* **2006**, *44*, 13–24.

**224.** M. J. Yanjarappa, K. V. Gujraty, A. Joshi, A. Saraph, R. S. Kane, *Biomacromolecules* **2006**, *7*, 1665–1670.

**225.** B. de Lambert, C. Chaix, M. T. Charreyre, A. Laurent, A. Aigoui, A. Perrin-Rubens, C. Pichot, *Bioconjugate Chem.* **2005**, *16*, 265–274.

**226.** M. Beija, P. Relogio, M. T. Charreyre, A. M. G. da Silva, P. Brogueira, J. P. S. Farinha, J. M. G. Martinho, *Langmuir* **2005**, *21*, 3940–3949.

**227.** E. T. A. van den Dungen, J. Rinquest, N. O. Pretorius, J. M. McKenzie, J. B. McLeary, R. D. Sanderson, B. Klumperman, *Aust. J. Chem.*, **2006**, *59*, 742–748.

**228.** Y. Shi, Z. F. Fu, W. T. Yang, *J. Polym. Sci., Part A: Polym. Chem.* **2006**, *44*, 2069–2075.

**229.** L. J. Shi, T. M. Chapman, E. J. Beckman, *Macromolecules* **2003**, *36*, 2563–2567.

**230.** C. Y. Hong, Y. Z. You, C. Y. Pan, *J. Polym. Sci., Part A: Polym. Chem.* **2004**, *42*, 4873–4881.

**231.** Z. Ma, P. Lacroix-Desmazes, *Polymer* **2004**, *45*, 6789–6797.

**232.** M. Hales, C. Barner-Kowollik, T. P. Davis, M. H. Stenzel, *Langmuir* **2004**, *20*, 10809–10817.

**233.** Y. Z. You, C. Y. Hong, W. P. Wang, W. Q. Lu, C. Y. Pan, *Macromolecules* **2004**, *37*, 9761–9767.

**234.** J. Rzayev, M. A. Hillmyer, *J. Am. Chem. Soc.* **2005**, *127*, 13373–13379.

**235.** S. L. Brown, C. M. Rayner, S. Perrier, *Macromol. Rapid Commun.* **2007**, *28*, 478–483.

**236.** J. Zhu, X. L. Zhu, Z. B. Zhang, Z. P. Cheng, *J. Polym. Sci., Part A: Polym. Chem.* **2006**, *44*, 6810–6816.

**237.** Y. Liu, J. P. He, J. T. Xu, D. Q. Fan, W. Tang, Y. L. Yang, *Macromolecules* **2005**, *38*, 10332–10335.

**238.** M. J. Monteiro, R. Bussels, S. Beuermann, M. Buback, *Aust. J. Chem.* **2002**, *55*, 433–437.

**239.** J. Rzayev, J. Penelle, *Angew. Chem., Int. Ed. Engl.* **2004**, *43*, 1691–1694.

**240.** T. Arita, M. Buback, O. Janssen, P. Vana, *Macromol. Rapid Commun.* **2004**, *25*, 1376–1381.

**241.** T. Arita, M. Buback, P. Vana, *Macromolecules* **2005**, *38*, 7935–7943.

**242.** C. L. McCormick, A. B. Lowe, *Acc. Chem. Res.* **2004**, *37*, 312–325.

**243.** A. B. Lowe, C. L. McCormick, *Aust. J. Chem.* **2002**, *55*, 367–379.

**244.** S. Perrier, T. P. Davis, A. J. Carmichael, D. M. Haddleton, *Chem. Commun.* **2002**, 2226–2227.

**245.** T. Arita, S. Beuermann, M. Buback, P. Vana, *e-Polymers* **2004**, 003.

**246.** T. Arita, S. Beuermann, M. Buback, P. Vana, *Macromol. Mater. Eng.* **2005**, *290*, 283–293.

**247.** Y. Z. You, C. Y. Hong, R. K. Bai, C. Y. Pan, J. Wang, *Macromol. Chem. Phys.* **2002**, *203*, 477–483.

**248.** L. Lu, H. Zhang, N. Yang, Y. Cai, *Macromolecules* **2006**, *39*, 3770–3776.

**249.** C. Y. Hong, Y. Z. You, R. K. Bai, C. Y. Pan, G. Borjihan, *J. Polym. Sci., Part A: Polym. Chem.* **2001**, *39*, 3934–3939.

**250.** Y. Z. You, R. K. Bai, C. Y. Pan, *Macromol. Chem. Phys.* **2001**, *202*, 1980–1985.

**251.** R. K. Bai, Y. Z. You, C. Y. Pan, *Macromol. Rapid Commun.* **2001**, *22*, 315–319.
**252.** R. K. Bai, Y. Z. You, P. Zhong, C. Y. Pan, *Macromol. Chem. Phys.* **2001**, *202*, 1970–1973.
**253.** J. F. Quinn, L. Barner, T. P. Davis, S. H. Thang, E. Rizzardo, *Macromol. Rapid Commun.* **2002**, *23*, 717–721.
**254.** L. Barner, J. F. Quinn, C. Barner-Kowollik, P. Vana, T. P. Davis, *Eur. Polym. J.* **2003**, *39*, 449–459.
**255.** Y. Zhou, J. A. Zhu, X. L. Zhu, Z. P. Cheng, *Radiat. Phys. Chem.* **2006**, *75*, 485–492.
**256.** P.-E. Millard, L. Barner, M. H. Stenzel, T. P. Davis, C. Barner-Kowollik, A. H. E. Müller, *Macromol. Rapid Commun.* **2006**, *27*, 821–828.
**257.** G. Chen, X. Zhu, J. Zhu, Z. Cheng, *Macromol. Rapid Commun.* **2004**, *25*, 818–824.
**258.** T. Otsu, *J. Polym. Sci.,Part A: Polym. Chem.* **2000**, *38*, 2121–2136.
**259.** P. Vana, L. Albertin, L. Barner, T. P. Davis, C. Barner-Kowollik, *J. Polym. Sci., Part A: Polym. Chem.* **2002**, *40*, 4032–4037.
**260.** G. Moad, E. Rizzardo, D. H. Solomon, S. R. Johns, R. I. Willing, *Makromol. Chem., Rapid Commun.* **1984**, *5*, 793–798.
**261.** P. Corpart, D. Charmot, S. Z. Zard, O. Biadatti, D. Michelet US6153705, **2000**, Rhodia.
**262.** D. Batt-Coutrot, J.-J. Robin, W. Bzdurcha, M. Destarac *Macromol. Chem. Phys.* **2005**, *206*, 1709–1717.

# 7
# RAFT Polymerization in Homogeneous Aqueous Media: Initiation Systems, RAFT Agent Stability, Monomers and Polymer Structures

*Andrew B. Lowe, and Charles L. McCormick*

## 7.1
## Introduction

Since its disclosure in the open literature in late 1998 [1], reversible addition–fragmentation chain transfer (RAFT) radical polymerization has evolved into arguably the most versatile and most readily executable of the controlled radical polymerization (CRP) techniques, at least with respect to monomer choice and polymerization conditions. One particularly redeeming feature of RAFT is the ease with which it can be employed under homogeneous aqueous conditions facilitating the direct synthesis of materials with high degrees of hydrophilic functionality, including the preparation of (co)polymers bearing anionic, cationic, zwitterionic and neutral groups [2, 3]. Such versatility under homogeneous aqueous conditions still presents some significant challenges for other common CRP techniques. Additionally, RAFT polymerizations are no more difficult to perform than conventional free radical polymerizations, the only difference being the requirement of addition of a suitable RAFT agent, or RAFT chain-transfer agent (CTA), to the polymerization system. RAFT operates on the principle of degenerative chain transfer and is thus fundamentally different from atom transfer radical polymerization (ATRP) and stable free radical polymerization (SFRP), both of which operate on the principle of reversible termination. The RAFT CTAs [4–6] are thiocarbonylthio compounds that are typically dithioesters [7–11], trithiocarbonates [12–18], xanthates [19–23] or dithiocarbamates [24–27], although some more specialized compounds have also been reported, such as the F-RAFT [28, 29] agents, the phosphoryl dithioesters [30], dithiocarbazates and the vinylologous thionothio [31] compounds. Of these four general families, the dithioesters and trithiocarbonates are the most versatile, not just for aqueous RAFT polymerizations but also in general, although all of these general families have been employed directly in aqueous media.

In this chapter we highlight the application of RAFT for the synthesis of water-soluble (co)polymers, with an emphasis on polymerizations conducted directly

*Handbook of RAFT Polymerization.* Edited by Christopher Barner-Kowollik

ISBN: 978-3-527-31924-4

in water under homogeneous conditions. It is to be noted, however, that water-soluble (co)polymers with certain, primarily nonionic, building blocks can also be prepared directly in organic media under homogeneous conditions, as we have recently reviewed [2].

## 7.2 Initiation Systems

Any discussion of the generally employed initiating systems/reagents requires an initial description of the RAFT mechanism in an effort to explain the structural requirements of the thiocarbonylthio-mediating species. The generally accepted RAFT mechanism is outlined in Scheme 7.1 [32].

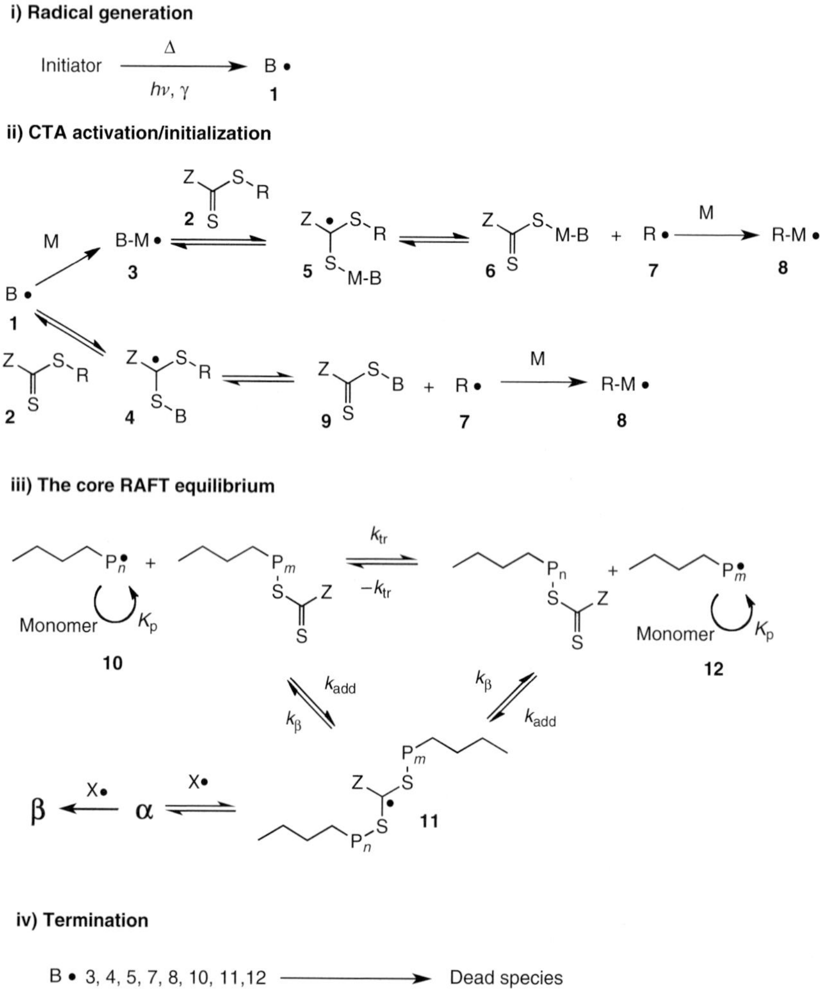


**Scheme 7.1** The generally accepted RAFT mechanism.

The initiation process in RAFT polymerizations is composed of steps (i) and (ii) shown in Scheme 7.1, which are often collectively referred to as the RAFT preequilibrium. As with any radical-mediated process, the first step requires the generation of primary radicals. In RAFT polymerizations this is accomplished via any of the traditional methods employed in conventional free radical, chain-growth polymerizations. Common examples include the use of azo initiators, gamma irradiation, redox initiation, as well as simple thermal initiation in the case of styrene. However, unlike conventional free radical polymerizations, the generated primary radicals, **1**, under RAFT conditions serve, ideally, only two specific purposes – to activate the RAFT CTA, i.e. liberate R• from ZC(=S)SR, and replenish radicals lost from the system due to undesirable side reactions. In contrast, in a traditional free radical polymerization these primary radicals are the true initiating species. After the generation of **1** several reactive pathways are available to the primary radicals. Neglecting cage reactions **1** can add either directly to monomer to yield **3** (as in a normal free radical polymerization; Scheme 7.1) or, preferably, add directly to CTA, **2**, to yield **4** (Scheme 7.1, step ii). Considering the first option, adduct **3** can likewise add monomer, i.e. propagate, or add to a RAFT agent. Given the inherently high chain-transfer constants of most RAFT agents, it is unlikely that more than a couple of monomers will add to a generated radical before adding to a RAFT agent. So, **3** is likely to add to **2**, forming the intermediate radical **5**. Note that this reaction is reversible and this species *may* fragment to regenerate **2** and **3**. Alternatively, and with correct choice of monomer/RAFT agent combination, the equilibrium for fragmentation of **5** to yield **6** and **7** can be favored. To achieve the favored fragmentation of **5** in the forward direction, R must be a better free radical leaving group than B—M. Here, one must give careful consideration to the structures of the radical fragments and factors affecting their intrinsic stability and fragmentation characteristics when considering a given RAFT agent with a specific monomer. Assuming the forward fragmentation is favored, a new RAFT agent, **6**, is formed as well as the radical **7**, R•. Again, this reaction is reversible and **7** may add back to **6** etc. Provided **7** is a good reinitiating species then there is a high probability that **7** will add to monomer, initiating a new chain **8**. The net result of this sequence of forward reactions is the generation of a new polymer (oligomer) chain, **8**, derived from the R fragment of the initial RAFT agent and the formation of a new thiocarbonylthio compound, **6**, which is itself capable of serving as a RAFT agent. This entire forward sequence of reactions represents one addition–fragmentation cycle. The second option for the primary radical **1** is to add directly to RAFT agent **2** to yield the intermediate radical **4** and is typically considered to occur preferentially to the first available pathway. This reaction is also reversible, and again, fragmentation in the forward direction is the preferred pathway. In order to favor fragmentation and the formation of the new RAFT agent **9** and the radical **7**, the R group must be a better free radical leaving group than **1**. It is this exact requirement that makes the choice of source of primary radicals an important consideration. The radical derived from the R group **7** will ideally add to monomer, initiating a new polymer chain. Clearly, given the reversibility of many of these steps and various reaction pathways available to generated radical species this part of the RAFT mechanism is fairly complex. However, the net result of this sequence of reactions, i.e.

Scheme 7.1, step (ii), is activation of all the initial RAFT agent molecules to new oligomeric-type RAFT agents. When we talk of activation we mean that all the R groups of the initial RAFT agent **2** are converted to oligo/macro-RAFT agents **6**. An important feature to bear in mind that is not apparent from the mechanism as outlined below, or in other common versions of the RAFT mechanism, is that there are, ideally, 2–10 molecules of RAFT agent **2** for every primary radical **1** generated (at least based on common RAFT stoichiometries, and assuming 100% initiator decomposition/efficiency). As such, the total number of radicals in the system is determined by the source of primary radicals whereas the number of polymer chains, including both dormant, i.e. thiocarbonylthio end-capped, and actively propagating species, is controlled, predominantly, by the RAFT agent. Given the reversibility of the addition–fragmentation steps it is therefore evident that a single radical species, such as **1**, can 'activate' many RAFT agent molecules, although the total number of 'active' radicals is low compared to the number of polymer chains, most of which are dormant.

Once this preequilibrium is complete, the polymerization enters the main RAFT equilibrium (Scheme 7.1, step iii). This is the stage that predominates in RAFT polymerizations, i.e. where the bulk of the monomer consumption occurs. This stage simply involves the degenerative chain transfer of the thiocarbonylthio species between polymer chains **10** and **12** via the intermediate radical species **11**. It is worth noting that since RAFT is a radical process, and given that all side reactions cannot be completely suppressed, it is, in principle, possible for any of the radical species to participate in a number of undesirable, non-RAFT radical reactions, including combination, disproportionation and conventional chain transfer. The occurrence of such reactions leads to the formation of dead chains/inactive species that end up as contaminants in the final material.

It should be noted that while the processes described above are generally accepted, there are aspects that are still not fully understood nor agreed upon. Much of this debate is focused on a reconciliation of certain proposed mechanistic reactions/pathways with experimentally determined kinetic characteristics for certain monomer/CTA combinations, and relate primarily to the nature and reactivity of the intermediate, carbon-centered radicals **4**, **5** and **11** (Scheme 7.1) and possible products derived from reactions of these radicals. The most common 'anomalous' features are the presence of exceedingly long inhibition periods (perhaps even complete inhibition) or significant rate retardation, and are most commonly observed for dithioester-mediated polymerizations of styrene and acrylates [32]. The explanations for such observations include slow fragmentation of the intermediate radical species **4**, **5** and **11** [33–35], reversible or irreversible coupling of the same intermediate radicals with some other species X• [36, 37], while more recently McLeary and coworkers [38] reported that a significant difference in the propagation rate constants of initiator and/or R-group-derived radicals vs monomer-derived radicals could also manifest itself as an apparent inhibition period in the case of methyl acrylate or styrene, and termed the phenomena *initialization* [39–41]. While high level ab initio molecular orbital calculations also point to slow fragmentation as the root cause of the experimentally observed inhibition/retardation phenomena [33, 41], the expected high concentration of intermediate radicals has not been observed by electron spin resonance (ESR) spectroscopy. Likewise the star polymers formed from the irreversible coupling of intermediate radicals with other polymer chains have been detected only under nontypical RAFT conditions [36, 43, 44], although a recent amendment to this proposed root cause is capable of explaining the apparent lack of detectable coupled products [45]. Further discussions of such kinetic/mechanistic features associated with the RAFT process are beyond the scope of this chapter and more detailed information may be found in Chapter 3, as well as in a recent review authored by an IUPAC committee [32].

Given the fine balance of reversible processes as outlined above for the RAFT process, it is evident that the key to conducting successful RAFT polymerizations is appropriate choice of RAFT CTA for a given monomer substrate since the nature of the substitution pattern about the thiocarbonylthio functional group (the nature of the Z and R groups) [4, 5] is responsible for controlling the balance of the addition–fragmentation pathways, although the structure of the monomer and the primary radical must also be borne in mind. At present, an impressively large and varied number of RAFT agents have been prepared from all the major thiocarbonylthio families, containing a broad range of functionality, including those designed specifically for homogeneous aqueous RAFT polymerizations [2, 6]. Figure 7.1 shows the generic structures of the common thiocarbonylthio RAFT agents, with Fig. 7.2 showing examples of the more common species employed in homogeneous aqueous media.

RAFT agents must be thiocarbonylthio compounds. Beyond this, the difference between the various species lies in the nature of the so-called Z and R groups. These two groups perform distinct functions. The Z group primarily controls the ease with which radical species add to the C=S bond, i.e. the addition step. As such, Z groups that have a stabilizing effect on any of the intermediate radicals **4**, **5** and **11** will have high addition rate constants. The most common of these stabilizing groups, at least within the dithioester family of RAFT agents, is the phenyl group. The presence of simple alkyl groups, such as methyl, is less favorable in this class of CTA. In the case of the xanthates and dithiocarbamates the presence of lone pairs of electrons on the heteroatoms bonded to the thiocarbonylthio group disfavors radical addition to the C=S bond by virtue of contributing zwitterionic canonical forms to the resonance hybrid. As such, we find that xanthates and dithiocarbamates are useful only for nonactivated monomers such as vinyl esters and vinyl amides, unless, in the case of dithiocarbamates, the heteroatom is part of an aromatic ring such as a

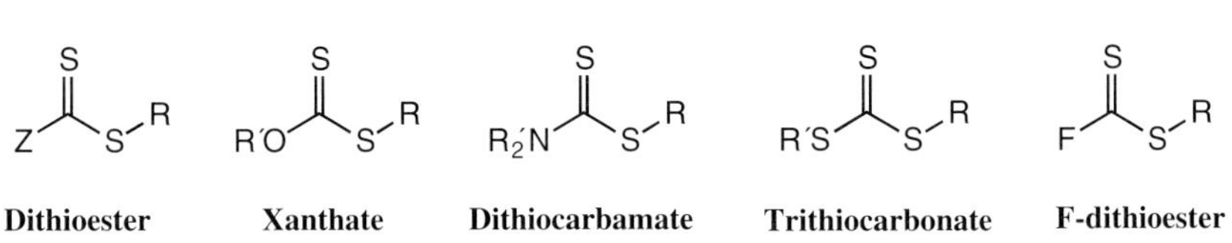


**Fig. 7.1** Generic structures of RAFT chain-transfer agents.

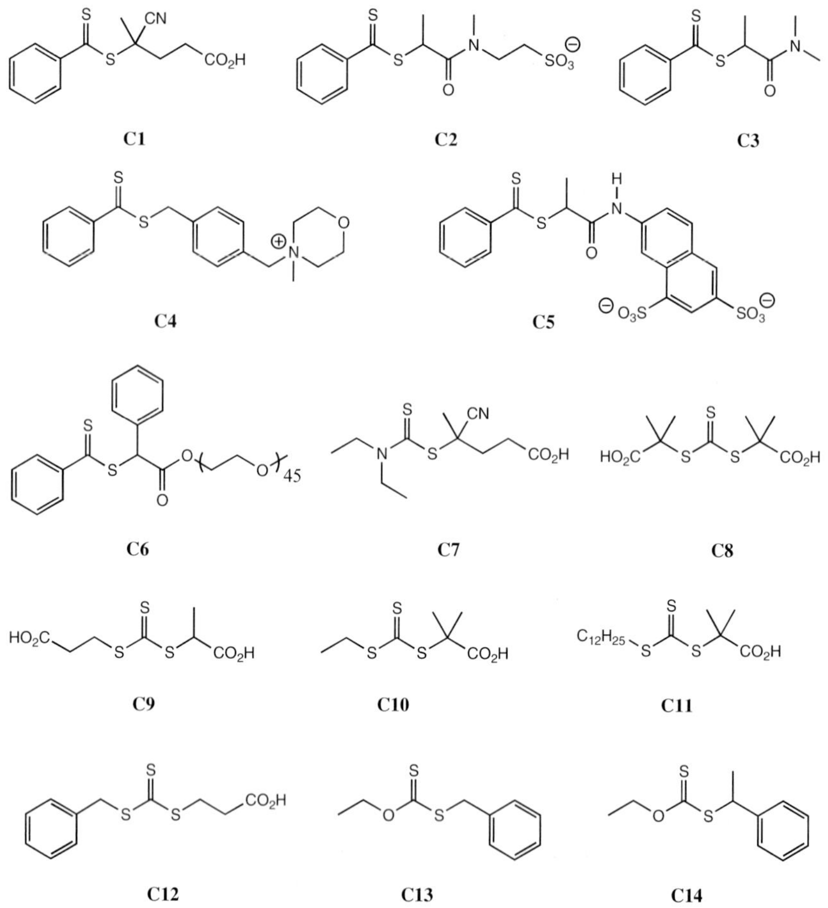


**Fig. 7.2** Examples of RAFT agents employed in the synthesis of water-soluble (co)polymers.

pyrrole. Intimately related to this function is clearly the resulting stability/lifetime of the intermediate radical. This is primarily controlled by the Z group but is R group influenced. The R group serves two important, but distinct, roles. Firstly, it must be a good homolytic leaving group, and secondly, the expelled radical R• must be able to add to monomer, i.e. initiate or propagate. The ease with which the fragmentation step (homolytic dissociation) occurs is dictated by the stability of the resulting radical, i.e. primary, vs secondary, vs tertiary, as well as polar and steric effects.

It could be argued that the major drawback, at present, of RAFT is the lack of commercially available RAFT agents. Fortunately, there are now many reliable and in many instances facile synthetic protocols for all the major families of RAFT agents [46–49]. The method employed is dictated primarily by the general family of RAFT agent being targeted since not all protocols are necessarily applicable to all general types of mediating agent. Recently, Moad et al. [46] summarized the major synthetic routes to RAFT agents:

1. The reaction of a carbodithioate salt (commonly prepared from the reaction of an appropriate Grignard reagent with $CS_2$) with a suitable alkylating agent: Such an approach works well for target RAFT agents bearing phenyl or benzyl Z groups and primary or secondary R groups.
2. Markovnikov addition of a dithio acid to an olefin: This approach is best exemplified by the synthesis of cumyl dithiobenzoate.
3. The reaction of a bis(thioacyl) disulfide with a radical species R•: This is the method of choice for the preparation of RAFT agents containing tertiary cyanoalkyl R groups, such as **C1** or **C7**. The experimental procedure for the preparation of **C1** via this method is given at the end of this chapter.
4. Radical-induced dithioester exchange.

While a wide range of water-soluble (co)polymers have been prepared both directly in water and in organic media, the majority of homogeneous aqueous-based RAFT polymerizations have been conducted with 4-cyanopentanoic acid dithiobenzoate, **C1** [2]. The choice of suitable CTA for conducting homogeneous polymerizations is dictated primarily by the nature of the monomeric substrate, i.e. (meth)acrylic vs (meth)acrylamido vs styrenic, since not all RAFT agent families are suitable for all monomers. Additionally, the nature of the R group must also be borne in mind since fragmentation to yield R• must be the preferred reaction pathway. Beyond these basic considerations, the introduction of other functionality such as poly(ethylene oxide), cyano groups, napthyl groups etc. (as highlighted above) as part of the R or Z groups must be considered in terms of conferring desired solubility.

RAFT polymerizations require a 'kick-start' – a radical species to initiate the cascade addition – fragmentation reaction to yield the oligomeric RAFT species **6**. Since RAFT is no more than a conventional free radical polymerization conducted simply in the presence of an appropriate thiocarbonylthio compound, all the traditional methods of radical generation can be employed. These include the use of thermally sensitive initiators, such as azo compounds (probably the most commonly employed method) [2], redox initiation, thermal initiation [50, 51], ultraviolet radiation [52–54], solar radiation [55], microwave radiation [56, 57] and gamma-ray radiation [35, 58–68].

The use of the various RAFT agents shown above, as well as others, will be discussed in more detail below when we consider the various hydrophilic monomers that have been successfully polymerized under homogeneous conditions via RAFT in water, mixed-aqueous media, as well as organic environments.

## 7.3 RAFT Agent Stability

Before we examine the types of hydrophilic monomers that have been successfully polymerized by RAFT under homogeneous conditions it is prudent to identify several potential drawbacks when considering such syntheses. While the economic and environmental advantages of performing RAFT polymerizations directly in water are evident, the hydrolysis of the thiocarbonylthio functional group is a concern, although other considerations must also be borne in mind. For example, RAFT agents are incompatible with some functional groups, specifically primary and secondary amines, thiols and reducing agents. Also, certain monomer substrates can themselves be problematic since they are susceptible to degradation in aqueous media yielding products that are capable of degrading RAFT CTAs.

### 7.3.1 RAFT Agent Hydrolysis

Since RAFT agents are simply the S analogs of esters, it is not surprising that they are susceptible to hydrolysis. As with esters, thiocarbonylthio compounds are prone to both acid- and base-catalyzed hydrolysis, although as a general remark it should be noted that thiocarbonylthio species are more stable under acidic conditions. Levesque et al. [61], as part of their protein thioacylation studies, examined the stability of several thiocarbonylthio compounds toward hydrolysis under mild conditions – conditions suitable for protein modifications. Specifically, reactions were monitored at 20–35 °C and at pH 7.5–8.5. As expected, both the pH and the temperature were found to have an effect on the rate and degree of hydrolysis. For example, the authors found that the rate of hydrolysis increases with temperature and decreases with decreasing pH. It was demonstrated that the degree of hydrolysis for a given thiocarbonylthio compound could be significant with up to ca. 40% loss at pH 8.5 after 4–5 h. These observations alone might imply that conducting wholly homogeneous aqueous RAFT polymerizations would be problematic given the often high temperatures and extended polymerization times required. Indeed, the impact of thiocarbonylthio hydrolysis in aqueous media is a particularly important issue since loss of these functional groups results in the loss of chain-end functionality, diminishes overall polymerization control and possibly causes an increase in the polydispersity index.

Besides these 'non-RAFT' studies, little has been done with respect to evaluating the general hydrolytic stability characteristics of particular RAFT CTAs. Thomas et al. [10] were the first to conduct a detailed study of the hydrolysis of both small molecule and macro-RAFT CTAs. In this study the hydrolytic stability of **C1** (Fig. 7.2) as a function of solution pH was examined alongside two examples of homopolymers prepared with **C1**. Given the large excess of water relative to thiocarbonylthio functionality the hydrolysis reaction can be considered to be zero order with respect to water. As such, the rate of CTA hydrolysis can be expressed in terms of an apparent rate constant of hydrolysis, $k_{\text{hyd}}$, and the CTA concentration (equation 7.1).

$$\frac{-d[CTA]}{dt} = k_{\text{hyd}}[\text{CTA}] \tag{7.1}$$

For example, Fig. 7.3 shows the pseudo-first-order rate plots for the hydrolysis of **C1** along with two macro-CTAs with differing degrees of polymerization (DP) derived from sodium 2-methyl-2-prop-2-enamidopropane-1-sulfonate (**M3**, Fig. 7.5).

It is evident that both solution pH and CTA molecular mass have a strong influence on the rate of thiocarbonylthio hydrolysis. For example, the magnitude of $k_{hyd}$ for **C1** increases by almost one order of magnitude from $2.5 \times 10^{-5}$ to $15 \times 10^{-5}$ $s^{-1}$ as the pH is increased from 7.0 to 10.0. In contrast, little change in $k_{hyd}$ is observed when the pH is lowered from 7.0 to 5.5, and finally to 2.0. Such observations are consistent with what is known about the pH-dependent stability of thiocarbonylthio functional groups. A comparison of the hydrolysis rates of **C1** with macro-CTAs derived from **M3** indicates that the polymeric CTAs hydrolyze at a significantly slower rate than the small-molecule RAFT agent, with the difference being attributed to the steric protection of the thiocarbonylthio end group in the case of the polymeric species.

The effect of added base on the homogeneous aqueous RAFT polymerization of 2-methacryloxyethyl glucoside (**M40**, Fig. 7.21) with **C1** has also been examined [62]. It was shown that when sodium carbonate or sodium bicarbonate were used to aid in the dissolution of **C1** at room temperature, an induction period of 60–90 min was observed coupled with an overall decrease in the polymerization rate. Additionally, experimentally determined molecular masses were found to be higher than expected based on the ratio of CTA to monomer for both added bases. Enhanced control was observed for sodium bicarbonate vs sodium carbonate with, for example, control essentially being lost after ca. 40% conversion in the case of the latter. This loss of control was attributed to the base hydrolysis of the thiocarbonylthio group and the formation of species (thiols and thionobenzoic acid) that were postulated to act as radical scavengers and/or retarders. Significantly better results were obtained when 10 vol % ethanol was added as opposed to base to aid in the dissolution of the CTA.

### 7.3.2
### Incompatibility with Primary and Secondary Amines

While thiocarbonylthio agents have demonstrated themselves to be tolerant of most functionality, they are not tolerant of all. Particularly problematic are amines, specifically primary and secondary amines. Just as thiocarbonylthio species are susceptible to hydrolysis, they are also prone to aminolysis. Indeed, dithioesters are well known to react with primary and secondary aliphatic amines at much faster rates than esters yielding thioamides and thiols. In the realms of RAFT chemistry this is detrimental and, for example, precludes the direct polymerization of monomers with such functional groups. Such reactions have been successfully employed, however, as a means of achieving facile protein modification. For example, Souppe et al. [63] described the modification of the lysine residues in both horseradish peroxidase and papain via reaction of the lysine $\epsilon$-$NH_2$ groups with carboxymethyl dithiotridecanoate in the case of the horseradish peroxidase and with carboxymethyl dithiobenzoate for papain (Scheme 7.2).

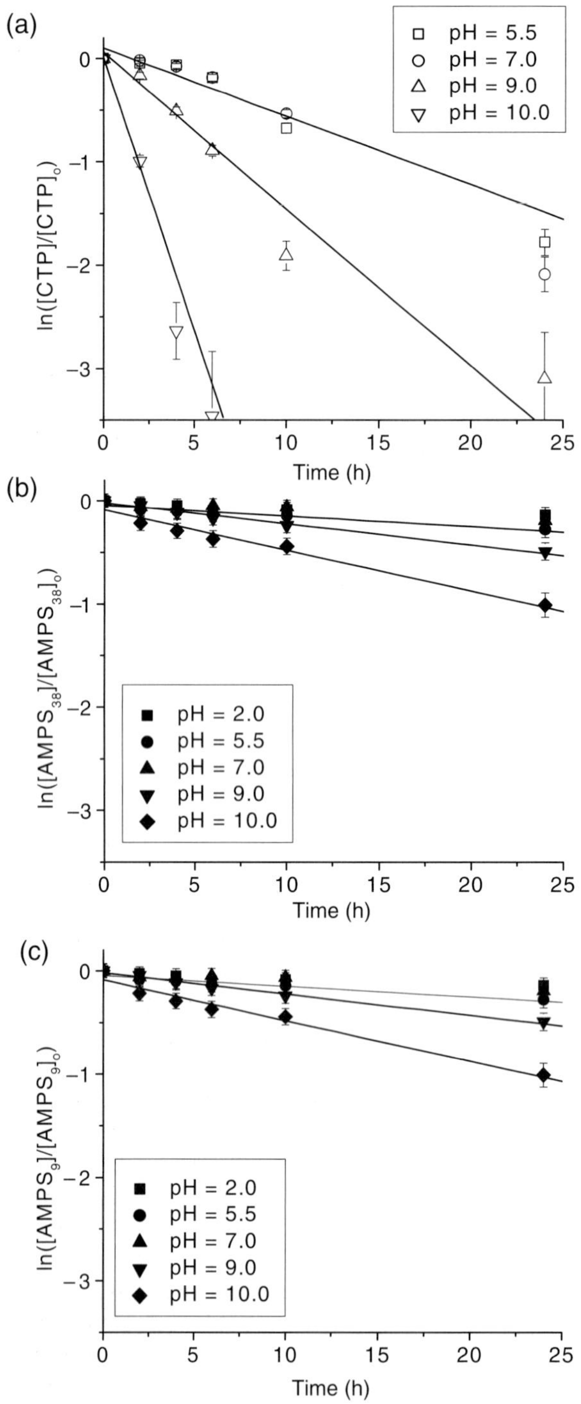


**Fig. 7.3** The pseudo-first-order rate plots for the hydrolysis of (a) **C1**, (b) P**M3**$_{38}$ and (c) P**M3**$_{9}$.

**Scheme 7.2** Thioacylation of ε-$NH_2$ groups in lysine.

Following this report, Deletre and Levesque [64] described the kinetics and mechanism of polythioamidation in solution. Reactions between mono and difunctional amines with dithioesters near room temperature were evaluated in the presence of a large excess of amine. The reactions were shown to have complex rate expressions with either second- or third-order dependence in dithioester and amine. Mechanistically, this was reported to involve an equilibrium between dithioester and amine and an intermediate followed by a bimolecular reaction of the intermediate with additional primary amine via prototropy.

Such facile reactions have been employed by RAFT researchers as a means of removing the thiocarbonylthio end group to yield —SH-terminated (co)polymers. For example, Qiu and Winnik [65] reported the facile end-group modification of trithiocarbonate-prepared **M33** polymers via a combination of aminolysis and Michael addition. The trithiocarbonate groups were reduced to thiol species by treatment with a fivefold excess of *n*-butylamine in the presence of a small quantity of tris(2-carboxyethyl)phosphine hydrochloride. Subsequent reaction with an activated ester, such as *n*-butyl acrylate or 2-hydroxyethyl acrylate, resulted in conjugate addition and the introduction of thioether/ester functional end groups.

## 7.3.3 Acrylamido Monomers as a Source of Primary and Secondary Amines

In addition to the inherent thermodynamic instability of the C=S bond in aqueous media and its susceptibility to hydrolysis, certain monomers can also be problematic in aqueous media unless the solution conditions are carefully controlled. Under homogeneous aqueous conditions, some monomers are themselves susceptible to hydrolysis, liberating species that are capable of consuming thiocarbonylthio compounds. (Meth)Acrylamido monomers are especially problematic in this respect since one product of monomer hydrolysis is either a primary or secondary amine. Since the concentration of monomer is significantly higher than that of CTA, very low levels of monomer hydrolysis can result in complete loss of CTA via an aminolysis reaction, as described above. For example, when conducting polymerizations of acrylamide, it is imperative to do so at low pH for two reasons. Firstly, RAFT agents are more stable under acidic conditions and secondly, acrylamide itself is susceptible to hydrolysis, yielding acrylic acid and ammonia. As noted above, such monomer hydrolysis reactions are undesirable since the liberation of ammonia will result in loss of RAFT agent. Thomas et al. [66, 67] examined the contribution of aminolysis to the failure of RAFT polymerizations for monomers such as acrylamide, which had been earlier postulated to be the primary cause of failure.

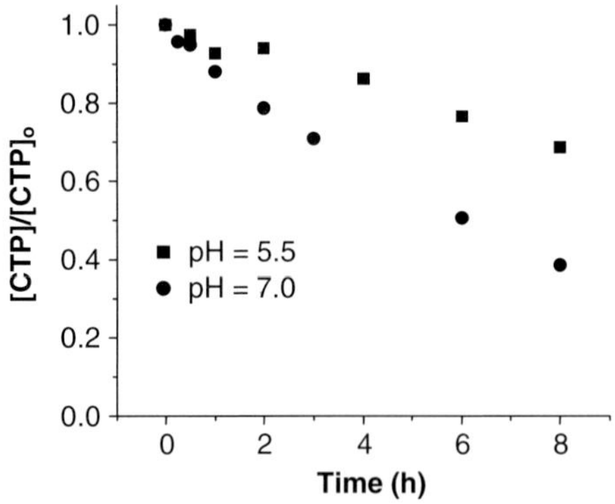


**Fig. 7.4** Fraction of **C1** remaining under conditions where both hydrolysis and aminolysis may be operative.

Aminolysis experiments with **C1** were conducted in buffered media with ammonium hydroxide to yield a final ammonia concentration of $5 \times 10^{-3}$. The fraction of **C1** remaining as a function of time at pH 5.5 and 7.0 was then determined (Fig. 7.4).

It is important to realize that the loss of **C1** is due to a combination of hydrolysis and aminolysis and may be described by

$$\frac{-d[C1]}{dt} = k_{\mathrm{hyd}}[C1] + k_{\mathrm{a}}[C1][\mathrm{NH}_3]^2 \tag{7.2}$$

where $k_\mathrm{a}$ is the third-order rate constant for aminolysis and $[\mathrm{NH}_3]$ is the concentration of ammonia in solution. Taking into account the time-dependent effect of both hydrolysis and aminolysis on the thiocarbonylthio concentration (either for small-molecule RAFT agents or macro-RAFT agents), the $M_{\mathrm{n,theory}}$ equations can be modified to take into account the loss of these end groups. It can be shown that:

$$M_{\mathrm{n,theory}} = \frac{M_{\mathrm{Mw}}([M]_0 - [M]_0 e^{-k_{\mathrm{p*}}(t-t_{\mathrm{ind}})})}{[CTA]_{\mathrm{ind}} e^{-(k_{\mathrm{hyd,macro}}+k_{\mathrm{a,macro}}[NH_3]^2)(t-t_{\mathrm{ind}})}} \tag{7.3}$$

Equation 7.3 was applied to the aqueous RAFT polymerization of acrylamide and was demonstrated to be effective for predicting the theoretical molecular mass as a function of polymerization time taking into account both hydrolysis and aminolysis side reactions.

### 7.3.4 Incompatibility with Reducing Agents

Thiocarbonylthio compounds are also susceptible to degradation by mild hydride reducing agents, such as sodium borohydride ($\mathrm{NaBH_4}$), and other hydride derivatives [68–72]. While such reagents are unlikely to be encountered during a typical

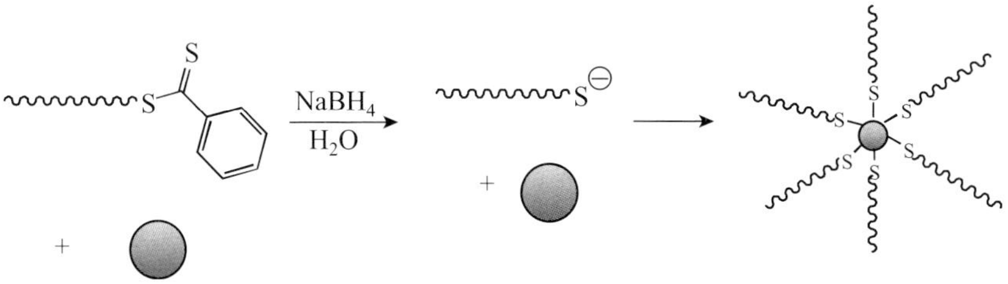


**Scheme 7.3** The stabilization of gold colloids via the in situ reduction of thiocarbonylthio-terminated polymers.

RAFT polymerization, reduction is a facile method for postpolymerization modification of the thiocarbonylthio end group. For example, Lowe et al. [68] reported the in situ reduction of thiocarbonylthio end groups in the presence of a gold sol to yield polymer-stabilized gold nanoparticles, a grafting-to approach (Scheme 7.3). This was subsequently extended to the functionalization of planar gold surfaces [71]. The reduction reaction is extremely facile and occurs near instantaneously in aqueous media at ambient temperature [68, 69, 71].

Other groups have adopted such a straightforward modification reaction. For example, Shan et al. [70] reported the synthesis of **M33** homopolymers using **C1** or cumyl dithiobenzoate that were used to prepare monolayer-protected clusters of gold nanoparticles. The thiocarbonylthio end groups were reduced in situ with lithium triethylborohydride. Most recently, this approach was utilized for the preparation of RAFT-synthesized, polymer-stabilized gold nanorods [73]. In addition to the grafting-to approach, gold substrates can be modified via grafting-from approach. Raula et al. [74] described the preparation of gold nanoparticles stabilized by 11-mercapto-1-undecanol. **C1** was subsequently coupled to the terminal —OH group via carbodiimide coupling with dicyclohexylcarbodiimide (DCC)/*N,N*-dimethylaminopyridine (DMAP) to yield RAFT-agent-functionalized gold nanoparticles. **M33** was subsequently polymerized directly from the surface in DMF at 60 °C.

### 7.3.5
### Susceptibility to Radical Reagents

The reactivity of thiocarbonylthio compounds toward radicals is well documented, with much of the early work having been done in small-molecule organic syntheses by Barton and coworkers [75, 76] and Zard [77]. The inherent reactivity toward radicals is the fundamental reason for the success of polymerizations mediated by such species. As such, radical reactions can be employed in a similar manner to that of hydride reduction, i.e. as a means of postcleaving/modifying the thiocarbonylthio end groups. For example, Perrier et al. [78] recently reported a straightforward approach for cleaving the thiocarbonylthio end group while simultaneously introducing a new functionality. This was achieved by postpolymerization reaction of various RAFT-synthesized (co)polymers with a variety of azo initiators. End-group modifications were near quantitative, allowing for a recycling of RAFT agents. Such end-group modifications are discussed in more detail in Chapter 12.

## 7.4 Suitable Monomers

As highlighted above, some of the most beneficial features of RAFT include its versatility with respect to monomer choice, functional group tolerance and general experimental conditions under which it can be successfully executed. Since RAFT is, in reality, no more than a conventional free radical polymerization conducted in the presence of a suitable thiocarbonylthio species, it possesses all the benefits of a conventional free radical polymerization process while simultaneously conferring many of those features associated with a living, or controlled, polymerization. As such, one advantage includes the ability to choose substrates from a broad range of monomer families bearing a wide range of functional groups. With specific reference to the preparation of water-soluble (co)polymers under homogeneous conditions in either aqueous or nonaqueous media, RAFT has so far been successfully employed with, for example, styrenic, (meth)acrylic, and the (meth)acrylamido general families of monomers. Additionally, within these general monomer groups, substrates containing neutral, anionic, cationic and zwitterionic, hydrophilic functionality have been successfully polymerized in a controlled manner. Below we will highlight those hydrophilic monomers that have been polymerized via RAFT to yield water-soluble polymers. As noted above, many of these monomers can be and have been polymerized in both organic and aqueous media [2].

### 7.4.1 Anionic/Acidic Monomers

Figure 7.5 shows examples of hydrophilic anionic/acidic monomers that have been polymerized via RAFT either directly in water or in organic media. The potential of RAFT in homogeneous aqueous media was hinted at in the very first literature report where sodium styrenesulfonate (**M1**, Fig. 7.5) was successfully polymerized in water at a concentration of 1.21 M, using **C1** at 70 °C and V-501 as a water-soluble azo initiator [1].

The ability to polymerize even this group of 13 monomers in a controlled fashion via RAFT serves to reiterate both the versatility and the functional group tolerance of the technique. While we will not discuss each of these in detail (readers are directed to a recent review by the chapter authors for a more thorough discussion) [2], pertinent features will be highlighted. The motivations behind the choice of these monomers as building blocks are varied. Aside from the inherent desire to demonstrate the versatility of RAFT by choosing monomers from various general families, the non-sulfonate-containing monomers, i.e. **M2**, **M4–M7** and **M10–M13**, have also been examined due to additional desirable features associated with the

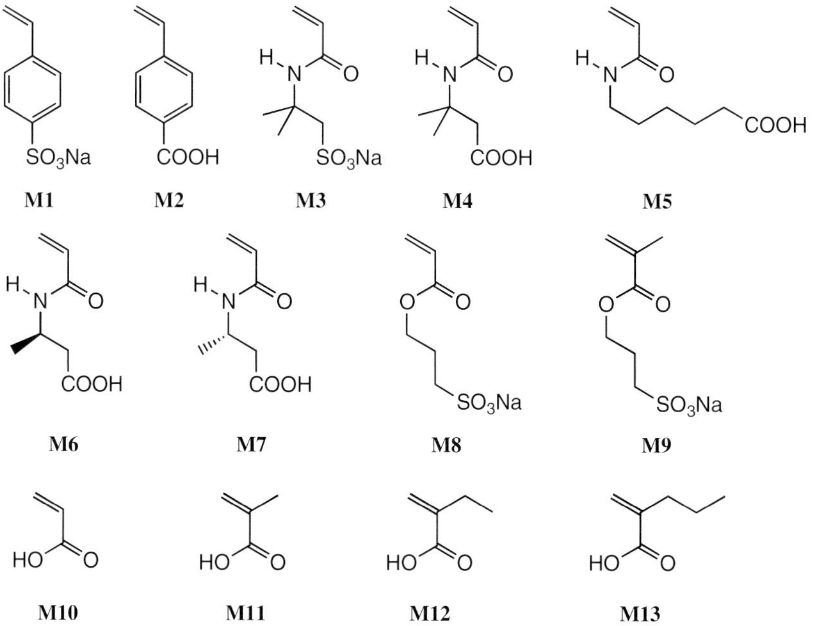


**Fig. 7.5** Chemical structures of anionic/acidic monomers polymerized via RAFT.

monomers. For example, **M2** and **M4–M7** represent examples of 'smart' building blocks. By 'smart' we mean that the monomer, and resulting (co)polymer, is tunably hydrophilic/hydrophobic by virtue of the readily accessible p$K_a$ of the carboxylic acid functional groups. As such, materials composed of these building blocks are capable of exhibiting interesting, and often complex, aqueous solution behavior. For example, we, and others, have shown that AB diblock copolymers with **M2**, **M4** or **M5** as one of the comonomers yield materials that are capable of undergoing reversible, pH-induced supramolecular self-assembly in aqueous media and will be discussed in more detail below [79–83].

The ability to prepare (co)polymers with well-defined, predetermined molecular characteristics from these building blocks under homogeneous conditions is limited only by the choice of a solvent capable of solubilizing the monomer/(co)polymer and by appropriate choice of a RAFT agent. The near-permanently hydrophilic nature of the sulfonate monomers limits the solvent choice as well as the application of a suitable RAFT agent. All of the sulfonate-containing monomers shown in Fig. 7.5 have been polymerized in water [79–81, 84, 85] only due to their lack of solubility in organic media. As such, only water-soluble RAFT agents have been applicable. **M1**, **M3**, **M8** and **M9** have all been homopolymerized with **C1** [79–81, 85], the most common RAFT agent employed for polymerizations conducted homogeneously in water. Additionally, **M9** has been polymerized using **C5** [85].

The carboxylic-acid-containing monomers can be readily polymerized in a broader range of solvents by virtue of their tunable hydrophilic/hydrophobic nature, as noted above. As such, a likewise larger pool of RAFT agents may also be employed. **M2**, for example, has been polymerized directly in water [79, 86] (as a second or third building block – vide infra), and also homopolymerized in DMF with the trithiocarbonate RAFT agent **C9**. The acrylamido substrates **M4–M7** [80, 81, 83, 87] have only been (co)polymerized in water using either **C1** or **C10** [87] in the case of **M6** and **M7** with an appropriate water-soluble azo initiator. Acrylic acid, **M10**, has been screened with more RAFT CTAs than any other monomer. The homopolymerization of **M10** was also disclosed in the first literature report. **M10** was polymerized in DMF, at 60 °C with 1-phenylethyl dithiobenzoate as the RAFT agent and 2,2′-azobis(isobutyronitrile) (AIBN) as the source of primary radicals to yield a homopolymer with an $M_n = 13\,800$ and $M_w/M_n = 1.23$ [1]. The most detailed evaluation regarding the general homopolymerizability of **M10** is that of Ladaviere et al. [88]. In this study, 17 different RAFT agents were screened with examples from the dithioester, dithiocarbamate, xanthate and trithiocarbonate families of thiocarbonylthio compounds. Polymerizations were conducted in ethanol with V-510 used as the source of primary radicals. In all instances, [**M10**]:[CTA]:[V-501] = 50:1:0.1. Best results were reported using phenoxyxanthate or trithiocarbonate RAFT agent derivatives. Gaillard et al. [89] have also reported the synthesis of **M10**–butyl acrylate block copolymers prepared using both xanthate and trithiocarbonate mediating agents. Various researchers have conducted more detailed evaluations regarding the effectiveness of trithiocarbonate CTAs. For example, Loiseau et al. [90] evaluated the homopolymerization of **M10** with dibenzyl trithiocarbonate and bis(1-phenylethyl)trithiocarbonate in methanol, ethanol, 2-propanol and dioxane. Polymerizations were generally controlled, but some transfer to solvent was observed for higher targeted molecular masses/conversions. Lai et al. [13] evaluated the carboxyl-functional trithiocarbonates **C8** and **C11** in both aqueous and organic (DMF) media at 70 or 80 °C. Both CTAs proved effective, with the resulting **M10** materials having narrow molecular mass distributions. More recently, Khousakoun et al. [91] reported the synthesis of ABA triblocks of poly(ethylene oxide) methyl ether methacrylate (A block) with **M10** using dibenzyl trithiocarbonate, V-501 in *n*-butanol at 90 °C. The aqueous solution properties of the ABA triblock copolymers were subsequently evaluated as a function of temperature and ionic strength. Schilli et al. [92] have also reported the synthesis of such doubly hydrophilic block copolymers of **M10** with **M33** in methanol. Additionally, Millard et al. [60] have reported the use of $\gamma$-irradiation and aqueous room temperature RAFT via the trithiocarbonate-mediated polymerization of **M10**. Morel et al. [93] have described the xanthate-mediated block copolymerization of **M10** with vinyl acetate, and Hoogenboom et al. [94] have reported the indirect synthesis of poly-**M10** employing the protected acrylic monomer 1-ethoxyethyl acrylate in toluene with AIBN and 2-cyano-2-butyl dithiobenzoate as the initiator/RAFT CTA pair.

In contrast to the fairly extensive RAFT studies of **M10**, relatively little has been done with respect to the RAFT polymerization of **M11**. The first report describing the polymerization of **M11** was that of Chong et al. [95], in which AB diblock copolymers with benzyl and methyl methacrylates were prepared in DMF with cumyl dithiobenzoate as the mediating agent. Following this, de Brouwer et al. [96] reported the RAFT miniemulsion synthesis of poly[(2-ethylhexyl) methacrylate-*block*-(methyl methacrylate-*co*-**M11**)] via a semicontinuous procedure. Sprong et al. [97, 98] reported the RAFT synthesis of model alkali-soluble rheology modifiers, in which **M11** was copolymerized with methyl methacrylate in dioxane with AIBN and bis(thiocarbonyl) disulfide. In addition to **M11**, two other α-substituted acrylic acids have been polymerized via RAFT, namely, α-ethylacrylic acid (**M12**) [99] and α-propylacrylic acid (**M13**) [100].

### 7.4.2
### Cationic Monomers

Like anionic monomers, cationic or potentially cationic, substrates are also of fundamental interest given their broad applications in areas as diverse as water purification, enhanced oil recovery, cosmetics and as antibacterial agents. As with the acidic/anionic substrates, various cationic/3° amine monomers have been examined. Figure 7.6 shows examples of such substrates. Also mirroring the acidic/anionic substrates, examples from all the major monomer families have been examined, as well as some 'unique' representatives of this class, namely, the vinyl pyridines, **M18** and **M19** [2]. As with the anionic/acidic monomers discussed above, the reason for choosing such substrates, at least with respect to the synthesis of water-soluble copolymers, is typically dictated by their solubility characteristics. For example, while **M15–M17** and **M23** are permanently hydrophilic, the remaining substrates exhibit reversible hydrophilicity/hydrophobicity either as a function of aqueous solution temperature (**M20** and **M22**) or aqueous solution pH (**M14**, **M18**, **M19** and **M21**).

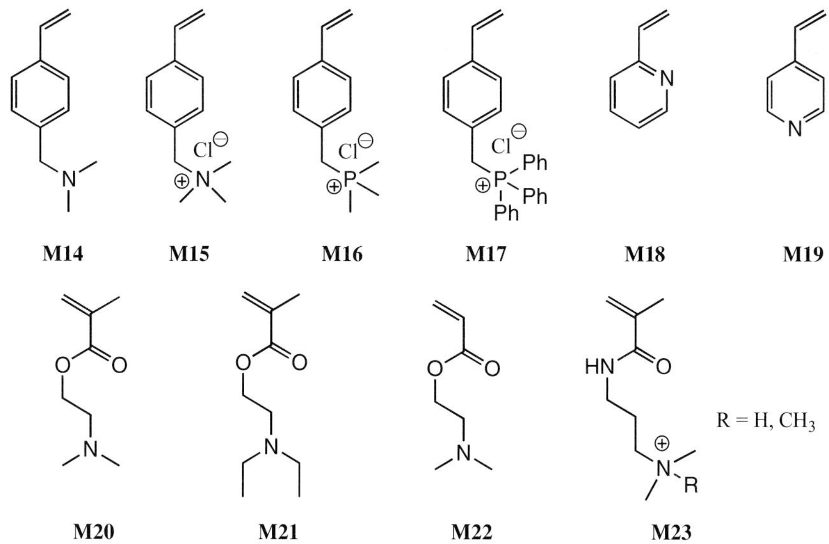


**Fig. 7.6** Chemical structures of permanently cationic and tunably cationic monomers that have been polymerized via RAFT.

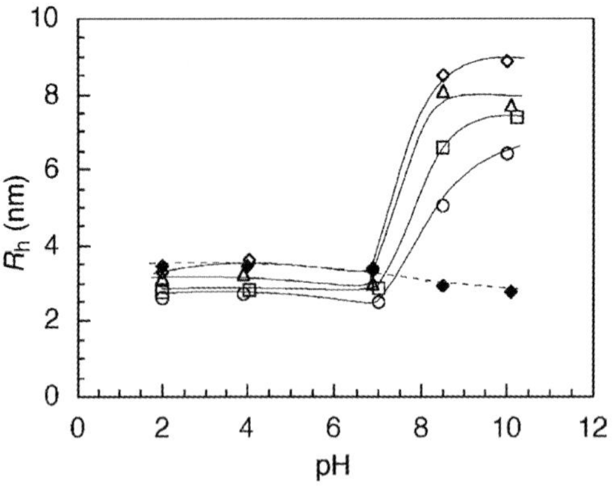


**Fig. 7.7** Hydrodynamic radius plotted as a function of solution pH for **M14–M15** block copolymers of varying composition – 57:11 (open circle), 57:22 (open square), 57:34 (open triangle), 57:50 (open diamond) and a random copolymer of **M14–M15** with a composition of 55:57 (solid diamond).

*N,N*-dimethylbenzylvinylamine, **M14**, is an excellent example of a 'smart' building block, being pH responsive. Specifically, polymers derived from **M14** are hydrophobic with neutral but become hydrophilic when the tertiary amine residues are protonated. **M14** has been employed as a 'smart' building block in several studies [79, 101, 102]. For example, Mitsukami and coworkers recently reported the synthesis and aqueous solution properties of a range of **M14–M15** AB diblock copolymers [101]. The block copolymers were prepared using **C1**/V-501 as the CTA/initiator pair directly in water to yield materials with controllable molecular masses and low polydispersity indices. The pH-induced, reversible self-assembly of such block copolymers was demonstrated using a combination of NMR spectroscopy, dynamic and static light scattering (SLS), as well as fluorescence spectroscopy.

For example, Fig. 7.7 shows the change in measured hydrodynamic radius ($R_h$), as determined by dynamic light scattering, for a series of **M14–M15** block copolymers, as well as a statistical copolymer for comparative purposes, as a function of aqueous solution pH. In all instances, sizes consistent with unimers are observed at pH $>$ ca. 7. Under such conditions, the **M14** block is expected to be predominantly protonated and thus hydrophilic, as is the permanently charged **M15** block. At pH $\sim$ 7, with the exception of the statistical copolymer, an abrupt change in the $R_h$ is observed with observed sizes in the range of 6.0–9.0 nm. Such changes are consistent with a hydrophilic-to-hydrophobic phase transition associated with the **M14** block and the onset of a self-assembly process. Such supramolecular self-assembly was verified by SLS and fluorescence spectroscopy. Sumerlin et al. have likewise taken advantage of the facile phase transition associated with **M14** as a function of pH in their studies of **M14**–*N,N*-dimethylacrylamide AB diblock copolymers [102]. As with the **M14–M15** copolymers, the AB diblocks were demonstrated to undergo pH-induced micellization. Amphiphilic AB diblock copolymers of styrene with structurally similar derivatives of **M15** have been prepared indirectly by the reaction of a tertiary amine such as $NEt_3$ with the vinylbenzyl chloride residues in styrene–vinylbenzyl chloride precursor copolymers [103]. The permanently cationic, hydrophilic phosphonium styrenic derivatives **M16** and **M17** have only recently been evaluated under RAFT conditions and will be discussed in more detail below [82].

The vinyl pyridine substrates **M18** and **M19** have likewise been little studied to date, although both are readily polymerized in a controlled manner. Yuan et al. [104] were the first to report on the RAFT polymerization of this particular monomer class. AB diblock copolymers of styrene with **M19** were prepared in DMF with dibenzyl trithiocarbonate and AIBN. Following this report, Convertine et al. [105] reported the synthesis of **M18** and **M19** homopolymers as well as AB diblock copolymers composed of the two monomers. Homopolymers of **M18** or **M19** were prepared with cumyl dithiobenzoate and AIBN at 60 °C. All experimental data indicate that the (co)polymerizations proceeded in a controlled manner (Figs. 7.8 and 7.9). Figure 7.8a–7.8c show the size-exclusion chromatograms for an **M18** homopolymerization coupled with the evolution of $M_n$ and $M_w/M_n$ with conversion (Fig. 7.8b) and the pseudo-first-order kinetic plot (Fig. 7.8c). The systematic shift to lower retention volume of the size-exclusion chromatograms with increasing polymerization time, the linear evolution of $M_n$, agreement with the theoretical $M_n$'s, the low $M_w/M_n$ values and the linear first-order kinetic plot are consistent with a controlled polymerization. The ability to prepare **M18–M19** in the 'forward' and 'reverse' directions was also demonstrated, as can be clearly seen in the size-exclusion chromatograms in Fig. 7.9.

The RAFT polymerization of the (meth)acrylic derivatives, **M20–M22**, has been examined by several research groups. The polymerization of **M20** has been reported in several papers, although typically as a part of a larger screening study [1, 95, 106]. Recently, Sahnoun et al. [106] conducted a detailed examination of the homopolymerization of **M20** in dioxane at 90 °C with 1-methyl-1-cyano-ethyl dithiobenzoate, with an emphasis on the effect of [M]:[CTA] and [CTA]:[I] with respect to control over $M_n$ and the molecular mass distribution, conversion and polymerization time.

Yusa et al. [107] reported the synthesis of **M21**-based AB diblock copolymers with **M23**. Consistent with previous reports regarding the aqueous solution behavior of **M21**-based block copolymers, the **M21–M23** block copolymers were shown to undergo pH-induced supramolecular assembly by virtue of the tunably hydrophilicity/hydrophobicity of the **M21** block.

To date, very little has been reported with respect to the synthesis of cationic acrylamido (co)polymers by RAFT. Vasilieva et al. [108] investigated the homo- and copolymerization of **M23** ($R = CH_3$) with **C1** and V-501 under buffered conditions (Fig. 7.10).

The acceptable correlation between the theoretical and the observed $M_n$'s coupled with the linearity of the $M_n$ vs conversion curve and the low $M_w/M_n$ values all indicate that **M23** polymerizes in a controlled fashion under buffered conditions. The use of buffer is an important requirement for achieving such controlled

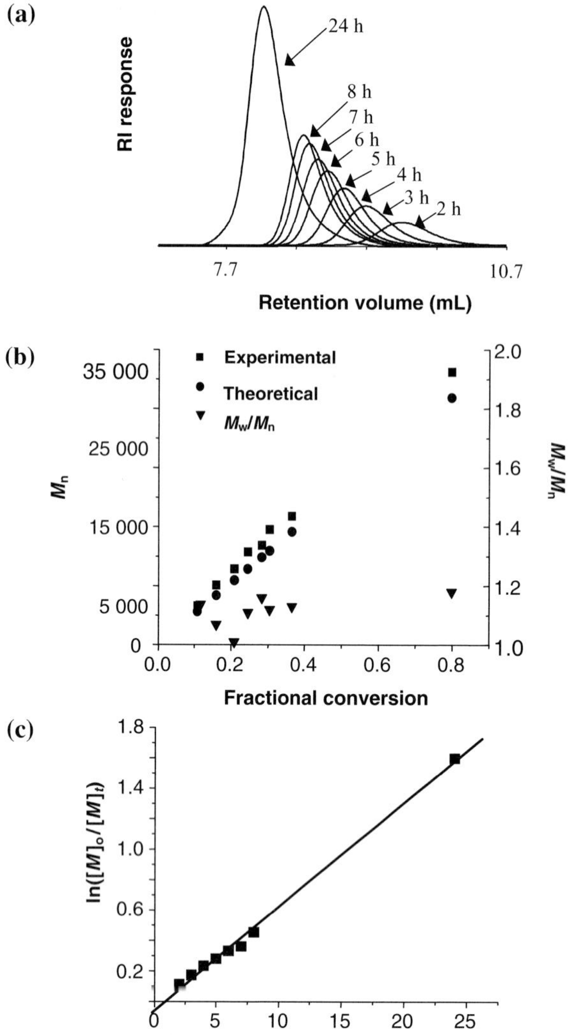


**Fig. 7.8** Size-exclusion chromatographic traces (a) for a **PM18** homopolymer, demonstrating the evolution of molecular mass with time, the plots of $M_n$ and $M_w/M_n$ vs conversion (b), and the corresponding pseudo-first-order rate plot (c).

polymerizations. As discussed above, acrylamido monomers are susceptible to hydrolysis in aqueous media, liberating primary, secondary or tertiary amines. In the case of **M23**, monomer hydrolysis would yield a primary amine that is capable of reacting with the thiocarbonylthio functional group of the CTA, resulting in loss of mediating agent. Additionally, the low-pH buffered environment is desirable since RAFT CTAs are more stable toward hydrolysis under acidic conditions and it also facilitates the protonation of any primary amine liberated via monomer hydrolysis [109].

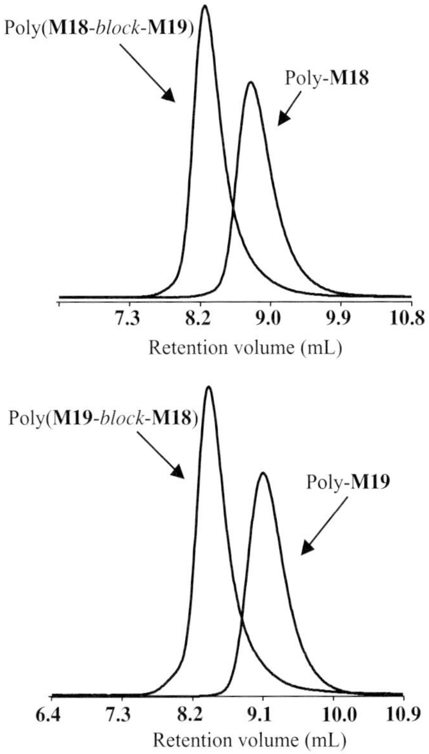


**Fig. 7.9** Size-exclusion chromatographic traces for **M18** and **M19** macro-RAFT agents and the corresponding **M18–M19** and **M19–M18** AB diblock copolymers.

### 7.4.3
### Zwitterionic Betaine Monomers

Betaine monomers represent a specialized family of functional substrates. Polymeric betaines are interesting for several reasons, including their *antipolyelectrolyte* characteristics in water [110], their well-documented bio/blood compatibility [109, 111, 112], as well as their less well-documented antimicrobial properties [113]. Synthetic polybetaines have been known since the 1950s; however, even today there are a very few examples of such materials that have been synthesized with well-defined molecular characteristics/architectures [114–117] and even fewer reports in which such materials have been prepared directly, i.e. via the direct, controlled polymerization of betaine monomers.

One of the reasons why it is difficult to polymerize such substrates in a direct manner is their extremely limited solubility in suitable solvents. Sulfobetaine monomers, and polymers derived thereof, in particular, i.e. **M24–M28**, Fig. 7.11 exhibit especially limited solubility properties. For example, materials derived from such zwitterionic building blocks are generally soluble *only* in aqueous salt solutions

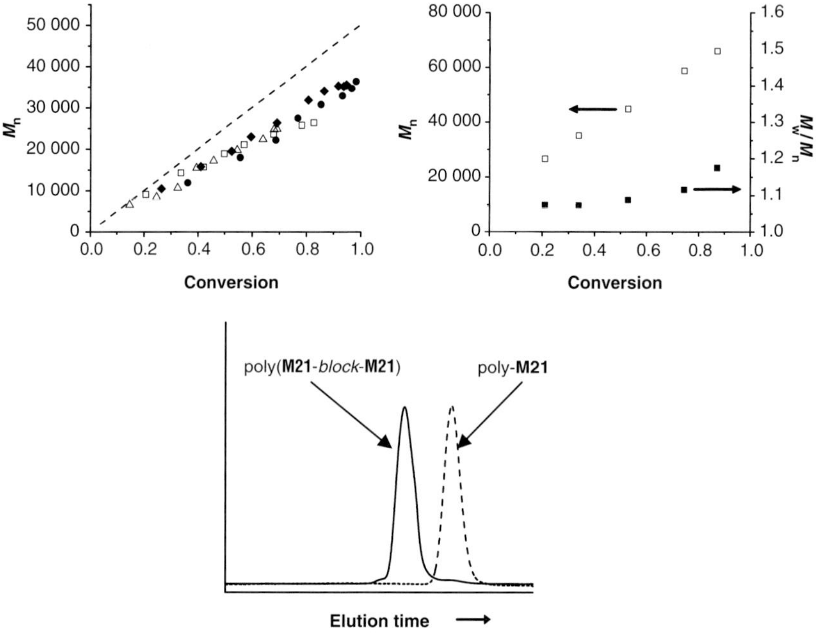


**Fig. 7.10** Dependence of $M_n$ on the monomer consumption for the homopolymerization of **M21** in an acetic acid/sodium acetate buffer at pH 5 and 70 °C; [CTA]:[I] = 1.5/1 (circles), 3/1 (diamonds) and 8/1 (triangles) (a), the dependence of $M_n$ (open symbols) and $M_w/M_n$ (closed symbols) on the monomer consumption in the polymerization of **M21** using an **M21** macro-RAFT agent (b), and aqueous size-exclusion chromatographic traces of a macro-**M21** RAFT agent and the successful chain extension with additional **M21**.

and certain fluorinated alcohols/acids [118–120]. Even in aqueous salt solutions the inherent solubility is dependent on the type (nature of the cation and whether it is mono vs divalent) and total concentration of low-molecular-weight electrolyte. The CSC, or critical salt concentration, defined as the concentration of low-molecular-weight electrolyte needed for dissolution, is one aqueous solution characteristic that is often examined for new polymeric betaines. 2,2,2-Trifluoroethanol has been previously demonstrated to be a thermodynamically better solvent for polymers derived from **M25** than 1.0 M NaCl [121]. The first disclosure of the RAFT polymerization of a betaine monomer was by Arotcarena et al. [122]. Unfortunately, no information was presented actually verifying that the betaine monomer (co)polymerized in a controlled fashion. Subsequently, Donovan et al. [123, 124] reported the polymerization of **M24**–**M26** directly in water in the presence of 0.5 M NaBr, employing **C1** as the RAFT agent and V-501 as the source of primary radicals at 70°C. In this instance all the polymerizations were demonstrated to possess the features associated with a controlled/living polymerization (see Fig. 7.12).

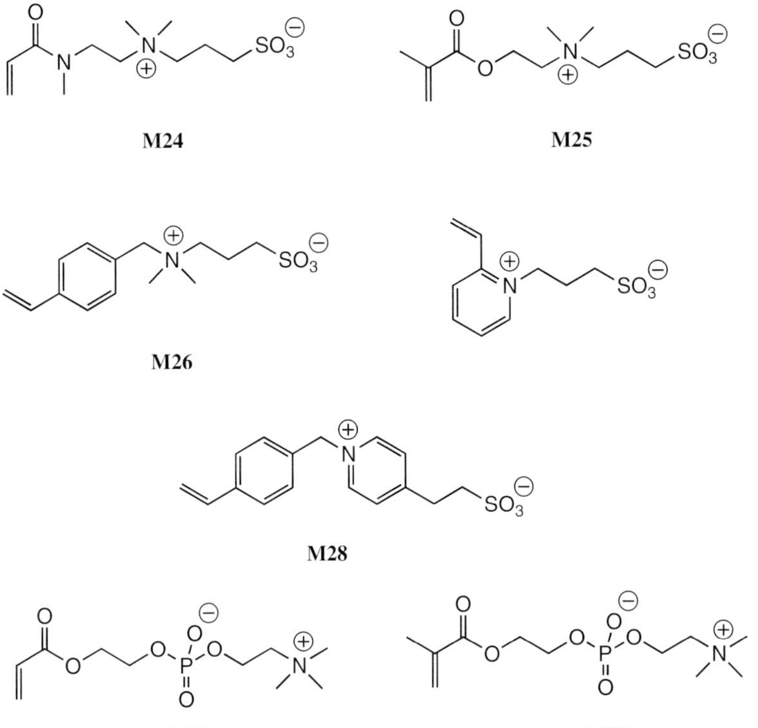


**Fig. 7.11** Chemical structures of betaine monomers (co)polymerized by RAFT.

**M27** and **M28** represent betaine monomers reported by Mertoglu et al. [85, 86]. The RAFT polymerization of **M27** has thus far proven to be elusive, although the reasons for this are not clear since the monomer will homopolymerize under conventional free radical conditions.

In addition to the sulfobetaines, several groups have reported the polymerization of phosphobetaine monomers. For example, the CAMD group in Australia disclosed the polymerization of 2-acryloyloxyethyl phosphorylcholine (**M29**, Fig. 7.11), with both butyl acrylate [125] and styrene [126]. In the case of the styrene-based block copolymers, a polystyrene macro-CTA was first prepared using benzyl dithiobenzoate. Block copolymers were subsequently prepared in a DMF/methanol solvent mixture, with AIBN employed as the free radical source. Copolymerizations were conducted at 60 °C for 24 h. Two polystyrene macro-CTAs of different $M_n$'s were prepared, thus allowing the synthesis of a wide range of block copolymers of varying composition and molecular mass. The resulting block copolymers were generally soluble in carbon disulfide and could be cast into films from this solvent to yield novel honeycomb-structured porous materials. In the case of the butyl-acrylate-based block copolymers, butyl acrylate homopolymers were first prepared

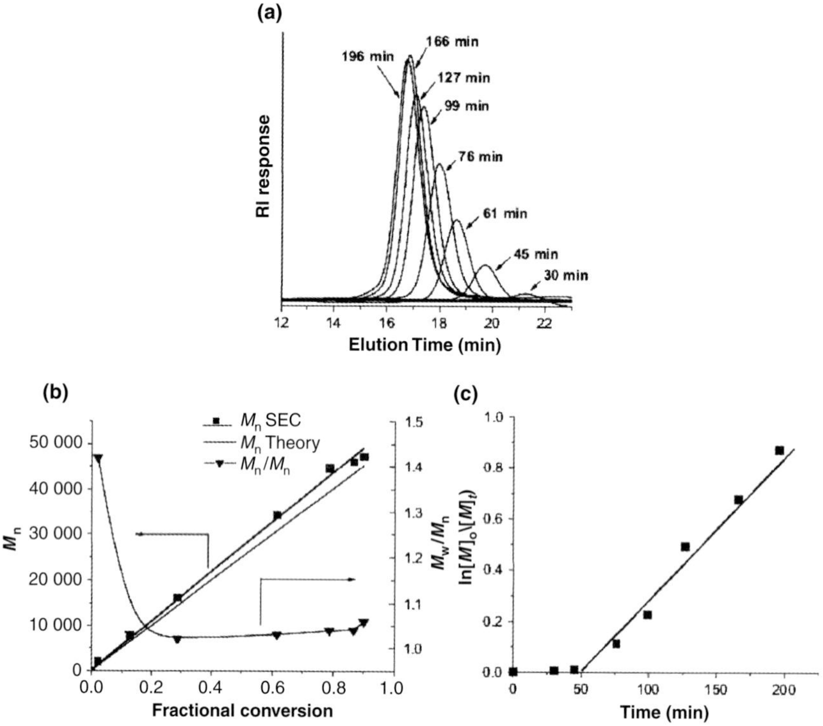


**Fig. 7.12** Aqueous size-exclusion chromatographic traces demonstrating the evolution of molecular mass (a), the $M_n$ vs conversion plot (b, and the pseudo-first-order rate plot (c) for the homopolymerization of **M26**.

using **C12** as the RAFT agent with AIBN in toluene. **M29** was block copolymerized with the macro-CTA in a methanol/dimethylacetamide solvent mixture with AIBN. While analysis of the resulting block copolymers proved difficult, the authors did demonstrate, via light scattering studies, that the AB diblock materials were capable of undergoing supramolecular self-assembly to form nanosized aggregates.

Yusa et al. [127] reported the aqueous RAFT polymerization of 2-methacryloyloxyethyl phosphorylcholine (**M30**) using **C1**, with V-501 at 70 °C for 2 h. After a small induction time of ca. 10 min, the polymerization proceeded in a controlled manner, as evidenced by the kinetics, the linear evolution of $M_n$ with conversion and the narrow molecular mass distributions. The only apparent anomaly was the poor agreement between the theoretical and experimental $M_n$ values, as determined by aqueous size-exclusion chromatography (SEC). However, the SEC system was calibrated with poly(sodium styrenesulfonate) standards that are not ideal for the poly-**M30** homopolymers. Significantly better agreement between the target and experimental $M_n$'s was observed in SLS studies where the measured value of 28 400 was in reasonable agreement with the theoretical value of 21 800. A poly-**M30** homopolymer with a DP of 96 was subsequently employed in the block copolymerization with *n*-butyl methacrylate (BMA) to form amphiphilic block copolymers. Block copolymerizations were performed in MeOH. The first-order kinetic plot for the polymerization of BMA with the poly-**M30** macro-CTA did exhibit some downward curvature, implying the loss of propagating radicals. Several block copolymers were prepared and their aggregation behavior in aqueous media examined. The formation of micelles was confirmed using fluorescence spectroscopy and *N*-phenyl-1-naphthylamine (PNA) as a probe. The emission maxima for PNA were plotted as a function of copolymer concentration. In the case of the $\mathbf{M30_{96}–BMA_{76}}$ block copolymer, the maxima were essentially constant at ca. 464 nm in the low-concentration regime but steadily decreased to a blue shift wavelength of 394 nm at concentrations >0.2 $g{\cdot}L^{-1}$. In the case of $\mathbf{M30_{96}–BMA_{22}}$, the blue shift was less pronounced, reaching saturation at 408 nm at concentrations >0.5 $g{\cdot}L^{-1}$. Such decreases in the emission maxima are consistent with the PNA being sequestered into a more hydrophobic environment, and given the block architecture of the materials, it is likely that multimeric micelles are being formed. The concentration at which the blue shift starts can be regarded as the critical micelle concentration (cmc). SLS measurements of the polymer aggregates allow a determination of the aggregate molecular mass and $N_{agg}$. The experimentally determined $N_{agg}$ for $\mathbf{M30_{96}–BMA_{76}}$ and $\mathbf{M30_{96}–BMA_{22}}$ were 224 and 39 respectively. The core-shell structure of multimeric polymer micelles gives nanoparticles with hydrophobic cores or potential 'cargo holds' capable of sequestering hydrophobic compounds; this forms the basis of the fluorescence experiment. However, the sequestering abilities of such nanoparticles are not limited to fluorescence probes. Indeed, polymeric micelles have been extensively evaluated as potential delivery vehicles for pharmaceutically important compounds. The $\mathbf{M30_m–BMA_m}$ block copolymers were briefly examined with respect to their ability to sequester paclitaxel (PTX), a hydrophobic drug commonly used for the treatment of ovarian and breast cancers. It was shown that the block copolymers were able to sequester significantly more PTX than analogous *statistical* **M30**–BMA copolymers, a feature attributed to the different types of structures formed in aqueous media.

Inoue et al. [128] reported the synthesis of copolymers with differing architectures composed of **M30** with 2,2,2-trifluoroethyl methacrylate. Copolymers with block, statistical and gradient architectures were prepared using **C1** and AIBN in various organic solvents and/or cosolvent combinations. SEC analysis, in 1,1,1,3,3,3-hexafluoroisopropanol, revealed that the copolymers possessed narrow molecular mass distributions. The thermal properties of the copolymers were examined to determine the effect, if any, of chain architecture on the bulk properties. Differences in the differential scanning calorimetry (DSC) thermograms were observed for the block vs gradient copolymers and were attributed to the different bulk morphologies, with the block copolymer most likely possessing a distinct phase-separated morphology, whereas the gradient copolymers were presumed not to exist in a distinct phase-separated state by virtue of the continuous change in monomer distribution in the copolymer.

### 7.4.4 Polyampholytic Materials

Related to polymeric betaines are polyampholytes, which are the second major group of polyzwitterions [129]. These differ from polybetaines in the nature of the location and number of cationic/anionic residues. Whereas polymeric betaines contain both the cationic and anionic/potentially anionic functionality on the same monomer unit, and as a result are charged balanced, polyampholytes have the ionic residues located on distinct monomer units. As a result, polyampholytes may be charged balanced, or may contain various ratios of anionic/cationic functionality [110]. Well-defined polyampholytes with either statistical or block architectures are, like polybetaines, difficult to prepare directly. However, such materials can be prepared indirectly via protecting group chemistry and/or postpolymerization modification. RAFT, given its versatility, would be expected to be the best currently available method for preparing such materials directly with well-defined molecular characteristics. Surprisingly, however, little has been reported to date utilizing RAFT for the synthesis of such specialized materials with only two papers describing their direct synthesis by this technique [82, 130]. The first report was by Xin et al. [130], in which homopolymers, with narrow molecular mass distributions, of **M20** were prepared in anisole with AIBN and *tert*-butyl dithiobenzoate. The subsequent use of these poly-**M20** homopolymers as macro-CTAs in the block copolymerization with sodium acrylate in a water/methanol mixture yielded the corresponding block polyampholytes. More recently, Wang and Lowe [82] described the synthesis of statistical and block copolymers of **M16** or **M17** with **M2**. **M16**/**M17** homopolymers were first prepared with **C9**/V-501 directly in water (Fig. 7.13).

Figure 7.13a shows a series of aqueous SEC traces of aliquots taken from a homopolymerization of **M16**. The systematic shift to lower retention time, i.e. higher molecular mass, is consistent with a controlled polymerization. Figure 7.13b shows plots of $M_n$ and $M_w/M_n$ vs conversion for the same **M16** homopolymerization. Importantly, and further confirming the controlled nature of the homopolymerization, the evolution of $M_n$ is linear and agrees almost perfectly with the theoretical values. Additionally, and as expected, there is no effect of [CTA]:[I] on the $M_n$ profile. In all instances the measured polydispersity indices are low and are typically $\leq 1.10$. Given the inherent difficulty in analyzing polyampholytes by aqueous SEC, the successful formation of the statistical and block copolymers was confirmed qualitatively by Fourier transform infrared (FT-IR) spectroscopy (Fig. 7.14).

Figures 7.14a and 7.14b show the FT-IR spectra for **M16** and **M2** homopolymers respectively, while Figs. 7.14c and 7.14d show the spectra for **M16** and **M2** statistical and block copolymers. The key signals associated with the two separate monomeric building blocks are clearly visible in Figs. 7.14c and 7.14d and specifically, the

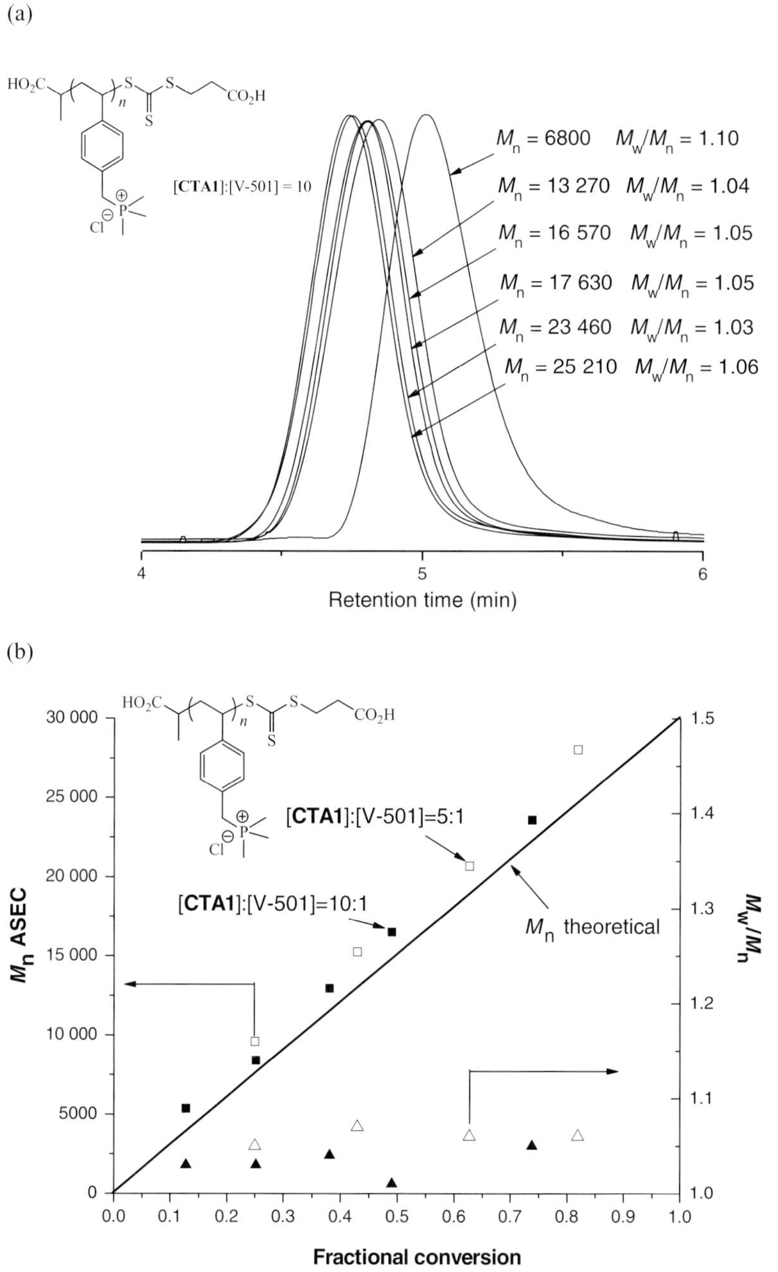


**Fig. 7.13** Aqueous size-exclusion chromatographic traces (RI signal) for the homopolymerization of **M16** with **C9** in water (a), and the evolution of $M_n$ vs conversion at [CTA]:[I] ratios of 5 and 10 (b).

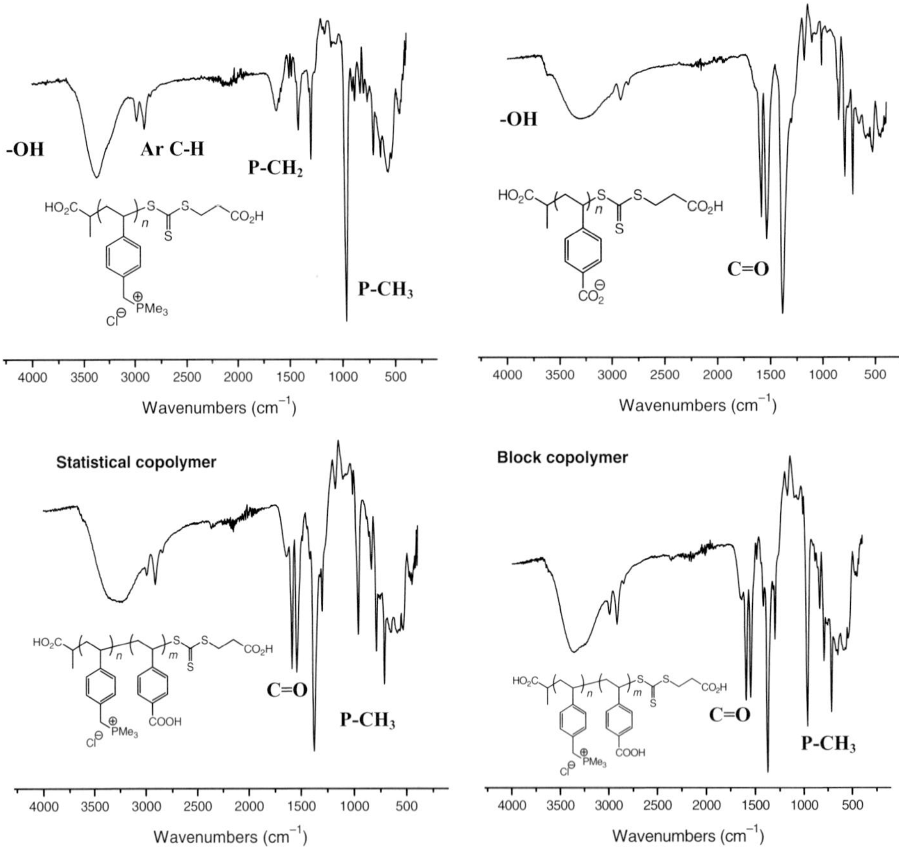


**Fig. 7.14** FT-IR spectra of (a) a poly-**M16** homopolymer, (b) a poly-**M2** homopolymer, (c) an **M16–M2** statistical copolymer and (d) an **M16–M2** block copolymer.

C=O and P—$CH_3$ signals associated with **M2** and **M16** respectively are clearly evident.

## 7.4.5 Nonionic Monomers

Of all the monomeric substrates so far examined by RAFT, nonionic hydrophilic species have been the most widely studied. Examples include various amide-based substrates (Fig. 7.15), glycomonomers (Fig. 7.21) and (meth)acrylic poly(ethylene glycol) derivatives (Fig. 7.23) [2]. Indeed, considering all RAFT polymerizable monomers, the ability to control the polymerization of the amide-based monomers, as shown in Fig. 7.15, has been, arguably, one of the most significant challenges and represents an important advance in controlled radical polymerization (CRP). While examples of such monomers such as *N*,*N*-dimethylacrylamide, **M32**, have

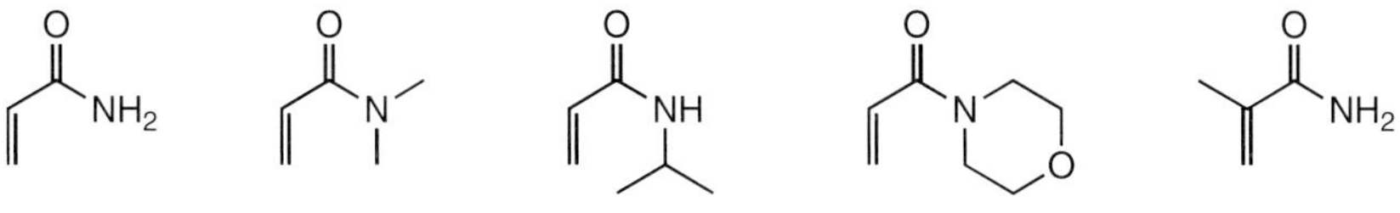

M31 M32 M33 M34 M35

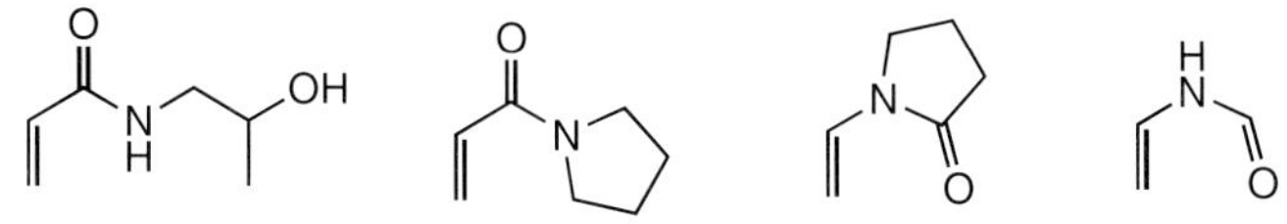

M36 M37 M38 M39

**Fig. 7.15** Chemical structures of examples of activated and nonactivated, nonionic amide-based monomers that have been polymerized via RAFT.

been polymerized in a controlled fashion by living anionic polymerization, CRP techniques other than RAFT have thus far proven to be of limited use for this general class of substrates.

RAFT is the only polymerization technique that reliably allows for the controlled polymerization of acrylamide, **M31**. However, careful attention must be paid to the choice of mediating agent and polymerization conditions [66]. The successful polymerization of **M31** has been achieved with **C2** in conjunction with 2,2′-azobis[2-methyl-*N*-(2-hydroxyethyl)propionamide] (VA-086) under buffered conditions ($pH = 5.0$). **M31** has also been successfully polymerized at room temperature in aqueous media, with the trithiocarbonate-based RAFT agent **C8** employing various azo initiators [131].

Poly(*N*,*N*-dimethylacrylamide), from **M32**, is a readily water-soluble polymer that does not exhibit any phase-transition behavior in water, at least not between 0 and 100 °C and 1 atm. It readily serves as a permanently hydrophilic building block in the synthesis of amphiphilic or stimuli-responsive materials. **M32** has been polymerized by RAFT in both aqueous and organic media under a range of conditions. For example, Donovan et al. have reported the polymerization of **M32** in water, benzene and a water/DMF solvent mixture [132, 133]. These studies evaluated several different CTAs, including **C1** and **C3**, as well as cumyl dithiobenzoate under a range of conditions. Of all the CTAs evaluated, **C3** was demonstrated to be particularly useful. Laschewsky and coworkers [86, 134] have reported the RAFT synthesis and aqueous solution studies of several different **M32**-based copolymers, including copolymers with **M28** and *n*-butyl acrylate. For example, homopolymers of **M32** were prepared with **C1** in water at 48 °C for 5 h, using V-545 as the source of primary radicals. The **M32** homopolymer was subsequently used for the block copolymerization with **M28**. The AB diblock copolymers were shown to undergo stimulus-induced self-assembly by virtue of the salt-responsive nature of the **M28** block. Li et al. [135] recently described the synthesis of ABC triblock copolymers of poly(ethylene oxide) (PEO), **M32** and **M33** with a small amount of *N*-acryloxysuccinimide using a PEO-functional RAFT CTA (Fig. 7.16).

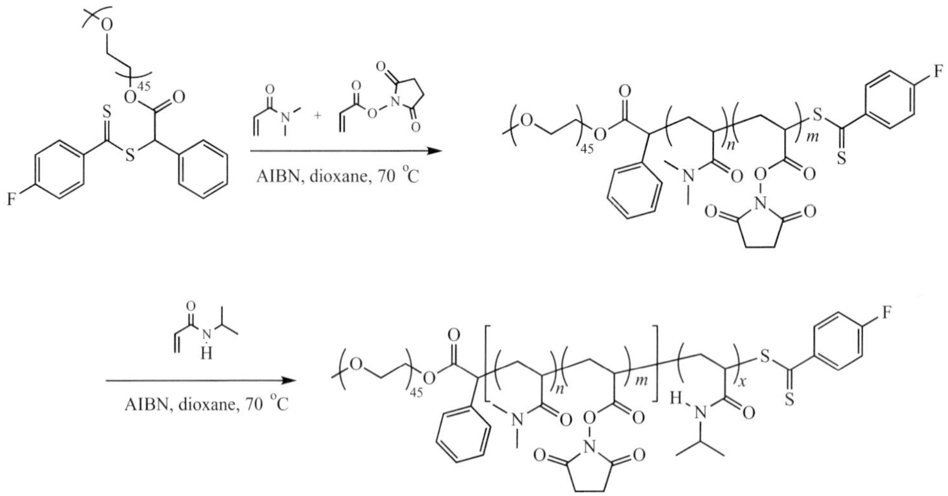


**Fig. 7.16** The synthesis of PEO-*block*-(**M32**-stat-*N*-acryloxysuccinimide)-*block*-**M33** copolymers employing a PEO-functional CTA.

It was shown that the triblock copolymers were molecularly dissolved at $T <$ LCST (lower critical solution temperature) of the **M33** block whereas the triblock copolymers underwent supramolecular self-assembly when the aqueous solution temperature was raised above the LCST of the **M33** block to form aggregates with a hydrophobic **M33** core, an outer PEO stabilizing corona and an inner hydrophilic corona of **M32** and *N*-acryloxysuccinimide. The presence of the highly reactive *N*-acryloxysuccinimide in the inner corona facilitated a facile, postassembly modification by reaction with ethylenediamine to yield shell-cross-linked aggregates.

*N*-Isopropylacrylamide, **M33**, is one of, if not the, most widely studied neutral hydrophilic monomer in polymer science. This is due in part to its readily accessible LCST in water (ca. 32 °C) and the fact that this value is close to physiological temperature. Such interest has been mirrored in RAFT studies. It is not possible to review all relevant literature here and readers are directed to a recent review article where the RAFT synthesis and characterization/properties of **M33**-based materials are discussed in more detail [2]. An early detailed study of the RAFT polymerization of **M33** was reported by Ganachaud et al. [136]. **M33** was polymerized in the presence of benzyl and cumyl dithiobenzoates with AIBN in benzene and dioxane respectively. Also, Schilli et al. [25] have polymerized **M33** with benzyl and cumyl dithiocarbamates in 1,4-dioxane. Given that **M33** is readily water soluble, it would be desirable to prepare (co)polymers directly in water. However, homogeneous aqueous conditions require that the solution temperature be kept below the LCST of the poly-**M33** to prevent phase separation during polymerization. Careful choice of reagents has recently enabled this to be achieved. As highlighted above, **M33** can be polymerized in organic media. The use of organic solvents has the added advantage of reagent compatibility; i.e. there is a much wider choice of reagents, such as initiators, and RAFT agents, which can be employed in organic media, under a broader range of polymerization conditions.

Recently [137], we described facile conditions for the homopolymerization of **M33**, using 2-dodecylsulfanylthiocarbonylsulfanyl-2-methyl propionic acid, **C11**, as the RAFT agent and 2,2′-azobis(4-methoxy-2,4-dimethylvaleronitrile) (V-70) as the initiator in DMF (33 wt % monomer) at room temperature. Key to success in this particular study was the use of an azo initiator (V-70) with an appropriate half-life at room temperature.

Figure 7.17a shows an example of the experimentally determined SEC traces (RI signal) for the homopolymerization of **M33** at $[CTA]_0$:$[I]_0$ (**C11**:V-70) of 20:1. The monotonic shift of the traces to lower elution volume (higher molecular mass) with increasing polymerization time is qualitatively indicative of a controlled polymerization. Additionally, the chromatograms show no evidence of high-molecular-mass impurities (most commonly visualized as a shoulder on the left of the trace), even at extended polymerization times. Figure 7.17b shows the plot of polydispersity index (PDI) vs conversion. It is clear that the PDI remains low ($M_w/M_n \leq 1.10$) throughout the course of the homopolymerizations. A distinguishing feature of controlled polymerizations is a linear evolution of $M_n$ with conversion. Figure 7.17c shows these plots at three different ratios of **C11**/V-70. In all instances the plots are linear and indicate that there is no particular favored **C11**/V-70 ratio, at least of the three examined. There is however, as expected, an effect of this ratio on the kinetics of homopolymerization (see Fig. 7.17d), with the higher ratio leading to slower polymerization. The pseudo-first-order kinetic plots are essentially linear, at least for the major part of the polymerization, indicating a constant number of active species, although some slight curvature is observed at extended polymerization times/higher conversions.

Convertine et al. [138] also described the synthesis and aqueous solution properties of thermally responsive **M33**/**M32** diblock copolymers as well as **M33**/**M32**/**M33** triblock materials. The ABA triblock copolymers were prepared using the difunctional trithiocarbonate RAFT agent **C8**, while the diblock copolymers were synthesized using the novel monofunctional, water-soluble species **C10**. All polymerizations were performed at 25 °C directly in water with 4,4′-azobis[2-(imidazolin-2-yl)propane] dihydrochloride (VA-044) as the source of primary radicals at a $[CTA]_0$:$[VA\text{-}044]_0$ of 3:1. This low ratio was employed due to the relatively long half-life of VA-044 at 25 °C. The AB and ABA block copolymers were prepared with fixed **M32** but variable **M33** block lengths so as to facilitate the systematic evaluation of the effect of the DP of **M33** on the aqueous solution properties. The block copolymers were characterized using a combination of SEC, $^1$H NMR spectroscopy, $T_2$ relaxation measurements, dynamic light scattering and SLS. The **M32** macro-CTAs were prepared in a controlled manner as judged from the kinetics, control over $M_n$ and the low PDIs, which ranged from 1.03 to 1.32. Likewise the formation of the block copolymers proceeded with a high degree of efficiency as likewise judged from the $M_n$ control, the low PDIs ($M_w/M_n = 1.03$–$1.21$) and the clean, symmetric nature of the block copolymers SEC traces. While poly-**M32** is permanently hydrophilic, i.e. it does not undergo a macroscopically observed phase transition as a function of aqueous solution temperature, as has already been noted, poly-**M33** is characterized by its readily accessible coil-to-globule phase transition

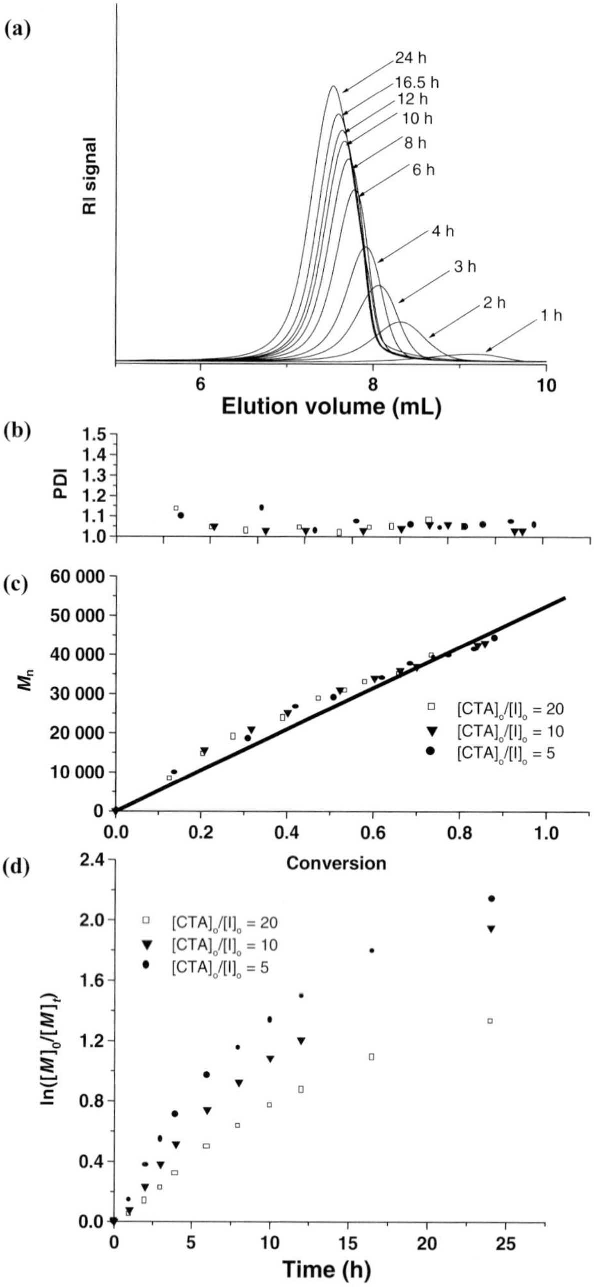


**Fig. 7.17** Experimental data for the room temperature polymerization of **M33** with **C11**/V-70 in DMF. (a) Size-exclusion chromatograms demonstrating the evolution of molecular mass with conversion, (b) the change in polydispersity with conversion, (c) the evolution of $M_n$ with conversion as a function of [**C11**]:[V-70] and (d) the pseudo-first-order kinetic plots as a function of [**C11**]:[V-70].

that occurs at ca. 32 °C. As such these di- and triblock copolymers are expected to undergo temperature-induced supramolecular self-assembly at temperatures above the LCST of the poly-**M33** blocks. To demonstrate this self-assembly process, as well as the reversibility of this process, Fig. 7.18a shows the z-average intensity size distribution for an $\mathbf{M33}_{460}$–$\mathbf{M32}_{100}$ diblock copolymer at 1.00 g·L$^{-1}$ at 25 and 45 °C, i.e. below and above the LCST of the poly-**M33** block. At 25 °C the z-average $D_h$ is ca. 10 nm, which is consistent with molecularly dissolved unimer chains, whereas at 45 °C, above the LCST of poly-**M33**, the $D_h$ is ca. 80 nm and has a measured aggregation number of ca. 213, as determined by SLS.

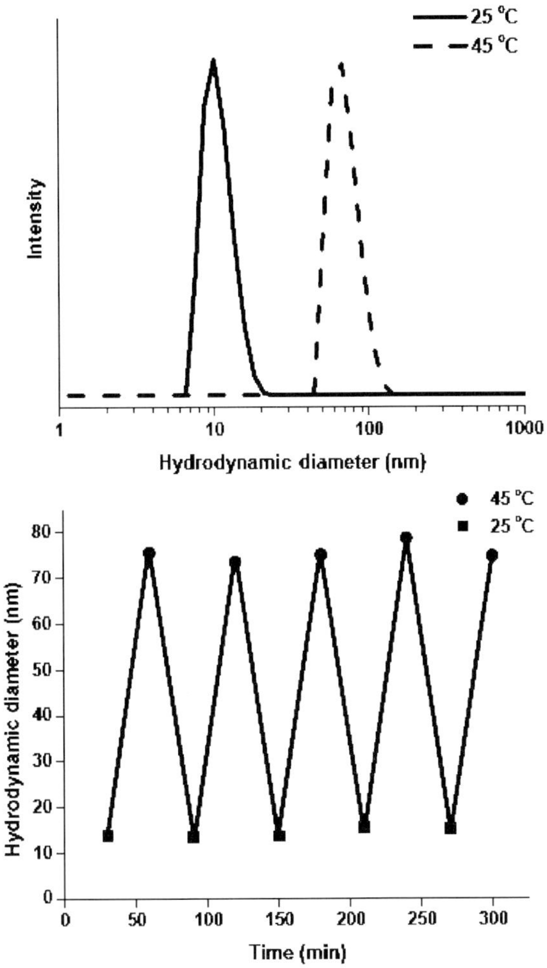


**Fig. 7.18** (a) The z-average size distribution for an $\mathbf{M33}_{460}$–$\mathbf{M32}_{100}$ diblock copolymer at 25 and 45 °C, and (b) the change in hydrodynamic diameter for the same block copolymer when subjected to five heating/cooling cycles.

The reversibility of this self-assembly process is demonstrated in Fig. 7.18b, which shows the change in $D_h$ for the same diblock copolymer when cycled through five heating/cooling cycles. Not only is this self-assembly process clearly reversible, it is also remarkably consistent. A complimentary method for monitoring the self-assembly process is NMR spectroscopy. The relative peak intensity of specific copolymer building blocks can be conveniently monitored as a function of temperature. For example, Fig. 7.19a shows a series of $^1H$ NMR spectra recorded in $D_2O$ of the **M33**$_{460}$–**M32**$_{100}$ block copolymer as a function of solution temperature. The key peaks associated with the poly-**M33** block (Figs. 7.19a, peaks labeled b and c) clearly decrease in intensity, at or around the LCST of the poly-**M33** block, relative

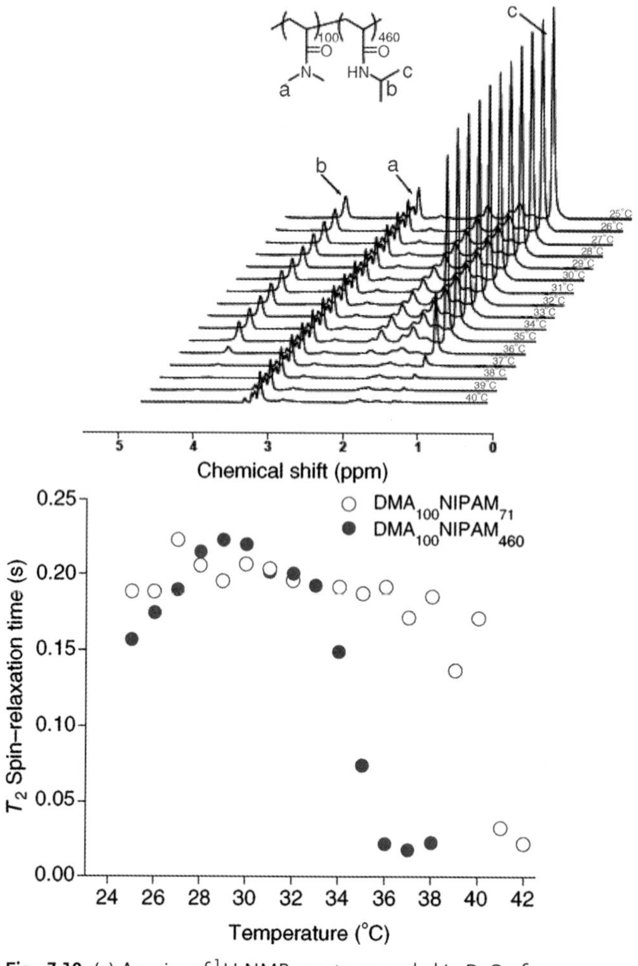


**Fig. 7.19** (a) A series of $^1H$ NMR spectra recorded in $D_2O$ of an **M32–M33** block copolymer as a function of temperature, and (b) the change in $T_2$ spin–spin relaxation times for two **M32–M33** block copolymers.

to the dimethyl protons of the poly-**M32** block (Fig. 7.19a, peaks labeled a). Such a relative decrease is consistent with dehydration/desolvation and restricted mobility of the poly-**M33** block. While such a straightforward experiment gives a qualitative indication of the dehydration/reduced mobility associated with the coil-to-globule transition of the poly-**M33** block, a more quantitative picture of this process can be obtained by monitoring the changes in proton spin–spin relaxations. The spin–spin relaxation ($T_2$) for polymers is heavily influenced by local rigidity with protons in more constrained environments exhibiting faster $T_2$ relaxations than identical protons in less constrained molecular environments. As such $T_2$ measurements afford a method for monitoring the change in segmental motion that accompanies the coil-to-globule phase transition of the poly-**M33** block and as such the self-assembly process.

Figure 7.19b shows the $T_2$ values observed at $d \sim 3.9$ ppm, which correspond to the pendent poly-**M33** methyne proton, for the AB diblock copolymers poly-**M33**$_{71}$–poly-**M32**$_{100}$ and poly-**M33**$_{460}$–poly-**M32**$_{100}$. Initially, there is gradual increase in the $T_2$ values due to increased thermal motion, after which the $T_2$ value stabilizes at around 0.20 s for both block copolymers. In the case of the poly-**M33**$_{460}$-based copolymer, a sharp decrease is observed at ca. 33 °C, with the $T_2$ value falling from 0.194 to 0.0213 s at 36 °C. A similar decrease is observed for the poly-**M33**$_{71}$-based copolymer, although the transition occurs at a slightly higher temperature. Such transitions are entirely consistent with the onset of supramolecular self-assembly.

Similar stimuli-responsive block copolymers with an **M33** as the 'smart' building block have been reported by Morishima and coworkers [139], Liu and Perrier [140], Yin et al. [100], Millard et al. [60], Zhang et al. [141] and Mertoglu et al. [86] under a range of experimental conditions with a variety of RAFT CTAs. In addition to linear block copolymers, **M33** has been used as a building block in the preparation of other novel materials including water-dispersible multiwalled carbon nanotubes via a grafting-from procedure using an immobilized dithioester CTA [142]; dendritic materials incorporating **M33** have been reported by Zheng and Pan and Ge and coworkers [143, 144] (Figs. 7.20a–7.20c); star copolymers have been reported by Zheng and Pan [145]; highly branched poly-**M33** prepared using the polymerizable CTA 3*H*-imidazole-4-carbodithioic acid 4-vinylbenzylester were reported by Carter et al. [146, 147] (Fig. 7.20d); and the synthesis of biotin–**M33** bioconjugates either via post-polymerization modification or via the use of a biotin-functional trithiocarbonate have been reported by Kulkarni et al. [148, 149] and Hong and Pan [150e] (Fig. 7.20).

*N*-acryloylmorpholine, **M34**, is a little studied water-soluble acrylamido derivative [7, 151–153]. To date, the RAFT polymerization of this monomer has been studied exclusively by Charreyre et al. The initial report focused on the effect of dithioester structure, with the researchers evaluating a range of RAFT agents. These included carboxymethyl dithiobenzoate, menthonyl dithiobenzoate, the difunctional species 1,3-bis(2-(thiobenzoylthio)prop-2-yl)benzene and *tert*-butyl dithiobenzoate [154]. Polymerizations were conducted in dioxane at 60 and 90 °C, with AIBN used as the source of primary radicals. A comparison of carboxymethyl dithiobenzoate with *tert*-butyl dithiobenzoate under identical conditions (at 60 °C) clearly highlighted the effect of RAFT agent structure, and specifically the nature of the

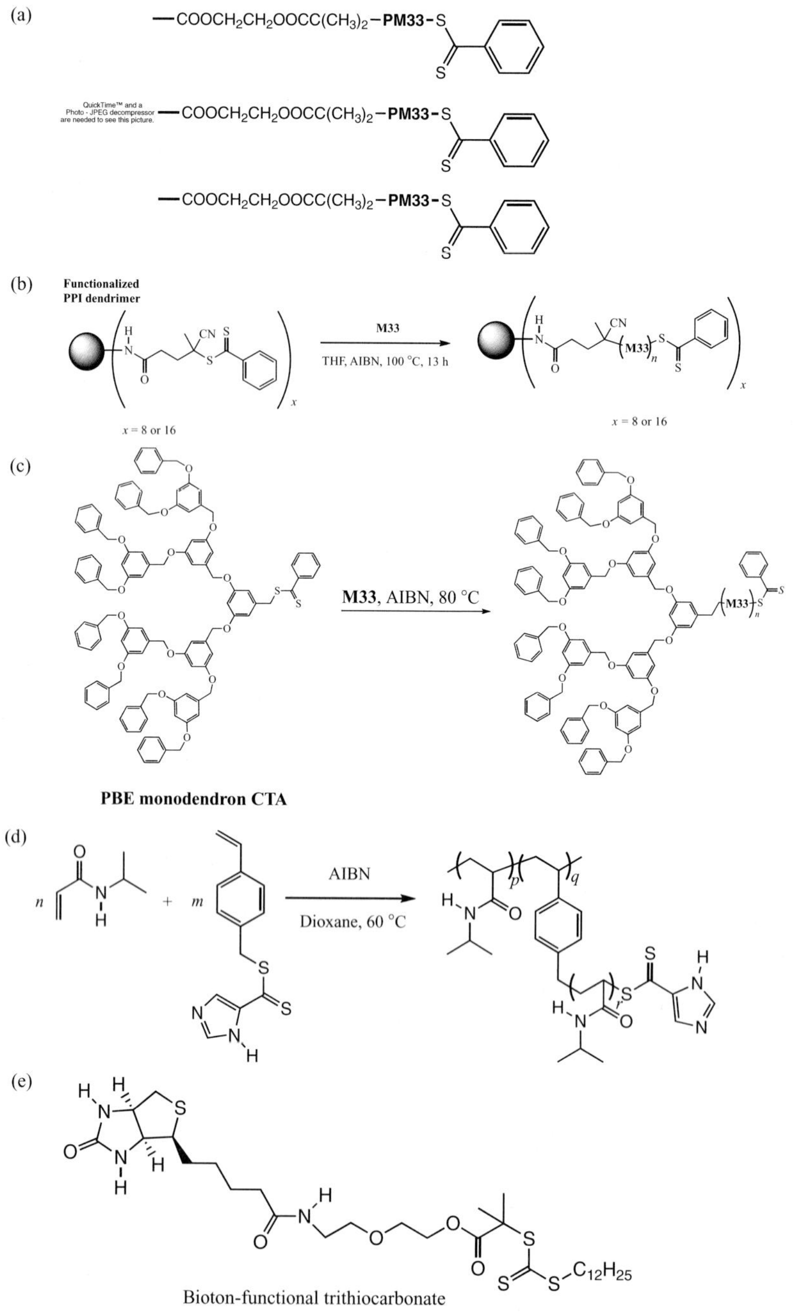


**Fig. 7.20** (a) Water-dispersible multi-walled carbon nanotubes prepared by the grafting-from of **M33**, (b) an **M33**-based dendrimer star prepared from a poly(propylene imine), RAFT CTA functionalized core, (c) the synthesis of poly(benzyl ether)–**M33** copolymers from a poly(benzyl ether)-monodendron functionalized dithioester, (d) the synthesis of highly branched **M33** materials via the application of a polymerizable dithioester RAFT agent and (e) the chemical structure of a biotin-functionalized trithiocarbonate RAFT agent.

R group, on the polymerization kinetics, polydispersity and molecular weight control, with *tert*-butyl dithiobenzoate (which possesses a tertiary R group vs a primary R group on carboxymethyl dithiobenzoate) proving to be overall more effective. A comparison of *tert*-butyl dithiobenzoate with menthonyl dithiobenzoate led to no distinctive differences in effectiveness as mediating agents. Both RAFT agents have bulky R groups that fragment to yield tertiary-carbon-centered radicals. While a small difference in expected fragmentation and reinitiation rates was proposed for the menthonyl derivative based on simple steric considerations, the effect was insufficient to manifest itself under the experimental conditions. The difunctional RAFT agent also proved to be an efficient mediating species. It was also noted that in some instances apparent side reactions led to a decrease in the overall control, given the observed decrease in $M_n$ and a rapid increase in polydispersity at high conversion. Along with the concomitant discoloration observed in these instances, the authors proposed reactions resulting in dithioester degradation to yield sulfides that are capable of acting as nondegradative irreversible transfer agents. More recently, Bathfield et al. [155] reported the homopolymerization of **M34** using a series of functional amide-based dithioester RAFT agents, including two examples with functional bio-related R groups derived from galactopyranose and biotin. Of particular note here is the ability to selectively react with the primary amine-containing molecules at the succinimidyl ester without any competing thiocarbonylthio degradation. Homopolymerizations were conducted in dioxane at 90 °C and in all instances were shown to proceed in a controlled manner.

Methacrylamide, **M35**, has recently been shown to polymerize in a controlled fashion in aqueous media provided appropriate attention is paid to the solution conditions [156]. Polymerizations of **M35** were performed with **C1** at 70 °C in both buffered and nonbuffered aqueous media. Similar to the observations made for other (meth)acrylamido monomers in nonbuffered conditions at elevated temperatures with a dithioester RAFT agent, it was found that **M35** homopolymerizations exhibited controlled characteristics for the first ~3 h after which the molecular mass distribution broadened significantly and a substantial deviation of the experimentally determined molecular mass from the theoretical value was observed. Such behavior was ascribed to the loss of CTA (or macro-CTA) presumably via a combination of hydrolysis and aminolysis. Conducting the polymerization in an acidic buffer circumvented the problem of CTA loss and led to conditions in which the polymerization was controlled over the entire course of the reaction.

*N*-(2-hydroxypropyl)methacrylamide, **M36**, is a well-established biocompatible species that has long-attracted attention in the biomedical field. At present there are only a few reports describing the RAFT synthesis of (co)polymers derived from **M36**. Scales and coworkers [157] described the direct polymerization of **M36** in water, using **C1** and V-501 in an acetic acid buffer at pH 5.2 and 70 °C. This report focused simply on demonstrating the controlled nature of the polymerization of **M36** and, aside from some deviations in the first-order kinetic plots at intermediate-to-high conversion, the polymerizations exhibited the expected characteristics for a controlled process. Subsequently, Yanjarappa et al. [158] reported the synthesis of statistical copolymers of **M36**, with *N*-methacryloyloxysuccinimide employing 1-methyl-1-cyanoethyl dithiobenzoate, and AIBN in *t*-BuOH/DMF at 80 °C for 8 h. The controlled-structure, activated copolymers were subsequently derivatized via peptide coupling with a species that had previously been demonstrated to inhibit the assembly of anthrax toxin. The modification was performed in anhydrous dimethyl sulfoxide (DMSO) at 50°C followed by quenching with 1-amino-2-propanol to yield a polyvalent inhibitor. This polyvalent inhibitor was shown to be at least 3 orders of magnitude more potent with respect to its inhibitory characteristics than the corresponding monovalent peptide. **M36** has also been homopolymerized and block copolymerized, with **M33** using a novel biotinylated RAFT CTA, as described by Hong and Pan [150]. Homopolymers of **M36** were readily prepared using this novel CTA in MeOH with AIBN as the source of primary radicals. Consistent with the use of this CTA for the homopolymerization of **M33**, the homopolymerization of **M36** proceeded in a controlled fashion. The authors subsequently used a poly-**M36** macro-CTA to prepare novel temperature-responsive block copolymers with **M33**. The novel biotin-functionalized block copolymers were shown to undergo supramolecular self-assembly to form novel core-shell structures, in which the surface is decorated with biotin, with average hydrodynamic diameters of ca. 40–50 nm. Recently, the Scales et al. reported the synthesis of AB diblock copolymers of **M36** with **M23** (R = H) and their ability to complex and stabilize small interfering ribonucleic acid (siRNA) [159]. It was shown that the block copolymers were able to effectively stabilize a model, 43 nucleotide siRNA under simulated physiological conditions. The ability of the block copolymers to protect the siRNA from enzymatic degradation was also evaluated. It was shown that two block copolymers in particular (of molar composition $\mathbf{M36}_{258}$-*b*-$\mathbf{M23}_{13}$ and $\mathbf{M36}_{258}$-*b*-$\mathbf{M23}_{23}$) were particularly effective in preventing degradation of the siRNA by ribonuclease A.

*N*-Acryloyl pyrrolidine (**M37**) is a well-known water-soluble amide monomer whose homopolymers exhibit LCST behavior with a cloud point of ca. 50 °C [86]. At present **M37** has only been employed as a hydrophilic/'smart' building block in RAFT-synthesized (co)polymers by the group of Laschewsky [86]. **M37** was successfully homopolymerized for 23 h with cumyl dithiobenzoate at 70 °C in toluene with AIBN to yield poly-**M37** with an experimentally determined $M_n$ of 15 000 at 78% conversion, although unfortunately no experimental evidence confirming the controlled nature was presented. In the same report, **M37** was successfully block copolymerized with *n*-butyl acrylate, **M32** and **M33**. Subsequently, the same group published a detailed report concerning the surfactant properties of a series of amphiphilic block copolymers with a poly(*n*-butyl acrylate) hydrophobic block including one example with **M37** as comonomer [160].

*N*-Vinylpyrrolidone, **M38**, is an extremely important water-soluble building block and has attracted significant attention from the biomedical field due to its biocompatibility and nontoxicity. Unfortunately, **M38** is a nonconjugated monomer in the sense that the alkene $\pi$-electrons are not conjugated with the C=O functional group and as such is structurally similar to vinyl acetate. As with vinyl acetate the controlled polymerization of **M38** represents a significant synthetic challenge. With respect to RAFT, the controlled polymerization of **M38** has only recently been reported and can only be achieved with select families of RAFT agents, and specifically with certain trithiocarbonate, xanthate and dithiocarbamate derivatives [161–163]. For example, Wan et al. [161] described the simultaneous control of both molecular weight and polymer tacticity in the polymerization of **M38** mediated with xanthates, **C13** and **C14**, under bulk conditions at 60 °C with AIBN. Of the two xanthates, **C13** was shown to be more effective in conferring controlled characteristics with, for example, $M_w/M_n$ values in the range 1.10–1.26 in contrast to 1.32–1.47 for **C13**. This difference in ability to control the polymerization of **M38** was attributed to the enhanced leaving group ability of the R group in **C14** compared to the benzylic species in **C13**. Also in this report the authors demonstrated that it is possible to induce the stereospecific polymerization of **M38** using fluoroalcohol solvents. Tacticity control in the free radical polymerization of **M38** has proven to be difficult, although there has been some success with vinyl acetate. The authors examined the use of $CF_3CH_2OH$, $(CF_3)_2CHOH$ and $(CF_3)_3COH$ under both conventional free radical and RAFT polymerization conditions. For normal free radical polymerization of **M38**, the authors found that the syndiotacticity of the resulting poly-**M38** increased with increasing bulkiness of fluoroalcohol, with $(CF_3)_3COH$ being the most effective. Additionally, lowering the polymerization temperature also enhanced the syndiotacticity (~59.0 % obtained with $(CF_3)_3COH$ at 20 °C). This observation was consistent with those made for vinyl acetate where bulkier, more acidic fluoroalcohols likewise resulted in enhanced syndiotacticity. The nature of the mechanism responsible for inducing such stereocontrol was probed using a combination of $^1$H and $^{13}$C NMR spectroscopy. Addition of fluoroalcohol to a $CDCl_3$ solution of **M38** indicated the formation of a 1:1 **M38**:fluoroalcohol, hydrogen-bonded complex, as evidenced by the downfield shift of the C=O resonance associated with **M38** in the $^{13}$C NMR spectrum (from ca. 173.3 to 175.6 ppm) and the much more significant downfield shift of the —OH resonance of the fluoroalcohol in the $^1$H NMR spectrum from 4.34 to 9.04 ppm (for $(CF_3)_3COH$). Finally, the authors demonstrated the ability to simultaneously control the polymerization as well as the tacticity by conducting a series of RAFT polymerizations of **M38** in various fluoroalcohols at 20 and 60 °C in the presence of **C14**.

Polymers derived from *N*-vinylformamide (**M39**) can serve as precursors to poly(vinylamine). The only report regarding the RAFT synthesis of **M39**-based materials is that of Shi et al. [164]. A poly(ethylene glycol)-functionalized xanthate RAFT agent was employed to prepare block copolymers with **M39** in DMSO using 1,1′-azobis(cyclohexanecarbonitrile) (VAZO-88) as the source of primary radicals at 100 °C. The polymerizations were stated to proceed in a controlled fashion, although the resulting polydispersity indices were large with $M_w/M_n = 1.7$–2.3.

### 7.4.6 Glycomonomers

Recently, researchers have extended their studies of nonionic water-soluble species to include the biologically significant sugar-containing family of monomers (glycomonomers). Glycopolymers are most commonly prepared by the polymerization

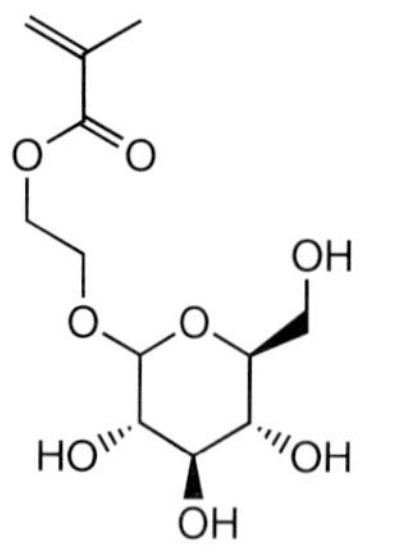

**M40**

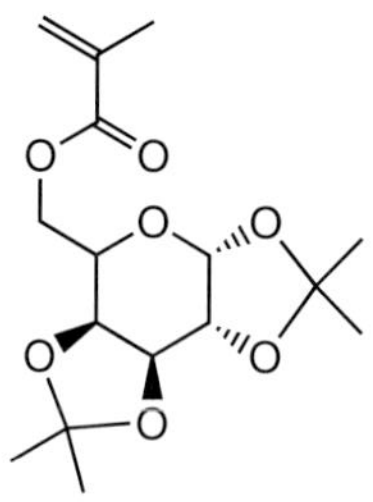

**M41**

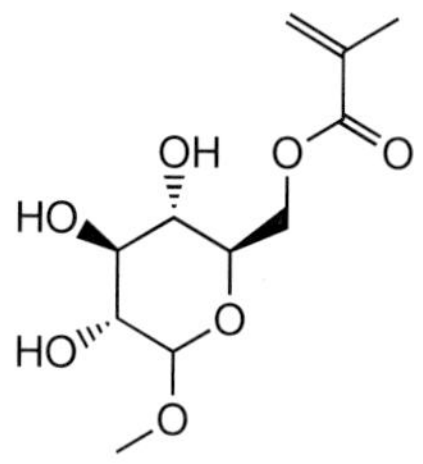

**M42**

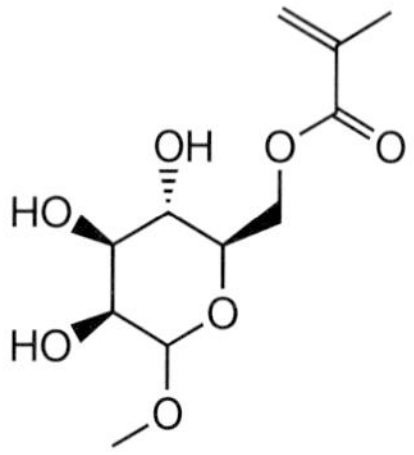

**M43**

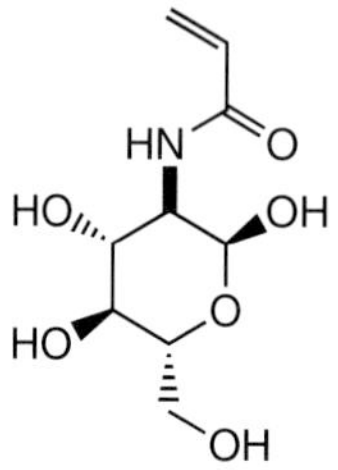

**M44**

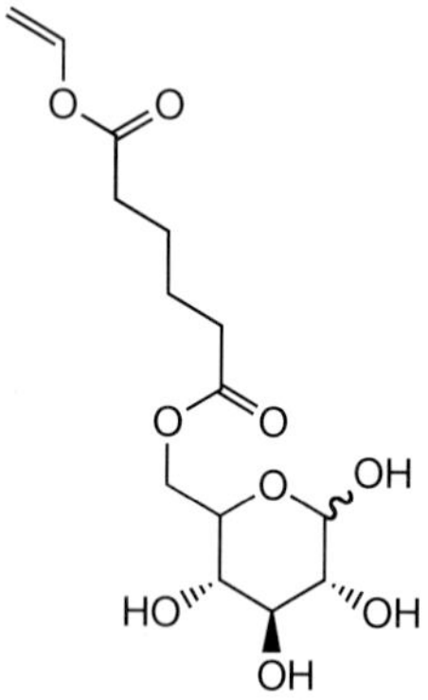

**M45**

**Fig. 7.21** Chemical structures of glycomonomers polymerized via RAFT.

of a protected precursor followed by postpolymerization conversion to the free sugar. There have, however, been several recent reports of the direct polymerization of glycomonomers without recourse to protecting group chemistries. Of particular note are the reports by Armes et al., describing the ATRP synthesis of several well-defined glycopolymers. Given the versatility of RAFT, it is not surprising that it is also possible to polymerize glycomonomers directly. For example, Lowe et al. demonstrated that 2-methacryloxyethyl glucoside (MAGlu, **M40**, Fig. 7.21) may be polymerized in water with **C1** as the RAFT agent and V-501 as source of free radicals, at 70 °C.

Under these conditions, the homopolymerization of **M40** proceeds cleanly (see Fig. 7.22) at least up to 40% conversion, after which deviations from the theoretical $M_n$ are observed. The upward curve in the experimentally determined $M_n$ values could well be due to thiocarbonylthio hydrolysis. The pseudo-first-order rate plot (see inset) is linear, indicating a first-order dependence on monomer. The PDI decreases with increasing conversion. All these features indicate that under these conditions the homopolymerization of **M40** is a controlled process. Additionally, the homopolymerization of the *protected* glycomonomer 3-*O*-methacryloyl-1,

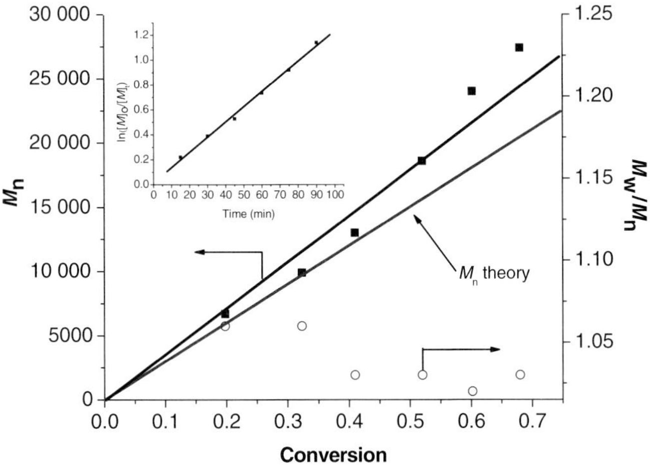


**Fig. 7.22** $M_n$ vs conversion and pseudo-first-order kinetic plot (inset) for the homopolymerization of **M40** directly in water at 70 °C with **C1**.

2:3,4-di-*O*-isopropylidene-D-galactopyranose (MAIpGP, **M41**) and its subsequent use as a macro-RAFT agent for the synthesis of AB diblock copolymers with 2-(dimethylamino)ethyl methacrylate in DMF have been reported [165]. **M41** is not commercially available, but can be prepared by a number of routes, including acylation of 1,2:3,4-di-*O*-isopropylidene-D-galactopyranose with methacrylic anhydride or reaction with methacryloyl chloride. **M41** polymerizes in a controlled fashion in DMF, employing either cumyl dithiobenzoate or 1-methyl-1-cyanoethyl dithiobenzoate in conjunction with AIBN. Homopolymers of **M41** are easily and quantitatively converted to the corresponding free sugar by treatment with a trifluoroacetic acid/water mixture at room temperature. Furthermore, homopolymers of **M41** may be employed as macro-RAFT agents for the preparation of block copolymers with other methacrylic monomers such as **M20**. The CAMD group in Australia has also reported the direct polymerization of free sugar monomers by dithioester-, dithiocarbamate- and xanthate-mediated RAFT. For example, they reported the synthesis of the vinyl ester monomer 6-*O*-vinyladipoyl-D-glucopyranose, **M45**, via a chemoenzymatic route, and its subsequent polymerization with **C7** and the xanthate derivative methyl 2-(ethoxycarbonothioylthio)propanoate [166]. The same group has also described the RAFT polymerization of methyl-6-*O*-methacryloyl-α-D-glucoside, **M42**, using **C1** with V-501 at 70 °C in aqueous media [167], and recently extended these studies to include the preparation of novel diblock glycopolymers [168]. Macro-RAFT agents were prepared via the homopolymerization of **M40** and **M42**, using **C1** in aqueous media to yield homopolymers with experimentally determined $M_n$'s of 21 200 and 25 600 for poly-**M40** and poly-**M42** respectively with corresponding $M_w/M_n$ values of 1.12 for both species. Poly-**M40** was subsequently chain extended with **M42** and poly-**M42** was chain extended with **M43** to yield well-defined glycol-block copolymers. Block copolymerizations were performed in water/ethanol mixtures (9/1 v/v) and proceeded with first-order kinetics up to ca. 50% conversion in both instances. The structure of the resulting copolymers was confirmed via

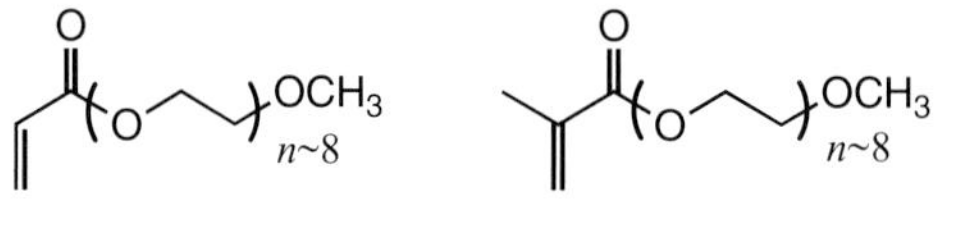

M46 M47

Fig. 7.23 Chemical structures of (meth)acrylic poly(ethylene glycol) monomers polymerized via RAFT.

$^1$H and $^{13}$C NMR spectroscopy. The thermal properties were briefly examined via DSC, in which a single $T_g$ was observed for all materials (homopolymers and block copolymers) in the range 105–180 °C and indicates the absence of phase separation of the block copolymers under bulk conditions. Bernard et al. described the homo and block copolymerization of acryloyl glucosamine, **M44**, using **C12** and V-501 in a water/ethanol mixture at 60 °C [169]. After confirming the controlled nature of the homopolymerization, a macro-CTA of **M44**, with a DP of 180, was employed for the subsequent block copolymerization with **M33** to yield new thermoresponsive AB diblock copolymers. Block polymerizations were conducted at 60 °C in a 1/1 $H_2O$/DMSO cosolvent mixture that was required to maintain the solubility of the growing poly-**M33** chains. Successful block copolymerization was confirmed by SEC with high blocking efficiency observed, although there was some evidence of residual homopolymer impurity.

Aside from the nonionic glycomonomers highlighted above, the other major class of nonionic acrylics are the poly(ethylene glycol) derivatives, poly(ethylene glycol) methyl ether acrylate, **M46**, and poly(ethylene glycol) methyl ether methacrylate, **M47** (Fig. 7.23). The RAFT (co)polymerization of these monomers has been successfully accomplished, as reported by Mertoglu et al. [85]. **M46** was polymerized in 0.5 M KCl at 48 °C with **C5** and 2,2′-azobis(2-methyl-*N*-phenylpropionamide) dihydrochloride as the initiator. After an induction period of ca. 60 min the monomer homopolymerized smoothly following first-order kinetics up to ca. 85% conversion, after which a negative deviation was observed. The evolution of molecular mass was monitored using SEC with an online light scattering detector and increased in a linear manner with conversion with the experimentally determined molecular masses being close to the theoretically targeted values. Similar experimental observations were made for **M47**, although **C1** was employed in conjunction with 2,2′-azobis(2-methylpropionamidine) dihydrochloride as the CTA/initiator pair and polymerizations were conducted in pure water.

## 7.5
## Examples of Experimental Procedures

Here we give examples of experimental procedures for the synthesis of suitable aqueous RAFT agents, as well as general conditions for performing RAFT polymerizations under homogeneous aqueous conditions.

## 7.5.1 Synthesis of 4-Cyano(Pentanoic Acid) Dithiobenzoate (C1) [79]

### 7.5.1.1 Synthesis of Dithobenzoic Acid

To a thoroughly dried 1.0-L, three-necked, round-bottomed flask equipped with a magnetic stir bar, addition funnel (250.0-mL capacity), thermometer and rubber septum for liquid transfers, was added sodium methoxide (30% solution in methanol, 180.0 g, 1.0 mol). Anhydrous methanol (250.0 g) was added to the flask via cannula, followed by the rapid addition of elemental sulfur (32.0 g, 1.0 mol). Benzyl chloride (63.0 g, 0.5 mol) was then added dropwise via the addition funnel over a period of 1 h, at room temperature under a dry nitrogen atmosphere. The reaction mixture was subsequently heated in an oil bath at 67 °C for 10 h. After this time, the reaction mixture was cooled to 7 °C using an oil bath. The precipitated salt was removed by filtration and the solvent removed in vacuo. To the residue was added deionized water (500.0 mL). The solution was filtered a second time and then transferred to a 2.0-L separatory funnel. The crude sodium dithiobenzoate solution was washed with diethyl ether (3 × 200.0 mL). Diethyl ether (200.0 mL) and 1.0 N HCl (500.0 mL) were added, and the dithiobenzoic acid was extracted into the ethereal layer. Deionized water (300.0 mL) and 1.0 N NaOH (600.0 mL) were added, and the sodium dithiobenzoate was extracted into the aqueous layer. This washing process was repeated two more times to finally yield a solution of sodium dithiobenzoate.

### 7.5.1.2 Synthesis of Di(Thiobenzoyl) Disulfide

The synthesis of di(thiobenzoyl) disulfide (DTBD) is an extremely important synthetic procedure since DTBD serves as a convenient precursor to many different RAFT CTAs, and not just those species suited for polymerization in water. The general procedure is shown in step 1 of Scheme 7.4.

Potassium ferricyanide (III) (32.9 g, 0.1 mol) was dissolved in deionized water (500.0 mL). Sodium dithiobenzoate solution (350.0 mL) was transferred to a 1.0-L conical flask equipped with a magnetic stir bar. The potassium ferricyanide solution was added dropwise to the sodium dithiobenzoate solution via an addition funnel over a period of ~1 h under vigorous stirring. The resulting red precipitate was

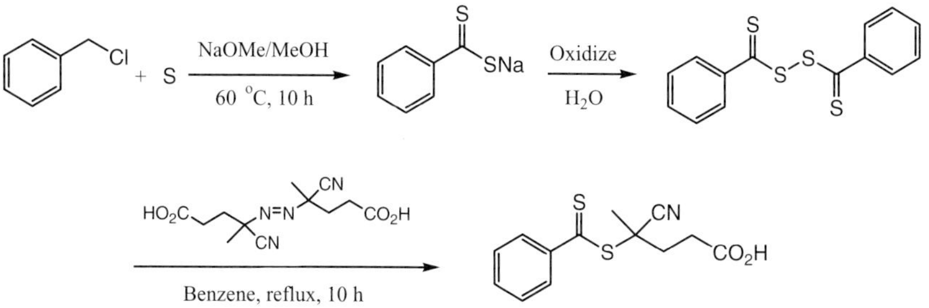


**Scheme 7.4** Synthesis of **C1** from dithiobenzoic acid.

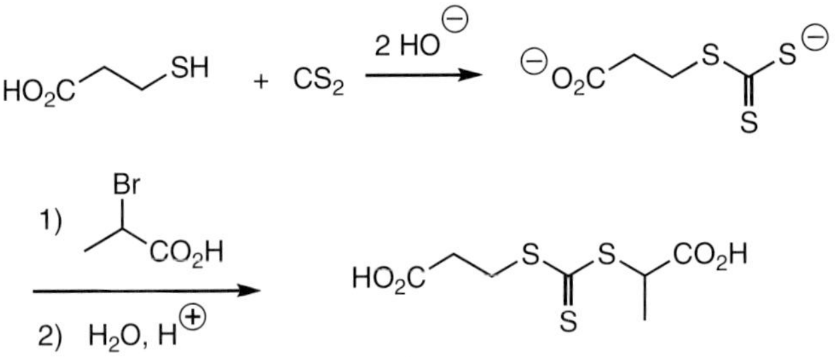

**Scheme 7.5** Synthetic outline for the preparation of **C9**.

filtered and washed with deionized water until the washings were colorless. The solid was dried in vacuo at room temperature overnight.

#### 7.5.1.3 Synthesis of C1

To a 250-mL, round-bottomed flask was added distilled ethyl acetate (80.0 mL). To the flask was added dry 4,4′-azobis(4-cyanopentanoic acid) (V-501) (5.84 g, 21.0 mmol) and DTBD (4.25 g, 14.0 mmol). The reaction solution was then heated at reflux for 18 h. After cooling, the ethyl acetate was removed in vacuo. The crude product was purified by column chromatography (silicagel 60 Å, 70–230 mesh), using ethyl acetate:hexane 2:3 as eluent. Fractions that were red in color were combined and dried over anhydrous sodium sulfate overnight. The solvent mixture was removed in vacuo, and the red oil residue placed in a freezer at −20 °C whereupon it crystallized. The target compound was recrystallized from benzene.

### 7.5.2 Synthesis of C9

The synthesis and application of **C9**, Scheme 7.5, was first reported by Wang et al. [14].

#### 7.5.2.1 Experimental

3-Mercaptopropionic acid (10.6 g, 0.1 mol), distilled/deionized water (100 mL) and 50 wt % NaOH solution (16.0 g, 0.2 mol) were added to a 250-mL, round-bottomed flask equipped with a magnetic stir bar. The mixture was stirred for ca. 30 min prior to the dropwise addition of carbon disulfide (6.0 mL, 0.1 mol). The resulting yellow solution was stirred overnight. 2-Bromopropionic acid (15.3 g, 0.1 mol) was then added to the solution dropwise and the mixture left to stir overnight. The reaction mixture was then acidified via the addition of concentrated hydrochloric acid, and the resulting precipitate isolated by Buchner filtration. The isolated solid was washed with deionized water and subsequently dried in vacuo overnight at ambient temperature.

### 7.5.3
### Homopolymerization of M16 with C9

To a 20.0-mL, round-bottomed flask equipped with a magnetic stir bar was added 4-vinylbenzyl(trimethylphosphonium) chloride (**M16**) (4.0 g, $1.751 \times 10^{-2}$ mol), 2-(2-carboxy-ethylsulfanylthiocarbonylsulfanyl) propionic acid (**C9**) (34.0 mg, $1.339 \times 10^{-4}$ mol), V-501 (4.0 mg, $1.429 \times 10^{-5}$ mmol) (target $M_n = 30\,000$, [CTA]:[I] = 10) and $D_2O$ (8.0 g). The flask was immersed in an ice bath and left to stir for ~1 h to ensure complete dissolution of all components. Subsequently, the contents were split equally between eight small vials that were sealed with rubber septa. Each vial was then purged with $N_2$ for ca. 30 min while immersed in an ice bath. After purging, all vials were immersed in a preheated oil bath at 80°C. Vials were removed from the oil bath at regular time intervals and polymerization terminated via immediate exposure to air and quenching with liquid nitrogen.

### 7.5.4
### Block Copolymerization of M16 with M14

Below is a typical procedure for the preparation of a poly(**M16**-*block*-**M14**) copolymer with a molar ratio of **M16**:**M14** = 1:1.

To a 25-mL, round-bottomed flask equipped with a magnetic stir bar were added the **M16** macro-CTA (0.5 g, $2.19 \times 10^{-3}$ mol), **M14** (0.35 g, $2.20 \times 10^{-3}$ mol), 10-mL pH 4 buffer solution (sodium acetate/acetic acid, 6.0 M, prepared in advance) and V-501 (2.0 mg, $7.14 \times 10^{-6}$ mol). The mixture was stirred while being purged with dry nitrogen for ca. 1 h before it was immersed in a preheated oil bath at 80 °C. After 1 h the polymerization was stopped by immediate exposure to air and quenching in liquid nitrogen. The mixture was dialyzed against distilled water for 2 days with change of water twice daily. Following this, the product was isolated via lyophilization.

## Acknowledgements

ABL and CLM thank the following for supporting parts of their research: Energizer, Genzyme, US Department of Energy (DE-FC26-01BC15317), the MRSEC program of the National Science Foundation (DMR-0213883) and Avery Dennison.

## References

1. J. Chiefari, Y. K. Chong, F. Ercole, J. Krstina, J. Jeffery, T. P. T. Le, R. T. A. Mayadunne, G. F. Meijs, C. L. Moad, G. Moad, E. Rizzardo, S. H. Thang, *Macromolecules* **1998**, *31*, 5559–5562.
2. A. B. Lowe, C. L. McCormick, *Prog. Polym. Sci.* **2007**, *32*, 283–351.
3. C. L. McCormick, A. B. Lowe, *Acc. Chem. Res.* **2004**, *37*, 312–325.

4. Y. K. Chong, J. Krstina, T. P. T. Le, G. Moad, A. Postma, E. Rizzardo, S. H. Thang, *Macromolecules* **2003**, *36*, 2256–2272.
5. J. Chiefari, R. T. A. Mayadunne, C. L. Moad, G. Moad, E. Rizzardo, A. Postma, M. A. Skidmore, S. H. Thang, *Macromolecules* **2003**, *36*, 2273–2283.
6. M.-T. Charreyere, A. Favier, *Macromol. Rapid Commun.* **2006**, *27*, 653–692.
7. A. Favier, M.-T. Charreyere, P. Chaumont, C. Pichot, *Macromolecules* **2002**, *35*, 8271–8280.
8. P. Lebreton, B. Ameduri, B. Boutevin, J. M. Corpart, *Macromol. Chem. Phys.* **2002**, *203*, 522–537.
9. C. L. McCormick, M. S. Donovan, A. B. Lowe, B. S. Sumerlin, D. B. Thomas, US Patent 6,855,840 B2, **2005**.
10. D. B. Thomas, A. J. Convertine, R. D. Hester, A. B. Lowe, C. L. McCormick, *Macromolecules* **2004**, *37*, 1735–1741.
11. M. Benaglia, E. Rizzardo, A. Alberti, M. Guerra, *Macromolecules* **2005**, *38*, 3129–3140.
12. R. T. A. Mayadunne, E. Rizzardo, J. Chiefari, J. Krstina, G. Moad, A. Postma, S. H. Thang, *Macromolecules* **2000**, *33*, 243–245.
13. J. T. Lai, D. Filla, R. Shea, *Macromolecules* **2002**, *35*, 6754–6756.
14. R. Wang, C. L. McCormick, A. B. Lowe, *Macromolecules* **2005**, *38*, 9518–9525.
15. X.-H. Liu, G.-B. Zhang, X.-F. Lu, J.-Y. Liu, D. Pan, Y.-S. Li, *J. Polym. Sci., Polym. Chem.* **2006**, *44*, 490–498.
16. Q. Wang, Y.-X. Li, J. Hong, Z.-Q. Fan, *Chin. J. Polym. Sci.* **2006**, *24*, 593–597.
17. P. Lei, Q. Wang, J. Hong, Y. Li, *J. Polym. Sci., Polym. Chem.* **2006**, *44*, 6600–6606.
18. P. Lei, Q. Wang, J. Hong, Y. Li, *J. Polym. Sci., Polym. Chem.* **2006**, *44*, 6600–6606.
19. M. Destarac, C. Brochon, J.-M. Catala, A. Wilczewska, S. Z. Zard, *Macromol. Chem. Phys.* **2002**, *203*, 2281–2289.
20. M. Destarac, W. Bzducha, D. Taton, I. Gauthier-Gillaizeau, S. Z. Zard, *Macromol. Rapid Commun.* **2002**, *23*, 1049–1054.
21. M. L. Coote, L. Radom, *Macromolecules* **2004**, *37*, 590–596.
22. D. A. Shipp, K. Vercoe, T. Zhang, M. Thopasridharan, *Polym. Prep.* **2005**, *46*, 389–390.
23. H. Mori, H. Ookuma, S. Nakano, T. Endo, *Macromol. Chem. Phys.* **2006**, *207*, 1005–1017.
24. R. T. A. Mayadunne, E. Rizzardo, J. Chiefari, Y. K. Chong, G. Moad, S. H. Thang, *Macromolecules* **1999**, *32*, 6977–6980.
25. C. Schilli, M. G. Lanzendorfer, A. H. E. Muller, *Macromolecules* **2002**, *35*, 6819–6827.
26. D. Zhou, X. Zhu, J. Zhu, H. Yin, *J. Polym. Sci., Polym. Chem.* **2005**, *43*, 4849–4856.
27. R. Bussels, C. Bergman-Goettgens, B. Klumperman, J. Meuldijk, C. Koning, *J. Polym. Sci., Polym. Chem.* **2006**, *44*, 6419–6434.
28. M. L. Coote, E. I. Izgorodina, G. E. Cavigliasso, M. Roth, M. Busch, C. Barner-Kowollik, *Macromolecules* **2006**, *39*, 4585–4591.
29. A. Thies, M. H. Stenzel, T. P. Davis, M. L. Coote, C. A. Barner-Kowollik, *Aust. J. Chem.* **2005**, *58*, 437–431.
30. M. Laus, R. Papa, K. Sparnacci, A. Alberti, M. Benaglia, D. Macciantelli, *Macromolecules* **2001**, *34*, 7269–7275.
31. M. Destarac, I. Gauthier-Gillaizeau, C. T. Vuong, S. Z. Zard, *Macromolecules* **2006**, *39*, 912–914.
32. C. Barner-Kowollik, M. Buback, B. Charleux, M. L. Coote, M. Drache, T. Fukuda, A. Goto, B. Klumperman, A. B. Lowe, J. B. McLeary, G. Moad, M. J. Monteiro, R. D. Sanderson, M. P. Tonge, P. Vana, *J. Polym. Sci., Polym. Chem.* **2006**, *44*, 5809–5831.
33. M. L. Coote, L. Radom, *J. Am. Chem. Soc.* **2003**, *125*, 1490–1491.
34. A. Feldermann, M. L. Coote, M. H. Stenzel, T. P. Davis, C. Barner-Kowollik, *J. Am. Chem. Soc.* **2004**, *126*, 15915–15923.
35. C. Barner-Kowollik, P. Vana, J. F. Quinn, T. P. Davis, *J. Polym. Sci., Polym. Chem.* **2002**, *40*, 1058–1063.
36. Y. Kwak, A. Goto, T. Fukuda, *Macromolecules* **2004**, *37*, 1219–1225.

**37.** Y. Kwak, A. Goto, Y. Tsujii, Y. Murata, K. Komatsu, T. Fukuda, *Macromolecules* **2002**, *35*, 3026–3029.

**38.** E. T. A. Van Den Dungen, J. Rinquest, N. O. Pretorius, J. M. McKenzie, J. B. McLeary, R. D. Sanderson, B. Klumperman, *Aust. J. Chem.* **2006**, *59*, 742–748.

**39.** J. B. McLeary, J. M. McKenzie, M. P. Tonge, R. D. Sanderson, B. Klumperman, *Chem. Commun.* **2004**, 1950–1951.

**40.** J. B. McLeary, M. P. Tonge, B. Klumperman, *Macromol. Rapid Commun.* **2006**, *27*, 1233–1240.

**41.** M. L. Coote, E. I. Izgorodina, E. H. Krenske, M. Busch, C. Barner-Kowollik, *Macromol. Rapid Commun.* **2006**, *27*, 1015–1022.

**42.** M. L. Coote, *Macromolecules* **2004**, *37*, 5023–5031.

**43.** Y. Kwak, A. Goto, K. Komatsu, Y. Sugiura, T. Fukuda, *Macromolecules* **2004**, *37*, 4434–4440.

**44.** R. Venkatesh, B. B. P. Staal, B. Klumperman, M. J. Monteiro, *Macromolecules* **2004**, *37*, 7906–7917.

**45.** P. Vana, M. Buback, *Macromol. Rapid Commun.* **2006**, *27*, 1299–1305.

**46.** G. Moad, E. Rizzardo, S. H. Thang, *Aust. J. Chem.* **2005**, *58*, 379–410.

**47.** S. H. Thang, Y. K. Chong, R. T. A. Mayadunne, G. Moad, E. Rizzardo, *Tetrahedron Lett.* **1999**, *40*, 2435–2438.

**48.** A. Sudalai, S. Kanagasabapathy, B. C. Benicewicz, *Org. Lett.* **2000**, *2*, 213–216.

**49.** M. R. Wood, D. J. Duncalf, S. P. Rannard, S. Perrier, *Org. Lett.* **2006**, *8*, 553–556.

**50.** Z. Zhang, X. Zhu, J. Zhu, Z. Cheng, S. Zhu, *J. Polym. Sci., Polym. Chem.* **2006**, *44*, 3343–3354.

**51.** Z. Zhang, X. Zhu, J. Zhu, Z. Cheng, *Polymer* **2006**, *47*, 6970–6977.

**52.** J. F. Quinn, L. Barner, C. Barner-Kowollik, E. Rizzardo, T. P. Davis, *Macromolecules* **2002**, *35*, 7620–7627.

**53.** L. Lu, N. Yang, Y. Cai, *Chem. Commun.* **2005**, *42*, 5287–5288.

**54.** S. Muthukrishnan, E. H. Pan, M. H. Stenzel, C. Barner-Kowollik, T. P. Davis, D. Lewis, L. Barner, *Macromolecules* **2007**, *40*, 2978–2980.

**55.** W. Jiang, L. Lu, Y. Cai, *Macromol. Rapid Commun.* **2007**, *28*, 725–728.

**56.** J. Zhu, X. Zhu, Z. Zhang, Z. Cheng, *J. Polym. Sci., Polym. Chem.* **2006**, *44*, 6810–6816.

**57.** S. L. Brown, C. M. Rayner, S. Perrier, *Macromol. Rapid Commun.* **2007**, *28*, 478–483.

**58.** J. F. Quinn, L. Barner, E. Rizzardo, T. P. Davis, *J. Polym. Sci., Polym. Chem.* **2001**, *40*, 19–25.

**59.** L. Barner, J. F. Quinn, C. Barner-Kowollik, P. Vana, T. P. Davis, *Eur. Polym. J.* **2003**, *39*, 449–459.

**60.** P.-E. Millard, L. Barner, M. H. Stenzel, T. P. Davis, C. Barner-Kowollik, A. H. E. Mueller, *Macromol. Rapid Commun.* **2006**, *27*, 821–828.

**61.** G. Levesque, P. Arsene, V. Fanneau-Bellenger, T.-N. Pham, *Biomacromolecules* **2000**, *1*, 400–406.

**62.** L. Albertin, M. H. Stenzel, C. Barner-Kowollik, L. J. R. Foster, T. P. Davis, *Polymer* **2006**, *47*, 1011–1019.

**63.** J. Souppe, M. Urrutigoity, G. Levesque, *Biochim. Biophys. Acta* **1988**, *957*, 254–257.

**64.** M. Deletre, G. Levesque, *Macromolecules* **1990**, *23*, 4733–4741.

**65.** X.-P. Qiu, F. M. Winnik, *Macromol. Rapid Commun.* **2006**, *27*, 1648–1653.

**66.** D. B. Thomas, B. S. Sumerlin, A. B. Lowe, C. L. McCormick, *Macromolecules* **2003**, *36*, 1436–1439.

**67.** D. B. Thomas, A. J. Convertine, L. J. Myrick, C. W. Scales, A. E. Smith, A. B. Lowe, Y. A. Vasilieva, N. Ayres, C. L. McCormick, *Macromolecules* **2004**, *37*, 8941–8950.

**68.** A. B. Lowe, B. S. Sumerlin, M. S. Donovan, C. L. McCormick, *J. Am. Chem. Soc.* **2002**, *124*, 11562–11563.

**69.** C. L. McCormick, A. B. Lowe, B. S. Sumerlin, US Patent *7,138,468*, **2006**.

**70.** J. Shan, M. Nuopponen, H. Jiang, E. Kauppinen, H. Tenhu, *Macromolecules* **2003**, *36*, 4526–4533.

**71.** B. S. Sumerlin, A. B. Lowe, P. A. Stroud, P. Zhang, M. W. Urban, C. L. McCormick, *Langmuir* **2003**, *19*, 5559–5562.

72. M. Nuopponen, H. Tenhu, *Langmuir* **2007**, *23*, 5352–5357.
73. J. W. Hotchkiss, A. B. Lowe, S. G. Boyes, *Chem. Mater.* **2007**, *9*, 6–13.
74. J. Raula, J. Shan, M. Nuopponen, A. Niskanen, H. Jiang, E. I. Kauppinen, H. Tenhu, *Langmuir* **2003**, *19*, 3499–3504.
75. D. H. R. Barton, S. W. McCombie, *J. Chem Soc., Perkin Trans. 1* **1975**, *1*, 1574–1585.
76. D. H. R. Barton, J. D. Ok, J. C. Jaszberenyi, *Tetrahedron Lett.* **1990**, *42*, 3991–3994.
77. S. Z. Zard, *Angew. Chem. Int. Ed. Engl.* **1997**, *36*, 672–685.
78. S. Perrier, P. Takolpuckdee, C. A. Mars, *Macromolecules* **2005**, *38*, 2033–2036.
79. Y. Mitsukami, M. S. Donovan, A. B. Lowe, C. L. McCormick, *Macromolecules* **2001**, *34*, 2248–2256.
80. B. S. Sumerlin, M. S. Donovan, Y. Mitsukami, A. B. Lowe, C. L. McCormick, *Macromolecules* **2001**, *34*, 6561–6564.
81. B. S. Sumerlin, A. B. Lowe, D. B. Thomas, C. L. McCormick, *Macromolecules* **2003**, *36*, 5982–5987.
82. R. Wang, A. B. Lowe, *J. Polym. Sci., Polym. Chem.* **2007**, *45*, 2468–2483.
83. S. Yusa, Y. Shimada, Y. Mitsukami, T. Yamamoto, Y. Morishima, *Macromolecules* **2003**, *36*, 4208–4215.
84. A. B. Lowe, B. S. Sumerlin, C. L. McCormick, *Polymer* **2003**, *44*, 6761–6765.
85. M. Mertoglu, A. Laschewsky, K. Skrabania, C. Wieland, *Macromolecules* **2005**, *38*, 3601–3614.
86. M. Mertoglu, S. Garnier, A. Laschewsky, K. Skrabania, J. Storsberg, *Polymer* **2005**, *46*, 7726–7740.
87. B. S. Lokitz, J. E. Stempka, A. W. York, Y. Li, H. K. Goel, G. R. Bishop, C. L. McCormick, *Aust. J. Chem.* **2006**, *59*, 749–754.
88. C. Ladaviere, N. Dorr, J. P. Claverie, *Macromolecules* **2001**, *34*, 5370–5372.
89. N. Gaillard, A. Guyot, J. Claviere, *J. Polym. Sci., Polym. Chem.* **2003**, *41*, 684–698.
90. J. Loiseau, N. Doerr, J. M. Suau, J. B. Egraz, M. F. Llauro, C. Ladaviere, J. Claverie, *Macromolecules* **2003**, *36*, 3066–3077.
91. E. Khousakoun, J.-F. Gohy, R. Jerome, *Polymer* **2004**, *45*, 8303–8310.
92. C. M. Schilli, M. Zhang, E. Rizzardo, S. H. Thang, Y. K. Chong, K. Edwards, G. Karlsson, A. H. E. Mueller, *Macromolecules* **2004**, *37*, 7861–7866.
93. A. Morel, H. Cottet, M. In, S. Deroo, M. Destarac, *Macromolecules* **2005**, *38*, 6620–6628.
94. R. Hoogenboom, U. S. Schubert, W. VanCamp, F. E. DuPrez, *Macromolecules* **2005**, *38*, 7653–7659.
95. Y. K. Chong, T. P. T. Le, G. Moad, E. Rizzardo, S. H. Thang, *Macromolecules* **1999**, *32*, 2071–2074.
96. H. de Brouwer, J. G. Tsavalas, F. J. Schork, M. J. Monteiro, *Macromolecules* **2000**, *33*, 9239–9246.
97. E. Sprong, D. De Wet-Roos, M. P. Tonge, R. D. Sanderson, *J. Polym. Sci., Polym. Chem.* **2002**, *40*, 223–235.
98. E. Sprong, D. De Wet-Roos, M. P. Tonge, R. D. Sanderson, *J. Polym. Sci., Polym. Chem.* **2004**, *42*, 2502–2512.
99. Y. Ren, Z. Zhu, J. Huang, *J. Polym. Sci., Polym. Chem.* **2004**, *42*, 3828–3835.
100. X. Yin, A. S. Hoffman, P. S. Stayton, *Biomacromolecules* **2006**, *7*, 1381–1385.
101. Y. Mitsukami, A. Hashidzume, S.-I. Yusa, Y. Morishima, A. B. Lowe, C. L. McCormick, *Polymer* **2006**, *47*, 4333–4340.
102. B. S. Sumerlin, A. B. Lowe, D. B. Thomas, A. J. Convertine, M. S. Donovan, C. L. McCormick, *J. Polym. Sci., Polym. Chem.* **2004**, *42*, 1724–1734.
103. M. Save, M. Manguian, C. Chassenieux, B. Charleux, *Macromolecules* **2005**, *38*, 280–289.
104. J.-J. Yuan, R. Ma, Q. Gao, Y.-F. Wang, S.-Y. Cheng, L.-X. Feng, Z.-Q. Fan, L. Jiang, *J. Appl. Polym. Sci.* **2003**, *89*, 1017–1025.
105. A. J. Convertine, B. S. Sumerlin, D. B. Thomas, A. B. Lowe, C. L. McCormick, *Macromolecules* **2003**, *36*, 4679–4681.
106. M. Sahnoun, M.-T. Charreyre, L. Veron, T. Delair, F. D'Agosto, *J. Polym. Sci., Polym. Chem.* **2005**, *43*, 3551–3565.

**107.** S. Yusa, Y. Konishi, Y. Mitsukami, T. Yamamoto, Y. Morishima, *Polym. J.* **2005**, *37*, 480–488.

**108.** Y. A. Vasilieva, D. B. Thomas, C. W. Scales, C. L. McCormick, *Macromolecules* **2004**, *37*, 2728–2737.

**109.** S. L. West, J. P. Salvage, E. J. Lobb, S. P. Armes, N. C. Billingham, A. L. Lewis, G. W. Hanlon, A. W. Lloyd, *Biomaterials* **2004**, *25*, 1195–1204.

**110.** A. B Lowe, C. L. McCormick, *Polyelectrolytes and Polyzwitterions*, Oxford University Press, Washington, DC, **2006**; ACS Symposium Series Vol *937*.

**111.** T. Nakaya, Y. J. Li, *Prog. Polym. Sci.* **1999**, *24*, 143–181.

**112.** A. L. Lewis, P. D. Hughes, L. C. Kirkwood, S. W. Leppard, R. P. Redman, L. A. Tolhurst, P. W. Stratford, *Biomaterials* **2000**, *21*, 1847–1859.

**113.** M. Ward, M. Sanchez, M. O. Elasri, A. B. Lowe, *J. Appl. Polym. Sci.* **2006**, *101*, 1036–1041.

**114.** A. B. Lowe, N. C. Billingham, S. P. Armes, *Chem. Commun.* **1996**, 1555–1556.

**115.** A. B. Lowe, N. C. Billingham, S. P. Armes, *Macromolecules* **1999**, *32*, 2141–2148.

**116.** Z. Tuzar, H. Pospisil, J. Plestil, A. B. Lowe, F. L. Baines, N. C. Billingham, S. P. Armes, *Macromolecules* **1997**, *30*, 2509–2512.

**117.** V. Bütün, C. E. Bennett, M. Vamvakaki, A. B. Lowe, N. C. Billingham, S. P. Armes, *J. Mater. Chem.* **1997**, *7*, 1693–1695.

**118.** J. C. Salamone, W. Volksen, A. P. Olson, S. C. Israel, *Polymer* **1978**, *19*, 1157–1162.

**119.** V. M. Monroy Soto, J. C. Galin, *Polymer* **1984**, *25*, 254–262.

**120.** D. N. Schulz, D. G. Peiffer, P. K. Agarwal, J. Larabee, J. J. Kaladas, L. Soni, B. Handwerker, R. T. Garner, *Polymer* **1986**, *27*, 1734–1742.

**121.** M. B. Huglin, M. A. Radwan, *Makromol. Chem.* **1991**, *192*, 2433–2445.

**122.** M. Arotcarena, B. Heise, S. Ishaya, A. Laschewsky, *J. Am. Chem. Soc.* **2002**, *124*, 3787–3793.

**123.** M. S. Donovan, B. S. Sumerlin, A. B. Lowe, C. L. McCormick, *Macromolecules* **2002**, *35*, 8663–8666.

**124.** M. S. Donovan, A. B. Lowe, T. A. Sanford, C. L. McCormick, *J. Polym. Sci., Polym. Chem.* **2003**, *41*, 1262–1281.

**125.** H. Stenzel Martina, C. Barner-Kowollik, P. Davis Thomas, H. M. Dalton, *Macromol. Biosci.* **2004**, *4*, 445–453.

**126.** H. Stenzel Martina, P. Davis Thomas, *Aust. J. Chem.* **2003**, *56*, 1035–1038.

**127.** S. Yusa, K. Fukuda, T. Yamamoto, K. Ishihara, Y. Morishima, *Biomacromolecules* **2005**, *6*, 663–670.

**128.** Y. Inoue, J. Watanabe, M. Takai, S.-I. Yusa, K. Ishihara, *J. Polym. Sci., Polym. Chem.* **2005**, *43*, 6073–6083.

**129.** A. B. Lowe, C. L. McCormick, *Chem. Rev.* **2002**, *102*, 4177–4190.

**130.** X. Xin, Y. Wang, W. Liu, *Eur. Polym. J.* **2005**, *41*, 1539–1545.

**131.** A. J. Convertine, B. S. Lokitz, A. B. Lowe, C. W. Scales, L. J. Myrick, C. L. McCormick, *Macromol. Rapid Commun.* **2005**, *26*, 791–795.

**132.** M. S. Donovan, A. B. Lowe, B. S. Sumerlin, C. L. McCormick, *Macromolecules* **2002**, *35*, 4123–4132.

**133.** M. S. Donovan, T. A. Sanford, A. B. Lowe, B. S. Sumerlin, Y. Mitsukami, C. L. McCormick, *Macromolecules* **2002**, *35*, 4570–4572.

**134.** S. Garnier, A. Laschewsky, *Colloid Polym. Sci.* **2006**, 1243–1254.

**135.** Y. Li, B. S. Lokitz, C. L. McCormick, *Macromolecules* **2006**, *39*, 81–89.

**136.** F. Ganachaud, M. J. Monteiro, R. G. Gilbert, A.-A. Dourges, S. H. Thang, E. Rizzardo, *Macromolecules* **2000**, *33*, 6738–6745.

**137.** A. J. Convertine, N. Ayres, C. W. Scales, A. B. Lowe, C. L. McCormick, *Biomacromolecules* **2004**, *5*, 1177–1180.

**138.** A. J. Convertine, B. S. Lokitz, Y. Vasileva, L. J. Myrick, C. W. Scales, A. B. Lowe, C. L. McCormick, *Macromolecules* **2006**, *39*, 1724–1730.

**139.** S. Yusa, Y. Shimada, Y. Mitsukami, T. Yamamoto, Y. Morishima, *Macromolecules* **2004**, *37*, 7507–7513.

**140.** B. Liu, S. Perrier, *J. Polym. Sci., Polym. Chem.* **2005**, *43*, 3643–3654.

**141.** P. Zhang, Q. Liu, A. Qing, J. Shi, M. Lu, *J. Polym. Sci., Polym. Chem.* **2006**, *44*, 3312–3320.

**142.** C. Y. Hong, Y. Z. You, C. Y. Pan, *Chem. Mater.* **2005**, *17*, 2247–2254.

**143.** Q. Zheng, C.-Y. Pan, *Eur. Polym. J.* **2006**, *42*, 807–814.

**144.** Z. Ge, S. Luo, S. Liu, *J. Polym. Sci., Polym. Chem.* **2006**, *44*, 1357–1371.

**145.** G. Zheng, C. Pan, *Polymer* **2005**, *46*, 2802–2810.

**146.** S. Carter, B. Hunt, S. Rimmer, *Macromolecules* **2005**, *38*, 4595–4603.

**147.** S. Carter, S. Rimmer, A. Sturdy, M. Webb, *Macromol. Biosci.* **2005**, *5*, 373–378.

**148.** S. Kulkarni, C. Schilli, A. H. E. Mueller, A. S. Hoffman, P. S. Stayton, *Bioconjug. Chem.* **2004**, *15*, 747–753.

**149.** S. Kulkarni, C. Schilli, B. Grin, A. H. E. Muller, A. S. Hoffman, P. S. Stayton, *Biomacromolecules* **2006**, *7*, 2736–2741.

**150.** C. Y. Hong, C. Y. Pan, *Macromolecules* **2006**, *39*, 3517–3524.

**151.** A. Favier, C. Ladaviere, M. T. Charreyre, C. Pichot, *Macromolecules* **2004**, *37*, 2026–2034.

**152.** F. D'Agosto, R. Hughes, M.-T. Charreyre, C. Pichot, R. G. Gilbert, *Macromolecules* **2003**, *36*, 621–629.

**153.** F. D'Agosto, M.-T. Charreyre, C. Pichot, R. G. Gilbert, *J. Polym. Sci., Polym. Chem.* **2003**, *41*, 1188–1195.

**154.** A. Favier, M.-T. Charreyre, P. Chaumont, C. Pichot, *Macromolecules* **2002**, *35*, 8271–8280.

**155.** M. Bathfield, F. D'Agosto, R. Spitz, M. T. Charreyre, T. Delair, *J. Am. Chem. Soc.* **2006**, *128*, 2546–2547.

**156.** Y. A. Vasilieva, C. W. Scales, D. B. Thomas, R. G. Ezell, A. B. Lowe, N. Ayres, C. L. McCormick, *J. Polym. Sci., Polym. Chem.* **2005**, *43*, 3141–3152.

**157.** C. W. Scales, Y. A. Vasilieva, A. J. Convertine, A. B. Lowe, C. L. McCormick, *Biomacromolecules* **2005**, *6*, 1846–1850.

**158.** M. J. Yanjarappa, K. V. Gujraty, A. Joshi, A. Saraph, R. S. Kane, *Biomacromolecules* **2006**, *7*, 1665–1670.

**159.** C. W. Scales, F. Huang, N. Li, Y. A. Vasilieva, J. Ray, A. J. Convertine, C. L. McCormick, *Macromolecules* **2006**, *39*, 6871–6881.

**160.** S. Garnier, A. Laschewsky, *Langmuir* **2006**, *22*, 4044–4053.

**161.** D. Wan, K. Satoh, M. Kamigaito, Y. Okamoto, *Macromolecules* **2005**, *38*, 10397–10405.

**162.** P. Bilalis, M. Pitsikalis, N. Hadjichristidis, *J. Polym. Sci., Polym. Chem.* **2005**, *44*, 659–665.

**163.** Z. Q. Hu, Z. C. Zhang, *Macromolecules* **2006**, *39*, 1384–1390.

**164.** L. Shi, T. M. Chapman, E. J. Beckman, *Macromolecules* **2003**, *36*, 2563–2567.

**165.** A. B. Lowe, R. Wang, *Polymer* **2007**, *32*, 283–351.

**166.** L. Albertin, C. Kohlert, M. H. Stenzel, L. J. R. Foster, T. P. Davis, *Biomacromolecules* **2004**, *5*, 255–260.

**167.** L. Albertin, M. Stenzel, C. Barner-Kowollik, L. J. R. Foster, T. P. Davis, *Macromolecules* **2004**, *37*, 7530–7537.

**168.** L. Albertin, M. H. Stenzel, C. Barner-Kowollik, L. J. R. Foster, T. P. Davis, *Macromolecules* **2005**, *38*, 9075–9084.

**169.** J. Bernard, X. Hao, T. P. Davis, C. Barner-Kowollik, M. H. Stenzel, *Biomacromolecules* **2006**, *7*, 232–238.

# 8
# RAFT-Mediated Polymerization in Heterogeneous Systems

*Carl N. Urbani, and Michael J. Monteiro*

## 8.1
## Introduction

### 8.1.1
### Background of Emulsion Polymerization

Emulsion (or 'latex') polymerization allows the formation of polymer colloids dispersed in water. This technology gained importance rapidly during the Second World War to produce synthetic polymers to replace natural rubber, and now has become a global industry that continues to expand from base commodity products for the building industry (e.g. paints, concrete additives, adhesives, sealants and tough plastics) to high-value-added biomedical products (e.g. drug-delivery devices, materials for diagnostic kits and assays) [1, 2]. The reason for the ubiquitous use of synthetic polymer colloids formed via emulsion polymerizations is the control of particle size distribution (PSD), particle morphology, molecular weight distribution (MWD) and polymer chemical composition, all of which control the final properties and function of the material. Polymer colloids can be constructed of two or many different types of polymers with very different chemical compositions, leading to the formation of a variety of 'latex' particle morphologies (see Fig. 8.1), ranging from core-shell, hemisphere, salami, raspberry to separated individual particles (each with different polymers) [2, 3]. Depending upon the morphology the physical properties can be tailored to the application. For example, core-shell morphologies are generally used for pressure-sensitive adhesives where the core imparts cohesive strength and the shell-adhesive strength. This morphology is also widely used in films to impede the propagation of microcracks, in which the crack will propagate until it reaches a hard polymer nanosphere embedded in the film. The surface of polymer colloids can also be chemically modified for the attachment of antibody molecules. These particles can serve as immunospecific markers for the corresponding antigens and can be fluorescently labeled for analytical detection of specific proteins.

*Handbook of RAFT Polymerization.* Edited by Christopher Barner-Kowollik

ISBN: 978-3-527-31924-4

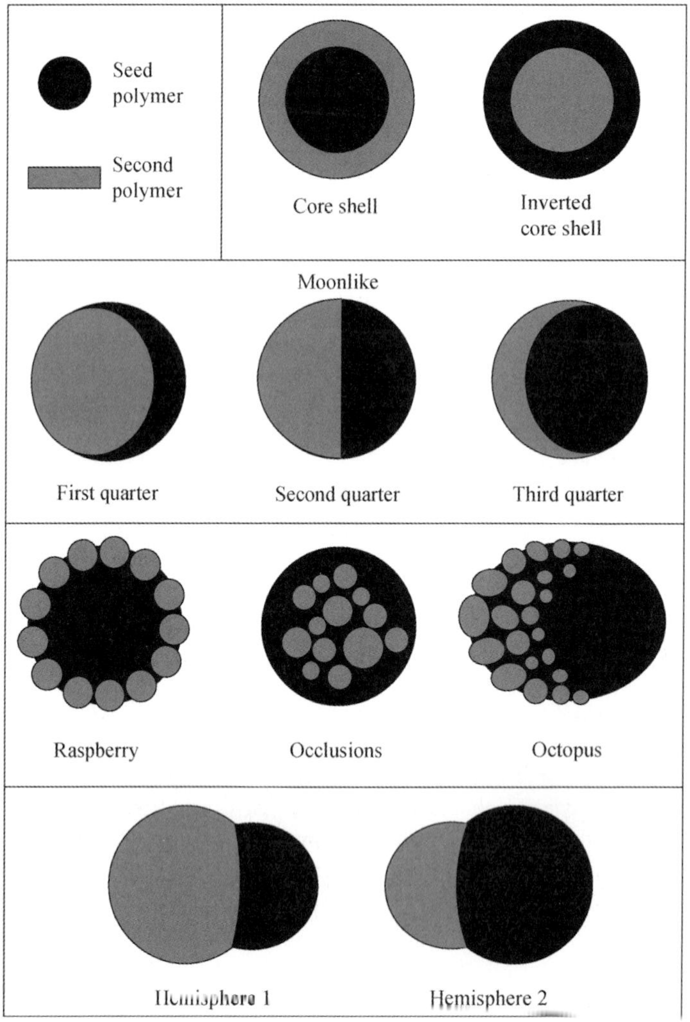


**Fig. 8.1** Types of polymer particle morphologies.

The major advantage of water over bulk monomer or organic solvents is the environment-friendly nature of the reaction medium. Waterborne processes are cheap, can be used for a broad range of monomers and a wide range of experimental conditions, the heat transfer is highly efficient, high conversions with low monomer residuals can be reached, there are no organic volatile compounds and one can obtain high polymer solids (~50 wt %) in a low viscosity environment, which means the polymer is easy to process. Another advantage in emulsion copolymerization is that the morphology of the particle can be controlled. Coupling the advantages of living radical polymerization (LRP) to prepare well-defined polymer architectures with the above advantages of carrying out the polymerization in water will provide a new class of specialty polymer materials for use in the coatings industry and in biomedical applications.

The end result of the emulsion polymerization process is the formation of polymer colloids stabilized with surfactant and dispersed in an environment-friendly medium, water. The kinetic processes of entry, exit and termination in the water phase and particles dictate the PSD and MWD [4]. The obvious advantage of using emulsion polymerization is that radicals are compartmentalized in each 'latex' particle, physically isolating polymeric radicals in one particle from polymeric radicals in all other particles. This leads to very fast rates of polymerizations and high conversion of monomer to polymer. Should such rapid polymerizations be found in solution or bulk reactions, the increase in temperature would most probably cause the reaction vessels to explode. In emulsions, this is overcome, as the water provides good heat transfer from the reactor to the environment.

The MWD, copolymer chemical composition, PSD and morphology can also be controlled by the type of polymerization process. In general, there are three types of processes [1, 2, 5–7]. The first is a batch polymerization, in which all ingredients are mixed together at the beginning of the polymerization, and thus the growth of particles occurs at the same time. The limitation of this process is that little can be done, apart from changing the temperature, to control the properties of the 'latex'. The second is a semicontinuous or semibatch process, which involves the initial addition of some of the ingredients and the slow feed of other ingredients over time. The mode of addition allows control of particle growth, copolymer composition and morphology. It also provides a process to control the heat of reaction through the feeding of initiator or redox activator, and allows reproducibility of the 'latex'. The third is the continuous addition of all ingredients over time into a stirred tank or tanks linked in series, while the 'latex' is removed at a similar rate as that of addition. This means that 'latexes' with high uniformity and reproducibility are made at high rates.

### 8.1.2
### Types of Heterogeneous Free-Radical Polymerizations

There are a wide variety of heterogeneous polymerization systems that have been developed. In this chapter, we will focus only on those specifically used to implement RAFT-mediated polymerizations. These include the following: ab initio and seeded emulsion polymerizations, miniemulsions, microemulsions, self-assembly and degassed polymerizations.

An *ab initio emulsion polymerization* [2–6, 8] begins with a reaction containing a water-insoluble vinyl monomer(s) in the presence of a free-radical aqueous soluble initiator, water and micelle-forming surfactant. When these ingredients are placed under shear, the system forms monomer-swollen micelles (approximately 15–30 nm in diameter) dispersed in water and large droplets of monomer stabilized by surfactant. Polymerization occurs when an initiator-derived radical reacts with the monomer in the water phase and propagates to a critical chain length (of z monomer units) whereby its solubility in water diminishes and it enters a monomer-swollen micelle. This oligomeric radical (or $z$-mer) is now in a monomer-rich environment and propagates rapidly to form a young 'latex' particle. This stage of the emulsion polymerization is termed interval I (i.e. the nucleation stage for particles), and its end is defined by the disappearance of micelles due to the drop in the surfactant concentration in the water phase below the critical micelle concentration. Particle growth continues via propagation of the polymeric free radical in the particle, and the monomer that is being converted into polymer is continuously replenished by monomer from the droplets, maintaining the concentration in the polymer particles high and constant – termed interval II. Interval III commences when there are no more monomer droplets present in the system, resulting in a concomitant decrease in monomer concentration in the growing particles with conversion (Scheme 8.1). To obtain a narrow PSD, a very high initiator concentration or high temperature is required to produce a high radical flux to nucleate and convert as many micelles as possible to particles.

In an emulsion polymerization, the propagating polymeric radical can undergo several fates at any stage of the polymerization. For example, the polymeric radical can transfer to monomer to form a monomeric radical and a dead polymer chain. The monomeric radical can either propagate with monomer present in the particle or exit the particle, via diffusion, and can either terminate in the water phase or enter another particle and terminate the propagating polymeric radical present in that particle via instantaneous termination. The end results of transfer, entry and exit events are an increase in the number of dead chains formed and ultimately a

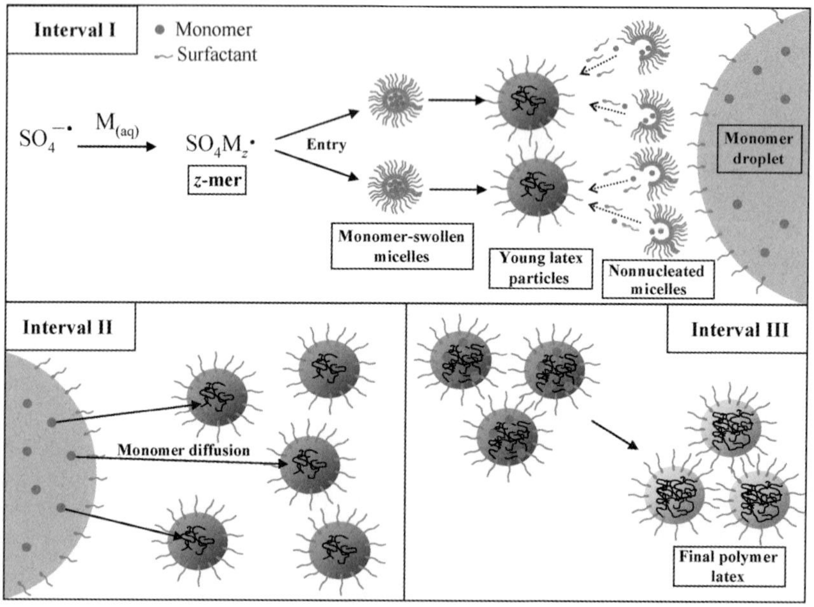


**Scheme 8.1** Mechanism of particle formation in a basic ab initio emulsion polymerization.

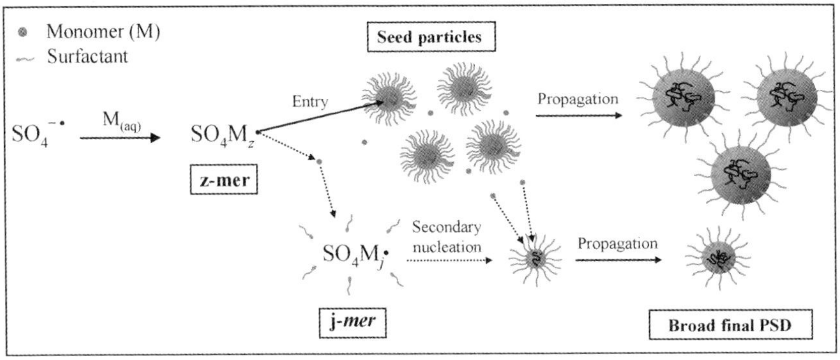


**Scheme 8.2** Mechanism of secondary particle formation (or homogeneous nucleation).

broadening of the MWD. It is the formation of dead polymer in both emulsion and bulk and solution free-radical polymerizations that limits the control over the main components of macromolecular structure and design.

A *seeded emulsion polymerization* [2–4, 9, 10] begins with a reaction containing a 'latex' that has been previously prepared with a well-defined particle size and PSD. Monomer, aqueous soluble initiator and some additional surfactant are then added and the polymerization initiated. The main locus of polymerization is in the seed particles, thus eliminating the nucleation period (or interval I in an ab initio polymerization), and therefore one can start the polymerization in interval II or III. In other words, the seed particles act as nanoreactors, and allow control over the rate of polymerization, the MWD and more importantly the final particle size and distribution. To avoid additional or a second crop of particles through secondary nucleation (or homogeneous nucleation), the initial number of seeded particles must be high to increase the probability of entry into particles rather than growth of the radicals in the water phase to a water-insoluble species (reaching a $j_{\text{crit-mer}}$, see Scheme 8.2). These species collapse, swell with monomer and grow in the same fashion as the other particles. Seeded particles in some cases can consist of a different polymer to that of the second polymer, and allow control of the particle morphology due to the chemical incompatibility of the seed polymer and the second polymer. However, for these polymerizations to be carried out successfully the second monomer should effectively swell the seeded particles. Seeded emulsion polymerizations are also an ideal system to study the kinetics of heterogeneous polymerizations and allow kinetic parameters, such as entry, exit and termination, to be determined.

A *miniemulsion polymerization* [2, 3, 11–14] begins with a reaction containing water-insoluble monomer, surfactant, cosurfactant (e.g. hexadecane), water-soluble initiator and water. The strict definition of miniemulsions is the formation of surfactant-stabilized monomer droplets of size ranging between 50 and 500 nm. The cosurfactant is an important ingredient as it can limit the diffusion of monomer from the smaller particles to the larger ones (i.e. Ostwald ripening), and can provide additional surface stability against droplet coagulation depending on its chemical

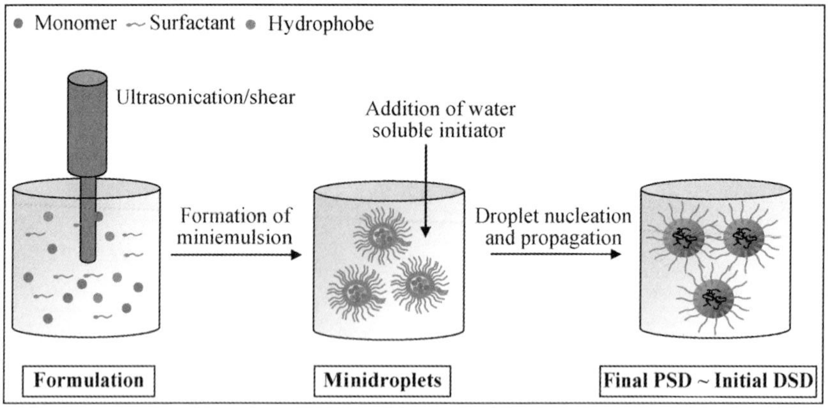


**Scheme 8.3** Method for the formation and polymerization of miniemulsions.

composition. The diffusion of monomer via the action of Ostwald ripening from small to larger droplets results in the growth of the larger droplets until they reach a critical size of usually greater than 500 nm, become buoyant and then rise to form a single monomer phase at the top of the reaction vessel. Miniemulsions are typically formed through high shear, by subjecting the system to ultrasonification, homogenizer or microfluidizer. The high shear produces small monomer droplets stabilized by the surfactant and cosurfactant. Polymerization of the system should result in each droplet becoming a particle (or one-to-one copy of droplets to particles, Scheme 8.3). In reality, there are fewer particles than droplets. This suggests that nucleation is an important consideration even in miniemulsions and as will be described later, has implications on RAFT-mediated miniemulsion polymerizations. Once again, the number of droplets must be high to avoid secondary particle (homogeneous) nucleation. The number of particles is therefore strongly dependent upon the initiator concentration and suggests that nucleation is primarily via the monomer droplets and that only a fraction of droplets become particles.

A *microemulsion polymerization* [2, 15, 16] begins with a reaction containing water, monomer (few wt % to water), surfactant, water-soluble initiator and cosurfactant that should be low-molecular-weight alcohol. Microemulsions are thermodynamically stable due to a low interfacial tension that compensates for the dispersion entropy. The droplet sizes are much smaller than miniemulsion droplets, and the reaction mixtures are usually transparent. Consequently, the number of droplets is very high and as such the resulting 'latex' contains only a few polymer chains per particle, sometimes equal or close to one. (Conventional ab initio emulsion polymerizations, on the other hand, produce particles that contain many polymer chains per particle.) The amount of surfactant and cosurfactant required to obtain microemulsions is usually extremely high (with monomer to surfactant ratios ranging from 0.3 to 1.0), thus limiting their use in biomedical applications or for the production of large quantities of pure well-defined polymer. However, very small 'latex' particles are formed, ranging between 20 and 40 nm. The mechanism of microemulsions is believed to occur via the continuous nucleation of droplets over time, and thus the number of particles increases with conversion. This is a result of the high surfactant and thus high micelle concentration, in which nucleation is preferential in the micelles rather than particles.

### 8.1.3
### Aim of the Chapter

LRP has emerged as the most versatile technique to produce polymers with controlled architecture and MWD. In solution and bulk systems, the most used living radical techniques are (i) reversible addition–fragmentation chain transfer (RAFT) [17–22], (ii) atom transfer radical polymerization (ATRP) [23–28], (iii) nitroxide-mediated polymerization (NMP) [29–33] and (iv) metal-catalyzed radical polymerization (e.g. single-electron-transfer LRP) [34–39], which have been used to make new, interesting and smart polymeric materials. The versatility of these polymerization techniques is attributed to the vast number of monomers capable of being polymerized effectively, a wide range of acceptable solvents, including water, and the ability to conduct polymerizations at ambient temperature. LRP has offered the polymer scientist an avenue for preparing polymers with effectively uniform chain lengths parallel to that of an organic chemist's synthesis of organic compounds. The various architectures that can be prepared in bulk or solution are now left up to the imagination, and moreover the applications for such architectures are slowly being realized (see Chapter 13).

The challenge is to prepare polymer colloids with these architectures in an environment-friendly media, water, and with controlled PSD, MWD, morphology and chemical composition. All living radical techniques have been applied under heterogeneous conditions in suspension, dispersion, ab initio emulsion, seeded emulsion and miniemulsion, using water as the reaction medium. There have been recent reviews of the techniques in dispersed media [40–42]. The purpose of this chapter is to survey the current literature on the use of RAFT-mediated polymerization in dispersed media, derive a mechanistic understanding and describe the advantages and limitations of RAFT in dispersed media.

There have been many claims of successful LRP via ab initio, seeded emulsion polymerizations or miniemulsion polymerizations. We define a successful emulsion polymerization as the ability to produce polymer with an MWD that would be found in bulk or solution polymerizations. For example, should the polymerization of monomer in the presence of RAFT agent give a polydispersity index (PDI) of 1.1 in solution, then a successful emulsion polymerization would produce polymer with close to the same PDI and number-average molecular weight ($M_n$). This will also apply to RAFT agents that have a low chain-transfer constant ($C_{tr,RAFT}$) to the polymeric radical, in which PDIs of close to 2 are found in bulk and solution.

## 8.2
## Effect of $C_{tr,RAFT}$ on $M_n$ and PDI in Homogeneous Systems

The mechanism for reversible chain transfer relies on an exchange reaction between the dormant and active species. The reaction mixture consists of a specific RAFT agent (CTA), monomer (solvent is optional) and initiator. The initiation step to produce active species is identical to conventional free-radical polymerization (e.g. thermal decomposition of initiator). The amount of initiator that has decomposed directly relates to the amount of dead polymer formed during the polymerization, and therefore the initiator concentration must be kept low compared to that of CTA to obtain a well-controlled number-average molecular weight ($M_n$) and MWD [43]. A compromise should be reached between MWD control and the speed of the reaction.

The reactivity of CTA toward the active species has a great influence on the $M_n$ and MWD evolution with monomer conversion ($x$) and is controlled by the value of $C_{tr,RAFT}$ ($=k_{ex}/k_p$), where $k_p$ is the rate coefficient for propagation. Analytical equations for the evolution of the degree of polymerization, $X_n$ ($=M_n/M_w$ of monomer) and PDI with $x$ are given as follows [44]:

$$x_n = \frac{\gamma_0 x}{1-(1-\alpha)(1-x)^{\beta}} \quad \text{or} \quad M_n = \frac{\gamma_0 x}{1-(1-\alpha)(1-x)^{\beta}} M_0 \tag{8.1}$$

$$\mathrm{PDI} = \frac{1}{\gamma_0 x} + \frac{1}{x}\left[2 + \frac{\beta-1}{\alpha-\beta}(2-x)\right] - \frac{2\alpha(1-\alpha)}{(\beta-\alpha)x}[1-(1-x)^{1+\beta/\alpha}] \tag{8.2}$$

where $M_0$ is the monomer molar mass, $\gamma_0 = [M]_0/[CTA]_0$, $x$ is fractional conversion, $\alpha = [P^{\cdot}]/[CTA]$ (with $P^{\cdot}$, the concentration of propagating radicals) and $\beta = C_{tr,RAFT}$.

These equations can be used for reversible chain transfer where termination, transfer to monomer and all other side reactions are neglected. Figure 8.2 shows the evolution of $M_n$ and PDI with conversion by varying $C_{tr,RAFT}$ over a range of values. At $C_{tr,RAFT} = 1$, $M_n$ reaches its maximum value early in the polymerization and remains constant at that value until all the monomer is consumed (i.e. at $x = 1$). Similarly, the PDI at early conversion reaches 2 and remains constant to $x = 1$, which is similar to what is found for the addition of a conventional CTA. As the value of $C_{tr,RAFT}$ is increased, the evolution of $M_n$ and PDI with conversion starts to resemble that of an 'ideal' living polymerization. The results show that for a $C_{tr,RAFT}$ value greater than 10, $M_n$ increases linearly with $x$, and there is no change in the $M_n$ evolution as $C_{tr,RAFT}$ becomes larger. In contrast, the PDI is more sensitive to the value of $C_{tr,RAFT}$, and the greater the value of $C_{tr,RAFT}$, the lower the PDI at early conversion. The analytical equations clearly show that the reversible chain-transfer technique allows polymers with controlled MWD to be made. This, coupled with the ability to make complex architectures, offers a wide range of new materials that can be synthesized. A full description of RAFT-mediated kinetic simulations has been reviewed [43].

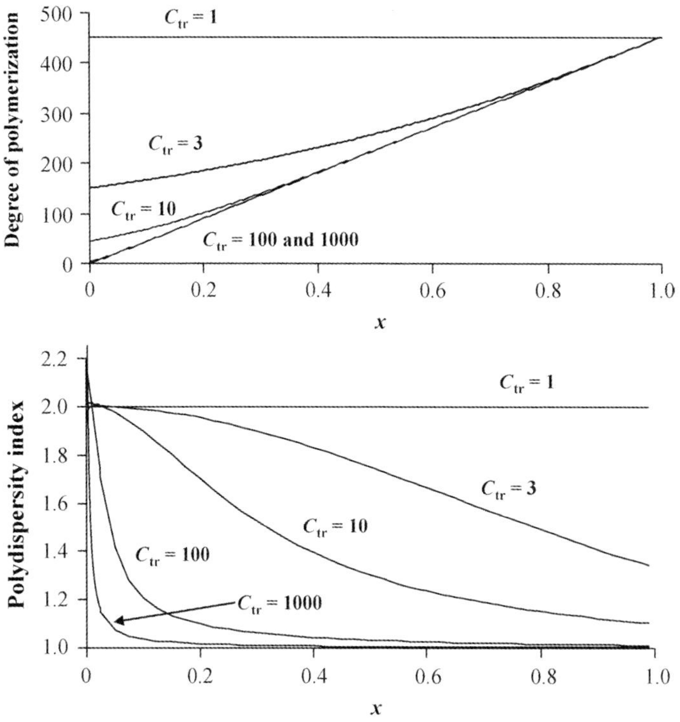


**Fig. 8.2** Effect of $C_{tr}$ on the degree of polymerization ($X_n$) and PDI vs conversion in RAFT.

## 8.3 Raft in Heterogeneous Systems

RAFT has been studied in many dispersion systems, including emulsion, miniemulsion, self-assembly and surfactant-free polymerizations. These will be discussed in this section. A summary of all the techniques and results are given in Table 8.1.

### 8.3.1 Emulsion Polymerization

#### 8.3.1.1 Ab Initio

The first ab initio emulsion polymerizations were reported by the CSIRO group in their patent [67]. However, their experimental procedures were designed to avoid the presence of monomer droplets. A typical example was the emulsion polymerization of butyl methacrylate (BMA) in the presence of cumyl dithiobenzoate (CDB, a highly reactive RAFT agent with $C_{tr,RAFT} > 1000$), in which BMA (1.7 g), CDB and water-soluble azo initiator were added into the reaction flask and then 18.7 g of BMA and 71 mg of CDB were added slowly over 72 min. The polymer conversion

**Table 8.1** Summary of RAFT-mediated ab initio emulsion, miniemulsion, microemulsion and nanoprecipitation polymerizations

| Emulsion type | RAFT agent | $C_{tr,RAFT}$ | Monomer(s) | Conditions | $M_n$ range | PDI range | Successful[a] | Ref. |
|---|---|---|---|---|---|---|---|---|
| Ab initio | 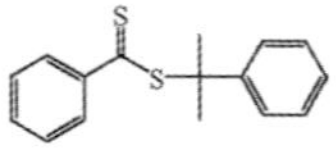 | 10 [45] | BMA | Starved-feed, SDS, 80 °C, ACP | 58 000 | 1.22 | × | [46] |
| Ab initio | 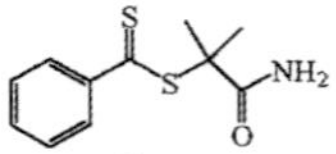 | — | STY | CTAB, V-50 | 60 000 | 1.7 | ×× | [47] |
| Ab initio | 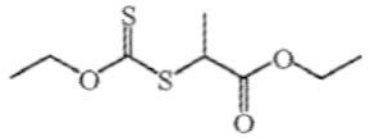 | 0.7 [48] | STY BA | SLS, 85 °C, NaPS | 17 000–100 000 | 1.4–2.7 | √√ | [49] |
| Ab initio | 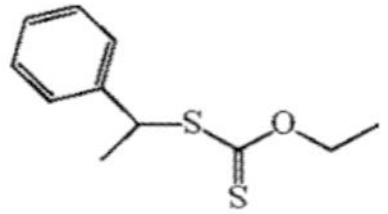 | 0.7 [48] | BA | SDS, 70 °C, NaPS | 40 000–100 000 | 1.6–2.4 | √ | [50] |
| Ab initio | 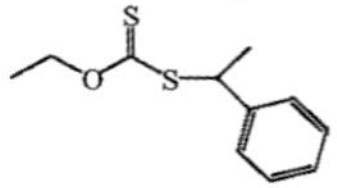 | 0.7 [48] | STY | SDS, 70 °C, NaPS | 100 000–100 000 | 1.9–2.6 | √√ | [51] |
| Ab initio | 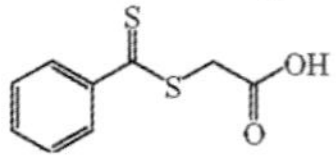 | — | 4-ASTY | SDS, 70 °C, KPS | 5500 | 1.6 | ×× | [52] |
| Ab initio | 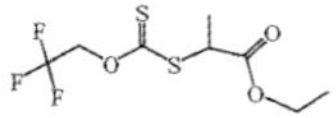 | 3.5 [48] | STY | SDS, 70 °C, NaPS | 5000–60 000 | 1.5–2 | √√ | [53] |

| Emulsion type | RAFT agent | $C_{tr,RAFT}$ | Monomer(s) | Conditions | $M_n$ range | PDI range | Successful[a] | Ref. |
|---|---|---|---|---|---|---|---|---|
| Ab initio | 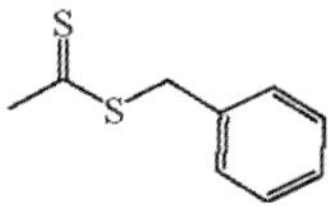 | 10 [45] | STY | SDS, 80 °C, KPS | 50 000 | 1.5 | ×× | [54] |
| Ab initio | 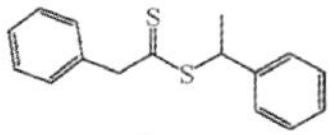 | 130 [55] | STY | Brij98, 70 °C, NaPS | 5000–9000 | 1.15–1.2 | √ | [56] |
| Ab initio | 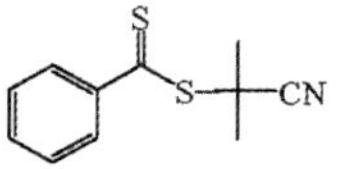 | 13 [45] | MMA | SDS, 70 °C, KPS | 100 000–40 000 | 1.2–1.4 | × | [57] |
| Seeded | 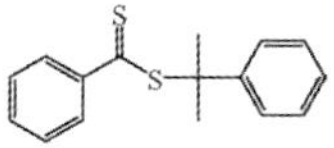 | 1030 [55] | STY | PMMA seed, interval II SDS, 60 °C, KPS | 5000–60 000 | 2.1–3.7 | ×× | [58] |
| Seeded | 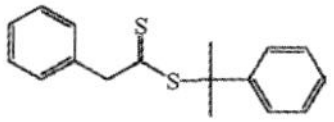 | 650 [55] | STY | PSTY seed, interval II SDS, 50 °C, NaPS | 10 000–50 000 | 1.1–1.4 | √ | [59] |
| Seeded | 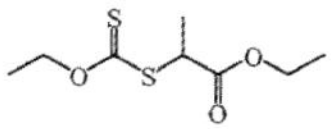 | 0.7 [48] | STY | PMMA seed, interval II SDS, 50 °C, NaPS | 15 000–80 000 | ~2 | √√ | [60] |
| Seeded | 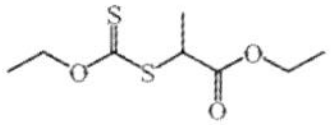 | 0.7 [48] | BA | PSTY-RAFT seed, interval II SDS, 60 °C, NaPS | 2000–15000 | ~2 | √√ | [61] |

(Continued)

Table 8.1 (*Continued*)

| Emulsion type | RAFT agent | $C_{tr,RAFT}$ | Monomer(s) | Conditions | $M_n$ range | PDI range | Successful[a] | Ref. |
|---|---|---|---|---|---|---|---|---|
| Miniemulsion | 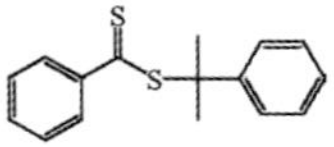 | 1030 [55] | STY | SDS, hexadecane, 75 °C, KPS | 10000–20000 | 1.5–2.8 | ×× | [62] |
| Miniemulsion | 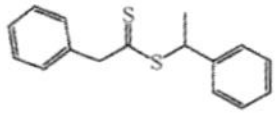 | 1030 [55]<br>130 [55] | STY | SDS, hexadecane, 75 °C, KPS | 2000–17000 | 1.7–2 | ×× | [55] |
| Miniemulsion | 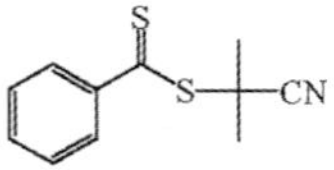 | 13 [45] | STY, EHMA, *i*-BMA, *n*-BMA | Brij 98 or Igepal 890, 70 °C, KPS | 5000–1000 | 1.07–1.25 | √√ | [63] |
| Miniemulsion | 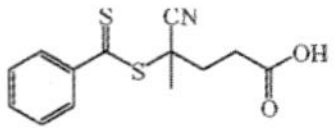 | — | STY | SDS, hexadecane, 70–85 °C, AIBN | 5000–20 000 | 1.2–1.5 | O | [64] |
| Microemulsion | 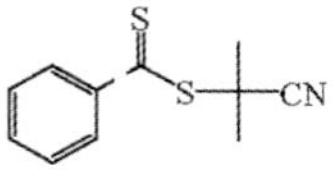 | 13 [45] | *n*-HMA | DTAB, 75 °C, V-50 | 3000–20 000 | 1.2–1.3 | O | [65] |
| Nanoprecipitation | 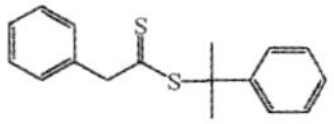 | 650 [55] | STY | PSTY-RAFT, 65 °C; or 75 °C, KPS; or BPO | 1500–8000 | 1.1–1.3 | √ | [66] |

[a] We define a successful emulsion polymerization as the ability to produce polymer with an MWD that would be found in bulk or solution polymerizations. Thus, in the above table a process gets a 'double tick' if it reproduces the bulk conditions almost exactly.

Note: STY = styrene; MMA = methyl methacrylate; BA = butyl acrylate; EHMA = ethyl hexylmethacrylate; BMA = butyl methacrylate; HMA = hexyl methacrylate; 4-ASTY = 4-acetoxy styrene.

reached 95% with a PDI of 1.22, with $M_{n,theory}$ being close to the experimental $M_n$ [46]. These results suggested that transportation of the RAFT from droplets to growing particles needs to be avoided. Uzulina et al. [47] carried out ab initio emulsion polymerizations with water-soluble and highly reactive dithio-RAFT agents (Z = phenyl and R = $CH_2CO_2H$ or $C(CH_3)_2CONH_2$) for three different monomers: styrene (STY), methyl methacrylate and vinyl acetate. They found for STY polymerizations that high ratios of initiator (V-50) to RAFT agent (where R = $CH_2CO_2H$) were required to get agreement between $M_{n,theory}$ and $M_{expt}$, but at the cost of high PDIs (>3) and high amounts of flocculation up to approximately 40%. They tried to overcome this problem by using a different RAFT agent (where R = $C(CH_3)_2CONH_2$); however, the PDI decreased to only 1.7 with little or no observed flocculation. The work here suggests that water-phase reactions are important. Regardless of the water solubility of the RAFT agent, control of the MWD was not ideal.

Charmot et al. [68] introduced xanthates as the controlling agent and named their invention MADIX (macromolecular design via interchange of xanthate). These RAFT agents were found to have a low reactivity ($C_{tr,RAFT} < 1$ in STY polymerizations) and thus should follow the $M_n$ and PDI curves where $C_{tr} = 1$, as shown in Fig. 8.2. The ab initio emulsion polymerizations for styrene (STY) and butyl acrylate (BA) showed that a similar MWD could be obtained as in bulk or solution; however, the PDIs were much greater than 1 and were close to 2 [49]. Monteiro et al. [50] showed that the evolution of $M_n$ and PDI for BA mediated by xanthates was very close to theory, and the greater the xanthate concentration, the lower the average particle size. These polymer particles could be used in a subsequent polymerization with STY to produce block copolymers of STY and BA. It was found that these block copolymers formed films, and the film surface as imaged by atomic force microscopy showed that the hard polystyrene (PSTY) spheres were segregated from each other by a soft poly-BA continuous phase. This suggests that the particle morphology was most probably core shell and formed during the polymerization where the phase separation of the polymers gave the most energetically favorable morphology.

RAFT-mediated ab initio emulsion polymerization was also successfully carried out for STY using the xanthate, 1-(*o*-ethylxanthyl)ethylbenzene, with a chain-transfer constant close to 0.8 [51]. The study looked at the effects of surfactant, initiator and RAFT-agent concentrations on the rate, particle size and MWD. It was found that with an increased concentration of surfactant, sodium dodecylsulfate (SDS), the average particle size decreased (higher number of particles) and the PSD became narrower. Similar results were found when the RAFT-agent concentrations were increased and it was postulated that the R radical from the xanthate produced from the fragmentation process resulted in a greater nucleation of micelles during interval I. Since this RAFT agent has a low chain-transfer constant, it was assumed that the R radical would play a role throughout the entire polymerization. The $M_n$ found from experiment was in general twice that of $M_{n,theory}$. A tentative explanation was put forward suggesting that this xanthate was surface active and resided on the surface of the particles. Transmission electron microscopy showed that after the polymerization of these PSTY latex particles with BA and acetoacetoxyethyl methacrylate (AAEMA, containing a reactive ketone) core-shell particle morphologies were indeed observed.

Kanagasabapathy et al. [52] carried out an emulsion polymerization of a STY-based monomer containing a reactive ketone group, 4-acetoxystyrene, mediated by the RAFT agent, S-thiobenzoylthioglycolic acid (a highly reactive agent). They found that the $M_n$ increased linearly with conversion but interestingly, so too the PDI (PDI = 1.61 at 70% conversion), which is opposite of what should be expected in bulk or solution (see curve where $C_{tr} = 100$, Fig. 8.2). Polymers with functional monomers such as AAEMA and acetoxystyrene provide a means of either cross-linking polymer chains together or attaching compounds with amine functionality. A new class of polymer latexes were prepared by placing the AAEMA monomer units in different locations of the polymer chain in a poly(styrene-*b*-BA) latex [69]. A diamine was then added to these latexes and immediately film formed. The mechanical properties of the film were assessed by stress–strain tensile measurements, and the film properties were found to be superior to conventional blends of similar polymer latexes even after cross-linking. Importantly, the authors were able to produce films from polymer latexes with the same chemical composition that showed mechanical behavior similar to polymers that are brittle, semicrystalline or rubbery. The reason for this is the location of the reactive ketone groups in the chain and their location either in the core or in the shell of the latex particle. It should also be noted that the continuous phase is coupled to the PSTY nanospheres, which will inherently provide different mechanical properties to composite blends of homopolymers.

All the successful RAFT-mediated ab initio emulsion polymerizations were carried out using low reactive agents ($C_{tr,RAFT} < 1$). Adamy et al. [48] found that when a xanthate contained trifluoroethyl moieties as the Z group, the $C_{tr,RAFT}$ for STY was close to 3.5. The authors then used this RAFT agent in an ab initio emulsion polymerization at 70 °C, using SDS as surfactant and sodium peroxydisulfate as initiator [53]. The $M_n$ and PDI could be accurately predicted and gave PDIs less than 1.5, and the particle size could be controlled by changing the concentration of the fluorinated xanthate; that is, an increase in xanthate concentration resulted in the average particle size decreasing from approximately 90-nm diameter to 70-nm diameter.

The use of highly reactive RAFT agents ($C_{tr,RAFT} > 10$) was studied by Nozari and Tauer [54] for STY polymerizations under ab initio conditions. They showed that the best control of molecular weight (in particular PDI) was from the least hydrophobic of the dithiobenzoate agents and proposed that this was probably due to the faster transportation of the more hydrophilic RAFT agents from monomer droplet to growing particles. However, the PDIs found from this study were much greater than those found from either bulk or solution polymerizations. The authors then showed [70] that the diffusion of the RAFT agent strongly depended on its water solubility, supporting their postulate of slow RAFT-agent transportation. They also found degradation of the RAFT agent with peroxide initiators both inside the particle and in the water phase.

Urbani et al. [56] carried out a comprehensive study on the ab initio emulsion polymerizations of STY in the presence of a reactive RAFT agent (1-phenylethyl phenyldithioacetate, PEPDTA). However, they used a nonionic surfactant (Brij 98) instead of the convention anionic SDS. The weight fraction of STY was 10 or 20 wt % and the Brij 98 surfactant was 5 wt %. Due to the high surfactant concentration it was found that from 'creaming' experiments at 10 wt % of STY, approximately 99.4% of monomer was contained in the surfactant micelles, and at 20 wt % approximately 86.4% of monomer was contained in the micelles. The nonionic surfactant can accommodate the excess monomer by increasing either in size or in number, which is very different to when SDS is used as surfactant. The results suggest that the system in many ways resembles a micro- or miniemulsion or an interval II emulsion polymerization. The evolution of $M_n$ with conversion for all targeted molecular weights (at 100% conversion) was similar to theoretically calculated $M_n$'s. Interestingly, the PDIs were low and below 1.2 when $M_n$'s below 9000 were targeted (at 100% conversion). At higher targeted $M_n$'s, the PDI would increase with conversion to values greater than 2 (at high conversions). These results suggest a fundamental mechanism at play that will be discussed later in this chapter.

All the ab initio emulsion experiments show that the transportation of the RAFT agent is crucial to obtaining good 'living' behavior. However, Urbani's et al. [56] data have shown that this alone is not the governing mechanism and molecular weight of the polymer also plays a dominant role. Can this latter mechanism be explained by thermodynamics or kinetics?

#### 8.3.1.2 Seeded Emulsions

Carrying out experiments using a preformed seed latex allows mechanistic data to be derived since such systems eliminate the nucleation process. The polymerization starts in interval II or III and parameters such as entry and exit can be readily determined. Monteiro et al. [58] were the first to study RAFT-mediated seeded emulsion polymerizations, using highly reactive dithiobenzoate RAFT agents starting in interval II. The polymerizations were problematic, as high levels of flocculation were found and a red monomer layer, containing high amounts of RAFT agent or RAFT-based oligomers, was observed. These oligomers would form a coagulant once interval III was reached. They proposed that exit of the R groups from the RAFT agent out of the growing particles was responsible for retardation, and the MWD did not resemble that found from either bulk or solution polymerizations with the same RAFT agents. The authors stated that the rate of RAFT-agent transportation from droplets to growing particles should be at the rate of its consumption in the particles. Therefore, for a successful 'living' emulsion to be carried out all the RAFT agent must be transported within the first few percent conversion. The results found from this work and others suggest that the diffusion rate of the RAFT agents examined is much slower than this criterion. This explains why the low reactive RAFT agents (xanthates) showed good 'living' behavior [49–51, 68, 69–] due to the fact that the xanthate needs transport only over the whole conversion range.

Prescott et al. [59] used acetone to transport the RAFT agent (PEPDTA) to the seed particles and then the residual acetone removed by rotovaporation and STY added. The polymerizations were then carried out at 50 °C, and the $M_n$ evolved linearly with conversion and the PDIs were below 1.4 after the polymer from the seed was subtracted. This shows that the transportation of the RAFT agent is important, and once inside the particles can act to mediate the 'living' behavior of the polymerization. With all emulsion polymerizations, the addition of RAFT agent usually results in retardation or even inhibition of rate. Work carried out by Prescott et al. [71, 72] and Peklak and Butte [73] has used simulations to show that inhibition of rate is most likely due to exit and termination of R radicals formed from fragmentation from the RAFT agent, in support of Monteiro et al.'s original work [58] on the subject. Retardation is more complex and has been proposed to be due to many contributing factors: size of the particles, probability of exit of R radicals, reactivity of R radicals toward monomer, termination of radicals within the particles and addition rate constants of entry and R radicals toward the S=C moiety. The general trend found was that a greater retardation in rate was observed with an increase in RAFT-agent concentration. Once the inhibition period has ceased and all the initial RAFT agent converted to small oligomeric RAFT agents, retardation would be governed by the probability of exit and the reactivity of $z$-mers (formed in the aqueous phase) to enter and react with the RAFT oligomers. Since the average molecular weight of the polymer chains is a function of conversion, termination will be dependent upon chain length. As such termination will be high at low conversions should the system be under pseudobulk conditions (where one or more radicals can reside in the same particle). The amount of retardation is dependent upon the size of the particles and the reactivity of the S=C bonds to the radicals.

To examine these parameters more closely, the kinetics of low reactive RAFT agents (xanthates) was studied by Smulders et al. [60]. They found that exit of R radicals from the particles occurred via a second-order loss rate with respect to the average number of radicals per particle, but the entry rate coefficient was surprisingly reduced with the increase in RAFT concentration. The most plausible explanation was that the xanthates studied were surface active due to their canonical structure. The surface-active moieties would therefore be preferentially located at the surface of the particles and create an inhomogeneous distribution of xanthate oligomer or polymeric species throughout the particle. The entry of $z$-mers into the particles would now have a high probability to react with the surface-active S=C moiety, resulting in exit of the R radicals – the process termed 'frustrated entry'. Smulders and Monteiro [61] used this knowledge to prepare block copolymers of STY and BA with a core-shell morphology using a polymethyl methacrylate (PMMA) seed latex. The importance of this research was not only to control the size and morphology of the particles, but also to obtain better control of the MWD by artificially increasing the $C_{tr,RAFT}$ value by feeding the monomer into the reaction vessel slowly over time. The PMMA seed was used to control the PSD, in which STY in the presence of the RAFT agent was polymerized under batch conditions to give polymer particles with an $M_n$ of 7000 and PDI of 2. BA was then polymerized under semibatch conditions into these particles to give an overall $M_n$ of 20 000 and PDI close to 1.3 for the block copolymer, in which greater than 90% block purity was observed. The calculated PDI of the second block (PBA) was close to 1.4, which is lower than the theoretically determined value under batch conditions (PDI = 1.6). The results show that by slow monomer addition feed into the reactor the PDI can be reduced and that by using the advantage of 'random coupling' between the two blocks the PDI of the final block copolymer is lower than either of the individual blocks.

#### 8.3.1.3 Miniemulsion and Microemulsion

**Miniemulsion RAFT-Mediated Polymerizations** As described above, the most important consideration for successful RAFT-mediated emulsion polymerizations is the success of transporting the RAFT agent from droplets to the growing particles. Miniemulsions provide the most ideal dispersion methodology to locate all the ingredients in the stabilized monomer droplets, thus avoiding the need for RAFT transportation. The size of these droplets are much smaller (50–500 nm) than monomer droplets found in ab initio emulsion (>1 μm) and thus have the disadvantage that the PSD is not well controlled and dependent upon the method to form the miniemulsion [2]. The other disadvantage of miniemulsion is that the hydrophobic compound used to stabilize the monomer droplets from Ostwald ripening is approximately 2 wt % relative to monomer [74].

Several phenomena were observed when using the RAFT-mediated miniemulsion polymerizations indicating a deviation from the idealized theory when the miniemulsion was stabilized by ionic surfactant [62]. Inefficient droplet nucleation, a steadily rising polydispersity over the reaction and the appearance of a separate organic phase after initiation were all indications of particle instability. A distinct difference between standard polymerizations and those that involve highly active RAFT agents comes from the fact that in RAFT polymerization there is a time interval early in the reaction where oligomers dominate the MWD. The presence of large quantities of oligomers is believed to be the major culprit behind the destabilization observed through a detrimental interaction with the ionic surfactant of the miniemulsion [62]. Conductivity measurements verified the increase of free surfactant in the aqueous phase over the course of reaction. Despite this, results showed clear indication of 'living' character with a linear evolution of molecular weight until roughly 40% monomer conversion, after which the molecular weight showed contributions from initiator-derived chains. Lansalot et al. [55] showed that they could find conditions where colloidal stability was observed when using SDS. However, the polydispersity found in their experiments ranged between 1.7 and 2.0, suggesting that these systems gave nonideal RAFT polymerization.

It was thought that destabilization could be similar to that observed for ATRP in the presence of SDS and therefore, ionic surfactants were substituted for a nonionic surfactant (e.g. Brij 98) [63]. This allowed well-defined polymer (PDI < 1.2) to be prepared with no stability problems. However, retardation compared to polymerization without a RAFT agent was found. This was prescribed to be due to termination of the intermediate radical species, which lowered the propagating radical concentration considerably [75]. Although the rate is significantly reduced by intermediate radical termination (IRT), it should have little or no effect on the MWD since the amount of RAFT-dormant chains lost through intermediate termination is less than 5%. It was also found that a wide range of monomers could be polymerized with control and could be further used to prepare block copolymers with low polydispersities (<1.2) [63]. One major advantage of the RAFT process is that acidic monomers can be used, which provide very efficient stability to the polymer latex particles. For example, using Brij 98 (nonionic surfactant) a block copolymer of poly-EHMA-*block*-poly(methyl methacrylate-*co*-methacrylic acid) was prepared.

A destabilization theory to explain the observations in miniemuslions was described by Luo et al. [74]. They argued that the growing particles in an LRP would have a lower chemical potential than nonnucleated droplets due to the 'superswelling' effect of small oligomers. This would result in monomer transfer from high to low chemical potential, where monomer would swell the growing particles until equilibrium was reached. The authors suggested that by simply increasing the costabilizer level the problems found by using ionic stabilizers would be eliminated. Further work [64] by carrying out miniemulsions with higher levels of surfactant and costabilizer (hexadecane) tentatively supported the Luo et al. [74] postulate. The 'superswelling' theory is becoming more widely accepted as the thermodynamic reason for poor control in RAFT-mediated dispersion polymerizations, and was used by Urbani et al. [56] to describe their results. They commented that by utilizing a nonionic surfactant such as Brij 98 'superswelling' could be avoided due to the surfactant's lower efficiency [76] (i.e. it can stabilize a larger surface area than, for example, SDS), and also hinder monomer transportation [74]. Luo and Cui [57] used the 'superswelling' theory to aid in the experimental design to produce well-defined polymer using RAFT in the presence of high amounts of SDS and initiator. However, deviation from theory was observed when high molecular weights were targeted analogous to Urbani et al.'s [56] results. In this case, although $M_n$ increased linearly with conversion and was close to that calculated, the PDI values ranged between 1.5 and 1.8. They suggested that these high PDI values were not a result of the colloidal stability but were most probably a result of a heterogeneous distribution of RAFT agent among the particles formed at different nucleation times. Therefore, according to the 'superswelling' theory, an increase in the nucleation rate should lead to lower PDIs, which was found by Urbani et al. [56].

**'Superswelling' Theory** The original theory of 'superswelling' was derived by Luo et al. [74] and was aimed at investigating the cause of the frustrating problems of latex instability and broad MWDs observed in RAFT-mediated ab initio and miniemulsion polymerizations. The theory is based on the work of Ugelstad et al. [77, 78], which describes the swelling capacity of latex particles consisting of polymer and oligomer and mixtures thereof using a modification of the Morton equation.

The Luo–Tsavalas–Schork 'superswelling' theory is centered on the superswelling equilibrium of monomer between the droplets and the particles in an emulsion polymerization consisting of monomer, molecular weight control agent (RAFT agent) surfactant, costabilizer, initiator and water. The diffusion of monomer in this system is governed by the monomer chemical potential difference between droplets and particles. After nucleation, the chemical potential of monomer in the particles is less than that in the droplets due to the presence of oligomeric polymer [64]. In order to minimize the chemical potential in the system there is diffusion of monomer from droplets to oligomeric particles. Experimental and theoretical data by Ugelstad et al. [77, 78] have demonstrated that oligomers are very effective swelling agents and can cause high swelling of polymer particles with monomer, whereas high-molecular-weight polymer swells monomer to a much lesser extent. In conventional free-radical polymerization, oligomers are formed throughout the polymerization, rapidly increasing in molecular weight becoming macromolecules. However, in a living free-radical polymerization the oligomers formed dominate the MWD early in the reaction. The large amount of monomer transferred from droplets (high monomer chemical potential) to oligomeric particles (low monomer chemical potential) would give rise to the colloidal instability and ultimately loss of molecular weight control observed experimentally. Luo–Tsavalas–Schork have also showed using Ugelstaad's equation that the superswelling equilibrium will be effected to some extent by the costabilizer concentration and length, initial droplet size, interfacial tension and molecular weight control agent (i.e. RAFT agent) concentration.

Considering the kinetic aspect of the polymerization system, the rate of nucleation and rate of monomer diffusion through the water phase will greatly influence the establishment of the superswelling state. If nucleation is slow and monomer diffusion through water is fast (relative to initiation) then superswelling would be increased due to the few number of particles being formed. If nucleation is fast, superswelling will be suppressed due to the formation of greater numbers of particles.

Luo et al. [79] also used miniemulsion to determine the rate constant for fragmentation and addition for CDB and PEPDTA. They determined values of 314 $L \cdot mol^{-1}$ for PS-CDB and 22 $L \cdot mol^{-1}$ for PS-PEPDTA, and concluded that the fragmentation rate constant was of the order between $10^4$ and $10^5$ $s^{-1}$. However, there is still a debate within the literature between slow fragmentation [80] and IRT [75], and has been reviewed extensively [81].

**Microemulsion RAFT-Mediated Polymerizations** Surprisingly, there have been only two publications on the RAFT-mediated microemulsion polymerizations by Kaler and coworkers [65]. They conducted the RAFT-mediated microemulsion polymerizations of hexyl methacrylate (2.64 wt %) using dodecyltrimethylammonium bromide (12 wt %) as surfactant and 2-cyano-2-yl dithiobenzoate as RAFT agent. The average size of the particles was small and ranged between 18 and 30 nm. The $M_n$ increased linearly with conversions and the PDI was low provided that the number of RAFT agents per particles was much greater than 1 [65]. They also found that when the RAFT-agent concentration was increased, the rate of polymerization decreased with long inhibition times. Hermanson et al. [82] carried out simulations to provide an explanation for the retardation in rate and the loss of molecular weight control at low RAFT-agent concentrations. In their simulations they used a slow fragmentation rate constant of 2.9 $s^{-1}$, no exit of the R radicals and other oligomeric radicals from the particles was implemented and it was assumed that the RAFT agent could transport freely between particles and micelles. They found a good correlation between simulation and experimental for the rate of polymerization, $M_n$, PDI and average particle size. However, their model assumptions are contrary to what has been found in the literature. Luo et al. [79] found that fast fragmentation and IRT were required to fit their miniemulsion data, and the numerous works by Monteiro and coworkers [50, 53, 58, 60, 61], Prescott et al. [59, 71, 72, 83–] and Butte and coworkers [73, 84] showed that exit of the R radicals provides an explanation to why the rate was inhibited and retarded.

We believe that the RAFT-mediated polymerizations in dispersed media are a combination between kinetic effects (exit, entry, 'frustrated entry' and monomer droplet nucleation) and thermodynamic effects ('superswelling' theory). However, the 'superswelling' theory does not fully explain why a red layer is found in ab initio, seeded and miniemulsion polymerizations. A qualitative study by Huang et al. [26] tried to elucidate the reason for this observation and proposed that the red layer most probably formed due to Ostwald ripening of small droplets to large ones where the amount of hydrophobe (hexadecane) was diluted and could not act to stop monomer diffusion to these droplets. They found that there was approximately 3.5 times more RAFT agent to monomer in these droplets than in the stabilized droplets. A quantitative description of the destabilization effect was described by Qi and Schork [85], who used a derivation of the Lifshitz–Slyozov–Wagner theory to explain the stability of miniemulsions in the presence of RAFT agents. They found that RAFT agents more hydrophobic than the cosurfactant (the hydrophobe) led to greater instability prior to the polymerization, but RAFT agents with similar hydrophobicity as the cosurfactant led to stability that was much better than miniemulsions without RAFT agent. The evidence given here is most probably why the use of a nonionic surfactant led to stable and well-controlled ('living') miniemulsion polymerizations [63].

**Nanoprecipitation RAFT-Mediated Polymerizations** Georges and coworkers [66] used a nanoprecipitation method to prepare 'latex' particles where the polymer was well defined (low PDIs). They overcame all the obstacles of destabilization found in conventional dispersion polymerizations by precipitating low- and narrow-molecular-weight polymer (dissolved in acetone) in water to form nanoparticles in the presence of polyvinyl alcohol. The acetone was removed and monomer added and polymerization initiated. The MWD was well controlled for RAFT-mediated polymerizations when initiated with benzoyl peroxide, with PDIs for one polymerization remaining constant at 1.14 from 19 to 31% conversion [66].

### 8.3.2 Surfactant-Free Dispersion Polymerizations

In all the previous emulsion polymerization systems described, surfactant is used to stabilize the initial emulsion and final 'latex' particles. A significant disadvantage of surfactants is that they can migrate upon film formation and create regions that are water sensitive. If such surfactant-stabilized 'latex' particles are used in biomedical applications, there is great concern that the surfactant may dissociate from the particles, reducing particle stability and resulting in particle coagulation. It is usually difficult and expensive to remove surfactants from the system postpolymerization. Now the challenge for researchers is to prepare stable surfactant-free 'latexes' with the polymer MWD controlled by 'living' radical polymerization. The next section will discuss two types of surfactant-free dispersion polymerizations. The first is the self-assembly of amphiphilic block copolymers to make stabilized 'latex' particles developed by Hawkett and coworkers, and the second is the novel degassing technique developed by the group of Pashley and implemented in RAFT-mediated dispersion polymerizations by the group of Monteiro.

#### 8.3.2.1 Self-Assembly

In 2001, Charleux and coworkers [87] showed that diblock amphiphilic copolymers could function to stabilize conventional ab initio emulsion polymerizations. This work led Hawkett's team [88] to develop the synthesis of 'latex' particle via a self-assembly process. The initiator (4,4′-azobis(4-cyanopentonoic acid)) was added to a solution of polyacrylic acid (PAA) macro-RAFT agent in water. The mixture was brought to the reaction temperature and BA fed into the reaction to maintain the monomer concentration below the saturation concentration of BA in the water phase. The BA units added to the PAA macro-RAFT agent to form a diblock copolymer, and when the number of BA units was great enough the diblock would self-assemble into small PAA-stabilized nanoparticles, where the core consisted of PBA. The monomer would then swell the hydrophobic PBA region and polymerization would continue with growth of the particles. There was a linear increase in $M_n$ with conversion, and the PDI increased from approximately 1.25 (10% conversion) to 1.5 (70% conversion).

They then used STY as the comonomer [89] and found that $M_n$ increased linearly with conversion, but the PDIs were low (below 1.2) only for molecular weights below 10 K, analogous to the work by Urbani et al. [56]. The increase in PDI with conversion for the higher molecular weight polymerizations could possibly be explained by the 'superswelling' phenomenon.

Other methods to form 'latex' core-shell particles are through the self-assembly of block copolymers in water. There has been a great deal of work on the self-assembly

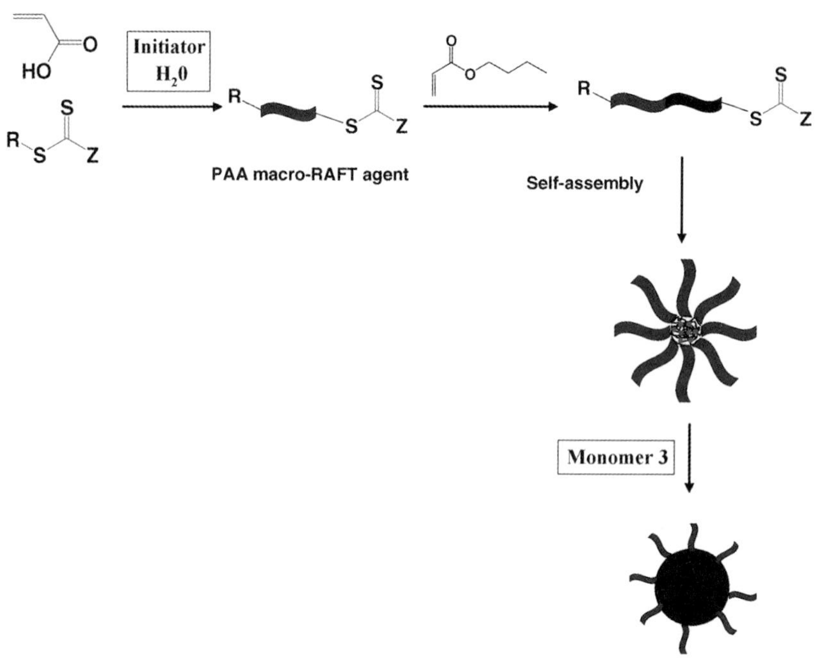


**Scheme 8.4** In situ polymerization of amphiphilic block copolymers that self-assemble into stabilized polymer particles.

of amphiphilic diblock copolymers in water, but of recent interest is the self-assembly of amphiphilic star diblock copolymers (Scheme 8.4) [20]. Amphiphilic four-arm star diblock copolymers consisting of styrene (STY) and acrylic acid (AA) were made using RAFT (Z-group approach with no star–star coupling), in which the PSTY was attached to the RAFT core moiety and PAA attached to the end of the PSTY chains. The size of the poly($AA_{132}$-$STY_m$)$_4$ stars in DMF was small and close to 7 nm, suggesting no star aggregation. Slow addition of water (pH = 6.8) to this mixture resulted in aggregates of 15 stars per micelle with core-shell morphology with an average diameter of 40 nm. The PSD was very narrow as determined from field flow fractionation. In this work the authors showed through the use of star amphiphilic polymers that the micelle size, aggregation number and morphology could be controlled [20].

The authors used an alternative and nonintuitive method to form tethered PAA loops in core-shell PSTY–PAA nanoparticles [90]. Instead of placing the hydrophilic polymer (i.e. PAA) on the exterior of the four-arm star, PAA was located at the interior of the star (close to the RAFT core moiety) and PSTY attached to the chain ends of the PAA. When these four-arm stars were aggregated in water, they formed core-shell particles with diameters ranging from 88 to 95 nm, in which the shell consists of tethered PAA loops. The entropic penalty for having such loops resulted in less densely packed PSTY core when compared to linear diblock copolymers of the same arm length. The surface of the shell is irregular due to the tethering points, but when cleaved the PAA chains extend to form a regular and relatively uniform corona.

#### 8.3.2.2 Degassing Technique

A brief theoretical description of how degassing can stabilize hydrophobic particles will be given, followed by the implementation of the degassing technique in conventional emulsion and RAFT-mediated emulsion polymerizations.

##### Theory and Historical Background

***Derjaguin–Landau–Verwey–Overbeek Theory*** Derjaguin–Landau–Verwey–Overbeek (DLVO) theory has long been used to describe the colloidal stability of electrostatically stabilized dispersions [1, 2]. The theory, developed in the 1940s by Derjaguin, Landau, Verwey and Overbeek [91, 92], describes the forces between charged surfaces in a liquid medium, which combines van der Waal's force of attraction and the electrostatic repulsive forces generated from the counterions in the electric double layer. The electric double layer can be generated by any ionic species absorbing at the surface of hydrophobic materials dispersed in water, but generally for emulsion polymerizations this is created by added ionic surfactant. The potential at the electric double layer is assumed to be distributed by the nonlinear Poisson Boltzmann equation for point ions. The theory suggests that the stability of electrostatically stabilized oil in water dispersions is dependent on the total energy function, $V_T$. Where $V_T$ is a contribution of the repulsive, $V_R$ and $V_B$ (the born repulsion) and attractive, $V_A$, forces (Fig. 8.3).

Using this theory, the interaction energy, $V_m$, between two charged particles in water can be calculated, where $V_m$ is a measure of the minimum energy required for coagulation of the dispersed particles or coalescence of the dispersed oil droplets. Generally, if $V_m$, for electrostatically charged particles dispersed in water, is greater than 10 kT, it is likely that the particles will remain dispersed and stable [2].

***The Hydrophobic Interaction*** In recent years the development of a plethora of new forces has emerged, which have been unintentionally grouped and labeled non-DLVO forces. These are suggested forces which DLVO theory has failed to account and have included the effects of hydration, specific ion effects, Sugami forces, ion fluctuation, protrusion, Helfrich and depletion [93].

Using DLVO theory alone, Pashley [94] calculated that the interaction energy between two surfactant-free dodecane droplets of 300 nm in water had an energy barrier of 800 kT. This suggests that once dispersed the significant surface electrostatic potential of the droplets will increase the energy required for coalescence greater than that for any other colloidal system, and they should remain dispersed and stable in water. The failure of DLVO to account for droplet coalescence in this system was prescribed to a new long-range hydrophobic attractive force [94]. The hydrophobic interaction has received much attention. Israelachvili and Pashley [95] showed that for two crossed cylinders of mica, surface coated with a monolayer of

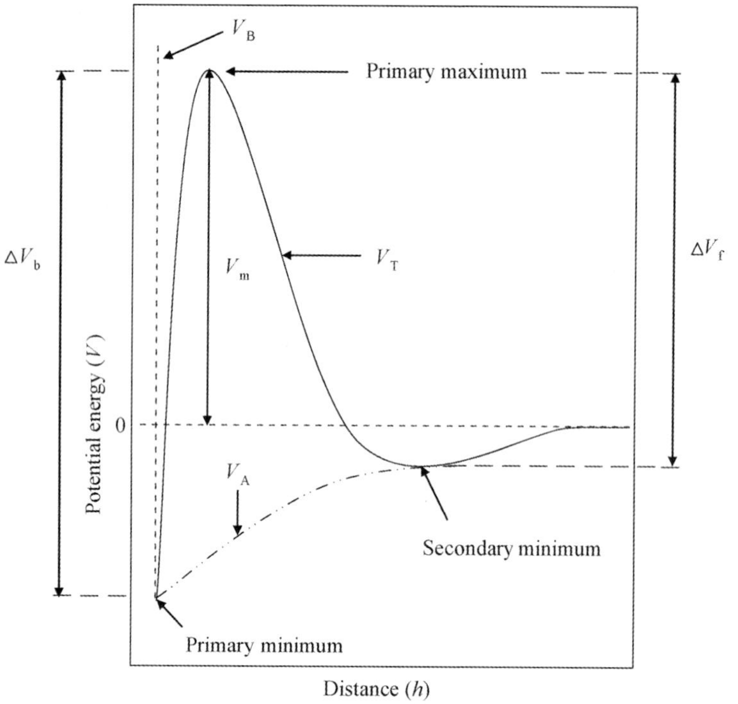


**Fig. 8.3** Schematic describing the potential energy ($V_T$) as a function of distance ($h$) between electrostatically stabilized dispersed particles.

cationic surfactant, in water, the contribution of the hydrophobic interaction was found to be an order of magnitude greater than van der Waal's forces and decayed exponentially in the range of 1–10 nm. For pure polypropylene and polystyrene surfaces in water the hydrophobic interaction extended up to 30 nm and was observed to be largely unaffected by the presence of electrolytes [96]. The attractive force has been observed to extend over very long ranges (up to 300 nm) between surfaces formed through nonequilibrium processes, such as LB deposition [97, 98], and silanation [99, 100]. The apparent range of these forces extends beyond the range for van der Waals and is up to one hundred times greater in attraction. There is currently no theoretical working model to explain the mechanism of the hydrophobic interaction. However various theories have been suggested. Initially, it has been proposed that changes in water structure at or near the surface of two hydrophobic species, due to enhanced hydrogen bonding, resulted in displacement of water molecules in the interlayer region to the bulk and a lowering of the free energy of the system [96]. This explanation, however, does not account for the long-range nature of the interaction. Another theory suggested that the interaction was driven by the metastability of water films between two hydrophobic surfaces with contact angles greater than 90 °C [97]. When the particles are at close range (<1 nm), an

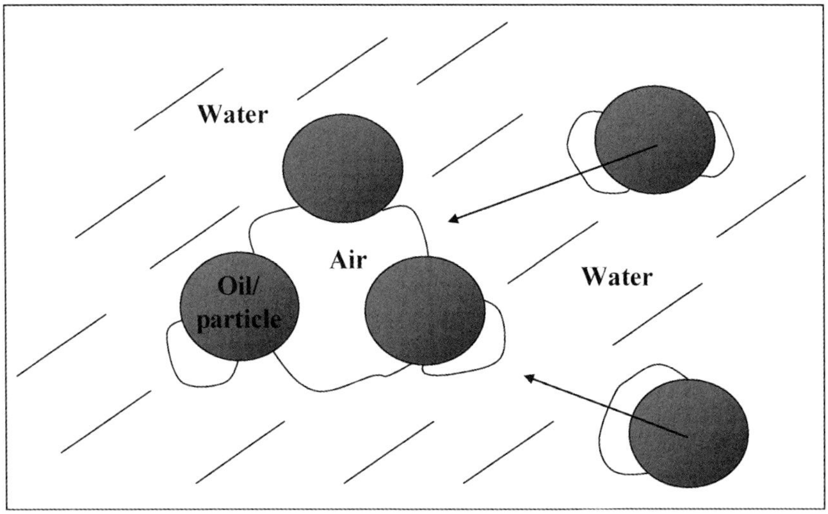


**Scheme 8.5** Mechanism for the gas adsorption bridge coagulation proposed by Pashley and coworkers.

attraction is predicted just prior to spinodal cavitation, which, as suggested, could be the microscopic origin of the hydrophobic interaction [101].

A theory that has received must attention and experimental agreement is 'bridging bubbles'. Wennerstroem [102] suggested that the attraction was due to the formation of a gaseous film between two hydrophobic surfaces in water. Dissolved gas, being hydrophobic, will accumulate to some extent close to the hydrophobic surface and the observed attractive force between the surfaces is the occurrence of gas bubbles bridging the surfaces [102]. Pashley [94] also suggests that the coalescence of oil droplets in water might occur due to absorption of dissolved gas on the droplets hydrophobic surfaces. Concentration of these gases may lead to cavitations when the two surfaces approach (Scheme 8.5 – transport of oil via air bridges).

A study conducted by Meagher and Craig [96] showed that removing the dissolved gases from dilute sodium chloride solutions containing polypropylene surfaces reduced the magnitude and range of the attractive force. Mahnke et al. [103] studied, using atomic force microscopy, the 'jump into contact' distance of methylated and dehydroxylated silica surfaces in water. The authors confirmed that the jump to contact distances was sufficiently large so as to not be attributed to van der Waal's attractive force. It was determined that for surfaces with large jump distances ($>25$ nm) the effects of degassing greatly reduced the jump distance, but at shorter distances, degassing had little effect. They attributed this reduction to the presence of bubbles on the hydrophobic solid surfaces stabilized by a combination of hydrophobicity and chemical heterogeneity. Later work by Meyer et al. [104] observed similar results for mica surfaces coated with a monolayer of surfactant. Removal of dissolved gases was seen to reduce long-range attraction, while short-range

(<250 Å) attraction remained unchanged. Gong et al. [105] also showed that completely removing the dissolved carbon dioxide from a suspension of methylated silica in water reduced the rate of aggregation of the silica but for dehydroxylated silica there was no difference.

Although there is great support for the bubble bridging theory, there is yet a resolution of the mechanism of the hydrophobic interaction. The observations that the magnitude and range of the attractive force varies with the type of hydrophobic surface supports dissolved gases and bubble absorption plays a role. Whether the case or not, oil-in-water emulsions can be stabilized by removal of the dissolved gases in the system. The purpose of this observation and of novel application to the polymer scientist is the preparation and polymerization of stable emulsions initially formed and stabilized by 'the degassing technique'.

**Implementation of the Degassing Technique** The general aim of the degassing technique is to prepare stable emulsions of monomer droplets dispersed in water without the use of added surfactant. The stable droplets (prepared by the complete removal of dissolved gas) could then be nucleated with free radicals and polymerized to afford stable surfactant-free polymer latex.

Early work by Karaman et al. [106] showed that stable emulsions of STY in water could be prepared using this technique and subsequent polymerization produced stable PSTY latex. A nondegassed mixture (control) showed clear phase separation of the STY and water prior to polymerization. The PSD was much more monodispersed for the degassed system with spherical particles of mean diameter 46 nm. The nondegassed polymerization latex particles were roughly spherical in appearance with broad PSD (35–200 nm diameter, average 38 nm). The narrow polydispersity of the final PSD in the degassed system suggests the initial droplet size distribution to be narrow.

The first RAFT-mediated surfactant-free emulsion prepared by the removal of dissolved gases was conducted by Hartmann et al. [107]. In this work an emulsion of STY and RAFT agent (xanthate or PEPDTA) in water was prepared by subjecting the mix to consecutive freeze-pump-thaw cycles (high vacuum degassing). After the degassing cycles, a turbid solution formed, with a thin monomer layer residing on the top of the mixture (Fig. 8.4). It was determined by dynamic light scattering that the initial droplet sizes were in the order of 3–4 μm. Further emulsions were prepared in this manner but included the hydrophobe hexadecane at 2 wt % concentration relative to STY. The initial droplet sizes in this system were much smaller as a result of the hexadecane minimizing the degree of coalescence of droplets caused by Oswald ripening. The STY emulsions formed in both these systems were then polymerized with persulfate at 70 °C. The polymerizations showed living character; however, secondary nucleation was attributed to the generation of high-molecular-weight dead polymer. *We believe that the stability of the droplets and particles is due to hydroxyl anions on the surface.* These preliminary data open up an exciting avenue for the preparation of nonstabilized polymer latexes with 'living' functionality.

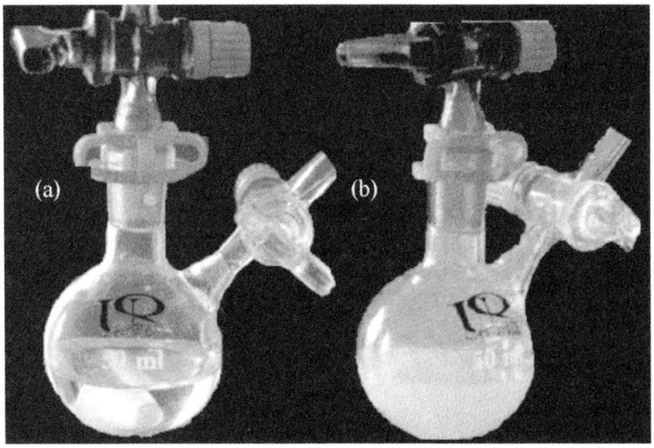


**Fig. 8.4** Photograph of the formation of a surfactant-free emulsion of styrene in water before (a) and after (b) three consecutive freeze-pump-thaw cycles. A turbid solution is formed after degassing, with a thin monomer layer residing on the top of the reaction mixture [107].

## 8.4 Conclusion

'Living' radical polymerization has opened a new field in polymer science of well-defined nanostructures. Implementation of LRP to dispersion polymerization is a complex interplay of both thermodynamic and kinetic effects. This is exemplified in RAFT-mediated emulsion polymerizations. We believe that the fundamental understanding gained from the years of work by many researchers will allow polymer nanostructures to be synthesized with precise size, chemical composition, morphology and chemical functionality. Such nanostructures will provide new products in the coatings area as well as biomedical applications. However, for us to reach this goal, more fundamental research is required to further unravel the mysteries of RAFT-mediated and other LRP systems in dispersion polymerizations.

## References

1. A. M. van Herk, *Chemistry and Technology of Emulsion Polymerization*, Blackwell Publishing, Oxford, UK, **2005**.
2. P. A. Lovell, M. S. El-Aasser, *Emulsion Polymerization and Emulsion Polymers*, John Wiley & Sons Ltd., Chichester, England, **1997**.
3. J. M. Asua, *Polymeric Dispersions: Principles and Applications*, Kluwer Academic Publishers, The Netherlands, **1997**.
4. R. G. Gilbert, *Emulsion Polymerization: A Mechanistic Approach*, Academic Press, London, **1995**.
5. C. S. Chern, *Prog. Polym. Sci.* **2006**, *31*(5), 443–486.
6. M. Nomura, *J. Ind. Eng. Chem. (Seoul, Repub. Korea)* **2004**, *10*(7), 1182–1216.
7. R. M. Fitch, *Polymer Colloids*, Academic Press, London, **1997**.
8. J. M. Asua, *J. Polym. Sci., Part A: Polym. Chem.* **2004**, *42*(5), 1025–1041.

**9.** B. S. Hawkett, D. H. Napper, R. G. Gilbert, *J. Chem. Soc., Faraday Trans 1: Phys Chem. Condensed Phases* **1980**, *76*(6), 1323–1343.

**10.** C. Hagiopol, V. Dimonie, M. Georgescu, I. Deaconescu, T. Deleanu, M. Marinescu, *Acta Polym.* **1981**, *32*(7), 390–397.

**11.** M. Antonietti, K. Landfester, *Prog. Polym. Sci.* **2002**, *27*(4), 689–757.

**12.** J. M. Asua, *Prog. Polym. Sci.* **2002**, *27*(7), 1283–1346.

**13.** K. Landfester, *Top. Curr. Chem. – Colloid Chem. II* **2003**, *227*, 75–123.

**14.** F. J. Schork, Y. Luo, W. Smulders, J. P. Russum, A. Butte, K. Fontenot, *Adv. Polym. Sci. – Polym. Part.* **2005**, *175*, 129–255.

**15.** P. Y. Chow, L. M. Gan, *Adv. Polym. Sci. – Polym. Part.* **2005**, *175*, 257–298.

**16.** X.-J. Xu, L. M. Gan, *Curr. Opin. Colloid Interface Sci.* **2005**, *10*(5, 6), 239–244.

**17.** G. Moad, Y. K. Chong, A. Postma, E. Rizzardo, S. H. Thang, *Polymer* **2005**, *46*(19), 8458–8468.

**18.** L. Barner, C. Barner-Kowollik, T. P. Davis, M. H. Stenzel, *Aust. J. Chem.* **2004**, *57*(1), 19–24.

**19.** J. Bernard, A. Favier, T. P. Davis, C. Barner-Kowollik, M. H. Stenzel, *Polymer* **2006**, *47*(4), 1073–1080.

**20.** M. R. Whittaker, M. J. Monteiro, *Langmuir* **2006**, *22*(23), 9746–9752.

**21.** R. T. A. Mayadunne, J. Jeffery, G. Moad, E. Rizzardo, *Macromolecules* **2003**, *36*(5), 1505–1513.

**22.** W.-p. Wang, Y.-z. You, C.-Y. Hong, J. Xu, C.-Y. Pan, *Polymer* **2005**, *46*(22), 9489–9494.

**23.** K. Matyjaszewski, *Mol. Cryst. Liq. Cryst.* **2004**, *415*, 23–34.

**24.** K. Matyjaszewski, *Curr. Org. Chem.* **2002**, *6*(2), 67–82.

**25.** M. R. Whittaker, C. N. Urbani, M. J. Monteiro, *J. Am. Chem. Soc.* **2006**, *128*(35), 11360–11361.

**26.** J. Huang, S. Jia, D. J. Siegwart, T. Kowalewski, K. Matyjaszewski, *Macromol. Chem. Phys.* **2006**, *207*(9), 801–811.

**27.** M. Li, N. M. Jahed, K. Min, K. Matyjaszewski, *Macromolecules* **2004**, *37*(7), 2434–2441.

**28.** S. G. An, G. H. Li, C. G. Cho, *Polymer* **2006**, *47*(11), 4154–4162.

**29.** C. J. Hawker, A. W. Bosman, E. Harth, *Chem. Rev. (Washington, D. C.)* **2001**, *101*(12), 3661–3688.

**30.** A. Studer, T. Schulte, *Chemical record, New York, N.Y.* **2005**, *5*(1), 27–35.

**31.** N. L. Hill, R. Braslau *Abstracts of Papers, 232nd ACS National Meeting, San Francisco, CA, United States* **10–14 Sept. 2006**, POLY-040.

**32.** F. Chauvin, P.-E. Dufils, D. Gigmes, Y. Guillaneuf, S. R. A. Marque, P. Tordo, D. Bertin, *Macromolecules* **2006**, *39*(16), 5238–5250.

**33.** O. Lagrille, N. R. Cameron, P. A. Lovell, R. Blanchard, A. E. Goeta, R. Koch, *J. Polym. Sci., Part A: Polym. Chem.* **2006**, *44*(6), 1926–1940.

**34.** V. Percec, T. Guliashvili, J. S. Ladislaw, A. Wistrand, A. Stjerndahl, M. J. Sienkowska, M. J. Monteiro, S. Sahoo, *J. Am. Chem. Soc.* **2006**, *128*(43), 14156–14165.

**35.** V. Percec, A. V. Popov, E. Ramirez-Castillo, M. Monteiro, B. Barboiu, O. Weichold, A. D. Asandei, C. M. Mitchell, *J. Am. Chem. Soc.* **2002**, *124*(18), 4940–4941.

**36.** M. Monteiro, T. Guliashvili, V. Percec, *J. Polym. Sci., Part A: Polym. Chem.* **2007**, *45*, 1835–1847.

**37.** V. Percec, A. V. Popov, E. Ramirez-Castillo, L. A. Hinojosa-Falcon, *J. Polym. Sci., Part A: Polym. Chem.* **2005**, *43*(11), 2276–2280.

**38.** V. Percec, E. Ramirez-Castillo, L. A. Hinojosa-Falcon, A. V. Popov, *J. Polym. Sci., Part A: Polym. Chem.* **2005**, *43*(10), 2185–2187.

**39.** V. Percec, E. Ramirez-Castillo, A. V. Popov, L. A. Hinojosa-Falcon, T. Guliashvili, *J. Polym. Sci., Part A: Polym. Chem.* **2005**, *43*(10), 2178–2184.

**40.** A. van Herk, M. J. Monteiro, *Handbook of Radical Polymerization*, K. Matyjaszewski and T. P. Davis, Eds., John Wiley and Sons, Chichester, England, **2002**, pp. 301–332.

**41.** M. J. Monteiro, B. Charleux, *Chemistry and Technology of Emulsion Polymerisation*, A. v. Herk Ed., Blackwell

Publishing Ltd., Oxford, UK, **2005**, pp. 111–139.
**42.** J. B. McLeary, B. Klumperman, *Soft Matter* **2006**, *2*(1), 45–53.
**43.** M. J. Monteiro, *J. Polym. Sci., Part A: Polym. Chem.* **2005**, *43*(15), 3189–3204.
**44.** A. H. E. Müller, R. Zhuang, D. Yan, G. Litvinenko, *Macromolecules* **1995**, *28*(12), 4326–4333.
**45.** G. Moad, J. Chiefari, Y. K. Chong, J. Krstina, R. T. A. Mayadunne, A. Postma, E. Rizzardo, S. H. Thang, *Polym. Int.* **2000**, *49*(9), 993–1001.
**46.** J. Chiefari, Y. K. Chong, F. Ercole, J. Krstina, J. Jeffery, T. P. Le, R. T. A. Mayadunne, G. F. Meijs, C. L. Moad, G. Moad, E. Rizzardo, S. H. Thang, *Macromolecules* **1998**, *31*, 5559–5562.
**47.** I. Uzulina, S. Kanagasabapathy, J. Claverie, *Macromol. Symp.* **2000**, *150*, 33–38.
**48.** M. Adamy, A. M. van Herk, M. Destarac, M. J. Monteiro, *Macromolecules* **2003**, *36*(7), 2293–2301.
**49.** D. Charmot, P. Corpart, H. Adam, S. Z. Zard, T. Biadatti, G. Bouhadir, *Macromol. Symp.* **2000**, *150*, 23–32.
**50.** M. J. Monteiro, M. Sjoberg, J. Van Der Vlist, C. M. Gottgens, *J. Polym. Sci., Part A: Polym. Chem.* **2000**, *38*(23), 4206–4217.
**51.** M. J. Monteiro, J. de Barbeyrac, *Macromolecules* **2001**, *34*(13), 4416–4423.
**52.** S. Kanagasabapathy, A. Sudalai, B. C. Benicewicz, *Macromol. Rapid Commun.* **2001**, *22*(13), 1076–1080.
**53.** M. J. Monteiro, M. M. Adamy, B. J. Leeuwen, A. M. van Herk, M. Destarac, *Macromolecules* **2005**, *38*(5), 1538–1541.
**54.** S. Nozari, K. Tauer, *Polymer* **2005**, *46*, 1033–1043.
**55.** M. Lansalot, T. P. Davis, J. P. A. Heuts, *Macromolecules* **2002**, *35*(20), 7582–7591.
**56.** C. N. Urbani, H. N. Nguyen, M. J. Monteiro, *Aust. J. Chem.* **2006**, *59*(10), 728–732.
**57.** Y. Luo, X. J. Cui, *Polym. Sci., Part A: Polym. Chem* **2006**, *44*, 2837–2847.
**58.** M. J. Monteiro, M. Hodgson, H. De Brouwer, *J. Polym. Sci., Part A: Polym. Chem.* **2000**, *38*(21), 3864–3874.
**59.** S. W. Prescott, M. J. Ballard, E. Rizzardo, R. G. Gilbert, *Macromolecules* **2002**, *35*(14), 5417–5425.
**60.** W. Smulders, R. G. Gilbert, M. J. Monteiro, *Macromolecules* **2003**, *36*(12), 4309–4318.
**61.** W. Smulders, M. J. Monteiro, *Macromolecules* **2004**, *37*, 4474–4483.
**62.** J. G. Tsavalas, F. J. Schork, H. de Brouwer, M. J. Monteiro, *Macromolecules* **2001**, *34*(12), 3938–3946.
**63.** H. de Brouwer, J. G. Tsavalas, F. J. Schork, M. J. Monteiro, *Macromolecules* **2000**, *33*(25), 9239–9246.
**64.** J. B. McLeary, M. P. Tonge, D. De Wet Roos, R. D. Sanderson, B. Klumperman, *J. Polym. Sci., Part A: Polym. Chem.* **2004**, *42*(4), 960–974.
**65.** S. Liu, K. D. Hermanson, E. W. Kaler, *Macromolecules* **2006**, *39*(13), 4345–4350.
**66.** A. R. Szkurhan, T. Kasahara, M. K. Georges, *J. Polym. Sci., Part A: Polym. Chem.* **2006**, *44*(19), 5708–5718.
**67.** T. P. Le, G. Moad, E. Rizzardo, S. H. Thang Int. Pat. WO 9801478 [*Chem. Abs. 128*, 115390f], **1998**.
**68.** D. Charmot, P. Corpart, D. Michelet, S. Z. Zard, T. Biadatti WO 9858974, **1998**.
**69.** M. J. Monteiro, J. De Barbeyrac, *Macromol. Rapid Commun.* **2002**, *23*(5, 6), 370–374.
**70.** S. Nozari, K. Tauer, A. M. Imroz Ali, *Macromolecules* **2005**, *38*, 10449–10454.
**71.** S. W. Prescott, M. J. Ballard, E. Rizzardo, R. G. Gilbert, *Macromolecules* **2005**, *38*(11), 4901–4912.
**72.** S. W. Prescott, M. J. Ballard, E. Rizzardo, R. G. Gilbert, *Macromol. Theory Simul.* **2006**, *15*(1), 70–86.
**73.** A. D. Peklak, A. Butte, *J. Polym. Sci., Part A: Polym. Chem.* **2006**, *44*(20), 6114–6135.
**74.** Y. Luo, J. Tsavalas, F. J. Schork, *Macromolecules* **2001**, *34*(16), 5501–5507.
**75.** M. J. Monteiro, H. de Brouwer, *Macromolecules* **2001**, *34*, 349–352.

76. K. Landfester, N. Bechthold, F. Tiarks, M. Antonietti, *Macromolecules* **1999**, *32*, 2679–2683.
77. J. Ugelstad, K. H. Kaggerud, F. K. Hansen, A. Berge, *Makromol. Chem.* **1979**, *180*(3), 737–744.
78. J. Ugelstad, P. C. Moerk, K. Herder Kaggerud, T. Ellingsen, A. Berge, *Adv. Colloid Interface Sci.* **1980**, *13*(1–2), 101–140.
79. Y. Luo, R. Wang, L. Yang, B. Yu, B. Li, S. Zhu, *Macromolecules* **2006**, *39*, 1328–1337.
80. C. Barner-Kowollik, J. F. Quinn, D. R. Morsely, T. P. Davis, *J. Polym. Sci., Part A: Polym. Chem.* **2001**, *39*, 1353–1357.
81. C. Barner-Kowollik, M. Buback, B. Charleux, M. L. Coote, M. Drache, T. Fukuda, A. Goto, B. Klumperman, A. B. Lowe, J. B. Mcleary, G. Moad, M. J. Monteiro, R. D. Sanderson, M. P. Tonge, P. Vana, *J. Polym. Sci., Part A: Polym. Chem.* **2006**, *44*, 5809–5831.
82. K. D. Hermanson, S. Liu, E. W. Kaler, *J. Polym. Sci., Part A: Polym. Chem.* **2006**, *44*, 6055–6070.
83. S. W. Prescott, M. J. Ballard, E. Rizzardo, R. G. Gilbert, *Aust. J. Chem.* **2002**, *55*(6, 7), 415–424.
84. A. Butte, G. Storti, M. Morbidelli, *Macromolecules* **2000**, *33*(9), 3485–3487.
85. G. Qi, J. Schork, *Langmuir* **2006**, *22*, 9075–9078.
86. A. R. Szkurhan, M. K. Georges, *Macromolecules* **2004**, *37*, 4776–4782.
87. C. Burguiere, S. Pascual, C. Bui, J. Varion, B. Charleux, K. A. Davis, K. Matyjaszewski, *Macromolecules* **2001**, *34*, 4439–4451.
88. C. J. Ferguson, R. J. Hughes, B. T. T. Pham, B. S. Hawkett, R. G. Gilbert, A. K. Serelis, C. H. Such, *Macromolecules* **2002**, *35*(25), 9243–9245.
89. B. T. T. Pham, D. Nguyen, C. J. Ferguson, B. S. Hawkett, A. K. Serelis, C. H. Such, *Macromolecules* **2003**, *36*(24), 8907–8909.
90. M. R. Whittaker, C. N. Urbani, M. J. Monteiro, *Langmuir*, **2007**, *23* 7887–7890.
91. B. V. Derjaguin, L. Landau, *Acta Physicochim. USSR* **1941**, *14*, 633.
92. E. J. W. Verwey, J. T. G. Overbeek, *Theory of Stability of Lyophobic Colloids*, Elsevier, Amsterdam, The Netherlands, **1948**.
93. B. W. Ninham, *Adv. Colloid Interface Sci.* **1999**, *83*(1–3), 1–17.
94. R. M. Pashley, *J. Phys. Chem. B* **2003**, *107*(7), 1714–1720.
95. J. Israelachvili, R. Pashley, *Lett. Nat.* **1982**, *300*, 341–342.
96. L. Meagher, V. S. J. Craig, *Langmuir* **1994**, *10*, 2736–2742.
97. H. K. Christenson, P. M. Claesson, *Science* **1988**, *239*, 390.
98. H. K. Christenson, J. Fang, B. W. Ninham, J. L. Parker, *J. Phys. Chem.* **1990**, *94*, 8004.
99. J. L. Parker, D. L. Cho, P. M. Claesson, *J. Phys. Chem.* **1989**, *93*, 6121.
100. J. L. Parker, P. M. Claesson, *Langmuir* **1994**, *10*, 635.
101. V. V. Yaminsky, B. W. Ninham, *Langmuir* **1993**, *9*, 3618.
102. H. Wennerstroem, *J. Phys. Chem. B* **2003**, *107*(50), 13772–13773.
103. J. Mahnke, J. Stearnes, R. A. Hayes, D. Fornasiero, J. Ralston, *Phys. Chem. Chem. Phys.* **1999**, *1*, 2793–2798.
104. E. E. Meyer, Q. Lin, J. N. Israelachvili, *Langmuir* **2005**, *21*, 256–259.
105. W. Gong, J. Stearnes, D. Fornasiero, R. A. Hayes, J. Ralston, *Phys. Chem. Chem. Phys.* **1999**, *1*, 2799–2803.
106. M. E. Karaman, B. W. Ninham, R. M. Pashley, *J. Phys. Chem.* **1996**, *100*(38), 15503–15507.
107. J. Hartmann, C. Urbani, M. R. Whittaker, M. J. Monteiro, *Macromolecules* **2006**, *39*(3), 904–907.

# 9
# Complex Architecture Design via the RAFT Process: Scope, Strengths and Limitations

*Martina H. Stenzel*

## 9.1
## Complex Polymer Architectures

The connection of polymer science with material science requests the synthesis of different polymer structures in order to produce material with novel properties. These novel properties arise from the ability of complex architectures to show significantly different solution behaviors as well as from their ability to self-assemble into structures of higher order.

A range of structures are depicted in Scheme 9.1. Many of these structures have been realized using anionic polymerization.

The intention of this chapter is to demonstrate that reversible addition–fragmentation chain transfer (RAFT) polymerization is a versatile and powerful tool to contribute to the development of further complex polymer architectures.

Key to the successful synthesis of complex polymer architectures is the presence of a thiocarbonylthio group as an end functionality at each polymer chain as a result of the RAFT process, which allows further chain extension or the formation of branching points after prior attachment of the RAFT agent to a multifunctional compound (Scheme 9.2).

## 9.2
## Block Copolymers

The synthesis of block copolymers via RAFT polymerization has drawn significant attention due to the range of potential applications of these structures (see Chapter 13). Block copolymer synthesis has a strong focus on diblock copolymers, but triblocks are also described. Block copolymers are most commonly obtained via chain extension of a macro-RAFT agent. In addition, other avenues are outlined, such as the combination of RAFT-made blocks with other polymers generated

*Handbook of RAFT Polymerization.* Edited by Christopher Barner-Kowollik

ISBN: 978-3-527-31924-4

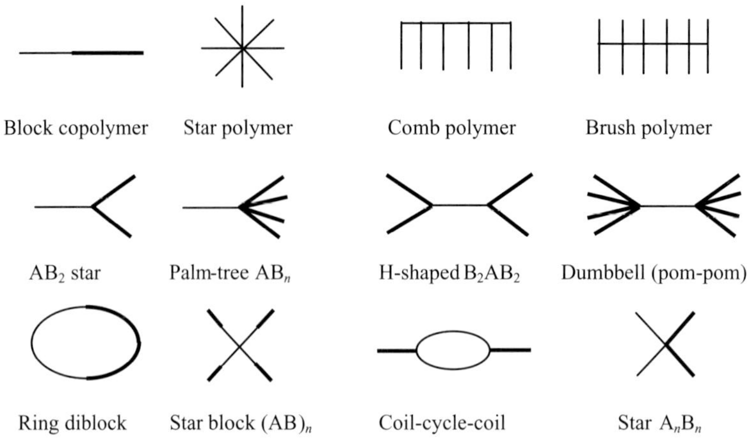


**Scheme 9.1** Complex architectures.

via other polymerization techniques. Lately, novel techniques emerge – such as click chemistry – allowing the connection of two homopolymers to generate block copolymers and other structures.

## 9.2.1
## Block Copolymers via Chain Extension of Macro-RAFT Agent

All polymer chains generated during the RAFT process carry a thiocarbonylthio end group, which now acts as a so-called macro-RAFT agent. The synthesis of block copolymers via chain extension is the simple process of starting a polymerization of a monomer in the presence of a macro-RAFT agent resulting in the formation of A—B diblock copolymers: The macro-RAFT agent takes on a similar role as the low-molecular-weight RAFT agent during homopolymerization.

### 9.2.1.1 Theoretical Considerations

A simplified view of the chain extension as presented above displays the desired outcome; it does not take into consideration the amount of reactions (and possible termination reactions) taking place during the reaction. Scheme 9.4 displays a more detailed summary of the process. As pointed out earlier, prerequisite is the utilization of a macro-RAFT agent. Similar to the homopolymerization via RAFT the process is initiated by radicals, which start the radical polymerization of monomer $M_2$, which will form the second block (Scheme 9.4, pathway I). The macroradical $M_2\cdot$ will then undergo chain transfer with the macro-RAFT agent, which has been obtained during RAFT polymerization of $M_1$. Considerations regarding addition and fragmentation are similar to those during the homopolymerization when using a low-molecular-weight RAFT agent. After fragmentation, a macroradical based on $M_1$ is formed, while a macro-RAFT agent – consisting of solely $M_2$ repeating units – is generated. At the early stages of the polymerization the occurrences

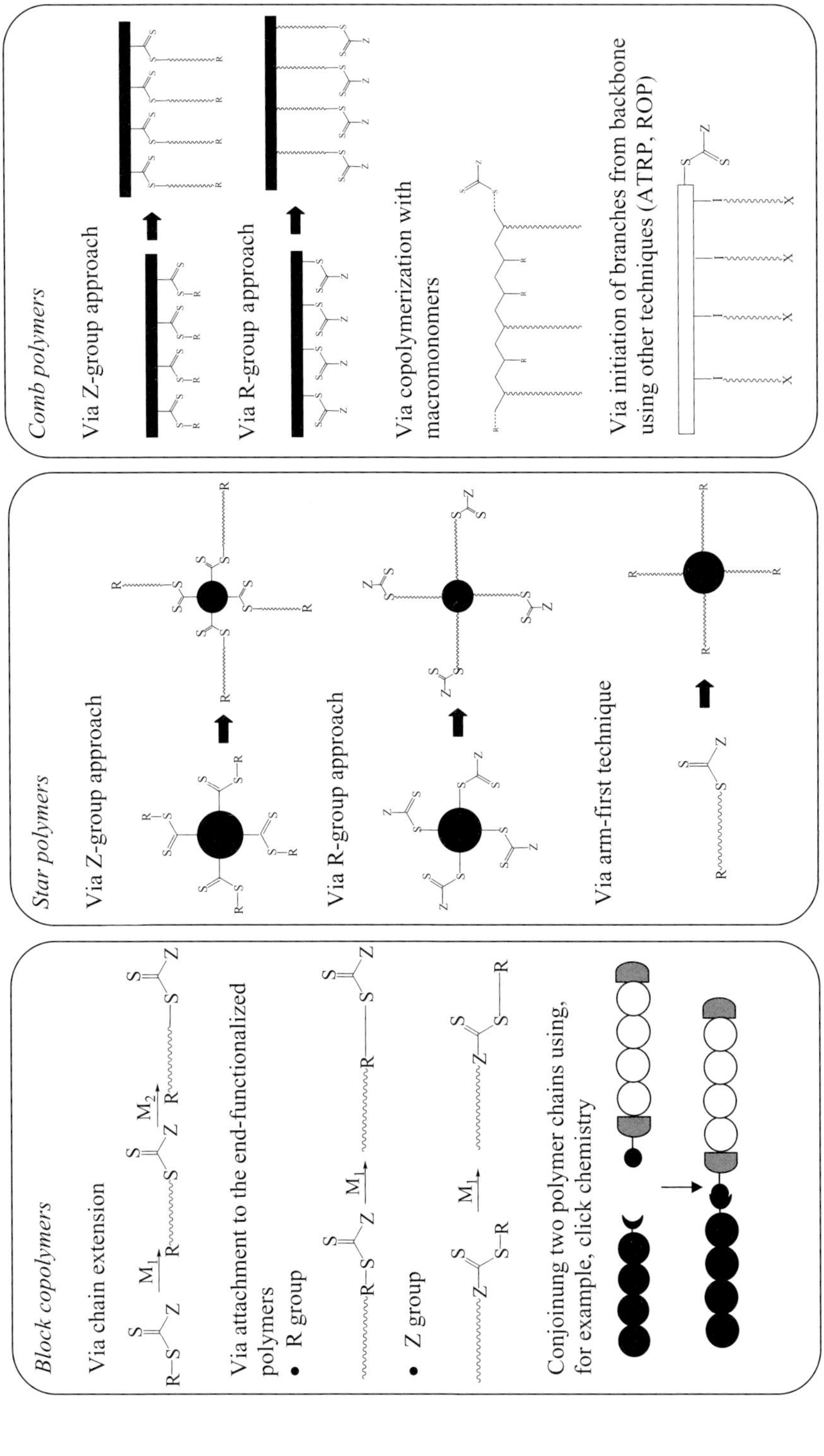


**Scheme 9.2** Selection of complex polymer architectures prepared via RAFT polymerization.

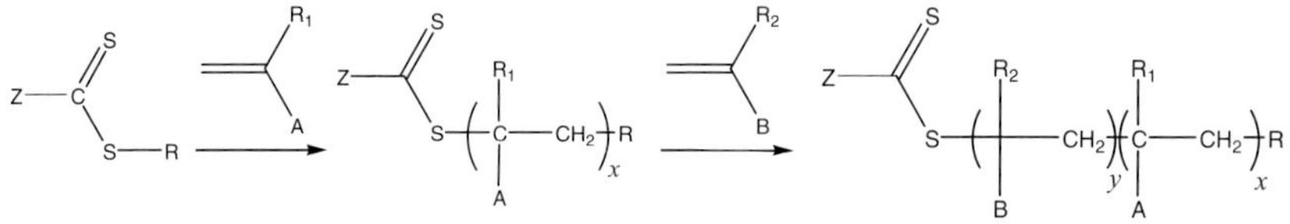

**Scheme 9.3** Synthesis of block copolymers via chain extension of a macro-RAFT agent.

of two different macro-RAFT agents can be expected (Scheme 9.4, pathway II). Block copolymers are formed only in the next step, when the $M_1$ macroradical reacts with monomer $M_2$ (Scheme 9.4, pathway III). The blocklike macroradical can then undergo chain transfer with two different types of macro-RAFT agents, as displayed in Scheme 9.4, pathway IV. However, reaction IV(a) will become less likely since all initial macro-RAFT agent should have reacted via Scheme 9.4, pathway II. The macroradical formed via reaction IV(a) will lead to further block structures (Scheme 9.4, pathway III), while the macroradical resulting from IV(b) can never contribute to the formation of block copolymer and will always remain as a homopolymer impurity in the system. Detailed inspection reveals that $P(M_2)$ is clearly a function of the radical concentration as introduced via an initiator. Similar to hompolymerization, termination reactions occur (Scheme 9.4, pathway V). Depending on the mode of termination, triblock copolymers can theoretically be expected.

Based on the theoretical mechanism, the development of the molecular weight has been modeled taking chain-transfer constants into account. Depending on the chain-transfer rate constant (the rate of addition to the macro-RAFT agent and the fragmentation of the intermediate radical) and the rate of propagation of the monomer, the evolution of the molecular-weight polydispersity can be predicted. Also the monomer concentration was found to play a crucial role in achieving well-defined products [1].

#### 9.2.1.2 **Practical Considerations and Experimental Results**

In principle, consideration regarding the block copolymer synthesis is similar to homopolymer synthesis. Prerequisites regarding the Z group and the leaving R group are equivalent. The design of the Z group should facilitate the controlled polymerization of both monomers. A requirement is therefore the identification of a common RAFT agent for both monomers regarding the Z-group design. The only difference is that the leaving group is a polymeric chain, which will form the first block of the block copolymer. The ability of the polymeric leaving group to fragment into a macroradical and restart the polymerization – here with the monomer forming the second block – is vital for the successful formation of block copolymers. Similar to low-molecular-weight compounds, not every leaving group can undergo fragmentation at the desired rate. If the second monomer forms a relatively stable radical – such as in methacrylates – it is advised to synthesize the macro-RAFT agent with a leaving group of higher or equivalent stability. Recommendations on preparation of the first blocks have been established already in the first publication

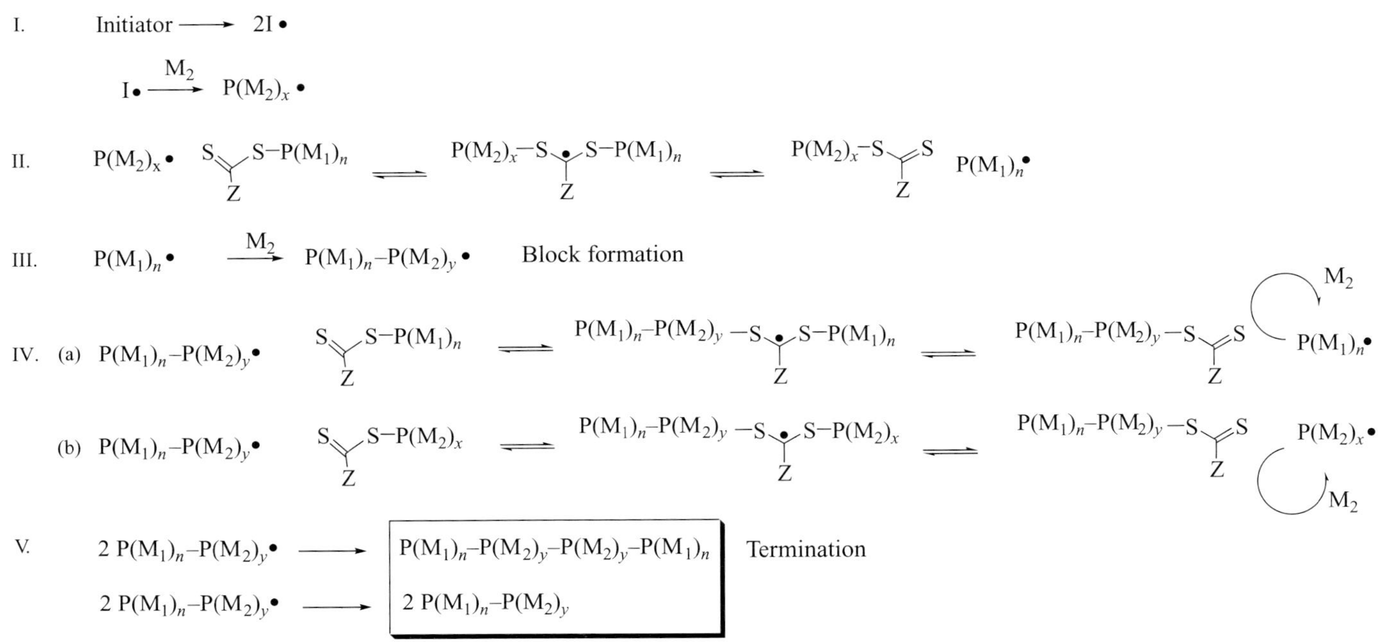


**Scheme 9.4** Formation of block copolymers via chain extension.

on block copolymers via RAFT [2]. Methacrylyl radicals have a greater leaving capability than styryl or acrylyl radicals. Looking at the chain equilibrium in Scheme 9.4, pathway II, the fragmentation toward $P(M_1)_n\cdot$ should be strongly favored, which is possible only when $P(M_1)_n\cdot$ has a better leaving ability. If fragmentation toward the starting material $P(M_2)_n\cdot$ is more pronounced, mainly homopolymer is produced. The general recommendation is therefore to prepare a methacrylate-type macro-RAFT agent first, followed by chain extension with styrene- or acrylate-type monomers.

These recommendations have been confirmed via experimental results. Methacrylate macro-RAFT agent could easily be chain extended via acrylates, styrene or acrylamide, while the opposite preparative strategy leads to broad distributions and unreacted macro-RAFT agent. While certain decision-making processes on the sequence of blocks are obvious and the estimation of the leaving ability of the macroradical is not intricate, other macroradical stabilities are established only during the actual experiment. Many block copolymers have been reported to allow synthesis in both ways. Butyl acrylate (BA) has successfully been polymerized in the presence of polystyrene (PS) macro-RAFT agents, but also the opposite pathway – using styrene to chain extend poly(butyl acrylate) – resulted in well-defined block copolymers [3]. Similar observations were made for other systems, such as styrene with 4-vinylbenzylchloride [4] or sodium-2-acrylamido-2-methylpropanesulfonate with sodium-3-acrylamido-3-methylbutanoate [5]. More common is, however, that a block copolymer can only be prepared according to a specific sequence. Even if the same monomer type is used for both blocks, i.e. acrylates, the side chain can influence the stability of the macroradical resulting in better defined block copolymers when the macro-RAFT agent is prepared from the more stable macroradical. Examples are (polystyrene) macro-RAFT with *N*-ethyl-3-vinylcarbazole [6], poly(acrylonitril) macro-RAFT with BA [7], poly(*N*-[3-(dimethylamino)propyl]methacrylamide) macro-RAFT with *N,N*-dimethylacrylamide (DMA) [8] or poly(*N*-*tert*-butyl acrylamide) macro-RAFT with *N*-acryloylmorpholine [9]. In certain cases it seems possible to use a more unusual sequence of block preparation. However, a broadening of the molecular-weight distribution is observed indicative of some remaining macro-RAFT agent. The block copolymer poly(*N*-isopropyl acrylamide)-*b*-poly($\gamma$-methacroyloxypropyltrimethoxysilane) had to be prepared employing an acrylamide macro-RAFT agent due to the insolubility of the homopolymer made from $\gamma$-methacroyloxypropyltrimethoxysilane [10]. Other examples include methyl methacrylate/*N*-isopropyl acrylamide (NIPAAm) [11] and methacrylic acid/NIPAAm [12], which were synthesized using either block first. Another thermoresponsive zwitterionic block copolymer was successfully prepared by adding 3-[*N*-(3-methacrylamidopropyl)-*N,N*-dimethyl]ammoniumpropane sulfonate onto poly(*N*-isopropyl acrylamide) (PNIPAAm) macro-RAFT agent [13].

The molecular-weight distribution of the block copolymer can only be as good as the initial first block, the so-called macro-RAFT agent. A significant effort in the synthesis of block copolymers should focus on the preparation of well-defined macro-RAFT agents. Consideration regarding the choice of RAFT agent for the first block as well as appropriate concentrations of monomer and initiator are prerequisite. High initiator concentrations during the macro-RAFT agent synthesis may result in a significant amount of termination products. These dead polymer chains cannot undergo chain extension, resulting in the formation of homopolymer impurity in the final product. A small amount of termination products can never be avoided, but under careful optimization be minimized [14]. The amount of termination products will increase during the course of the polymerization, meaning that more dead polymer is formed when the monomer conversion is high. In order to obtain a better defined block copolymer with less remaining homopolymer, the synthesis of the macro-RAFT agent via a low conversion process may be advantageous. This has been demonstrated in the synthesis of poly(6-*O*-methacryloyl-a-D-glucoside)-*b*-poly(2-hydroxyethyl methacrylate). The macro-RAFT agent based on poly(6-*O*-methacryloyl-α-D-glucoside) was isolated after having been polymerized to a conversion of 98%. During the subsequent chain extension some initial polymer was not initiated, which has been assigned to termination events during the macro-RAFT synthesis [15]. Despite careful optimization of the reaction conditions, some systems will always result in the formation of better defined macro-RAFT agents (meaning less terminated polymers) than other polymers. Very slowly propagating monomers such as hindered monomers will naturally result in a higher fraction of polymer without thiocarbonylthio end groups, which cannot be chain extended and will contaminate the final block copolymer with homopolymer [16]. In addition, it has to be considered that some polymers such as hindered monomers will undergo a large fraction of transfer to monomer, resulting in polymers without thiocarbonylthio end groups.

An important issue is the stability of the thiocarbonylthio end group. While the macro-RAFT agent may have been prepared in an optimized manner from a kinetic point of view, the destruction of the RAFT end group can result in additional dead polymer, which cannot be chain extended. Several investigations demonstrate the loss of the RAFT activity under the influence of heat [17], light [18], certain pH values [19] and solvents that are known to contain oxidizing species such as dioxane or tetrahydrofuran (THF) [20]. The influences can potentially lead to the destruction of the thiocarbonylthio end group during the polymerization, during storage or during purification via precipitation or dialysis.

Care should be taken during the synthesis in aqueous solution since the RAFT end group is especially prone to hydrolysis. It has been demonstrated that depending on the pH value of the solution the hydrolysis can be accelerated. The hydrolysis was significantly delayed or suppressed below pH = 7 [21]. The effect of the pH value on the chain extension has been demonstrated, comparing the block copolymerization carried out at pH = 5 and pH = 7. Excellent control over the molecular-weight evolution and distribution up to high conversions was observed at low pH values [8]. It was also confirmed that hydrolysis not only occurs during polymerization, but also during purification. For example: isolation of the macro-RAFT agent by dialysis against nonbuffered water resulted in the loss of thiocarbonylthio end groups [8].

Thiocarbonylthio end groups are not only sensitive against hydrolysis via water. Aminolysis, in addition, can cause similar devastating effects and will as well result in the cleavage of active RAFT groups [21]. The possibility of aminolysis has to be considered when polymerizing monomers with alkaline amino functionalities. A strong color change – typically from red/pink to orange or yellow – can indicate the occurring aminolysis reaction [22]. Several possible pathways can suppress or prevent this side reaction, such as solvent adjustments or the recourse to protective chemistry. Alternatively, the increase of the rate of polymerization by decreasing the amount of solvent can lead to faster propagation, thus restraining the competing aminolysis process [22].

Peroxides can be responsible for the destruction of the RAFT end group, especially in solvents, which are known to generate peroxides such as dioxane and THF. A RAFT-generated polymer left in a THF solution for certain period of time will eventually loose its color due to the exchange of a sulfur atom within the thiocarbonylthio end group by oxygen resulting in an inactive end group [20]. Indeed, block copolymerizations carried out in dioxane report frequently on incomplete chain extension as evidenced by residual macro-RAFT agent observed during size-exclusion chromatography (SEC) measurements [23, 24].

It can therefore be concluded that there are several obstacles in the way of the synthesis of a well-defined macro-RAFT agent. Termination events or the destruction of the thiocarbonylthio end group can prevent the complete chain extension, leading to a mixture of homopolymer based on the first block and the desired block copolymer. It is certainly recommended to optimize reaction conditions and quantify the amount of dead polymer prior to chain extension, using either mass spectroscopy [20], NMR or UV analysis [25]. A small amount of residual dead polymer from the macro-RAFT agent synthesis was observed in many block copolymers' synthesis attempts. Depending on the system, dead polymer is evidenced either as a distinguishable second peak or as low-molecular-weight tailing in the SEC diagram [26–28]. Small amount of residual polymer is in many cases not immediately evident upon inspection of SEC curves. Alternative displays of SEC curves taking concentration gradients into account can reveal incomplete chain extension, hence confirming the amount of inactive or dead homopolymer.

Figure 9.1 displays the conventional way of presenting SEC curves compared to concentration-modified SEC curves. The alternative way takes into account that with increasing conversion of the second block the intensity of the homopolymer is naturally suppressed [29]. This way of presenting SEC data could frequently confirm the occurrence of homopolymer next to block copolymer [15, 30–33].

Finally, the block copolymer synthesis can be carried out after considering the choice of the first block and the careful synthesis of the macro-RAFT agent. The block copolymer synthesis can be challenged by the suitable choice of solvent, especially in the synthesis of amphiphilic block copolymers. To circumvent this problem, some block copolymers were obtained using protective chemistry or using a precursor of the side group. This alternative pathway opens up a range of solvents, which may facilitate the synthesis by reducing problems caused by inhomogeneities or by chain transfer to certain solvents. Acetal protected acrylic acid

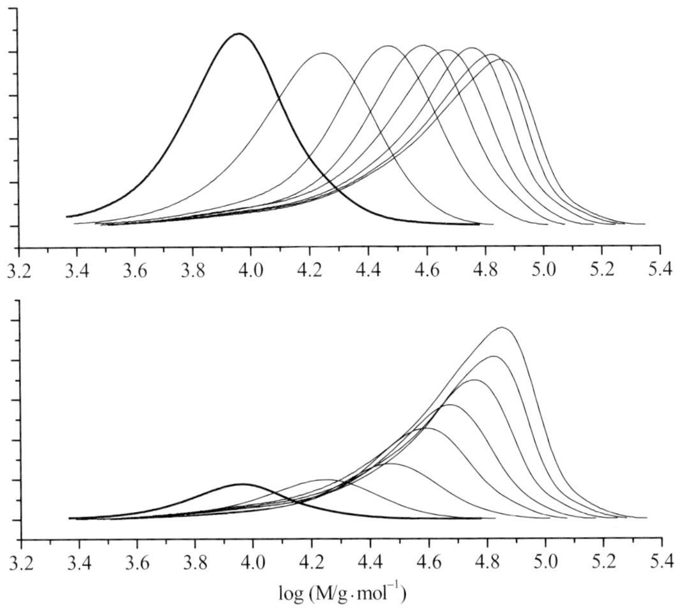


**Fig. 9.1** Size-exclusion chromatography (SEC) traces of polystyrene-*b*-poly(*N*,*N*-dimethyl methacrylate) (PS-*b*-PDMA) block copolymers prepared from polystyrene macro-RAFT agent ($M_n$ = 8000 g·mol$^{-1}$, RAFT agent is 3-benzylsulfanylthio-carbonylsufanylpropionic acid) with increasing PDMA block size using the normalized response (top) and obtained by multiplying the normalized response with the conversion and the maximum ratio at 100% conversion of ($N_{PDMA}$ + $N_{PS}$) and $N_{PS}$ (here = 7.5) (bottom) [32].

allowed the synthesis of poly(methyl methacrylate)-*b*-poly(acrylic acid) in solvents such as toluene [34], while amphiphilic block copolymers based on acrylic acid such as polystyrene-*b*-polyacrylic acid are typically polymerized in polar solvents such as *N*,*N*-dimethyl acetamide (DMAc) [35]. A cationic block copolymer was prepared by quaternizing polystyrene-*b*-poly(4-vinylpyridine) after polymerization to avoid the search for a suitable solvent for PS and a charged monomer [36]. A similar attempt using postquaternization was described elsewhere [4]. In another example, the chain extension of poly(4-acetoxystyrene) macro-RAFT agent with styrene could be carried out in bulk to afford poly(4-hydroxystyrene)-*b*-polystyrene after hydrolysis [37]. Even if no suitable solvent can be found that dissolves both blocks, the RAFT process has been proven to be highly robust even under heterogeneous conditions. The search for a common solvent for the synthesis of block copolymers based on an ionic block and a hydrophobic block can be impossible. However, suspension- or emulsionlike systems can still successfully be employed to prepare these structures. In the initial state of the polymerization, the macro-RAFT agent is insoluble in the solvent employed. With the onset of the chain extension and the formation of block copolymer a homogenous solution is formed [38, 39].

A successful chain extension leading to block copolymers is confirmed by the shift of the SEC curve to higher conversions, while the molecular-weight distribution remains narrow. In an ideal case the molecular-weight evolution increases linear with conversion and can be predicted using the ratio between the (second) monomer concentration and the macro-RAFT agent concentration and the monomer conversion:

$$M_{\text{block copolymer}} = \frac{[M]}{[\text{macro-RAFT}]} \times \text{conversion} \times M_{\text{M}} + M_{\text{macro-RAFT}}$$

where $[M]$ and [macro-RAFT] are the monomer and macro-RAFT agent concentrations, and $M_{\text{M}}$ and $M_{\text{macro-RAFT}}$ are the molecular weights of the monomer of the second block and the macro-RAFT agent, respectively.

This linear relationship can indeed be found in most cases despite reported difficulties in determining the real molecular weight of block copolymers using SEC. However, a slight broadening of the molecular-weight distribution – usually in the form of low-molecular-weight tailing – can frequently be observed. This can be assigned to the above-discussed impurities of dead polymer found in the macro-RAFT agent. However, the mechanism of the block copolymer should be inspected upon in detail again at this point (Scheme 9.4).

Mechanistic considerations valid for the macro-RAFT agent synthesis are also true for the block copolymer synthesis. As outlined above and displayed in Scheme 9.4, the amount of radicals to start the block copolymer formation will determine the amount of termination events as well as the amount of homopolymer (generated from the second monomer). It is therefore recommended to keep the amount of radicals as small as possible. Typically, the broadening of the molecular-weight distribution was observed with increasing initiator concentration. The formation of dead chains also led to the loss of control over the molecular-weight evolution [4]. Similar to homopolymerization, a high ratio of macro-RAFT agent to initiator concentration can suppress these unwanted side reactions. The amount of dead chains can be estimated taking the rate of initiator decomposition as well as the concentration of macro-RAFT agent and initiator into account [9]:

$$\%_{\text{dead chains}} = \frac{2f[I]_0 \times (1 - e^{-k_{\text{d}}t})}{[\text{macro-RAFT}] + 2f[I]_0 \times (1 - e^{-k_{\text{d}}t})}$$

where $f$ is the initiator efficiency, $k_{\text{d}}$ the initiator decomposition rate coefficient and $[I]$ and [macro-RAFT] are the concentrations of initiator and macro-RAFT agent, respectively.

In fact, depending on reaction conditions, dead polymer can be found and removed from the block copolymer by purification. In many cases, homopolymers have significantly different solubility properties than the corresponding block copolymers. Careful precipitation allows the removal of homopolymer generated during the chain extension. The SEC curves before and after precipitation confirm the formation of homopolymer [13, 27]. Impurities such as homopolymers will result in deviations from the calculated molecular weight. To accommodate the amount of dead polymer formed, the molecular-weight calculation has been modified to take the amount of polymer chains into account, which have been initiated by initiator only and not by the polymeric leaving group of the macro-RAFT agent. However, it should be considered that this equation does not take termination events into account, which can result in broadening or the formation of distinct high-molecular-weight shoulders:

$$M_{\text{block copolymer}} = \frac{[M]}{[\text{macro} - \text{RAFT}] + 2f[I]_0 \times (1 - e^{-k_d t})} \times \text{conversion} \times M_M + M_{\text{macro-RAFT}}$$

Despite theoretical considerations regarding the optimization of all concentrations including radical concentrations, the process can in reality result in additional complications, which are derived by solvent and diffusion effects. It has been found that the length of the macro-RAFT agent can have an effect on the block copolymer formation, as observed during the block copolymer synthesis of polystyrene-*b*-poly(*N*,*N*-dimethacrylamide) [32]. While a low-molecular-weight macro-RAFT agent resulted in complete chain extension with the complete disappearance of the macro-RAFT agent, as seen in SEC, a high-molecular-weight macro-RAFT agent can lead to bimodal distributions (Fig. 9.2). Radical concentrations were kept constant in both cases and can therefore not be responsible for the broadening. Solvent effects or potentially chain-length-dependent addition rate coefficients to the RAFT end group can be responsible for this observation. More investigations are necessary in this area to identify the origin of this behavior.

Worth mentioning is the effect of the macro-RAFT agent on the rate of polymerization. Similar to homopolymer synthesis increasing amounts of macro-RAFT

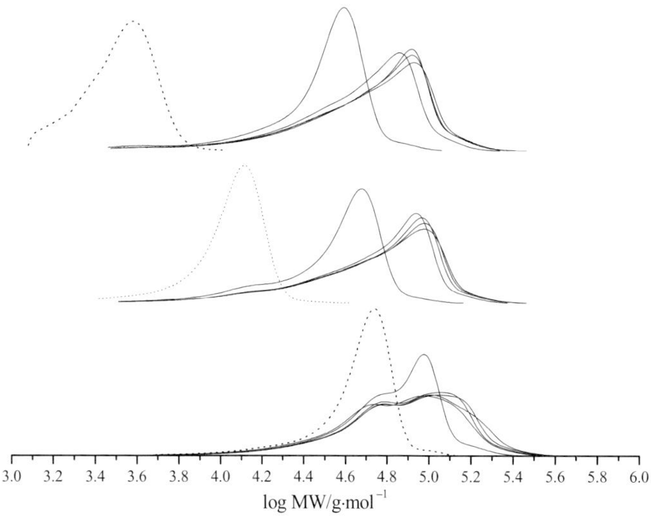


**Fig. 9.2** SEC curves (DMAc) of PS-*b*-PDMA block copolymers using polystyrene macro-RAFT agent with molecular weights 3000 g·mol$^{-1}$ (top), 9000 g·mol$^{-1}$ (middle) and 38 000 g·mol$^{-1}$ (bottom); [macro-RAFT] = $4 \times 10^{-3}$ mol·L$^{-1}$, [*N*,*N*-dimethylacrylamide] = 2 mol·L$^{-1}$, [AIBN] = $4 \times 10^{-4}$ mol·L$^{-1}$ in *N*,*N*-dimethylacetamide at 60 °C [32].

agent can introduce retardation decreasing the rate of polymerization. In addition, similar macro-RAFT agent concentrations but with varying molecular weight can influence the rate of polymerization significantly. A significant decrease of the rate of polymerization with increasing chain length was observed [23, 32, 35]. The pseudo-first-order kinetic block was, however, always found to be linear, indicating a constant radical concentration. Retardation effects caused by the equilibrium of the RAFT agent cannot be made responsible, since the nature of the leaving group is similar. The cause must lie in the length of the leaving group (=first block) resulting in a range of solvent effects. Other observations regarding the rate of polymerization include a significant deviation from the first-order kinetic plot. Polymerizations were found to slow down, without having consumed all available initiator [40–42].

The synthesis of block copolymers via chain extension using the RAFT process is subject to a range of possible side reactions. However, careful optimization allows the synthesis of well-defined block copolymers with polydispersity indices of 1.1. A range of block copolymers were successfully synthesized using the RAFT process. Structures include pH- and thermoresponsive sequences, glycopolymers, ionic blocks including cationic, anionic and zwitterionic blocks and many more (Table 9.1). Only one limitation in the synthesis of block copolymer should be highlighted. Only block copolymer with monomer having comparable reactivities can be prepared. Monomers with vastly disparate monomer reactivities, for example, styrene and vinyl acetate, cannot be copolymerized, since both monomers require significantly different Z groups.

RAFT agents and monomers used to prepare a macro-RAFT agent and synthesize the second block as well as investigated block copolymer systems are summarized in Table 9.1.

### 9.2.2 Triblock Copolymers via Chain Extension of Macro-RAFT Agent

Triblock copolymers are prepared either by further chain extension of a diblock copolymer (I), by utilization of a difunctional RAFT agent (II) or by employing a trithiocarbonate RAFT agent with two leaving groups (III) (Scheme 9.5).

The implications regarding the synthesis via pathway (I) (Scheme 9.5) are similar to the generation of diblock copolymers. Termination reactions, however, do now result in the formation of A–B–C–B–A pentablock copolymers. A typical A–B–C–triblock was reported employing NIPAAm, *N*-acryloylalanine and DMA as building blocks [63]. In a similar approach, a range of stimuli-responsive charged A–B–C triblocks were obtained [46]. This pathway was also employed to prepare pH- and stimuli-responsive A–B–A triblock copolymers [12].

Pathway II requires the linkage of two RAFT agents, which can be attached to a bifunctional linker via either the R group [28, 70] or the Z group. The mechanism and side reactions arising during the polymerization are similar to the synthesis of star polymers via RAFT and are discussed in the appropriate section below.

**Table 9.1** Examples of block copolymers including RAFT agent employed to generate the macro-RAFT agent.

**RAFT agents**

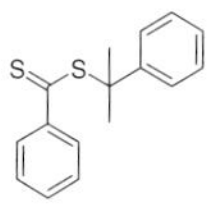
$R_1$

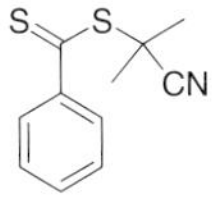
$R_2$

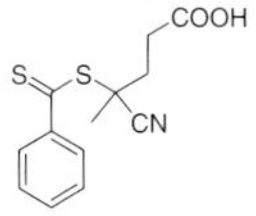
$R_3$

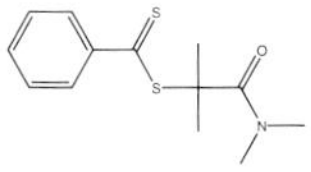
$R_4$

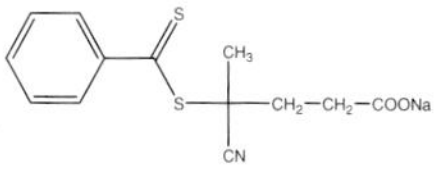
$R_5$

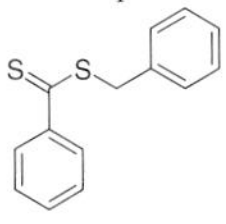
$R_6$

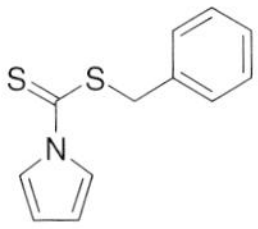
$R_7$

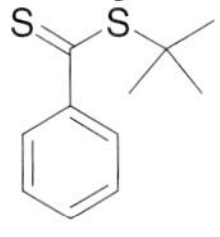
$R_8$

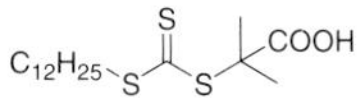
$R_9$

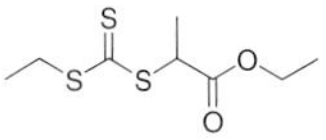
$R_{10}$

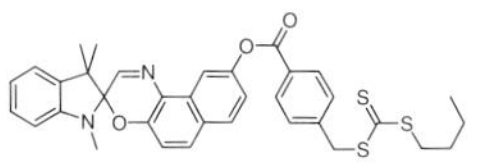
$R_{11}$

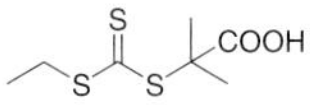
$R_{12}$

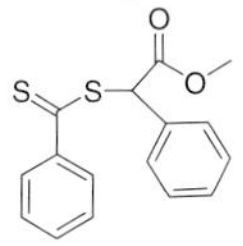
$R_{13}$

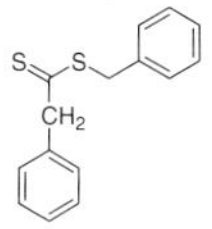
$R_{14}$

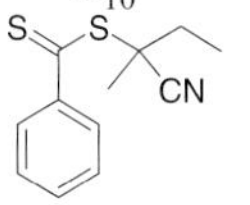
$R_{15}$

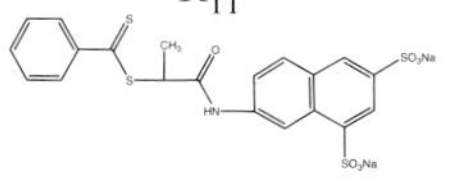
$R_{16}$

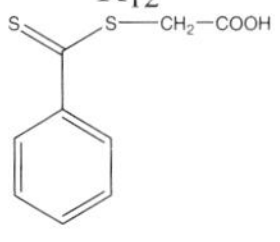
$R_{17}$

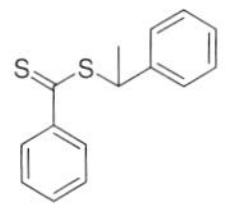
$R_{18}$

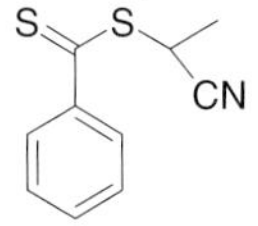
$R_{19}$

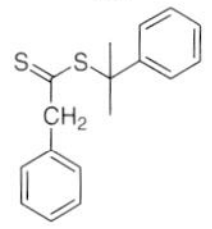
$R_{20}$

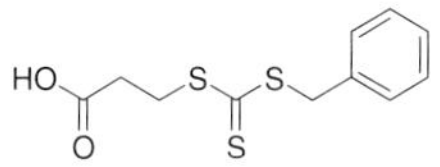
$R_{21}$

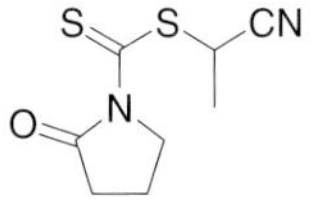
$R_{22}$

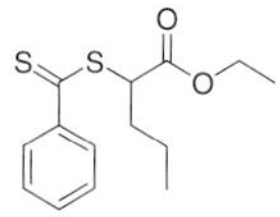
$R_{23}$

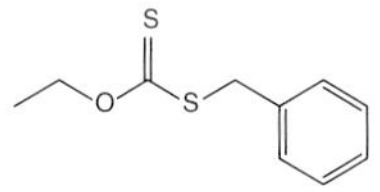
$R_{24}$

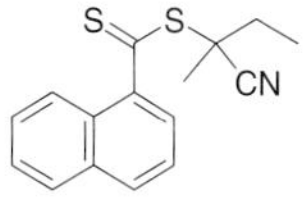
$R_{25}$

**Neutral and stimuli-responsive monomers**

The numbers in bracket indicate the table entry:

AA (**13, 43, 56**), AN (**12, 18**), BA (**1, 8, 9, 10, 11, 13, 18, 23, 34, 38, 39**), BMA (**1**), BzMA (**13, 50**), DEA (**4**), DEAEMA (**57, 58**), DMA (**9, 11, 13, 16, 24, 32, 35, 46, 49, 50, 54**), DMAEMA (**9, 12, 13, 42, 62**), DMAPMA (**63**), GMA (**12, 26**), HEA (**23, 52**), HEMA (**21**), Isoprene (**7**), MAA (**13, 55**), MAn (**12, 15, 19**), MMA (**1, 4, 10, 12, 13, 48**), NAM (**4, 6**), NAP (**9, 11, 47**), NIPAAm (**1, 45–57, 61, 64**), NVP (**5, 25**), PEGA (**9, 35, 37**), PEGMA (**11, 26**), STY (**2, 7, 8, 12, 13, 14, 15, 17, 19, 20, 22, 24, 33, 41, 43, 44, 51, 52**), t-BA (**7**), t-BAm (**6**), VAc (**25**)

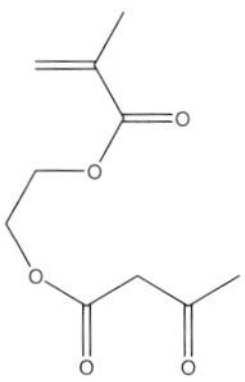
$M_1$ (**1**)

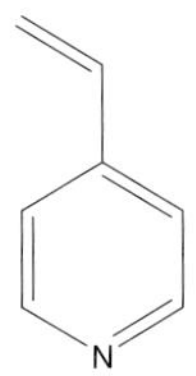
$M_2$ (**41, 44**)

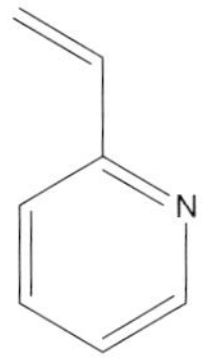
$M_3$ (**5**)

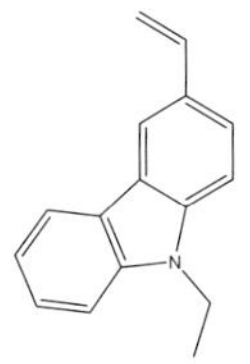
$M_4$ (**2**)

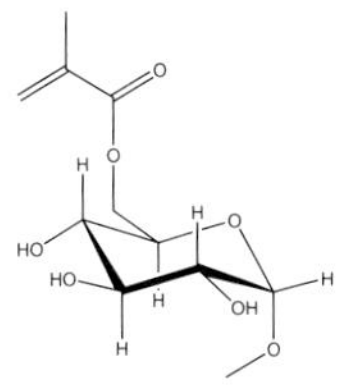
$M_5$ (**3**)

(*Continued*)

Table 9.1 (*Continued*)

| | | | | |
|---|---|---|---|---|
| 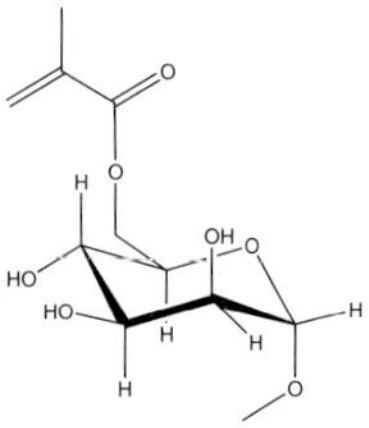 | 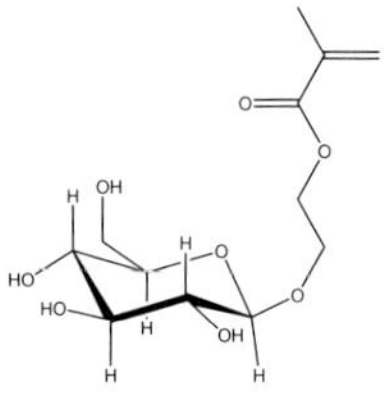 | 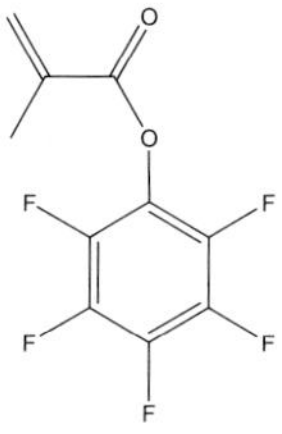 | 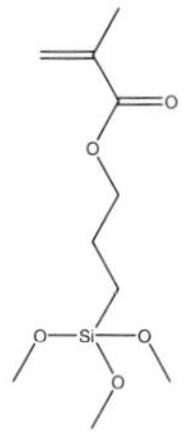 | 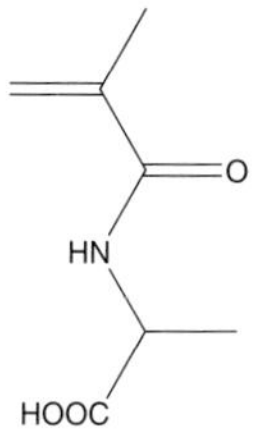 |
| $M_{6}$ (**3**) | $M_{7}$ (**3**) | $M_{8}$ (**4**) | $M_{9}$ (**45**) | $M_{10}$ (**54**) |
| 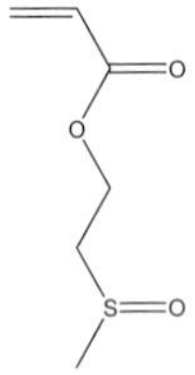 | 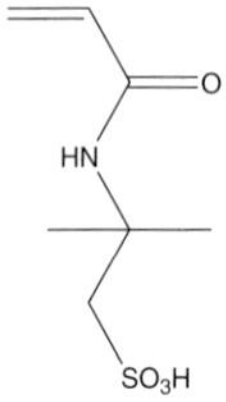 | 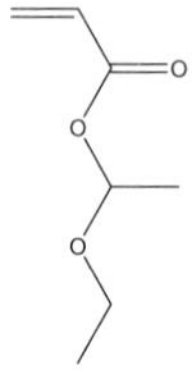 | 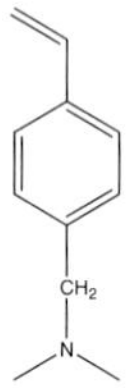 | 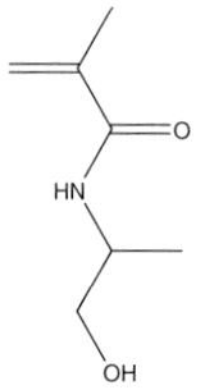 |
| $M_{11}$ (**9**) | $M_{12}$ (**9**) | $M_{13}$ (**10**) | $M_{14}$ (**59**) | $M_{15}$ (**36, 40**) |
| 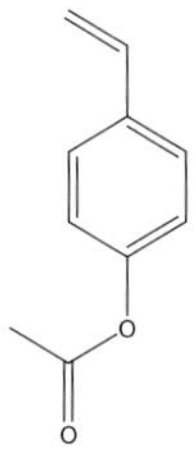 | 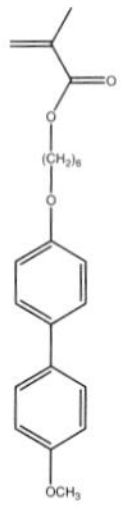 | 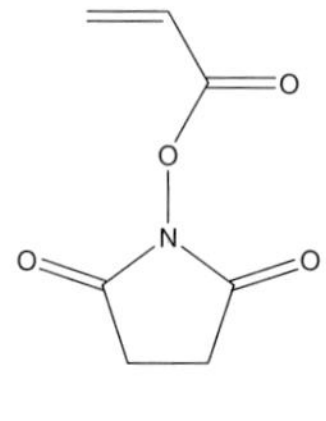 | 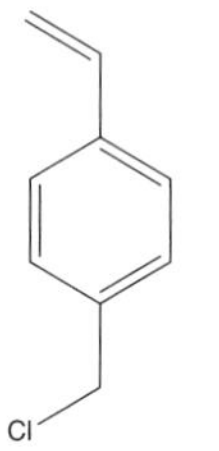 | 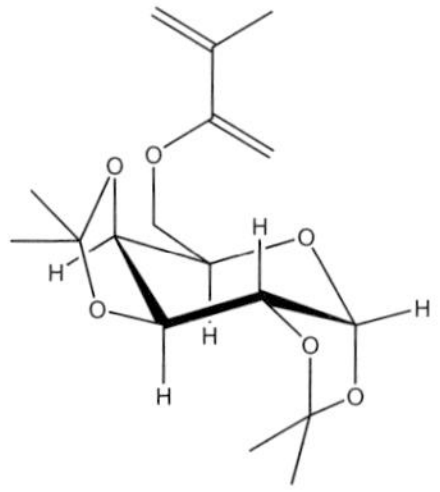 |
| $M_{16}$ (**14**) | $M_{17}$ (**15**) | $M_{18}$ (**16**) | $M_{19}$ (**17**) | $M_{20}$ (**42**) |
| 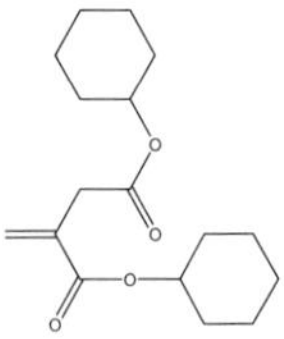 | 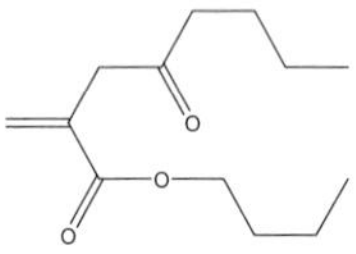 | 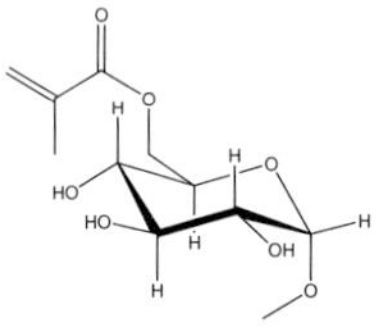 | 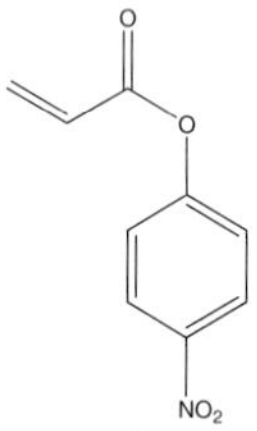 | 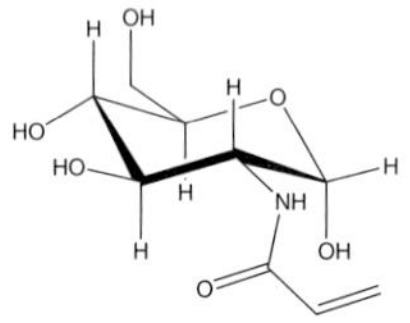 |
| $M_{21}$ (**20**) | $M_{22}$ (**20**) | $M_{23}$ (**21**) | $M_{24}$ (**22**) | $M_{25}$ (**43**) |

**Table 9.1** (*Continued*)

| Ionic monomers | | | | |
|---|---|---|---|---|
| 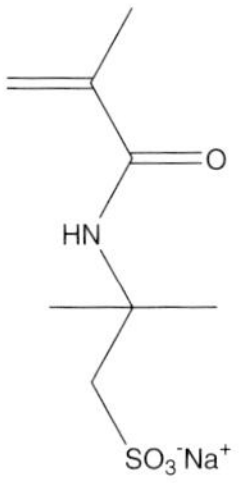 | 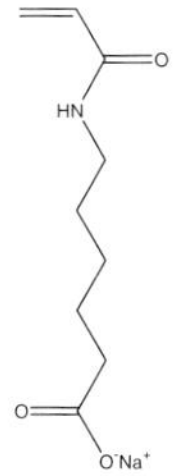 | 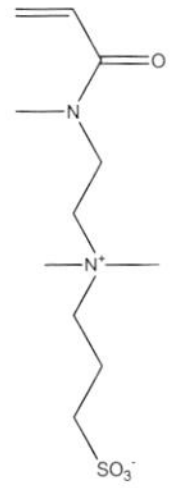 | 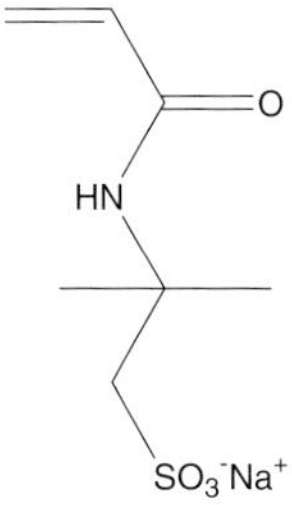 | 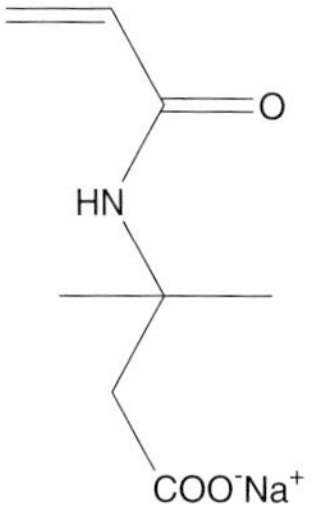 |
| $CM_1$ (**27**) | $CM_2$ (**27**) | $CM_3$ (**28, 32**) | $CM_4$ (**31**) | $CM_5$ (**28, 31**) |
| 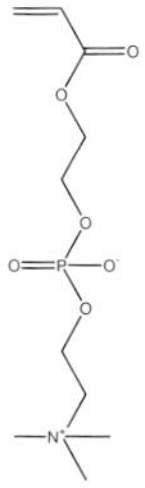 | 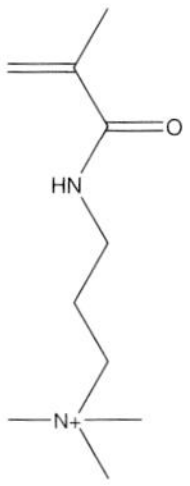 | 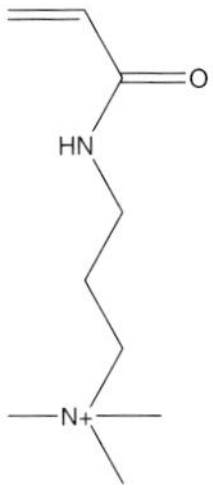 | 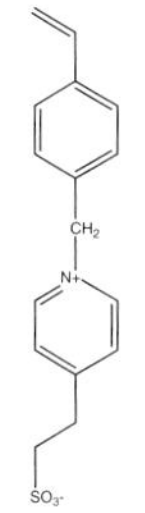 | 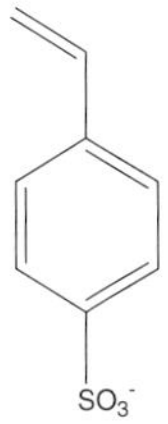 |
| $CM_6$ (**33, 39**) | $CM_7$ (**36, 40, 58, 60**) | $CM_8$ (**34**) | $CM_9$ (**29, 35**) | $CM_{10}$ (**29, 30**) |
| 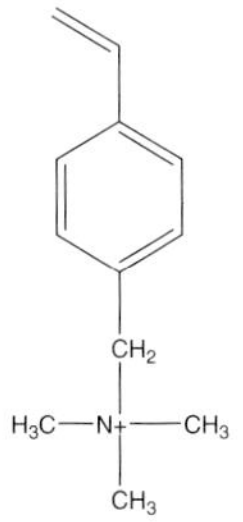 | 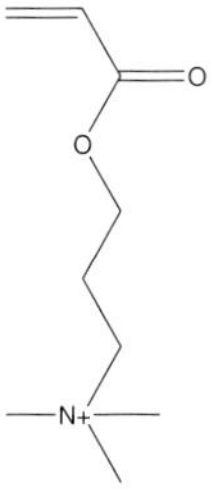 | 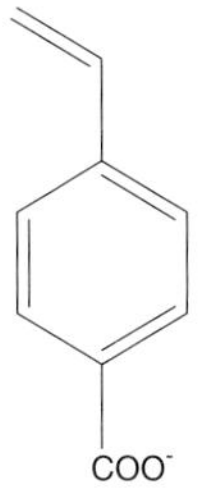 | 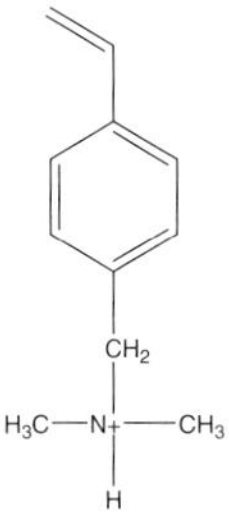 | 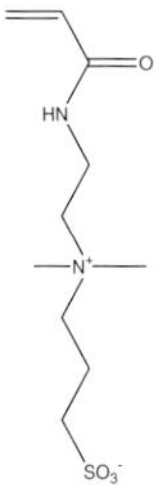 |
| $CM_{11}$ (**30, 59, 63**) | $CM_{12}$ (**37**) | $CM_{13}$ (**30**) | $CM_{14}$ (**30**) | $CM_{15}$ |
| 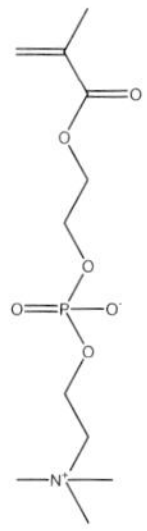 | 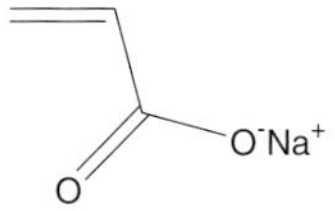 | 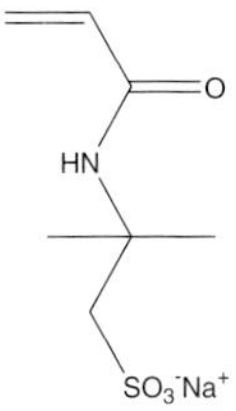 | | |
| $CM_{16}$ (**38**) | $CM_{17}$ (**62**) | $CM_{18}$ (**64**) | | |

(*Continued*)

**Table 9.1** (*Continued*)

| Entry | First block | Second block | RAFT agent | T (°C) | Solvent | Ref. |
|---|---|---|---|---|---|---|
| | | **Neutral block copolymers** | | | | |
| 1 | $M_1$ | MMA | $R_1$ | 60 | Ethyl acetate | [26] |
| | | BMA | $R_2$ | | | |
| | | BA | | | | |
| | | NIPAAm | | | | |
| 2 | $M_4$ | STY | $M_7$ | 60 | Toluene | [6] |
| | STY | $M_4$ | | | | |
| 3 | $M_5$ | $M_6$ | $R_3$ | 70 | Water/ethanol | [30] |
| | $M_7$ | $M_7$ | | | | |
| 4 | $M_8$ | MMA | $R_1$ | 90 | Dioxane | [43] |
| | | NAM | $R_3$ | | | |
| | | DEA | | | | |
| 5 | NVP | $M_3$ | $R_9$ | 75 | DMF | [44] |
| 6 | *t*-BAm | NAM | $R_8$ | 90 | Dioxane | [9] |
| | NAM | *t*-BAm | | | | |
| 7 | STY | Isoprene | $R_{10}$ | 115 | Bulk | [45] |
| | *t*-BA | | | | | |
| 8 | BA | STY | $R_{11}$ | 110 | Benzene | [3] |
| | STY | BA | | 60 | | |
| 9 | BA | NAP | $R_{14}$ | 66 | Dioxane | [38] |
| | | DMA | | | Dioxane | |
| | | $M_{11}$ | | | DMF | |
| | | PEGA | | | THF | |
| | | $M_{12}$ | | | NMP | |
| 10 | MA | $M_{13}$ | $R_{15}$ | 70 | Toluene | [34] |
| | BA | | | | | |
| | MMA | | | | | |
| | DMAEMA | | | | | |
| 11 | BA | NAP | $R_{14}$ | 66 | Dioxane | [46] |
| | PEGMA | BA | $R_3$ | 55 | Water | |
| | NAP | DMA | $R_1$ | 70 | Toluene | |
| 12 | MMA | STY/AN | $R_1$ | 60 | Benzene | [31] |
| | GMA | | | | Benzene | |
| | DMAEMA | | | | Dioxane/MeOH | |
| | STY/MAn | | | | Dioxane | |
| 13 | STY | DMA | $R_1$ | | Benzene | [2] |
| | STY | MeSTY | $R_3$ | | Benzene | |
| | BA | AA | $R_6$ | | DMF | |
| | MA | EA | $R_{18}$ | | Benzene | |
| | MMA | MAA | | | DMF | |
| | MMA | STY | | | Bulk | |
| | BzMA | DMAEMA | | | EtOAc | |
| | BzMA | MAA | | | DMF | |
| 14 | $M_{16}$ | STY | $R_{17}$ | 90 | Bulk | [37] |
| 15 | $M_{17}$ | STY/MAn | $R_3$ | 60 | Dioxane | [23] |
| 16 | DMA | $M_{18}$ | $R_8$ | 90 | Dioxane | [24] |
| 17 | STY | $M_{19}$ | $R_6$ | 90 | Bulk | [4] |
| | $M_{19}$ | STY | | | | |

**Table 9.1** (*Continued*)

| Entry | First block | Second block | RAFT agent | T (°C) | Solvent | Ref. |
|---|---|---|---|---|---|---|
| 18 | AN | BA | $R_{19}$ | 60 | Ethylene carbonate | [7] |
| | BA | AN | $R_1$ | | | |
| 19 | STY/MAn | STY | $R_9$ | 60 | Dioxane | [47] |
| 20 | $M_{21}$ | STY | $R_1$ | 65 | Bulk | [16] |
| | $M_{22}$ | | $R_{20}$ | | | |
| 21 | $M_{23}$ | HEMA | $R_3$ | 60 | Water/ethanol | [15] |
| 22 | $M_{24}$ | STY | $R_{23}$ | 70 | DMSO | [27] |
| 23 | BA | HEA | $R_{21}$ | 60 | DMAc | [48] |
| 24 | STY | DMA | $R_{21}$ | 60 | DMAc | [32] |
| 25 | NVP | VAc | $R_{24}$ | 60 | MeOH | [49] |
| 26 | GMA | PEGMA | $R_{25}$ | 85 | DMF | [50] |
| | | **Block copolymers with two ionic blocks** | | | | |
| 27 | $CM_1$ | $CM_2$ | $R_3$ | 70 | Water | [40] |
| 28 | $CM_3$ | $CM_5$ | $R_5$ | 70 | Water | [51] |
| 29 | $CM_{10}$ | $CM_9$ | $R_{16}$ | 55 | Water | [46] |
| 30 | $CM_{10}$ | $CM_{13}$ | $R_3$ | 70 | Water | [52] |
| | $CM_{11}$ | $CM_{14}$ | | | | |
| 31 | $CM_4$ | $CM_5$ | $R_3$ | 70 | Water | [5] |
| | | **Block copolymers with one ionic block** | | | | |
| 32 | DMA | $CM_3$ | $R_4$ | 70 | Water | [28] |
| | | | $R_5$ | | | |
| 33 | STY | $CM_6$ | $R_6$ | 60 | DMF/methanol | [53] |
| 34 | BA | $CM_8$ | $R_{14}$ | 66 | NMP | [38] |
| | | | | | DMF | |
| 35 | PEGA | $CM_9$ | $R_{16}$ | 55 | Water | [46] |
| | DMA | $CM_9$ | $R_3$ | 53 | Water | |
| 36 | $M_{15}$ | $CM_7$ | $R_3$ | 70 | Water | [54] |
| | $CM_7$ | $M_{15}$ | | | | |
| 37 | $CM_{12}$ | PEGA | $R_{16}$ | 55 | Water | [25] |
| 38 | $CM_{16}$ | BA | $R_3$ | 70 | Methanol | [41] |
| 39 | BA | $CM_6$ | $R_{21}$ | 60 | Methanol/NMP | [39] |
| | | **pH-responsive block copolymers** | | | | |
| 40 | $CM_7$ | $M_{15}$ | $R_1$ | 60 | Bulk | [55] |
| | $M_{15}$ | $CM_7$ | | | DMF | |
| 41 | STY | $M_2$ | $R_6$ | 80 | DMF | [36] |
| 42 | $M_{20}$ | DMAEMA | $R_1$ | 60 | DMF | [56] |
| | | | $R_2$ | | | |
| 43 | STY | AA | $R_{21}$ | 60 | DMAc | [35] |
| 44 | STY | $M_2$ | $R_{21}$ | 60 | DMAc | [22] |
| | | | $R_6$ | | | |
| | | **Thermoresponsive block copolymers** | | | | |
| 45 | NIPAAm | $M_9$ | $R_6$ | 80 | Dioxane | [10] |
| 46 | NIPAAm | DMA | $R_{13}$ | 60 | Dioxane | [57] |
| 47 | NIPAAm | NAP | $R_1$ | 70 | Toluene | [46] |

(*Continued*)

**Table 9.1** (*Continued*)

| Entry | First block | Second block | RAFT agent | T (°C) | Solvent | Ref. |
|---|---|---|---|---|---|---|
| 48 | MMA | NIPAAm | $R_8$ | 70 | Toluene | [11] |
| | NIPAAm | MMA | $R_{13}$ | | | |
| 49 | DMA | NIPAAm | $R_{12}$ | 25 | Water | [58] |
| 50 | BzMA | NIPAAm/DMA | $R_2$ | 70 | Benzene | [59] |
| 51 | NIPAAm | STY | $R_3$ | 60 | Dioxane | [60] |
| | *t*-BMA | NIPAAm | | | | |
| 52 | NIPAAm | STY | $R_{21}$ | 75 | THF | [61] |
| | | HEA | | | | |
| 53 | $M_{25}$ | NIPAAm | $R_{21}$ | 60 | Water/DMSO | [62] |
| | | **Thermo- and pH-responsive block copolymers** | | | | |
| 54 | DMA | NIPAAm | $R_3$ | 25 | Water, pH = 4.8 | [63] |
| | | $M_{10}$ | $R_{12}$ | 70 | Water, pH = 6.5 | |
| 55 | NIPAAm | MAA | $R_{17}$ | 60 | Methanol | [12] |
| 56 | NIPAAm | AA | $R_{22}$ | 60 | Methanol | [64] |
| 57 | NIPAAm | DEAEMA | $R_2$ | 80 | Dioxane | [65] |
| | | **Block copolymers with ionic and stimuli-responsive block** | | | | |
| 58 | $CM_7$ | DEAEMA | $R_3$ | 70 | Water, pH=6.5 | [66] |
| 59 | $CM_{11}$ | $M_{14}$ | $R_3$ | 70 | Water | [67] |
| 60 | $CM_7$ | NIPAAM | $R_3$ | 70 | Dioxane/Water | [68] |
| 61 | NIPAAm | $CM_{15}$ | $R_6$ | | Methanol | [13] |
| 62 | DMAEMA | $CM_{17}$ | $R_1$ | 80 | Methanol/water | [69] |
| 63 | DMAPMA | $CM_{11}$ | $R_3$ | 70 | Water | [8] |
| | $CM_{11}$ | DMAPMA | | | | |
| 64 | $CM_{18}$ | NIPAAm | $R_3$ | 70 | Methanol/water | [42] |

AA = acrylic acid; AN = acrylonitril; BA = butyl acrylate; BMA = butyl methacrylate; BzMA = benzyl methacrylate; DEA = *N,N*-diethylacrylamide; DEAEMA = 2-(diethylamino) ethyl methacrylate; DMA = *N,N*-dimethyl acrylamide; DMAEMA = 2-(dimethylamino) ethyl methacrylate; DMAPMA = 2-(dimethylamino) propyl methacrylate; GMA = glycidyl methacrylate; HEA = 2-hydroxyethyl acrylate; HEMA = 2-hydroxyethyl methacrylate; MAA = methacrylic acid; MAn = maleic anhydride; MMA = methyl methacrylate; NAM = *N*-acryloylmorpholine; NAP = *N*-acryloylpyrrolidone; NIPAAm = *N*-isopropyl acrylamide; NVP = *N*-vinyl pyrrolidone; PEGA = poly(ethylene glycol) acrylate; PEGMA = poly(ethylene glycol) methacrylate; STY = styrene; *t*-BA = *tert*-butyl acrylate; *t*-Bam = *tert*-butyl methacrylate; VAc = vinyl acetate.

A popular way to generate A–B–A triblock copolymer is via a trithiocarbonate-based RAFT agent, which carries two leaving groups (Scheme 9.5, pathway III). A conventional Z group is absent and the leaving group as well as later the polymer chain takes on the role of the Z group as well as the R group. Examples for such triblock copolymers are described using styrene/*n*-BA [2, 71], styrene/4-vinylpyridine [72], styrene/maleic anhydride [47], Poly(ethylene glycol) methacrylate (PEGMA)/styrene [73] or dicyclohexyl itaconate/styrene [16].

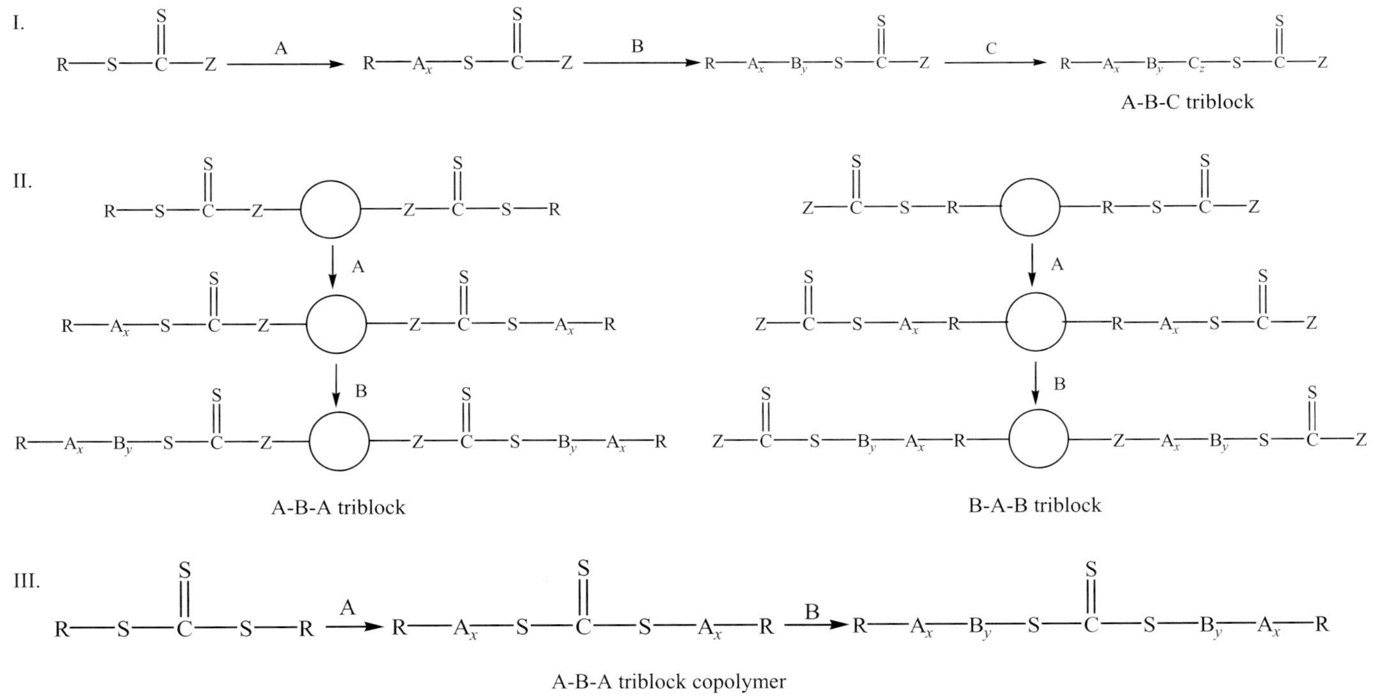


**Scheme 9.5** Possible synthetic pathways to triblock copolymers.

I. Covalent attachment of a RAFT agent to an end-functionalized non-RAFT polymer ($P_A$), followed by RAFT polymerization ($P_B$):

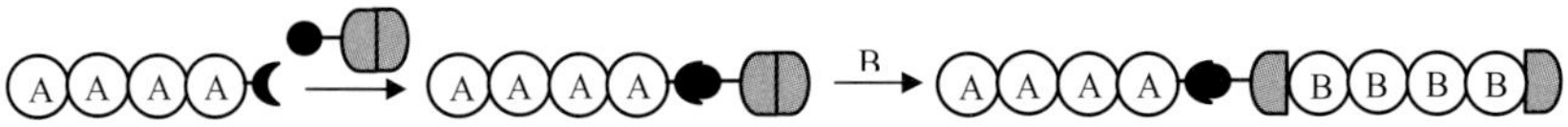

II. Initiation of polymerization of non-RAFT polymer ($P_A$) using a functional RAFT agent, followed by RAFT polymerization ($P_B$):

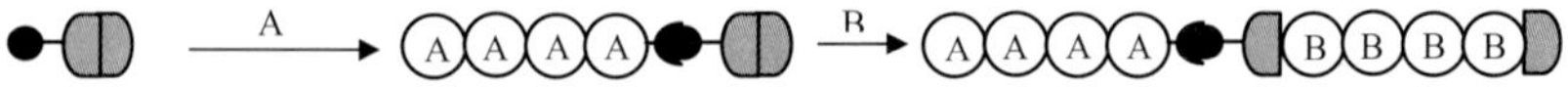

III. RAFT polymerization using a functional RAFT agent ($P_B$) followed by attachment to non-RAFT polymer ($P_A$):

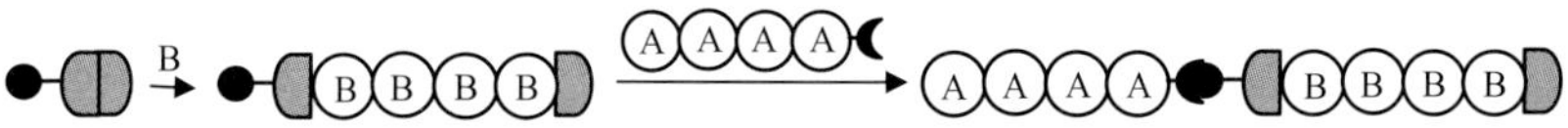

IV. RAFT polymerization using functional RAFT agent ($P_B$) followed by initiation of polymerization of non-RAFT polymer ($P_A$):

V. Simultaneous polymerization of ($P_A$) and ($P_B$):

**Scheme 9.6** Theoretical approaches to block copolymers based on a block made by RAFT polymerization and a block composed of non-RAFT polymers.

### 9.2.3
### Block Copolymers Based on Polycondensates and Other Polymer Chains

The combination of polymers made from other techniques such as polycondensation or ring-opening polymerization with polymers generated via a radical process is increasingly popular. The RAFT agent is hereby linked to another polymer via end-group functionality. Several potential approaches are conceivable (Scheme 9.6):

#### 9.2.3.1 **Theoretical Consideration**

While the synthesis of non-RAFT-made polymers follows its own kinetics, only the RAFT process is the focus of attention here. While the five possible pathways to combine two blocks of a different nature have all their own implications, a main concern is, however, the direction of attachment of the RAFT agent. Prior to the reaction, a RAFT agent has to be designed in a suitable way to undergo reactions with the non-RAFT polymer such as a condensation reaction or act as an initiator for a polymerization, for example, a ring-opening polymerization. This functional group can be located either on the Z group or on the R group of the RAFT agent. With the attachment via the R group (R-group approach) or Z group (Z-group approach), two different types of polymeric RAFT agents (poly-RAFT) are obtained, both with their own characteristic mechanism.

#### 9.2.3.2 **Poly-RAFT Agent via R-Group Approach**

The polymer is covalently bound to a RAFT agent via R group; thus, the first block is part of the leaving group. Consequently, this polymer block is detached from the RAFT agent after the initial transfer step (Scheme 9.7, pathway II). Typical features of this type of block copolymer formation are the possibility of the formation of homopolymer composed of the monomer of the second block (Scheme 9.7, pathway IV) and the formation of triblock copolymers via termination by combination. Both reactions are caused by the initiation step or the initiator concentration, respectively, and can be adjusted accordingly (Scheme 9.7).

#### 9.2.3.3 **Poly-RAFT Agent via Z-Group Approach**

The RAFT agent is covalently attached to the polymer via the Z group. Therefore, the thiocarbonylthio group is constantly bound to the polymer. Termination reactions

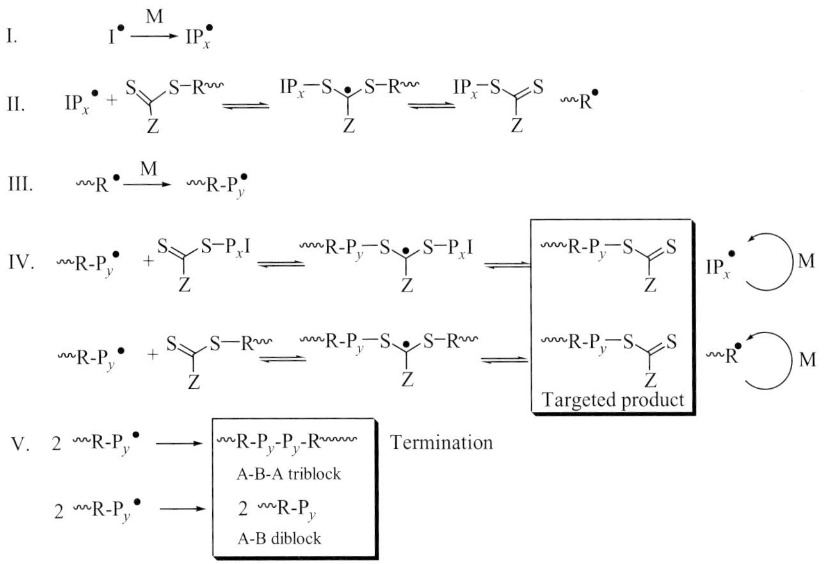


**Scheme 9.7** Block copolymer synthesis using a RAFT agent, which is covalently attached to an end-functionalized polymer chain in an R-group approach.

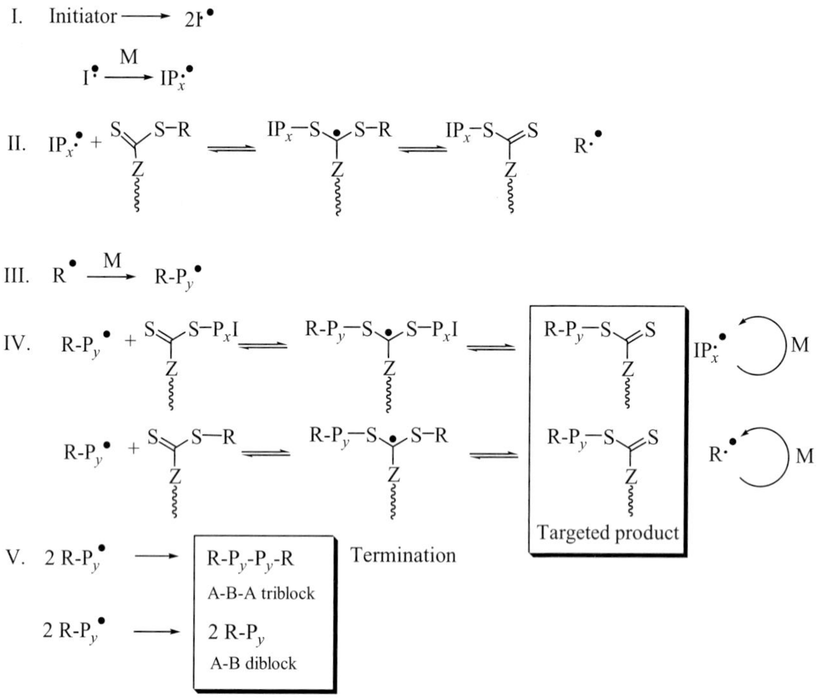

**Scheme 9.8** Block copolymer synthesis using a RAFT agent, which is covalently attached to an end-functionalized polymer chain in a Z-group approach.

lead now to homopolymers of the second monomer, while the formation of triblocks is absent (Scheme 9.8).

### 9.2.3.4 **Practical Considerations and Experimental Results**

All the five possible pathways (Scheme 9.6) suggested above have been employed to generate complex polymer architectures based on polymers, which have been prepared with other techniques. The most common one in literature is the technique 1, where a preformed end-functionalized polymer is converted into a poly-RAFT agent by covalent attachment of a thiocarbonylthio compound. Technique 4 – a macro-RAFT agent starting a reaction to add a block of different nature – is usually not explored to generate block copolymers, but to produce other complex architectures, such as comb polymers (see Section 9.4).

The types of non-RAFT polymers employed are summarized in Table 9.2. The polymers range from poly(ethylene oxide) [2, 75–79, 84], kraton [80], polyethylene [81, 82], poly(dimethylsiloxane) (PDMS) [83], poly(lactide) (PLA) [85, 86], poly(butadiene) [87], to poly($\varepsilon$-caprolactone) [90].

Type 1 poly-RAFT agents (Scheme 9.6, pathway I) have mostly been obtained by reaction of hydroxyl-end-functionalized polymers with RAFT agents containing carboxy groups to form an ester. An alternative route has been described for the

**Table 9.2** Block copolymers prepared from RAFT and non-RAFT polymers

| Entry | poly-RAFT agent | Monomer | A[a] | Solvent | T (°C) | Ref. |
|---|---|---|---|---|---|---|
| 1 | 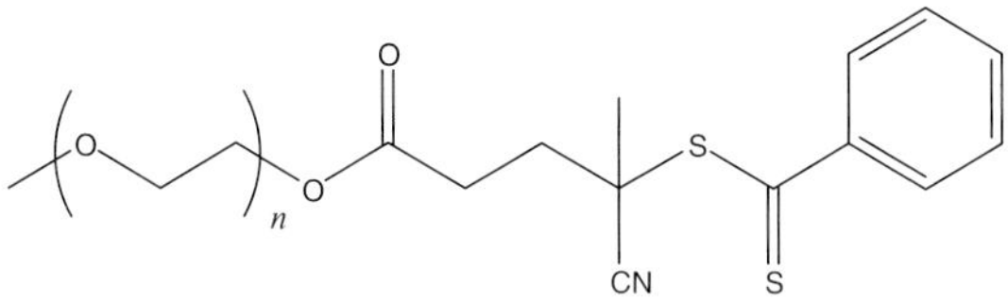 | STY BzMA | 1-R | Bulk benzene | 60 | [2] |
| 2 | 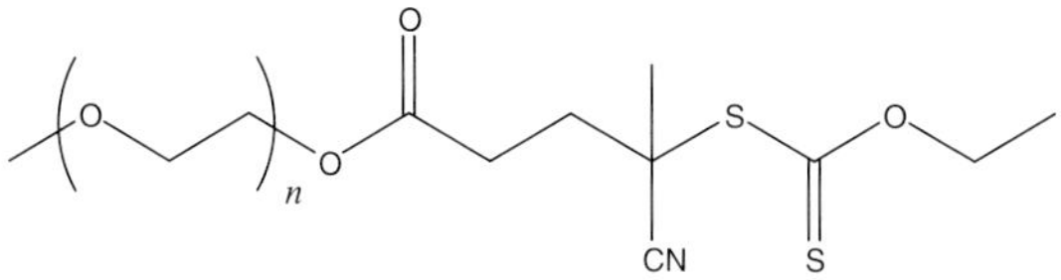 | *N*-vinyl formaide | 1-R | DMSO | 100 | [75] |
| 3 | 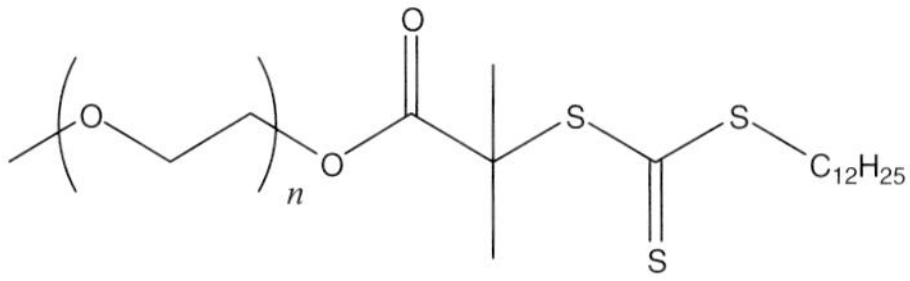 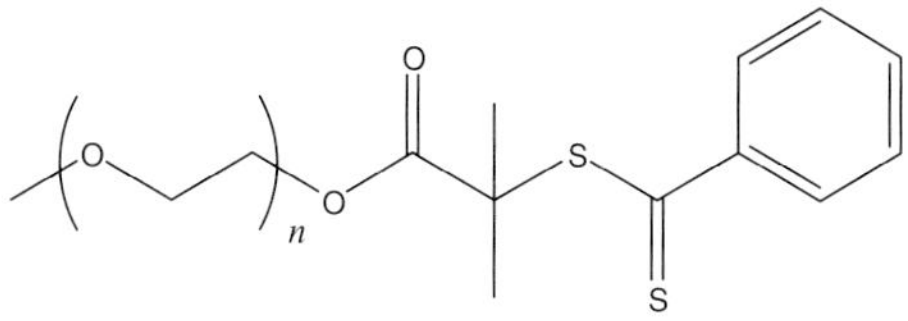 | BA FDA | 1-R | TFT | 65 | [76 77] |
| 4 | 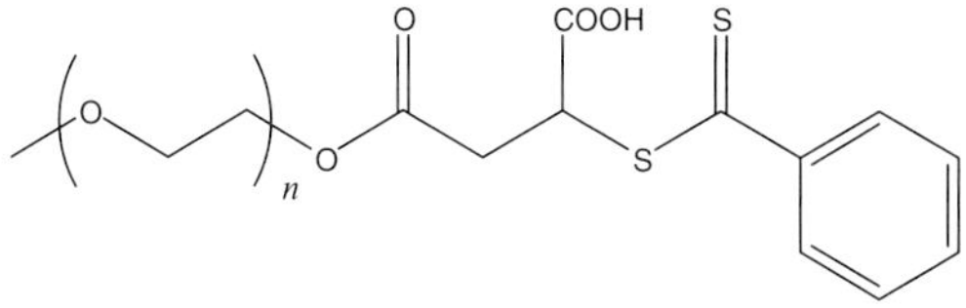 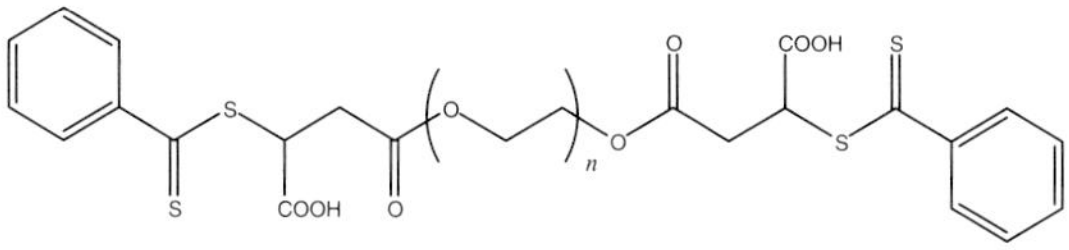 | NIPAAm | 1-R | THF | 100 | [78] |
| 5 | 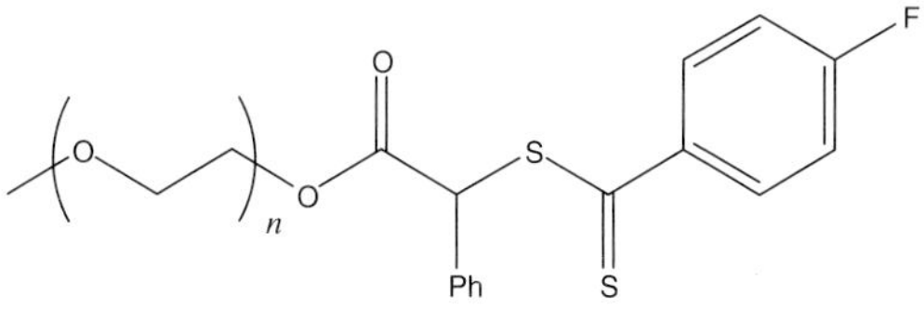 | DMA NAS | 1-R | Dioxane | 70 | [79] |

(*continued*)

**Table 9.2** (*Continued*)

| Entry | poly-RAFT agent | Monomer | A[a] | Solvent | T (°C) | Ref. |
|---|---|---|---|---|---|---|
| 6 | 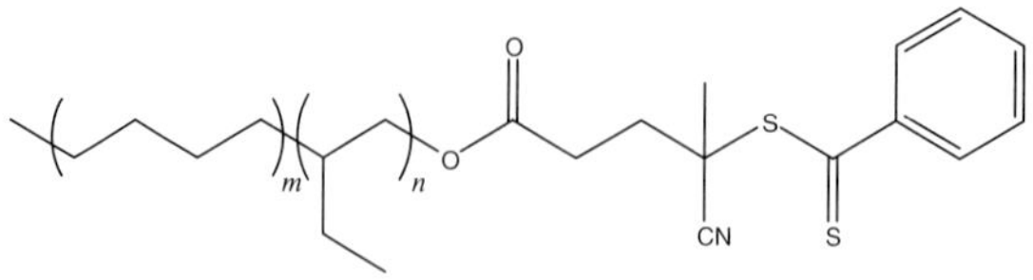 | STY/MAn | 1-R | Xylene BuAc | 60 | [80] |
| 7 | 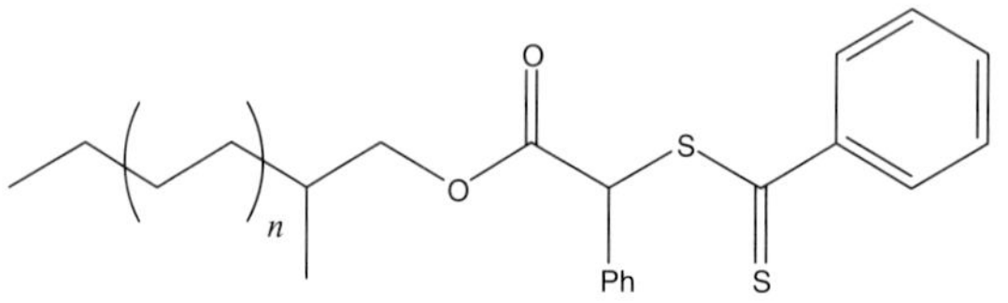 | MMA | 1-R | Toluene | 60 | [81] |
| 8 | 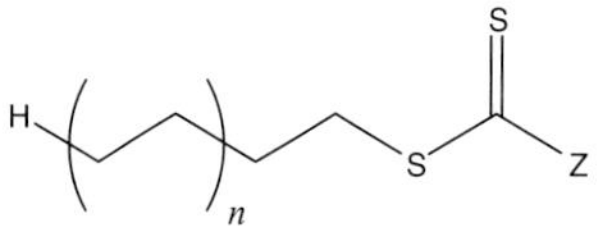 | N/A | 1-R | N/A | N/A | [82] |
| 9 | 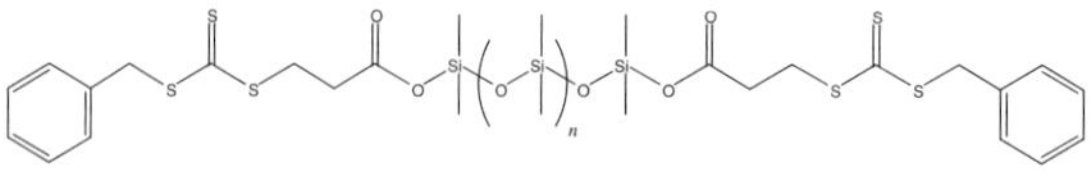 | DMA/FA | 1-Z | TFT | 60 | [83] |
| 10 | 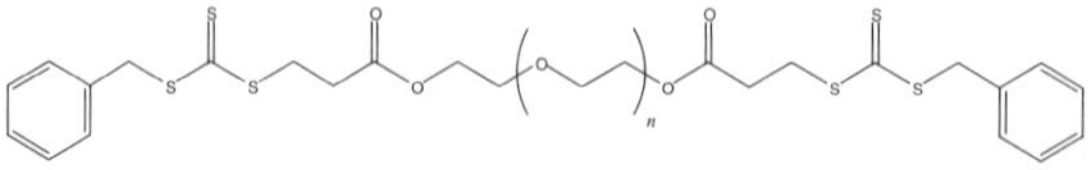 | STY/ HEMA | 1-Z | DMF | 60 | [84] |
| 11 | 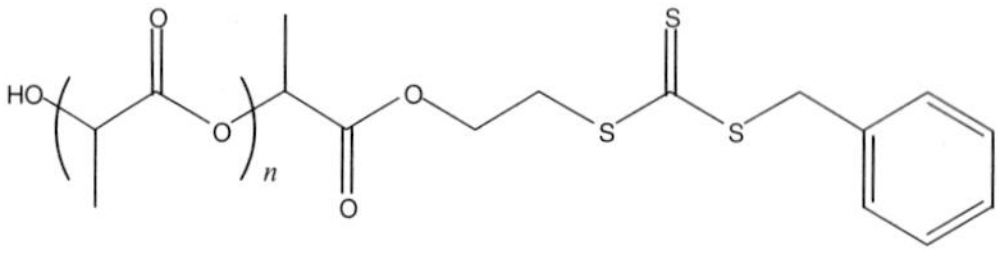 | NIPAAm | 2-Z | DMAc | 60 | [85] |
| 12 | 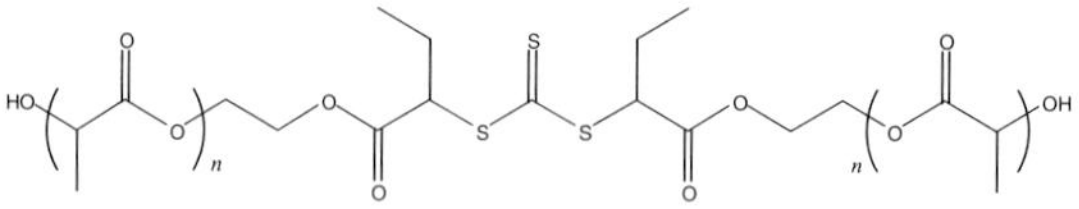 | NIPAAm | 2-Z | THF | 100 | [86] |
| 13 | 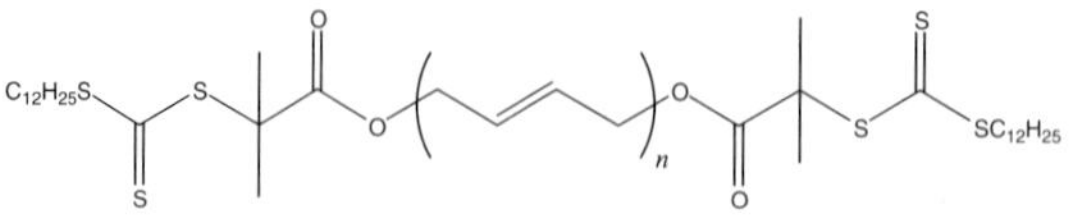 | *t*-BA | 2-R | Benzene | 60 | [87] |
| 14 | 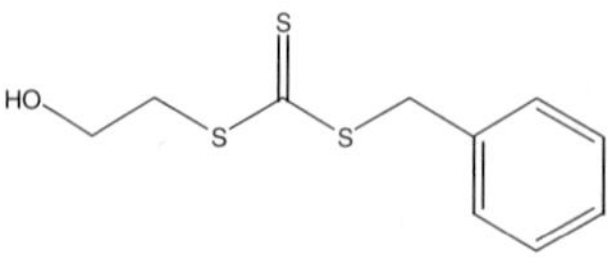 | ε-Caprol actone/ STY | 5-Z | sc-$CO_2$ | 65 | [88] |

[a]Approach referring to Scheme 9.6 in combination with either the R-group approach (Scheme 9.7) or the Z-group approach (Scheme 9.8).

synthesis of polyethylene-based poly-RAFT agents, which were obtained by reacting thiocarbonylated disulfide compounds Z—(CS)—S—S—(CS)—Z with $Mg(PE)_2$ via Grignard reaction [82]. The completeness of the reaction is usually confirmed using NMR studies. However, residual poly-RAFT agent was occasionally observed in the subsequent polymerization. The origin of this remaining poly-RAFT agent may be either an incomplete chain transfer or the incomplete functionalization of the polymer with RAFT agent. Since NMR is subject to uncertainties, especially at high molecular weight, UV spectroscopy has been shown to be a suitable technique to quantify the amount of RAFT agent attached [84]. An elegant way to enumerate the amount of unfunctionalized polymer and the amount of poly-RAFT agent that did not undergo chain extension has been demonstrated via a dual detector system. In the described case, an evaporative light-scattering detector was combined with a UV detector in order to obtain the absolute amount of polymer next to the amount of polymer carrying thiocarbonylthio groups [80].

The RAFT agents were in most cases attached via the R group when employing the type 1 pathway (Table 9.2). As mentioned above, R-group approaches are always subject to termination reactions resulting in the formation of A–B–A triblock copolymers, assuming the monomer will terminate via combination. Indeed, high-molecular-weight shoulders are commonly observed, especially at higher conversions. The formation of these triblock copolymers can even be forced using UV radiation: Decomposition of the RAFT end group into radicals can cause additional termination reactions, as evidenced by a significant increase of high-molecular-weight products, especially triblock copolymers [80]. High-molecular-weight shoulders were reported to be absent when using the Z-group approach, confirming the mechanism suggested in Scheme 9.3.

The block copolymer formation evolved mostly in a controlled manner. However, it has to be considered that the resulting block copolymer can only be as well defined as its underlying non-RAFT polymer. Poly(ethylene oxide) has usually a narrow-molecular-weight distribution [2, 75–79, 84] and the resulting block copolymers were observed to have low polydispersity indices as well despite the occurrence of some termination reactions. In contrast, the synthesis of polyethylene-poly(methyl methacrylate) (PE-PMMA) block copolymers could only result in block copolymers with polydispersity indices of above 2, which is mainly attributed to a relatively broad-molecular-weight distribution of the underlying polyethylene [81].

The kinetics of the polymerization was found to be slightly influenced by the presence of the polymeric chain and often proceeded in a fashion different to the RAFT polymerization employing the equivalent low-molecular-weight RAFT agent. Since the structure surrounding the thiocarbonylthio end group is equivalent, this may be assigned to steric aspects or altered polarities adjacent to the RAFT group. A decline of inhibition time – which is usually explained by the slow fragmentation of the intermediate radical in the preequilibrium – was observed when comparing the polymerization of DMA in the presence of a low-molecular-weight RAFT agent and in the presence of the same RAFT agent, which is now attached to a PDMS chain [83]. The poly-RAFT agent was additionally observed to influence apparent reactivity ratios. Investigations revealed that the RAFT process does not significantly alter reactivity ratios in contrast to random copolymerization carried out via conventional free-radical polymerization [74]. Attachment of a polymer chain to a RAFT agent can alter the microenvironment. During the subsequent copolymerization, the monomer with a similar polarity to the poly-RAFT agent is preferably built in. This bootstrap effect was observed to have a significant effect on the apparent reactivity ratio [83]. The microenvironment was also thought to be responsible for a slow consumption of the poly-RAFT agent. While the thiocarbonylthio group employed may have appropriate chain-transfer efficiency for a certain monomer, the block copolymer formation was found to be incomplete. Only with increasing monomer conversion, all poly-RAFT agents are eventually involved in block copolymer formation [80].

Type 2 pathway (Scheme 9.6, pathway II) to generate block copolymers uses a RAFT agent to initiate the polymerization of the non-RAFT polymer. RAFT agents carrying hydroxy groups were successfully employed in the ring-opening polymerization to generate poly(lactate) end functionalized with a RAFT agent [85, 86]. The amount of RAFT agent starting a ring-opening process was then determined via UV spectroscopy, confirming the successful polymerization despite some remaining RAFT agents, which were not involved in the process and had to be removed by precipitation [85]. Ring-opening metathesis polymerization of 1,5-cyclobutadiene using suitably designed RAFT agents was demonstrated successful to obtain poly(butadiene)-poly(*t*-butyl acrylate) triblock copolymers [87].

An elegant way (Scheme 9.6, pathway V) of preparing these block copolymers in a one-step process has been employed using the simultaneous process of RAFT polymerization of styrene and ring-opening polymerization of ε-caprolactone in supercritical $CO_2$. While the polymerization of styrene was found to be very well controlled, the ring-opening polymerization – catalyzed by enzymes – was reported to result in blocks of rather broad-molecular-weight distribution [88].

Multiblock copolymers were obtained in a similar manner by polymerizing a bifunctional RAFT agent via a polycondensation/addition step. The resulting polymer acts as a multifunctional RAFT agent for the subsequent polymerization.

A multifunctional poly(ethylene oxide) (PEO)-based RAFT agent was synthesized for the subsequent RAFT polymerization. The condensation process between PEO and the diacyloyl chloride RAFT agent followed condensation kinetics, resulting in rather broad-molecular-weight distributions. The following RAFT polymerization of styrene, however, was well controlled (Scheme 9.9) [89].

Alternatively, a polymer was constructed from alternating RAFT agents and thiourethane building blocks [90]. The RAFT process using styrene leads to the insertion of PS resulting in the increased spacing between two thiourethane groups forming a sequentially ordered polymer (Scheme 9.9).

### 9.2.4
### Block Copolymers via Click Chemistry

A drawback of the RAFT process is that only block copolymers based on monomers with similar reactivities can be generated. Macroradicals with highly

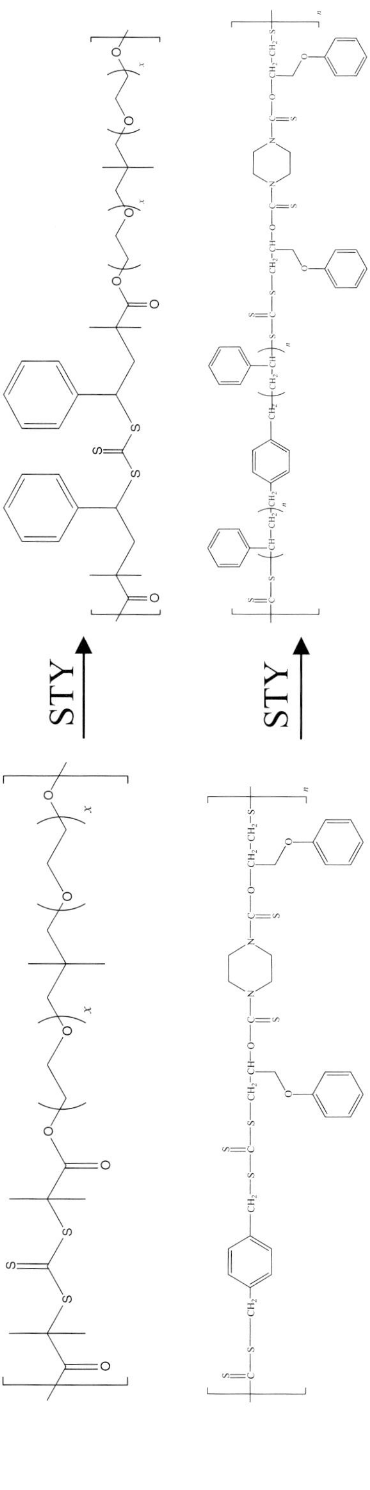


**Scheme 9.9** Synthesis of multiblock copolymers via RAFT polymerization [89, 90].

disparate reactivities require substantially different RAFT agents to achieve a good polymerization control. For example, vinyl acetate RAFT polymerization can only be mediated by a xanthate agent [91], whereas styrene is polymerized in the presence of dithiobenzoate compounds [92]. The only potential exception to the above dilemma is the use of universal RAFT agents, such as F-RAFT (fluorodithioformates) [93], which holds great promise for sequential block copolymer formation of monomers with disparate reactivities, such as styrene and vinyl acetate.

One possible alternative strategy to circumvent the dilemma is to separately prepare two homopolymers, which are suitably functionalized, to enable their efficient combination. This approach is similar to type 3, as displayed in Scheme 9.6. Prerequisite for such a pathway is the highly efficient reaction between two homopolymers. This reaction should be preferably robust against the presence of other functional groups. An elegant pathway to block copolymer synthesis via conjunction of two homopolymers is pericyclic [2 + 3] reactions ('click chemistry'). Coupling of two polymers is usually thermodynamically unfavorable. The steric hindrance of the polymer chains acts as a shield, preventing the molecular reaction between polymer end groups. However, the coupling reaction was found to be achievable using click reactions, which was coined by Sharpless and coworkers [94] to encompass all reactions of high yields, modularity and stereospecificity, such as copper-catalyzed alkyne and azide 1,3-dipolar cycloaddition (CuAAC) or the Diels Alder (DA) reactions. The successful combination of click reactions with a range of polymerization techniques including the RAFT process opens the doors to a range of complex polymer architectures.

A new block copolymer – polystyrene-*block*-poly(vinyl acetate) – was obtained by the combination of RAFT and click chemistry [95]. Two complementary RAFT agents were prepared to entail the functionalities required for click chemistry – one with azide functionality and one with acetylene functionality. Prior to polymerization, the acetylene functionality is required to be protected in order to preserve the chemically and thermally instable acetylene functionality during polymerization (Scheme 9.10).

To ensure the success of the click reaction and enable the formation of well-defined polymeric material, it is paramount that equimolar amounts of each of the functional homopolymers are employed, as otherwise unreacted material will remain. This dilemma has been faced when conjoining RAFT-made poly(vinyl acetate) (PVAc) and glycopolymer via click chemistry [96]. Low-molecular-weight tailing indicates some remaining homopolymer, which can be assigned to the loss of functionality either via destruction during the polymerization or by the formation of inactive homopolymer caused by a high radical concentration, as discussed earlier (Scheme 9.4).

Block copolymers – polystyrene-*block*-poly(*N*,*N*-dimethylacrylamide) – were prepared via the RAFT process, using chain extension of a macro-RAFT agent carrying an azide functionality. The resulting well-defined block copolymers were clicked to a range of acetylene-based compounds signifying the versatility of click chemistry to generate an array of end-functionalized polymers [97].

An alkynyl-containing RAFT agent was directly employed without the recourse to protective chemistry in the synthesis of block copolymers to afford well-defined

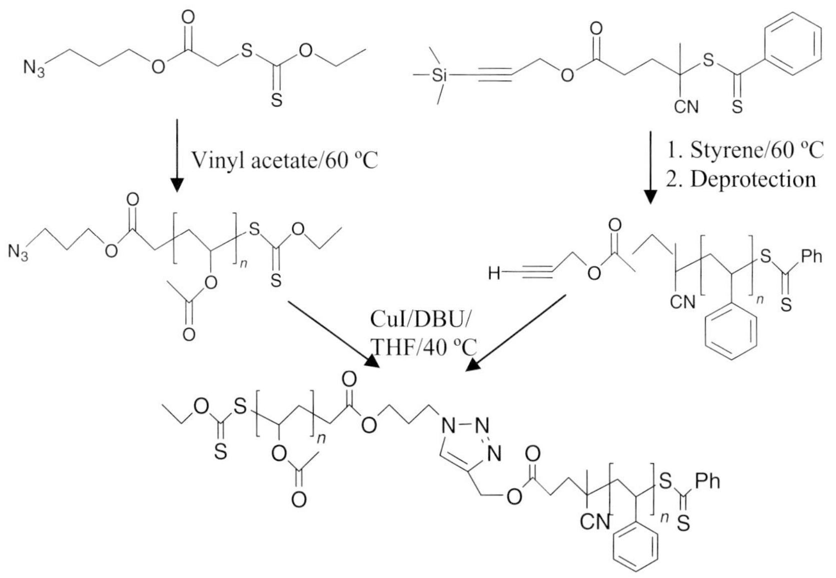


**Scheme 9.10** Schematic approach to polystyrene-*block*-poly(vinyl acetate) via combination of RAFT process and click chemistry [95].

poly(acrylic acid-*block*-polystyrene) structures with terminal acetylene functionalities [98].

Instead of using azide- or acetylene-functionalized RAFT agent, these moieties can be introduced into block copolymers by employing clickable monomers. By preparing an amphiphilic block copolymer using a mixture of styrene and 4-(trimethylsilylethynyl)styrene protected with a trimethylsilyl protective group for the hydrophobic block and acrylic acid for the hydrophilic sequence, a highly reactive block copolymer was generated. Upon self-assembly into micelles, the click functionalities located in the core were confirmed to be available for further reactions, such as azido dyes (Scheme 9.11) [99].

## 9.3 Star Polymers via RAFT Polymerization

RAFT polymerization offers several avenues to generate star polymers. Similar to other synthesis pathways, core-first and arm-first techniques can be employed. Unique to the RAFT process is, however, the subdivision of the core-first technique into R-group and Z-group approach, which describe the direction of attachment of the RAFT agent onto a multifunctional core. R-group and Z-group approach have different implementation regarding the mechanism and the outcome [100].

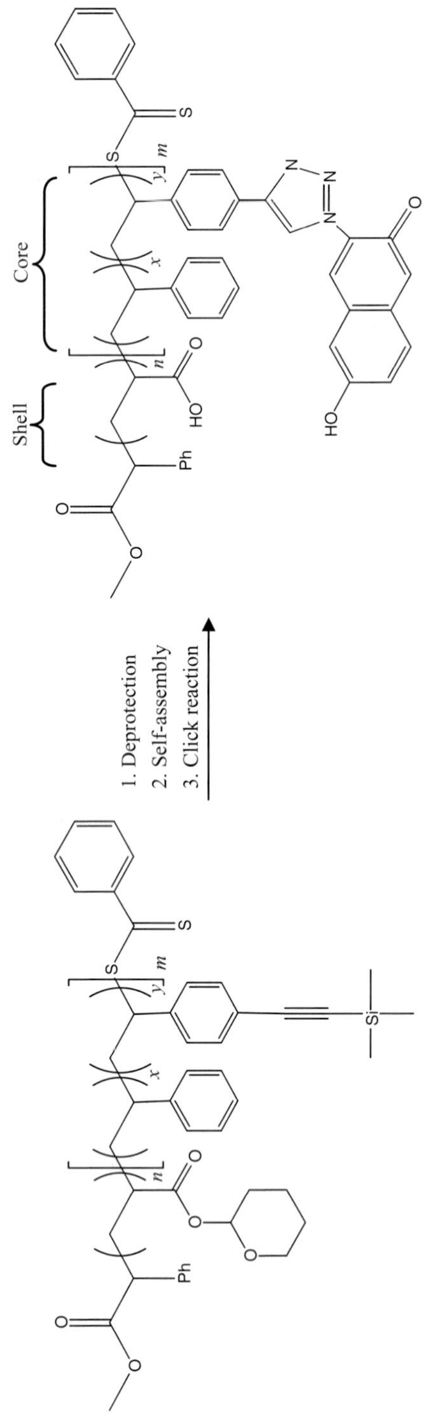


**Scheme 9.11** Preparation of micelles with clickable moieties in the core [99].

## 9.3.1
## Star Polymers via Core-First Technique

The core-first technique employs a multifunctional initiator. The number of arms is directly given by the number of initiating sites on the core. The RAFT agents are attached via the Z group or R group (Scheme 9.12). Both techniques have been described in the literature employing a variety of monomers (Table 9.3). For the core-first technique, RAFT agents were covalently bound via R- and Z group to the core, which were based on (aromatic) hydrocarbons [62, 101, 102, 117] to cyclodextrin [103], hyperbranched polyesters (ethers) [104, 121], metal complexes [105, 112, 113] and well-defined dendrimers [106, 107, 114–116, 123]. The resulting number of arms is determined by the number of RAFT agents attached to the core. A range of star polymers were synthesized using monomers such as styrene, acrylates, acrylamide, vinyl ester and vinyl pyrrolidone, but also functionalized polymers such as glycopolymers and light-harvesting polymers were generated using the RAFT process (Table 9.3).

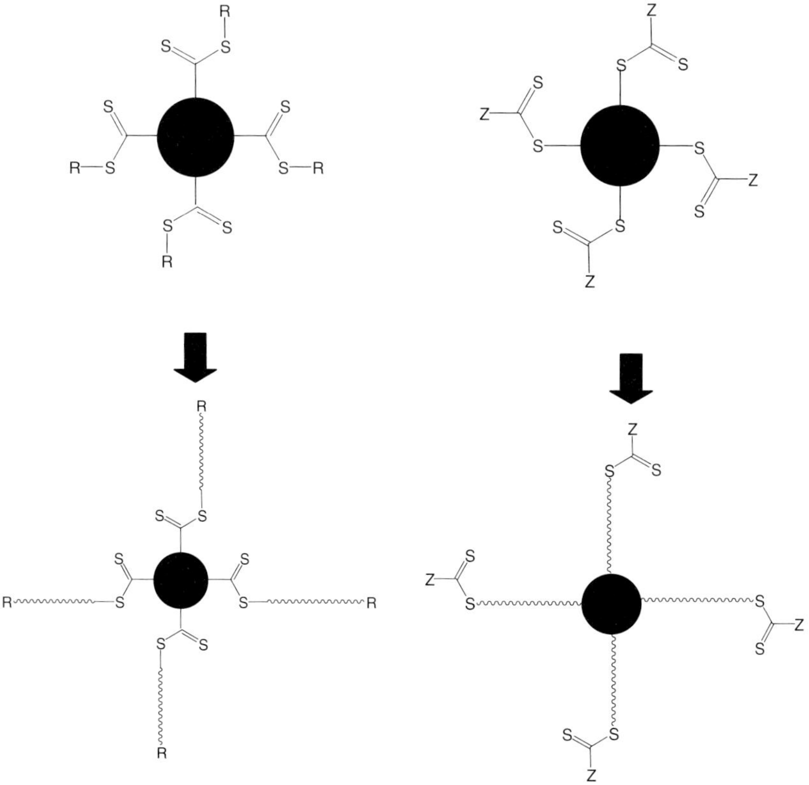

**Scheme 9.12** Comparison of star synthesis using the core-first technique via Z-group (left) and R-group (right) approach.

**Table 9.3** Star polymers prepared via RAFT process using R-group and Z-group approach

| A[a] | Type of core | RAFT group | N[b] | Monomer | Solvent | T (°C) | Ref. |
|---|---|---|---|---|---|---|---|
| R | Ruthenium tris(bipyridyl) complex | 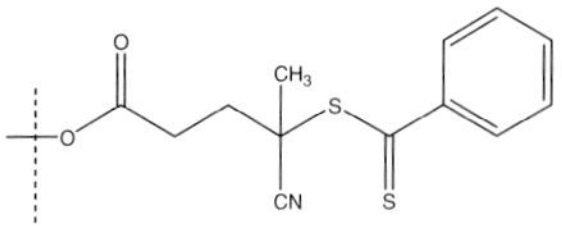 | 6 | NIPAAm<br>Styryl coumarin | Chlorobenzene | 70 | [105, 112] |
| R | Europium tris(β-diketenate) complex | 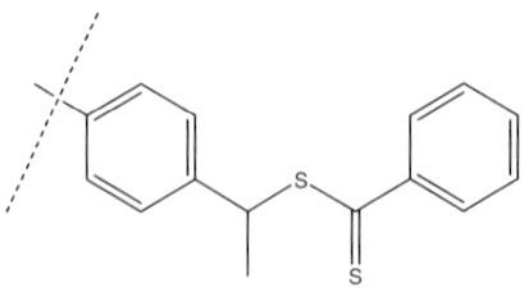 | 6 | MMA<br>STY | Toluene | 70 | [113] |
| R | Benzene | 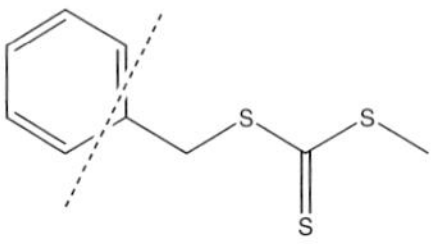 | 4 | STY<br>MA | Bulk | 110<br>60 | [102] |
| R | Dendrimer | 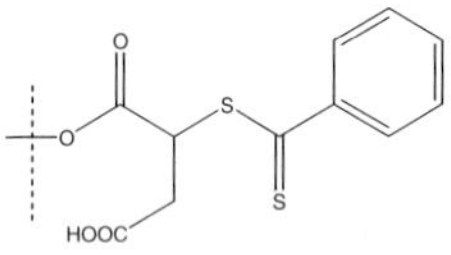 | 16 | STY<br>MA | THF | 100 | [114] |
| R | Dendrimer | 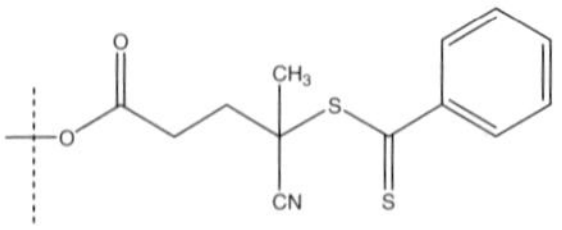 | 8<br>16 | NIPAAm | THF | 100 | [115] |
| R | Benzene | 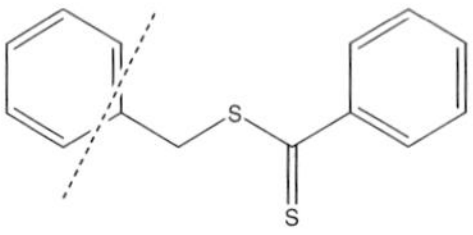 | 6 | STY | Bulk | 80<br>100<br>120 | [101] |
| R | Dendrimer | 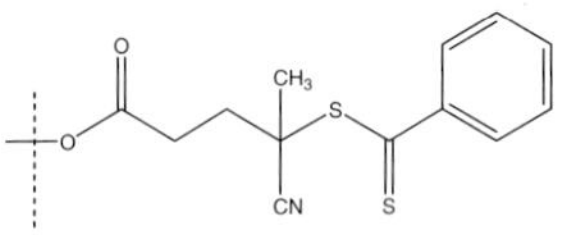 | 8<br>16 | STY<br>MA | THF | 120 | [116] |
| R | 1,1,1-tris(hydroxy methyl)ethane, pentaerythriol | 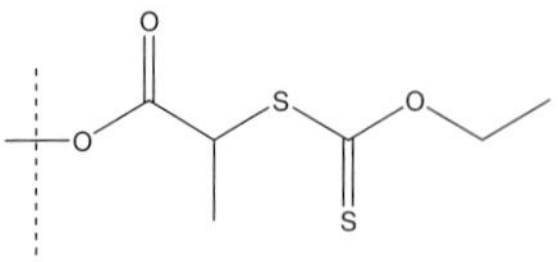 | 3<br>4 | VAc<br>VPi<br>VND<br>VAG | Bulk | 60 | [117] |

Table 9.3 (*Continued*)

| A[a] | Type of core | RAFT group | N[b] | Monomer | Solvent | T (°C) | Ref. |
|---|---|---|---|---|---|---|---|
| R | 1,1,1-tris (hydroxy methyl)ethane, pentaerythriol | 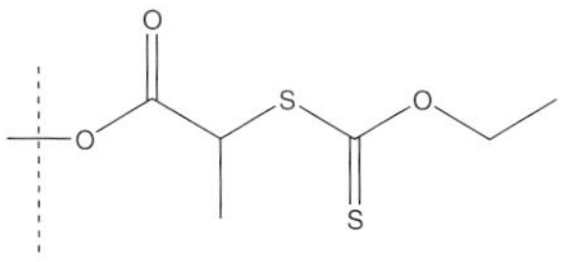 | 3 4 | VAc | Bulk | 60 | [118] |
| R | Benzene | 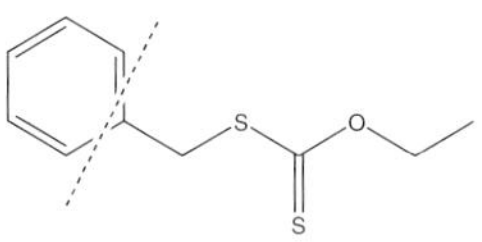 | 4 | NVP | Bulk | 60 | [49] |
| Z | hyperbranched polyglycerol | 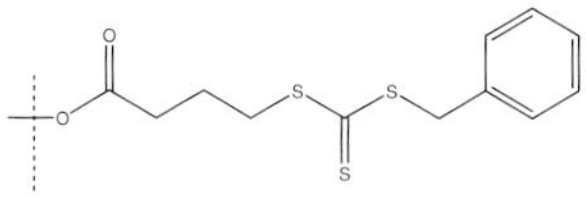 | 17 | NIPAAm<br>DMAEA | Dioxane | 65<br>70 | [119] |
| Z | Pentaerythritol | 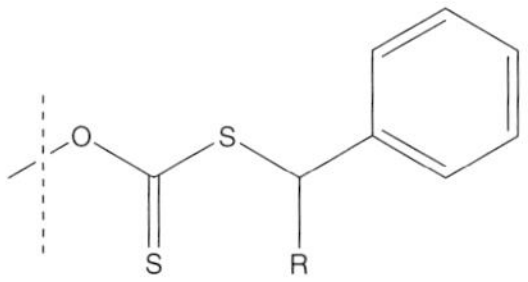 | 4 | VAc<br>VPr | Bulk | 60<br>90 | [120] |
| Z | Pentaerythritol | 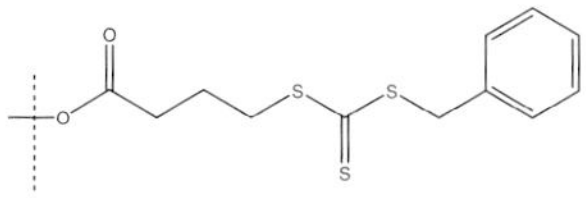 | 4 | STY<br>MA | Bulk | 110<br>60 | [102] |
| Z | Hyperbranched polyglycerol | 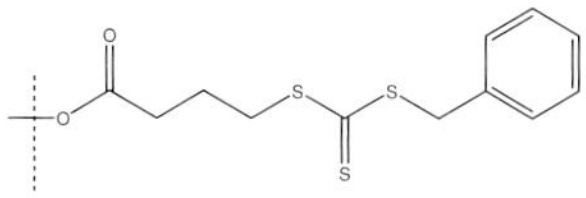 | 17 | EA | Bulk | 80 | [121] |
| Z | Thiourethane-isocyanurate | 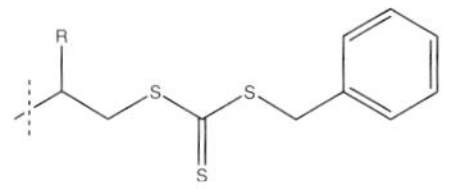 | 3 | STY | Bulk | 60 | [122] |
| Z | β-Cyclodextrine | 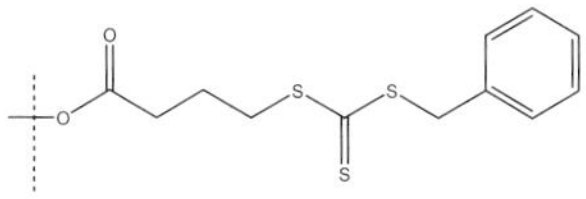 | 7 | STY<br>EA | Bulk | 60 100<br>120 | [103] |
| Z | Hyperbranched polyester | 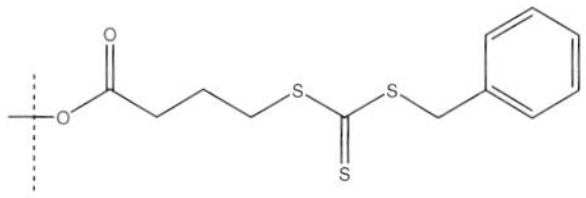 | 12 | STY<br>BA | Bulk | 60 | [104] |

(*Continued*)

Table 9.3 (*Continued*)

| A[a] | Type of core | RAFT group | *N*[b] | Monomer | Solvent | T (°C) | Ref. |
|---|---|---|---|---|---|---|---|
| Z | Dendrimer | 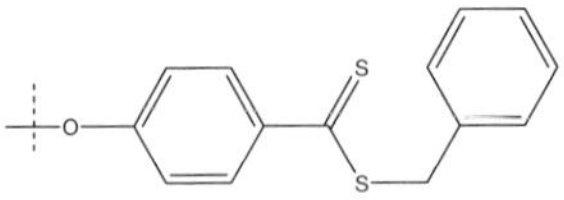 | 12 | STY | Bulk | 110 | [106] |
| Z | 1,1,1-tris(hydroxy methyl) ethane, pentaerythriol | 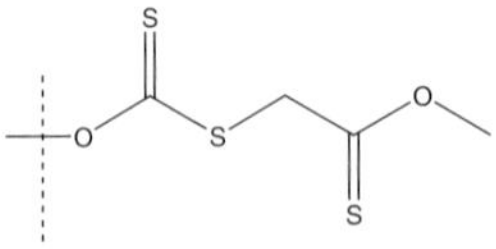 | 3<br>4 | VAc | Bulk | 60 | [117] |
| Z | Dendrimer | 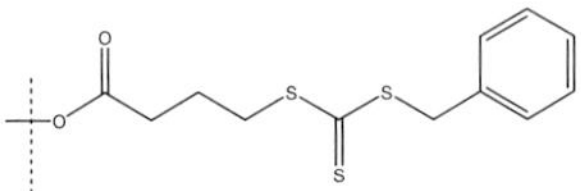 | 3<br>6<br>12 | STY<br>BA | Bulk | 60 | [118] |
| Z | Dendrimer | 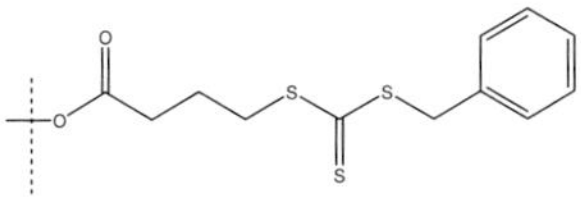 | 3<br>6<br>12 | STY<br>BA | Bulk | 60 | [123] |
| Z | 1,1,1-tris(hydroxy methyl)ethane | 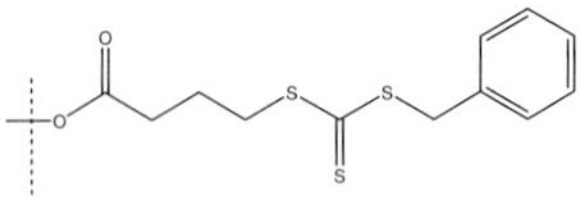 | 3 | AGA | Water | 60 | [62] |
| Z | Pentaerythriol | 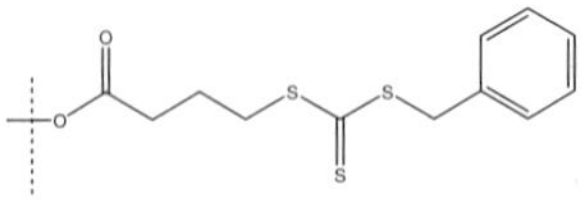 | 4 | *t*-BA<br>STY | Toluene | 60 | [124] |
| Z | Pentaerythriol | 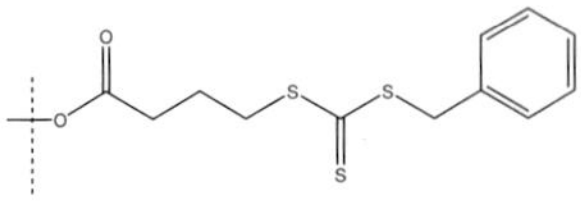 | 4 | NIPAAM | DMF | 60 | [125] |
| Z | Dipentacrithriol | 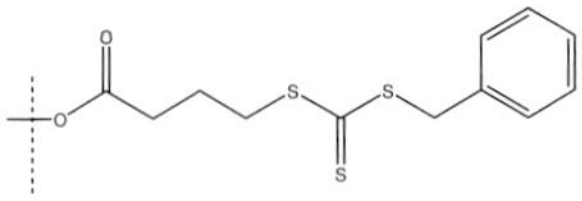 | 6 | MA<br>BA<br>DA | Bulk | 60 | [126] |

[a] A refers to type of approach.
[b] *N* refers to number of arms.

#### 9.3.1.1 Star Polymers via R-Group Approach

**Theoretical Considerations.** In this approach, the RAFT agent is attached to the core via R group. After the initial transfer the radical is located on the core, leading subsequently to the growth of arms. Additionally, a linear macro-RAFT agent is formed, which can undergo further transfer with a macroradical. The transfer of the star macroradical with a thiocarbonylthio group of another star or a linear macro-RAFT agent results in a branch with the RAFT end group, which is growing away from the core. Next to this main reaction, significant side reactions can occur in the synthesis of stars using the R-group approach. Termination reactions via star–star coupling can be considered to occur as well as termination via star–chain coupling leading to the loss of an active arm (Scheme 9.13).

It is evident that next to the main star product a range of side product may appear, broadening the distribution of the polymerization. Considering the mechanism, it seems to be crucial to keep the concentration of free radicals as low as possible since all side reactions mentioned above are the result of bimolecular termination reactions. In fact, the amount of linear macro-RAFT agents and termination products should be directly correlated to the amount of radicals generated by initiation of the polymerization. The amount of radicals can be expressed by two different determining parameters: the radical concentration at any given time during the polymerization and the sum of radicals generated during the course of the polymerization. Lowering the concentration of active radicals can possibly be achieved either by a low initiator concentration or by a high RAFT agent content, especially when a highly retarding RAFT agent is employed [108]. In addition, an increasing rate of polymerization, such as with fast propagating monomers, is anticipated to suppress termination reactions since less radical are usually generated considering the shorter reaction time.

Computational modeling has been employed to correlate radical concentration, propagation rate coefficients and addition–fragmentation constant of the RAFT equilibrium and other parameter against the amount of termination reactions and thus the molecular-weight distribution. PREDICI® coupled with high-level ab initio quantum chemical calculations were employed to foresee the occurrence of side products in dependency of a range of parameters. A list of recommendations can be established in order to optimize the polymerization, meaning a high fraction of star polymers while suppressing termination reactions [109, 110]:

- *Ratio RAFT agent to initiator*: A high RAFT agent to initiator concentration reduces the amount of radicals and thus the amount of termination reactions.
- *Number of arms*: The amount of termination products increases with increasing number of arms.
- *RAFT equilibrium*: A high addition rate of the macroradical to the RAFT agent $k_\beta$ and a high transfer of the linear macroradical to a star-bound RAFT group $k_{trStar}$ are as beneficial for a small-molecular-weight distribution as is a strongly retarding RAFT agent.
- *Propagation rate coefficient*: Fast-propagating monomers are predicted to lead to more narrow-molecular-weight distributions.

**Practical Considerations and Experimental Results.** Theoretical considerations supported by modeling results predict a significant broadening of the molecular-weight distribution in dependency of the reaction parameters. It seems that optimization of these parameter may suppress termination reaction increasing the fraction of the desired star polymers, as depicted in Scheme 9.13.

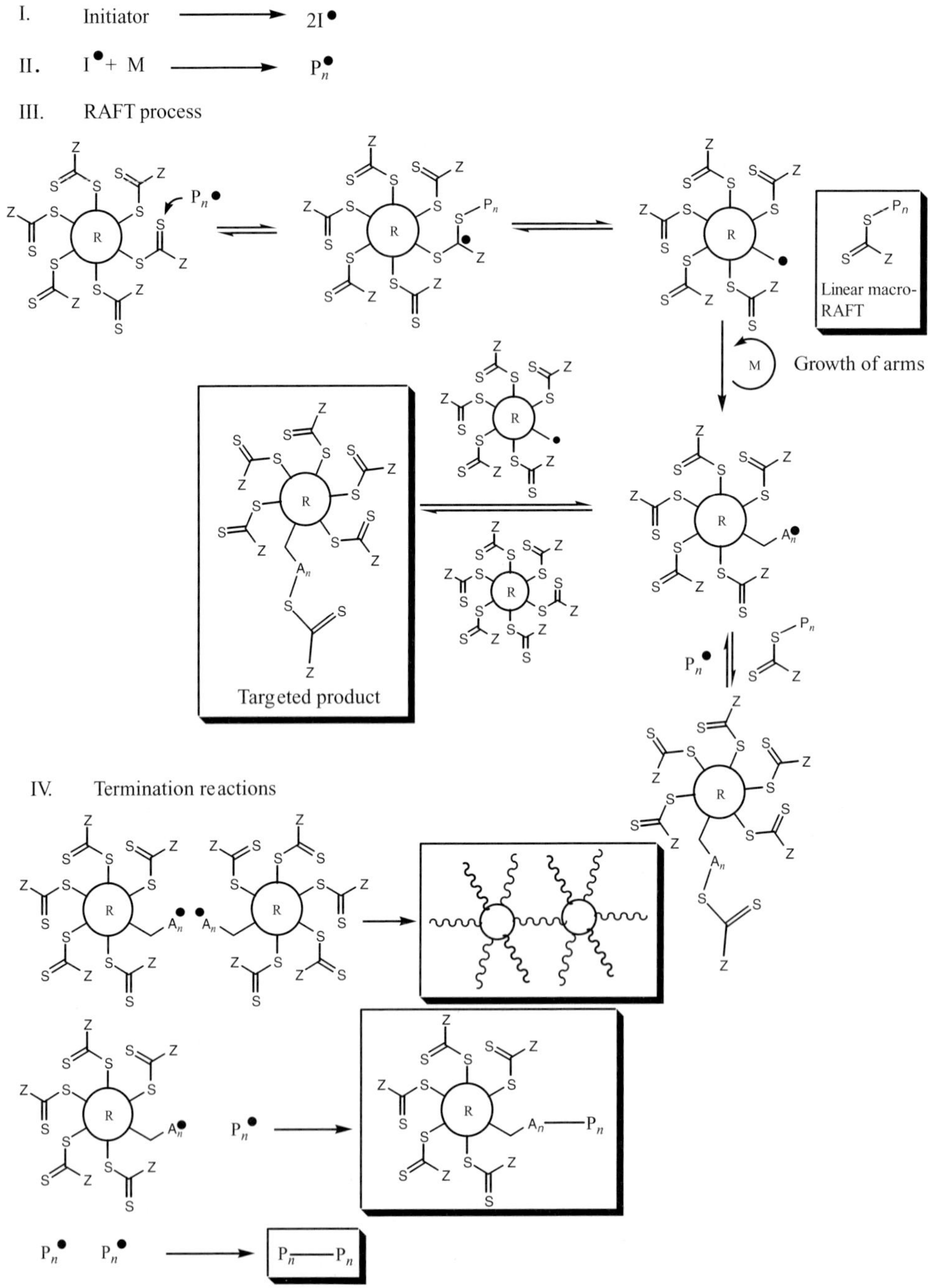


**Scheme 9.13** Schematic drawing of the synthesis of star polymers via R-group approach.

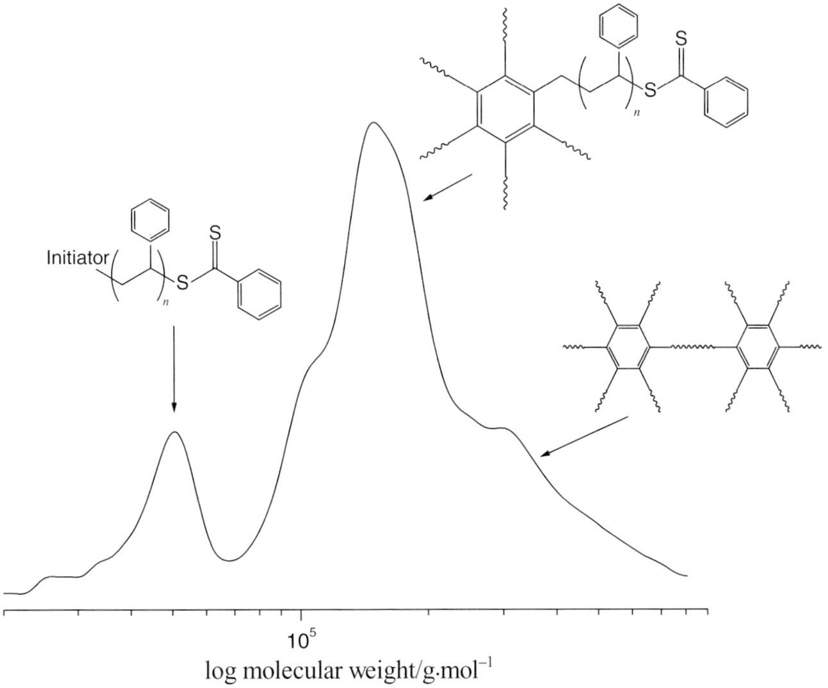


**Fig. 9.3** SEC curve of a polystyrene six-arm star obtained by the polymerization of styrene at 100 °C in the presence of hexakis(thiobenzoylthiomethyl)benzene [100].

In practice, a range of systems have been explored using varying core structures and different monomers (Table 9.3). The multimodality of many SEC curves confirms indeed the suggested mechanism with the appearance of star–star coupling termination products and the formation of a linear macro-RAFT agent that increases in molecular weight linear with conversion. The molecular weight of the targeted star polymer increases typically linear, with conversion close to the expected molecular weight (Fig. 9.3) [101]

$$M_n = \frac{[M]}{[\text{star-RAFT}]} \times x \times M_M + M_{\text{star-RAFT}} = \frac{[M] \times n}{[\text{RAFT}]} \times x \times M_M + M_{\text{star-RAFT}}$$

where $M_n$ is the number-average molecular weight of the star, [star-RAFT] is the concentration of the multifunctional RAFT agent, $n$ the number of arms, $x$ the monomer conversion, [RAFT] the concentration of thiocarbonyl groups and $M_{\text{star-RAFT}}$ the molecular weight of the multifunctional RAFT agent.

Especially the synthesis of PS star polymers has led to broad-molecular-weight distributions similar to the multimodal distribution portrayed in Fig. 9.3 [101, 102, 114, 115]. In contrast, vinyl acetate resulted in well-defined star polymers having a single monomodal molecular-weight distribution, with the visible absence of linear chains and termination products [117, 118]. Therefore, termination reactions and the formation of linear chains could indeed be minimized by an increased rate of polymerization caused by the faster propagation of vinyl acetate as predicted by the computational approach [110]. Indeed, a range of stars prepared from monomers with a high propagation rate coefficient were reported to have polydispersity indices well below 1.3 and a monomodal molecular-weight distribution. Examples are PNIPAAm eight-arm stars [115], PVAc three- and four-arm stars [117, 118] and poly(vinyl pyrrolidone) four-arm stars [49] as well as a range of acrylate stars (Table 9.3).

The number of arms is expected to play a vital role regarding the occurrence of star–star coupling products. It has to be considered that upon chain transfer the radical is located on the core. With increasing number of arms – which is equivalent to the number of attached RAFT agents – the likelihood of a radical being located on any given core is increased resulting in the increased occurrence of star–star coupling. A PS four-arm star was found to have only negligible amounts of these side products [102]. The opposite structure is a polymer chain with a multitude of RAFT agents attached, which is utilized to generate comb structures with a vast amount of branches along a backbone. As discussed in detail later, an exceedingly intensive amount of star–star (comb–comb) coupling by-products were observed, leading to significant broadening of the molecular-weight distribution in addition to the formation of cross-linked gel [111].

While the formation of well-defined stars from slowly propagating monomers or from cores with a significantly higher number of branches may be intricate, several ways of improving the outcome have been demonstrated. In general, these better defined star polymers were obtained using high temperatures and high RAFT agent concentrations [102, 114, 116], which is in agreement with the model requesting a high RAFT agent/initiator ratio and a high rate of propagation.

#### 9.3.1.2 Star Polymers via Z-Group Approach

**Theoretical Considerations.** Star synthesis via Z-group approach differs from the R-group approach by the mode of attachment of the RAFT agent to the multifunctional core: The RAFT agent is connected to the core via the Z group. The direction of RAFT-agent attachment results in consequences for the mechanism, significantly different to the R-group approach. Upon initiation, the growing macroradical undergoes its addition–fragmentation step, resulting in a linear macroradical. Consequently, the radical is never located on the core; termination reaction resulting in star–star coupling should be absent, unless chain transfer to monomer interferes. In addition, no RAFT-end-capped linear chains are generated (Scheme 9.14). Products generated during the star synthesis via the Z-group approach should – next to the star product – only be linear termination products, which are in correlation with the amount of radicals generated in the system.

As outlined in Scheme 9.14, the RAFT agent does not grow from the core. Therefore, the chain transfer takes place adjacent to the core independent of the length of the polymer chain. This, however, may cause complications, regarding the accessibility of the RAFT group. With increasing chain length, the star branches

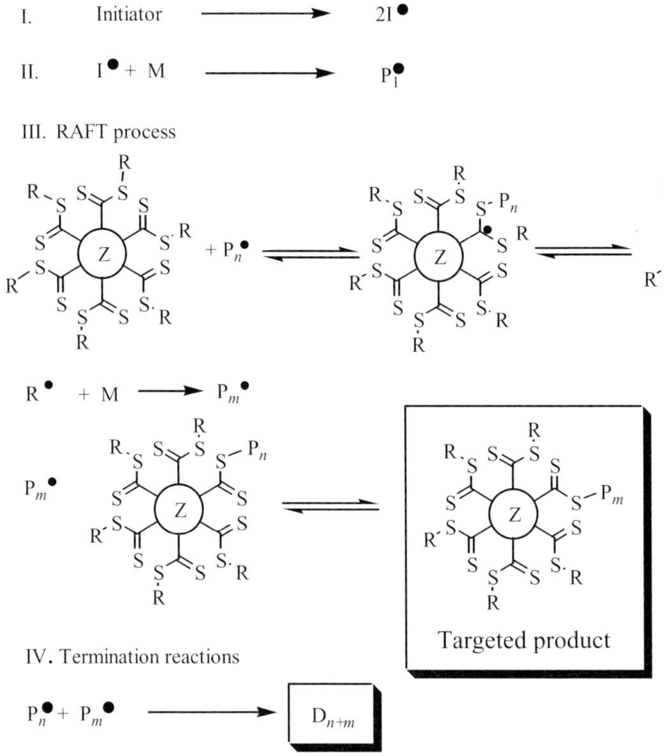


**Scheme 9.14** Schematic drawing of the synthesis of star polymers via Z-group approach.

may prevent more and more the diffusion of macroradicals into the core acting as an obstacle for the chain transfer to the RAFT agent.

**Practical Considerations and Experimental Results.** The proposed mechanism – depicted in Scheme 9.3 – suggests only the occurrence of the desired star product. Termination reactions occurring should only be derived by termination between two linear macroradicals. Their incidence can be influenced – similar to all RAFT processes – by a high ratio between thiocarbonylthio functionality and initiator concentration as well as other parameter such as rate of propagation.

Experiments using a range of cores and monomer (Table 9.3) indeed confirm the observation of a single product, while multimodal molecular-weight distributions are typically absent [62, 102–104, 107, 117, 119–123]. The molecular-weight shifts indicate increasing molecular weights of the star polymer with conversion. A more detailed inspection of some SEC curves reveals, however, that significant low-molecular-weight tailing occurs, broadening the molecular-weight distribution (Fig. 9.4). Occasionally, it has been observed that the molecular weight does not experience any significant shifts at high conversions. Figure 9.4 represents a typical example of a PS polymerization in the presence of a multifunctional RAFT agent.

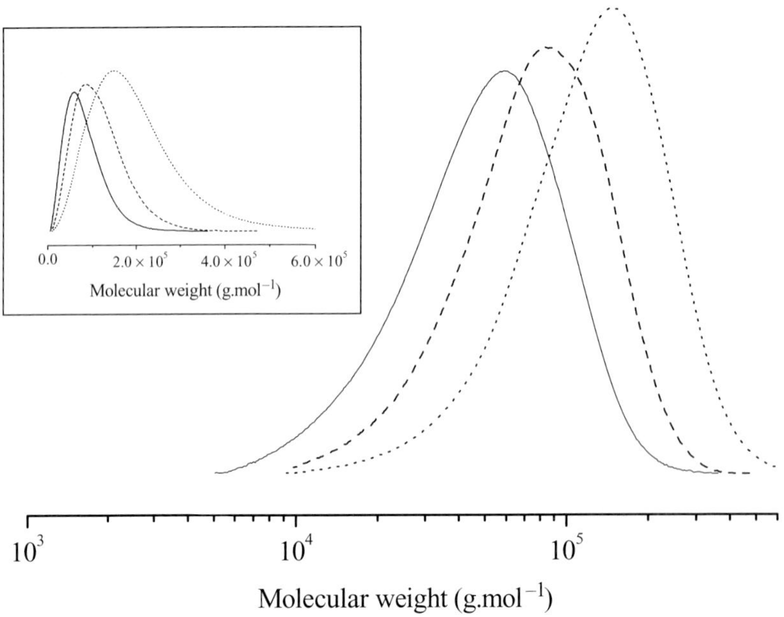


**Fig. 9.4** SEC traces of polystyrene, synthesized at 100 °C in the presence of a 7-arm cyclodextrine-based RAFT agent [103] (trithiocarbonate). The concentration of thiocarbonate groups is $1 \times 10^{-2}$ mol·L$^{-1}$. Samples were taken after 6, 13 and 23 h. The inserted figure shows the molecular-weight distribution without the logarithmic scale [100].

The molecular-weight distribution seems to broaden significantly with higher conversion. This effect becomes even more obvious when plotting the traces against the absolute molecular weight instead of employing the usual logarithmic scale.

Comparing theoretical molecular weights with measured values (disregarding the fact that some SEC systems do not provide absolute molecular-weight values) a significant deviation between the measured and the theoretical molecular weight was observed. The difference was observed to be dependent on conversions, RAFT agent concentration and the type of monomer used. A linear evolution between molecular weight and conversion close to the theoretical value was usually observed at low conversions, especially when using low monomer/RAFT ratios [103]. The deviation from the expected value became more pronounced at higher conversion when the growth of the molecular weight seemed to slow down until no further increase in molecular weight was measured [103, 104]. Closer inspection of the SEC curves reveal that the delay of the molecular-weight growth is accompanied by the broadening of the molecular-weight distribution and the appearance of a low-molecular-weight tail. Employing dithioesters as chain-transfer agents – which have usually a better chain-transfer efficiency – even reveals a distinct low-molecular-weight peak instead of low-molecular-weight tailing [106]. The tailing and/or the formation of bimodal distributions indicate the significant occurrence of side reactions. The cleavage of arms from the core by hydrolysis was

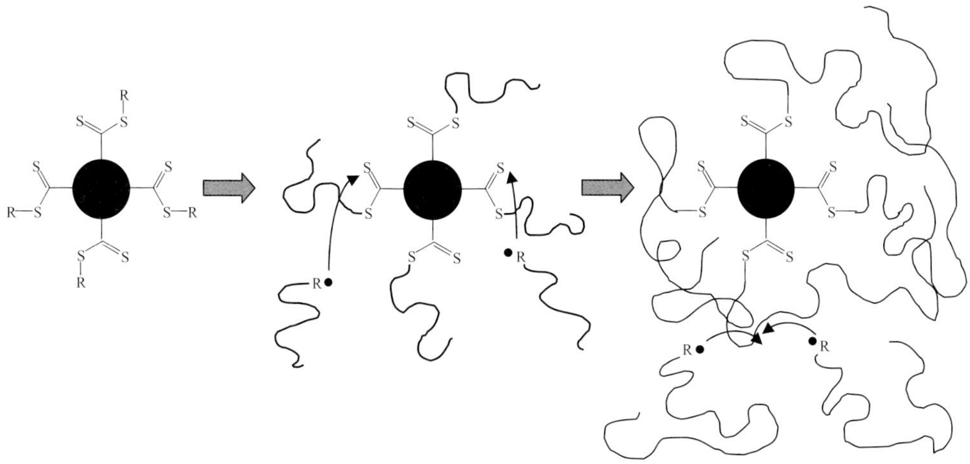


**Scheme 9.15** Schematic drawing of the hindered accessibility of the RAFT group during the synthesis of star polymers using the Z-group approach caused by the shielding effect of the growing polymer arms [100].

demonstrated to be an essential reaction step to further analyze the products [103]. The molecular weight of each arm is a vital indicator of the star growth process, revealing that indeed the star polymer stops growing in molecular weight at high conversions. In addition, temperature and monomer/RAFT ratios were observed to strongly influence this deviation [103]. One possible explanation for this behavior may be the increased occurrences of termination reactions between two linear macroradicals, as depicted in Scheme 9.14. Access of the macroradical to the RAFT group – which is adjacent to the core – may be prevented by a shielding effect of the star branches surrounding the core (Scheme 9.15). Reaching the core may become increasingly difficult; thus, the macroradicals rather undergo termination. This possible side reaction as illustrated in Scheme 9.15 is therefore indeed observed especially during the polymerization of styrene. In general, the shielding effect was found to be more or less pronounced depending on the system employed:

- *Number of arms*: The lower the number of arms, the better defined the resulting star polymers since the accessibility of the RAFT agent in the core is enhanced. Consequently, in the synthesis of three- [122] and four-arm [102] PS stars termination reactions were almost absent. While a 12-arm BA star possessed a narrow-molecular-weight distribution [107], a 17-arm star polymer contained termination products as confirmed when cleaving off the arms [121]. A high number of branches connected in one point (core) can immensely affect the conformation of each arm up to an entropic unsuitable change. As a result, the shielding effect is increased with increasing number of arms leading to the broadening of the molecular-weight distribution.
- *Type of core*: Interestingly, even the type of core was found to be responsible for a better accessibility. A hyperbranched core [104] with the same number of arms

as a well-defined dendritic [118] core did prevent further growth of the arm after a monomer conversion of 20%, while the dendritic core allowed the controlled growth of the star polymer to higher conversions.

- *Monomer to RAFT-agent concentration and conversion*: As a rule of thumb, the delay of branch growth is clearly a function of the length of the branch. High conversions or long branches caused by a high monomer/RAFT ratio are unfavorable for a well-defined growth.
- *Type of monomer*: Styrene was in general observed to lead to rather broad-molecular-weight distributions with significant low-molecular-weight tailing. The synthesis of stars based on acrylates was reported to proceed according to a living process [102, 104, 107, 123]. Other monomers such vinyl acetate [117, 120], NIPAAm [119, 125] and *N,N*-dimethylaminoethyl acrylate [119] were successfully employed in the synthesis of star polymers using the Z-group approach. This is easily understandable considering that every monomer has unique solution properties providing an enhanced or worsened contact to the center of the star.
- *Temperature*: High temperature can sometimes enhance diffusion of the polymer segments possibly leading to a better accessibility of the core [103].

High-molecular-weight shoulders were occasionally observed, especially when using acrylates. Detailed studies revealed that the observed star–star coupling was caused by intermolecular chain transfer to polymer. Kinetic simulation could indeed correlate the amount of star–star coupling to the rate coefficient of intermolecular transfer of the radical of several acrylates to the polymer [126].

### 9.3.2
### From Star Polymers via Arm-First Technique to Cross-Linked Micelles

The arm-first technique utilizes the simple approach to carry out the chain extension of the preformed arm, a macro-RAFT agent. Therefore, the thiocarbonylthio end groups are reactivated in a polymerization using divinyl compounds. This leads to the formation of star-shaped structures or microgels with a branch length determined by the size of the macro-RAFT agent and a cross-linked core (Scheme 9.16).

The synthesis of stars using the arm-first technique has been applied to obtain PS microgels, using PS macro-RAFT agents of varying molecular weights followed by cross-linking with divinyl benzene. The versatility of this approach is restricted by the difficulty to obtain narrow-molecular-weight distribution. Cross-linker concentration and reaction time have to be carefully adjusted to avoid the formation of a broad range of products with varying number of branches [127–129].

The broad-molecular-weight distribution of these microgels can be improved by employing a selective solvent to force the system into a self-assembled aggregate. Prerequisite is a different polarity of the arms forming the polymer and the gel-like core. For example, a PEO-based poly-RAFT agent was employed to generate a

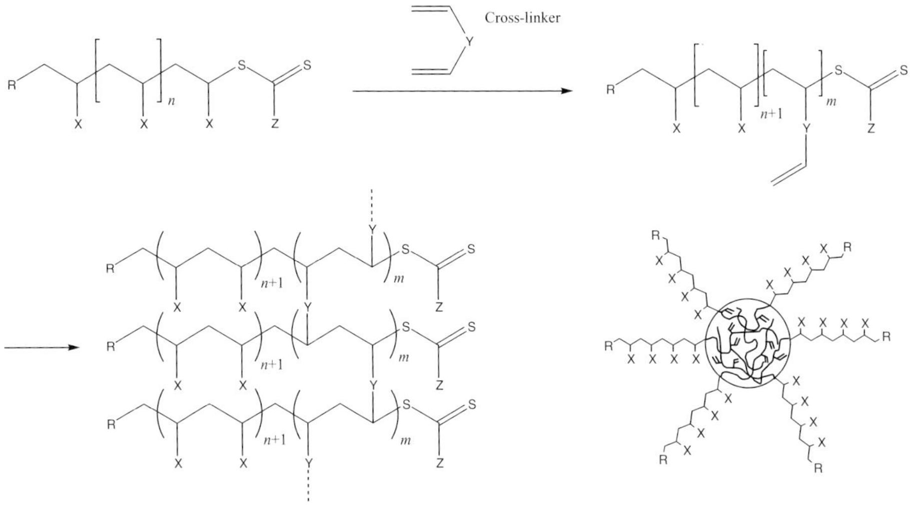


**Scheme 9.16** Schematic drawing of the synthesis of star polymers via arm-first approach.

microgel with PEO branches and a cross-linked styrene/divinyl benzene core (Table 9.4). When ethanol/THF (5:1 v/v) – a good solvent for PEO, but a nonsolvent for styrene and divinyl benzene – was utilized, self-assembly facilitated the formation of well-defined starlike structures [130]. Further similar approaches have been pursued and are summarized in Table 9.4.

The approach described is typically a one-pot reaction, leading to a fully cross-linked core. Cross-linked micelles can in addition be obtained in a subsequent reaction after the self-assembly of a block copolymer in a reactive solvent. Depending on the synthesis technique of the underlying block copolymer, the thiocarbonylthio group can be located either in the core, on the surface of the micelle or at the nexus of the two blocks (Scheme 9.17).

The successful synthesis of core-cross-linked micelles was demonstrated using either a poly(2-hydroxyethyl) acrylate-*block*-poly(*n*-butyl acrylate) RAFT [48] or a poly(*N*-isopropyl) acrylamide-*block*-polystyrene RAFT [134] block copolymer. Upon self-assembly in a selective solvent, micelles were obtained with the RAFT group being located in the core. Subsequent addition of cross-linking agent results in the fixation of the aggregate. The block copolymers of the cross-linked micelle is – in contrast to the one-pot technique described above – connected only in the center of the core.

With preparation of the core-forming block first, followed by the chain extension with the shell-forming polymer, a suitable system has been generated that allows further stabilization along the shell. The cross-linking process requires, however, significant optimization of the reaction conditions to prevent intermicellar cross-linking [135].

In addition, RAFT polymerization offers the opportunity to stabilize aggregates via cross-linking along the interface between shell and core. Prerequisite is the

**Table 9.4** Formation of microgels (or core cross-linked micelles) via arm-first technique

| Macro-RAFT agent | Solvent | Core | Ref. | |
|---|---|---|---|---|
| 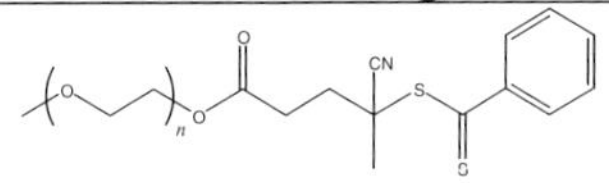 | Ethanol/THF | 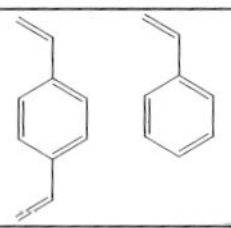 | [130] | 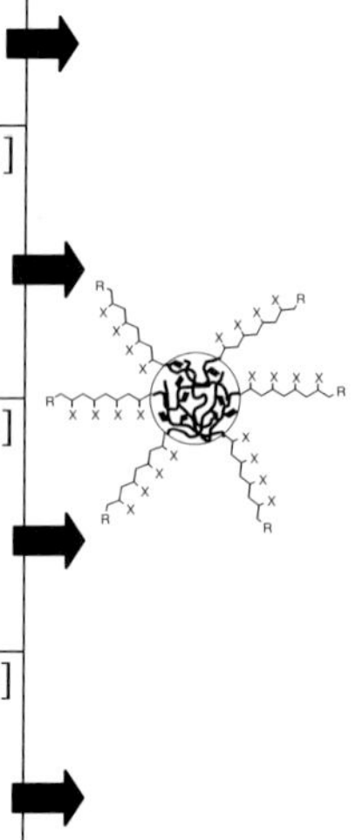 |
| 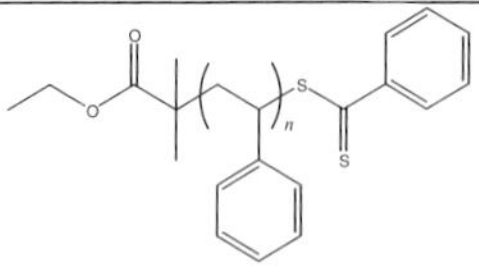 | Benzene | 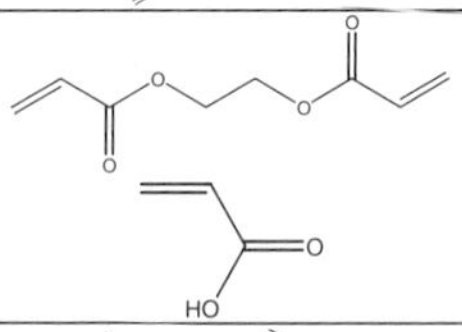 | [131] | |
| 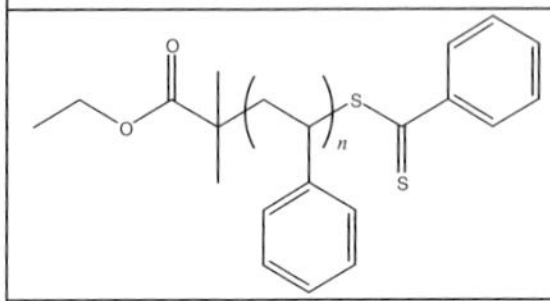 | Cyclohexane | 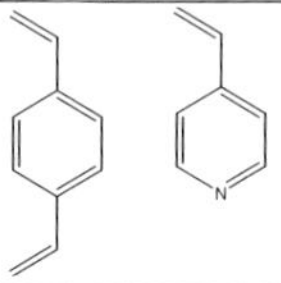 | [132] | |
| 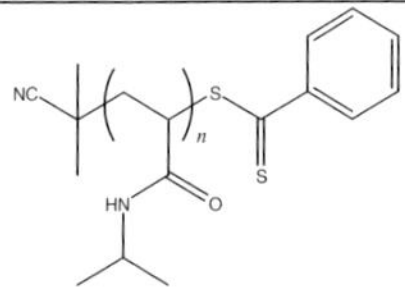 | Benzene | 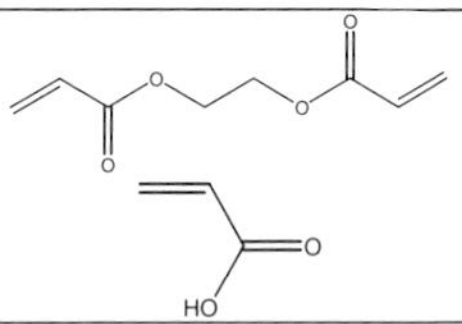 | [134] | |

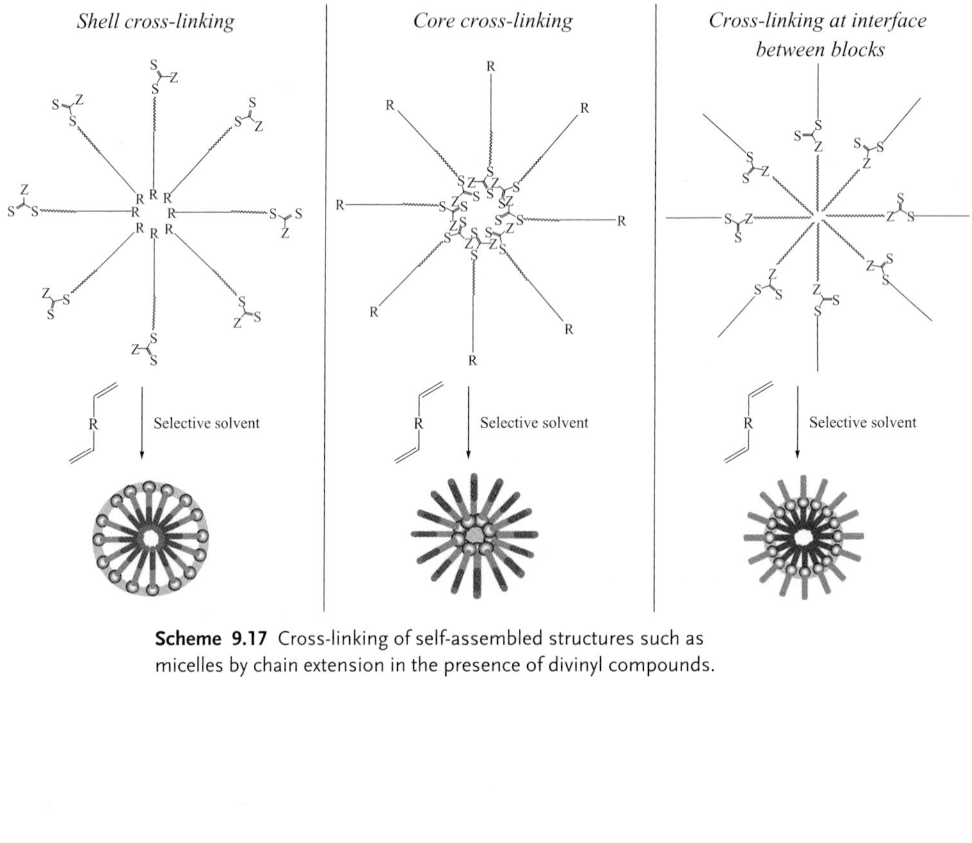

**Scheme 9.17** Cross-linking of self-assembled structures such as micelles by chain extension in the presence of divinyl compounds.

utilization of a poly-RAFT agent (generated by covalent attachment of a RAFT agent to a non-RAFT polymer), with the connection of the RAFT agent via Z group (Scheme 9.17) [85]. The resulting cross-linked aggregate was observed to behave only like the core-forming polymer – here PNIPAAm – but not like the underlying block copolymer.

### 9.3.3
### Other Techniques

Three-arm star polymers were prepared based on A–B–C triblock copolymers, which the second block B consisting only of one reactive repeating unit. Monomers, which cannot undergo homopolymerization such as maleic anhydride [136] or hydroxyethylene cinnamate [137], were employed to achieve $A_n$–$B_1$–$C_m$ structures. The reactive B block underwent reaction with an end-functionalized polymer, leading to three-arm star polymers with three chemically different arms – so-called miktoarm star polymers.

## 9.4
## Comb Polymers

Comb polymer – polymers with a linear backbone and a number of branches along this chain – can be prepared in a similar fashion to star polymers, using the attachment of a RAFT agent along a linear polymer chain – the backbone. The branches grow from the RAFT-agent anchor points in a controlled fashion. Branches were also prepared using other polymerization techniques by combing the RAFT process with a non-RAFT technique. A one-step approach is in contrast the random copolymerization of a macromonomer, which will form the branch with another (backbone-forming) monomer.

### 9.4.1
### Comb Polymers via Attachment of RAFT Agent to Backbone

Comb polymers via attachment of RAFT agent via R- or Z group have similar inferences regarding the mechanism and the resulting outcome as discussed above in the synthesis of star polymers. The RAFT agents can be attached either via R group (R-group approach) or via Z group (Z-group approach) (Scheme 9.19). The mechanism is similar to the approach to synthesize star polymers.

Comb polymers via R-group approach are subject to side reactions such as the formation of comb–comb termination products. The number of arms was proposed as a significant influence on the occurrence of these products. Indeed, comb polymers based on PS show a significant broadening caused by termination products. With increasing conversion, the fraction of termination products was observed to be so significant in order for gelation to occur [111]. In addition,

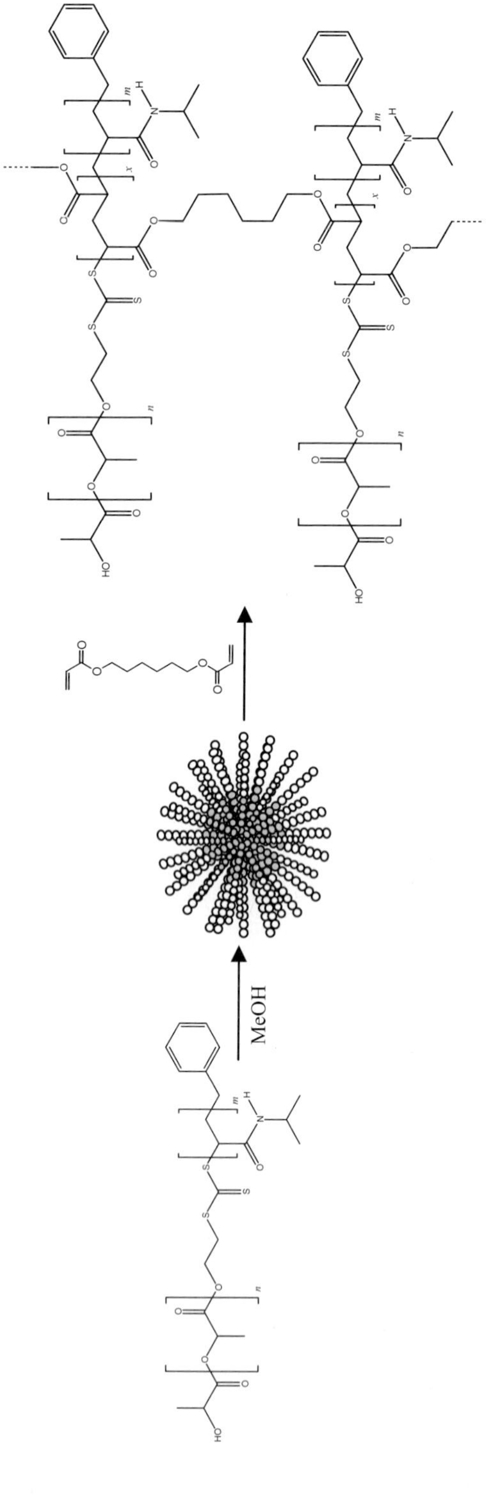


**Scheme 9.18** Cross-linking of micelles along the interface between core and shell [85].

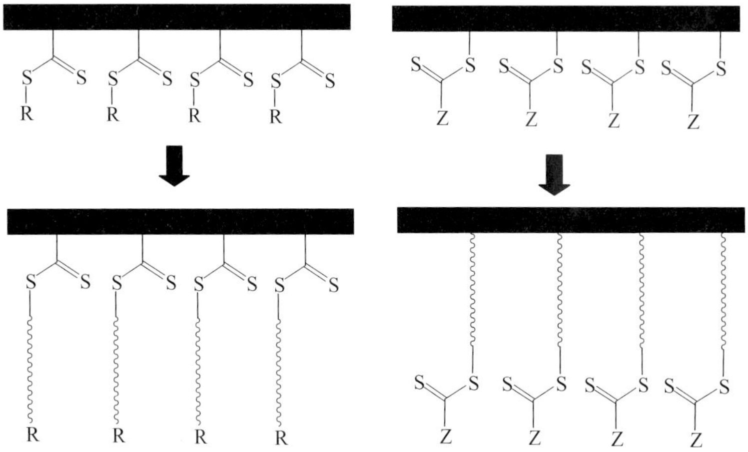


**Scheme 9.19** Synthesis of comb polymers via RAFT, employing the Z-group (left) or the R-group (right) approach.

termination reactions in comb polymers were found to be beyond the prediction via kinetic modeling and should possibly include solution parameter. PVAc star polymers with three- and four-arm stars followed the expected behavior since the fast-propagating monomer suppressed side reactions. Increasing the number of arm such as in comblike polymers should theoretically increase termination events. Multifunctional RAFT agents with 20, 100 and 200 dithioxanthate groups (Fig. 9.5) confirm the increased amount of comb–comb termination products with increasing number of arms. In parallel, the amount of linear macro-RAFT agents increased substantially with increasing number of branches, a trend against theoretical calculations (Fig. 9.5). Solution effects and the locally high concentration of RAFT groups along the backbone were suggested as origin of this effect [100, 138].

However, following the recommendations concerning the star synthesis via the R-group approach, better defined products can be obtained. Poly(butyl methacrylate) (poly-BMA) branches were grown from RAFT agents immobilized on a poly(butyl methacrylate) in a controlled manner by increasing the amount of RAFT groups considerably resulting in the formation of short well-defined branches [139].

Attachment of RAFT agents via the Z group attracts similar difficulties, such as the increased occurrence of termination reactions of two linear macroradicals. The hydrolysis of arms also revealed that the growth of each branch deviates even further from the expected value than during the star synthesis. The high local concentration of RAFT agents along a backbone results in an uneven distribution of RAFT agents in the solution. The chain transfer was found to be delayed, resulting in an uneven growth of branches. While some branches already reached a significant length, other RAFT agents still remained inactivated [140, 141].

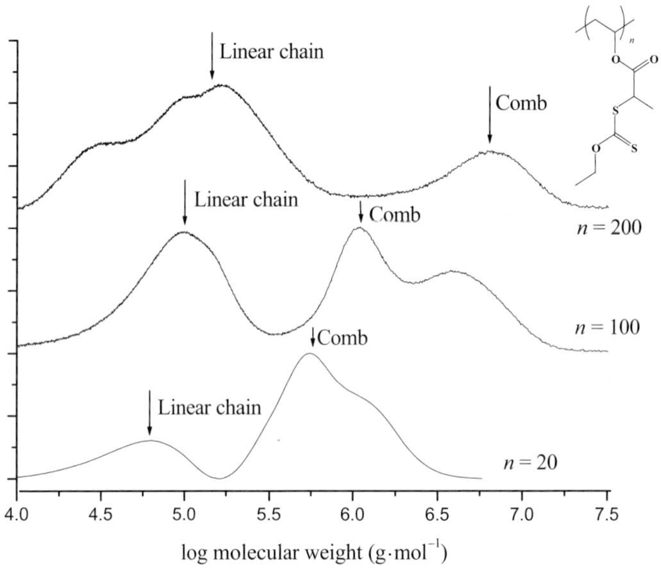


**Fig. 9.5** SEC curves of the bulk polymerization of VAc in the presence of varying sized multifunctional RAFT/MADIX agents at 60 °C with [xanthate groups] = $1.1 \times 10^{-2}$ mol·L$^{-1}$ and [AIBN] = $2.2 \times 10^{-3}$ mol·L$^{-1}$. The conversion is between 30 and 40%. Gel products were removed via filtration when using $n = 200$ [100].

### 9.4.2
### Comb Polymers via Attachment of Initiator to Backbone

Grafted chains were also obtained by attaching the initiator to the backbone instead of a RAFT agent. The length of the grafted chain is then controlled by the added RAFT agent (Scheme 9.20). The growth of the branch is subject to the decomposition rate of the initiator. Potential side products are linear macro-RAFT agents generated from the leaving group R of the RAFT agent preceding chain transfer.

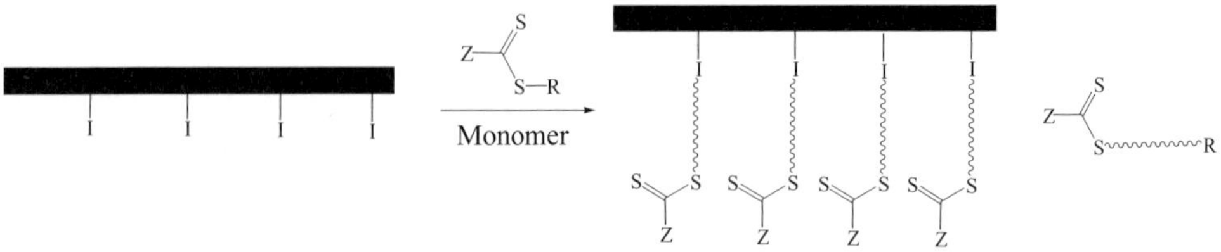


**Scheme 9.20** Synthesis of comb polymers via immobilization of radical initiator along the backbone and added RAFT agent to control the branch length.

PEGMA branches have been generated along a fluorinated polyimide [142] and a poly(vinylidene fluoride) backbone [143]. The initiator was in both cases generated by ozone treatment. Reasonably well-defined products were obtained, considering that the underlying polymers had already a broad-molecular-weight distribution. The distribution was improved by precipitation of the product in ethanol to remove linear PEGMA macro-RAFT agents.

### 9.4.3
### Comb Polymers Using Macromonomers

An alternative pathway to comb polymers is the controlled polymerization of macromonomers (Scheme 9.21).

A common macromonomer, which has already been described earlier, is a monomer based on polyethylene oxide, such as poly(ethylene glycol) (meth)acrylate (PEGMA) and poly(ethylene glycol) acrylate (PEGA). However, the molecular weight of PEGMA and PEGA was typically below 600 g·mol$^{-1}$. A PEO-based macromonomers with molecular weights of $M_n$ = 2000 g·mol$^{-1}$ was utilized to chain extend a PS macro-RAFT agent in order to prepare real toothbrushlike structures, with a PS block forming the handle and poly(PEGA) creating the brush [144]. The molecular weight of the PS macro-RAFT agent was found to influence the rate of polymerization considerably. Similar to the synthesis of block copolymers, a decrease in the rate of polymerization of the macromonomer with increasing molecular weight of PS macro-RAFT agent was noted [144].

PDMS-based methacrylate [145, 146] was copolymerized with methyl methacrylate, leading to comb polymers with a random branch distributions. The apparent reactivity ratio during the RAFT process was shown to derivate from the values observed using free-radical polymerization. Additionally, higher temperatures of the RAFT polymerization seem to influence the apparent reactivity ratios in favor of the macromonomers [145]. This was ascribed to the chemical structure of the macromonomer, which affects diffusion associated with the large molecular weight of the macromonomer and possible incompatibility effects between macromonomer and macroradical [146].

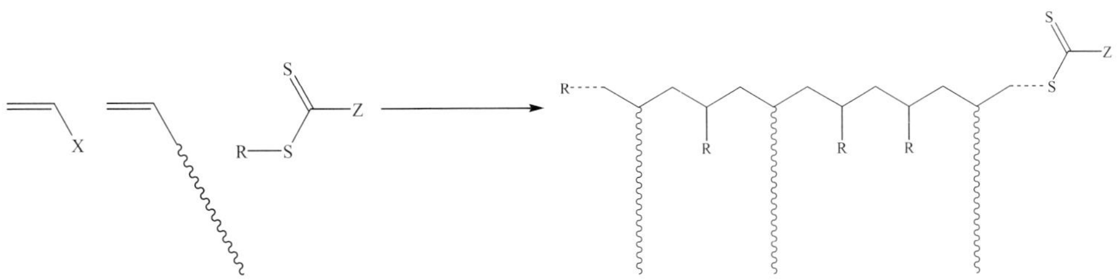


**Scheme 9.21** Synthesis of comb polymers via RAFT polymerization of macromonomers.

### 9.4.4
### Comb Polymers via Combination of RAFT and Other Techniques

The growth of branches from a backbone can be carried out by combining RAFT polymerizations with a range of other polymerization techniques.

Hydroxy groups in the backbone were utilized to initiate ring-opening polymerization. A random copolymer prepared via RAFT polymerization from styrene and 2-hydroxyethyl methacrylate functioned as the backbone and as initiator for the ring-opening polymerization of $\varepsilon$-caprolactone [147]. This work was then extended to prepare toothbrushlike structures by employing PEO-modified poly-RAFT agent, leading to a structure with a hydrophilic PEO handle and a brush consisting of poly($\varepsilon$-caprolactone) (Scheme 9.22) [84, 148]. The initiating efficiency was found to be very high with almost all hydroxy group leading to the growth of a branch.

A similar approach was applied to generate poly(tetramethylene oxide) branches by ring-opening polymerization of THF initiated by a well-defined backbone of randomly copolymerized styrene and chloromethyl styrene [149].

Chloromethyl styrene can not only initiate ring-opening polymerization, but also act as an initiator for the controlled polymerization via atom-transfer radical polymerization (ATRP). An A—B—C triblock copolymer – prepared by subsequent chain-extension steps using the RAFT process – composed of glycidyl methacrylate, chloromethyl styrene and polyethylene glycol methacrylate was utilized as a macroinitiator in the controlled polymerization of styrene via ATRP, resulting in well-defined PS brushes [150]. The same pathway was pursued to generate poly(methyl methacrylate) brushes growing from a PS backbone [151].

All the techniques suggested above are designed to obtain comb polymers with a high branching density, which resembles brushlike structures, or to generate comb polymers with a low branching density, but with the distribution of branching points being subject to reactivity ratio of the two monomers forming the backbone.

A controlled distance between two branching points is achieved by generating a multiblock copolymer containing a multitude of RAFT agents. Reaction with a functional monomer that cannot homopolymerize introduces anchor points for the branch in order to obtain regular comb polymers (Scheme 9.23) [152].

## 9.5
## Other Complex Architectures

As outlined in Scheme 9.1, a range of architectures are theoretically possible. However, not all these structures have been realized. However, RAFT polymerization in combination with a range of other polymerization techniques such as polycondensation and ring-opening polymerization do facilitate access to these structures. H-shaped polymers based on PS, PEO and PLA [153] or based on PS and poly(1,3-dioxepane) [154] were reported as well as $\pi$-shaped polymers [155] (Scheme 9.24). Access to novel architectures was also demonstrated by combining dendrimer chemistry with the RAFT approach [156].

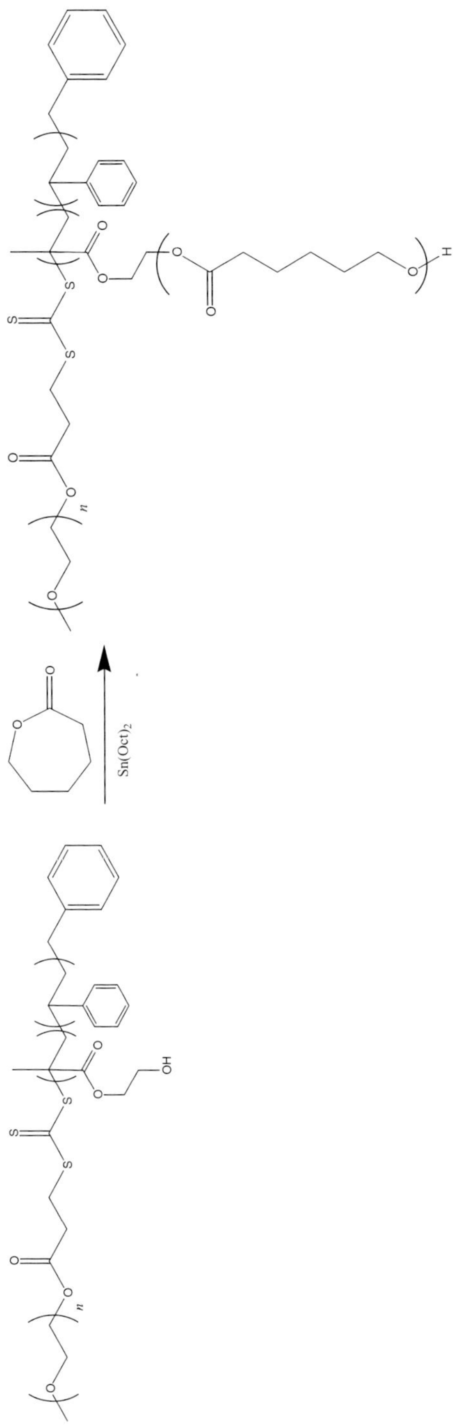


**Scheme 9.22** Synthesis of comb polymers via ring-opening polymerization from a backbone prepared via RAFT polymerization [84, 148].

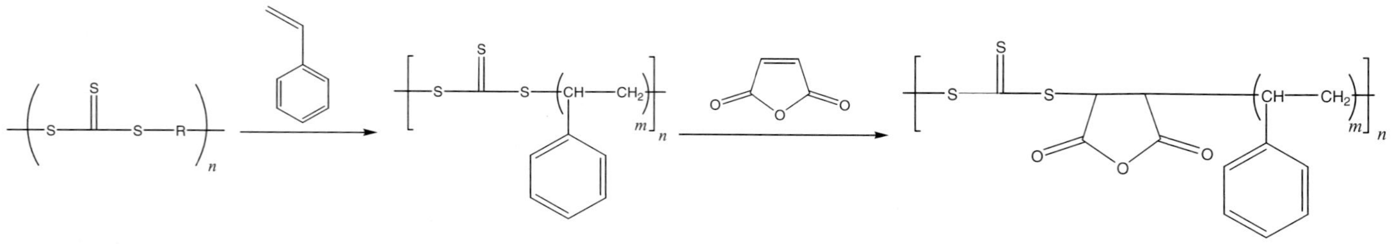

**Scheme 9.23** Synthesis of a backbone for comb polymers with equidistant branching points [152].

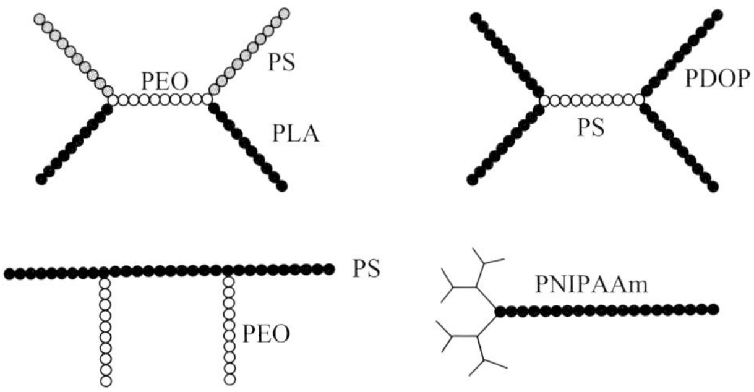


**Scheme 9.24** Various complex architectures prepared by combination of RAFT with other techniques.

## 9.6 Conclusions

Despite certain difficulties such as the occurrence of termination reactions, RAFT polymerization was shown to be a versatile tool to access a range of different complex architectures. Especially the broader range of monomers that can be polymerized in a controlled manner as well as the robustness of the process in the presence of a range of functional groups makes the RAFT process unique and advantageous.

Therefore, RAFT polymerization is a powerful way to obtain a range of structures with control over architecture as well as molecular weight. These complex polymer architectures were already successfully employed in a range of applications to general novel materials. These applications include polymer–protein conjugates [157], polymer–peptide conjugates [158] and polymer–oligonucletide conjugates [159]. Honeycomb-structured porous films – films with a highly regular hexagonal order of pores – were obtained using a variety of complex architectures [160]. Especially, amphiphilic block copolymers were investigated regarding their self-assembly properties in order to produce core-shell nanoparticles [39–41, 47, 48, 54, 57, 59, 63, 68, 79, 85, 96, 99].

Details of the application of these complex polymer architectures can be found in Chapter 13.

### Abbreviations

| | |
|---|---|
| ATRP | Atom-transfer radical polymerization |
| BA | *n*-Butyl acrylate |
| BMA | *n*-Butyl methacrylate |
| CuAAc | Copper-catalyzed alkyne and azide 1,3-dipolar cycloaddition |
| DA | Diels Alder |
| DMA | *N*,*N*-dimethyl acrylamide |
| MA | Methyl acrylate |

| | |
|---|---|
| NIPAAm | *N*-isopropyl acrylamide |
| PDMS | Poly(dimethylsiloxane) |
| PEGA | Poly(ethylene glycol) acrylate |
| PEGMA | Poly(ethylene glycol) methacrylate |
| PEO | Poly(ethylene oxide) |
| PLA | Poly(lactide) |
| PNIPAAm | Poly(*N*-isopropyl acrylamide) |
| PS | Polystyrene |
| PVAc | Poly(vinyl acetate) |
| RAFT | Reversible addition–fragmentation chain transfer |
| SEC | Size-exclusion chromatography |
| VAc | Vinyl acetate |

## References


1. M. J. Monteiro, *J. Polym. Sci., Part A: Polym. Chem.* **2005**, *43*, 5643–5651.
2. B. Y. K. Chong, T. P. L. Le, G. Moad, E. Rizzardo, S. H. Tang, *Macromolecules* **1999**, *32*, 2071–2074.
3. G. K. Such, R. A. Evans, T. P. Davis, *Macromolecules* **2006**, *39*, 9562–9570.
4. M. Save, M. Manguian, C. Chassenieux, B. Charleux, *Macromolecules* **2005**, *38*, 280–289.
5. B. S. Sumerlin, M. S. Donavan, Y. Mitsukami, A. B. Lowe, C. L. McCormick, *Macromolecules* **2001**, *34*, 6561–6564.
6. H. Mori, S. Nakano, T. Endo, *Macromolecules* **2005**, *38*, 8192–8201.
7. C. Tang, T. Kowalewski, K. Matyjaszewski, *Macromolecules* **2003**, *36*, 8587–8590.
8. Y. A. Vasilieva, D. B. Thomas, C. W. Scales, C. L. McCormick, *Macromolecules* **2004**, *37*, 2728–2737.
9. B. de Lambert, M.-T. Charreyre, C. Chaix, C. Pichot, *Polymer* **2007**, *48*, 437–447.
10. Y. Zhang, S. Luo, S. Liu, *Macromolecules* **2005**, *38*, 9813–9820.
11. T. Tang, V. Castelletto, P. Parras, I. W. Hamley, S. M. King, D. Roy, S. Perrier, R. Hoogenboom, U. S. Schubert, *Macromol. Chem. Phys.* **2006**, *207*, 1718–1726.
12. C. Yang, Y.-L. Cheng, *J. Appl. Polym. Sci.* **2006**, *102*, 1191–1201.
13. M. Arotçaréna, B. Heise, S. Ishaya, A. Laschewsky, *J. Am. Chem. Soc.* **2002**, *124*, 3787–3793.
14. P. Vana, T. P. Davis, C. Barner-Kowollik, *Macromol. Theory Simul.* **2002**, *11*, 823–835.
15. L. Albertin, M. H. Stenzel, C. Barner-Kowollik, L. J. R. Foster, T. P. Davis, *Macromolecules* **2004**, *37*, 7530–7537.
16. Z. Szablan, A. Ah Toy, M. H. Stenzel, T. P. Davis, C. Barner-Kowollik, *J. Polym. Sci. Polym. Chem.* **2004**, *42*, 2432–2443.
17. J. Xu, J. He, D. Fan, W. Tang, Y. Yang, *Macromolecules* **2006**, *39*, 3753–3759.
18. J. F. Quinn, L. Barner, C. Barner-Kowollik, E. Rizzardo, T. P. Davis, *Macromolecules* **2002**, *35*, 7620–7627.
19. L. Albertin, M. H. Stenzel, C. Barner-Kowollik, T. P. Davis, *Polymer* **2006**, *47*, 1011–1019.
20. P. Vana, L. Albertin, T. P. Davis, L. Barner, C. Barner-Kowollik, *J. Polym. Sci. Polym. Chem.* **2002**, *40*, 4032–4037.
21. D. B. Thomas, A. J. Convertine, R. D. Hester, A. B. Lowe, C. L. McCormick, *Macromolecules* **2004**, *37*, 1735–1741.
22. K. H. Wong, T. P. Davis, C. Barner-Kowollik, M. H. Stenzel, *Aust. J. Chem.* **2006**, *59*, 539–543.
23. X. Hao, M. H. Stenzel, C. Barner-Kowollik, T. P. Davis,

E. Evans, *Polymer* **2004**, *45*, 7401–7415.
**24.** P. Relógio, M.-T. Charreyre, J. P. S. Farinha, J. M. G. Martinho, C. Pichot, *Polymer* **2004**, *45*, 8639–8649.
**25.** A. Laschewsky, M. Mertoglu, S. Kubowicz, A. F. Thünemann, *Macromolecules* **2006**, *39*, 9337–9345.
**26.** T. Krasia, R. Soula, H. G. Börner, H. Schlaad, *Chem. Commun.* **2003**, 538–539.
**27.** Y.-C. Hu, C.-Y. Pan, *J. Polym. Sci. Polym. Chem.* **2004**, *42*, 4862–4872.
**28.** M. S. Donavan, A. B. Lowe, T. A. Sanford, C. L. McCormick, *J. Polym. Sci., Part A: Polym. Chem.* **2003**, *41*, 1262–1281.
**29.** L. Barner, T. P. Davis, M. H. Stenzel, C. Barner-Kowollik, *Macromol. Rapid Commun.* **2007**, *28*, 539–559.
**30.** L. Albertin, M. H. Stenzel, C. Barner-Kowollik, L. J. R. Foster, T. P. Davis, *Macromolecules* **2005**, *38*, 9075–9084.
**31.** D. Fan, J. He, J. Xu, W. Tang, Y. Liu, Y. Yang, *J. Polym. Sci., Part A: Polym. Chem.* **2006**, *44*, 2260–2269.
**32.** K. H. Wong, T. P. Davis, C. Barner-Kowollik, M. H. Stenzel, *Polymer*, **2007**, *48*, 4950–4965.
**33.** D. Fan, J. He, J. Xu, W. Tang, Y. Liu, Y. Yang, *J. Polym. Sci., Part A: Polym. Chem.* **2006**, *44*, 2260–2269.
**34.** R. Hoogenboom, U. S. Schubert, W. van Camp, F. Du Prez, *Macromolecules* **2005**, *38*, 7653–7659.
**35.** D. Beattie, K. H. Wong, C. Williams, L. A. Poole-Warren, T. P. Davis, C. Barner-Kowollik, M. H. Stenzel, *Biomacromolecules* **2006**, *7*, 1072–1082.
**36.** B.-Q. Zhang, G.-D. Chen, C.-Y. Pan, B. Luan, C.-Y. Hong, *J. Appl. Polym. Sci.* **2006**, *102*, 1950–1958.
**37.** S. Kanagasabapathy, A. Sudalai, B. C. Benicewicz, *Macromol. Rapid Commun.* **2001**, *22*, 1076–1080.
**38.** S. Garnier, A. Laschewsky, *Macromolecules* **2005**, *38*, 7580–7592.
**39.** M. H. Stenzel, C. Barner-Kowollik, T. P. Davis, H. M. Dalton, *Macromol. Biosci.* **2004**, *4*, 445–453.
**40.** S. Yusa, Y. Shimada, Y. Mitsukami, T. Yamamoto, Y. Morishima, *Macromolecules* **2003**, *36*, 4208–4215.
**41.** S.-I. Yusa, K. Fukuda, T. Yamamoto, K. Ishihara, Y. Morishima, *Biomacromolecules* **2005**, *6*, 663–670.
**42.** S.-I. Yusa, Y. Shimada, Y. Mitsukami, T. Yamamoto, Y. Morishima, *Macromolecules* **2004**, *37*, 7507–7513.
**43.** M. Eberhardt, P. Theato, *Macromol. Rapid Commun.* **2005**, *26*, 1488–1493.
**44.** P. Bilalis, M. Pitsikalis, N. Hadjichristidis, *J. Polym. Sci., Part A: Polym. Chem.* **2006**, *44*, 659–665.
**45.** V. Jitchum, S. Perrier, *Macromolecules* **2007**, *40*, 1408–1412.
**46.** M. Mertoglu, S. Garnier, A. Laschewsky, K. Skrabania, J. Storsberg, *Polymer* **2005**, *46*, 7726–7740.
**47.** S. Harrisson, K. L. Wooley, *Chem. Commun.* **2005**, 3259–3261.
**48.** L. Zhang, K. Katapodi, T. P. Davis, C. Barner-Kowollik, M. H. Stenzel, *J. Polym. Sci. – Polym. Chem.* **2006**, *44*, 2177–2194.
**49.** L. T. U. Nguyen, K. Eagles, T. P. Davis, C. Barner-Kowollik, M. H. Stenzel, *Polym. Sci. Polym. Chem.* **2006**, *44*, 4372–4383.
**50.** Z. Cheng, X. Zhu, E. T. Kang, K. G. Neoh, *Langmuir* **2005**, *21*, 7180–7185.
**51.** B. S. Sumerlin, A. B. Lowe, D. B. Thomas, C. L. McCormick, *Macromolecules* **2003**, *36*, 5982–5987.
**52.** Y. Mitsukami, M. S. Donavan, A. B. Lowe, C. L. McCormick, *Macromolecules* **2001**, *34*, 2248–2256.
**53.** M. H. Stenzel, T. P. Davis, *Aust. J. Chem.* **2003**, *56*, 1035–1038.
**54.** C. W. Scales, F. Huang, N. Li, Y. A. Vasilieva, J. Ray, A. J. Convertine, C. L. McCormick, *Macromolecules* **2006**, *39*, 6871–6881.
**55.** A. J. Convertine, B. S. Sumerlin, D. B. Thomas, A. B. Lowe, C. L. McCormick, *Macromolecules* **2003**, *36*, 4679–4681.
**56.** B. Lowe, R. Wan, *Polymer* **2007**, *48*, 2221–2230.
**57.** B. Liu, S. Perrier, *J. Polym. Sci., Part A: Polym. Chem.* **2005**, *43*, 3643–3654.
**58.** A. J. Convertine, B. S. Lokitz, Y. Vasileva, L. J. Myrick, C. W. Scales, A. B. Lowe, C. L. McCormick, *Macromolecules* **2006**, *39*, 1724–1730.
**59.** M. Nakayama, T. Okano, *Biomacromolecules* **2005**, *6*, 2320–2327.

**60.** M. Nuopponen, J. Ojala, H. Tenhu, *Polymer* **2004**, *45*, 3643–3650.

**61.** P. Zhang, Q. Liu, A. Qing, J. Shi, M. Lu, *J. Polym. Sci., Part A: Polym. Chem.* **2006**, *44*, 3312–3320.

**62.** J. Bernard, X. Hao, T. P. Davis, C. Barner-Kowollik, M. H. Stenzel, *Biomacromolecules* **2006**, *7*, 232–238.

**63.** B. S. Lokitz, A. J. Convertine, R. G. Ezell, A. Heidenreich, Y. Li, C. McCormick, *Macromolecules* **2006**, *39*, 8594–8602.

**64.** C. M. Schilli, M. Zhang, E. Rizzardo, S. H. Tang, B. Y. K. Chong, K. Edwards, G. Karlsson, A. H. E. Müller, *Macromolecules* **2004**, *37*, 7861–7866.

**65.** Z. Ge, Y. Cai, J. Yin, J. Rao, S. Liu, *Langmuir* **2007**, *23*, 1114–1122.

**66.** S.-I. Yusa, Y. Konishi, Y. Mitsukami, T. Yamamoto, Y. Morishima, *Polym. J.* **2005**, *37*, 480–488.

**67.** Y. Mitsukami, M. S. Donavan, A. B. Lowe, C. L. McCormick, *Macromolecules* **2001**, *34*, 2248–2340.

**68.** Y. Li, B. S. Lokitz, C. L. McCormick, *Angew. Chem. Int. Ed.* **2006**, *45*, 5792–5795.

**69.** X. Xin, Y. Wang, W. Liu, *Eur. Polym. J.* **2005**, *41*, 1539–1545.

**70.** R. Bussels, C. Bergman-Goettgens, B. Klumperman, J. Meuldijk, C. Koning, *J. Polym. Sci., Part A: Polym. Chem.* **2006**, *44*, 6419–6434.

**71.** R. T. A. Mayadunne, E. Rizzardo, J. Chiefari, J. Krstina, G. Moad, A. Postma, S. H. Tang, *Macromolecules* **2000**, *33*, 243–245.

**72.** J.-J. Yuan, R. Ma, Q. Gao, Y.-F. Wang, S.-Y. Cheng, L.-X. Feng, Z.-Q. Fan, L. Jiang, *J. Appl. Polym. Sci.* **2003**, *89*, 1017–1025.

**73.** D.-H. Han, C.-Y. Pan, *Macromol. Chem. Phys.* **2006**, *207*, 836–843.

**74.** A. Feldermann, A. Ah Toy, H. Phan, M. H. Stenzel, T. P. Davis, C. Barner-Kowollik, *Polymer* **2004**, *45*, 3997–4007.

**75.** L. Shi, T. M. Chapman, E. J. Beckman, *Macromolecules* **2003**, *36*, 2563–2567.

**76.** Z. Ma, P. Lacroix-Desmazes, *J. Polym. Sci., Part A: Polym. Chem.* **2004**, *42*, 2405–2415.

**77.** Z. Ma, P. Lacroix-Desmazes, *Polymer* **2004**, *45*, 6789–6797.

**78.** C.-Y. Hong, Y.-Z. You, C.-Y. Pan, *J. Polym. Sci., Part A: Polym. Chem.* **2004**, *42*, 4873–4881.

**79.** Y. Li, B. Lokitz, C. L. McCormick, *Macromolecules* **2006**, *39*, 81–89.

**80.** H. De Brower, M. A. J. Schellekens, B. Klumperman, M. J. Monteiro, A. L. German, *J. Polym. Sci., Part A: Polym. Chem.* **2000**, *38*, 3596–3603.

**81.** N. Kawahara, S.-I. Kojoh, S. Matsuo, H. Kaneko, T. Matsugi, J. Saito, N. Kashiwa, *Polym. Bull.* **2006**, *57*, 805–812.

**82.** R. Godoy-Lopez, C. Boisson, F. D'Agosto, R. Spoutz, F. Boisson, D. Gigmes, D. Bertin, *Macromol. Rapid Commun.* **2006**, *27*, 173–181.

**83.** T. S. C. Pai, C. Barner-Kowollik, T. P. Davis, M. H. Stenzel, *Polymer* **2004**, *45*, 4383–4383.

**84.** X. Xu, Z. Jia, R. Sun, J. Huang, *J. Polym. Sci., Part A: Polym. Chem.* **2006**, *44*, 4396–4408.

**85.** M. Hales, C. Barner-Kowollik, T. P. Davis, M. H. Stenzel, *Langmuir* **2004**, *20*, 10809–10817.

**86.** Y. You, C. Hong, W. Wang, W. Lu, C. Pan, *Macromolecules* **2004**, *37*, 9761–9761.

**87.** M. K. Mahanthappa, F. S. Bates, M. A. Hillmeyer, *Macromolecules* **2005**, *38*, 7890–7894.

**88.** K. J. Thurecht, A. M. Gregory, S. Villarroya, J. Zhou, A. Heise, S. M. Howdle, *Chem. Commun.* **2006**, 4383–4385.

**89.** Z. Jia, X. Xu, Q. Fu, J. Huang, *J. Polym. Sci., Part A: Polym. Chem.* **2006**, *44*, 6071–6082.

**90.** S. Motokucho, A. Sudo, F. Sanda, T. Endo, *Chem. Commun.* **2002**, 1956–1957.

**91.** M. Stenzel, L. Cummins, G. E. Roberts, T. P. Davis, P. Vana, C. Barner-Kowollik, *Macromol. Chem. Phys.* **2003**, *204*, 1160–1168.

**92.** C. Barner-Kowollik, J. F. Quinn, D. R. Morsley, T. P. Davis, *J. Polym. Sci., Part A: Polym. Chem.* **2001**, *39*, 1353–1365.

**93.** A. Theis, M. H. Stenzel, T. P. Davis, M. L. Coote, C. Barner-Kowollik, *Aust. J. Chem.* **2005**, *58*, 437–441.

**94.** H. C. Kolb, M. G. Finn, K. B. Sharpless, *Angew. Chem. Int. Ed.* **2001**, *40*, 2004–2021.

**95.** D. Quemener, T. P. Davis, C. Barner-Kowollik, M. H. Stenzel, *Chem. Commun.* **2006**, 5051–5053.

**96.** S. R. Simon Ting, A. Granville, T. P. Davis, M. H. Stenzel, C. Barner-Kowollik, *Aust. J. Chem.*, **2007**, *60*, 405–409.

**97.** S. R. Gondi, A. P. Vogt, B. S. Sumerlin, *Macromolecules* **2007**, *40*, 474–481.

**98.** R. K. O'Reilly, M. J. Joralemon, C. J. Hawker, K. L. Wooley, *J. Polym. Sci., Part A: Polym. Chem.* **2006**, *44*, 5203–5217.

**99.** R. K. O'Reilly, M. J. Joralemon, C. J. Hawker, K. L. Wooley, *Chem. Eur. J.* **2006**, *12*, 6776–6786.

**100.** C. Barner-Kowollik, T. P. Davis, M. H. Stenzel, *Aust. J. Chem* **2006**, *59*, 719–727.

**101.** M. Stenzel-Rosenbaum, T. P. Davis, A. G. Fane, V. Chen, *J. Polym. Sci., Part A: Polym. Chem.* **2001**, *39*, 2777–2783.

**102.** R. T. A Mayadunne, J. Jeffery, G. Moad, E. Rizzardo, *Macromolecules* **2003**, *36*, 1505–1513.

**103.** M. H. Stenzel, T. P. Davis, *J. Polym. Sci., Part A: Polym. Chem.* **2002**, *40*, 4498–4512.

**104.** M. Jesberger, L. Barner, M. H. Stenzel, E. Malmström, T. P. Davis, C. Barner-Kowollik, *J. Polym. Sci., Part A: Polym. Chem.* **2003**, *41*, 3847–3861.

**105.** M. Chen, K. P. Ghiggino, A. Launikonis, A. W. H. Mau, E. Rizzardo, W. H. F. Sasse, S. H. Thang, G. J. Wilson, *J. Mater. Chem.* **2003**, *13*, 2696–2700.

**106.** V. Darcos, A. Duréault, D. Taton, Y. Gnanou, P. Marchand, A. M. Caminade, J. P. Majoral, M. Destarac, F. Leising, *Chem. Commun.* **2004**, *18*, 2110–2111.

**107.** X. Hao, C. Nilsson, M. Jesberger, M. H. Stenzel, E. Malmström, T. P. Davis, E. Östmark, C. Barner-Kowollik, *J. Polym. Sci., Part A: Polym. Chem.* **2004**, *42*, 5877–5890.

**108.** A. Feldermann, M. L. Coote, M. H. Stenzel, T. P. Davis, C. Barner-Kowollik, *J. Am. Chem. Soc.* **2004**, *126*, 15915–15923.

**109.** H. Chaffey-Millar, M. Busch, T. P. Davis, M. H. Stenzel, C. Barner-Kowollik, *Macromol. Theory Simul.* **2005**, *14*, 143–157.

**110.** H. Chaffey-Millar, M. H. Stenzel, T. P. Davis, M. L. Coote, C. Barner-Kowollik, *Macromolecules* **2006**, *39*, 6406–6419.

**111.** J. F. Quinn, R. P. Chaplin, T. P. Davis, *J. Polym. Sci., Part A: Polym. Chem.* **2002**, *40*, 2956–2966.

**112.** M. Chen, K. P. Ghiggino, S. H. Tang, G. J. Wilson, *Polym. Int.* **2005**, *55*, 757–763.

**113.** E. Southard, K. A. van Houten, G. M. Murray, *Macromolecules* **2007**, *40*, 1395–1399.

**114.** C.-Y. Hong, Y.-Z. You, J. Liu, C.-Y. Pan, *Polym. Sci., Part A: Polym. Chem.* **2005**, *43*, 6379–6393.

**115.** Q. Zheng, C.-Y. Pan, *Eur. Polym. J.* **2006**, *42*, 807–814.

**116.** Q. Zheng, C.-Y. Pan, *Macromolecules* **2005**, *38*, 6841–6848.

**117.** J. Bernard, A. Favier, L. Zhang, A. Nilasaroya, C. Barner-Kowollik, T. P. Davis, M. H. Stenzel, *Macromolecules* **2005**, *38*, 5475–5484.

**118.** H. Stenzel, T. P. Davis, C. Barner-Kowollik, *Chem. Commun.* **2004**, *13*, 1546–1548.

**119.** D. C. Wan, Q. Fu, J. Huang, *J. Polym. Sci., Part A: Polym. Chem.* **2005**, *43*, 5652–5660.

**120.** D. Boschmann, P. Vana, *Polym. Bull.* **2005**, *53*, 231–242.

**121.** J. Huang, D. Wan, J. Huang, *J. Appl. Polym. Sci.* **2006**, *100*, 2203–2209.

**122.** A. Suzuki, D. Nagai, B. Ochiai, T. Endo, *Polym. Sci., Part A: Polym. Chem.* **2005**, *43*, 5498–5505.

**123.** X. Hao, E. Malmström, T. P. Davis, M. H. Stenzel, C. Barner-Kowollik, *Aust. J. Chem.* **2005**, *58*, 483–491.

**124.** M. R. Whittaker, M. J. Monteiro, *Langmuir* **2006**, *22*, 9746–9752.

**125.** R. Plummer, D. J. T. Hill, A. K. Whittaker, *Macromolecules* **2006**, *39*, 8379–8388.

**126.** D. Boschmann, P. Vana, *Macromolecules* **2007**, *40*, 2683–2693.

**127.** H. T. Lord, J. F. Quinn, S. D. Angus, M. R. Whittaker, M. H. Stenzel, T. P. Davis, *J. Mater. Chem.* **2003**, *13*, 2819–2824.

**128.** L. Zhang, Y. Chen, *Polymer* **2006**, *47*, 5259–5266.

**129.** G. Zheng, C.-Y. Pan, *Polymer* **2005**, *46*, 2802–2810.

**130.** Q. Zheng, G.-H. Zheng, C.-Y. Pan, *Polym. Int.* **2006**, *55*, 1114–1123.

**131.** G. Zheng, Q. Zheng, C. Pan, *Macromol. Chem. Phys.* **2006**, *207*, 216–223.

**132.** G. Zheng, C.-Y. Pan, *Macromolecules* **2006**, *39*, 95–102.

**133.** L. P. Yang, C.-Y. Pan, *Aust. J. Chem.* **2006**, *59*, 733–736.

**134.** P. Zhang, Q. Liu, A. Qing, J. Shi, M. Lu, *J. Polym. Sci., Part A: Polym. Chem.* **2006**, *44*, 3312–3320.

**135.** L. Zhang, L. T. U. Nguyen, J. Bernard, T. P. Davis, C. Barner-Kowollik, M. H. Stenzel, *Biomacromolecules* **2007**, *8*, 2404–2415.

**136.** S. Feng, C. Y. Pan, *Macromolecules* **2002**, *35*, 4888–4893.

**137.** Y. G. Li, Y. M. Wang, C. Y. Pan, *J. Polym. Sci., Part A: Polym. Chem.* **2003**, *41*, 1243–1250.

**138.** J. Bernard, A. Favier, T. P. Davis, C. Barner-Kowollik, M. H. Stenzel, *Polymer* **2006**, *47*, 1073–1080.

**139.** J. Vosloo, M. P. Tonge, C. M. Fellows, F. D'Agosto, R. D. Sanderson, R. G. Gilbert, *Macromolecules* **2004**, *37*, 2371–2382.

**140.** M. H. Stenzel, T. P. Davis, A. G. Fane, *J. Mater. Chem.* **2003**, *13*, 1–10.

**141.** M. Hernández-Guerrero, C. Barner-Kowollik, T. P. Davis, M. H. Stenzel, *Eur. Polym. J.* **2005**, *41*, 2264–2277.

**142.** Y. Chen, L. Chen, H. Nie, E. T. Kang, R. H. Vora, *Mat. Chem. Phys.* **2005**, *94*, 195–201.

**143.** Y. Chen, L. Ying, W. Yu, E. T. Kang, K. G. Neoh, *Macromolecules* **2003**, *36*, 9451–9457.

**144.** Y. G. Li, P. J. Shi, Y. Zhou, C.-Y. Pan, *Polym. Int.* **2004**, *53*, 349–354.

**145.** H. Shinoda, K. Matyjaszewski, *Macromol. Rapid Commun.* **2001**, *2*, 1176–1181.

**146.** H. Shinoda, K. Matyjaszewski, L. Okrasa, M. Mierzwa, T. Pakula, *Macromolecules* **2003**, *36*, 4772–4778.

**147.** X. Xu, J. Huang, *J. Polym. Sci., Part A: Polym. Chem.* **2004**, *42*, 5523–5529.

**148.** X. Xu, J. Huang, *J. Polym. Sci., Part A: Polym. Chem.* **2006**, *44*, 467–476.

**149.** W.-P. Wang, Y.-Z. You, C.-Y. Hong, J. Xu, C.-Y. Pan, *Polymer* **2005**, *46*, 9494–9494.

**150.** Z. Cheng, X. Zhu, G. D. Fu, E. T. Kang, K. G. Neoh, *Macromolecules* **2005**, *38*, 7187–7192.

**151.** C. Li, Y. Shi, Z. Fu, *Polym. Int.* **2006**, *55*, 25–30.

**152.** Y.-Z. You, C. Y. Hong, W.-P. Wang, P.-H. Wang, W.-Q. Lu, C. Y. Pan, *Macromolecules* **2004**, *37*, 7140–7145.

**153.** D.-H. Han, C.-Y. Pan, *J. Polym. Sci., Part A: Polym. Chem.* **2007**, *45*, 789–799.

**154.** Liu, C.-Y. Pan, *Polymer* **2005**, *46*, 11133–11141.

**155.** H. Han, C.-Y. Pan, *Eur. Polym. J.* **2006**, *42*, 507–515.

**156.** Z. Ge, S. Luo, S. Liu, *J. Polym. Sci., Part A: Polym. Chem.* **2006**, *44*, 1357–1371.

**157.** S. Kulkarni, C. Schilli, A. H. E. Mueller, A. S. Hoffman, P. S. Stayton, *Bioconjug. Chem.* **2004**, *15*, 747–753.

**158.** G. J. tenCate, H. Retting, K. Bernardt, H. G. Börner, *Macromolecules* **2005**, *38*, 10643–10649.

**159.** B. de Lambert, C. Chaix, M.-T. Charreyre, A. Laurent, A. Aigoui, A. Perrin-Rubens, C. Pichot, *Biocongug. Chem.* **2005**, *16*, 265–274.

**160.** M. H. Stenzel, *Aust. J. Chem.* **2002**, *55*, 239–243; M. H. Stenzel, C. Barner-Kowollik, T. P. Davis, *J. Polym. Sci., Part A: Polym. Chem.* **2006**, *44*, 2363–2375.

# 10
# Macromolecular Design by Interchange of Xanthates: Background, Design, Scope and Applications

*Daniel Taton, Mathias Destarac, and Samir Z. Zard*

## 10.1
## Introduction

Free-radical polymerization is extensively utilized industrially, with about 50% of polymeric materials being produced by this chain process [1]. Compared to ionic polymerizations, free-radical processes offer the advantage of being applicable to a wider variety of monomers and are not as demanding regarding the purity of the reagents used. Radical processes can be implemented in emulsion, suspension, solution or in bulk. In addition, radical growing species are highly tolerant of many functional groups, including acid, hydroxyl, amino, epoxide, etc.; hence, functional monomers can undergo radical polymerization without the help of protection chemistry. However, chain breakings occurring by irreversible terminations in 'conventional' free-radical polymerization seriously limit its relevance in macromolecular engineering. Substantial research has thus been devoted to controlled/living radical polymerization (C/LRP) [2] methodologies that combine the advantages of truly 'living' systems for the quality of the polymers formed with the easiness inherent to radical processes. Among C/LRP systems, nitroxide-mediated polymerization (NMP) [3], atom-transfer radical polymerization (ATRP) [4, 5], reversible addition–fragmentation chain transfer (RAFT) [6–9], macromolecular design by interchange of xanthates (MADIX) [10, 11] and iodine degenerative transfer polymerization (IDTP) [12, 13] have been extensively investigated, paving the way for a use in macromolecular engineering [14, 15].

An important development in this area is based on the use of thiocarbonylthio compounds of general structure Z—C(=S)SR, including dithioesters, dithiocarbamates, trithiocarbonates or xanthates in RAFT and MADIX processes [6–11]. From a mechanistic viewpoint, MADIX and RAFT processes are actually strictly identical and eventually only differ by the nature of the chain-transfer agent (CTA): The RAFT terminology prevails for CTAs Z—C(=S)—S—R in general, whereas MADIX refers to xanthates exclusively (with Z = OZ′). However, some authors using xanthates as CTAs refer to the more general RAFT terminology rather than MADIX.

*Handbook of RAFT Polymerization.* Edited by Christopher Barner-Kowollik

ISBN: 978-3-527-31924-4

The CSIRO in Australia [16] and Rhodia [17] were pioneers in the development and use of RAFT/MADIX polymerizations. Significant developments have then been directed toward RAFT and MADIX processes, including the design of new CTAs, kinetic and mechanistic investigations, direct synthesis of water-soluble materials and their implementation in dispersed media.

This review is preliminary intended to cover the literature on the particular use of xanthates in MADIX/RAFT polymerization, since the first peer-reviewed article by Charmot et al. on this topic in 2000 [10]. As far as we are aware, the number of contributions citing the particular use of xanthate in RAFT/MADIX polymerization is somehow limited (<200), as compared to the total number of reports on RAFT/MADIX polymerization in its general sense (>500 since 1998). The scope and limitations of MADIX polymerization will be discussed with respect to important criteria that include the variation of both R- and Z groups of xanthates to reach optimal control, monomers that can be polymerized and experimental conditions to be applied. Synthetic routes to xanthates that can operate in MADIX are also briefly reviewed. This will be followed by a short overview of the various polymer architectures achievable by MADIX. Practical ways to remove xanthate end groups in polymer chains after polymerization, which is an important matter for an industrial development of MADIX-derived (co)polymers, is also presented. Finally, some potential applications of such materials described in the recent patent literature are mentioned.

## 10.2
## History of MADIX Polymerization

The invention of RAFT/MADIX originates in two distinct types of works initiated in the mid-1980s. According to the CSIRO group, RAFT is an extension of their own works regarding the use of addition–fragmentation chain-transfer agents (AFCTA) in radical polymerization, which enable control of the molar mass and terminal functionality of the prepared polymers. A first generation of irreversible AFCTA agents was reported [18], including allylic compounds, vinyl ethers and thionoesters. The CSIRO took further step toward RAFT with the use of noncopolymerizable methacrylate macromonomers in the polymerization of methacrylics by a mechanism of addition–fragmentation, thus retaining a terminal olefinic group over the course of the polymerization [19]. In spite of the low transfer constant of macromonomers ($C_{tr} = 0.2$–$0.3$) in the polymerization of methyl methacrylate (MMA) [19], all methacrylic diblock copolymers of low polydispersity index (PDI) were obtained in semibatch emulsion polymerization [20]. However, RAFT/MADIX methodologies also find their origin in works in France by Zard and coworkers who exploited the degenerative transfer of xanthates in their addition to alkenes for the formation of 1:1 adducts [21, 22]. This addition reaction was designed to shed light on the mechanism of the exceedingly useful Barton–McCombie deoxygenation, which allows the reduction of an alcohol **1** into the corresponding alkane **4** via the corresponding xanthate **3** or a related thiocarbonyl derivative, as summarized in Scheme 10.1 [23]. The alcohol is converted into the *S*-methyl

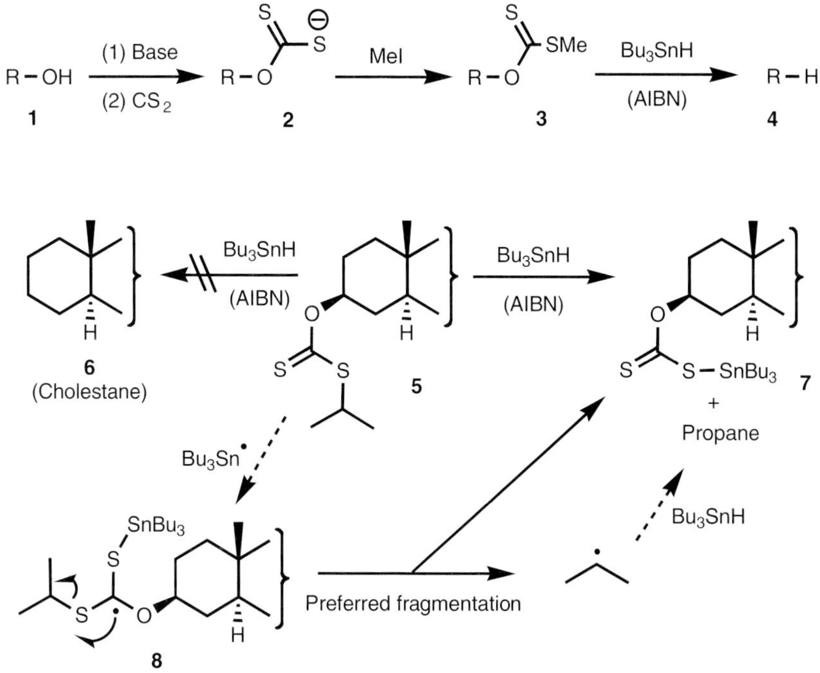


Scheme 10.1 An abnormal result in the Barton–McCombie reaction.

xanthate 3 by methylation of the intermediate xanthate salt 2 with methyl iodide. The placement of a methyl group on the sulfur atom was done initially out of sheer convenience, without much forethought as to its importance. When *O*-(cholestan-3-yl)-*S*-isopropyl xanthate 5 was reacted with tributyl stannane and a small amount of 2,2′-azobis(isobutyronitrile) as the initiator, none of the expected cholestane 6 was obtained: the reaction produced propane and tributyltin xanthate 7 [24]. This meant that the reaction intermediate 8, when given a choice of generating radicals of similar stabilities by rupture of either the C—O or C—S bond, will preferentially undergo scission of the latter.

By placing a primary group on the oxygen of the xanthate such as an ethyl, as in 9, cleavage of the C—O bond to give the high-energy ethyl radical becomes even more difficult. An incoming carbon-centered radical R′• will therefore add to the thiocarbonyl group and will expel radical R• from the other sulfur atom, if radical R• is more stable than attacking radical R′• (Scheme 10.2). A transfer of the xanthate group thus takes place from R to R′ on going from 9 to 11 via adduct 10, or vice versa depending on the relative stabilities of the two corresponding radicals.

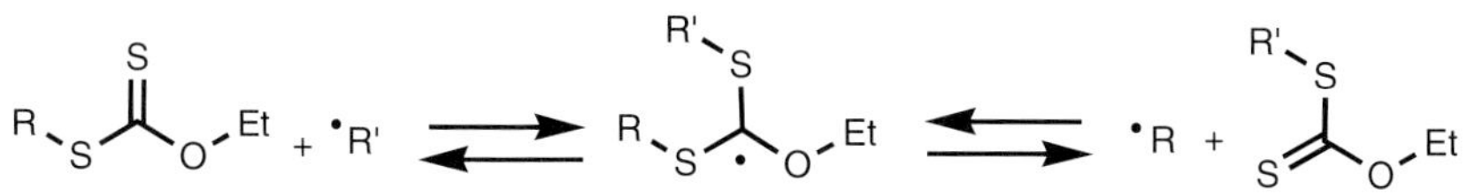

9 10 11

Scheme 10.2 Reversible exchange of a xanthate group between two radicals.

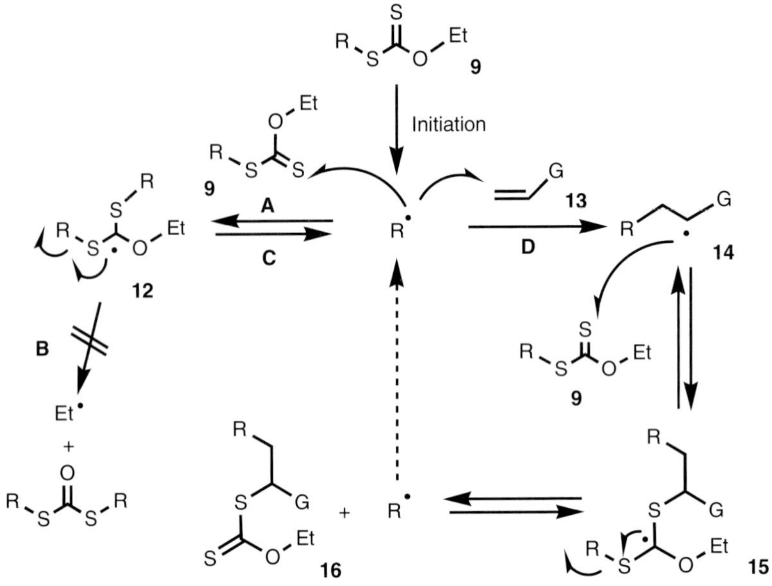


**Scheme 10.3** Radical addition of a xanthate to an olefin.

This RAFT process can be incorporated into a reaction scheme where radical R′• arises by addition of radical R• to an olefin, as outlined in Scheme 10.3 (i.e. R′ = **14**) [21, 22]. Thus, in a first initiation step, a radical R• is generated from the starting xanthate **9** and is rapidly captured by its precursor to give intermediate **12** (path A). This reaction is very fast but degenerate, because scission of the C—O bond to give an ethyl radical (path B) is disfavored, and the other possible fragmentation (path C) returns the starting materials. Interception of R• by the olefinic trap **13** (path D), although normally slower, is in fact not hampered by another major competing process. Radical **14**, arising from the addition to the olefin, is in turn rapidly captured by xanthate **9** to produce intermediate **15**, an unsymmetrical entity that can fragment in two ways: either to give back radical **14** or to proceed to product **16** and radical R•, which propagates the chain. This equilibrium is the same as the one in Scheme 10.2 and will be shifted in the direction of the more stabilized radical: If R• is more stable than radical **14**, the process will be driven forward and a clean addition to the olefin is observed.

This is indeed the case, as illustrated by the smooth addition of the cyclopropyl containing xanthate **19** to protected *N*-allylhydroxylamine **19** to give the expected adduct **20** in high yield [25]. This is one example from hundreds that has been performed over two decades [21, 22].

The example of addition to an olefin deployed in Scheme 10.4 embodies many of the advantages of this radical process:

- The starting xanthate is often trivial to prepare, as shown by the essentially quantitative formation of **19** from bromoketone **17**. Potassium *O*-ethyl xanthate, **18**, is

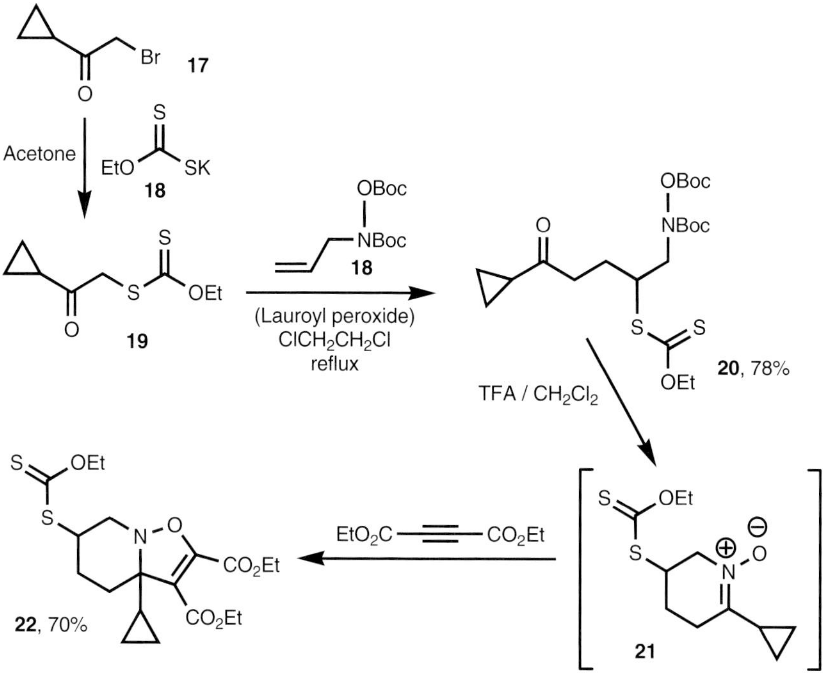


**Scheme 10.4** Example of an addition to an olefin and further transformation.

simply prepared by adding carbon disulfide to ethanolic potassium hydroxide; it is a very cheap commodity chemical that is extensively used as a flotation agent in the mining industry.

- The experimental procedure is exceedingly simple and the reaction can be performed under very high concentrations, typically 1–4 M. In the case of liquid reactants, no solvent is actually needed. Nearly equimolar quantities of the xanthate and olefin can be used, less initiator is needed and the reaction mixture is cleaner, resulting in a simplified purification, diminished waste and can be easily scaled up.
- Even though 1,2-dichloroethane (DCE) was most often used, because of its high solubilizing power, convenient boiling point and little tendency to act as a hydrogen donor, many other solvents can be used, including water. The choice of solvent depends on the substrate, the scale and, perhaps most importantly, the temperature at which the reaction is to be carried out, as this factor will also condition the half-life of the initiator.
- The experimental conditions are mild and neutral; numerous functional groups are tolerated and many of the polar groups that need protection in the case of ionic or organometallic reactions can be left in their native form.
- The absence of a major competing process gives the intermediate radical R• in Scheme 10.3 an effective long lifetime, allowing it to add even to nonactivated alkenes or, more generally, to undergo comparatively slow radical processes that

are often difficult to accomplish with other radical methods (e.g. ring closure to aromatic or heteroaromatic rings). The xanthate transfer addition has emerged as perhaps the most powerful process for creating C—C bonds on nonactivated alkenes in an *intermolecular* fashion. By modifying the starting xanthate and the olefinic trap a virtually infinite number of possible combinations can be made.

- The system is self-regulating, since reactive radicals are rapidly intercepted by the xanthate to furnish intermediates such as **12** and **15**. These are stabilized radicals that cannot disproportionate and are for all intents and purposes totally innocuous. They act in fact as storage entities for the reactive radicals and fragment selectively to liberate the most stable radical. This is a very subtle but important feature of the system with some interesting synthetic consequences (see Chapter 5 of this book for a thorough discussion).
- The process is convergent and atom economical, since all the components end up in the densely functionalized product. Indeed, one of the major strengths of this technology is that it allows the swift assembly, under neutral conditions, of complex structures containing a combination of functional groups. A change in the pH or the addition of an external reagent can then bring about a reaction between these various entities. This is illustrated by the further transformation of adduct **20** in Scheme 10.4: Addition of trifluoroacetic acid (TFA) results in the liberation of the terminal hydroxylamine followed by a fast internal condensation onto the ketone to give nitrone **21**, which can then be intercepted by diethyl acetylenedicarboxylate, a strong dipolarophile, to give finally bicyclic derivative **22** in good yield. Such structure would be accessible with some difficulty using more traditional routes.
- The reaction manifold depicted in Scheme 10.3 is applicable to other thiocarbonylthio derivatives, of general formula Z—C(=S)—SR, since the —OEt group in the xanthate is just a spectator, with only an indirect influence. An identical process can therefore be implemented with dithiocarbamates ($Z = NR'R''$), trithiocarbonates ($Z = SR'$), dithioesters ($Z = R'$), as long as the appropriate choice of substituents is made. The nature of the Z group influences the rate of radical addition onto the thiocarbonyl and the stability of the ensuing adduct [Z—C•(SR)SR′] and hence its propensity to fragment. Other properties such as polarity, solubility (in water for example), color or crystallinity can also be modified by selecting the appropriate substituent on oxygen.
- The product of the reaction is itself a xanthate, so that another radical transformation can be envisaged. Alternatively, the xanthate can act as a very convenient entry into the exceedingly rich and varied chemistry of sulfur, opening the way to limitless opportunities for subsequent modifications.

The fact that the product is itself a xanthate, or more generally a thiocarbonylthio derivative, it is therefore capable, in principle, of undergoing a second radical addition. If the alkene trap is a polymerizable monomer, then the addition process can be easily perpetuated until there is no more monomer in the medium. The RAFT and MADIX processes benefit from all the advantages listed above. This concept of degenerative transfer of xanthate group has thus been extrapolated by the Rhodia group [10, 17] and has given birth to the MADIX process. This new C/LRP method was patented [17] at about the same time that the CSIRO group in Australia discovered the RAFT process [16]. The first patent of the CSIRO deals with the use of dithioesters and trithiocarbonates as RAFT agents; xanthates and dithiocarbamates were not claimed in this invention [16]. In the literature, the first peer-reviewed article on RAFT was published in 1998 by the CSIRO [26], while Charmot et al. from Rhodia published the first article on MADIX in 2000 [10].

## 10.3 Mechanism of MADIX Polymerization

A MADIX polymerization consists of the simple introduction of tenths to few molar percent of a xanthate in a conventional free-radical system. Like for RAFT, the key feature in MADIX is a sequence of reversible addition–fragmentation reactions on the basis of the degenerative transfer of xanthate described above, as shown in Scheme 10.5 [10, 11]. The alkene plays the role of the monomer and is therefore used in large excess relative to the xanthate. Typical experimental conditions of MADIX polymerization are very similar to those used in conventional radical polymerization in the presence of a transfer agent. Propagating species $P_n\bullet$ are first produced from a conventional radical source (e.g. azo or peroxo compounds) and the monomer (Scheme 10.5, path I). These oligomeric radicals $P_n\bullet$ add onto the C=S double bond ($k_{add}$) of the xanthate, the transient radical thus formed subsequently undergoing a β-scission ($k_\beta$) to form a new thiocarbonylthio species and the expelled R• (Scheme 10.5, path II). The R• radical is chosen so that it is capable of reinitiating the polymerization ($k_{re\text{-}ini}$) and, as in conventional radical polymerization, active species propagate (Scheme 10.5, path III).

The so-called preequilibrium (II) in Scheme 10.5 consists in the sole consumption of the MADIX agent.

The faster the preequilibrium, the closer the experimental molar mass to the theoretical value will be. Ideally, as in any C/LRP where the controlling agent is fully converted in the first instants of the polymerization, the molar masses increase linearly with monomer conversion and can be predicted by equation (10.1):

$$\overline{M}_{n,\text{theo}} = \frac{[M]_0}{[X]_0} \times M_{\text{MU}} \times \text{Conv.} + M_{\text{X}} \qquad (10.1)$$

where $\overline{M}_{n,\text{theo}}$ is the theoretical number-average molar mass, $[M]_0$ and $[X]_0$ are the initial monomer and xanthate concentrations, respectively, $M_{\text{MU}}$ and $M_{\text{X}}$ are the molar masses of the monomer unit and the xanthate, respectively, and Conv. is the fractional conversion of the monomer. The above expression assumes that all growing chains arise from the sole R group of the CTA.

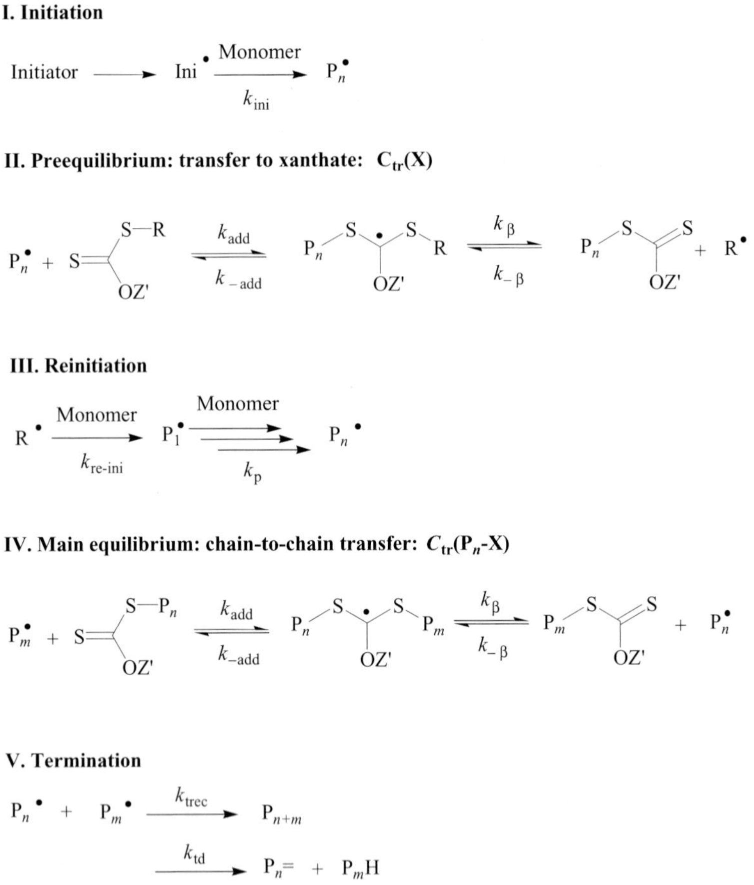


**Scheme 10.5** General mechanism of MADIX polymerization.

As for other CTAs used in RAFT/MADIX polymerization, the radical attack of propagating species onto the C=S bond of the xanthate RS(C=S)OZ′ strongly depends on the activation brought by the activating group Z = OZ′, whereas the β-scission is favored when R• has a marked stability and steric hindrance. As illustrated in Scheme 10.5, the chain-transfer constant $C_{tr}(X)$ accounts for both addition and fragmentation steps according to equation (10.2):

$$C_{tr}(X) = \left(\frac{k_{add}}{k_p}\right)\left[\frac{k_\beta}{(k_{-add} + k_\beta)}\right] \tag{10.2}$$

This period of consumption of the initial RAFT/MADIX agent that is sometimes called the 'initialization period' [27] can be conveniently probed by in situ $^1$H NMR spectroscopy, as reported by Klumperman and coworkers [27, 28]. Once the xanthate has been consumed, the macro-CTA agent enters in the so-called main equilibrium (IV) that is the chain-to-chain transfer of the $\omega$-xanthate groups between the dormant chains and the propagating species. The kinetics of the main equilibrium is expressed through a chain-transfer constant, noted $C_{tr}(P_n\mathbf{X})$, which can be different from the $C_{tr}(\mathbf{X})$ value of the preequilibrium. Since the two polymer fragments in the intermediate radical are considered as almost identical, one can consider that $k_{-add} = k_\beta$; hence, the $C_{tr}(P_n\mathbf{X})$ can be expressed according to equation (10.3):

$$C_{tr}(P_nX) = \frac{k_{add}}{2k_p} \tag{10.3}$$

Ideally, a rapid exchange ensures an equal probability for all chains to grow and a narrow molar mass distribution. After complete polymerization, the vast majority of the chains theoretically retain the dithiocarbonate group at one end and the R group arising from the initial xanthate at the other. Under appropriate conditions, the proportion of dead chains formed by irreversible chain breakings (by disproportionation and/or recombination) (Scheme 10.5, path V) can be minimized. Increasing the radical concentration will increase the overall rate of polymerization but at the same time chain breakings will be more pronounced, which will broaden the molar mass distribution of the final polymers.

If intermediate radicals derived from dithioesters formed in pre- and main equilibria could be observed via electron spin resonance [7], their inspection and quantification in xanthate-mediated polymerization has not been reported yet in the literature. The expected structures of MADIX-derived polymers could be confirmed by different characterization techniques, such as $^1$H NMR, UV or IR spectroscopy, as well as by electrospray ionization mass spectrometry (ESI MS) and matrix-assisted laser desorption ionization – time-of-flight mass spectrometry (MALDI-TOF MS). The xanthate-terminated polymers can be isolated or reactivated in chain extension experiments or for block copolymer synthesis.

In summary, the two following conditions must be fulfilled to reach an optimal control of a MADIX polymerization: (1) rapid exchanges of dithiocarbonate end groups between propagating species and all CTAs (the initial xanthate and the macro-CTA), meaning that both $C_{tr}(X)$ and $C_{tr}(P_n\mathbf{X})$ must be high; (2) use of a much lower concentration of initiator than that of xanthate (typically $<10$ mol %) to minimize the presence of dead chains; however, this concentration should be high enough to ensure completion of the polymerization. Obviously, the effectiveness of a MADIX polymerization varies as a function of the monomer/xanthate combination.

The MADIX process can be well described by the model reported by Müller et al. in 1995 for group transfer polymerization of alkyl (meth)acrylates [29], which is another notorious degenerative transfer process. This model assumes that irreversible terminations by coupling can be neglected and that $C_{tr}(X) = C_{tr}(P_n\mathbf{X})$, and allows one to predict the evolution of $M_n$ and that of PDI as a function of monomer conversion. The number-average degree of polymerization, $DP_n$, could be expressed as follows (equation (10.4)):

$$DP_n = \frac{[M]_0 \times \text{Conv.}}{[X]_0(1 - (1 - \text{Conv.})^{C_{tr}(X)}} \tag{10.4}$$

The final PDI in such polymerizations equals $1 + 1/C_{tr}(P_n\mathbf{X})$ [29]. Both variations of $M_n$ and PDI could thus be plotted for different $C_{tr}$ values. Figure 10.1a shows that the greater the $C_{tr}$ value, the closer the $DP_n$ versus monomer conversion profile to the theoretical straight line is. A linear increase of molar masses is achieved for $C_{tr}$ values greater than 10, meaning that the CTA is much more rapidly consumed than the monomer. As shown in Fig. 10.1b, it also clearly appears that the evolution of PDI versus conversion strongly varies with $C_{tr}$; values greater than 10 permit the obtainment of polymers with PDI around 1.2. Experimental $C_{tr}$(X) values can be

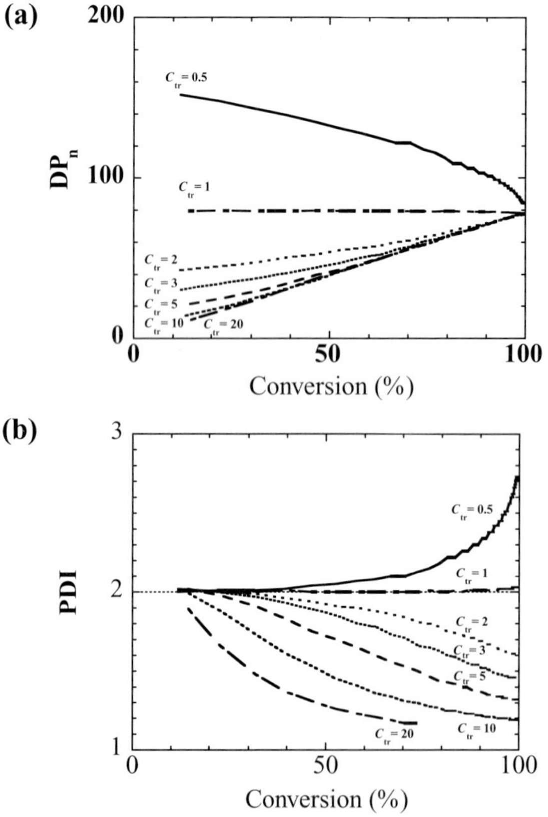


**Fig. 10.1** Theoretical evolution of the (a) number-average degree of polymerization $DP_n$ and (b) PDI = $M_w/M_n$ versus monomer conversion for different $C_{tr}$ in a polymerization following a degenerative transfer mechanism.

estimated from this model with the measured $DP_n$ at a given monomer conversion, provided that the simplifying conditions for the model to be valid are fulfilled and that the experimental $DP_n > \text{Conv.} \times [M]_0/[X]_0$.

Before the entire consumption of X, $DP_n > \text{Conv.} \times [M]_0/[X]_0$. This profile is sometimes referred to as the 'hybrid behavior', that is new polymer chains produced from the CTA late in the polymerization broadening of the molar mass distribution of the final polymer [27, 28]. The possible causes of the 'hybrid behavior' are linked to (i) an inappropriate selection of the Z group, (ii) the R group that can be a poor leaving group to fragment ($k_\beta$) relative to the oligomer/polymer chain from the intermediate radical or (iii) the fact that R• group may exhibit poor reinitiating ability ($k_{\text{re-ini}}$).

More than 50 xanthates have already been described in the literature to be used in MADIX polymerization. Main xanthates discussed in this chapter are represented in Figs. 10.2 and 10.8.

## 10.4 Kinetics of MADIX Polymerization

Chain-transfer reactions usually have no effect on the overall polymerization rate. As a consequence, kinetics of a RAFT/MADIX polymerization should be identical to those of a conventional free-radical polymerization, irrespective of the CTA. However, RAFT polymerizations utilizing some dithioesters (mainly dithiobenzoates) have been reported as being slower than conventional free-radical polymerization systems conducted under similar conditions [30]. Polymerization can even be completely inhibited for particular CTA/monomer combination (e.g. dithioester/vinyl acetate). As discussed later, some retardation was also evidenced in xanthate-mediated polymerization of vinyl acetate (VAc) [31]. Likewise the overall rate of polymerization can be decreased by increasing the concentration of the CTA. The existence of this retardation/inhibition effect can be related to a significant concentration and stabilization of radical intermediates appearing in pre- and main equilibria (II) and (IV) in Scheme 10.5. This has been the subject of an intense and ongoing debate in the literature [6–8, 30]. Some authors have suggested that the intermediate radical $P_nS{-}(Z)C^{\cdot}{-}SP_m$ undergoes a slow fragmentation and may terminate reversibly, the CTA and macro-CTAs acting therefore as radical sinks. Other groups have counterproposed that radical intermediates are irreversibly captured, for instance by propagating species $P_n^{\cdot}$ in irreversible terminations thereby producing three-arm stars, or may undergo conventional radical reactions such as transfer or propagation. However, inhibition can also be explained by a slow reinitiation of $R^{\cdot}$ during the 'initialization period'. In this context, the case of xanthates as MADIX agent can be put forward. Little retardation is generally observed in MADIX polymerization, except for fast-propagating monomers, such as vinyl acetate (VAc) or *N*-vinyl pyrrolidone (NVP). For instance, the Australian group from the Centre for Advanced Macromolecular Design (CAMD) designed a series of eight xanthates (**X**$_{20}$–**X**$_{27}$, Fig. 10.2) to investigate the kinetics of MADIX polymerization of VAc

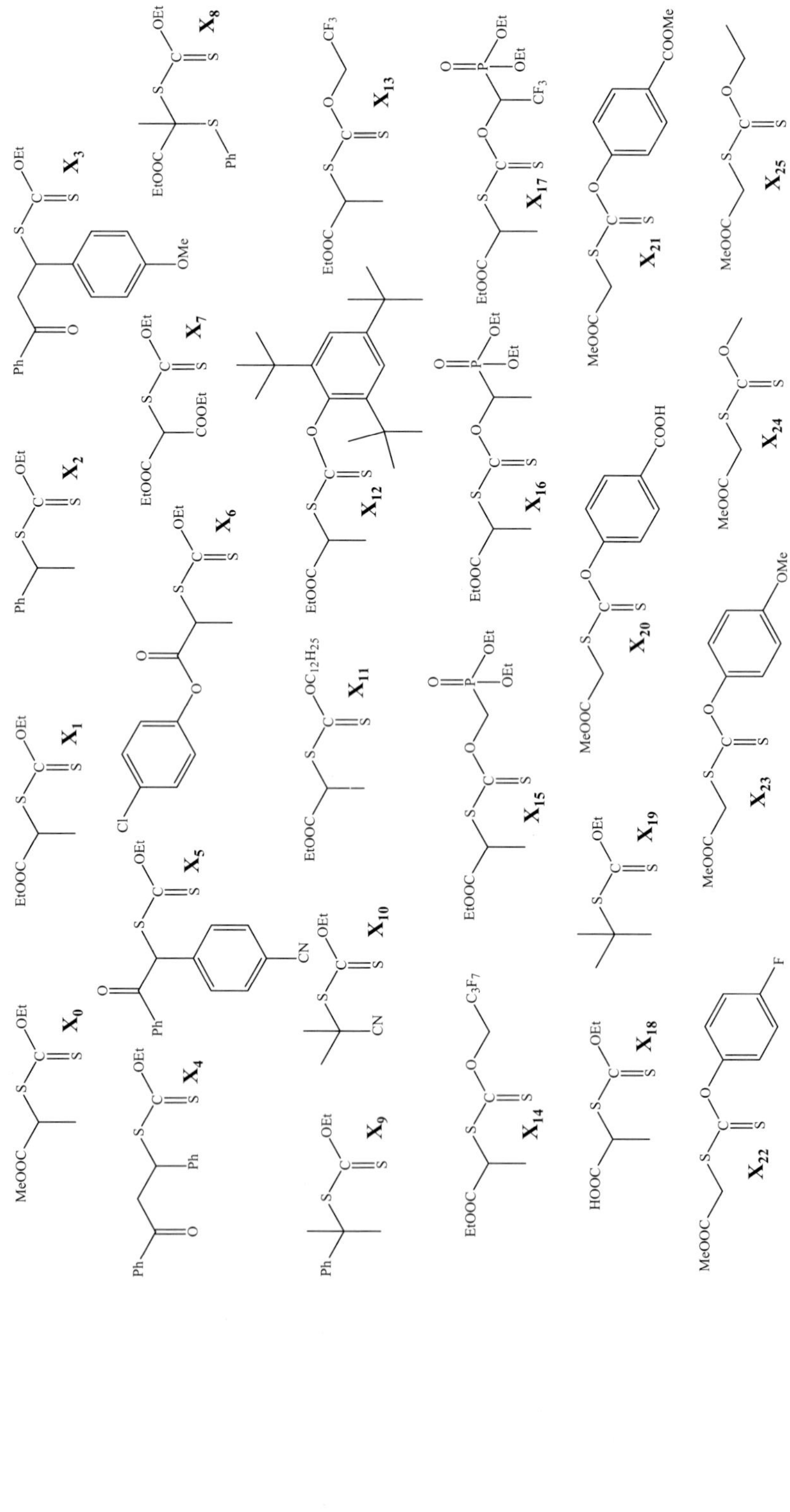

**Fig. 10.2** Xanthates used in MADIX polymerization.

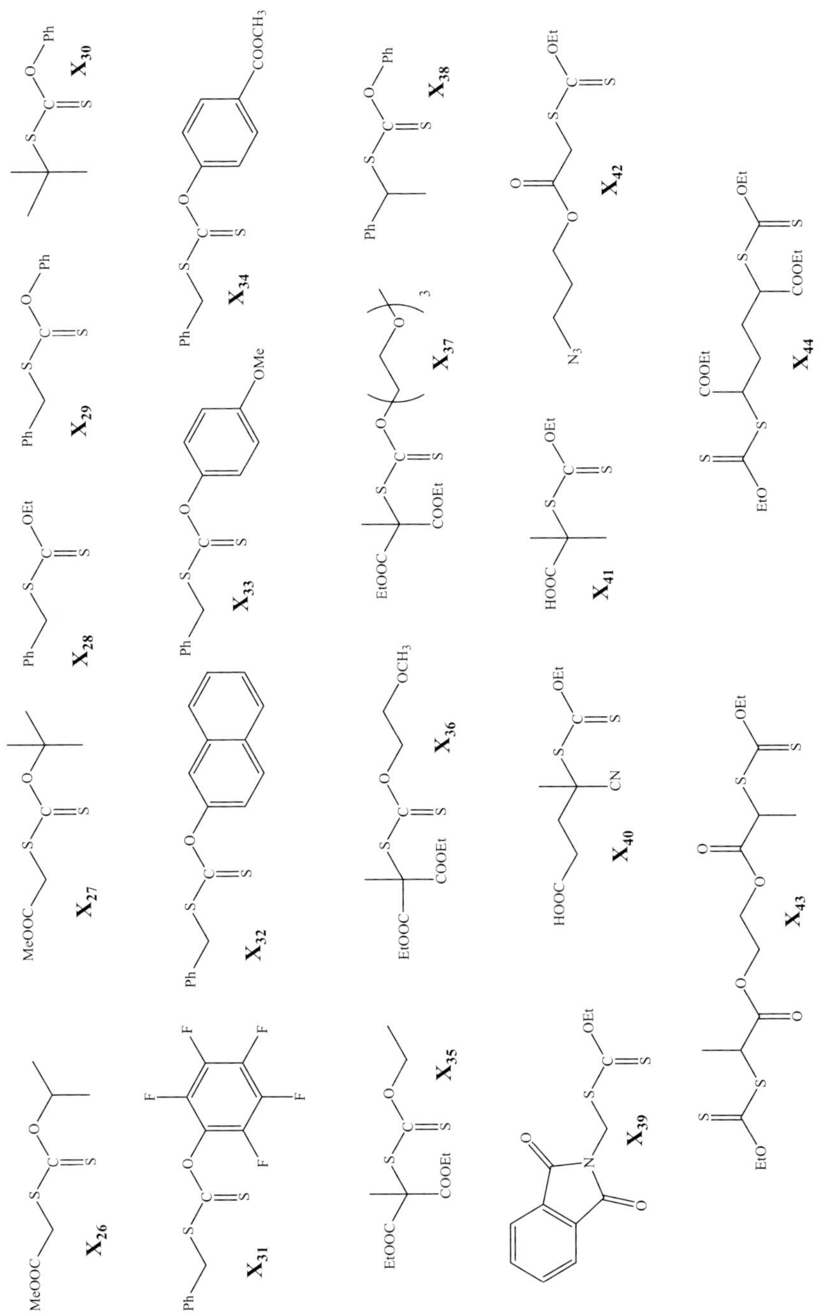


**Fig. 10.2** *(Continued)*

[31]. All MADIX agents exhibited extended periods of inhibition (0.3 h $< t_{inh} <$ 10 h) and moderate rate retardation. This was attributed to slow fragmentation of intermediate MADIX radicals appearing in pre- and main equilibria, though possible termination reactions were not ruled out. More recently, Klumperman and coworkers studied the MADIX polymerization of both VAc and NVP by in situ $^1$H NMR spectroscopy, as a convenient and complementary means to molar mass distribution characterization, to follow the concentrations of xanthate, monomer and oligomeric adduct of the CTA involved [28]. For instance, addition of $\mathbf{X_{10}}$ to NVP is highly selective during the first 275 min, with no significant polymerization before the complete xanthate conversion into the single monomer adduct. Afterward, the rate of monomer consumption significantly increased. Here, the nature of the leaving group R$^{\cdot}$, in particular its reinitiating ability ($k_{re\text{-}ini}$) during initialization process, was the key parameter in the observation of a retardation effect. Indeed, it was observed that the single monomer adduct was formed very slowly, while the concentration of the cyanoisopropyl recombination by-product was abnormally high. In contrast, changing the cyanoisopropyl for a 2-carboxyethyl ($\mathbf{X_{18}}$) or a *tert*-butyl ($\mathbf{X_{19}}$) leaving group led to the simultaneous formation of oligomeric adducts [28]. This was ascribed to the monomer-derived radicals exhibiting better leaving group ability than the R group. Consequently, monomer is polymerized before complete consumption of the CTA, which broadens the molar mass distribution. A 'selective initialization' leads to polymers with narrowly distributed molar masses – system which is referred to as an 'ideal' RAFT/MADIX polymerization – whereas polymers with higher PDIs are formed in the 'absence of selective initialization'.

The inhibition phenomenon was also studied by the CAMD group by in situ Fourier transform near-infrared and off-line $^1$H NMR spectroscopy, for MADIX bulk polymerization of VAc performed in the presence of the *O*-isopropyl xanthate $\mathbf{X_{26}}$ [32]. The presence of traces of impurities (oxygen or residual stabilizer) in the reaction mixture was put forward to explain the strong and variable inhibition periods observed during polymerization. This susceptibility of the poly(vinyl acetate) (PVAc) propagating radical to these impurities is obviously due to its high reactivity.

The overall kinetics of MADIX polymerization is thus dependent on both addition and fragmentation rate coefficients ($k_{add}$ and $k_{\beta}$), which in turn can be varied as a function of the R- and the Z groups. For instance, $k_{\beta}$ depends not only on stabilization of the intermediate radicals (mainly affected by the Z group) but on the weakness of the S—C bond (mainly related to the R group) as well.

Coote and Radom performed high-level ab initio calculations for MADIX polymerization of VAc, considering a series of xanthates ($\mathbf{X_{24}}$–$\mathbf{X_{27}}$) [33]. Computational studies confirmed that rate retardation can be induced by a slow fragmentation if substitution within the activating OZ′ group is increased. This, indeed, stabilizes the MADIX radical intermediate and hence reduces $k_{\beta}$. The authors also found that rate reduction for bulkier groups (isopropyl and *tert*-butyl compared to methyl and ethyl) is also accompanied by a sterically induced conformational change in the transition structures. In the particular case where *O*-*tert*-butyl xanthate was considered ($\mathbf{X_{27}}$), rate retardation was explained by an unexpected side reaction that was the fragmentation of the O—C bond in the *tert*-butoxy group of the MADIX adduct.

## 10.5
## Choice of MADIX Agents

A wide variety of xanthates have been designed and tested in radical polymerization of different monomers (Figs. 10.2 and 10.8). The structural variation of the Z- and R groups and its effect on the MADIX agent reactivity can be qualitatively and quantitatively predicted using low-level molecular orbital calculations and high-level ab initio calculations, respectively, as reported by the group of Coote and coworkers [33–37]. Current computational techniques allow calculation of energy barriers and enthalpies with a good accuracy. For instance, radical stabilization energies of the RAFT/MADIX-adduct radicals and values of the enthalpies of the fragmentation reactions were calculated for various combinations of the Z- and R substituents [34]. These investigations allowed the authors to examine the influence of the substituents on the stability of both the RAFT/MADIX-adduct radicals and the thiocarbonyl-thio compounds formed by fragmentation. It was confirmed that the stability of the intermediate radical is enhanced by electron donation from the two sulfur groups. Interestingly, lone-pair donor substituents in Z such as in *O*-alkyl xanthates have a smaller effect on radical stability. The R group, although having a minor effect on the stability of intermediate radical, can interact with the other sulfur substituent, affecting the strength of the S—R bond. In addition, bulky R- and Z groups were found to have a more significant effect on the destabilization of RAFT/MADIX agent than on the corresponding radical intermediate. Although this should be compared with experimental findings, computational chemistry can be used as a first-reference guide for selecting the R- and Z groups for a given monomer/xanthate combination [35–37].

### 10.5.1
### R Effect of Xanthates

In addition to be a good leaving group relative to the attacking propagating radical, the R$^{\cdot}$ radical should also efficiently reinitiate polymerization. For instance, benzyl, phenylethyl or cumyl groups as R groups in *O*-ethyl xanthates would be poor reinitiating groups for MADIX polymerization of VAc. In contrast, xanthate $\mathbf{X_0}$ proved to be a very efficient CTA for MADIX polymerization of this monomer [11]. The R group has a lesser contribution than the activating Z group toward the stabilization of the radical intermediate. To date, cumyl, cyanoisopropyl or propionyl groups proved to be the most efficient for reinitiation of RAFT/MADIX polymerization in general [6–11].

A series of xanthates containing the same Z = OEt activating group ($\mathbf{X_1}$–$\mathbf{X_{10}}$) were evaluated for the polymerization of styrene [38] (S) and ethyl acrylate (EA) [11, 38]. In this series, the rate constant of addition, $k_{add}$ (Scheme 10.5), is mainly influenced by the activating group Z = OEt. The results obtained were therefore directly correlated to the leaving R-group ability, that is to the probability for the transient radical to undergo a β-scission in the desired direction that is expressed by

$k_\beta/(k_{\text{-add}} + k_\beta)$, preferentially forming R$^\cdot$ rather than fragmenting back to reform the attacking $P_n^\cdot$ radical. It was thus found that the R group of *O*-ethyl xanthates had a marked influence on the $M_n$ evolution profile due to a change of the $C_{tr}$ value. For instance, the $M_n$ value obtained at high conversion with **X$_1$** is slightly higher than that predicted by the [S]/[**X$_1$**] feed ratio. This was ascribed to a slow and incomplete consumption of **X$_1$** over the course of the polymerization. This *hybrid behavior* was supported by the $C_{tr}$ value of **X$_1$** lower than unity ($C_{tr}(\mathbf{X_1}) = 0.89$). The chain-transfer activity then increased in the following order: $\mathbf{X_1} \sim \mathbf{X_2} < \mathbf{X_3} < \mathbf{X_4} \sim \mathbf{X_6} < \mathbf{X_5} \sim \mathbf{X_7} < \mathbf{X_8} < \mathbf{X_9} < \mathbf{X_{10}}$. It turns out that incorporation of electron-withdrawing groups increased xanthate reactivity: $C_{tr}(\mathbf{X_6}) = 1.65 > C_{tr}(\mathbf{X_1}) = 0.89$ and $C_{tr}(\mathbf{X_7}) = 2 > C_{tr}(\mathbf{X_1})$. Finally, *O*-ethyl xanthates with tertiary R groups further improve the control in S polymerization, the cyanoisopropyl group of **X$_{10}$** proving the best leaving group in this series: $C_{tr}(\mathbf{X_8}) = 3 < C_{tr}(\mathbf{X_9}) = 3.8 < C_{tr}(\mathbf{X_{10}}) = 6.8$. Consistently with findings for dithioesters, the more substituted and stabilized the R$^\cdot$ leaving radical, the higher the transfer constant [6–11]. It is noteworthy that an excellent correlation between the experimental $M_n$ evolution profiles and those predicted by Müller's model was observed with $\mathbf{X_1}$–$\mathbf{X_{10}}$, taking into account the $C_{tr}$ values determined by the Mayo method. On the other hand, little influence of the R group on the molar mass distributions was noted: PDIs were typical of those obtained in a xanthate-free polymerization (1.9–2.4). This could be ascribed to a slow interchange of the xanthate end groups between polymer chains, that is to a low $C_{tr}(P_n\mathbf{X})$. Catala and coworkers later calculated the $C_{tr}$ value of an *S*-polystyryl-*O*-ethyl xanthate and found $C_{tr}(P_n\mathbf{X}) = 0.8$ [39], a value that was very close to $C_{tr}(\mathbf{X_2}) = 0.82$. This demonstrated that the phenylethyl group exhibited the same leaving ability as the polystyryl chains. The observation of final PDIs around 2 was consistent with the expression $(1 + 1/C_{tr}(P_n\mathbf{X}))$ proposed by Müller et al. [29]. Despite rather high PDI values, good control of chain structures could be achieved, as evidenced by NMR analysis and MALDI-TOF MS: No chains derived from thermally generated radicals have been detected [38]. Importantly, the nature of the R group seemed to have no significant influence on the overall rate of polymerization of S and EA for this series of xanthates [11, 38]. Significant retardation was observed only with **X$_9$**, presumably due to degradative transfer.

### 10.5.2
### Z Effect of Xanthates

A comprehensive investigation of the effect of the Z activating group (Z effect) on RAFT/MADIX polymerization in its general sense was reported by the CSIRO group [40]. The effect of the OZ′ activating group of xanthates on the quality of control of MADIX was investigated by our group [41]. This was accomplished using a series of xanthates (**X$_1$** and $\mathbf{X_{11}}$–$\mathbf{X_{17}}$) carrying the same R leaving group, namely a (1-ethoxycarbonyl)ethyl group, for S polymerization. Both electron density and steric hindrance of the Z = OZ′ group were varied. The difference of reactivity of the C=S double bond toward growing radicals ($k_{add}$) could be put forward. First,

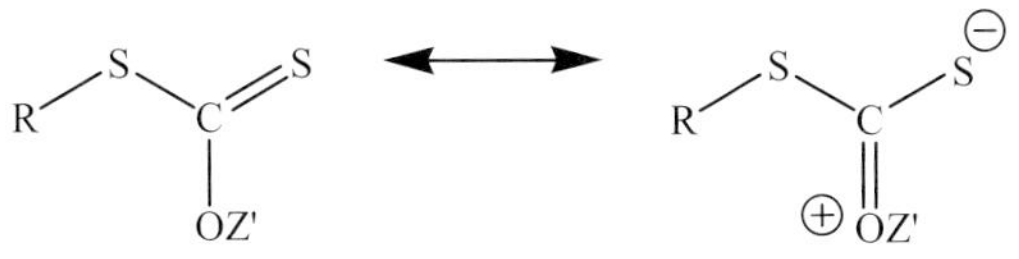

**Scheme 10.6** Canonical forms of xanthates.

no influence of the *O*-alkyl chain length on the level of control was noted, $\mathbf{X_1}$ and $\mathbf{X_{11}}$ giving roughly the same results. Introduction of a bulky group such as tris(*tert*-butylphenyl) group in $\alpha$-position to the oxygen prevents the growing radicals from accessing the C=S double bond of xanthate $\mathbf{X_{12}}$, which proved ineffective as MADIX agent. For MADIX polymerization of S and EA, the chain-transfer activity decreases in the following order: $\mathbf{X_{17}} > \mathbf{X_{13}} \sim \mathbf{X_{14}} > \mathbf{X_{15}} > \mathbf{X_1} \sim \mathbf{X_{11}} > \mathbf{X_{16}} \gg \mathbf{X_{12}}$. It appears that the perfluoroalkyl chain length has no influence on the activity of xanthates, $\mathbf{X_{13}}$ and $\mathbf{X_{14}}$ giving similar results. As for $\mathbf{X_{17}}$, it is fully consumed before 10% of the monomer is converted, resulting in a linear increase of the molar masses as a function of monomer conversion from the early stages of the polymerization, with $M_n$ values perfectly matching the theoretical ones based on the $[\mathrm{S}]_0/[\mathbf{X_{17}}]_0$ ratio.

The Z′ group was also found to have a dramatic impact on PDIs [41]. The moderate reactivity of *O*-ethyl xanthates was attributed to the conjugation of the lone pairs of electrons on the oxygen atom with the C=S bond, resulting in low $k_{add}$ values (Scheme 10.6).

Using high-level ab initio calculations, Coote and Henry confirmed that RAFT agents are strongly stabilized by the lone-pair donor Z substituent but can be destabilized by electron-withdrawing groups (CN and $CF_3$) in the R- and Z-positions [34]. From a practical viewpoint, the enhanced capability for transfer (increase of both $C_{tr}$(X) and $C_{tr}(P_n X)$) leading to a significant decrease in the PDI could be achieved, indeed, through the use of a fluoroalkyl substituents in the Z′ moiety [11, 41]. This can be rationalized by the fact that the conjugation effect mentioned above is considerably reduced with electron-withdrawing substituents. The CSIRO group reported similar findings; that is substituents rendering the oxygen lone pair less available for delocalization with the C=S can enhance the effectiveness of xanthates in MADIX (using $\mathbf{X_{29}}$ and $\mathbf{X_{31}}$) [20, 40]. Finally, the 1-diethoxyphosphonyl and 2,2′,2″-trifluoromethyl groups on the $\alpha$-carbon bonded to the oxygen atom seemed to have a cooperative effect since $\mathbf{X_{17}}$ further activates the chain-transfer process, as compared to $\mathbf{X_{13}}$ and $\mathbf{X_{14}}$ [41]. However, the introduction of the diethoxyphosphonyl group alone was not sufficient to enhance the reactivity of these xanthates: The substitution of the $CF_3$ group for a methyl group ($\mathbf{X_{16}}$) or a hydrogen group ($\mathbf{X_{15}}$) resulted in a moderate control.

Monteiro and coworkers reported a complementary study on the Z′ effect for three xanthates ($\mathbf{X_1}$, $\mathbf{X_2}$ and $\mathbf{X_{13}}$) on the control of S polymerization [42]. $C_{tr}$ values were determined using both the Mayo and chain-length distribution methods. Consistently with the above discussion, electron-withdrawing groups in Z increased xanthate reactivity toward propagating radicals: $C_{tr}(\mathbf{X_1}) = 0.69 < C_{tr}(\mathbf{X_{13}}) = 3.5$, the *O*-trifluoroethyl xanthate allowing the preparation of polystyrene (PS) with better

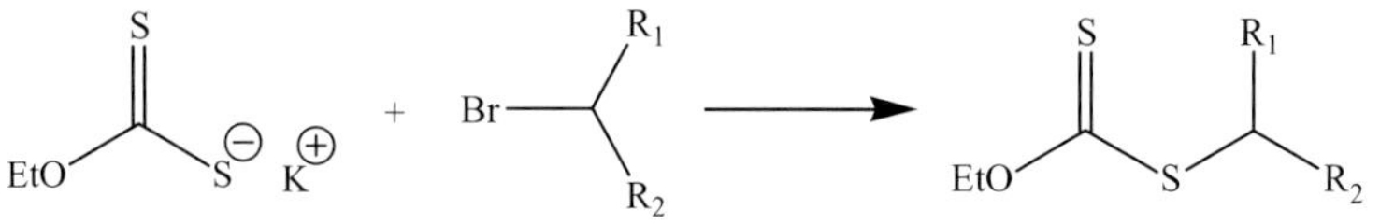

**Scheme 10.7** Synthesis of xanthate using potassium salt of xanthic acid.

controlled molar masses and lower polydispersities (PDI ∼ 1.6) compared to the use of $\mathbf{X}_1$ (PDI ∼ 2).

## 10.6
## Synthesis of MADIX Agents

Methods for synthesizing RAFT/MADIX agents in general have been reviewed by different groups [6–9]. Specifically, synthetic methodologies to xanthates from established organic synthetic procedures are well documented [21, 22]. Similar methods as those described for the synthesis of dithioesters can be used. However, it is noteworthy that the less electron-withdrawing alkoxy group of xanthates makes them more stable toward nucleophiles such as water or amines. Of particular interest also, the potassium salt of xanthic acid, $EtO(C{=}S)S^-K^+$, is commercially available and is a convenient source for *O*-ethyl xanthates. For instance, $EtO(C{=}S)S^-K^+$ can be alkylated using a primary or secondary alkyl halide (Scheme 10.7) [11, 17, 21, 22]. In this regard, many of the primary and secondary alkyl halides used as ATRP initiators are potential precursors for xanthates, following this route.

Xanthates bearing tertiary R leaving groups can be synthesized following a free-radical pathway, as reported by Zard and coworkers [43], from the reaction of a diazo compound as a radical source for R group and a xanthogen disulfide (Scheme 10.8), the latter being readily obtained from oxidative coupling of a xanthate salt.

Carbon disulfide ($CS_2$), although highly toxic, flammable and volatile, is an alternative precursor to achieve thiocarbonylthio in general. When treated in the

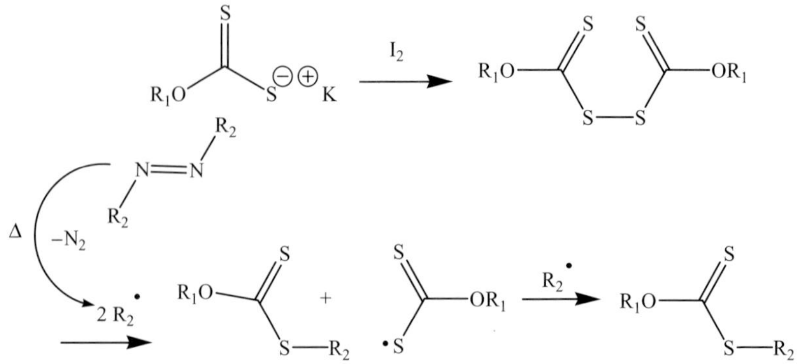

**Scheme 10.8** Free-radical synthesis of a xanthate from an azo compound and an *O*-alkyl xanthogen disulfide.

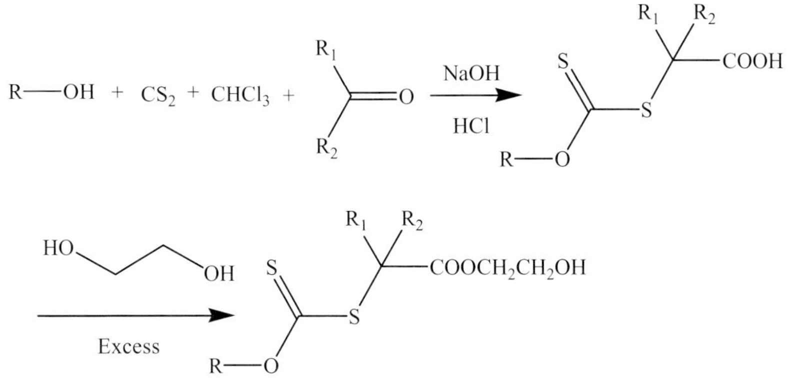


**Scheme 10.9** Synthesis of carboxyl- and hydroxyl-terminated xanthates.

presence of an alkoxide, followed by the addition of an alkyl halide, xanthates are readily obtained [6–11]. A variation of the latter procedure was reported by researchers from Noveon for making mono- and di-*S*-*t*-alkanoic acid terminated dithiocarbamates and xanthates (Scheme 10.9) [44, 45]. In this method, the potassium salt of *O*-alkyl xanthate is reacted with chloroform and a ketone (acetone or cyclohexanone) under basic conditions, followed by acidic treatment to form an *O*-alkyl-*S*-α,α′-disubstituted acetic acid dithiocarbonate. These COOH-terminated xanthates were further reacted with excess diol to derive OH-terminated MADIX agents.

In a recent contribution, Perrier and coworkers developed a new strategy based on the use of 1,10-thiocarbonyl diimidazole as a precursor to react with primary and secondary alcohols (or thiols or amines) to synthesize xanthates (or trithiocarbonates or dithiocarbamates), as depicted in Scheme 10.10 [46]. One advantage of this approach is to avoid the use of $CS_2$ and to perform a one-pot reaction.

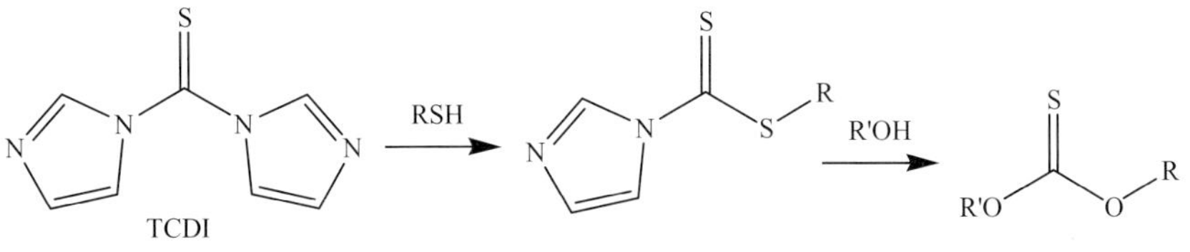


**Scheme 10.10** Synthesis of xanthate using TCDI as precursor.

## 10.7 Experimental Conditions in MADIX

This section will not cover the specific case of MADIX polymerizations performed in aqueous dispersed media, which is discussed later. As already emphasized, one main advantage of RAFT/MADIX polymerizations is their compatibility with various reaction conditions, as these methods only require the introduction of a CTA to an otherwise conventional free-radical polymerization. Xanthate-mediated polymerization has been performed in a variety of processes, including bulk, solution, emulsion and miniemulsion.

Rhodia researchers also showed that MADIX can be easily set up in an automated parallel synthetic approach, using high-throughput equipments, for the synthesis of a library of well-defined (co)polymers [47]. As a reminder, we discuss below typical reaction conditions that are used in MADIX polymerizations, although the influence of a few important parameters has not been examined yet. For instance, there is no detailed investigation reported on temperature effect or concentration effect on the use of xanthates in MADIX or on the possibility to perform polymerization in supercritical $CO_2$ or at high pressure, in contrast to dithioesters used in RAFT [6–8].

### 10.7.1
### Source of Radicals

As in conventional free-radical polymerization, MADIX polymerization is triggered by a source of radicals that should be carefully balanced between a reasonable rate of polymerization and an acceptable level of dead chains. As already mentioned, a too high concentration of radicals may (i) generate dead chains and broaden the molar mass distribution and (ii) affect the overall rate of polymerization. The following three methods for generating radicals have been reported in RAFT/MADIX polymerization in its general sense: thermal decomposition of initiators (e.g. azo-type initiators or peroxides), initiation by UV–vis or $\gamma$ radiation, generally at room temperature, or thermal initiation (no initiator added) [6–9]. The three methods have been reported in xanthate-mediated polymerization, though the former one is essentially employed.

The total number of polymer chains generated in RAFT/MADIX is equal to the number of R leaving groups of the CTA, plus the number of radicals derived from the initiator which is equal to$\{2f([I]_0 - [I]_t)$, where $f$ is the initiator efficiency, $[I]_0$ is the initial initiator concentration and $[I]_t$ is the initiator concentration at time $t$. $[I]_t$ can be expressed as follows: $[I]_t = [I]_0 \exp(-k_d t)$, where $k_d$ is the rate constant of decomposition of the initiator. Since the radical initiator produces polymer chains that do not carry the thiocarbonylthio moiety, its amount should be balanced to minimize the proportion of dead chains. As a consequence, the theoretical molar mass given in equation (10.1) in MADIX should be rewritten as follows:

$$\overline{M}_{\text{n,theo}} = \frac{[M]}{[X] + fd[I]_0(1 - e^{-k_d t})} \times M_{\text{MU}} \times \text{Conv.} + M_{\text{X}} \qquad (10.5)$$

where $d$ is the average number of chains produced by irreversible terminations. Thus, the higher the xanthate concentration, the lower the proportion of dead chains is.

Pan and coworkers investigated MADIX polymerization of three different monomers (S, methyl acrylate (MA) and methyl methacrylate (MMA)) under $^{60}$Co $\gamma$-ray irradiation [48]. Five xanthates with variation in Z and R groups ($\mathbf{X_{28}}$, $\mathbf{X_{29}}$ and $\mathbf{X_{32}}$–$\mathbf{X_{34}}$) were examined in this process. Polymerizations of MA was controlled in the presence of *O*-aryl xanthates, provided the aryl group was not too large otherwise the rate of polymerization was decreased. In contrast, the use of the *S*-benzyl-*O*-ethyl xanthate (**X**$_{28}$) led to a very poor control of the polymerization, with a broad molar mass distribution of the final polymers, some being even cross-linked. MADIX polymerization of S using *S*-benzyl *O*-2-naphthyl xanthate (**X**$_{32}$) was controlled, though very slow. As expected, polymerization of MMA was not controlled under these conditions.

### 10.7.2
### Solvent Effect

In principle, any solvent employed for free-radical polymerization can be used for MADIX, provided the chain-transfer constant to solvent is very low, otherwise the proportion of dead chains produced by irreversible transfer will increase. Using different analytical means, researchers from Rhodia showed for instance that numerous dead chains are obtained during xanthate **X**$_0$-mediated polymerization of butyl acrylate (BA) performed in solution in ethanol [49]. This was attributed to transfer to ethanol that mainly occurred at the end of the process. Under these conditions, a mass percentage of dead chains between 17 and 69 wt % was determined for samples of targeted molar mass $M_n$ between 4000 and 30 000 g·mol$^{-1}$. Of particular importance, MADIX can be performed in waterborne media and even in pure water to polymerize hydrophilic monomers [50]. Access to water-soluble (co)polymers by RAFT/MADIX polymerization has been recently reviewed by McCormick and coworkers [9].

### 10.7.3
### Additives

Kamigaito and coworkers reported that MADIX polymerization can be conducted in the presence of Lewis acids [51]. The use of such additives in free-radical polymerization aims at modifying the tacticity of the polymers or forcing the tendency for alternation in copolymerization. Thus, poly(*N*-vinyl pyrolidone) (PNVP) with well-controlled molar masses and slightly improved syndiotacticity were obtained from two different xanthates (**X**$_2$ and **X**$_{28}$), in the presence of fluoroalcohols (e.g. $(CF_3)_3COH$). Investigations by $^1$H NMR showed that the formation of a 1:1 hydrogen-bonding complex between the monomer and the fluoroalcohols was responsible for the change of tacticity of the polymers.

## 10.8
## Monomers Polymerizable by MADIX

As emphasized above, xanthates afford variable control over molar masses and PDIs depending on the Z and R substituents/monomer combination. In other words, there is no MADIX agent (neither RAFT agent) exhibiting a universal character. It turns out that MADIX is particularly suited for highly reactive propagating radicals, such as those deriving from vinyl esters and *N*-vinyl monomers. We recently brought evidence that *O*-ethyl xanthates like $\mathbf{X_0}$ promote the MADIX polymerization of vinyl phosphonate monomers [52]. In particular, MADIX enabled the first C/LRP of the challenging vinyl phosphonic acid monomer. In contrast, the *O*-ethyl MADIX agents exhibit a moderate reactivity toward polystyryl radicals and a slightly better one toward polyacrylyl and polyacrylamidyl radicals [11]. However, polymerization of both acrylics and styrenics can be well controlled by introducing electron-withdrawing substituents in the Z group. To date, the polymerization of alkyl methacrylates and methacrylamido monomers could not be controlled. A list of monomers that were efficiently polymerized under MADIX conditions is given in Fig. 10.3.

### 10.8.1 Styrenics

Characteristic examples of MADIX polymerization of S have been discussed in the Z- and R-effect sections. As just indicated, *O*-ethyl xanthates EtO(C=S)SR are moderately efficient in MADIX polymerization of S, leading to polymers with PDI around 2 and experimental molar mass higher than theoretical values before completion of the reaction. The fact that both measured $C_{tr}(X)$ and $C_{tr}(P_n\mathbf{X})$ are low mirrors the poor reactivity of such xanthates [11, 41, 42]. Addition ($k_{add}$) of polystyryl radicals onto xanthates is not favored because of the stabilization effect of these CTAs (see canonical forms in Scheme 10.6). However, PS with PDIs as low as 1.2 could be obtained from xanthate $\mathbf{X_{17}}$, possessing a *O*-diethoxyphosphonyl-2,2,2-trifluoroethyl substituents in the Z group, which considerably enhanced the xanthate reactivity [41]. Other styrenic monomers such as divinyl benzene serving as cross-linker or the water-soluble sodium styrene sulfonate (NaSS) [53] follow the same trends when polymerized by MADIX. Other *p*-substituted styrenic monomers with acid-cleavable groups, for example *p*-acetoxystyrene (pAcS) and *p*-(*tert*-butoxycarbonyloxy)styrene (*t*-BOCOS) were block copolymerized with S in the presence of $\mathbf{X_{17}}$ [54]. Also, first boronated polymers with controlled architectures based on *p*-vinylphenylboronic acid (VPBA) were synthesized by the MADIX process [55].

### 10.8.2 Alkyl Acrylates

Polyacrylyl-based radicals exhibit relatively low steric hindrance and a higher reactivity than polystyryl radicals. As for styrenics, efficient control of MADIX polymerization of acrylates can be achieved, provided that the xanthate is properly selected [11]. For instance, *O*-ethyl xanthates are fairly effective toward polyacrylate propagating chains, although they can provide good control over molar masses versus monomer conversion profiles, in particular with a tertiary R leaving group. Compared to PS, *O*-ethyl xanthates lead to polyacrylates with lower PDIs ($1.5 < \mathrm{PDI} <$

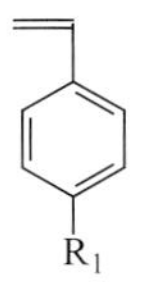

| $R_1$ | Abbreviation |
|---|---|
| H | S [10,11,38,41,42,80,83] |
| OAc | pAcS [54] |
| $SO_3^{\ominus}Na^{\oplus}$ | NaSS [53,59] |
| $B(OH)_2$ | VPBA [55] |
| t-BOC-O | tBOCOS [54] |

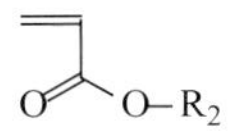

| $R_2$ | Abbreviation |
|---|---|
| H | AA [50,63] |
| $CH_3$ | MA [48] |
| $C_2H_5$ | EA [11,41,47] |
| $C_4H_9$ | BA [47,49,98] |
| $(CH_3)_3$ | t-BA [112] |
| $C_8H_{17}$ | 2-EHA [57] |
| $C_2H_4OH$ | 2-HEA [56] |
| $C_2H_4N(CH_3)_2$ | 2-DMAEA [58] |
| $C_2H_4\overset{\oplus}{N}(CH_3)_3,CH_3SO_3O^{\ominus}$ | 2-DMAEA MS QUAT [58,59] |
| $C_2H_4R_f$ | FA [60] |

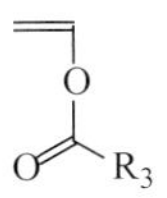

| $R_3$ | Abbreviation |
|---|---|
| $CH_3$ | VAc [10-11,31-32,107] |
| $CH_2CH_3$ | VProp [109] |
| $(CH_3)_3$ | VPiv [107] |
| $C_9H_{19}$ | VneD[11,107] |
| $C_{18}H_{37}$ | VSte [11] |

6-O-vinyladipoyl-D-glucopyranose (VAGluc) [62]

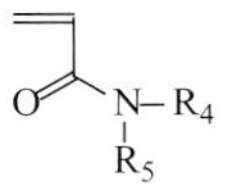

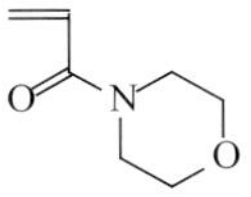

AcMor

| $R_4$ | $R_5$ | Abbreviation |
|---|---|---|
| H | H | Am [50] |
| H | $CH(CH_3)_2$ | NiPAm [64] |
| H | $C(CH_3)_2CH_2SO_3H$ | AMPS [59] |

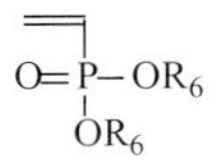

| $R_6$ | Abbreviation |
|---|---|
| H | VPA [52] |
| $CH_3$ | DMVP [52] |

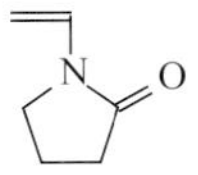

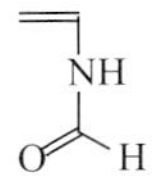

N-VP[28,51,65-67]NVF[100]

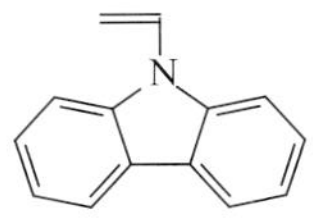

N-VCbz [68]

**Fig. 10.3** List of monomers compatible with the MADIX process.

1.8), suggesting a more rapid interchange of the dithiocarbonate moieties – higher $C_{tr}(P_n\mathbf{X})$ – during polymerization. In this case also, control is significantly improved by introducing electron-withdrawing groups in Z′ [41]. A very recent report by researchers from Rhodia provided quantitative analyses about xanthate ($\mathbf{X_0}$)-mediated polymerization of both di(ethylene glycol) ethyl ether acrylate (DEGA) and butyl acrylate (BA) [49]. The chain-transfer constant was 1.5 and 2.7 for DEGA and BA, respectively, attesting to the controlled character of the process, although such values do not permit to achieve very low PDIs. Many other acrylate monomers were polymerized via a MADIX process in the presence of $\mathbf{X_0}$, including 2-hydroxyethyl acrylate(2-HEA) [56], 2-ethylhexyl acrylate (2-EHA) [57], 2-dimethylaminoethyl acrylate (2-DMAEA) [58], [2-(acryloyolxy)ethyl]trimethylammonium methyl sulfate (2-DMAEA MS QUAT) [59] and fluorinated acrylates [60]. Their polymerization is described in Rhodia's patent literature.

### 10.8.3
### Alkyl Methacrylates, Methacrylamides and Other 1,1-disubstituted Alkenes

To date, MADIX polymerization of methacrylic and methacrylamido-type monomers is ineffective [6–11]. It is worth mentioning that the examples of efficient RAFT agents for 1,1-disubstituted monomers are scarce, that is few dithioesters and dithiocarbamates with properly chosen Z group combined with relevant tertiary R groups like cyanoisopropyl [6–9]. For a similar reason, attempts to control the polymerization of a captodative monomer, namely ethyl-α-acetoxyacrylate also proved unsuccessful by MADIX using $\mathbf{X_0}$ [61]. However, its MADIX copolymerization with different acrylic monomers, such as butyl acrylate (BA), acrylic acid (AA), 2-dimethylaminoethyl acrylate (2-DMAEA) or *N,N′*-dimethyl acrylamide (DMA) gave well-defined statistical copolymers [61].

### 10.8.4
### Vinyl Esters

Whereas the polymerization of vinyl esters (vinyl acetate (VAc), vinyl neodecanoate (VneD) and vinyl stearate (VSt)) is completely inhibited in the presence of dithioesters as RAFT agents [6–8], excellent control over molar masses and PDIs is achieved with *O*-ethyl xanthates [10, 11, 31, 32]. This means that a MADIX agent such as $\mathbf{X_0}$ is entirely consumed in the early stages of the polymerization, resulting in a linear increase of molar masses versus conversion of the monomer. A MALDI-TOF MS of a PVAc sample prepared with $\mathbf{X_0}$ showed a perfect agreement between the expected chain structure and the experimental one [11]. In 2000, MADIX was the first reported C/LRP technique to allow the polymerization of VAc under controlled conditions [10]. PVAc radicals are highly reactive and readily add the C=S double bond of *O*-ethyl xanthate. This is explained by the destabilization of the intermediate radical in the pre- and the main equilibrium. These MADIX-radical adducts undergo a much faster β-scission than when more reactive RAFT agents are employed, since re-forming a xanthate-capped polymer chain which is stabilized by the conjugation effect mentioned above. Inhibition periods may be associated with the preequilibrium, whereas the rate retardation effect is related to the main equilibrium. The CAMD group correlated the ability of xanthates $\mathbf{X}_{20}$–$\mathbf{X}_{27}$ to control MADIX polymerization of VAc with the electron density on the central carbon atom of the xanthate [31]. ESI MS analysis was also helpful to analyze the MADIX-derived polymers and showed an excellent agreement between the theoretical and experimental molar masses. The same group designed a glycomonomer of vinyl ester type, namely 6-*O*-vinyladipoyl-D-glucopyranose, by lipase-catalyzed transesterification of divinyladipate with α-D-glucopyranose [62]. Its $\mathbf{X}_1$-mediated polymerization was performed in methanol, affording a poly(vinyl ester)-like glycopolymer with $M_n = 19\,000\ \text{g·mol}^{-1}$ and PDI = 1.10.

### 10.8.5
### Acrylic Acid and Acrylamido-Type Monomers

Among the major advantages of the MADIX process, one can remind the possibility to directly polymerize hydrophilic monomers in aqueous media without resorting to protection/deprotection chemistry, making these methodologies environmentally friendly with a high potential of transfer into industrially viable processes. Earlier studies by Rhodia showed that controlled polymerizations of hydrophilic monomers such as acrylic acid (AA) and acrylamide (Am) could be performed in aqueous solution by MADIX [50]. Radicals deriving from the two monomers have little steric bulk and are highly reactive; hence, xanthates are again well suited to control their MADIX polymerization. For instance, both the monofunctional and difunctional xanthates, $\mathbf{X}_0$ and $\mathbf{X}_{43}$, respectively, were used to prepare well-defined homopolymers that were successfully chain extended for the synthesis of double hydrophilic block copolymers (DHBC) [50]. Beyond the example of Am, MADIX allows the controlled polymerization of a range of acrylamido monomers, among which *N*-isopropyl acrylamide (NiPAm) [63], *N*,*N*′-dimethyl acrylamide (DMA) (M. Destarac, unpublished results) and 4-acryloylmorpholine (AcMor) (M. Destarac, unpublished results). A MALDI-TOF mass spectrum of a P(AcMor)-$\mathbf{X}_0$ sample is shown in Fig. 10.4 to illustrate the excellent control of the polymer structure.

In 2001, Claverie and coworkers reported that MADIX polymerization of AA is controlled in aqueous or alcoholic solution, in the presence of *O*-phenoxy and *O*-alkyl xanthates ($\mathbf{X}_2$, $\mathbf{X}_9$, $\mathbf{X}_{28}$–$\mathbf{X}_{30}$ and $\mathbf{X}_{35}$–$\mathbf{X}_{38}$): Linear increase of molar masses versus conversion and PDIs lower than 1.3 were obtained [64].

### 10.8.6
### *N*-Vinyl Monomers: *N*-Vinyl Pyrrolidone and *N*-Vinyl Carbazole

PNVP is employed in a variety of applications, especially in the biomedical sector and in the cosmetic industry. This is due to its solubility in aqueous and nonaqueous

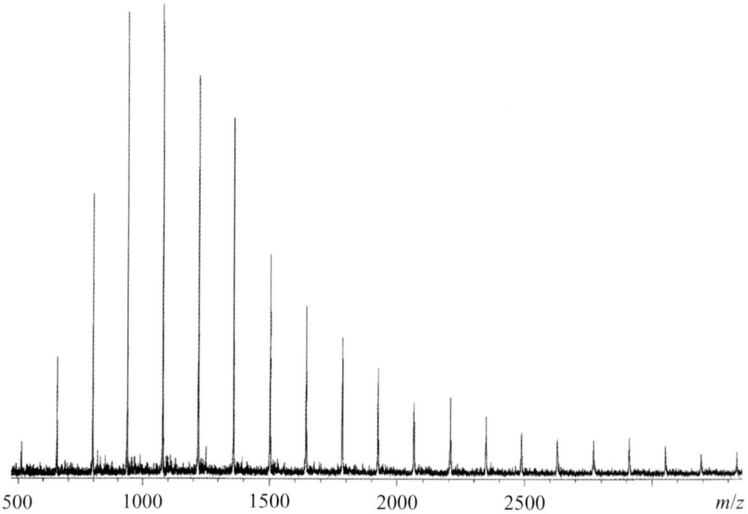


**Fig. 10.4** MALDI-TOF mass spectrum of poly(4-acryloylmorpholine) synthesized with $\mathbf{X_o}$. $M_n = 950$ g·mol$^{-1}$. The measured $m/z$ values correspond to the following controlled chain structure: $CH_3O_2C(CH_3)CH$—$(AcMor)_n$-S(C=S)$OCH_2CH_3$.

solvents as well as to its excellent biotolerance. Until recently, little was published on C/LRP of NVP. MADIX polymerization of that monomer follows the same trends than that of VAc: xanthates are particularly suited for the corresponding propagating radicals that are highly reactive, provided the R leaving group is properly selected [28, 51, 65–67]. The same trend applies for other *N*-vinyl monomers such as *N*-vinyl carbazole (*N*-VCbz) [68]. For instance, the CAMD group showed that *S*-benzyl or *S*-phenylethyl as the R group in *O*-ethyl xanthates ($\mathbf{X_2}$ and $\mathbf{X_{28}}$) allowed the synthesis of well-defined architectures based on PNVP, including linear, starlike and block copolymer compounds [65]. Likewise, xanthate $\mathbf{X_2}$ was used by Mori et al. to efficiently control the MADIX polymerization of *N*-VCbz: Polymers with $M_n$ values up to 48 000 g·mol$^{-1}$ and PDIs $< 1.2$ could be obtained [68]. Retention of the dithiocarbonate polymer end groups was demonstrated by chain-extension experiments. In contrast, with R = 2-carboxyethyl ($X_{18}$) as the leaving group, polymerization of NVP exhibited a 'hybrid behavior', but the same MADIX agent showed fast and selective initialization for VAc, although such a difference was not fully explained [28, 66].

As already mentioned, Kamigaito and coworkers reported that xanthates $\mathbf{X_2}$ and $\mathbf{X_{28}}$ gave well-controlled PNVPs, following inhibition periods of 6 and 1 h, respectively. (The inhibition period might be attributed to the poor reinitiating ability of the benzylic radicals arising from the R group.) [51]. Finally, the CSIRO group described the successful synthesis of PNVP using the *S*-phthalimidomethyl xanthate $\mathbf{X_{39}}$ [67].

## 10.9 MADIX Polymerization in Waterborne Dispersed Media

It is now generally accepted that emulsion RAFT polymerization with highly reactive CTAs like dithioesters or trithiocarbonates cannot be controlled in a conventional manner, that is under simple *ab initio* conditions comprising water, surfactant, monomer and RAFT agent [6–8]. The implementation of such a process is accompanied by a loss of control of the molar mass distribution, a strong decrease of the rate of polymerization together with a degradation of the colloidal stability of the latex [69]. These problems resulting from the lack of control of the nucleation step could be circumvented in different manners, including the following:

- the use of a phase-transfer agent to transport the RAFT agent in a preformed seed latex [70],
- the help of self-assembled RAFT-terminated amphiphilic oligomeric micelles used as nucleation sites [71, 72],
- the implementation of miniemulsion polymerization [73–75] or
- the use of AA-rich RAFT-ended PS seed generated by a spontaneous phase-inversion process [76].

Most of these systems, however, suffer from a broadening of the molar mass distribution over the course of the polymerization and from an increasing retardation upon increasing the RAFT-agent concentration. These approaches thus present severe limitations with a view of developing a cost-attractive industrial process. Gilbert and coworkers recently proposed an interesting model of an emulsion RAFT polymerization in order to optimize the rates of reaction [77]. It was proposed that the use of an oligomeric adduct to the RAFT agent, a less water-soluble RAFT reinitiating group and a less active RAFT agent would help to minimize the inhibition and retardation arising from the exit or desorption of radicals from the particles. As discussed below, these criteria can be fulfilled using xanthates as MADIX agents.

### 10.9.1 Ab Initio Emulsion Polymerization

As early as 1998, the first Rhodia patent on the MADIX process mentioned examples of *ab initio* xanthate ($X_1$)-mediated emulsion polymerization of S initiated by ammonium persulfate, using sodium lauryl sulfate as surfactant, with the monomer added semicontinuously. Under conditions typical of a standard emulsion process, fast polymerizations with $M_n$ increasing with monomer conversion were observed, together with high PDIs ($M_w/M_n \sim 1.8$–$1.9$), which were later explained by the slow reversible transfer of the *O*-ethyl xanthate groups [38]. The same research group reported these results in the first peer-reviewed article about MADIX [10]. The latter contribution was extended to the control of emulsion polymerization of BA using

**$X_1$**. At that time, it was believed that the low intrinsic reactivity of $X_1$ (compared to the far more reactive dithioester and trithiocarbonate RAFT agents) combined with its well-chosen solubility were responsible for the controlled polymerization. During the same time period, Rhodia R&D teams discovered that xanthate-capped hydrophilic (based on AA or Am) or amphiphilic (made of AA and BA) oligomers could be advantageously added to *ab initio* emulsion polymerization to produce styrene-acrylic latexes with controlled $M_n$ and surface chemistry [78]. Compared to reference latexes synthesized in the presence of comparable amounts of hydrophilic stabilizing monomer, the dried latexes exhibited better redispersibility and the corresponding films showed enhanced scrub resistance, thus opening new perspectives for applications in the fields of paints and coatings, adhesives and construction materials.

In collaboration with Rhodia, Monteiro and coworkers thoroughly studied *ab initio* emulsion polymerization of S and BA with xanthates **$X_1$** and **$X_2$** [79–83]. A preliminary kinetic study of solution MADIX polymerization of S and BA with **$X_1$** revealed that the transfer constant to xanthate was low for both monomers ($C_{tr} \sim 0.8$ for S [42] and $\sim 1.7$ for BA) [81]. Based on a model for entry and exit of radicals, $C_{tr}$ values and water solubility of **$X_1$** and **$X_2$**, the authors demonstrated that a negligible amount of aqueous-phase radicals undergo transfer to xanthate, which allows nucleation to occur without the perturbations encountered with other classes of too reactive RAFT agents.

In agreement with the kinetic model developed by Müller et al. [29], PDIs of 2.0 and 1.6 were obtained for S [84] and BA [85], respectively, under batch conditions. The PDI value for poly(n-butyl acrylate) (PBA) could be decreased down to 1.3–1.4 by slowly adding BA in a semicontinuous process [82]. As a result, styrene-acrylic block copolymer core-shell latex nanoparticles could be synthesized with controlled particle size and molar mass distributions [82]. The same authors took advantage of this synthetic strategy to incorporate the reactive acetoacetoxyethyl methacrylate in the acrylate block of a styrene–butyl acrylate core-shell diblock latex [86], to obtain latex films with remarkable mechanical properties when cross-linked with a diamine.

With the aim of synthesizing PS latexes of reduced PDI, emulsion polymerization of S was carried out using the xanthate **$X_{13}$** containing the trifluoroethyl Z′ group [83, 87].

Emulsion polymerization of S was carried out at 70 °C with **$X_{13}$**, using sodium dodecyl sulfate as surfactant and sodium persulfate as initiator [83]. The resulting PS exhibit $M_n$ and PDI evolution profiles that are in perfect agreement with the Müller's solution-based kinetic model: [29] nonlinear increase of $M_n$ versus conversion and PDIs close to 1.5. In order to optimize the control of both $M_n$ and PDI, our group carried out the polymerization of S that was added semicontinuously. Under these conditions, the intrinsic reactivity of **$X_{13}$** combined with the very low instantaneous monomer concentration throughout polymerization allowed for a fast consumption of **$X_{13}$** in the very early stages of the reaction, as shown by the linear increase of $M_n$ with conversion from the beginning of the polymerization (Fig. 10.5). PDIs are comparable to those obtained in batch ($1.4 < \text{PDI} < 1.6$, as shown in Fig. 10.5).

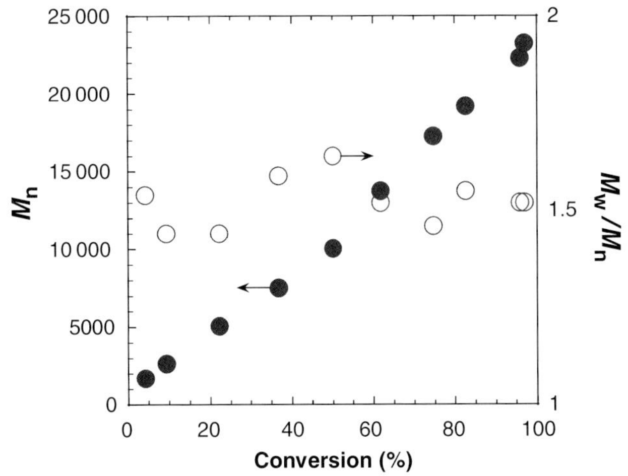


**Fig. 10.5** Evolution of $M_n$ and $M_w/M_n$ with conversion during semibatch emulsion polymerization of St with $\mathbf{X}_{13}$.

The PS-$\mathbf{X}_{13}$ latex ($M_n = 23250$ g·mol$^{-1}$; PDI = 1.52) was used as a seed for the polymerization of EA at 85 °C with ammonium persulfate as initiator, at a solid content of 30 wt %. EA was introduced during 1 hour and the reaction was kept at this temperature for an extra hour after the end of EA addition. A stable PS-*b*-PEA diblock copolymer latex was thus formed, with an EA conversion higher than 99%. $M_n$ was remarkably well controlled ($M_n = 43\,900$ g·mol$^{-1}$ and $M_{n\text{-theo}} = 43\,500$ g·mol$^{-1}$); however, a slight broadening of the molar mass distribution during EA polymerization was noteworthy (final PDI = 1.79). The controlled character of the diblock synthesis is illustrated in Fig. 10.6, showing the overlay of the size-exclusion chromatographic (SEC) traces of the PS first block and the corresponding diblock copolymer.

To conclude, the judicious choice of the xanthate MADIX agent of appropriate reactivity and solubility combined with monomer addition profiles adapted for a high instantaneous monomer conversion helped to strongly increase the *ab initio* MADIX polymerization of S and EA for the fast production of stable PS-*b*-PEA copolymer latexes with high monomer conversion (>99%), predetermined $M_n$ and relatively low PDIs. These advantages undoubtedly position MADIX as the most straightforward C/LRP technique to develop complex macromolecular architectures in waterborne emulsion under industrially relevant conditions.

### 10.9.2
### Miniemulsion Polymerization

Although MADIX emulsion polymerization could be implemented using standard conditions with xanthates $\mathbf{X}_1$, $\mathbf{X}_2$ and $\mathbf{X}_{13}$, miniemulsion polymerization was

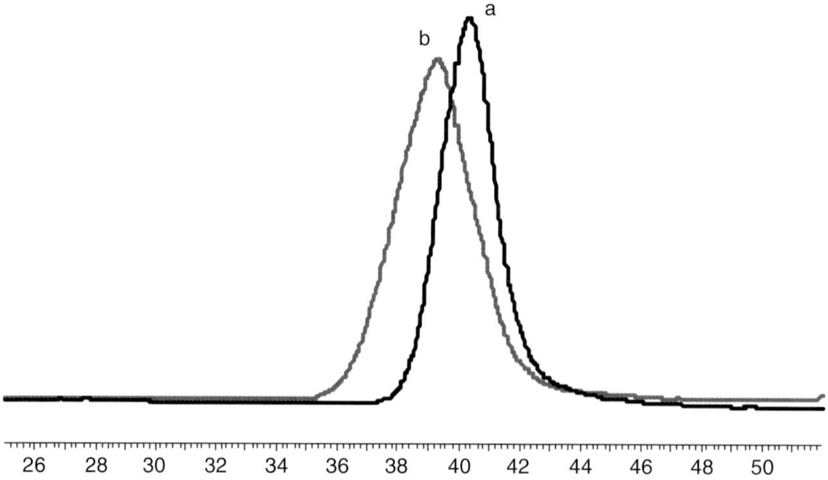


**Fig. 10.6** SEC chromatograms of (a) P(St)-$\mathbf{X}_{13}$ ($M_n = 23250$ g·mol$^{-1}$, PDI = 1.52) and (b) P(St)-*b*-P(AEt)-$\mathbf{X}_{13}$ ($M_n = 43\,900$ g·mol$^{-1}$, $M_w/M_n$ = 1.79). $M_{nth}$ (diblock) = 43 500 g·mol$^{-1}$.

considered for the two-step synthesis of PS-PBA-PS triblock copolymers with the difunctional $\mathbf{X}_{44}$ [88]. Polymers obtained from batch solution MADIX polymerization of BA with $\mathbf{X}_{44}$ exhibited expected characteristics when using an *O*-ethyl xanthate with a low $C_{tr}$: final $M_n$ close to theoretical value but nonlinear increase of $M_n$ versus conversion, $1.5 < \text{PDI} < 2$. Surprisingly, *ab initio* emulsion polymerization led polymers with higher $M_n$ values than expected, unreacted $\mathbf{X}_{44}$ at the end of the polymerization and quite high PDIs (~6 at final conversion) that increased during polymerization [79]. This was attributed to a slow reaction of $\mathbf{X}_{44}$ and slow nucleation of latex particles, probably due to a limited diffusion of $\mathbf{X}_{44}$ across the aqueous phase. This disadvantage was circumvented by directly polymerizing BA with $\mathbf{X}_{44}$ in nanosized droplets in the presence of a small amount of hydrophobe [88]. The resulting PBA latexes with controlled final $M_n$ (~$10^5$ g·mol$^{-1}$) and PDIs close to 2 were subsequently used as seeds for S polymerization. To the best of our knowledge, this miniemulsion approach represented the first example of hard-soft-hard triblock copolymer synthesis in waterborne dispersed media, with potential interest for thermoplastic elastomer properties.

More recently, two groups reported the successful miniemulsion MADIX polymerization of VAc using the monofunctional xanthate $\mathbf{X}_{25}$ [89, 90]. The level of control of $M_n$ and PDI was generally close to that obtained in bulk polymerization, with kinetics indicating both inhibition and retardation. Surprisingly, the possibility to perform emulsion MADIX polymerization of VAc under classical conditions was not questioned in these two reports.

## 10.10 Macromolecular Engineering by MADIX

MADIX now appears as a powerful and versatile synthetic tool to tailor-made polymers and 'complex' macromolecular architectures, including end-functionalized polymers, block copolymers, starlike polymers, graft copolymers and combs. The following sections review the potential of MADIX for macromolecular engineering.

### 10.10.1 End-Functionalized Polymers

A special section below is dedicated to the synthetic strategies to remove the xanthate end groups of (co)polymers made by MADIX. These methods generally replace the thiocarbonyl-thio moiety into either a thiol or a hydrogen atom. As emphasized above, MADIX produces (co)polymers possessing the R group at one end and the dithiocarbonate Z′O(C=S)S— at the other (see Scheme 10.5). MADIX being tolerant of many functional groups – except of primary and secondary amines that readily react with xanthate groups – a specific functionality might be incorporated either in R or in Z′. However, this has been little exploited in the literature. As an illustration, one can cite works by the Noveon group to synthesize COOH- and OH-containing xanthates, as shown in Scheme 10.9 [44, 45]. Related polymers thus contained the functional group in $\alpha$-position, which was confirmed by $^1$H NMR and MALDI-TOF MS.

### 10.10.2 Block Copolymers

Block copolymers enter in widespread applications as the result of their self-assembly properties, either in the solid state or in a selective solvent of one block, which provide a great variety of morphologies in the submicron size range [91, 92]. From a synthetic viewpoint [93], block copolymers are generally prepared by sequential addition of monomers, method (i). However, other methods can be used, including (ii) combination of two different modes of polymerization, (iii) coupling of preformed polymer segments possessing antagonist end groups and (iv) one-pot initiation from dual ('double-headed') initiators. Out of these four methods, it is interesting to note that the first three ones were used to synthesize block copolymers by MADIX. In method (i), it is essential to sequentially polymerize the two monomers in a certain order to access the targeted block. The golden rule is to comply with the scale of reactivity of propagating species. In RAFT/MADIX polymerization, one should start by the polymerization of the monomer that forms the higher reactive propagating radicals and then polymerizes the other monomer. Efficient crossover thus requires that the first block provides the better leaving radical.

In collaboration with different partners, the Rhodia group looked into two kinds of block copolymers, including DHBC [50], that is possessing one stabilizing block in water and a polyelectrolyte block [50, 94], and amphiphilic block copolymers (ABC) [49, 95–97]. For instance, the synthesis of a series of diblock and triblock copolymers based on Am and AA units, using the mono- and difunctional xanthates, $\mathbf{X}_0$ and $\mathbf{X}_{43}$, respectively, was described in 2001. Furthermore, copolymerizations of the two monomers yielded well-defined statistical copolymers that were chain extended for the synthesis of P(AA-*stat*-Am)-*b*-PAm DHBC, that is block copolymers with one block being a statistical one. In a subsequent contribution, Gérardin et al. applied these DHBC and similar PAA-*b*-P(2-HEA) copolymers to sterically stabilize metal hydrous oxide nanoparticles during their mineralization process [94]. In these DHBC, the metal-complexing block acts as a growth-control agent, while the neutral block promotes colloidal stabilization. Other anionic-neutral DHBC like PNaSS-*b*-PAm and PAMPS-*b*-PAm as well as P(2-DMAEA MS QUAT)-*b*-PAm cationic-neutral DHBC were synthesized in the presence of xanthate $\mathbf{X}_0$ and applied to the formation of homogeneous and stable monophasic compositions of complex coacervates made of the charged-neutral DHBC, a polyelectrolyte of the same sign and an oppositely charged surfactant [59].

In a series of papers, Ponsinet and coworkers investigated the self-assembly properties of PS-*b*-PAA and of P(S-*stat*-AA)-*b*-PAA ABC, both in bulk and in water solution [95, 96]. These diblock copolymers were synthesized by MADIX, following route (i) described above. Despite quite high PDIs ($\sim$2), these PS-*b*-PAA copolymers exhibited a microphase separation in bulk, with spherical domains made of the minority PS block embedded in a continuous PAA matrix [95]. The design of P(S-*stat*-AA)-*b*-PAA via the introduction of AA units in the first PS block induces a decrease of the Flory–Huggins parameter ($\chi$) between the two constitutive blocks, resulting in a decrease of the P(S-*stat*-AA)-core radius and a decrease of the chain density at the interfaces [96]. An order–disorder transition was observed in bulk only when the effective $\chi$ parameter was small enough: for a content in AA $>50\%$ in the first block, the ABC did not self-assemble any more. The behavior in water of a series of P(S-*stat*-AA)-*b*-PAA diblock copolymers was also investigated by small-angle scattering [97]. For a molar fraction of AA ($ø_{AA}$) in the first block less than 0.25, structures were found to be out-of-equilibrium micellelike objects, with no reorganization made possible upon dispersion. In contrast, for $ø_{AA} > 0.50$, the dispersion in water of the diblock copolymers was at equilibrium and for high values of $ø_{AA}$, the diblocks were entirely soluble in water. For $ø_{AA}$ close to 0.50, the diblocks gave a micellelike structure with a water-swollen core formed by the P(S-*stat*-AA) block and a swollen brush based on PAA.

Other researchers from Rhodia investigated the molecular features of both PBA-*b*-PAA and PDEGA-*b*-PAA amphiphilic diblock copolymers prepared by MADIX [49]. Several analytical techniques, including NMR, MALDI-TOF MS, SEC, liquid chromatography at the point of exclusion adsorption transition (LC-PEAT) and capillary electrophoresis were set up to analyze the chemical purity of the obtained ABC. For instance, the LC-PEAT technique used under the critical conditions of the

PBA block revealed that hydrophobic homopolymer was present in the final diblock compound, which could not be detected by regular SEC or by NMR. This also indicated that PAA chains were not grown from all the xanthate-ended PBA precursor chains. These studies permitted to reveal that MADIX, as any C/LRP method, is sensitive to experimental conditions and that special analytical techniques are needed to access clear-cut information about the structure of MADIX-derived (co)polymers.

Claverie and coworkers previously reported the synthesis of PAA-*b*-PBA diblock copolymers by MADIX using xanthate $\mathbf{X}_2$ in ethanol [98] on the basis of previous works on the synthesis of PAA by MADIX in protic media [64]. In contrast to the synthesis of PBA-*b*-PAA mentioned above, they performed the MADIX polymerization of AA before growing the PBA block. Characterization by different techniques, including static and dynamic light scattering, tensiometry and SEC showed that these ABC self-assembled in aqueous media, forming starlike micelles. Contrary to PS-*b*-PAA ABC that form frozen micelles, such PBA-*b*-PAA copolymers with a soft PBA block had diffusion coefficients in water that were high enough to adsorb onto growing particles. These ABC thus proved efficient polymeric stabilizing surfactants for BA and MMA emulsion polymerization with solid contents up to 50%. The same group showed that the phenylethyl end group near the hydrophilic block was beneficial for interaction with associative thickeners in paint formulations [99].

Other examples of block copolymers obtained by sequential MADIX polymerization were reported. For instance, Robin and coworkers in collaboration with Rhodia synthesized statistical copolymers consisting of captodative ethyl-α-acetoxyacrylate and BA units (see the Monomers section) that were chain extended with VAc [61].

Use of a preformed polymer for the growth of the second block by another mechanism is an alternative synthetic access to block copolymers, method (ii) [93]. One resort to this 'switch of mechanism' strategy is when monomers to pair in a diblock structure do not polymerize by the same mechanism. This strategy can be applied for MADIX, but it requires the chemical transformation of the end group of the prepolymer into a xanthate moiety. This has been applied for the synthesis of a poly(ethylene oxide) (PEO) macro-MADIX agent, by esterification between a COOH-containing xanthate ($\mathbf{X}_{40}$) and an OH-ended PEO [100]. The xanthate-ended PEO was then used to polymerize *N*-vinyl formamide (NVF), affording PEO-*b*-PNVF DHBC. Likewise, α,ω-dihydroxytelechelic poly(dimethyl siloxane) (PDMS) as well as OH-terminated poly(ethylene-*co*-butylene) precursors could be modified in a two-step sequence so as to obtain the corresponding xanthate-terminated polymers serving as macro-MADIX agents to obtain hybrid ABA-type triblock [101] and AB diblock copolymers [102], respectively. The latter case is illustrated in Fig. 10.7.

Finally, the covalent coupling of two polymeric chains at their respective ends also results in a diblock copolymer, method (iii). Noveon thus prepared block copolymers through the condensation of OH-functional polymers (e.g. PEO or PDMS) with COOH-terminated polymers derived by MADIX using $\mathbf{X}_{41}$ [44, 45].

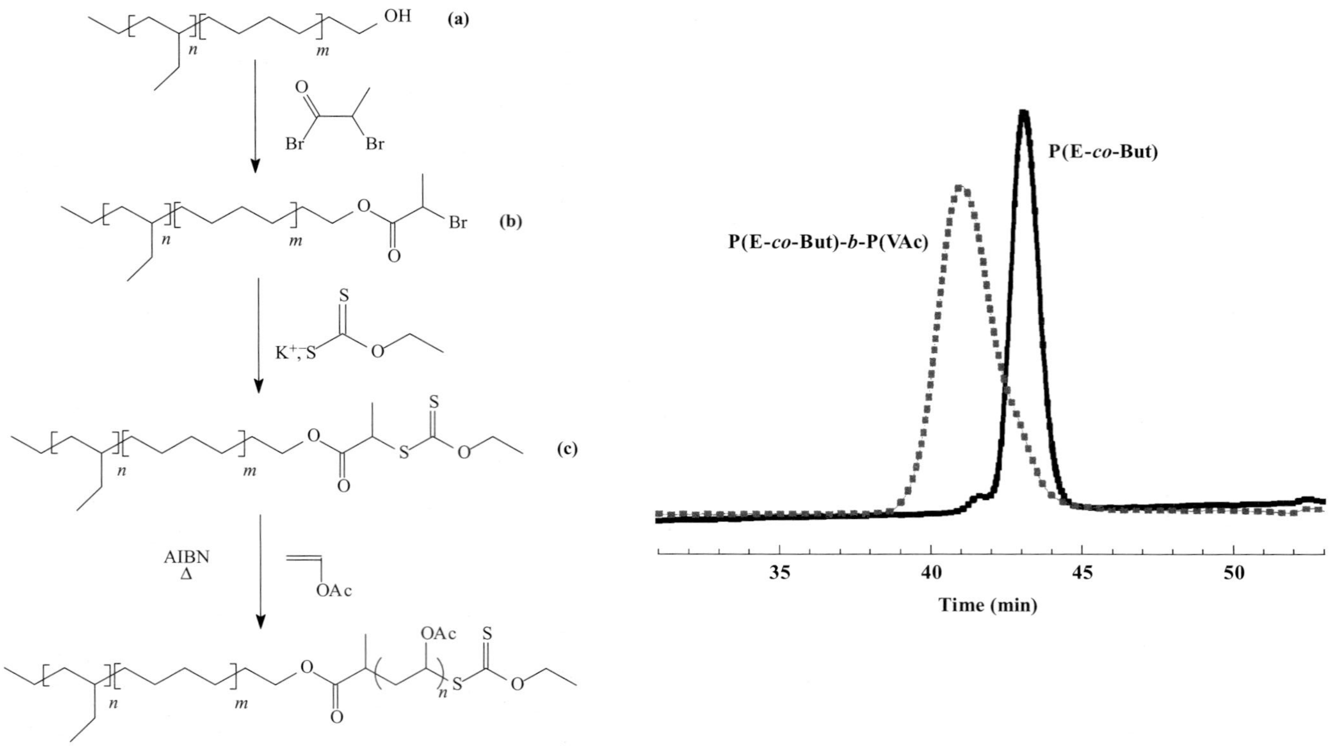


**Fig. 10.7** (a) Process for the preparation of xanthate-terminated P(E-*co*-But) and subsequent synthesis of P(E-*co*-But)-*b*-PVAc by MADIX. (b) SEC/RI traces of P(ethylene-*co*-butylene) ($M_n = 7250$, $M_w/M_n = 1.05$) and resulting P(E-*co*-But)-*b*-PVAc ($M_n = 15\,600$, $M_w/M_n = 1.32$) [102].

Recently, a synthetic coupling strategy to block copolymers based on the Huisgen's 1,3-dipolar cycloaddition ('click chemistry') was successfully developed by the CAMD group [103]. Well-defined PS-*b*-PVAc copolymers were obtained by coupling an azido-terminated PVAc prepared by MADIX and an alkyne-terminated PS prepared by RAFT. To this end, both an alkyne-containing RAFT agent and an azido-containing MADIX agent $\mathbf{X}_{42}$ were designed and used to synthesize the homopolymeric precursors. Block copolymers were then successfully obtained by copper(I)-catalyzed 1,3-dipolar cycloaddition, as evidenced by different analytical means (SEC, FT-IR and NMR).

### 10.10.3
### Star Polymers

Among all branched architectures, star polymers correspond to the simplest possible arrangement of macromolecular chains in a branched structure, since stars involve only one central branching point per macromolecule. In the last decade, there have been an increasing number of reports on the preparation of star polymers by C/LRP. Synthesis of starlike polymers using either the core-first or the arm-first approach has been reviewed recently [104]. In the context of RAFT/MADIX polymerization, two different types of multifunctional CTAs can be employed for star-polymer synthesis: those implying an outward growth of arms from the core (core-first approach referred to as 'the R-group approach') and those involving the reaction of linear chains with the functional core (arm-first approach also referred to as 'the Z-group approach'). Main of the multifunctional MADIX agents used either in the R-group approach or in the Z-group approach that have been reported so far are displayed in Fig. 10.8. Alternatively, coupling of linear chains still containing their thiocarnonyl-thio end groups onto multifunctional comonomers also leads to star polymers composed of a microgel-like core. This convergent ('arm-first') strategy is referred to as the 'nodulus approach'.

In the R-group approach, R leaving groups generated after fragmentation steps are part of the core of the stars. Star polymer synthesis following this route may be complicated by star–star couplings or by irreversible couplings between stars and/or star–linear chain couplings. However, for highly reactive radicals like those deriving from PAA [105], or PVAc [106] (high $k_p^2/k_t$ ratio) the probability for stars to get coupled can be minimized. A few examples of multifunctional MADIX agents consisting of various *S*-substituted R leaving groups and *O*-ethyl or *O*-trifluoroethyl activating groups aimed to grow star polymers by the R-group approach have been designed in a two-step sequence; these are shown in Fig. 10.8 ($\mathbf{X}_{45}$–$\mathbf{X}_{48}$) [66, 105–109]. This has been achieved in particular for VAc, NVP, NVCbz or AA monomers for which MADIX polymerization is highly efficient. For instance, we reported that well-defined three-arm PAA stars could be directly synthesized in solution in dimethylformamide at 70 °C from the trifunctional xanthates $\mathbf{X}_{45}$ and $\mathbf{X}_{46}$, minimizing side reactions such as star–star couplings owing to the very high reactivity of AA under such conditions [105]. The control of the polymerization was evidenced

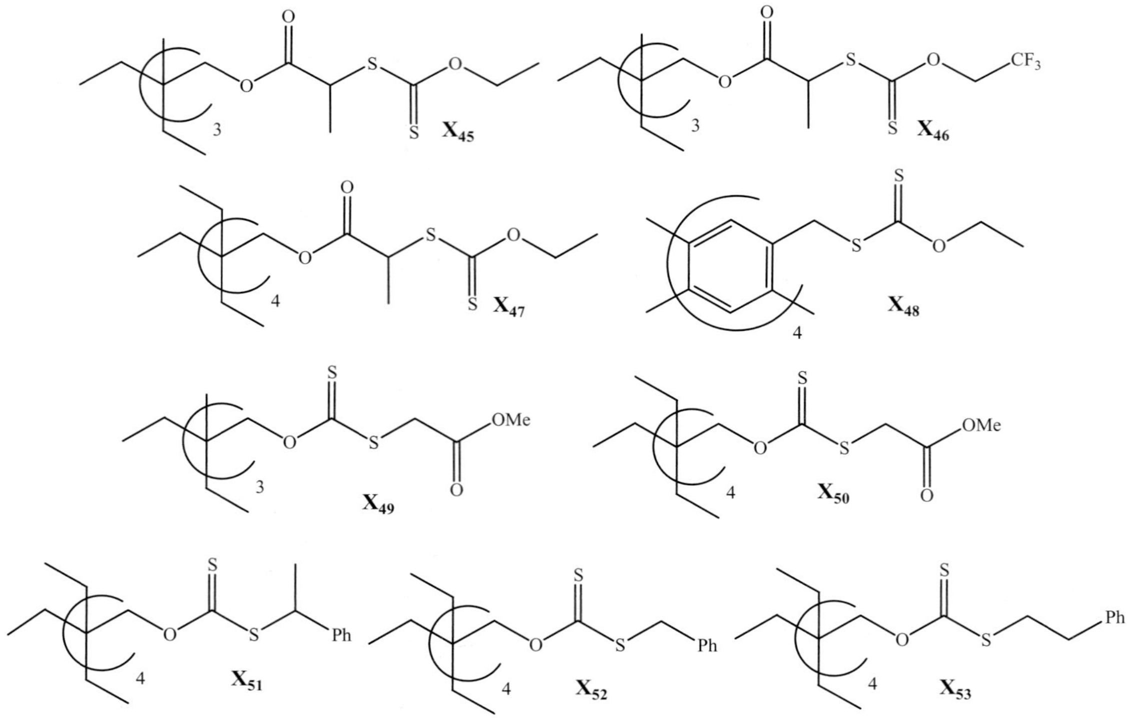


**Fig. 10.8** Representative multifunctional xanthates used in star polymer synthesis by MADIX. (**$X_{45}$–$X_{48}$** are used for the R-group approach, whereas **$X_{49}$–$X_{53}$** serve in the Z-group approach.)

by the increase of the molar masses with monomer conversion, whereas PDIs were in the range 1.2–1.3. Fidelity of the xanthate chain ends was demonstrated by their reactivation in chain-extension experiments in pure water with Am, thus affording double hydrophilic star-block copolymers, with neutral PAm blocks outside and pH-responsive PAA blocks inside. Following the same strategy, three- and four-arm PVAc stars were prepared by the CAMD group, using the tri- and the tetrafunctional MADIX agents (**X**$_{45}$ and **X**$_{47}$) [106]. At high conversions, however, a strong broadening of the molar mass distribution consistent with the occurrence of side reactions was noted. The subsequent methanolysis (use of KOH in $CH_3OH$) of these parent stars led to the formation of stars based on poly(vinyl alcohol) (PVA). In a more detailed study, the same group reported that the R-group approach yielded well-defined stars based on PVAc, poly(vinyl pivalate) (P(VPiv)) and P(VneD) with PDI $< 1.4$, using the same xanthates [107]. Application of this star-polymerization strategy to the vinylester-type glycomonomer already described was more problematic (limiting conversion, experimental molar masses higher than theoretical values), likely due to side reactions [107]. Based on previous works on MADIX polymerization of NVP, the CAMD group applied the R-group approach to design four-arm stars based on PNVP, using the xanthate **X**$_{48}$ [65]. A linear increase of $M_n$ versus monomer conversion was observed, whereas the PDIs remained $<1.3$. When used as stabilizers in suspension polymerization to prepare cross-linked P(VneD/ethylene glycol dimethacrylate) microspheres, PNVP stars gave smaller particles as compared to linear homologues.

In the Z-group approach mentioned above, the core corresponds to the activating group; chains are grown away from the core and are attached when undergoing transfer reactions. Addition of an oligomer chain ($P_n^{\cdot}$) onto one of the thiocarbonylthio groups results in an intermediate radical that releases the $R^{\cdot}$ leaving group upon fragmentation, allowing initiation of a new linear chain, the latter becoming one of the arms of a star after addition–fragmentation transfer reaction to its core. In other words, such arm-first stars contain branches only in a dormant form. One main advantage of this Z-group strategy by RAFT is that complications often seen in the core-first star synthesis, such as star–star and star–linear chain couplings, can be minimized. A potential problem from such an approach, however, is the accessibility to the thiocarbonylthio groups carried by the core (shielding effect), which can increase the probability of termination, thus increasing the concentration of dead polymers. For instance, molar mass distribution of PVAc stars prepared by the Z-group approach with **X**$_{49}$ or **X**$_{50}$ was found to be higher than stars obtained by the R-group approach using **X**$_{45}$ or **X**$_{47}$ [107]. A similar study by Vana and coworker gave slightly different results [109]. These authors reported the synthesis of both four-arm PVAc and poly(vinyl propionate) stars, using tetrafunctional xanthates **X**$_{51}$–**X**$_{53}$. Star polymers PDIs around 1.2 and apparent $M_n$ around 50 000 $g{\cdot}mol^{-1}$ were obtained in this way. The expected structure of these star polymers was confirmed by ESI MS. Kinetic results indicated that side reactions mentioned above, if occurring, could be neglected. As expected, however, an increasing shielding effect was observed, resulting in a progressive loss of control of the polymeriztion. The R-group approach was also found superior regarding the methanolysis step [107]. The overall architecture of stars generated by the Z-group approach, indeed, was destroyed during methanolysis, since the xanthate linkages between the arms and the core were also cleaved. Finally, Mori et al. reported the synthesis of well-defined four-arm PNVCbz following a Z-group approach, using the tetrafunctional xanthate $\mathbf{X}_{51}$ [108]. The authors noted that the multifunctional core had an effect on the polymerization kinetics when compared to the monofunctional xanthate $\mathbf{X}_2$, but no significant influence on the controlled character of the polymerization.

Alternatively, the Rhodia team developed the 'nodulus approach' to access a wide range of water-soluble star (co)polymers by MADIX [105]. This was achieved by adding the *N,N'*-methylenebisacrylamide (MBA) playing the role of a cross-linking agent onto preformed xanthate-capped linear polymers, which resulted in the formation of a microgel at the core stabilized by hydrophilic arms after couplings of linear chains (Scheme 10.11). This 'nodulus' method could be applied to various hydrophilic precursors, including not only homopolymers composed of Am or AA but also statistical and block copolymers of these two monomers. Interestingly, the cross-linking reaction could be simply performed in water at 70 °C. In the latter 'nodulus' arm-first method, key parameters having a dramatic influence on the number of chains attached to the core and the yield of star formation not only include the feed molar ratio ($r$) between the cross-linker and the linear precursor ($r$ = [cross-linker]$_0$/[P$_n$−X]$_0$), but also the nature of the linking agent, the size ($M_n$) of the precursor, the overall concentration of the reaction mixture and solvent nature. It was found, for instance, that the suitable range of $r$ ratio for star formation

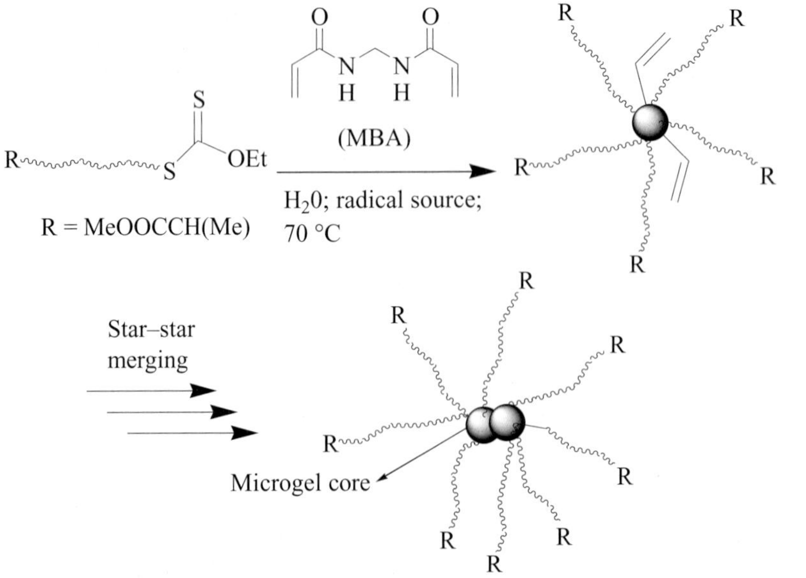


**Scheme 10.11** Synthesis of hydrophilic star polymers by MADIX by 'the nodulus approach'.

was $r = 5$–15 for a linear precursor with $M_n < 20\,000$ g·mol$^{-1}$. Above $r = 15$, gel formation generally occurred and below $r = 5$, poor yield of star formation was observed. Based on findings from SEC characterization, the apparent number of arms for these hydrophilic star (co)polymers was estimated in the range 20–70, depending on the $r$ value. The yield of star polymer formation could finally be increased by adding MBA with a monovinylic comonomer onto the preformed polymer during the cross-linking reaction. By doing so, the size of the core can be increased, thus decreasing the steric hindrance around the core.

### 10.10.4 Combs

Combs are graft copolymers where polymeric grafts are of the same chemical nature as that of the macromolecular backbone. The CAMD group reported the preparation of PVAc combs via MADIX [110]. To this end, well-defined PVA macromolecular backbones were first synthesized by MADIX polymerization of VAc, followed by methanolysis. Chemical modification of the pendant OH groups of PVA via either an R-group or a Z-group approach resulted in the formation of macromolecular MADIX agents. The R-group approach proved more efficient for growing the PVAc grafts by MADIX, the Z-group approach inhibiting the polymerization of VAc. At low conversions, however, the growth of the grafts was contaminated by the formation of both linear polymer chains as a result of the constant initiation and side populations formed by intermolecular couplings. The proportions of these side products were found to increase with the degree of polymerization of the macromolecular MADIX agents, broadening the molar mass distribution. The PVAc grafts could be subsequently treated under basic conditions without affecting the ester linkages between the backbone and the pendant chains, affording the targeted combs based on PVA.

### 10.10.5 Polymeric Nanogels

'Microgels' also called 'polymeric nanogels' are soluble intramolecularly cross-linked polymer chains in the submicron size range [104]. A method generally employed for synthesizing microgels is the radical cross-linking copolymerization of a vinylic monomer with a cross-linker, using one of the three following processes: highly diluted solution, emulsion and precipitation/dispersion polymerizations. We recently reported the one-pot batch solution RCC of AA in the presence of MBA and xanthate $\mathbf{X_0}$ [105, 111]. Highly branched and soluble copolymers referred to as polymeric nanogels were obtained under such conditions. Both the cross-linker and the xanthate concentrations were shown to have an important effect on the build-up of molar masses of the branched copolymers: (i) the higher the concentration of MBA, the larger the molar masses and PDIs for a given xanthate

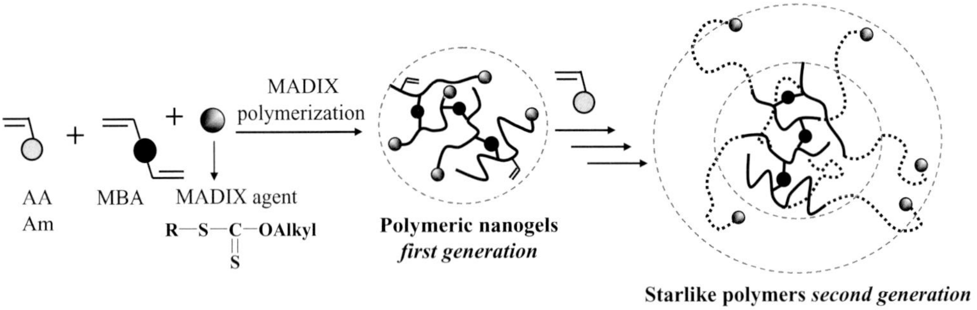


**Scheme 10.12** Polymeric nanogels synthesized by xanthate-mediated RCC.

concentration, and (ii) the lower the xanthate concentration, the larger the molar masses for a fixed MBA concentration. The use of MADIX agent thus permitted high monomer conversion – if not quantitative – higher solid content than in regular microgel synthesis, as well as a much higher content in cross-linker (up to 15% molar) without occurrence of gelation. Another advantage of this synthetic strategy is the possibility to carry out chain extensions from the multiple xanthates of the parent polymeric nanogel serving as macromolecular multifunctional MADIX agent. In this way, starlike structures composed of a polymeric nanogel-based core are obtained divergently, following an R-group approach, as shown in Scheme 10.12. For instance, nanogels based on PAA were chain extended with AA in pure water, with no visible formation of macrogel. This confirmed the low probability of intermolecular couplings in the R-group approach in the case of highly reactive propagating radicals.

## 10.11
## Methodologies to Remove the Dithiocarbonate End Groups

In spite of the numerous advantages offered by RAFT/MADIX polymerizations, one potential drawback lies in the fact that the thiocarbonylthio terminal group, if not deactivated, is likely to be degraded during the lifetime of the polymer or under specific application conditions. Our group paid considerable attention to the development of several appropriate treatments in order to prevent an uncontrolled degradation over time that may eventually generate low molar mass malodorous, potentially toxic sulfur-based by-products [112–116].

In this section, we specifically refer to xanthates since a chapter of this handbook is dedicated to the general methods proposed to irreversibly deactivate thiocarbonylthio terminal groups.

Three main kinds of chemical modification of a xanthate group were reported in the literature: oxidation, reduction and ionic cleavage (Scheme 10.13). For instance, several procedures for the cleavage of a xanthate group into the corresponding thiol are known. Aminolysis using amines [117] and ammonia [118] and reduction with

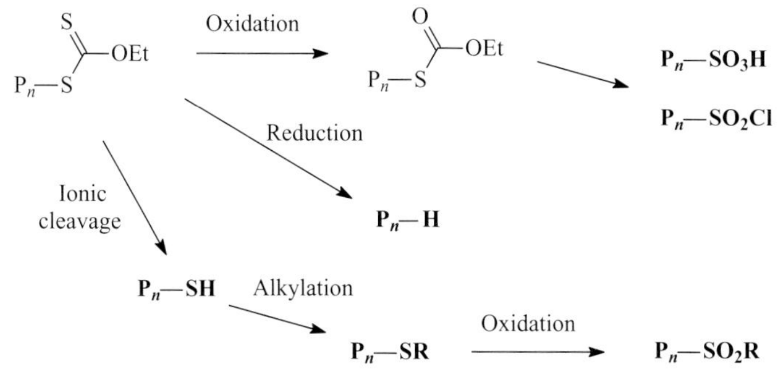


**Scheme 10.13** Modification of the xanthate group: a summary.

$LiAlH_4$ [119] were successfully implemented on xanthate derivatives. Reductive cleavage of the carbon—sulfur bond is often needed to obtain a sulfur-free target molecule. One first approach was described by Zard and collaborators [120], who found that exposure of a xanthate RS(C=S)OZ to stoichiometric amounts of peroxide in refluxing 2-propanol resulted in the complete replacement of the xanthate group by a hydrogen atom to give the corresponding alkane R—H. This method is ecologically superior to the traditionally employed Raney Nickel [121] and $Bu_3SnH$ [122], in the sense that it avoids the use of expensive and toxic heavy metal residues that are difficult to remove. More recently, Boivin et al. [123] reported the reductive cleavage of an *O*-ethyl xanthate group either with diethyl phosphite and a radical initiator or with an ammonium salt of hypophosphorus acid and a radical initiator. The latter approach was later applied to RAFT polymers by Farnham et al. [124]. The thermal stability of xanthate $\mathbf{X_0}$ was questioned in a recent paper of Legge et al. [125]. Thermogravimetric analysis revealed that the temperature at which 50% weight of $\mathbf{X_0}$ is lost and the breakdown temperature are low (131 and 75 °C, respectively) compared to other classes of RAFT agents. This strongly contrasts with earlier studies from our group where $\mathbf{X_0}$ was either efficiently added to various olefins at 160 °C in *o*-dichlorobenzene (S. Z. Zard, unpublished results) or successfully used as MADIX agent in self-initiated S bulk polymerization at 130 °C for 24 h (M. Destarac, unpublished results).

As part of our continuing work in this area, we have explored practical ways of usefully modifying the xanthate group on polymer structures. Two main methods have been explored, involving, on the one hand, the complete reductive removal of the xanthate motif with 2-propanol as hydrogen-atom donor [112, 113] and, on the other, its conversion into the corresponding thiol using the Chugaev fragmentation [112–114]. We have found that the terminal xanthate group present in PDMS and PAA prepared by the MADIX technique can be efficiently removed by a combination of peroxide and secondary alcohol, as summarized in Scheme 10.14 [112].

The second modification is based on a thermolysis process known as the Chugaev reaction [126]. It consists in the thermal cleavage of a xanthate of general structure $R_1R_2HC—CR_3R_4—O—(C=S)—SR$ to give olefin $R_1R_2=R_3R_4$, carbon oxysulfide and

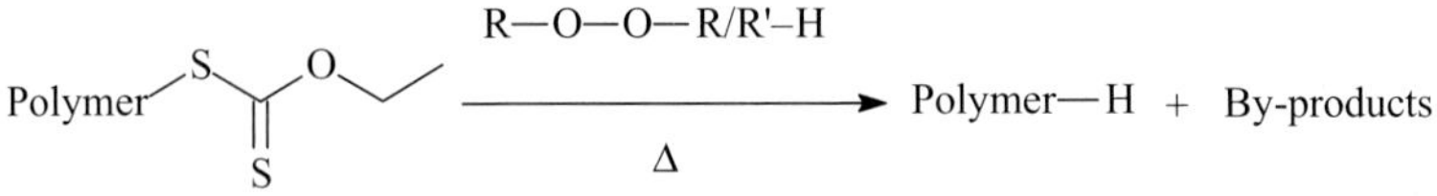


**Scheme 10.14** Peroxide-induced radical reduction of MADIX polymers.

thiol RSH (Scheme 10.15). Interestingly, this unimolecular elimination reaction generates volatile by-products, which may be fully removed from the reaction mixture by evaporation, simply by heating the polymer.

It was shown that the xanthate terminal group can be eliminated simply by heating solutions of MADIX polymers, without the need for additional reagents [112]. This economical and convenient technique was applied to xanthate-terminated PS and poly(*t*-butyl acrylate), as well as to MADIX-derived PDMS and related triblock copolymers with *t*-BA.

These chemical treatments, in addition to other approaches like aminolysis [115] and ozone treatment [116], strongly reduce the risk of possible undesired chemical ageing during the lifetime of the MADIX (co)polymer.

## 10.12 Applications of MADIX (co)polymers

As this was presented in this chapter, the MADIX technology toolbox offers the possibility of designing a nearly infinite array of functional complex polymer architectures, among which double hydrophilic and amphiphilic copolymers have shown to exhibit original interfacial properties in liquid formulations. In the following section, we illustrate the potential of MADIX (co)polymers for four selected types of properties:

- rheology modification,
- emulsion stabilization,
- surface modification, and
- preparation of organic–inorganic nanocolloids.

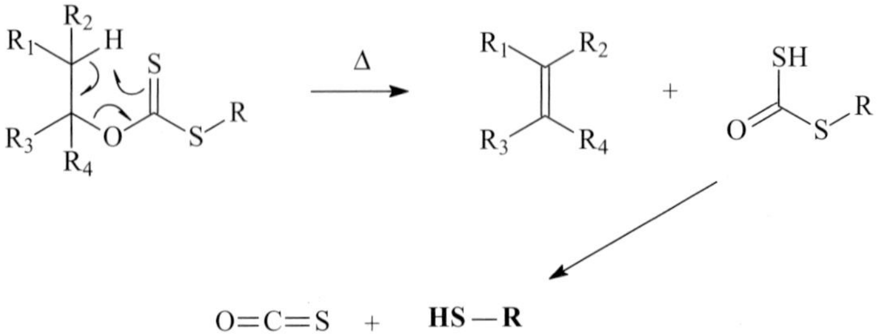


**Scheme 10.15** The Chugaev reaction.

### 10.12.1
### Rheology Modification

Amphiphilic copolymers and DHBC were reported to develop thickening properties in water, either alone or in the presence of coadditives. Viscoelactic aqueous gels were obtained with amphiphilic diblock and triblock copolymers based on styrene and sodium acrylate, with hydrophobic and hydrophilic blocks potentially comprising hydrophilic and hydrophobic monomers, respectively [127]. This range of products was successfully tested as hydraulic fracturing fluids in oil field application. [128] Star-block copolymers with either hydrophobic or cationic outer block were also designed for viscoelastic aqueous gel formation [129]. Well-architectured copolymers comprising boronated VPBA units associated with a polymeric ligand (e.g. a hydrocolloidlike guar) were developed for rheology control of water-based formulations [55]. These polymers were combined for use in the exploitation of oil and gas deposits [130]. Gelled aqueous compositions comprising an architectured copolymer with at least two ionically charged blocks (e.g. double hydrophilic star-block copolymers based on AA and Am) and an oppositely charged component were described in a Rhodia patent [131].

### 10.12.2
### Emulsion Stabilization

It was shown that MADIX amphiphilic diblock copolymers are candidates of choice for stabilizing emulsions, either during latex synthesis [132] or for emulsifying various actives in liquid formulations [133–138]. PDMS-based MADIX hybrid block and graft copolymers were efficiently tested as o/o [133] and o/o/w [134] emulsion stabilizers, in which one of the oil phases is a silicone. Various applications of hydrophilic–hydrophobic diblock copolymers were recently reported by Rhodia: emulsifiable concentrates [135], dried emulsions [136], low hydrophilic–lipophilic balance (HLB) diblock copolymers to stabilize w/o inverse emulsions [137] and the stabilization of dispersions in media of high ionic strength for phytosanitary applications [138].

### 10.12.3
### Surface Modification

Rhodia developed several kinds of hydrophilic–hydrophilic copolymers with tunable amphiphilic character in order to temporarily or permanently modify colloid or flat surfaces. For instance, water-soluble amphiphilic diblock copolymers synthesized by MADIX were advantageously added to preformed emulsion polymers in order to modify their surface chemistry and improve their colloidal stability [139]. Micellar solutions of amphiphilic diblock copolymers were efficiently envisaged to promote the adhesion of latex paints on plastic surfaces [140, 141]. Also, phosphonated copolymers with controlled architectures were found to behave like adhesion promoters for latex paints on metal surfaces [142, 143].

Rhodia patented hard-surface-cleaning compositions, comprising coacervate systems made of a DHBC and an oppositely charged component having soil antiadhesion and antideposition properties on hydrophilic hard surfaces [144]. More recently, a similar strategy was employed for rendering a surface antifouling and/or protein resistant [145].

### 10.12.4 Organic–Inorganic Nanoassemblies

Two approaches were considered to design organic–inorganic nanohydrids via MADIX-based materials: (i) a mineral (or polymer) synthetic step assisted by the polymer (or mineral) counterpart and (ii) the self-assembly of a charged mineral nanoparticle with a block copolymer comprising an oppositely charged block that anchors to the mineral surface and a neutral block soluble in the continuous phase. Following the first strategy, we patented a method for preparing mineral colloidal nanoparticles of controlled size and shape in aqueous dispersion via the basic hydrolysis of a metal cation mineral salt (like lanthanum nitrate $La(NO_3)_3$) in the presence of a DHBC comprising an anionic block (e.g. PAA) and a neutral block (e.g. PAm or P(2-HEA)). The obtained mineral hydroxide dispersions are transparent and stable over a very broad range of pH and ionic strength [56]. Another approach consisted in the grafting of alcoxysilyl-functional xanthate compounds [146] to mineral oxide nanoparticles (e.g. $SiO_2$) in order to control the growth of polymer chains from the mineral surface [147]. Organic–inorganic hybrid nanoparticles were thus obtained.

Without resorting to chemistry, a method for controlling the aggregation of preformed rare earth base nanoparticles like yttrium hydroxyacetate with oppositely charged–neutral block copolymers (e.g. PNaSS–PAm diblock copolymers) was reported to form complex of rare earth aggregates with remarkable stability over time [53]. Alternatively, organosols of mineral nanoparticles stabilized by ABC were recently described by our group [57]. The claimed organosols were prepared by direct-phase transfer of the nanoparticles from the water phase to the organic solvent. The main examples deal with the use of low-molecular-weight PAA–P(alkyl acrylate) diblock copolymers of appropriate HLB to extract nano-$CeO_2$ from a water solution to a broad range of solvents with various polarities.

## 10.13 Conclusion

The MADIX process relies on the interchange of xanthate groups at the polymer chain ends. A proper design of the xanthate RS—(C=S)OZ′ used as a reversible CTA is crucial for an optimal control of molar masses and polydispersities of the final polymer. Research on MADIX has come to maturity roughly in 10-year time, which has led to an advanced technology allowing a vast range of monomers to be (co)polymerized under 'living'/controlled and mild conditions. MADIX can be implemented in industrially viable processes, including waterborne homogeneous solution and dispersed media (emulsion or miniemulsion). It is a powerful synthetic tool for block copolymer synthesis, permitting the association of blocks with antagonist and/or complementary properties. Products obtained by MADIX – in particular block copolymers – may target segments of the specialty polymer markets for some specific functions (rheology control, emulsion stabilization, surface modification, etc.), but they may also find higher value in niche applications (for instance in industrial segments requiring surface modification of inorganic materials at the nanoscale level). A more comprehensive and systematic investigation of the structure/properties relationship of the corresponding materials should be addressed in the future, and these new materials should be benchmarked with analog products prepared by conventional polymerization. This will undoubtedly contribute to better express the competitive advantages of the MADIX technology on the way to industrial development and commercialization of MADIX (co)polymers.

## References


1. K., Matyjaszewski, T., Davis, Eds., *Handbook of Radical Polymerization*, John Wiley & Sons, Inc., Hoboken, NJ, USA, **2002**.
2. W. A., Braunecker, K., Matyjaszewski, *Prog. Polym. Sci.* **2007**, *32*, 93–146.
3. C. J., Hawker, A. W., Bosman, E., Harth, *Chem. Rev.* **2001**, *101*, 3661–3688.
4. K., Matyjaszewski, J., Xia, *Chem. Rev.* **2001**, *101*, 2921–2990.
5. M., Kamigaito, T., Ando, M., Sawamoto, *Chem. Rev.* **2001**, *101*, 3689–3745.
6. G., Moad, E., Rizzardo, S., Thang, *Aust. J. Chem.* **2005**, *58*, 379–410.
7. S., Perrier, P., Takolpuckdee, *J. Polym. Sci. A: Polym. Chem.* **2005**, *43*, 5347–5393.
8. A., Favier, M.-T., Charreyre, *Macromol. Rapid Commun.* **2006**, *27*, 653–692.
9. A. B., Lowe, C. L., McCormick, *Prog. Polym. Sci.* **2007**, *32*, 283–351.
10. D., Charmot, P., Corpart, H., Adam, S. Z., Zard, T., Biadatti, G., Bouhadir, *Macromol. Symp.* **2000**, *150*, 23–32.
11. M., Destarac, D., Taton, S. Z., Zard, Y., Saleh, *I. Six ACS Series 854*, In *Advances in Controlled/Living Radical Polymerization*, Vol. *37*, K., Matyjaszewski, Ed., American Chemical Society, Washington DC., USA, **2003**, pp. 536–550.
12. K., Matyjaszewski, S. G., Gaynor, J.-S., Wang, *Macromolecules* **1995**, *28*, 2093–2095.
13. G., David, C., Boyer, J., Tonnar, B., Ameduri, P., Lacroix-Deamazes, B., Boutevin, *Chem. Rev.* **2006**, *106*, 3936–3962.
14. Y., Gnanou, D., Taton, In *Handbook of Radical Polymerization*, Vol. *14*, K., Matyjaszewski, T., Davis, Eds., John Wiley & Sons, Inc., Hoboken, NJ, USA, **2002**, pp. 775–844.
15. K., Matyjaszewski, *Prog. Polym. Sci.* **2005**, *30*, 858–875.
16. T. P., Le, G., Moad, E., Rizzardo, S. H., Thang, E. I. Dupont deNemous and Company, WO 9801478, **1998**.
17. D., Charmot, P., Corpart, D., Michelet, S., Zard, T., Biadatti, Rhodia Chimie, WO 98/58974, **1998**.
18. J., Chiefari, E., Rizzardo, In *Handbook of Radical Polymerization*, Vol. *12*, K., Matyjaszewski, T., Davis, Eds., John Wiley & Sons, Inc., Hoboken, NJ, USA, **2002**, pp. 629–690.

19. J., Krstina, G., Moad, E., Rizzardo, C. L., Winzor, C. T., Berge, M., Fryd, *Macromolecules* **1995**, *28*, 5381–5385.
20. G., Moad, C. L., Moad, J., Krstina, E., Rizzardo, C. T., Berge, T. R., Darling, E. I. Dupont deNemours and Company, WO 96/15157, **1996**.
21. S. Z., Zard, *Angew. Chem. Int. Ed. Engl.* **1997**, *37*, 672–685.
22. B., Quiclet-Sire, S. Z., Zard, *Top. Curr. Chem.* **2006**, *264*, 201–236.
23. D. H. R., Barton, S. W., McCombie, *J. Chem. Soc. Perkin Trans. 1* **1975**, 1574–1585; D. H. R., Barton, *Half a Century of Free Radical Chemistry*, Cambridge University Press, Cambridge, **1993**; W., Hartwig, *Tetrahedron* **1983**, *39*, 2609–2645; D., Crich, L., Quintero, *Chem. Rev.* **1989**, *89*, 1413–1432.
24. D. H. R., Barton, D., Crich, A., Löbberding, S. Z., Zard, *J. Chem. Soc., Chem. Commun.* **1985**, 646–647; *Tetrahedron* **1986**, *42*, 2329–2338 ; M. D., Bachi, E., Bosch, *J. Chem. Soc., Perkin Trans.* **1988**, *1*, 1517–1519; (d) M. D., Bachi, E., Bosch, D., Denenmark, D., Girsh, *J. Org. Chem.* **1992**, *57*, 6803–6810.
25. B., Quiclet-Sire, B., Sortais, S. Z., Zard, *Synlet* **2002**, 903–906.
26. J., Chiefari, B. Y. K., Chong, F., Ercole, J., Krstina, J., Jeffery, T. P. T., Le, R. T. A., Mayadunne, G. F., Meijs, C. L., Moad, G., Moad, E., Rizzardo, S. H., Thang, *Macromolecules* **1998**, *31*, 5559–5562.
27. G., Pound, J. B., McLeary, J. M., McKenzie, R. F. M., Lange, B., Klumperman, *Macromolecules* **2006**, *39*, 7796–7797.
28. J. B., McLeary, F. M., Calitz, J. M., McKenzie, M. P., Tonge, R. D., Sanderson, B., Klumperman, *Macromolecules* **2004**, *37*, 2383–2394.
29. A. H. E., Müller, R., Zhuang, D., Yan, G., Litvinenko, *Macromolecules* **1995**, *28*, 4326–4333.
30. C., Barner-Kowollik, M., Buback, B., Charleux, M. L., Coote, M., Drache, T., Fukuda, A., Goto, B., Klumperman, A. B., Lowe, J. B., McLeary, G., Moad, M. J., Monteiro, R. D., Sanderson, M. P., Tonge, P., Vana, *J. Polym. Sci. A: Polym. Chem.* **2006**, *44*, 5809–5831.
31. M. H., Stenzel, L., Cummins, G. E., Roberts, T. P., Davis, P., Vana, C., Barner-Kowollik, *Macromol. Chem. Phys.* **2003**, *204*, 1160–1168.
32. A., Favier, C., Barner-Kowollik, T. P., Davis, M. H., Stenzel, *Macromol. Chem. Phys.* **2004**, *205*, 925–936.
33. M. L., Coote, L., Radom, *Macromolecules* **2004**, *37*, 590–596.
34. M., Coote, D. J., Henry, *Macromolecules* **2005**, *38*, 1415–1433.
35. M., Coote, D. J., Henry, *Macromolecules* **2005**, *38*, 5774–5779.
36. E. H., Krenske, E. I., Ezgorodina, M. L., Coote, ACS Symposium Series 944, ACS, In *Controlled/Living Radical Polymerization: From Synthesis to Materials*, K., Matyjaszewski, Ed., American Chemical Society, Washington DC., USA, **2006**, pp. 406–420.
37. M. L., Coote, E. H., Krenske, E. I., Ezgorodina, *Macromol. Rapid Commun.* *2006*, *27*, 473–497.
38. M., Destarac, C., Brochon, J.-M., Catala, S. Z., Zard, *Macromol. Chem. Phys.* **2002**, *203*, 2281–2289.
39. C., Brochon, J.-M., Catala, *ACS Boston, Polym. Prepr.* **2002**, *43* (2), 303–304.
40. J., Chiefari, R. T. A., Mayadunne, C. L., Moad, G., Moad, E., Rizzardo, A., Postma, M. A., Skidmore, S. H., Thang, *Macromolecules* **2003**, *36*, 2273–2283.
41. M., Destarac, W., Bzducha, D., Taton, I., Gauthier-Gillaizeau, S. Z., Zard, *Macromol. Rapid Commun.* **2002**, *23*, 1049–1054.
42. M., Adamy, A. M., van Herk, M., Destarac, M. J., Monteiro, *Macromolecules* **2003**, *36*, 2293–2301.
43. G., Bouhadir, N., Legrand, B., Quiclet-Sire, S. Z., Zard, *Tetrahedron Lett.* **1999**, *40*, 277–280.
44. J. T., Lai, R., Shea, *J. Polym. Sci. Part A: Polym. Chem.* **2006**, *44*, 4298–4316.
45. J., Lai, D., Egan, R., Hsu, C., Lepilleur, A., Lubnin, A., Pajerski, R., Shea, *ACS Symposium Series 944*, American Chemical Society, in *Controlled/Living Radical Polymerization: From Synthesis to Materials*, K., Matyjaszewski, Ed.,

American Chemical Society, Washington DC., USA, **2006**, pp. 547–563.
46. M. R., Wood, D. J., Duncalf, S. P., Rannard, S., Perrier, *Org. Lett.* **2006**, *8*, 553–556.
47. P., Chapon, C., Mignaud, G., Lizarraga, M., Destarac, *Macromol. Rapid Commun.* **2003**, *24*, 87–91.
48. D., Hua, J., Xiao, R., Bai, W., Lu, C., Pan, *Macromol. Chem. Phys.* **2004**, *205*, 1793–1799.
49. M., Jacquin, P., Muller, G., Lizarraga, C., Bauer, H., Cottet, O., Théodoly, *Macromolecules* **2007**, *40*, 2672–2682.
50. D., Taton, A.-Z., Wilczewska, M., Destarac, *Macromol. Rapid Commun.* **2001**, *22*, 1497–1503.
51. D., Wan, K., Satoh, M., Kamigaito, Y., Okamoto, *Macromolecules* **2005**, *38*, 10397–10405.
52. M., Destarac, Rhodia Chimie, WO 2006/125892, **2006**.
53. K., Yokota, J.-F., Berret, B., Tolla, M., Morvan, Rhodia Inc., WO 2005/074631, **2005**.
54. E., Prat, M., Destarac, Rhodia Chimie, WO 01/095034, **2001**.
55. M., Destarac, B., Bavouzet, W., Bzducha, E., Fleury, Rhodia Chimie, WO 03/095502, **2003**.
56. O., Anthony, J.-Y., Chane-Ching, M., Destarac, C., Gérardin, Rhodia Chimie, WO 01/094263, **2001**.
57. M., Destarac, B., Pavageau, B., Tolla, Rhodia Chimie, WO 2006/117476, **2006**.
58. P., Hervé, M., Destarac, B., Bavouzet, Rhodia Chimie, WO 03/050184, **2003**.
59. P., Hervé, M., Destarac, O., Anthony, B., Bavouzet, M., Joanicot, A., Wilczewska, Rhodia Chimie, WO 03/050185, **2003**.
60. J.-N., Bousseau, J.-C., Castaing, M., Dorget, M.-P., Labeau, Rhodia Terres Rares, WO 01/062857, **2001**.
61. D., Batt-Coutrot, J.-J., Robin, W., Bzducha, M., Destarac, *Macromol. Chem. Phys.* **2005**, *206*, 1709—-1717.
62. L., Albertin, C., Kohlert, M., Stenzel, L. J. R., Foster, T. P., Davis, *Biomacromolecules* **2004**, *5*, 255–260.
63. C., Ladavière, N., Dörr, J. P., Claverie, *Macromolecules* **2001**, *34*, 5370–5372.
64. M., Destarac, B., Bavouzet, Rhodia Chimie, WO 2007/012763, **2007**.
65. T. L. U., Nguyen, K., Eagles, T. P., Davis, C., Barner-Kowollik, M. H., Stenzel, *J. Polym. Sci., Part A: Polym. Chem.* **2006**, *44*, 4372–4383.
66. B., Klumperman, J. B., McLeary, E. T. A., Van Den Dungen, G., Pound, *Macromol. Symp.* **2007**, *248*, 141–149.
67. A., Postma, T. P., Davis, G., Li, G., Moad, M. S., O'Shea, *Macromolecules* **2006**, *39*, 5307–5318.
68. H., Mori, H., Ookuma, S., Nakano, T., Endo, *Macromol. Chem. Phys.* **2006**, *207*, 1005–1017.
69. H., de Brouwer, J. G., Tsavalas, F., Schork, M. J., Monteiro, *Macromolecules* **2000**, *33*, 9239–9246.
70. S. W., Prescott, M. J., Ballard, E., Rizzardo, R. G., Gilbert, *Macromolecules* **2002**, *35*, 5417–5425.
71. C. J., Ferguson, R. J., Hughes, B. T. T., Pham, B. S., Hawkett, R. G., Gilbert, A. K., Serelis, C. H., Such, *Macromolecules* **2002**, *35*, 9243–9245.
72. C. J., Ferguson, R. J., Hughes, D., Nguyen, B. T. T., Pham, R. G., Gilbert, A. K., Serelis, C. H., Such, B. S., Hawkett, *Macromolecules* **2005**, *38*, 2191–2204.
73. H., de Brouwer, J. G., Tsavalas, F., Schork, M. J., Monteiro, *Macromolecules* **2000**, *33*, 9239–9246; M., Lansalot, T. P., Davis, J. P. A., Heuts, *Macromolecules* **2002**, *35*, 7582–7591.
74. B. T. T., Pham, D. N., Nguyen, C. J., Ferguson, B. S., Hawkett, A. K., Serelis, C. H., Such, *Macromolecules* **2003**, *36*, 8907–8909.
75. S. E., Shim, H., Lee, S., Choe, *Macromolecules* **2004**, *37*, 5565–5571.
76. S., Fréal-Saison, M., Save, C., Bui, B., Charleux, S., Magnet, *Macromolecules* **2006**, *39*, 8632–8638.
77. S. W., Prescott, M. J., Ballard, E., Rizzardo, R. G., Gilbert, *Macromol. Theory Simul.* **2006**, *15*, 70–86.
78. W., Bett, J.-C., Castaing, J.-F., d'Allest, Rhodia Chimie, WO 01/42325, **2001**.
79. M. J., Monteiro, M., Sjoberg, C. M., Gottgens, J., Van Der Vlist, *J. Polym. Sci., Part A: Polym. Chem.* **2000**, *38*, 4206–4217.

**80.** M. J., Monteiro, J., de Barbeyrac, *Macromolecules* **2001**, *34*, 4416–4423.

**81.** W., Smulders, *Macromolecular architecture in aqueous dispersions, 'lining' free-radical polymerization in emulsion*, Thesis, Technische Universiteit Eindhoven, Eindhoven, **2002**.

**82.** W., Smulders, R. G., Gilbert, M. J., Monteiro, *Macromolecules* **2003**, *36*, 4309–4318; W., Smulders, M. J., Monteiro, *Macromolecules* **2004**, *37*, 4474–4483.

**83.** M. J., Monteiro, M., Adamy, B. J., Leeuwen, A. M., van Herk, M., Destarac, *Macromolecules* **2005**, *38*, 1538–1541.

**84.** M. J., Monteiro, J., de Barbeyrac, *Macromolecules* **2001**, *34*, 4416–4423.

**85.** M. J., Monteiro, M., Sjoberg, C. M., Gottgens, J., Van Der Vlist, *J. Polym. Sci., Part A: Polym. Chem.* **2000**, *38*, 4206–4217.

**86.** M. J., Monteiro, J., de Barbeyrac, *Macromol. Rapid Commun.* **2002**, *23*, 370–374.

**87.** M. J., Monteiro, M., Adamy, B., Leeuwen, A., Van Herk, M., Destarac, Polym. Prepr. (Am. Chem. Soc., Div. Polym. Chem.*)* **2005**, *46* (2), 353–354.

**88.** M., Destarac, W., Bzducha, Rhodia Chimie, WO 03/002614, **2003**.

**89.** R. W., Simms, T. P., Davis, M. F., Cunningham, *Macromol. Rapid Commun.* **2005**, *26*, 592–596.

**90.** J. P., Russum, N. D., Barbre, C. W., Jones, F. J., Schork, *J. Polym. Sci., Part A: Polym. Chem.* **2005**, *43*, 2188–2193.

**91.** I. W., Hamley, *The Physics of Block Copolymers*, Oxford Science Publication, Oxford, **1998**.

**92.** N., Hadjichristidis, S., Pispas, G. A., Floudas, *Block Copolymers: Synthetic Strategies, Physical Properties, and Applications*, Wiley Interscience, John Wiley & Sons, Inc., **2003**.

**93.** D., Taton, Y., Gnanou, In *Block Copolymers in Nanoscience, Vol. 2*, M., Lazzari, G., Liu, S., Lecommandoux, Eds., Wiley-VCH Verlag GmbH & Co., KGaA Weinheim, Germany, **2006**, pp. 9–38.

**94.** C., Gérardin, N., Sanson, F., Bouyer, F., Fajula, J.-L., Putaux, M., Joanicot, T., Chopin, *Angew. Chem. Int. Ed.* **2003**, *42*, 3681–3685.

**95.** D., Bendejacq, V., Ponsinet, M., Joanicot, L., Loo, R., Register, *Macromolecules* **2002**, *35*, 6645–6649.

**96.** D., Bendejacq, V., Ponsinet, M., Joanicot, A., Vacher, M., Airiau, *Macromolecules* **2003**, *36*, 7289–7295.

**97.** D., Bendejacq, V., Ponsinet, M., Joanicot, *Langmuir* **2005**, *21*, 1712–1718.

**98.** N., Gaillard, A., Guyot, J., Claverie, *J. Polym. Sci., Part A: Polym. Chem.* **2003**, *41*, 684–698.

**99.** N., Gaillard, J., Claverie, A., Guyot, *Prog. Org. Coatings* **2006**, *57*, 98–109.

**100.** L., Shi, T. M., Chapman, E. J., Beckman, *Macromolecules* **2003**, *36*, 2563–2567.

**101.** M., Destarac, G., Mignani, S., Zard, B., Sire, C., Kalai, Rhodia Chimie, WO 02/008307, **2002**.

**102.** M., Destarac, W., Bzducha, S. Z. Zard. Polym. Prepr. (Am. Chem. Soc., Div. Polym. Chem.) **2005**, *46* (2), 387–388.

**103.** D., Quémener, T. P., Davis, C., Barner-Kowollik, M. H., Stenzel, *Chem. Commun.* **2006**, *48*, 5051–5053.

**104.** D., Taton, In *Macromolecular Engineering, Vol. 2*, Y., Gnanou, K., Matyjaszewski, L., Leibler, Eds., Wiley-VCH Verlag GmbH & Co., KGaA Weinheim, Germany, **2007**, Chap. 8, pp. 1007–1056.

**105.** D., Taton, J.-F., Baussard, L., Dupayage, Y., Gnanou, M., Destarac, C., Mignaud, C., Pitois, In *ACS Series, Advances in Controlled/Living Radical Polymerization, Vol. 39*, K., Matyjaszewski, Ed., **2006**, pp. 578–594.

**106.** M. H., Stenzel, T. P., Davis, C., Barner-Kowollik, *Chem. Commun.* **2004**, *13*, 1546–1547.

**107.** J., Bernard, A., Favier, L., Zhang, A., Nilasaroya, T. P., Davis, C., Barner-Kowollik, M. H., Stenzel, *Macromolecules* **2005**, *38*, 5475–5484.

**108.** H., Mori, H., Ookuma, T., Endo, *Macromol. Symp.* **2007**, *249–250*, 406–411.

**109.** D., Boschmann, P., Vana, *Polym. Bull.* **2005**, *53*, 231–242.

**110.** J., Bernard, A., Favier, T. P., Davis, C., Barner-Kowollik, M. H., Stenzel, *Polymer* **2006**, *47*, 1073–1080.

**111.** D., Taton, J.-F., Baussard, L., Dupayage, J., Poly, Y., Gnanou, V., Ponsinet, M.,

Destarac, C., Mignaud, C., Pitois, *Chem. Commun.* **2006**, *18*, 1955–1957.
**112.** M., Destarac, C., Kalai, A., Wilczewska, L., Petit, E., Van Gramberen, S. Z., Zard, In ACS Symposium Series 944, *Controlled/Living Radical Polymerization: From Synthesis to Materials*, K., Matyjaszewski, Ed., American Chemical Society, Washington DC., USA, **2006**, pp. 564–577.
**113.** A., Wilczewska, M., Destarac, S. Z., Zard, C., Kalai, G., Mignani, H., Adam, Rhodia Chimie, WO 02/090397, **2002**; M., Destarac, C., Kalai, L., Petit, A., Wilczewska, G., Mignani, S. Z., Zard, *Polym. Prepr. (Am. Chem. Soc., Div. Polym. Chem.)* **2005**, *46*(2), 372–373.
**114.** A., Wilczewska, M., Destarac, S. Z., Zard, C., Kalai, G., Mignani, H., Adam, Rhodia Chimie, WO 02/090424, **2002**; M., Destarac, C., Kalai, A., Wilczewska, G., Mignani, S. Z., Zard, *Polym. Prepr. (Am. Chem. Soc., Div. Polym. Chem.)* **2005**, *46*(2), 213–214.
**115.** H., Adam, W.-L., Liu, Rhodia Chimie, WO 03/070780, **2003**.
**116.** S. Z., Zard, B., Quiclet-Sire, P., Jost, Rhodia Chimie, WO 2005/040233, **2005**.
**117.** M.-F., Chan, M. E., Garst, *J. Chem. Soc. Chem. Commun.* **1991**, *7*, 540–541.
**118.** B.-C., Chen, M. S., Bednarz, O. R., Kocy, J. E., Sundeen, *Tetrahedron Asym.* **1998**, 9 (10), 1641–1644.
**119.** D. M., Mulvey, H. J., Jones, *Heterocycl. Chem.* **1978**, *15*, 233–235.
**120.** A., Liard, B., Quiclet-Sire, S. Z., Zard, *Tetrahedron Lett.* **1996**, *37*, 5877–5880.
**121.** D. H. R., Barton, M. V., George, M., Tomoeda, *J. Chem. Soc.* **1962**, 1967.
**122.** J. H., Udding, J. P. M., Giesselink, H., Hiemstra, W. N., Speckamp, *J. Org. Chem.* **1994**, 59 (22), 6671.
**123.** J., Boivin, R., Jrad, S., Juge, V. T., Nguyen, *Org. Lett.* **2003**, *5*, 1645–1648.
**124.** W. B., Farnham, M., Fryd, G., Moad, S. H., Thang, E., Rizzardo, E. I. Dupont de Nemours and Company, WO 2005/113612, **2005**.
**125.** T. M., Legge, A. T., Slark, S., Perrier, *J. Polym. Sci., Part A: Polym. Chem.* **2006**, *44*, 6980–6987.
**126.** H. R., Nace, *Org. React.* **1962**, *12*, 57.
**127.** M., Destarac, R., Reeb, M., Joanicot, Rhodia Chimie, WO 01/16187, **2001**.
**128.** C., Heitz, M., Joanicot, R. J., Tillotson, Rhodia Chimie, WO 02/070861, **2002**.
**129.** D., Bendejacq, C., Pitois, K., Karagianni, Rhodia Chimie, WO 2006/067325, **2006**.
**130.** M., Destarac, B., Bavouzet, W., Bzducha, E., Fleury, Rhodia Chimie, WO 03/095502, **2003**.
**131.** B., Bavouzet, M., Destarac, Rhodia Chimie, WO 2004/106457, **2004**.
**132.** H., Adam, W.-L., Liu, Rhodia Chimie, WO 02/090392, **2002**.
**133.** H., Lannibois-Dréan, J.-M., Ricca, M., Destarac, P., Olier, Rhodia Chimie, WO 03/000396, **2003**.
**134.** H., Lannibois-Dréan, J.-M., Ricca, M., Destarac, P., Olier, Rhodia Chimie, WO 03/002636, **2003**.
**135.** S., Deroo, M., Morvan, M., Destarac, Rhodia Chimie, WO 03/090916, **2003**.
**136.** S., Deroo, A., Sénéchal, J.-M., Mercier, N., Martin, Rhodia Chimie, WO 2005/100454, **2005**.
**137.** S., Deroo, M., Morvan, Rhodia Chimie, WO 03/068848, **2003**.
**138.** M., Morvan, A., Sénéchal, Rhodia Chimie, WO 03/002242, **2003**.
**139.** J.-C., Castaing, J.-F., d'Allest, W., Bett, Rhodia Chimie, WO 02/22735, **2002**.
**140.** L., Queval, C., Bonnet-Gonnet, M., Destarac, Rhodia Chimie, WO 02/068550, **2002**.
**141.** L., Queval, C., Bonnet-Gonnet, M., Destarac, Rhodia Chimie, WO 02/068487, **2002**.
**142.** M., Destarac, C., Bonnet-Gonnet, A., Cadix, Rhodia Chimie, WO 03/076531, **2003**.
**143.** M., Destarac, C., Bonnet-Gonnet, A., Cadix, Rhodia Chimie, WO 03/076529, **2003**.
**144.** A., Schreiner, M., Morvan, Rhodia Inc., US 03/0276371, **2006**.
**145.** M. A., Cohen Stuart, S., Van Der Burgh, G. R., Fokkink, A., de Keizer, Rhodia Chimie and Wageningen Universiteit, WO 2005/030282, **2005**.
**146.** M., Destarac, S. Z., Zard, F., Rivals, B., Quiclet-Sire, G., Mignani, Rhodia Chimie, Fr 2889702, **2007**.
**147.** M., Destarac, C., Mignaud, Rhodia Chimie, Fr 2889704, **2007**.

# 11
# Surface and Particle Modification via the RAFT Process: Approach and Properties

*Yu Li, Linda S. Schadler, and Brian C. Benicewicz*

## 11.1
## Introduction

Surface modification of materials is of great importance, as it can alter the properties of the surface dramatically and thus control the interaction between materials and their environment. Due to the wide applications of polymers in many areas, for example adhesion, lubrication, friction and wear, composites, microelectronics and biotechnology [1–5], surface modification by polymers is gaining increasing attention [6–8]. Generally, there are two ways to achieve the surface modification of materials with polymers: physisorption and covalent attachment. Compared with the physisorption method, covalent attachment can avoid the desorption issue and provide a robust linkage between the introduced polymer chains and material surfaces.

Polymer grafting techniques provide a versatile tool to covalently modify the surface of materials. These techniques can be categorized into 'grafting to' and 'grafting from'. In the 'grafting-to' technique, the polymer, bearing an appropriate functional group, reacts with the material surfaces to form chemically attached chains. However, due to the steric hindrance imposed by the already-grafted chains, it becomes increasingly difficult for the incoming polymer chains to diffuse to the surface, which intrinsically results in low surface graft densities. In the 'grafting-from' technique, the initiators are initially anchored on the surface and then subsequently used to initiate the polymerization of monomer from the surface. Because the diffusion of monomer is not strongly hindered by the existing grafted polymer chains, this technique is more promising to achieve high graft densities.

The recent development of controlled polymerization techniques including cationic, anionic, ring-opening metathesis and controlled radical polymerizations (CRP) makes it possible to provide considerable control over both the structure of the polymer to be grafted onto the materials surface and surface graft densities. The combination of these polymerization methods with polymer grafting techniques has been successfully used as an approach to modify various surfaces with

*Handbook of RAFT Polymerization.* Edited by Christopher Barner-Kowollik

ISBN: 978-3-527-31924-4

a variety of functional polymers. As a relatively newer CRP technique, reversible addition–fragmentation chain transfer (RAFT) polymerization has been successfully applied to the controlled polymerization of various monomers under a wide range of conditions to prepare polymer materials with predetermined molecular weights, narrow polydispersities and advanced architectures [9–12]. RAFT polymerization is performed under mild conditions, is applicable to a wide range of monomers and does not require a catalyst. Due to these advantages, the RAFT technique has recently received substantial attention in the area of surface modification with polymers. Since the first report of applying this technique to surface-initiated graft polymerization on a solid surface in 2001 [13], the RAFT technique has been utilized in the surface modification of various substrates, including inorganic/organic particles [13–37], flat silicon wafers [38–43], clay [44–46], flat gold surfaces [47, 48], gold nanorods [49], glass slides [50], carbon nanotubes [51–57], cellulose [58–61], rigid plastic [62, 63] and polymer films [64–68]. This chapter focuses on the approaches that have been used to modify various surfaces via the RAFT process as well as the physical and molecular properties of the resulting materials.

## 11.2 Approach

### 11.2.1 'Grafting-to' Approach

The 'grafting-to' approach provides a convenient way to modify the surface of materials by utilizing an end-functionalized polymer chain reacting with an appropriately treated substrate. As the grafted chains are preformed in this technique, their types and structures can be carefully designed via various polymerization methods. As a versatile CRP technique, RAFT is compatible with almost all of the conventional radical polymerization monomers, which allows for the preparation of a wide range of polymers with well-defined structure. Because RAFT polymerization follows a degenerative chain-transfer mechanism in which thiocarbonylthio compounds act as chain-transfer agents (CTAs), polymers prepared by this technique usually bear dithioester or trithiocarbonate end groups that can be easily reduced to thiols. The high affinity of thiols for the surfaces of metals, in particular gold, makes it possible to modify various metal substrates with well-defined polymer chains prepared via RAFT.

Lowe et al. [31] developed a facile one-step process to prepare (co)polymer-stabilized transition metal nanoparticles based on Au ($HAuCl_4$ sol), Ag ($AgNO_3$), Pt ($Na_2PtCl_6{\bullet}6H_2O$) and Rh ($Na_3RhCl_6$). In this process, the (co)polymers employed as stabilizers, including poly(sodium 2-acrylamido-2-methyl propane sulfonate) (PAMPS), poly[(*ar*-vinylbenzyl)trimethylammonium chloride] (PVBTC), poly(*N*,*N*-dimethylacrylamide) (PDMA) and poly[3-(2-*N*-methylacrylamido]-ethyl dimethyl ammonio propane sulfonate-*b*-*N*,*N*-dimethylacrylamide) (PMAEDAPS-*b*-PDMA),

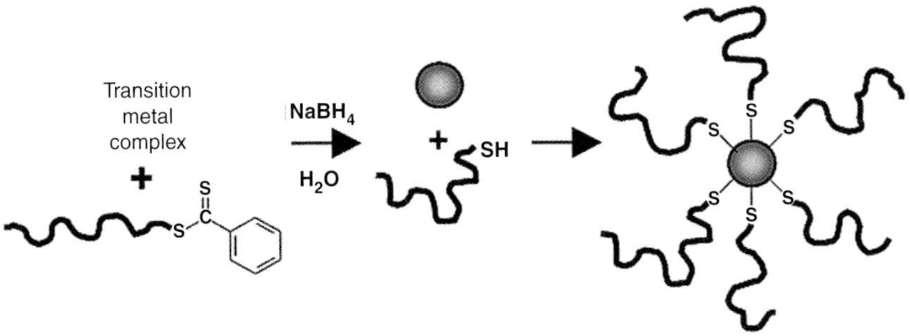


**Scheme 11.1** Preparation of polymer-stabilized transition metal nanoparticles. (Reproduced with permission from [31]. Copyright 2002 American Chemical Society.)

were synthesized by aqueous RAFT polymerization. The subsequent reduction of the dithioester end groups of these (co)polymer chains and a metal complex or metal solid occur simultaneously in aqueous media, giving a series of polymer-stabilized transition metal nanoparticles (Scheme 11.1). Transmission electron microscopy (TEM) was used to examine the metal nanoparticles after stabilization. Compared with those nanoparticles obtained by the reduction performed in the absence of the RAFT-synthesized (co)polymer, the polymer-stabilized metal nanoparticles were extremely stable. Sumerlin et al. [47] further extended this work to the modification of gold films, in which the reduction of the dithioester end-capped (co)polymers was performed in the presence of the gold substrates. Attenuated total reflectance Fourier transform infrared spectroscopy and atomic force microscopy (AFM) confirmed the presence of the monolayer (co)polymers on the surface of gold films. Chemical bonding of the thiol end groups to the surface of gold films was evidenced by the fact that the (co)polymers remained immobilized after thorough rinsing with solvent.

Using a similar approach, Spain et al. [32] prepared biologically active gold nanoparticles stabilized with multivalent neoglycopolymers synthesized via RAFT. Shan et al. [34] prepared amphiphilic gold nanoparticles grafted with a mixture of RAFT-prepared poly(*N*-isopropylacrylamide) (PNIPAM) and polystyrene (PS) chains with two different ratios. These amphiphilic gold nanoparticles showed different behaviors at the air–water interface in Langmuir monolayer experiments, and the contact-angle measurements revealed that PS and PNIPAM chains grafted on the surface of the gold cores appeared to be phase separated.

Shan et al. [17] also employed three methods in the preparation of PNIPAM-monolayer-protected clusters (PNIPAM-MPC) of gold nanoparticles (Scheme 11.2), in which three types of PNIPAMs were used: RAFT-prepared PNIPAMs bearing dithiobenzoate end groups, RAFT-prepared PNIPAM end capped with a thiol group obtained through hydrazinolysis and thiol-functionalized PNIPAM obtained through a conventional radical polymerization and subsequent modification. It was found that the one-step method was facile in controlling the sizes of gold clusters

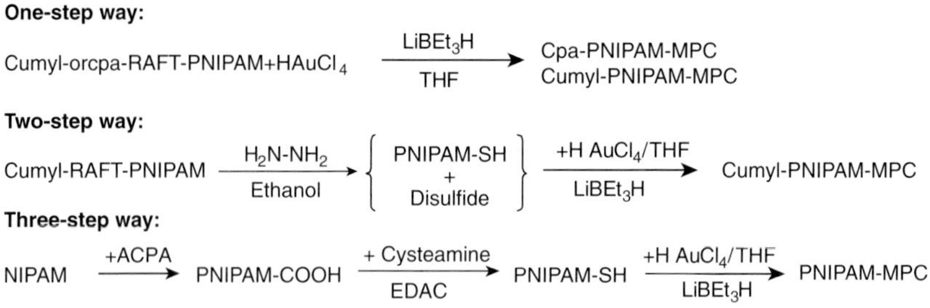


**Scheme 11.2** Schematic representation of three ways to prepare PNIPAM-MPCs. (Reproduced with permission from [17]. Copyright 2003 American Chemical Society.)

with reasonably narrow size distributions compared to the other two methods. The presence of the PNIPAM disulfide caused a broad size distribution of MPCs in the two-step method, and separate gold clusters could not be prepared by the three-step method due to a certain amount of dithiolated PNIPAM that acted as a cross-linking agent.

Recently, an interesting study by Duwez et al. [48] showed that dithioesters or trithiocarbonates can be directly chemisorbed on gold substrates (Scheme 11.3) without the need for reduction into thiols. Polystyrenes prepared by RAFT with two different CTAs, benzyldithiobenzoate (BDTB) and dibenzyl trithiocarbonate (DBTTC), were successfully grafted onto gold substrates via the chemisorption of the dithioester and trithiocarbonate end groups. This strategy simplifies the conventional procedures and avoids the formation of disulfides, resulting from the coupling between two thiol-functionalized polymer chains, which may cause a broad size distribution of the grafted chains [17].

In a more recent study, Hotchkiss et al. [49] modified the surface of gold nanorods by RAFT-prepared polymers, including poly[2-(dimethylamino)ethyl methacrylate] (PDMAEMA), poly(acrylic acid) (PAA) and polystyrene (PS), with or without the use of reducing agents (Scheme 11.4). TEM and UV–vis spectroscopy results confirmed that both reduced and nonreduced RAFT-prepared polymers were covalently

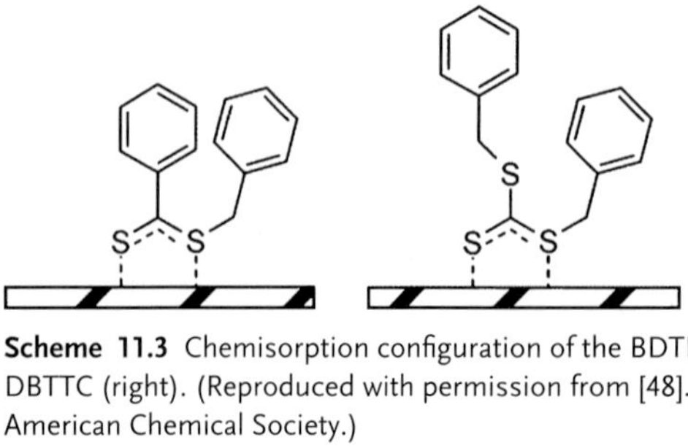

**Scheme 11.3** Chemisorption configuration of the BDTB (left) and DBTTC (right). (Reproduced with permission from [48]. Copyright 2006 American Chemical Society.)

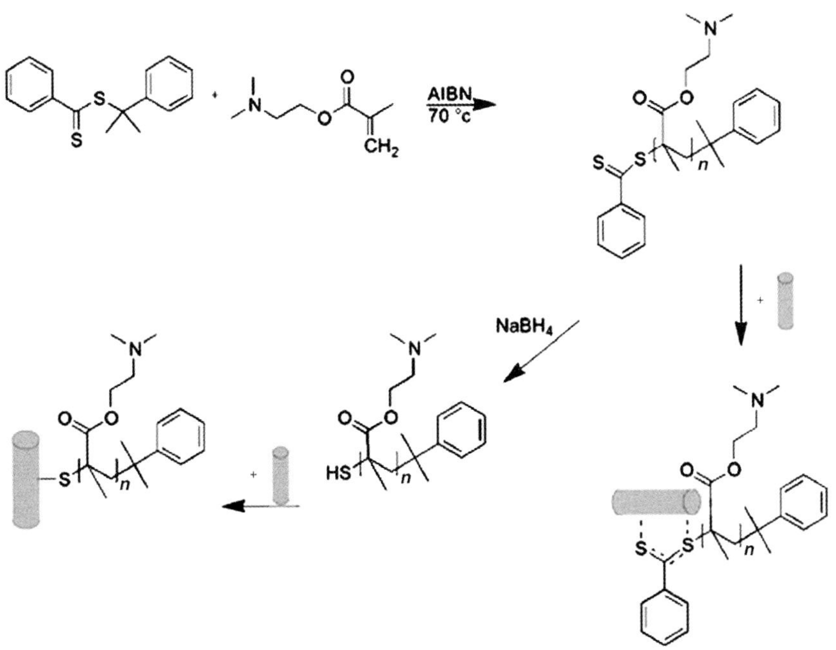


**Scheme 11.4** Proposed mechanism describing synthesis, reduction and immobilization of RAFT-prepared PDMAEMA on a gold surface. (Reproduced with permission from [49]. Copyright 2007 American Chemical Society.)

attached to the gold nanorods, and the thickness of the grafted polymer varied from 3 to 14 nm, depending on the polymer and grafting conditions used.

Instead of grafting polymer chains onto the substrate surfaces via the sulfur-metal bond, Guo et al. [24] used a different approach to prepare glycopolymer-modified silica gel particles. The surface-attached monomers were first prepared by modifying the surface of silica gel particles with γ-methacryoxypropyltrimethoxysilane, which was reacted with RAFT-prepared glycopolymers via radical exchange in the presence of 2,2′-azobisisobutyronitrile (AIBN) (Scheme 11.5). The subsequent cleavage of the acetyl groups of the grafted polymer resulted in silica gel particles modified with well-defined lactose-carrying polymer.

Although the 'grafting-to' approach provides a convenient way to modify the substrate surface with well-defined RAFT-prepared polymers, the inherent problem associated with this approach is the limitation of surface graft density. The diffusion barrier established by the already-grafted polymer chains makes it difficult for the new polymer chains to access the reactive sites on the substrate. Thus the amount of the grafted polymer chains was limited, which usually resulted in low grafting densities and film thickness. To overcome this problem, great attention has been paid to the modification of material surfaces via surface-initiated RAFT polymerization.

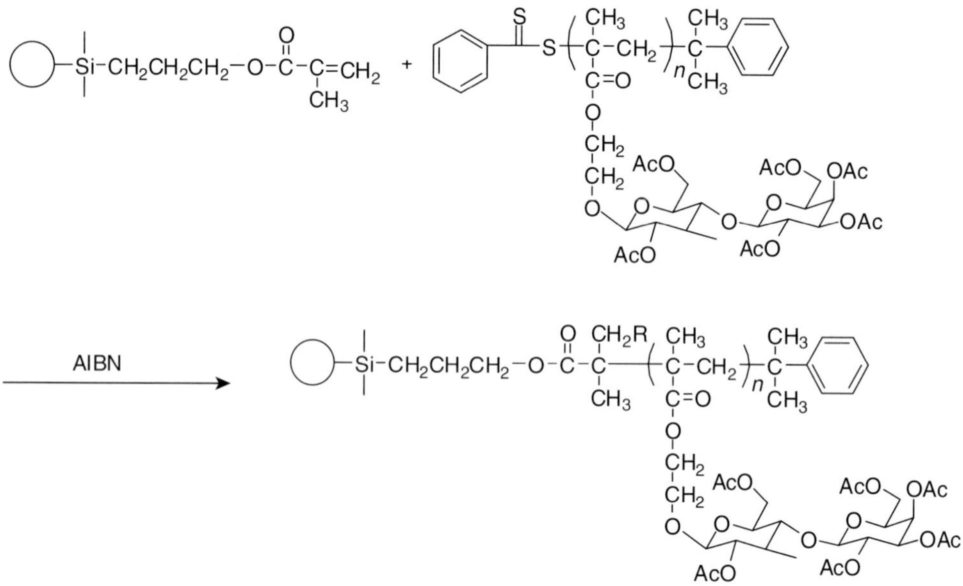


**Scheme 11.5** Schematic illustration of grafting glycopolymers onto silica gel particles. (Reproduced with permission from [24]. Copyright 2006 American Chemical Society.)

### 11.2.2
### Surface-Initiated RAFT Approach

Surface-initiated RAFT polymerization has been widely explored as an approach to modify the material surfaces due to its ability to precisely control the structure of the grafted polymer chains with a low-to-high range of graft densities. In this approach, there are two general routes to prepare surface-grafted polymer chains, including using (1) a surface-anchored initiator with free CTA in solution and (2) a surface-anchored CTA with appropriate initiation method. In both cases, the polymer chains are able to grow from the surface of materials rather than diffuse to the surface against the concentration gradient of the existing grafted polymers. Thus compared to the 'grafting-to' approach, surface-initiated RAFT polymerization is a more promising approach to construct dense and thick polymer layers on the surface of materials.

#### 11.2.2.1 **Grafting-From Surface-Anchored Initiators**

The immobilization of initiators on the material surfaces can be achieved by various techniques, including chemical reaction, plasma discharge and high-energy irradiation. The subsequent polymerization from these surface-anchored initiators in the presence of free CTA can generate surface-grafted polymer chains with uniform structure and adjustable length.

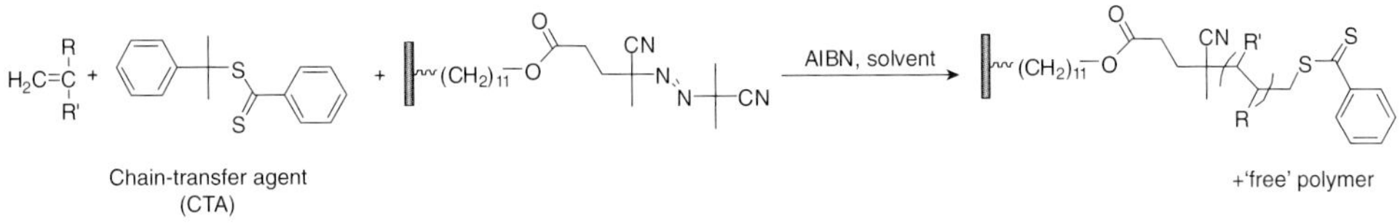


**Scheme 11.6** A general process of surface-initiated RAFT polymerizations from a surface-anchored azo initiator. (Reproduced with permission from [38]. Copyright 2002 American Chemical Society.)

Baum and Brittain [38] utilized RAFT to graft PS, polymethylmethacrylate (PMMA), PDMA and their copolymers from silica substrates using a surface-anchored azo initiator (Scheme 11.6). A silane coupling agent was used to immobilize the azo initiator on the silicate surfaces. 2-Phenylprop-2-yl dithiobenzoate was used as a free CTA in solution to control the graft polymerization. It was found that addition of free initiators was needed to achieve an effective polymerization rate, and increasing the concentration of free initiators produced a thicker polymer layer but decreased the control over the polymerization. A linear increase of film thicknesses with sequential monomer additions was observed, indicating the living characteristics of the grafted polymer chains prepared by this surface-initiated RAFT approach. Both $M_n$ and polydispersity index (PDI) of the PS and PMMA homopolymers cleaved from the surface of silica gel were comparable to those of the corresponding free polymers generated in solution, suggesting that the properties of the surface-grafted polymer chains can be estimated by analyzing the free polymer. Tensiometry tests showed that compared to a typical PS overlayer, PS homopolymer brushes made by RAFT showed a lower water contact angle, which was attributed to the dithioester end group.

Zhai et al. [39] used a similar azo initiator to prepare polybetaine brushes from the surface of hydrogen-terminated Si(100) substrates by surface-initiated RAFT polymerization. The azo initiator was immobilized on the Si—H surface in three steps. An alkyl ester was first immobilized on the surface under UV irradiation, which was reduced to a hydroxyl group and then coupled with a carboxylated azo initiator by esterification. A free initiator was also used in solution. The thickness of the polymer films increased linearly with the time of polymerization. Yu et al. [40] further expanded this approach by synthesizing poly(4-vinylbenzyl chloride) (PVBC) brushes from the same substrate. The free polymer generated in solution was analyzed to estimate the molecular weight of the surface-grafted polymer. A linear relationship between the film thickness and $M_n$ of the free polymer was observed, and the PDI of the free polymer was approximately 1.2–1.3. The surface-grafted PVBC was further functionalized to give the Si-g-viologen surface with redox-responsive properties. Chen et al. [67] used a similar strategy to graft polymer brushes of PMMA and poly[poly(ethylene glycol) monomethacrylate] (PPEGMA) from poly(vinylidene fluoride) (PVDF) surfaces. The azo-initiator coverage on the PVDF surface was determined by reaction with 2,2-diphenyl-1-picrylhydrazyl (DPPH) and estimated to be approximately 0.68 units $nm^{-2}$. The molecular weight of the grafted PMMA brushes was calculated from the thickness of polymer brushes and the estimated initiator coverage, which was comparable to the molecular weight of free polymer formed in solution. Compared to the native PVDF film, a more hydrophilic surface was generated after surface grafting of PEGMA and PMMA.

Bae et al. [68] applied microwave plasma to modify the surface of poly-(dimethylsiloxane) (PDMS) substrates with maleic anhydride. The introduced carboxylic groups were used to immobilize the azo initiators by condensation reactions, followed by surface-initiated RAFT polymerizations of *N,N*-dimethylacrylamide (DMA), styrenesulfonate (SS), and (*ar*-vinylbenzyl) trimethylammonium chloride (VBTC) on PDMS surfaces. Subsequently, a layer-by-layer process of alternately depositing RAFT-prepared PSS and PVBTAC homopolymers on the surface of PVBTAC-grafted PDMS was applied to create stable and highly hydrophilic surfaces on the PDMS substrates.

Xu et al. [43] utilized an interesting strategy to micropattern spatially well-defined binary polymer brushes on the Si(100) surface via a combination of surface-initiated atom-transfer radical polymerization (ATRP) and RAFT (Scheme 11.7). The ATRP initiator was first immobilized on the Si(100) surface via UV-induced hydrosilylation through a photomask, which was used to prepare sodium 4-styrenesulfonate (NaSS) polymer (PNaSS) brushes. The azo initiator for RAFT polymerization was immobilized to the unhydrosilylated $SiO_2$ domains using a silane coupling agent, which was then used to prepare poly(2-hydroxyethyl methacrylate) (PHEMA)

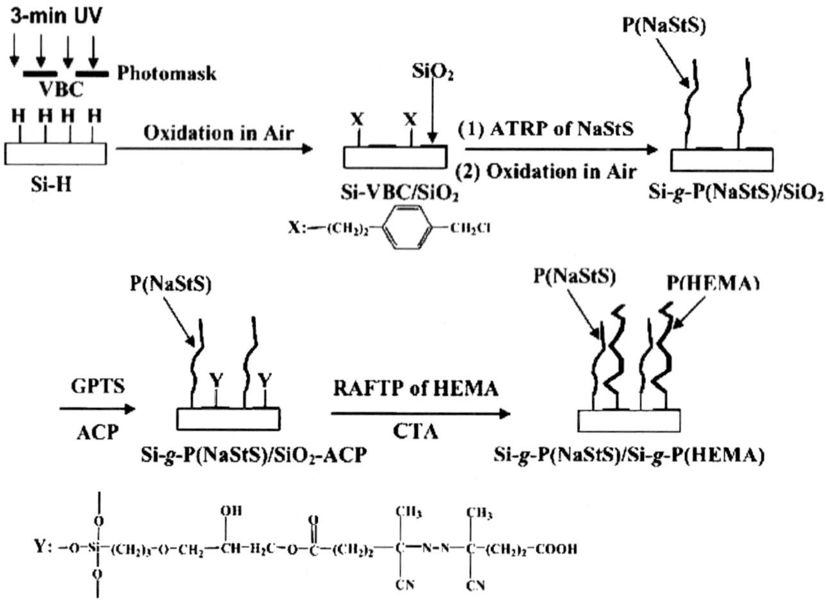


Scheme 11.7 Schematic diagram illustrating the process of nonlithographic micropatterning of a silicon surface by a combination of surface-initiated ATRP and RAFT. (Reproduced with permission from [43]. Copyright 2006 Royal Society of Chemistry.)

brushes. The thicknesses of PNaSS and PHEMA brushes, determined by AFM, were estimated to be 21.3 and 25 nm, respectively.

Rather than using a surface-anchored azo initiator, Pirri et al. [50] prepared polymer brushes of *N,N*-dimethylacrylamide (DMA), glycidyl methacrylate (GMA) and poly(DMA-*b*-GMA) from the surface of glass slides and silica beads by depositing a thiol-bearing organosilane. With a radical initiator in solution, S• radicals were formed on the surface by radical exchange, which were able to initiate the RAFT polymerization of monomers in the presence of a free CTA. Surfaces modified by poly(DMA-*b*-GMA) brushes bearing oxirane groups showed superior performance in oligonucleotide hybridization experiments compared to those coated with nonpolymeric self-assembled monolayers containing the same functional group. Barner et al. [15, 36] grafted PS from cross-linked poly(divinylbenzene) (PDVB) core microspheres by thermally induced RAFT polymerization. The core microspheres were prepared by precipitation polymerization. The residual double bonds located at the surface of core microspheres facilitated the growth of the polymer chains from the surface by radical capture of oligomers and monomers. 1-Phenylethyl dithiobenzoate was used as a free CTA in solution. A rapid increase in the average particle volume was found during the early stages of polymerization, which was attributed to polymer chains that grow from both the surface and the outer layer of the microspheres. After this initial stage the particle volume increase was slower and linear with reaction time. The PDIs of the free polymer in solution were below 1.2 for all reaction times. Using the same approach, Joso et al. [21] grafted poly(*n*-butyl acrylate) (PBuA) and poly(*N,N*-dimethyl acrylamide) from PDVB core microspheres. Cumyl dithiobenzoate was used as the RAFT agent.

In addition to chemical deposition, high-energy radiation and plasma are also convenient and powerful tools to generate initiating sites on the surface of substrates. Barner et al. [62] applied $\gamma$-initiated RAFT polymerization to graft PS from a polypropylene solid phase at ambient temperature. Since initiating radicals can be generated both on the polypropylene (PP) surface and in the PS chains by $\gamma$ radiation, two distinct grafting regimes were observed. In the first regime, polymer chains grew in a grafting layer, in which the surface was not completely covered by polymer chains. In the second regime, the surface was completely covered with polymer chains, and new polymer chains grew from radicals generated in the already-grafted polymer chains. It was believed that the growing surface-grafted polymer chains are in a dynamic equilibrium with free polymer chains in the solution. The free polymers in the solution were analyzed to estimate the chemistry of surface-grafted polymer chains. The results showed that $M_n$ of the free polymers increased linearly with conversion, and the PDI remained below 1.2 throughout the entire polymerization. In a later report, Barner et al. [63] expanded this approach by grafting a comonomer system of styrene (St) and *m*-isopropenyl-$\alpha,\alpha'$-dimethylbenzyl isocyanate (TMI) from a PP solid phase. Two different grafting regimes were also observed.

By utilizing an $O_2$-plasma treatment, Yoshikawa et al. [65] were able to prepare high-density PHEMA brushes on the surface of poly(tetrafluoroethylene-*co*-hexafluoropropylane) (FEP) films (Scheme 11.8). Peroxides were first introduced

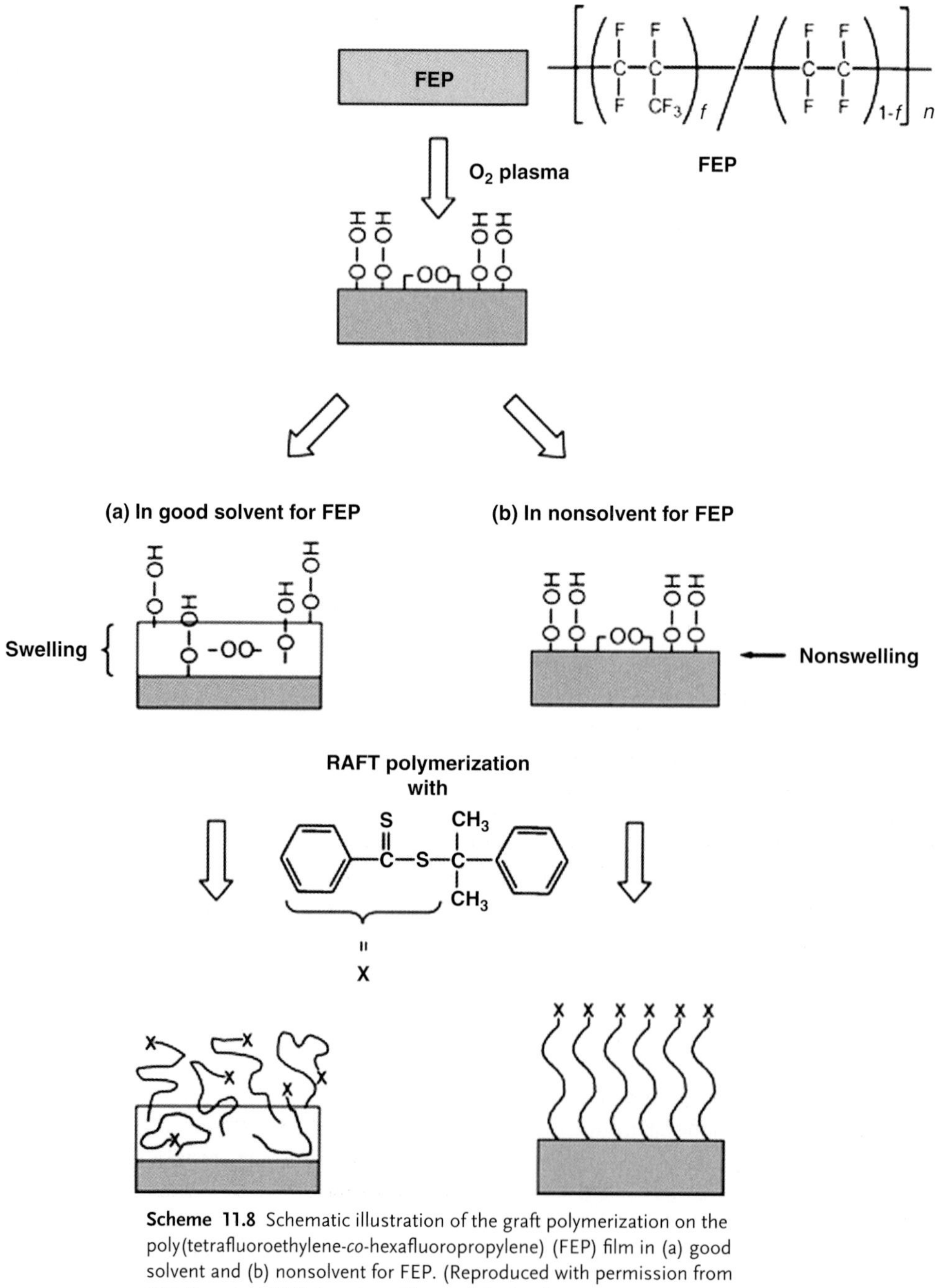


**Scheme 11.8** Schematic illustration of the graft polymerization on the poly(tetrafluoroethylene-*co*-hexafluoropropylene) (FEP) film in (a) good solvent and (b) nonsolvent for FEP. (Reproduced with permission from [65]. Copyright 2005 American Chemical Society.)

onto the surface as initiating moieties by the $O_2$-plasma treatment. The density of peroxides on the surface was determined by a radical-scavenging method, which increased with increasing plasma-treatment time and reached a constant value of about 10 peroxides $nm^{-2}$. Surface-initiated RAFT polymerization of HEMA from the plasma-treated FEP film was conducted in a nonsolvent for FEP at 40 °C to avoid the growth of polymer chains from deep within the swollen FEP film surface. *N,N*-dimethylaniline was added to accelerate the decomposition of the peroxides via a redox process. V-70 was used as a free initiator in solution. The graft density was estimated to be about 0.3 chains $nm^{-2}$. The results of the contact angle and electron spectroscopy for chemical analysis measurements indicated that the PHEMA chains were densely grafted to a considerably thin ($\leq$10 nm) boundary layer. To intentionally swell the FEP film surface, a graft polymerization was also attempted at 80 °C, which resulted in the growth of polymer chains from the inner part of the swollen film as well as the surface, giving high-density polymer brushes with ill-defined structure.

Yu et al. [66] reported the synthesis of comb copolymer brushes from plasma-treated poly(tetrafluoroethylene) (PTFE) films via a combination of surface-initiated RAFT with ATRP. The PTFE film was subjected to 90 s of radio frequency Ar plasma pretreatment to introduce peroxides on the surface with a density of about 0.3 units $nm^{-2}$. Poly(glycidyl methacrylate) (PGMA) brushes were first synthesized by surface-initiated RAFT polymerization from the immobilized peroxides. The ATRP initiators were then introduced by reacting 2-bromo-2-methylpropionic acid with the epoxy groups in the PGMA side chains. The subsequent surface-initiated ATRP of hydrophilic monomers, including poly(ethylene glycol) methyl ether methacrylate and sodium 4-styrenesulfonate, produced comb copolymer brushes on the surface of PTFE films. In a recent report by Wang et al. [27], plasma irradiation was also used to introduce peroxides on the surface of $Fe_3O_4$ magnetic nanoparticles (MNP). The subsequent surface-initiated RAFT polymerization of St and AA on the plasma-treated MNP produced core-shell $Fe_3O_4$-g-PS and $Fe_3O_4$-g-AAc nanoparticles. These surface-modified nanoparticles showed excellent dispersibility and stability in organic solvents.

Generally, in addition to the initiator immobilized on the surface of substrates, a free initiator is also added in the above investigations. Because of the low concentration of initiating sites on the surface, which can be terminated by trace amounts of impurities present in the reaction mixture, it has been found that the added free initiator can act as a scavenger for the impurities to facilitate the growth of the grafted polymer chains [38]. The addition of free initiator can also result in the formation of ungrafted polymer in solution. Researchers have shown that the $M_n$ and PDI of the grafted polymer chains agreed closely with those of the ungrafted polymer chains [38]. In these cases, the characterization of the ungrafted polymer provides a convenient method to estimate the properties of the grafted polymer, especially those polymer chains that are difficult to separate from the surface of substrates. However, the disadvantage, which is also derived from the presence of the large amount of ungrafted polymer in the final product, is the requirement of additional isolation and purification procedures after polymerization.

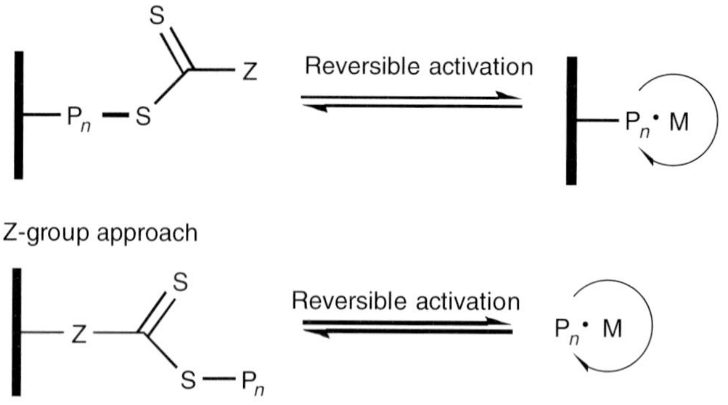


**Scheme 11.9** Comparison of R-group and Z-group approaches for surface-initiated RAFT polymerizations.

#### 11.2.2.2 **Grafting-From Surface-Anchored CTAs**

An alternative way to modify the surface of materials via surface-initiated RAFT polymerization is grafting-from surface-anchored CTAs, which generally can be accomplished through either the R-group or Z-group approach (Scheme 11.9). In the R-group approach, the RAFT agent is attached to the substrate surface via its leaving and reinitiating R group. The solid substrate acts as part of the leaving R group, and thus the propagating radicals are located on the terminal end of the surface-grafted polymer, which facilitates the growth of grafted polymer chains. This approach resembles a 'grafting-from' approach. In the Z-group approach, the RAFT agent is attached to the surface via its stabilizing Z group. Because the RAFT agent is permanently attached to the surface, this approach resembles a 'grafting-to' approach. The polymeric radicals always propagate in solution before they attach to the surface of substrate via the chain-transfer reactions with attached RAFT agents.

11.2.2.2.1 **R-group Approach** Tsujii et al. [13] reported the first application of surface-initiated RAFT polymerization in the modification of silica particles via an R-group approach. An ATRP macroinitiator was first prepared on the surface of silica particles, which was subsequently converted to a terminal RAFT moiety by reacting with 1-phenylethyl dithiobenzoate in the presence of CuBr via an atom-transfer addition (ATA) reaction. The conversion of this reaction was estimated to be 70% by UV–vis absorption spectroscopy. The surface-initiated RAFT polymerization of St from the immobilized RAFT moiety was carried out at 110 °C with a free RAFT agent in solution. The addition of the free RAFT agent in solution not only controlled the free polymerization in the bulk phase but also kept the graft polymerization under control at high conversions. After polymerization, the grafted PS chains were cleaved from silica particles by treating with HF and analyzed by GPC. The results revealed that the grafted radicals predominantly undergo bimolecular termination at an unusually high rate. The enhanced recombination was attributed to the fast migration of radicals on the surface by sequential

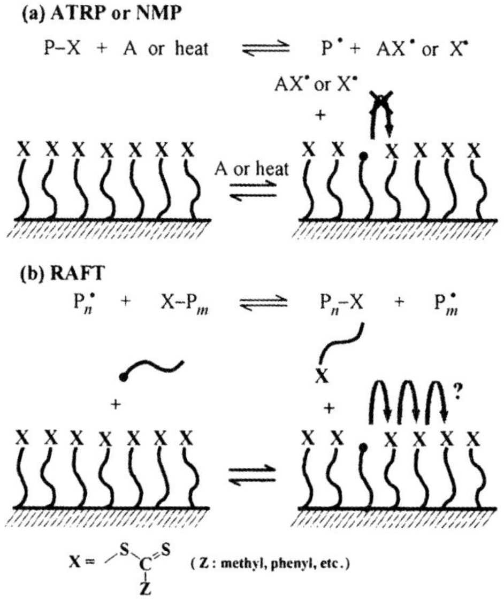


**Scheme 11.10** Comparison of the key processes in (a) the ATRP- or NMP-mediated and (b) RAFT-mediated graft polymerizations. (Reproduced with permission from [13]. Copyright 2001 American Chemical Society.)

chain-transfer reactions, which is not observed in an ATRP system (Scheme 11.10). It was also observed that the surface graft density had a critical value of about 0.08 chains $nm^{-2}$below which the surface migration of radicals hardly occurred. Rowe-Konopacki and Boyes [30] used a similar strategy to prepare a series of diblock copolymer brushes, including PMMA-*b*-PDMAEMA, PMMA-*b*-PS and PS-*b*-PMMA on the surface of flat silicon substrates. A modified ATA reaction was applied to convert a surface-immobilized ATRP initiator to a RAFT agent. The addition of Cu(0) was critical to achieve effective conversion of the ATA reaction. Diblock copolymer brushes were then synthesized from the surface-immobilized RAFT agent via sequential surface-initiated RAFT polymerization. The addition of free CTA in solution was also required to control the growth of polymer brushes. Due to the low concentration of polymer on the surface the grafted polymer was unable to be degrafted for direct characterization. The free polymer in solution was isolated and analyzed by GPC to estimate the properties of the grafted polymer, which showed a narrow polydispersity and predictable molecular weight.

Although well-defined polymer brushes were successfully prepared on the surface of silicate substrates in the above studies, large amounts of free polymer were also produced in the final products due to the free RAFT agents required in the polymerization, which required laborious purification steps. To overcome this problem, Li and Benicewicz [22] used a different strategy to synthesize polymer brushes on the surface of silica nanoparticles. A RAFT-silane agent was first prepared in three steps, which was then reacted with the surface of silica

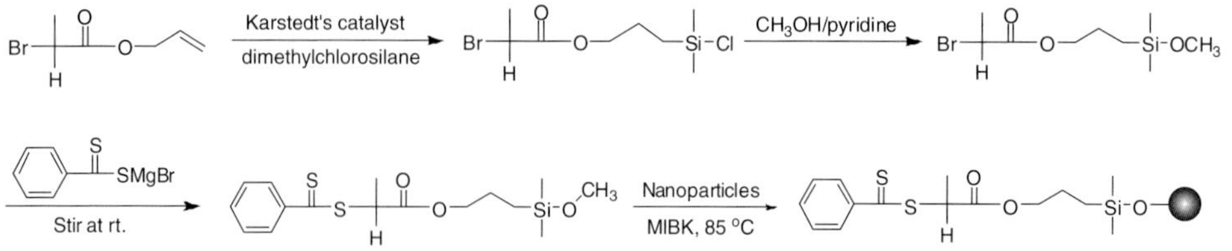


**Scheme 11.11** Synthesis procedures for attaching RAFT agent onto silica nanoparticles. (Reproduced with permission from [22]. Copyright 2005 American Chemical Society.)

nanoparticles, providing a surface-immobilized RAFT agent (Scheme 11.11). The amount of RAFT agent on the modified silica nanoparticles was determined quantitatively by UV–vis spectroscopy. By varying the silane concentration utilized for deposition, various graft densities of RAFT agent on silica nanoparticles, ranging from 0.15 to 0.68 units $nm^{-2}$ were prepared. Using these surface-immobilized RAFT agents, homopolymer and block copolymer brushes of PS and PBuA were prepared on the silica nanoparticle surfaces via surface-initiated RAFT polymerization without the addition of free RAFT agents in solution. A low AIBN:CTA ratio (<0.1) was used to minimize radical recombination and the amount of free polymer derived from the radicals formed by AIBN decomposition. The polymerizations were conducted at low conversion range (<20%) to avoid possible gelation or interparticle polymeric radical coupling. After polymerization, the grafted polymer chains were cleaved from the silica particles by treating with HF and characterized. The results showed that the grafted polymer chains had narrow polydispersities and predictable molecular weights, indicating that the surface-immobilized RAFT agents participated in the polymerization with a high activity. Polymerization retardation was observed for the surface-initiated RAFT polymerization of both PS and PBuA, which was ascribed to the localized high RAFT-agent concentration. Preliminary work with PS-modified fumed silica prepared at 16.6% monomer conversion using the same approach showed that the fraction of ungrafted polymer estimated by TGA was only about 9%. However, due to the lack of tertiary R-group structure, the surface-immobilized RAFT agent used in this work could not be used in controlling the polymerization of methacrylate monomers. Hence, in a subsequent investigation by Li et al. [25], a more versatile RAFT agent containing a 4-cyanopentanoic acid dithiobenzoate (CPDB) moiety was immobilized on the surface of silica nanoparticles and used to prepare both PS- and PMMA-grafted silica nanoparticles (Scheme 11.12). Amino-group-functionalized silica nanoparticles were first prepared by reacting 3-aminopropyldimethylethoxysilane with silica particles. An initial attempt of directly reacting CPDB with amino-group-functionalized silica nanoparticles via condensation failed due to the aminolysis of the dithiobenzoate group of CPDB. Therefore, the carboxyl group of CPDB was first activated by reacting with 2-mercaptothiazoline. Due to the ability of mercaptothiazoline-activated amide bond to selectively consume the amino groups in the presence of dithiobenzoate groups, the subsequent reaction of activated CPDB with amino-group-functionalized silica

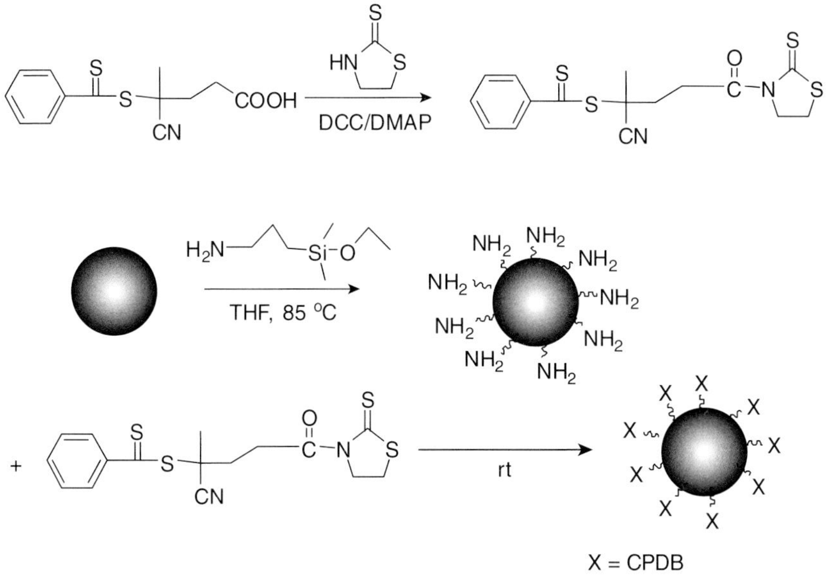


**Scheme 11.12** Synthesis procedures for anchoring CPDB moieties onto silica nanoparticles. (Reproduced with permission from [25]. Copyright 2006 American Chemical Society.)

nanoparticles successfully produced CPDB-anchored silica nanoparticles with variable graft density. Surface-initiated RAFT polymerizations of methyl methacrylate (MMA) and styrene (St) were mediated by CPDB-anchored silica nanoparticles without the addition of free CTA in solution, producing surface-grafted polymers with narrow polydispersities and predictable molecular weights. It was found that the rate of surface-initiated RAFT polymerization of MMA from the CPDB-anchored silica nanoparticles was much higher than that of the MMA polymerization mediated by free CPDB, and polydispersities of the PMMA cleaved from the silica nanoparticle surfaces were much narrower than those of the PMMA prepared using free CPDB as the RAFT agent. These effects were attributed to the unique structure and steric environment of the surface-anchored intermediate macro-RAFT-agent radical and also the localized high RAFT-agent concentration effect. Polymerizations mediated by a hybrid CPDB system, consisting of both free RAFT agent and surface-anchored RAFT agent, showed that the free polymer had a higher initial molecular weight than the grafted polymer and that they converged at high conversions. HPLC equipped with a $C_{18}$-coated silica column was used to isolate and quantitatively characterize the ungrafted PMMA polymer. The results showed that the amount of ungrafted polymer was generally very low, only around 5 wt % of the total polymer prepared up to 16% conversion and less than 15 wt % of the total polymer prepared at 22% monomer conversion.

Zhang et al. [46] grafted PS chains from the surface of layered silicates by RAFT polymerization. A RAFT agent, 10-carboxylic acid-10-dithiobenzoatedecyltrimethylammonium bromide, was first synthesized and intercalated into montmorillonite (MMT) via electrostatic attraction. The resulting CDDA-intercalated MMT was used to mediate the RAFT polymerization of St at 110 °C with AIBN in solution. GPC results revealed that the grafted PS chains had predicable molecular weights and narrow polydispersities. Also, the molecular weights and polydispersities of the free PS chains in solution were similar to those of the grafted PS chains, indicating that the intercalated MMT did not restrict the diffusion of propagating radicals and dormant chains. The obtained PS/MMT nanocomposites had an exfoliated structure and higher thermal stability compared to neat PS.

Many other substrate surfaces in addition to silicate substrates can be modified using surface-initiated RAFT polymerization by attaching the appropriate CTA through the R group. Skaff and Emrick [20] grafted a series of homopolymers and copolymers from the surface of CdSe nanoparticles via surface-initiated RAFT polymerization. A phosphine oxide ligand containing a trithiocarbonate moiety was first prepared, which was then anchored to a conventional tri-*n*-octylphosphine oxide (TOPO)-covered CdSe nanoparticle through ligand-exchange chemistry. Graft polymerizations of various monomers from the CTA-functionalized nanoparticles were achieved at 70 °C. It was found that common free-radical initiators including AIBN and benzoyl peroxide could induce the degradation of nanoparticles quickly at 70 °C, which could be attributed to the susceptibility of CdSe nanoparticles to free-radical degradation. Therefore, di-*tert*-butylperoxide was selected as a free-radical initiator, which has lower radical yield. The number-average molecular weight of the grafted polymer ranged from 9000 to 49 000 $g \cdot mol^{-1}$ and PDIs were generally below 1.3. The unique optical properties of the CdSe nanoparticles were well maintained after graft polymerization. TEM analysis of a composite thin film cast from the PS-grafted CdSe nanoparticles revealed that the CdSe nanoparticles were uniformly dispersed throughout the matrix.

Hu et al. [14] grafted linear thermally sensitive PNIPAM chains onto a spherical PNIPAM/hydroxyethyl acrylate (HEA) copolymer microgel. The hydroxyl group bearing NIPAM/HEA microgel was first prepared by dispersion polymerization. Then, α-butyl acid dithiobenzoate was immobilized on the surface of NIPAM/HEA microgel by esterification. The subsequent RAFT polymerization of PNIPAM from the CTA-immobilized microgel was conducted with AIBN in solution, resulting in a core-shell nanostructure. It was observed that the thickness of the grafted PNIPAM layer first decreased in the low-temperature range 25–32 °C and then increased in the high-temperature range 32–35 °C, which was related to a coil–globule–brush transition of linear grafted PNIPAM chains. Using a similar strategy, Raula et al. [16] synthesized gold nanoparticles grafted with PNIPAM by surface-initiated RAFT polymerization. CPDB was attached to gold nanoparticles by reacting with the 11-mercapto-1-undecanol ligands on the surface. The resulting CPDB-anchored nanoparticles were used to mediate the RAFT polymerization of NIPAM. The grafted PNIPAM was removed from the particle surfaces by treating with $I_2$ in $CH_2Cl_2$/ethanol. The molar mass and polydispersity of the grafted PNIPAM chains determined by GPC were 21 000 and 1.17, respectively. The optical properties of

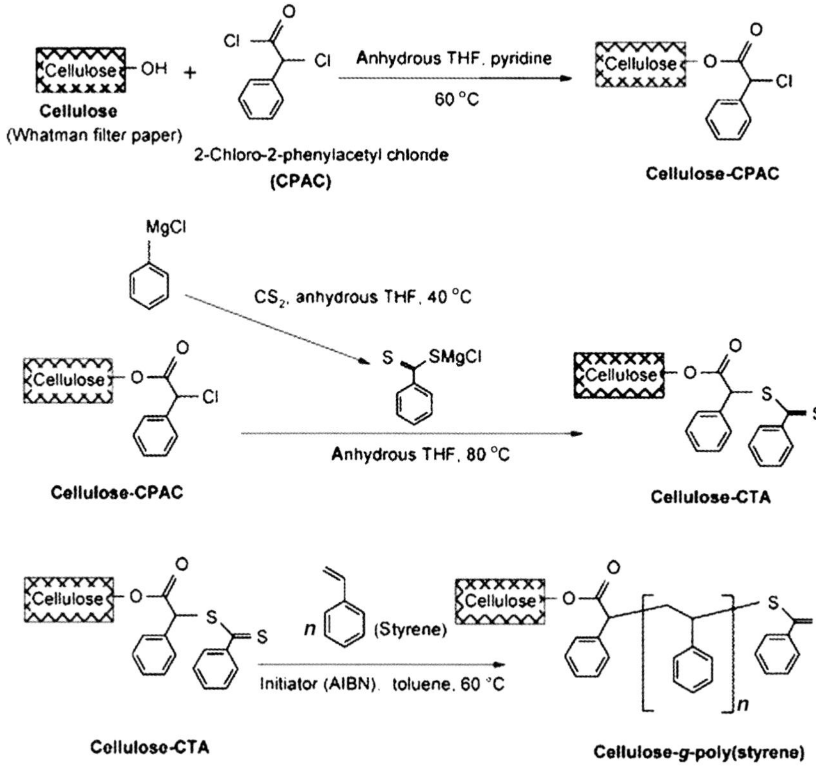


**Scheme 11.13** Synthesis of cellulose CTA for reversible addition–fragmentation chain transfer polymerization and their use to mediate St polymerization. (Reproduced with permission from [59]. Copyright 2005 American Chemical Society.)

these PNIPAM-modified gold nanoparticles varied with changes in environmental temperature and particle concentration.

Roy et al. [59, 69] utilized surface-initiated RAFT polymerization to graft PS from a cellulose substrate (Scheme 11.13). The hydroxyl groups of the cellulose were first treated with 2-chloro-2-phenylacetyl chloride and then converted to a thiocarbonylthio RAFT agent using a Grignard reagent. Based on the results of elemental analysis, the average loading of RAFT agent on the cellulose substrates was calculated to be 1.9 mmol·g$^{-1}$. From this cellulose-bound RAFT agent, St was polymerized in the presence of AIBN in solution. The graft ratio ranged from 11 to 28 wt %, depending on polymerization conditions. The grafted PS was cleaved from the cellulose backbone by treating with HCl solution. The cleaved PS chains from a sample with 28 wt % graft ratio were analyzed by size-exclusion chromatography, which gave values of $M_n = 21\,000$ g·mol$^{-1}$ and PDI = 1.1. Contact-angle measurements revealed a dramatic increase in hydrophobicity of the PS-modified cellulose surface compared to the untreated cellulose surface. Using the same strategy, they also grafted poly[2-(dimetylaminoethyl) methacrylate] from cellulose fiber. They found that the addition of a free RAFT agent in solution could increase the graft ratio. The maximum graft ratio was obtained when the molar ratio of free RAFT agent to cellulose-bound RAFT agent was 1.5:1.0 [58].

Using a surface-immobilized RAFT agent, Cui et al. [51] were able to graft PS from the surface of multiwalled carbon nanotubes (MWNTs). Carboxylic acid groups were first introduced to the surface of MWNTs by treating with nitric acid. The carboxylic acid group functionalized MWNTs were then treated with thionyl chloride and 2-hydroxyethyl-2′-bromoisobutyrate to form bromoisobutyrate-functionalized MWNTs, which were further reacted with a Grignard reagent to produce a surface-immobilized RAFT agent attached to the MWNT substrate via its R group (Scheme 11.14). Polymerization of St was conducted in THF at 100 °C, using AIBN as a free-radical initiator. In later investigations, the same strategy was used to modify the surface of MWNTs with PMMA-*b*-PS block copolymer [56] and a series of aqueous soluble polymers [52, 53, 55, 57]. In a graft polymerization of PNIPAM mediated by the RAFT-agent-functionalized MWNTs [52], the molecular weight of the grafted PNIPAM increased linearly with monomer conversion and the molecular-weight

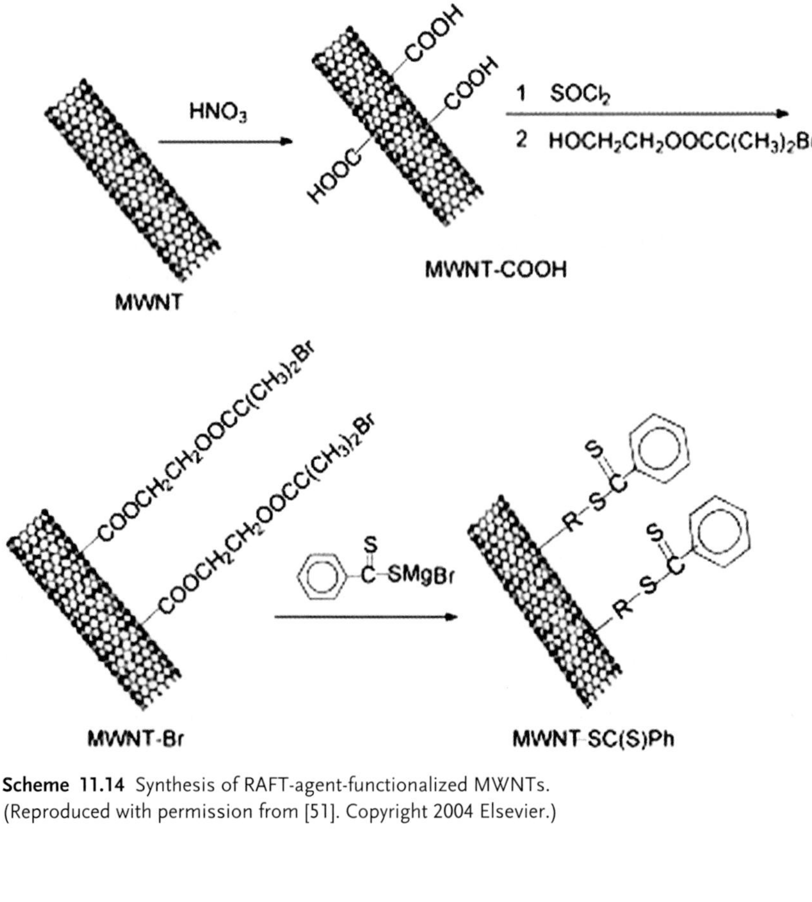


**Scheme 11.14** Synthesis of RAFT-agent-functionalized MWNTs. (Reproduced with permission from [51]. Copyright 2004 Elsevier.)

distribution was around 1.3. On the basis of this strategy, Hong et al. [54] also reported a new method to graft functional polymers and small molecules onto MWNTs, without significantly altering their surface structure. MWNTs were first slightly functionalized with RAFT agents to maximally maintain the surface structure. The graft density of RAFT agents was estimated to be approximately 1.5 RAFT agents per 1000 carbon atoms. From these RAFT-agent-immobilized MWNTs, an alternating copolymer of poly(styrene-*alt*-maleic anhydride) was grafted. The introduced highly reactive maleic anhydride groups could further react with various hydroxyl or amino groups that were contained in functional polymers and small molecules to achieve the functionalization of MWNTs.

11.2.2.2.2 **Z-group Approach** In comparison with the substantial application of the R-group approach to the surface modification of various substrates, relatively little attention has been paid to the Z-group approach. In the Z-group approach, the RAFT agent is located close to the surface throughout the polymerization and chain-transfer reactions between propagating polymer radicals and attached RAFT agents must occur near the surface of the substrates. Thus, these reactions can be severely hampered due to the steric hindrance of the neighboring attached polymer chains, which makes it difficult to prepare high-density grafted polymer. However, since the propagation of the polymer chains occurs only in solution in the Z-group approach, the polymer chains attached on the surface are always dormant, which excludes the bimolecular termination of grafted radicals often observed in the R-group approach. Recently, a few reports showed that this approach holds a unique advantage in constructing well-defined homopolymers and block copolymers grafted on the surface of solid substrates [23, 28, 33].

Perrier et al. [23] utilized a Merrifield-supported RAFT agent s-methoxycarbonylphenymethyl dithiobenzoate (Mer-MCPDB) and silica-supported MCPDB (Si-MCPDB) to mediate the RAFT polymerization of methyl acrylate (MA). MCPDB was attached to the surface of Merrifield resin via its Z group in two steps. The chlorobenzyl functional groups on the resin were first reacted with sodium methoxide and elemental sulfur to form a sodium dithiobenzoate salt, which was then converted to Mer-MCPDB by treating with methyl-$\alpha$-bromophenylacetate. The same procedures were used to prepare Si-MCPDB except that a silane was first used to introduce a chlorobenzyl functional group to the surface of the resin. The RAFT polymerizations of MA were conducted at 60 °C in the presence of AIBN. After polymerization, excess AIBN was added to cleave the grafted polymer chains via radical exchange. In a polymerization of MA mediated by a Mer-MCPDB, GPC was used to characterize both the grafted polymethylacrylate (PMA) chains and free PMA chains. The results revealed that the grafted PMA chains had lower polydispersity compared to the free PMA chains and did not show the GPC hump at high molecular weights due to termination by combination. It was found that the use of free RAFT agents in solution could not only help to increase the control over polymerization but also help to reduce the amount of free polymer chains in solution. When a free CTA (MCPDB) was used with a ratio free CTA:supported CTA = 1:1, the fraction of free polymer chains decreased from

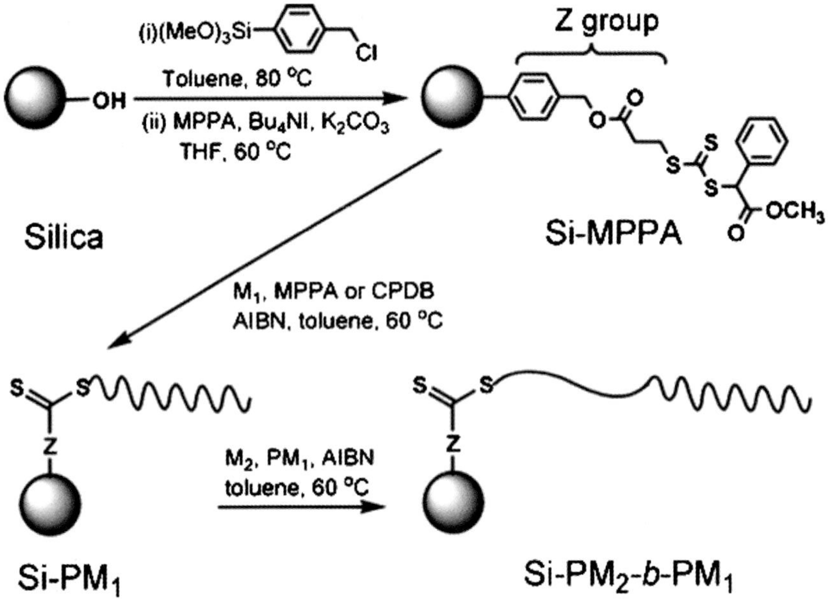


**Scheme 11.15** Synthetic route to polymer-grafted silica particles by Z-supported RAFT polymerization. (Reproduced with permission from [28]. Copyright 2006 American Chemical Society.)

51 to 38% in the Mer-MCPDB-mediated polymerization of MA and from 90 to 48% in the Si-MCPDB-mediated polymerization of MA. As the free polymer chains can be removed by simple filtration and all grafted chains are attached to the solid substrates via the RAFT agent, this strategy provides a way to modify the surface of substrates with polymer chains that can be considered 'truely' living. Zhao and Perrier [28] further extended this approach to the modification of silica particles with various homopolymers and block copolymers. Silica-supported 3-(methoxycarbonylphenylmethylsulfanylthiocarbonysulfanyl)-propionic acid (Si-MPPA) was first prepared by a two-step reaction (Scheme 11.15). The surface density of RAFT agent was estimated to be 0.388 molecule of CTA $nm^{-2}$. Si-MPPA was then used to mediate the RAFT polymerization of various monomers with AIBN and free MPPA in solution. The grafted polymer chains were cleaved by treating with *n*-hexylamine and analyzed by GPC. In a RAFT polymerization of MA mediated by Si-MPPA, it was found that the molecular weights of free and grafted polymers were similar at low conversion but differed with increasing conversion. At high conversion (>40%), the molecular weights of free polymers in solution were much higher than those of grafted polymers, which was attributed to the increased shielding effect of the grafted polymer chains with increasing conversion. Compared to free polymers produced in solution, the grafted polymers had better defined structure with a lack of GPC shoulders and obvious tailings corresponding to irreversible termination. A different free RAFT agent (CPDB) was also utilized to

mediate the RAFT polymerization, which afforded better defined grafted polymer than using MPPA due to the slower polymerization rate of CPDB-mediated RAFT polymerization in solution. The homopolymer-grafted silica particles were also subjected to a RAFT chain-extension polymerization of a second monomer, which produced well-defined diblock copolymer-grafted silica particles. Using a similar strategy, Nguyen and Vana [37] attached a cumyl dithiobenzoate to the surface of silica particles via its Z group. The resulting silica-supported RAFT agent was used to mediate the RAFT polymerization of MMA and St.

Stenzel et al. [42] used the Z-group approach to grow temperature-responsive glycopolymer brushes from flat silicon wafers. 3-Aminopropyl trimethoxysilane was first used to introduce amino functional groups to the surface of silicon wafers. The resulting amino-functionalized surface was reacted with 3-benzylsulfanylthiocarbonyl sulfanylpropanyl chloride to produce a surface-immobilized RAFT agent. The surface-initiated RAFT polymerizations of *N*-acryloyl glucosamin (AGA) and NIPAM were conducted at 60 °C with a radical initiator and free RAFT agent in solution. The RAFT process in solution was monitored to estimate the chemistry of the RAFT polymerization on the surface. The molecular weight of the polymer in solution increased linearly with conversion close to the theoretical value and the molecular-weight distributions remained narrow throughout the polymerization. Both the PNIPAM and PAGA layer thicknesses were found to increase linearly with monomer conversion. The PAGA-grafted surface was subjected to a chain-extension polymerization with NIPAM to prepare block copolymers grafted onto the silicon wafers. They found that the PNIPAM block grew approximately at the same rate independent of the initial size of the RAFT agent or the size of the PAGA block. This result was contrary to that from earlier studies using the Z-group approach to synthesize star polymers, in which the growth of the second block was found to be very difficult due to the increasing steric hindrance [70]. A possible explanation for this result may involve a highly entangled macroradical within the brush layer that cannot diffuse away from the reactive site. Consequently, to allow further brush growth, monomer must diffuse into the brush, which is not severely affected by the increasing steric hindrance.

In the above studies, a silane agent was used to introduce functional groups onto the surface that facilitated the subsequent attachment of the RAFT agent via its Z group. Peng et al. [41] explored a slightly different approach to immobilize the RAFT agent on the surface of silicon wafers (Scheme 11.16). A UV-induced hydrosilylation was used to immobilize a 4-vinylbenzyl chloride (VBC) monolayer on the Si—H

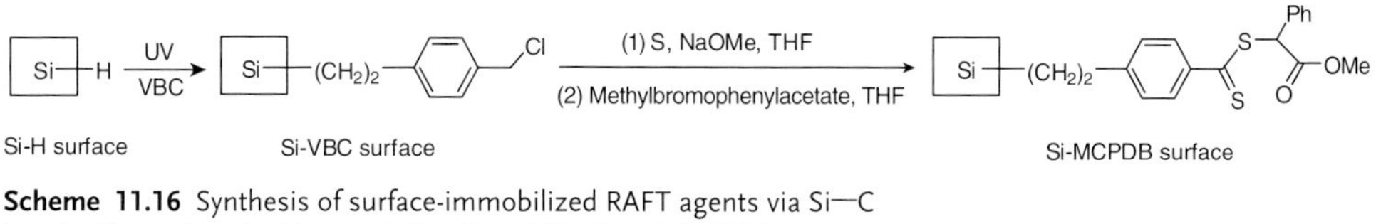


**Scheme 11.16** Synthesis of surface-immobilized RAFT agents via Si—C bonds. (Reproduced with permission from [41]. Copyright 2006 American Chemical Society.)

surface via robust Si—C linkages. The resulting Si—VBC surface was coupled with a RAFT agent in a two-step reaction. The RAFT polymerizations of MMA and HEMA were conducted at 70 °C with AIBN and a free RAFT agent in solution. An approximately linear increase in thickness of the grafted PMMA and PHEMA brushes on the surfaces with polymerization time was observed. The free polymers recovered from the reaction solution showed narrow polydispersities. The PHEMA and PMMA brushes were also subjected to a chain-extension polymerization of (2-dimethyl-amino)ethyl methylacrylate. The formation of a diblock copolymer-grafted surface was confirmed by X-ray photoelectron spectroscopy and time-of-flight secondary-ion mass spectrometry.

## 11.3
## Properties

### 11.3.1
### Surface Structure and Properties

The RAFT technique provides a versatile way to modify the surface of different materials with surface-attached polymers. By controlling the type, structure and graft density of the surface-attached polymers, surface hydrophobicity/hydrophilicity of substrate materials can be significantly modified, which further affects other properties including adhesion, wettability, compatibility and solubility [71, 72]. Due to the great promise of stimuli-responsive materials in many fields including nanotechnology, biochemistry and materials science, substantial attention has been given to the modification of material surfaces with stimuli-responsive polymers via the RAFT process. Typically, the surface-attached stimuli-responsive polymers can be mixed polymers, block copolymers or functional homopolymers (especially PNIPAM). Under external stimuli such as pH, temperature or solvency, the properties of these polymers can be affected through either structure rearrangement or conformational change, which is very useful for controlling the surface properties of materials.

Baum and Brittain [38] prepared diblock copolymer brushes attached on silicate substrates using RAFT polymerizations. The rearrangement behavior of diblock copolymer brushes upon treatment with different solvents was observed via tensiometry. Upon treatment with methylcyclohexane at 35 °C, the advancing water contact angle of a PS-*b*-PDMA brush (where the PS block is adjacent to the silicate surface) increased from 42 to 65°. Treatment of the same sample with tetrahydrofuran (THF)/ $H_2O$ (1/1, v/v) at 35 °C reversed the contact angle back to the original value. For the PDMA-*b*-PMMA brush, the advancing water contact angle decreased from 66 to 58° after the sample was treated with THF/$H_2O$ (1/1, v/v) at 35 °C and returned to the original contact-angle value after treatment with dichloromethane. These results were reproducible over several cycles of solvent treatment.

Sumerlin et al. [47] modified gold surfaces with a PMAEDAPS-*b*-DMA copolymer via a grafting-to approach where the DMA block is adjacent to the gold surface.

Water contact-angle measurement results showed that the contact angle for the PMAEDAPS-*b*-DMA sample was 29.1°, which was nearly identical to that of the PDMA-modified gold (30.5°). This result was interesting because the outer block is expected to be PMAEDAPS. To gain further insight, the contact angle was measured for a gold film modified with MAEDAPS homopolymer, which was determined to be 41.5°, indicating the surface was more hydrophobic than the gold modified with the block copolymer. By combining these two results, it was concluded that the relatively hydrophobic PMAEDAPS block causes the block copolymer to adopt a conformation such that the more hydrophilic PDMA block is exposed to the aqueous environment. Rearrangement of the blocks most likely occurred when the sample was treated with deionized water during the rinsing step that immediately followed the immobilization procedure.

Yoshikawa et al. [65] prepared a high-density PHEMA brush on the surface of FEP film by surface-initiated RAFT polymerization. The contact angle θ on the film surface was measured both in water with an air bubble and in air with a water droplet (Fig. 11.1). In the former measurement, it was found that all the PHEMA-grafted samples gave a θ of 26°, which was equal to the value for the pure PHEMA surface, indicating the uppermost surface was totally coved with PHEMA. In the later measurement, when $L > 20$ nm ($L$ is defined as the layer thickness), samples showed the same θ value as the pure PHEMA film. For samples with $L < 20$ nm, θ increased with the decrease of $L$, suggesting the surface was not totally covered with a pure PHEMA layer. This was ascribed to the arrangement on structure of FEP and PHEMA chains at the surface. In air (hydrophobic environment), a rearrangement occurred to allow the more hydrophobic FEP chains to move to the outer surface. When $L \geq 20$ nm, this surface rearrangement was completely suppressed.

PNIPAM is one of the most studied stimuli-responsive polymers and exhibits a lower critical solution temperature (LCST) in water at approximately 32 °C. At the LCST, PNIPAM undergoes a volume-phase transition, causing a change from a hydrophilic to a hydrophobic state, which results in dramatic changes in physical properties [73]. Hu et al. [14] grafted linear PNIPAM chains onto a spherical PNIPAM/HEA copolymer microgel. In the low-temperature range 25–32 °C, they found that the grafted PNIPAM layer thickness decreased, which was related to the coil–globule transitions of linear grafted PNIPAM chains. While in the high-temperature range 32–35 °C, the layer thickness increased linearly with the graft density. This was attributed to a strong steric repulsion among the chains grafted on the shrunken MG core, which forced the tethered PNIPAM chains to stretch into a brushlike conformation on the surface (Fig. 11.2). Raula et al. [16] polymerized NIPAM from the surface of gold nanoparticles using the RAFT technique. An aqueous PNIPAM-attached gold nanoparticles solution was gradually heated above the LCST of PNIPAM. They found that both the absorption maximum of the surface plasmon ($\lambda_{max}$) and the absorbance intensity ($I_{abs}$) at 650 nm decreased strongly during the collapse of PNIPAM around 34 °C (Fig. 11.3). The shift of the $\lambda_{max}$ to lower wavelengths indicated that the surroundings of the surface of the gold core became less hydrophilic, as the surface was covered by the collapsed PNIPAM. The decrease in the surface plasmon band was due to the shielding of the gold surface

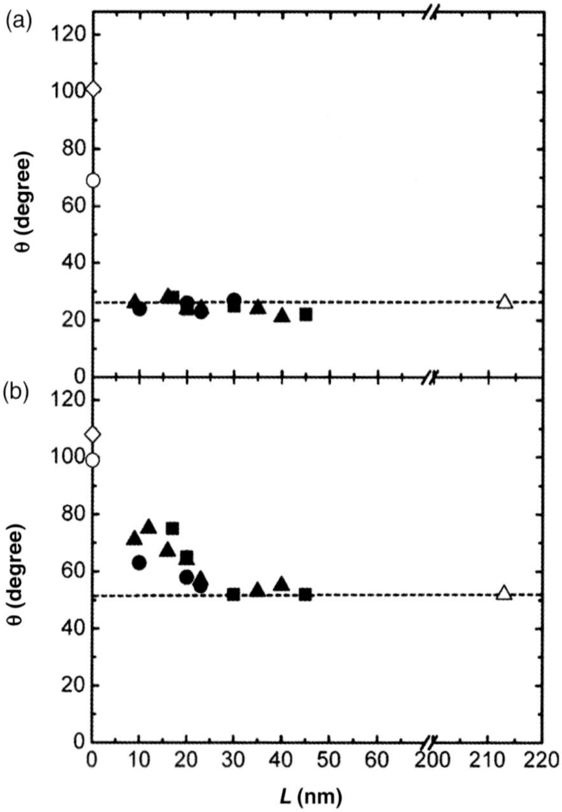


**Fig. 11.1** Plot of θ versus $L$ (a) in water with air bubbles and (b) in air with water droplets: (◊) pristine FEP film, (○) 120-s plasma-treated FEP film, (Δ) PHEMA-spin-cast FEP film and PHEMA-grafted FEP film with plasma-treatment time of (■) 60 s, (▲) 80 s and (•) 120 s. The broken lines show the contact angles of the pure PHEMA surface. (Reproduced with permission from [65]. Copyright 2005 American Chemical Society.)

by collapsed PNIPAM, which was observed to be reversible as the temperature was decreased back to 20 °C.

### 11.3.2
### Interfacial Properties in Polymer Nanocomposites

Nanoparticles have been used extensively as fillers in polymer nanocomposites to improve the mechanical, thermal, electric and optical properties. The large surface area of nanoparticles has the ability to affect a large volume fraction of the matrix polymer. Therefore, interfacial interactions between nanoparticles and the matrix polymer are especially important in determining the properties of

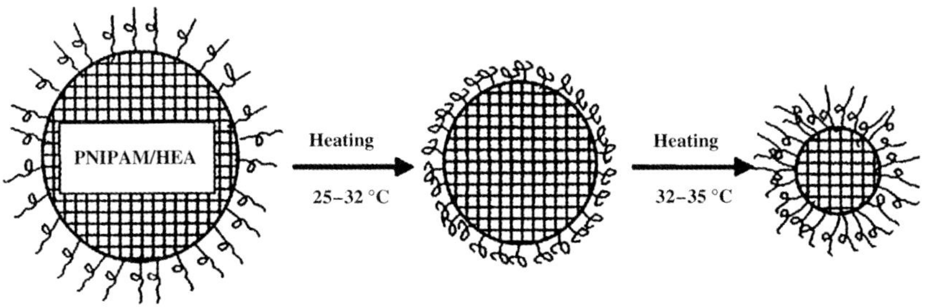


Fig. 11.2 Schematic of the coil–globule–brush transition of linear PNIPAM chains grafted on a thermally sensitive microgel. (Reproduced with permission from [14]. Copyright 2002 American Chemical Society.)

polymer nanocomposites. The RAFT technique provides a versatile tool to tailor the surface of nanoparticles and thus enables researchers to design interfaces between nanoparticles and the matrix polymers with several levels of control over chemistry, chain length, chain density and layer thickness. Li and coworkers [22, 25] developed a RAFT polymerization method capable of growing PS with molecular weights ($M$) of up to 150 000 g·mol$^{-1}$, polydispersity of less than 1.2 and graft densities ranging from 0.05 to 0.8 chains nm$^{-2}$ on the surfaces of silica nanoparticles. Using these nanoparticles with controlled interfaces as nanofillers in polymer nanocomposites, Bansal et al. [74, 75] studied the relationship between local interface behavior and thermomechanical properties of polymer nanocomposites. To study the wetting of $SiO_2$-g-PS surfaces by PS matrices with various molecular weights, films of $SiO_2$-g-PS nanoparticles were spin-cast onto Si wafers, which were previously

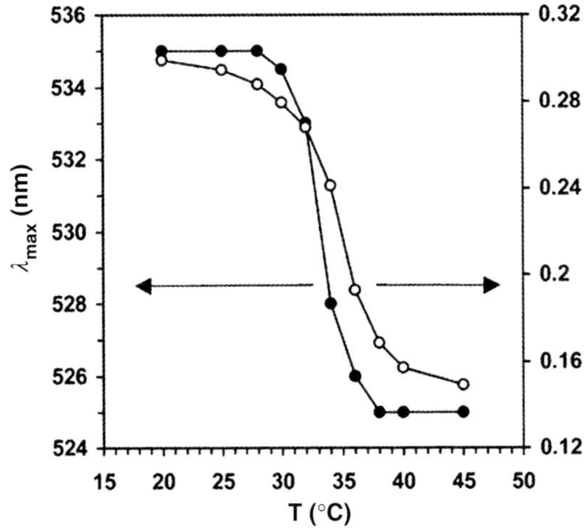


Fig. 11.3 Changes in the $\lambda_{max}$ of the surface plasmon and the $I_{abs}$ at 650 nm of the aqueous MPC-PNIPAM solution as a function of temperature. (Reproduced with permission from [16]. Copyright 2003 American Chemical Society.)

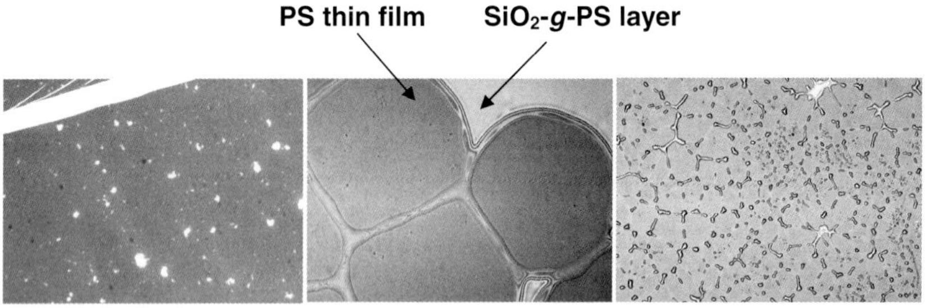


**Fig. 11.4** Optical micrographs of the surface of PS films atop a layer of $SiO_2$-g-PS nanoparticles. The matrix molecular weight is varied from
(a) 44 200 g·mol$^{-1}$, where complete wetting is seen, to
(b) 92 000 g·mol$^{-1}$, where partial dewetting is seen and finally to
(c) 252 000 g·mol$^{-1}$, where the PS film completely dewets and forms islands [74].

etched with an oxygen plasma for 15 min to remove organic impurities. A layer of narrow-polydispersity 'free' PS was then placed on top of this layer. Figure 11.4 shows optical micrographs of PS films of various molecular weights sitting atop a film of $SiO_2$-g-PS nanoparticles ($M = 110\,000$ g·mol$^{-1}$, the graft density ($\sigma$) was about 0.27 chains nm$^{-2}$). It can be seen that the PS films with a lower molecular weight ($M = 42\,000$ g·mol$^{-1}$ had a homogeneous surface that indicated that the PS film wets the surface of the $SiO_2$-g-PS nanoparticles. As the $M$ of PS film increased to 92 000 g·mol$^{-1}$, partial dewetting was observed. Complete dewetting of the PS films was apparent when the $M$ of the PS films increased to 250 000 g·mol$^{-1}$, evidenced by the many small PS islands on the $SiO_2$-g-PS surface. These results suggested that the wetting behavior of nanoparticles can be controlled by varying the ratio of the $M$ of free polymer to that of the grafted polymer on the surface of nanoparticles. To further evaluate the effect of polymer-nanoparticle wetting behavior on the thermomethanical properties of polymer nanocomposites, they studied the glass transition temperature ($T_g$) of $SiO_2$-g-PS/PS nanocomposites. As shown in Fig. 11.5, the crossover from wetting to autophobicity was observed when $M^{matrix}/M^{graft} \sim 0.7$. For $M^{matrix}/M^{graft} < 0.7$, the polymer matrix wetted the nanoparticles and the $T_g$ of the nanocomposites increased. For $M^{matrix}/M^{graft} > 0.7$, dewetting was observed and the $T_g$ decreased with increasing $SiO_2$ concentration. A possible explanation for these phenomena was that the low $M$ matrix, which strongly wets the particles, interdigitates with the brush, creating a strong interface. The brush chains extend further and thus, the grafted PS not only loses conformational entropy but also occupies a greater volume fraction of the nanocomposite. Both of these effects lead to reduced mobility of the polymer chains in the nanocomposite and could explain the observed increase in $T_g$. For the high $M$ case, where dewetting occurs, the interface acts akin to a free polymer surface thereby resulting in a reduced $T_g$.

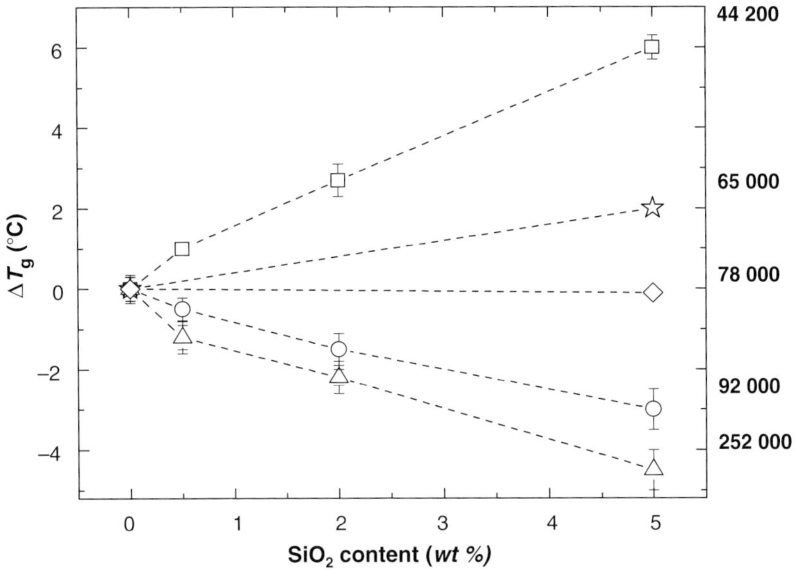


**Fig. 11.5** Change in glass transition temperature ($\Delta T_g$) of $SiO_2$-g-PS (110 000 molecular weight) in PS nanocomposites as a function of $SiO_2$ concentration for various molecular weights (in $g{\cdot}mol^{-1}$) of the matrix. Dashed lines are a guide for the eye [74].

## 11.4 Conclusions

The employment of the RAFT process to modify the surface of materials has attracted increasing interest in recent years. The versatility of the RAFT technique provides the ability to modify the material surfaces via various grafting approaches. By varying the combinations of grafting approach with substrate, surface modification can be achieved by grafting a wide range of polymer chains with considerable control over chain structure, chain length and graft density.

The increasing demand for materials with novel surface properties will continuously direct researchers to design sophisticated polymer structures on the surface of materials, which requires a further understanding of the polymerization mechanism and kinetics. Although the kinetics of RAFT polymerization in solution has been extensively studied, it has been found that the kinetics of RAFT polymerization on the surface of solid substrates can be quite different from that in solution, which was ascribed in early studies to the unique steric environment of the intermediate macro-RAFT-agent radical on the substrate surfaces. Further investigations are still needed to understand how the experimental factors involved, such as RAFT-agent surface density and monomer steric structure, affect the polymerization kinetics. Also, while the 'toolbox' of synthetic methods has expanded in the last several years, improvements in the levels of control (e.g. degree of polymer attached to the surface vs free polymer) are still needed.

Finally, while substantial attention has been given to the synthetic approaches via the RAFT process, extensive studies of the properties of the interface and resulting nanocomposite materials are still at an early stage. A clear understanding of the structure and properties will help to develop many potential applications in various fields and may also, in turn, guide researchers to design novel structures by selecting the most efficient approaches [76].

## 11.5 Abbreviations

| | |
|---|---|
| AFM | Atomic force microscopy |
| AGA | *N*-acryloyl glucosamin |
| AIBN | 2,2′-Azobisisobutyronitrile |
| ATRP | Atom-transfer radical polymerization |
| BDTB | Benzyldithiobenzoate |
| CPDB | 4-Cyanopentanoic acid dithiobenzoate |
| CRP | Controlled radical polymerizations |
| CTA | Chain-transfer agent |
| DBTTC | Dibenzyl trithiocarbonate |
| DMA | *N*,*N*-dimethylacrylamide |
| DPPH | 2,2-Diphenyl-1-picrylhydrazyl |
| FEP | Poly(tetrafluoroethylene-*co*-hexafluoropropylane) |
| GMA | Glycidyl methacrylate |
| HEA | Hydroxyethyl acrylate |
| HEMA | 2-Hydroxyethyl methacrylate |
| LCST | Lower critical solution temperature |
| MA | Methyl acrylate |
| Mer-MCPDB | *s*-Methoxycarbonylphenymethyl dithiobenzoate |
| MMA | Methyl methacrylate |
| MMT | Montmorillonite |
| MNP | Magnetic nanoparticles |
| MWNT | Multiwalled carbon nanotube |
| NaSS | Sodium 4-styrenesulfonate |
| NIPAM | *N*-isopropylacrylamide |
| RAFT | Reversible addition–fragmentation chain transfer |
| PAA | Poly(acrylic acid) |
| PAMPS | Poly(sodium 2-acrylamido-2-methyl propane sulfonate) |
| PBuA | Poly(*n*-butyl acrylate) |
| PDI | Polydispersity index |
| PDMA | Poly(*N*,*N*-dimethylacrylamide) |
| PDMAEMA | Poly[2-(dimethylamino)ethyl methacrylate] |
| PDMS | Poly(dimethylsiloxane) |

| | |
|---|---|
| PDVB | Poly(divinylbenzene) |
| PGMA | Poly(glycidyl methacrylate) |
| PHEMA | Poly(2-hydroxyethyl methacrylate) |
| PMA | Polymethylacrylate |
| PMMA | Polymethylmethacrylate |
| PNIPAM | Poly(*N*-isopropylacrylamide) |
| PNIPAM-MPC | PNIPAM-monolayer-protected clusters |
| PP | Polypropene |
| PPEGMA | Poly[poly(ethylene glycol) monomethacrylate] |
| PS | Polystyrene |
| PTFE | Poly(tetrafluoroethylene) |
| PVBC | Poly(4-vinylbenzyl chloride) |
| PVBTC | Poly[(*ar*-vinylbenzyl)trimethylammonium chloride] |
| PVDF | Poly(vinylidene fluoride) |
| Si-MCPDB | Silica-supported MCPDB |
| Si-MPPA | Silica-supported 3-(methoxycarbonylphenylmethylsulfanylthiocarbonysulfanyl)-propionic acid |
| SS | Styrenesulfonate |
| St | Styrene |
| TEM | Transmission electron microscopy |
| THF | Tetrahydrofuran |
| TOPO | Tri-*n*-octylphosphine oxide |
| VBC | 4-Vinylbenzyl chloride |
| VBTC | (*ar*-Vinylbenzyl)trimethylammonium chloride |

## 11.6 Acknowledgment

The authors gratefully acknowledge support through the Nanoscale Science and Engineering Initiative of the National Science Foundation under NSF Award Number DMR-0642573.

## References


1. Y. Ikada, *Adv. Polym. Sci.* **1984**, *57*, 103–140.
2. E. Raphael, P. G. Degennes, *J. Phys. Chem.* **1992**, *96* (10), 4002–4007.
3. S. L. Kaplan, E. S. Lopata, J. Smith, *Surf. Interface Anal.* **1993**, *20* (5), 331–336.
4. R. S. Parnas, Y. Cohen, *Rheol. Acta* **1994**, *33* (6), 485–505.
5. J. Klein, *Annu. Rev. Mater. Sci.* **1996**, *26*, 581–612.
6. R. C. Advincula, W. J. Brittain, K. C. Caster, J. Rühe, Eds., *Polymer Brushes*, Wiley-VCH, Weinheim, **2004**.
7. Y. Tsujii, K. Ohno, S. Yamamoto, A. Goto, T. Fukuda, In *Surface-Initiated Polymerization I*, **2006**, pp. 1–45.
8. B. Zhao, W. J. Brittain, *Prog. Polym. Sci.* **2000**, *25* (5), 677–710.
9. G. Moad, E. Rizzardo, S. H. Thang, *Aust. J. Chem.* **2005**, *58* (6), 379–410.

10. A. B. Lowe, C. L. McCormick, *Aust. J. Chem.* **2002**, *55* (6–7), 367–379.
11. G. Moad, E. Rizzardo, S. H. Thang, *Aust. J. Chem.* **2006**, *59* (10), 669–692.
12. J. Chiefari, Y. K. Chong, F. Ercole, J. Krstina, J. Jeffery, T. P. T. Le, R. T. A. Mayadunne, G. F. Meijs, C. L. Moad, G. Moad, E. Rizzardo, S. H. Thang, *Macromolecules* **1998**, *31* (16), 5559–5562.
13. Y. Tsujii, M. Ejaz, K. Sato, A. Goto, T. Fukuda, *Macromolecules* **2001**, *34* (26), 8872–8878.
14. T. J. Hu, Y. Z. You, C. Y. Pan, C. Wu, *J. Phys. Chem. B* **2002**, *106* (26), 6659–6662.
15. L. Barner, *Aust. J. Chem.* **2003**, *56* (10), 1091–1091.
16. J. Raula, J. Shan, M. Nuopponen, A. Niskanen, H. Jiang, E. I. Kauppinen, H. Tenhu, *Langmuir* **2003**, *19* (8), 3499–3504.
17. J. Shan, M. Nuopponen, H. Jiang, E. Kauppinen, H. Tenhu, *Macromolecules* **2003**, *36* (12), 4526–4533.
18. K. Matsumoto, R. Tsuji, Y. Yonemushi, T. Yoshida, *J. Nanopart. Res.* **2004**, *6* (6), 649–659.
19. K. Matsumoto, R. Tsuji, Y. Yonemushi, T. Yoshida, *Chem. Lett.* **2004**, *33* (10), 1256–1257.
20. H. Skaff, T. Emrick, *Angew. Chem. Int. Ed. Engl.* **2004**, *43* (40), 5383–5386.
21. R. Joso, M. H. Stenzel, T. P. Davis, C. Barner-Kowollik, L. Barner, *Aust. J. Chem.* **2005**, *58* (6), 468–471.
22. C. Li, B. C. Benicewicz, *Macromolecules* **2005**, *38* (14), 5929–5936.
23. S. Perrier, P. Takolpuckdee, C. A. Mars, *Macromolecules* **2005**, *38* (16), 6770–6774.
24. T. Y. Guo, P. Liu, J. W. Zhu, M. D. Song, B. H. Zhang, *Biomacromolecules* **2006**, *7* (4), 1196–1202.
25. C. Li, J. Han, C. Y. Ryu, B. C. Benicewicz, *Macromolecules* **2006**, *39* (9), 3175–3183.
26. M. M. Titirici, B. Sellergren, *Chem. Mater.* **2006**, *18* (7), 1773–1779.
27. W. C. Wang, K. G. Neoh, E. T. Kang, *Macromol. Rapid Commun.* **2006**, *27* (19), 1665–1669.
28. Y. L. Zhao, S. Perrier, *Macromolecules* **2006**, *39* (25), 8603–8608.
29. J. F. Quinn, R. P. Chaplin, T. P. Davis, *J. Polym. Sci. Part A: Polym. Chem.* **2002**, *40* (17), 2956–2966.
30. M. D. Rowe-Konopacki, S. G. Boyes, *Macromolecules* **2007**, *40* (4), 879–888.
31. A. B. Lowe, B. S. Sumerlin, M. S. Donovan, C. L. McCormick, *J. Am. Chem. Soc.* **2002**, *124* (39), 11562–11563.
32. S. G. Spain, L. Albertin, N. R. Cameron, *Chem. Commun.* **2006**, (40), 4198–4200.
33. P. Takolpuckdee, C. A. Mars, S. Perrier, *Org. Lett.* **2005**, *7* (16), 3449–3452.
34. J. Shan, M. Nuopponen, H. Jiang, T. Viitala, E. Kauppinen, K. Kontturi, H. Tenhu, *Macromolecules* **2005**, *38* (7), 2918–2926.
35. D. H. Nguyen, P. Vana, *Polym. Adv. Technol.* **2006**, *17* (9–10), 625–633.
36. L. Barner, C. Li, X. J. Hao, M. H. Stenzel, C. Barner-Kowollik, T. P. Davis, *J. Polym. Sci. Part A: Polym. Chem.* **2004**, *42* (20), 5067–5076.
37. D. H. Nguyen, P. Vana, *Polym. Adv. Technol.* **2006**, *17* (9–10), 625–633.
38. M. Baum, W. J. Brittain, *Macromolecules* **2002**, *35* (3), 610–615.
39. G. Q. Zhai, W. H. Yu, E. T. Kang, K. G. Neoh, C. C. Huang, D. J. Liaw, *Ind. Eng. Chem. Res.* **2004**, *43* (7), 1673–1680.
40. W. H. Yu, E. T. Kang, K. G. Neoh, *Ind. Eng. Chem. Res.* **2004**, *43* (17), 5194–5202.
41. Q. Peng, D. M. Y. Lai, E. T. Kang, K. G. Neoh, *Macromolecules* **2006**, *39* (16), 5577–5582.
42. M. H. Stenzel, L. Zhang, W. T. S. Huck, *Macromol. Rapid Commun.* **2006**, *27* (14), 1121–1126.
43. F. J. Xu, E. T. Kang, K. G. Neoh, *J. Mater. Chem.* **2006**, *16* (28), 2948–2952.
44. N. Salem, D. A. Shipp, *Polymer* **2005**, *46* (19), 8573–8581.
45. P. Ding, M. Zhang, J. Gai, B. J. Qu, *J. Mater. Chem.* **2007**, *17* (11), 1117–1122.
46. B. Q. Zhang, C. Y. Pan, C. Y. Hong, B. Luan, P. J. Shi, *Macromol. Rapid Commun.* **2006**, *27* (2), 97–102.
47. B. S. Sumerlin, A. B. Lowe, P. A. Stroud, P. Zhang, M. W. Urban, C. L. McCormick, *Langmuir* **2003**, *19* (14), 5559–5562.
48. A. S. Duwez, P. Guillet, C. Colard, J. F. Gohy, C. A. Fustin, *Macromolecules* **2006**, *39* (8), 2729–2731.

**49.** J. W. Hotchkiss, A. B. Lowe, S. G. Boyes, *Chem. Mater.* **2007**, *19* (1), 6–13.

**50.** G. Pirri, M. Chiari, F. Damin, A. Meo, *Anal. Chem.* **2006**, *78* (9), 3118–3124.

**51.** H. Cui, W. P. Wang, Y. Z. You, C. H. Liu, P. H. Wang, *Polymer* **2004**, *45* (26), 8717–8721.

**52.** C. Y. Hong, Y. Z. You, C. Y. Pan, *Chem. Mater.* **2005**, *17* (9), 2247–2254.

**53.** C. Y. Hong, Y. Z. You, C. Y. Pan, *J. Polym. Sci. Part A: Polym. Chem.* **2006**, *44* (8), 2419–2427.

**54.** C. Y. Hong, Y. Z. You, C. Y. Pan, *Polymer* **2006**, *47* (12), 4300–4309.

**55.** G. Y. Xu, W. T. Wu, Y. S. Wang, W. M. Pang, P. H. Wang, G. R. Zhu, F. Lu, *Nanotechnology* **2006**, *17* (10), 2458–2465.

**56.** G. Y. Xu, W. T. Wu, Y. S. Wang, W. M. Pang, Q. R. Zhu, P. H. Wang, Y. Z. You, *Polymer* **2006**, *47* (16), 5909–5918.

**57.** Y. Z. You, C. Y. Hong, C. Y. Pan, *Nanotechnology* **2006**, *17* (9), 2350–2354.

**58.** D. Roy, J. T. Guthrie, S. Perrier, *Aust. J. Chem.* **2006**, *59* (10), 737–741.

**59.** D. Roy, J. T. Guthrie, S. Perrier, *Macromolecules* **2005**, *38* (25), 10363–10372.

**60.** D. Roy, *Aust. J. Chem.* **2006**, *59* (3), 229–229.

**61.** S. Perrier, P. Takolpuckdee, J. Westwood, D. M. Lewis, *Macromolecules* **2004**, *37* (8), 2709–2717.

**62.** L. Barner, N. Zwaneveld, S. Perera, Y. Pham, T. P. Davis, *J. Polym. Sci. Part A: Polym. Chem.* **2002**, *40* (23), 4180–4192.

**63.** L. Barner, S. Perera, S. Sandanayake, T. P. Davis, *J. Polym. Sci. Part A: Polym. Chem.* **2006**, *44* (2), 857–864.

**64.** M. Grasselli, N. Betz, *Nucl. Instrum. Meth. B* **2005**, *236*, 201–207.

**65.** C. Yoshikawa, A. Goto, Y. Tsujii, T. Fukuda, K. Yamamoto, A. Kishida, *Macromolecules* **2005**, *38* (11), 4604–4610.

**66.** W. H. Yu, E. T. Kang, K. G. Neoh, *Langmuir* **2005**, *21* (1), 450–456.

**67.** Y. W. Chen, W. Sun, Q. L. Deng, L. Chen, *J. Polym. Sci. Part A: Polym. Chem.* **2006**, *44* (9), 3071–3082.

**68.** W. S. Bae, A. J. Convertine, C. L. McCormick, M. W. Urban, *Langmuir* **2007**, *23* (2), 667–672.

**69.** D. Roy, *Aust. J. Chem.* **2006**, *59* (3), 229–229.

**70.** M. H. Stenzel, T. P. Davis, *J. Polym. Sci. Part A: Polym. Chem.* **2002**, *40* (24), 4498–4512.

**71.** S. G. Boyes, A. M. Granville, M. Baum, B. Akgun, B. K. Mirous, W. J. Brittain, *Surf. Sci.* **2004**, *570* (1–2), 1–12.

**72.** R. C. Advincula, *J. Disper. Sci. Technol.* **2003**, *24* (3–4), 343–361.

**73.** H. G. Schild, *Prog. Polym. Sci.* **1992**, *17* (2), 163–249.

**74.** A. Bansal, H. C. Yang, C. Li, B. C. Benicewicz, S. K. Kumar, L. S. Schadler, *J. Polym. Sci. Part B: Polym. Phys.* **2006**, *44* (20), 2944–2950.

**75.** A. Bansal, H. C. Yang, C. Li, K. W. Cho, B. C. Benicewicz, S. K. Kumar, L. S. Schadler, *Nat. Mater.* **2005**, *4* (9), 693–698.

**76.** L. S. Schadler, S. K. Kumar, B. C. Benicewicz, S. L. Lewis, S. L. Harton, *MRS Bull.* **2007**, *32*, 335–340.

# 12
# Polymers with Well-Defined End Groups via RAFT – Synthesis, Applications and *Post*modifications

*Leonie Barner, and Sébastien Perrier*

## 12.1
## Introduction

The control over the chain-end functionality of a polymeric chain produced by controlled/'living' radical polymerization is inherent to the mechanism of the reaction. Indeed, the final product contains a majority of polymeric chains showing an $\omega$-functional end group which is used to control the molecular weight growth. The growth of the molecular weight can be mediated either via the reversible homolytic cleavage of the covalent bond between the terminal carbon and the chain-end group (e.g. halogen for transition-metal-mediated living radical polymerization/atom transfer radical polymerization (ATRP), nitroxide for nitroxide-mediated polymerization (NMP)) or via the degenerative transfer of chain-end groups between propagating radicals and dormant species (e.g. thiocarbonylthio groups for reversible addition–fragmentation chain transfer (RAFT)). Furthermore, $\alpha$-functional end groups can also be introduced via the initiator/mediator of the polymerization, for example, halogen alkyls (ATRP), alkyl nitroxides (NMP) or dithioesters (RAFT).

The versatility of the RAFT process toward functional groups allows for the introduction of a wide range of chain-end functionalities. Moreover, the structure of the thiocarbonylthio compounds, mostly used to mediate RAFT polymerization, provides three approaches to control the functionality of a polymeric chain end (see Scheme 12.1):

1. $\alpha$-Functional groups can be introduced via the R group of the chain transfer agent (CTA). The R group of the RAFT agent is a free radical leaving group that must also be able to reinitiate the polymerization. Although there is always a proportion of polymeric chains formed during the RAFT process that are initiated by radicals derived from the free radical initiator, these are usually kept very low due to the high CTA-to-initiator ratios utilized. It follows that most polymeric

*Handbook of RAFT Polymerization.* Edited by Christopher Barner-Kowollik

ISBN: 978-3-527-31924-4

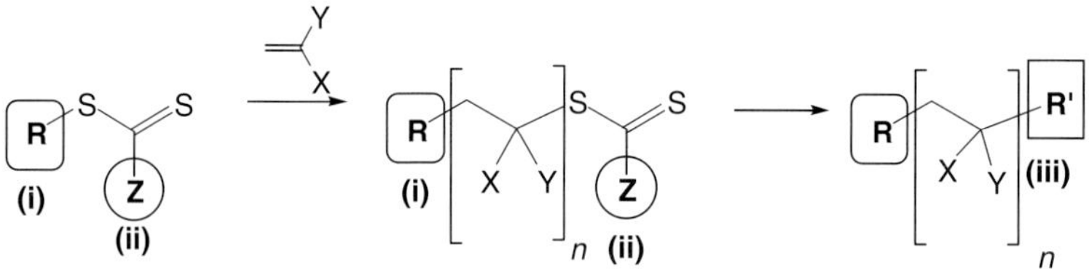


**Scheme 12.1** Three routes to introduce end functionalities in polymers synthesized via the RAFT process.

chains produced via RAFT will show an α-functionality provided by the R group of the initial CTA.

2. ω-Functional groups can be introduced via the Z group. The Z group is responsible for the stabilization of the intermediate radicals that are produced during the polymerization. Although the chains that terminate during polymerization will lose the thiocarbonylthio end group and therefore the Z functionality, the ratio of living chains (with Z group) to dead chains is usually kept very high (typically above 90%), thus ensuring that most chains retain their Z-group functionality. This approach is unique to the RAFT process and cannot be achieved by other living radical polymerization techniques. Problems arise, however, for applications in which the terminal C—S bond may be broken under certain conditions, thus leading to the loss of the ω-functionality.
3. ω-Functional groups can also be introduced via the modification of the thiocarbonylthio group *post*polymerization. The modification of thiocarbonylthio groups is well known, and a few methodologies have been applied to RAFT-synthesized polymers to remove this group and modify it into an alternative functional group.

In this chapter, we review the various methodologies that have been used to produce polymers with well-defined end groups via the RAFT process. We highlight the introduction of chain-end groups via the R and Z groups of CTAs (section 12.2). A specific section (12.3) is devoted to the use of functionalities in the CTA and/or the telechelic polymer that can initiate polymerizations proceeding via an alternative mechanism to radical polymerization (e.g. ring-opening polymerization (ROP)). Another part of the chapter is concerned with the removal of the thiocarbonylthio groups, its effect on the stability of RAFT-synthesized polymers and its use to introduce ω-functionality *post*polymerization (section 12.4).

## 12.2
## Terminal Functionalities Introduced via the CTA

The number of RAFT agents that could be used to introduce α- and/or ω-functionality in polymers is already sizeable. In this chapter, we describe the most important functional groups and their corresponding RAFT agents. We have not included RAFT agents that show latent functionalities, accessible via further reactions, but that have not yet been exploited.

## 12.2.1
## Carboxyl and Hydroxyl End-Functional Polymers

The facile synthesis of mono- and dicarboxyl-terminated trithiocarbonate RAFT agents was reported by Lai et al. [1]. *S,S'*-Bis(α,α'-dimethylacetic acid) trithiocarbonate (**1**), a dicarboxyl-terminated trithiocarbonate RAFT agent, and *S*-1-dodecyl-*S'*-(α,α'-dimethylacetic acid) trithiocarbonate (**2**), a monocarboxyl-terminated trithiocarbonate RAFT agent (see Table 12.1), mediate the polymerization of ethyl acrylate (EA), acrylic acid (AA), butyl acrylate (BA), 2-hydroxyethyl acrylate (HEA), *tert*-butyl acrylamide (tBAm) and styrene (St) [1]. The synthesis of block copolymers – including amphiphilic copolymers – mediated via these RAFT agents is also possible. However, narrow-disperse poly(methyl methacrylate) (PMMA) could not be achieved using RAFT acids **1** and **2**. Lai and Shea also reported the synthesis of carboxyl- and hydroxyl-terminated dithiocarbamates and xanthates, which can mediate the RAFT polymerization of various monomers [28].

Ferguson et al. [2] reported the synthesis and application of an amphipathic RAFT agent (**3**) which can mediate polymerization in both aqueous and organic phases. The RAFT acid **3** was applied for ab initio emulsion polymerization of AA and BA.

Moad et al. [3] designed an acid end-functionalized trithiocarbonate RAFT agent (**4**) which is able to control the polymerization of methyl methacrylate (MMA). This RAFT acid has a tertiary cyanoalkyl R group, which is a good radical leaving group with respect to the PMMA propagating radical.

Stenzel et al. [4] synthesized a monocarboxyl-terminated trithiocarbonate RAFT acid (**5**) which is able to mediate the polymerization of BA [5], 2-acryloylethyl phosphorylcholine [6], AA and *N*-isopropyl acrylamide (NIPAM) [7]. The RAFT acid **5** was also applied to modify α-D-glucose, β-cyclodextrine and cellulose and to facilitate the subsequent polymerization of St [4]. The RAFT acid **5** has – due to the hydrophobic R group – a reduced solubility in aqueous solutions. Wang et al. [5] synthesized water-soluble dicarboxyl-terminated trithiocarbonate RAFT acids (**6** and **7**) whose structures were similar to that of the RAFT acid **5**. These RAFT acids were applied to synthesize telechelic poly(*n*BA)s [5].

α,ω-Dihydroxyl-terminated telechelic polymers of St and MA with predetermined molecular weight and low polydispersity can be synthesized using the RAFT agent *S,S'*-bis(2-hydroxyethyl-2'-butyrate) trithiocarbonate (BHBT, **8**; see Table 12.1) [8]. HO–polystyrene–OH (HO-PSt-OH) and HO–PMA–OH macro-RAFT polymers could subsequently be used for the synthesis of triblock copolymers with the second polymer in the center of the block copolymer chains and two terminal hydroxyl groups.

Functional dithioester RAFT agents have also been reported for the synthesis of telechelic polymers. D'Agosto et al. [10] synthesized two dithioesters (**9** and **10**; see Table 12.1) having a propanoic acid group as R group and phenyl or benzyl group as Z group. These RAFT agents control the polymerization of *N*-acryloylmorpholine (NAM, **11**; see Scheme 12.2), a water-soluble bisubstituted acrylamide derivative. Successful formation of AB block copolymers, poly(NAM-*b*-PSt), has also been reported [10].

**Table 12.1** Selected RAFT agents for the synthesis of end-functional polymers

| Functional group | RAFT agent | Monomers |
|---|---|---|
| | *Trithiocarbonates* | |
| α,ω-Dicarboxyl | 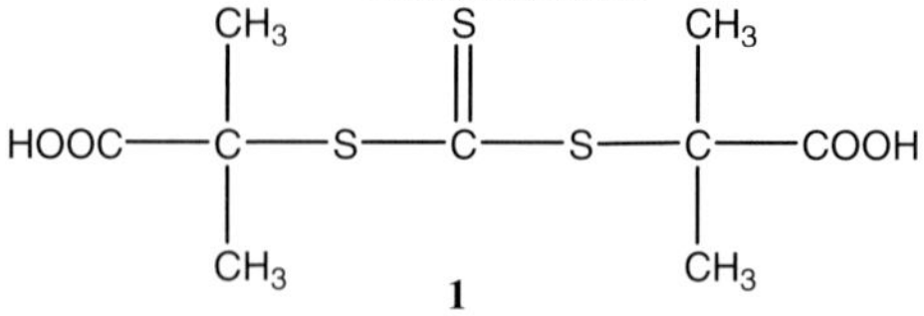 **1** | [1] EA, AA, BA, HEA, tBAm, St |
| α-Carboxyl | 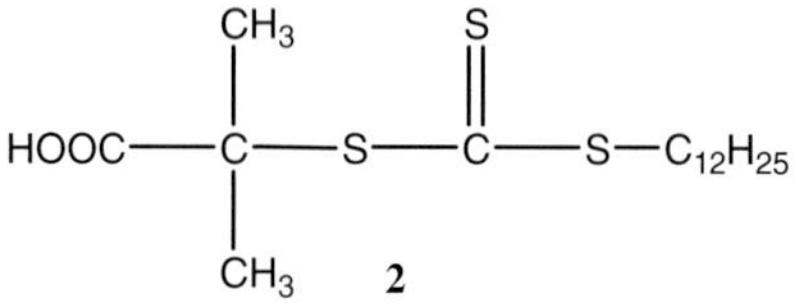 **2** | [1] EA, AA, BA, HEA, tBAm, St |
| α-Carboxyl | 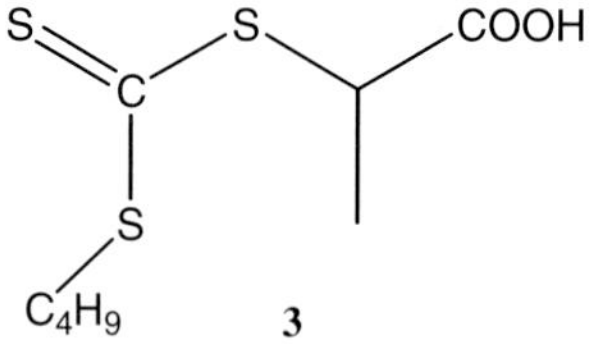 **3** | [2] AA, BA |
| α-Carboxyl | 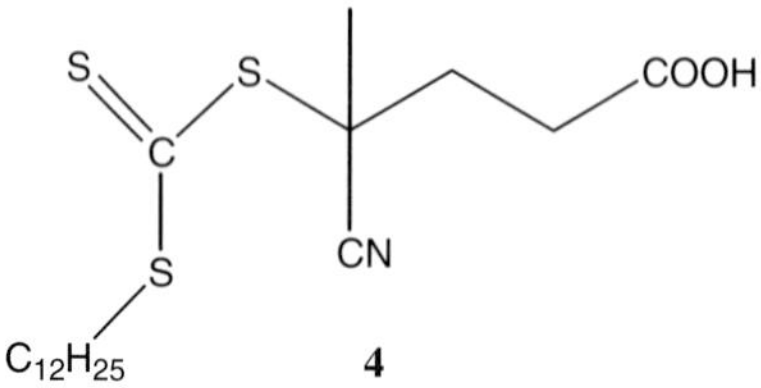 **4** | [3] MMA |
| ω-Carboxyl | 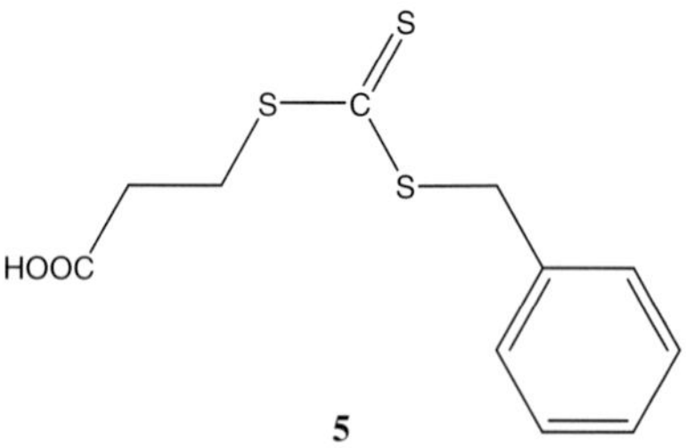 **5** | [4, 5] BA, [6] 2-acryloylethyl phosphoryl-choline, [7] AA, NIPAM |

**Table 12.1** (*Continued*)

| Functional group | RAFT agent | Monomers |
|---|---|---|
| α,ω-Dicarboxyl | 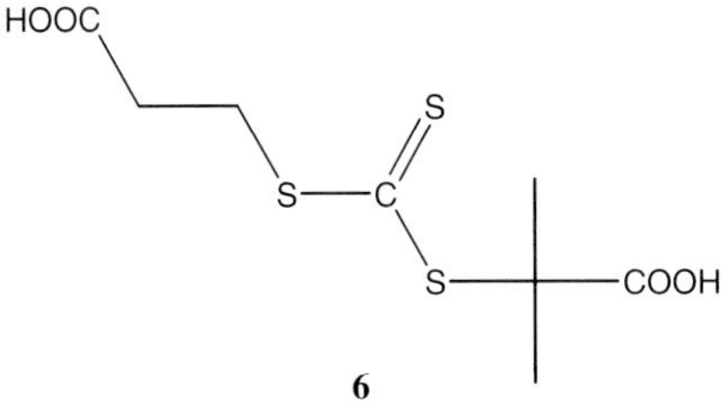 **6** | [5] BA |
| α,ω-Dicarboxyl | 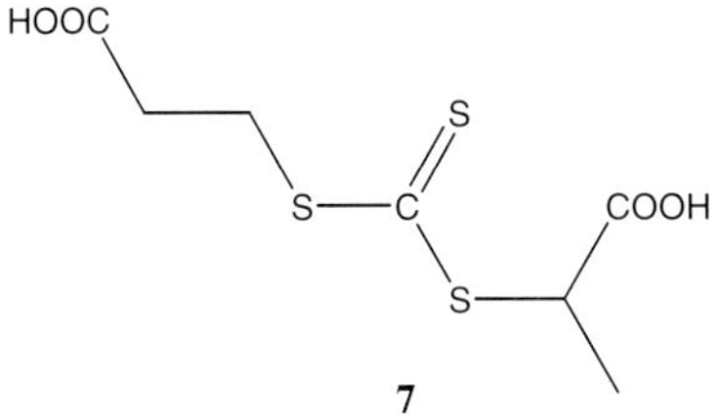 **7** | [5] BA |
| α,ω-Dihydroxyl | 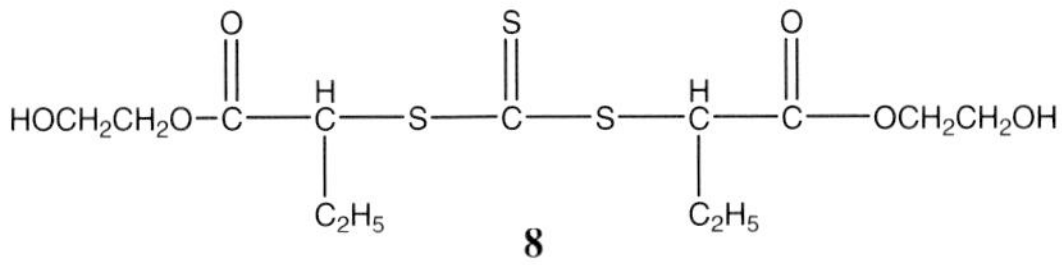 **8** | [8] MA, St, [9] NIPAM |
| | *Dithioesters* | |
| α-Carboxyl | 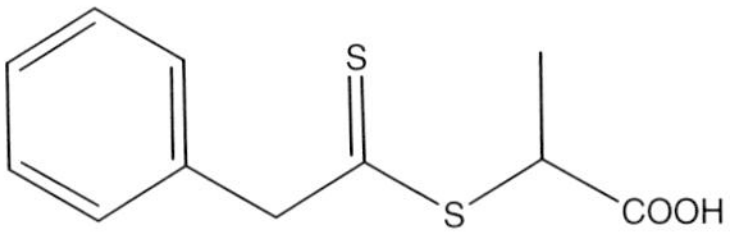 **9** | [10] NAM |
| α-Carboxyl | 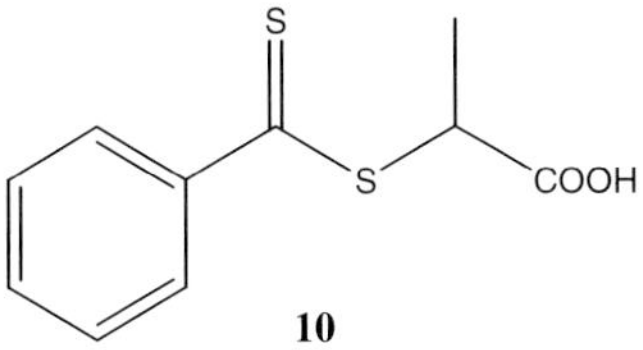 **10** | [10] NAM |

(*Continued*)

**Table 12.1** (*Continued*)

| Functional group | RAFT agent | Monomers |
|---|---|---|
| α-Carboxyl | 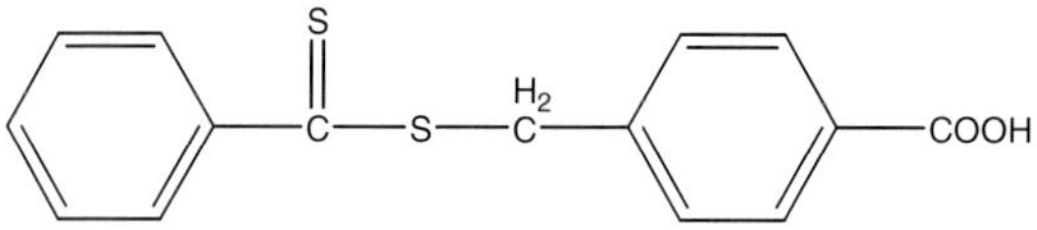 **12** | [11, 12] MMA, [13] St |
| α-Hydroxyl α,ω-Dihydroxyl | 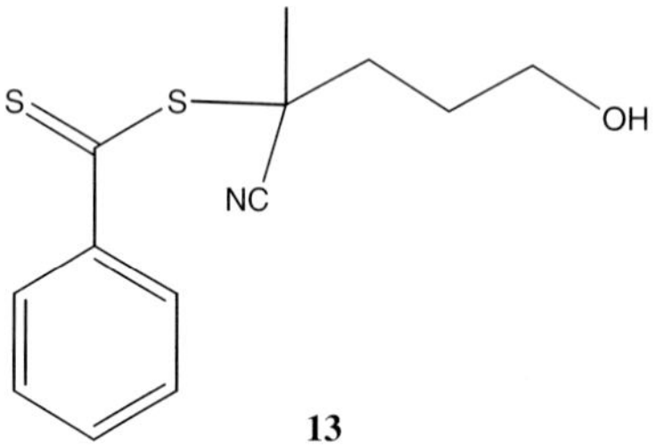 **13** | [14] MA |
| α-Amide | 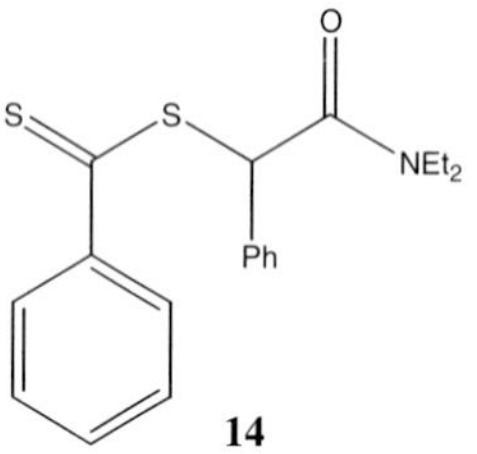 **14** | [15] St, MA, DMA |
| α-Primary amine | 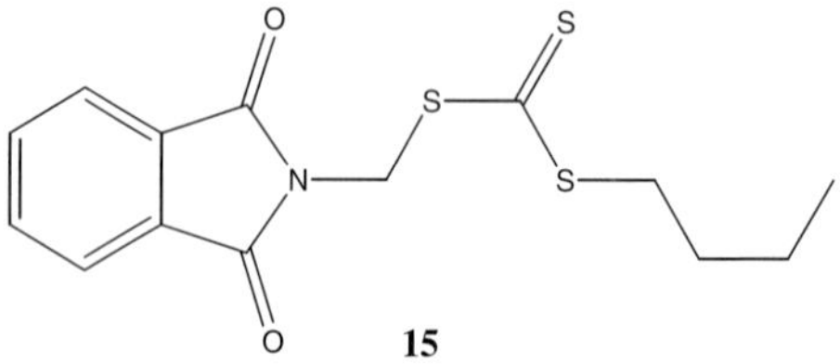 **15** | [16] St, [17] BA, NIPAM |
| α-Primary amine | 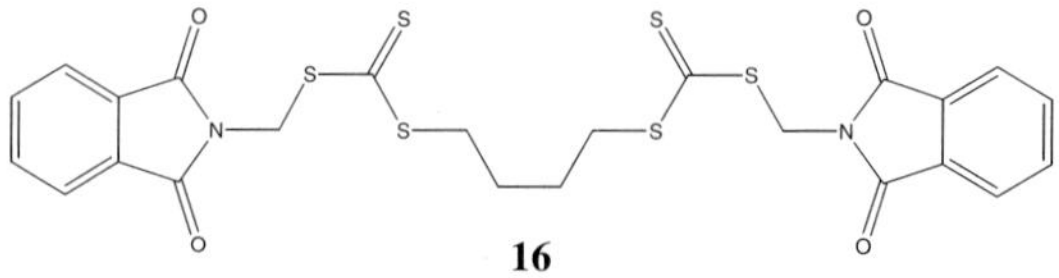 **16** | [16] St, [17] BA, NIPAM |

Table 12.1 (*Continued*)

| Functional group | RAFT agent | Monomers |
|---|---|---|
| α-Primary amine | 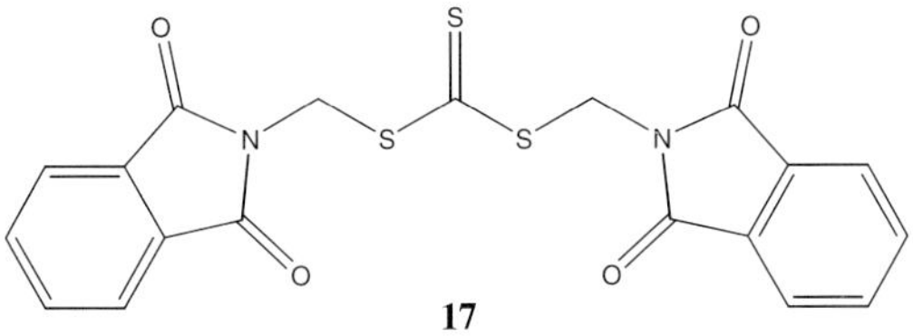 **17** | [16] St, [17] BA, NIPAM |
| α-Primary amine | 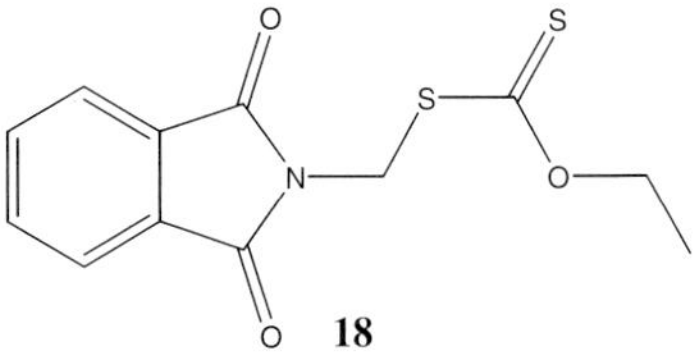 **18** | [17] NVP, VAc |
| α-Fluorescent | 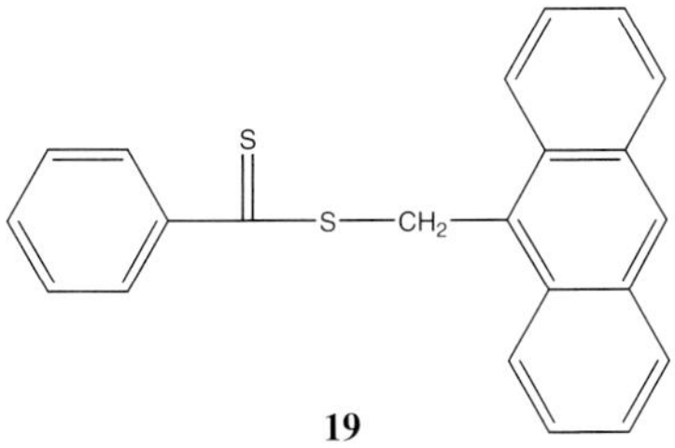 **19** | [18] St, MA |
| α,ω-Di-1,3-benzodioxole | 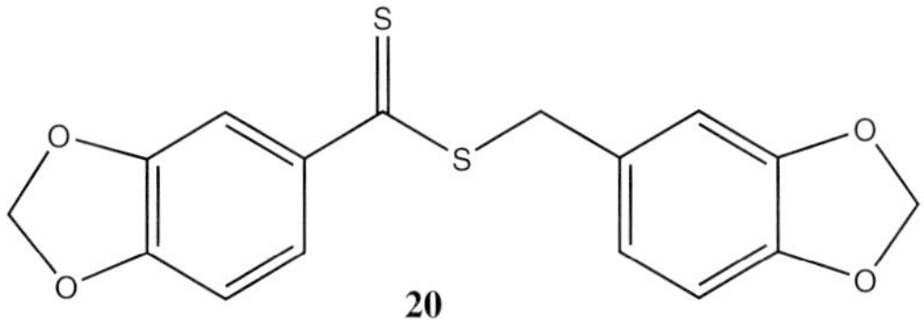 **20** | [19] St |
| | 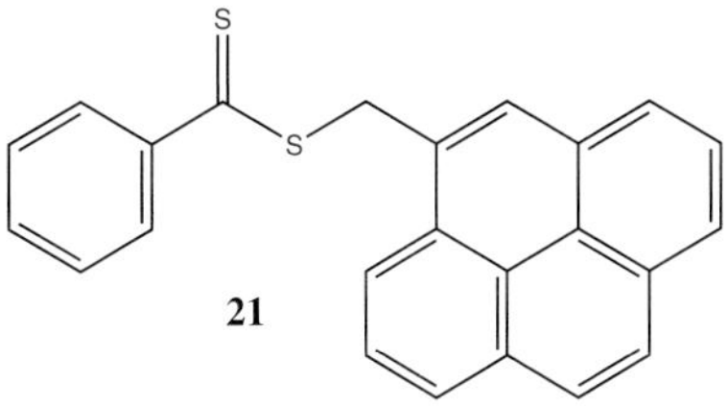 **21** | [20] St |

(*Continued*)

Table 12.1 (Continued)

| Functional group | RAFT agent | Monomers |
|---|---|---|
| α-Terpyridine | 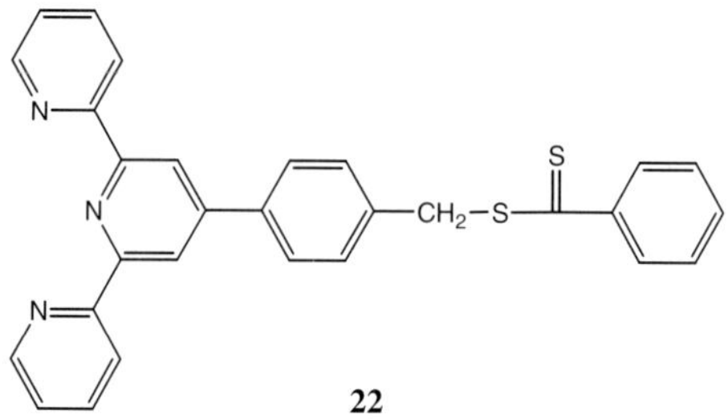 22 | [29] St, NIPAM |
| α,ω-Bisterpyridine | 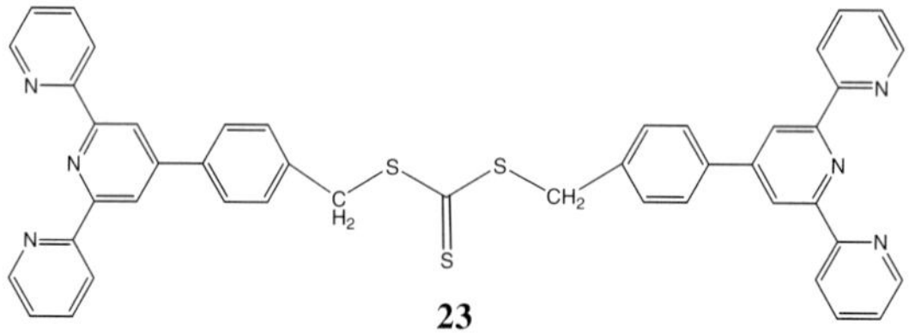 23 | [21] St, BA |
| α,ω-Bisallyl | 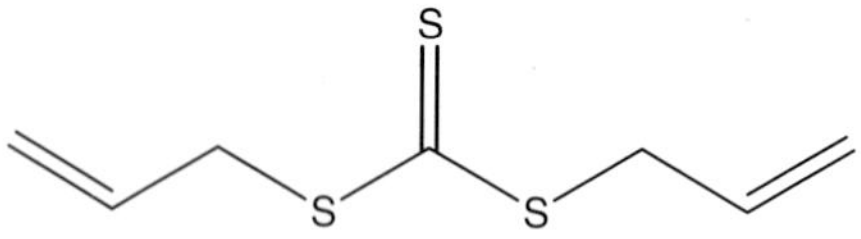 24 | [22] St, BA |
| α-Norbornenyl | 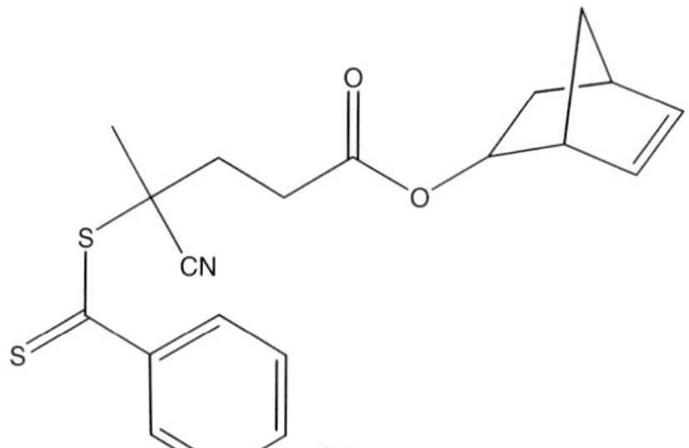 25 | [23] St, MMA, MA |

**Table 12.1** (*Continued*)

| Functional group | RAFT agent | Monomers |
|---|---|---|
| α-Vinyl | 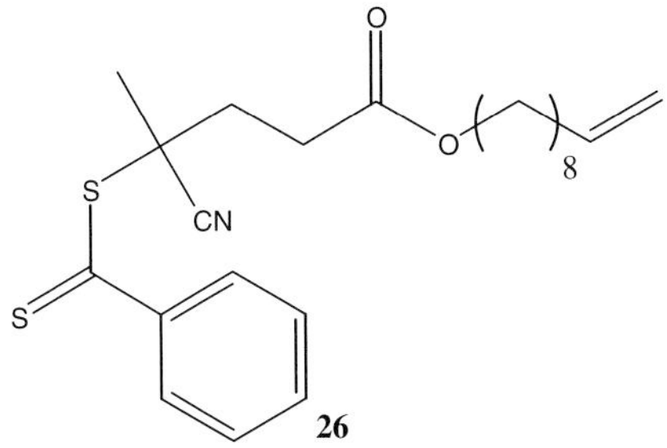 **26** | [23] St, MMA, MA |
| α-Cinnamyl | 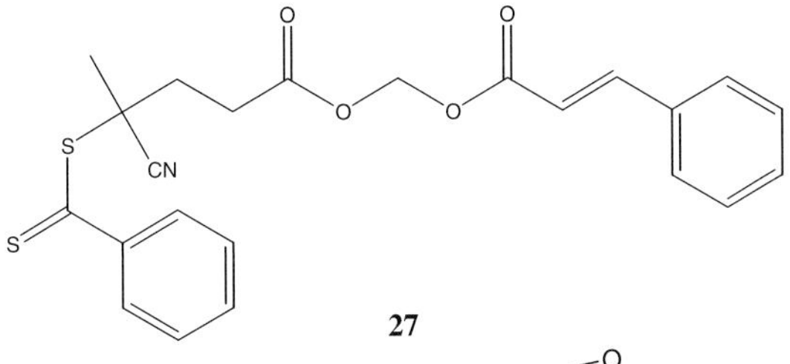 **27** | [23] St, MMA, MA |
| α-Carbohydrate derivative | 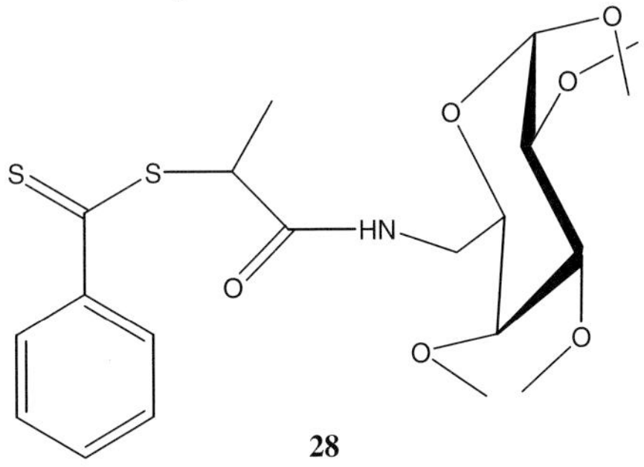 **28** | [24] NAM |
| α-Biotin derivative | 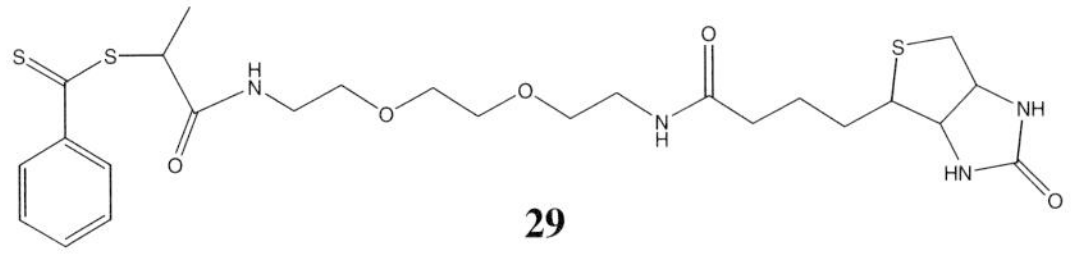 **29** | [24] NAM |
| α-Biotin derivative | 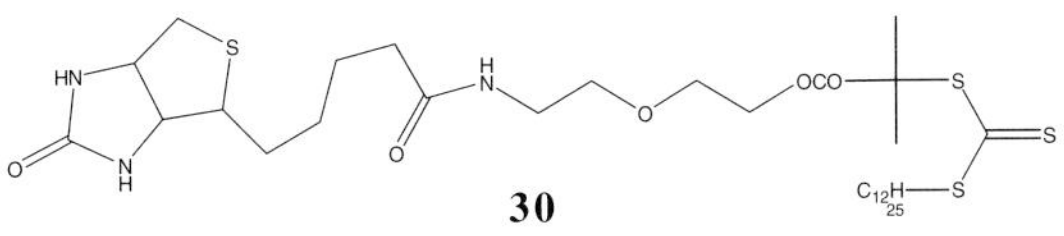 **30** | [25] NIPAM, HPMA |

(*Continued*)

Table 12.1 (*Continued*)

| Functional group | RAFT agent | Monomers |
|---|---|---|
| ω-BSA | 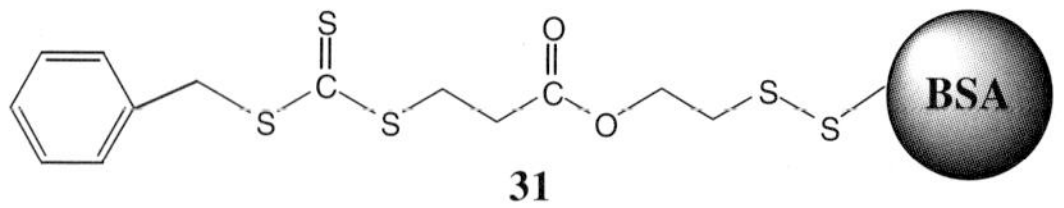 **31** | [26] PEG-A |
| ω-Pyridyl disulfide | 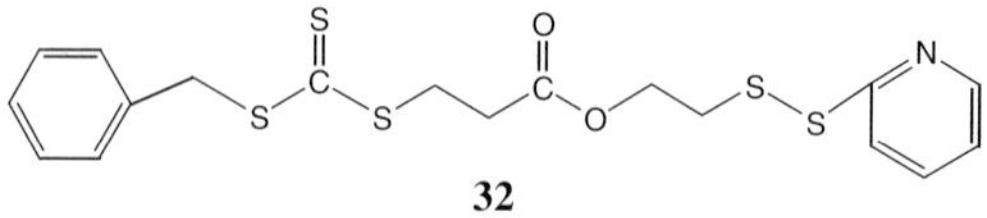 **32** | [26, 27] PEG-A, BA |

Another carboxylic acid end-functionalized dithioester has been described by Choe and coworkers [11]. Thiobenzoyl sulfanylmethylbenzoic acid (**12**; see Table 12.1) can be applied in the RAFT miniemulsion polymerization of MMA [12] and St [13] to yield carboxylic acid-functionalized nanoparticles.

Functional dithioesters can also be applied to synthesize hydroxyl-terminated polymers. Lima et al. [14] utilized a monofunctional hydroxyl RAFT agent (**13**; see Table 12.1) to synthesize α,ω-telechelic PMA. First, the polymerization of MA was mediated by **13**, resulting in α-hydroxyl-terminated PMA. The dithioester end group was subsequently modified by an aminolysis followed by a Michael addition on the resulting thiol, leading to α,ω-dihydroxyl-terminated PMA.

## 12.2.2
## Primary Amine and Amide End-Functional Polymers

α-Amide-terminated polymers can be directly synthesized using the RAFT agent **14** (*S*-diethylcarbamoylphenylmethyl dithiobenzoate) containing an amide functionality in the R group (see Table 12.1) [15]. The RAFT agent **14** mediates the controlled polymerization of styrene, acrylate and acrylamide derivatives. Block copolymers can also be synthesized using the RAFT agent **14**.

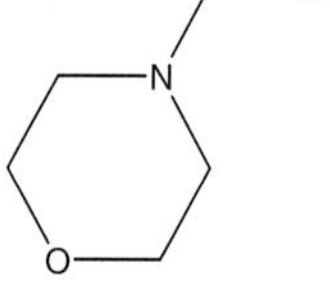

**11**

**Scheme 12.2** *N*-Acryloylmorpholine (NAM).

Primary amine end-functionalized polymers are not directly accessible through RAFT polymerization as unprotected primary and secondary amine groups react rapidly with the thiocarbonylthio group of the RAFT agent [3]. However, primary amine functionality can be introduced via phthalimido RAFT agents [16, 17]. RAFT agents **15–17** (see Table 12.1) are effective mediators for the living polymerization of St, BA and NIPAM. After RAFT polymerization the trithiocarbonate functionality in the polymers is transformed into inert chain ends by radical-induced reduction with tributylstannane and subsequently converted into primary amine end groups by hydrazinolysis. The xanthate **18** controls the polymerization of *N*-vinyl pyrrolidone (NVP) and vinyl acetate (VAc) [17].

### 12.2.3
### Other End-Functional Polymers

Fluorescent end-labeled polymers can be synthesized via the RAFT technique by mediation of a RAFT agent with a methyl anthracene R group [18]. Anthracene-9-yl methyl benzodithioate (AMB, **19**; see Table 12.1) controls the RAFT polymerization of St. The resulting polymers showed enhanced fluorescence properties compared to AMB in *N,N*-dimethylformamide.

The synthesis of $\alpha,\omega$-1,3-benzodioxole end-functionalized PSt via mediation of the RAFT agent **21** (see Table 12.1) was reported by Zhou et al. [19].

The RAFT agent pyrenylmethyl benzodithioate (**21**) also mediates the polymerization of St [20]. The resulting $\alpha$-fluorescent end-labeled polymers exhibit enhanced fluorescence properties in chloroform.

RAFT polymerization can also be applied for the synthesis of supramolecular metallopolymer complexes. Zhou and Harruna synthesized a terpyridine-functionalized dithioester (**22**; see Table 12.1) and proved that this RAFT agent controls the polymerization of St and NIPAM, yielding $\alpha$-terpyridine-functionalized polymers [29]. The terpyridine end-functionalized PSt can be utilized to generate supramolecular dimeric polystyrene ruthenium complexes and – in combination with terpyridine end-functionalized PNIPAM – amphiphilic diblock metallopolymers. This concept was adopted by Zhang et al. [21] who synthesized the bisfunctional terpyridine RAFT agent **23** which mediates the polymerization of St and BA, yielding $\alpha,\omega$-terpyridine functional polymers. Chelating interaction between the terpyridine end groups and $Ru^{II}$ ions results in supramolecular metal polymers.

Another attractive functional group is the allyl group as it can be utilized in various addition reactions leading to other functional groups. In addition, allyl end-functional polymers can be used as macromonomers to prepare graft copolymers [30]. Zhang and Chen [22] synthesized a bisallyl trithiocarbonate (**24**; see Table 12.1) which is able to mediate the polymerization of St and BA, resulting in bisallyl-functional telechelic homo- and triblock copolymers. They also showed that the allyl groups can be transformed into bromides and subsequently into azido groups. In addition, reaction of the bisallyl PSt with divinyl benzene resulted in PSt stars with allyl end-functional arms.

The concept of synthesizing macromonomers via RAFT polymerization was also investigated by Patton and Advincula [23]. They developed dithioesters (see Table 12.1) that carry norbornenyl- (**25**), vinyl- (**26**) and cinnamyl- (**27**) functionalities in the R groups of the RAFT agents. Macromonomers with norbornene and vinyl end groups can be applied for the construction of complex architectures via ROMP and metathesis polymerization. The cinnamate group can be utilized in the preparation of photo-cross-linkable polymer networks and stabilized polymer nanostructures. Patton and Advincula showed that the three RAFT agents **25–27** mediate the polymerization of St, MMA and MA.

### 12.2.4
### Toward Biomolecule–Polymer Conjugates via End-Functional RAFT Polymers

Recently, several research groups have started to employ the RAFT technique toward the preparation of biomolecule–polymer conjugates [24–27]. The combined properties of polymer and biomolecules may lead to unique properties and may subsequently be employed in applications such as affinity separations, immunoassays, enzyme recovery and bioengineering [25].

Bathfield et al. [24] developed two novel functional RAFT agents (see Table 12.1): one with a carbohydrate derivative R group (**28**) and one with a biotin derivative R group (**29**). These functional RAFT agents were synthesized via a precursor RAFT agent bearing a succinimidyl-activated ester group which reacts with amino derivatives without competitive degradation of the thiocarbonylthio function. The RAFT polymerization of NAM was successfully mediated via RAFT agents **28** and **29**.

Hong and Pan [25] developed a biotinylated trithiocarbonate RAFT agent (**30**; see Table 12.1) which efficiently mediates the polymerization of NIPAM and *N*-(2-hydroxypropyl) methacrylamide (HPMA). Biotin-ended diblock copolymers of PHPMA and PNIPAM were also successfully synthesized.

Liu et al. synthesized a biohybrid RAFT agent (**31**), that is, a bovine serum albumin RAFT agent (BSA-RAFT agent), which is able to control the $\gamma$-radiation-initiated polymerization of poly(ethylene glycol) acrylate (PEG-A) at room temperature (see Scheme 12.3) [26]. A novel RAFT agent (**32**) consisting of a trithiocarbonate with a pyridyl disulfide-modified Z group and a benzyl R group is reactive toward the selective exchange reaction with free thiol-tethered molecules under mild conditions. Reaction of **32** with one free thiol group bearing BSA forms the BSA-RAFT agent.

The $\omega$-pyridyl disulfide-functionalized RAFT agent **32** also mediates the polymerization of PEG-A and BA resulting in pyridyl disulfide-functionalized homo- and amphiphilic block copolymers [27]. Subsequently, the pyridyl disulfide end groups were reacted with a thiol-containing model compound, that is, 11-mercapto-1-undecanol. It should be possible to apply this technique to the development of tailor-designed biomolecule–polymer conjugates.

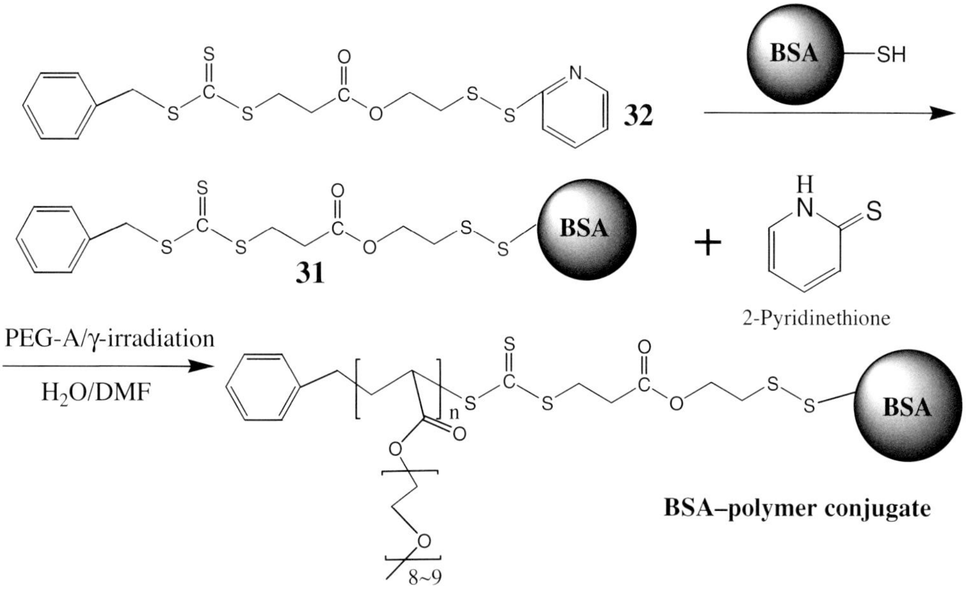


**Scheme 12.3** Site-specific modification of BSA with a pyridyl disulfide-terminated RAFT agent and the polymerization in situ of oligo(ethylene glycol) acrylate [26].

## 12.3 RAFT in Combination with Other Polymerization Techniques

Synthetic strategies that combine RAFT with other (controlled/living) polymerization techniques are highly attractive for the preparation of block copolymers of mechanistically incompatible monomers or monomers of extremely disparate reactivities (e.g. St and VAc) and polymers of complex architecture as star, brush and graft copolymers. To date, reports of combining RAFT polymerization with ROP, ATRP and highly orthogonal [2+3] cycloaddition (i.e. 'click chemistry') were reported.

### 12.3.1 RAFT in Combination with ROP

In 2003, Pan and coworkers reported the combination of RAFT and ROP for the synthesis of poly(MMA)/poly(1,3-dioxepane)/PSt ABC miktoarm star copolymers [31]. In their synthetic strategy they used a PSt with two terminal chain-transfer groups: one group which is able to mediate RAFT polymerization of MMA and one group which functions as CTA for the cationic ROP of 1,3-dioxepane. Pan and coworkers also synthesized poly(ethylene oxide) methyl ether/PSt/poly(L-lactide) ABC miktoarm star copolymers via this synthetic strategy [32].

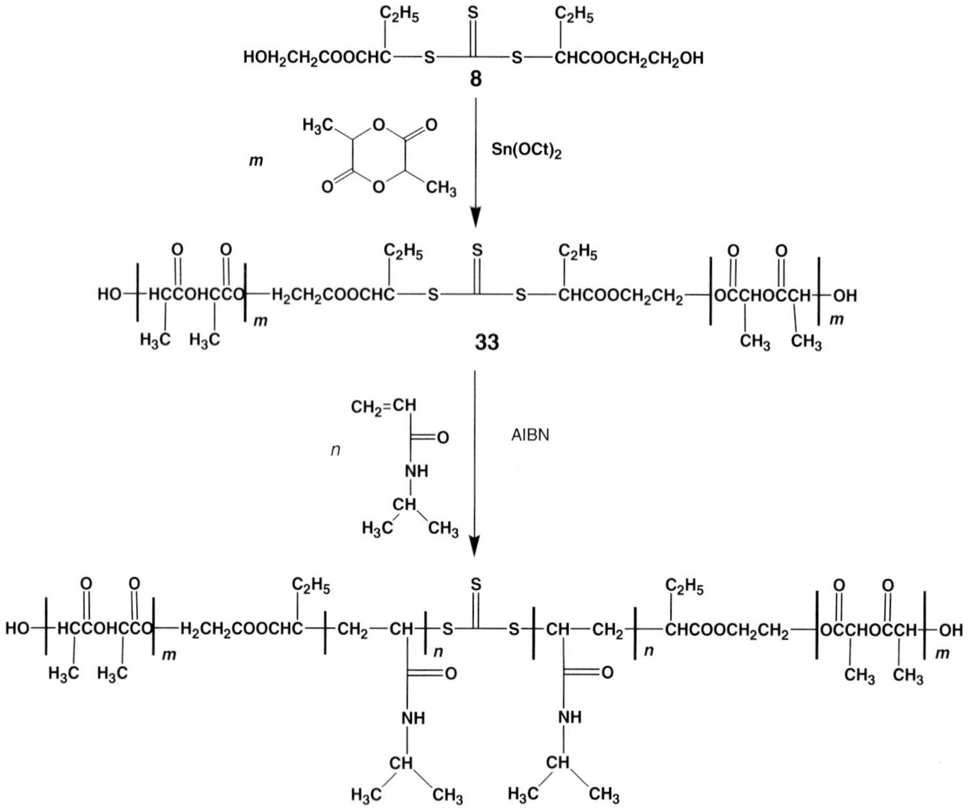

**34**

**Scheme 12.4** Synthetic strategy for thermally responsive and biodegradable block copolymers via the combination of ROP and RAFT. Reprinted with permission from Ref. [9]. Copyright 2004 American Chemical Society.

In 2004, the group of Pan also developed a new RAFT agent (BHBT, **8**) with two terminal hydroxyl groups [9]. On the one hand, the terminal hydroxyl groups of BHBT can act as initiating centers in ROP of L-lactide. On the other hand, BHBT is a RAFT agent and can be used to mediate, for example, the polymerization of NIPAM. First, BHBT was used to synthesize PLA with a centered trithiocarbonate (**8**; see Scheme 12.4). Second, the RAFT polymerization of NIPAM was performed using **33**, resulting in thermally sensitive and biodegradable PLLA–*b*-PNIPAM–*b*-PLLA block copolymers **34**.

H-shaped copolymers and heteroarm H-shaped terpolymers were also synthesized via a combination of RAFT polymerization and (cationic) ROP by Pan and coworkers [33, 34].

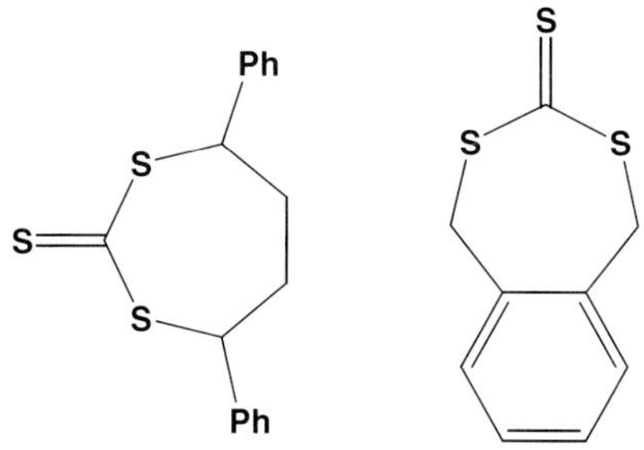

35 36

**Scheme 12.5** Cyclic trithiocarbonates (CTTC).

Xu and Huang applied sequential controlled polymerization (i.e. RAFT polymerization followed by ROP) to synthesize well-defined graft copolymers [35]. First, they prepared poly(2-hydroxyethylmethacrylate-*co*-styrene) (poly(HEMA-*co*-St)) via RAFT and subsequently used the hydroxyl groups of the poly(HEMA-*co*-St) for the ROP of ε-caprolactone in the presence of stannous octoate. Huang and coworkers also reported the synthesis of well-defined amphiphilic copolymers [36, 37] and well-defined, brush-type copolymers [38] via sequential controlled polymerization.

Hillmyer and coworkers developed a synthetic strategy where they first performed a controlled ROP, then modified the resulting polymer into a macro-RAFT agent and subsequently performed a RAFT polymerization [39, 40]. They named this synthetic strategy 'tandem ROMP-RAFT' and reported the synthesis of ABC triblock and ABA triblock copolymers of mechanistically incompatible monomers.

Hales et al. [41] prepared poly(D,L-lactide)-*b*-poly(NIPAM) using a two-step approach. First they utilized a RAFT agent, 2-(benzylsulfanylthiocarbonylsulfanyl) ethanol, as a coinitiator for the ROP of 3,6-dimethyl-1,4-dioxane-2,5-dione, resulting in a polylactide (D,L-PLA) macro-RAFT agent. This macro-RAFT agent was subsequently used to polymerize NIPAM, resulting in amphiphilic block copolymers.

Hong et al. reported the synthesis of cyclic trithiocarbonates (CTTC) **35** and **36** (see Scheme 12.5), which can be used for the polymerization of St [42, 43].

CTTC **35** was 'copolymerized' with St via an ROP (see Scheme 12.6), incorporating trithiocarbonate moieties. Subsequently, a RAFT polymerization of St via the incorporated trithiocarbonate moieties was performed. After cleaving the trithiocarbonate moieties by a method suggested by Wang et al. [42], low-polydispersity polymers with thiol end groups were obtained.

### 12.3.2
### RAFT in Combination with ATRP

Pan and coworkers successfully synthesized branched copolymers with PLLA-*b*-$PSt_2$ branches by a combination of RAFT, ROP and ATRP [44]. They first prepared the copolymer poly(MA-*co*-HEA) via RAFT and subsequently performed the ROP

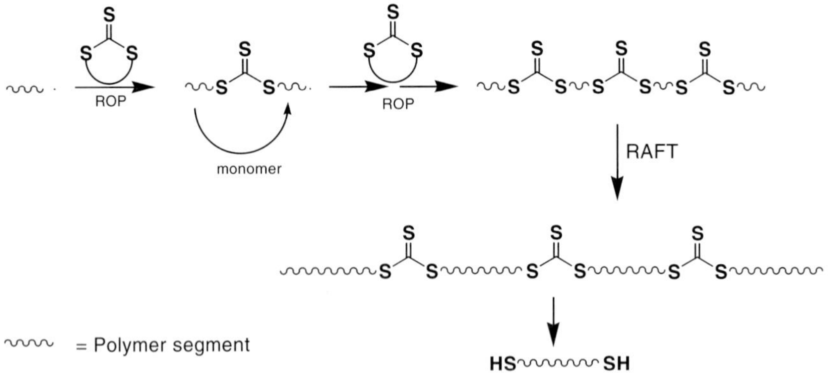


**Scheme 12.6** Integrated process of ring-opening and RAFT polymerization involving CTTC [43].

of L-lactide by using the side hydroxyl group of HEA as initiator and $Sn(Oct)_2$ as catalyst. 2,2-Bis(methylene α-bromoisobutyrate) propionyl chloride was subsequently used to transform one hydroxyl group of the PLLA branch into two ATRP-initiating sites, followed by an ATRP polymerization of St, yielding poly(MA-*co*-HEA)-*g*-(PLLA-*b*-$PSt_2$). Neoh and coworkers synthesized a dual-brush-type amphiphilic triblock copolymer with intact epoxide functional groups by consecutive RAFT polymerization and ATRP [45]. First they synthesized a diblock copolymer of poly(glycidyl methacrylate)-*b*-poly(4-vinylbenzyl chloride) (poly(GMA)-*b*-poly(VBC)) via a two-step RAFT polymerization. A third block, consisting of poly(ethylene glycol) methyl ether methacrylate (PEGMA) units, was synthesized via RAFT. The poly(ethylene glycol) side chains of PEGMA form the hydrophilic side chains of the triblock copolymer. Subsequently, a heterogeneous dual-brush-type amphiphilic triblock copolymer was synthesized by using the benzyl chloride group of the VBC units of the poly(VBC) block as an ATRP macroinitiator for the polymerization of St [46]. Neoh and coworkers also applied a consecutive ATRP/RAFT approach to graft a micropatterned, binary polymer brush system from a Si(100) chip [46].

Bon and coworkers reported an interesting strategy to convert macromolecular ATRP initiators into macro-RAFT agents [47]. Poly(MMA), poly(DMAEMA) (i.e. poly(*N,N*-dimethylethylaminoethyl methacrylate)) and PEG ATRP initiators were activated using $Cu^{I}Br$ under modified ATRP conditions. The activated polymer chains were subsequently reacted with bis(thiobenzoyl) disulfide lead to yield RAFT end functionality (see Scheme 12.7).

### 12.3.3
### RAFT in Combination with 'Click Chemistry'

Recently, RAFT agent-mediated polymerization has also been combined with highly orthogonal [2+3] cycloadditions (also called 'click chemistry' [48]). The combination

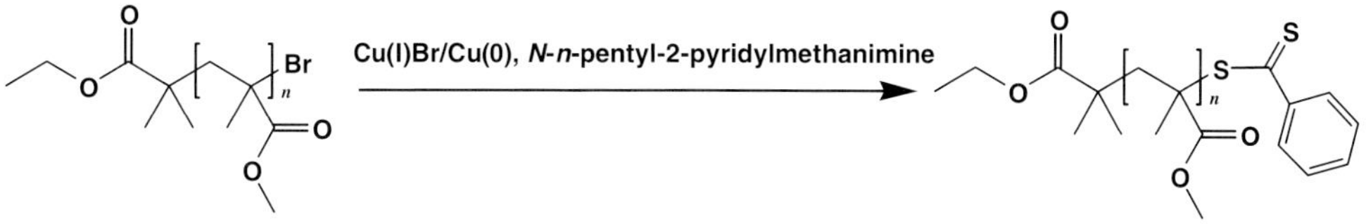


**Scheme 12.7** Conversion of macromolecular ATRP initiators into macro-RAFT agents. Reprinted with permission from Ref. [47]. Copyright 2004 Elsevier Ltd.

of these two techniques allows for the synthesis of well-defined block copolymers of monomers with extremely disparate reactivities [49]. Quémener et al. synthesized two novel RAFT agents carrying azide or acetylene functions (see Scheme 12.8) which allow subsequent click reactions. They showed that this synthetic route can be applied to synthesize very narrow polydispersity PSt-*b*-PVAc block copolymers.

Gondi et al. synthesized two novel azido-functionalized RAFT agents (see Scheme 12.9) and used them to mediate the RAFT polymerization of St and DMA (*N*,*N*-dimethylacrylamide) [50]. Subsequently, the resulting α-azido-terminated polymers were reacted with acetylene species, synthesizing a range of telechelics.

## 12.4
## Stability of the Thiocarbonylthio End Group and Its Modification *Post*polymerization

One of the characteristic features of the RAFT process is the incorporation of the thiocarbonylthio moiety from the CTA at the chain end of all the living polymeric chains. Such an end group provides a fantastic handle for further chemical modification, allowing for the incorporation into the polymeric chains of chain-end functionalities *post*polymerization. This end group usually confers color to the polymer (ranging from purple for dithiobenzoate to yellow for trithiocarbonates, xanthates or dithiocarbamates; see Figure 12.1), which might not be desirable

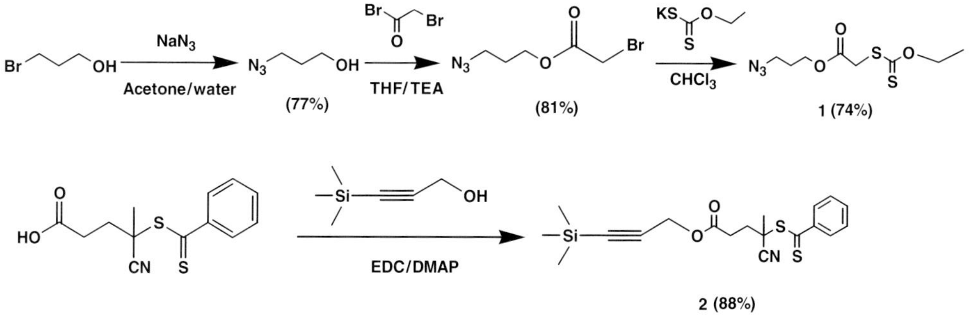


**Scheme 12.8** Synthesis of 'clickable' RAFT agents. Ref. [49] – Reproduced with permission of the Royal Society of Chemistry.

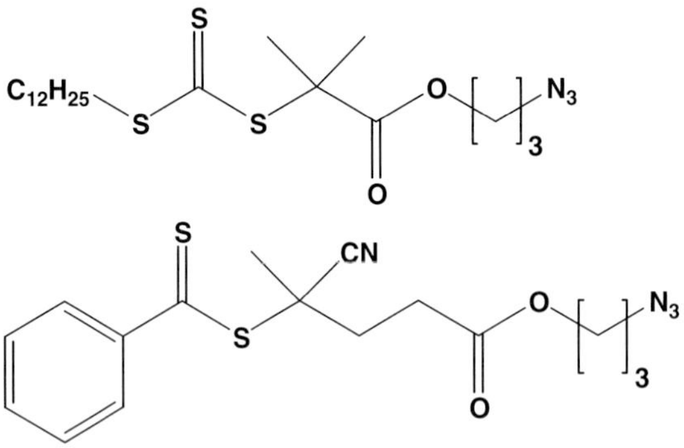


**Scheme 12.9** Azido-functionalized RAFT agents. Reprinted with permission from Ref. [50]. Copyright 2007 American Chemical Society.

depending on the applications. The thiocarbonylthio end group can weaken the thermal stability of the polymeric chains. There are a number of options to remove/transform the thiocarbonylthio end group from polymers produced by RAFT (Scheme 12.10).

### 12.4.1
### Thermolysis

The thermal decomposition of RAFT agents during polymerization has been studied. It was shown that cumyl dithiobenzoate undergoes thermal decomposition above 120 °C to yield dithiobenzoic acid and α-methyl styrene [52]. When used in the polymerization of MMA, the cumyl dithiobenzoate is converted into a polymeric dithiobenzoate, which also yields an unsaturated group at the polymeric chain end and dithiobenzoic acid at temperatures above 120 °C. However, surprisingly, 2-(ethoxycarbonyl)prop-2-yl dithiobenzoates, for which the R group mimics the repeating unit of a PMMA chain, do not yield an unsaturated group as a product

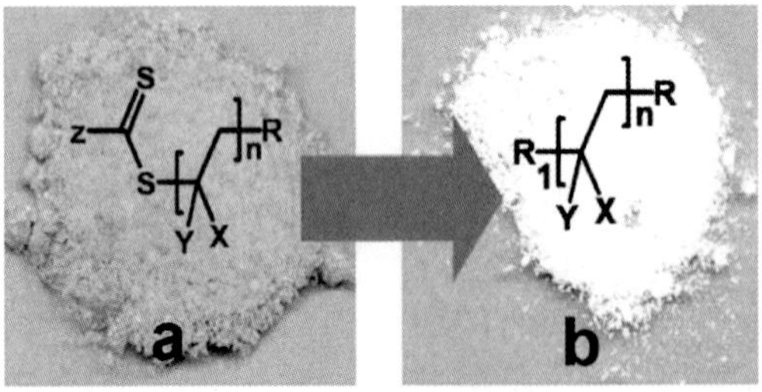


**Fig. 12.1** Typical physical aspect of a polymer synthesized by the RAFT process before (a) and after (b) removal of the thiocarbonyl\thio end group. Here a PMMA was synthesized by the RAFT process utilizing isocyano dithiobenzoate and treated with an excess of *N*,*N*′-azobis(isobutyronitrile) (AIBN) at 80 °C for 2.5 h [51].

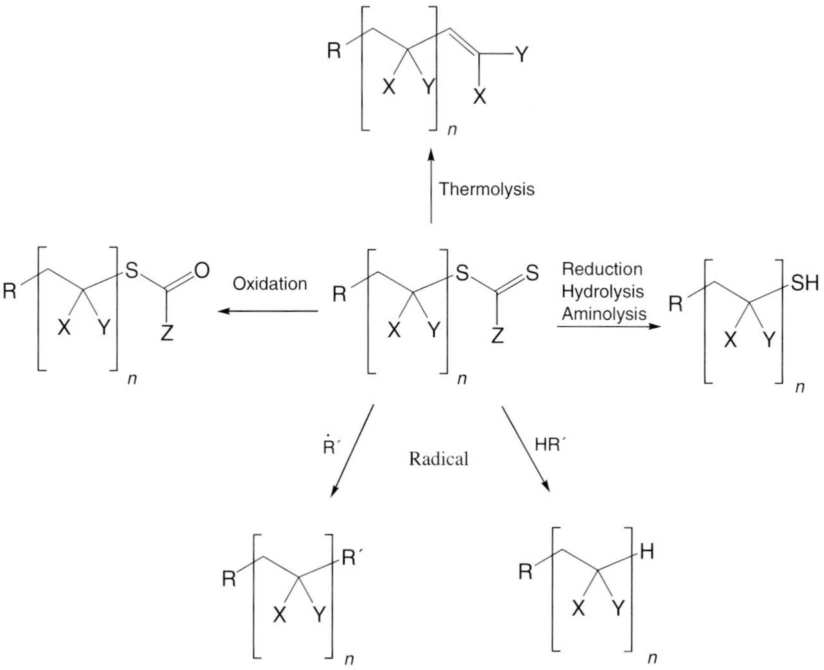


**Scheme 12.10** Various paths for the modification of the thiocarbonylthio end group of a RAFT-synthesized polymeric chain.

of thermolysis at these temperatures [53]. Nevertheless, the decomposition of the thiocarbonylthio group during MMA polymerization affects the control of the polymerization and yields to retarded reaction rates and broadened molecular weight distributions [53]. However, the St polymerization mediated by cumyl dithiobenzoate at high temperatures (120, 150 and 180 °C) was not affected by thermal decomposition of the CTA [52, 54]. This was attributed to the fast conversion of the initial CTA into a more stable polymeric CTA [52]. This observation is confirmed by the fact that 1-phenylethyl dithiobenzoate (for which the R group mimics the repeating unit of a PSt chain) and benzyl dithiobenzoate do not decompose when heated at 120 °C [52].

Legge et al. demonstrated that the Z groups also affect the thermal stability of thiocarbonylthio groups, which varies greatly across the classes, with most being stable above 150 °C [55]. In general, the order of thermal stability was found to be dithiobenzoates > trithiocarbonates > xanthates. The authors showed that when Z is an aromatic group (phenyl, carbazole, etc.), the thermal stability is much improved when compared to analogous trithiocarbonates or xanthates. R groups based on acrylates were found to confer relative instability to the thiocarbonylthio group, with the trithiocarbonate and xanthate derivatives decomposing below 100 °C. However, substitution of the thiocarbonylthio group with a phenylacetate R group was shown to improve the stability of both trithiocarbonates and dithiobenzoates. As a general

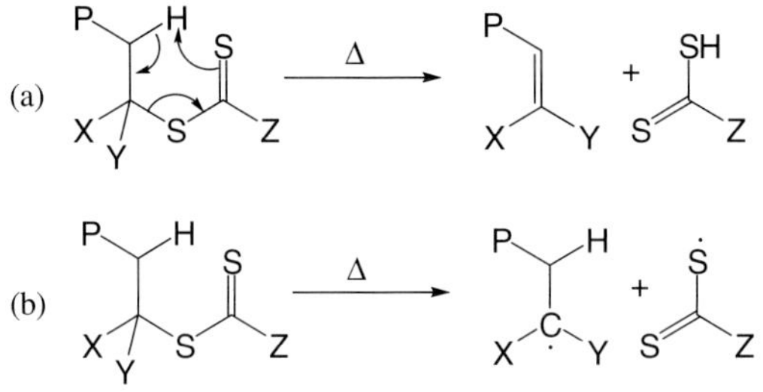

Scheme 12.11 Thermolysis of thiocarbonylthio group via (a) a concerted elimination (Chugaev reaction [58, 59]) or (b) C—S bond homolysis.

rule, the carbon–sulfur single bond is the most labile bond and factors that effect its strength include strongly electron-donating Z groups (e.g. dithiobenzoates) and high-energy penalty for the formation of breakdown products (e.g. methylbenzene carbocations).

The thermal lability of the thiocarbonylthio group at the chain end of a RAFT-synthesized polymer also offers a simple and efficient way of removing the end group after polymerization. In the case of PMMA, it has been shown that the decomposition of the thiocarbonylthio group depends strongly on the CTA used during polymer synthesis [14, 53, 56, 57]. The loss of the end group in the melt occurs around 180 °C and involves homolysis of the terminal C—S bond and subsequent depropagation in the case of a trithiocarbonate end group, while a polymer showing a dithiobenzoate end group seems more stable, and solely the thiocarbonylthio moiety is removed via a concerted elimination (Chugaev reaction [58, 59]; Scheme 12.11) [56].

In the case of PSt terminated by a trithiocarbonate group [16, 60, 61], the concerted elimination reaction is preponderant, while in the case of PBA terminated by a trithiocarbonate [17], C—S bond homolysis seems to be at the origin of the thiocarbonylthio group loss. In the case of polymers showing a xanthate as end group, the mechanism of degradation also depends on the structure of the polymer. The reported mechanism for the thermolysis of a PSt and poly(*tert*-butyl acrylate) terminated by an *O*-isobutyl xanthate involves selective elimination to form 2-butene and a polymer with a thiol end group [62], while in the case of a PVAc showing an *O*-ethyl xanthate as end group, it is suggested that homolysis of the C—S bond triggers the degradation [17].

### 12.4.2
### Transformation to Thiol

Hydrolysis, under basic or acidic conditions, is a commonly used reaction for the conversion of thioesters into thiols. Such reaction must be taken into account when undertaking RAFT polymerization in an aqueous environment, as hydrolysis of the terminal thiocarbonylthio group during reaction leads to elimination of

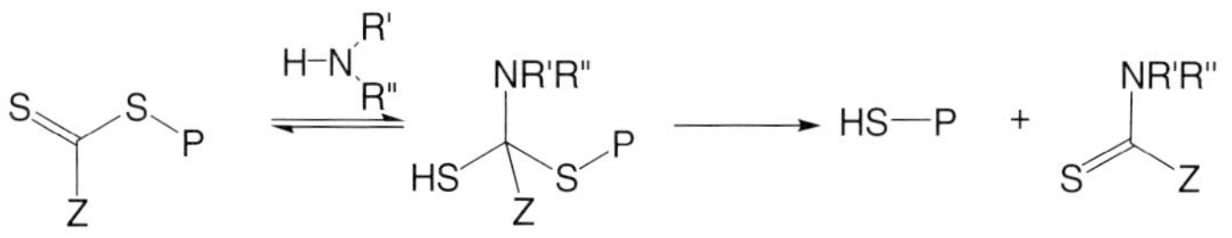


**Scheme 12.12** Aminolysis of pendant dithioester groups on a polymeric chain.

the active end groups necessary for maintaining livingness of the polymerization. Thomas et al. have determined the rate constants of hydrolysis and aminolysis for a dithiobenzoate derivative (cyanopentanoic acid dithiobenzoate) and the macro-chain-transfer agents of poly(sodium 2-acrylamido-2-methylpropanesulfonate) and poly(acrylamide) at selected pH values [63]. The authors found that the rates of hydrolysis and aminolysis both increase with increasing pH and decrease with increasing the molecular weight of the dithioester derivatives.

In addition, hydrolysis is a powerful method to remove the thiocarbonylthio group of RAFT-synthesized polymers to yield thiol-terminated polymeric chains. Generally, a strong base such as sodium hydroxide is used [64–66]. However, this approach leads to side reactions, with the formation of disulfide bridges; base-catalyzed elimination to form vinyl end groups; and cyclization to form cyclic lactones or thiolactones [65]. Alternatively, the hydrolysis can be catalyzed by an acid, as in the work reported by Hruby et al. who showed the acid hydrolysis of dithiocarbonate pendant groups at 90 °C using 35% hydrochloric acid [67].

An alternative route to the transformation of thiocarbonylthio groups into thiols is aminolysis. Either primary or secondary amines, acting as nucleophiles, can convert a thiocarbonylthio moiety to a thiol, and this has been applied to a variety of RAFT-synthesized polymers (Scheme 12.12) [14, 57, 60, 68–73].

The transformation of the polymeric end groups into thiols often sees side reactions through the oxidation of thiols into disulfides (for instance if traces of oxygen are present in the reaction medium). Such reactions can lead to bimolecular molecular weight distributions due to oxidative coupling. This approach was used to reversibly produce monocyclic and linear multiblock polystyrene after reduction of the thiocarbonylthio end groups of a polymer prepared via RAFT polymerization of St mediated by a difunctional CTA [74]. A similar approach was utilized to produce multiblock copolymers of PBA and PMMA from telechelic poly(BA-*b*-MMA-*b*-BA) via sequential RAFT polymerization mediated by a difunctional CTA, followed by reduction–oxidation reactions [75]. In the case of methacrylate derivatives (PMMA, poly(*N*,*N*-dimethylaminoethyl methacrylate) and poly(lauryl methacrylate)), however, Xu et al. have shown experimentally that the thiol end groups may cyclize through backbiting to form a thiolactone [76].

To prevent the oxidation of thiols into sulfide, reducing reagents can be introduced into the system, such as sodium bisulfite [14, 57], Zn/acetic acid [60, 71] or tris(2-carboxyethyl) phosphine·HCl (TCEP·HCl) [77]. Reduction of the thiocarbonylthio group into a thiol can also be achieved in the presence of metal hydride compounds (e.g. $LiAlH_4$ or $NaBH_4$). This route was illustrated with a range of water-soluble polymers synthesized via aqueous RAFT and their reaction with $NaBH_4$

at ambient temperature for 1 h. The resulting thiol end-functionalized polymers were used for the stabilization of gold nanoparticles [78–80] or were further reacted with a maleimide moiety [77]. Postmodification of thiol-terminated polymers was also illustrated by the reaction of a telechelic PNIPAM synthesized via RAFT with butylamine in the presence of TCEP. Further modification of the resulting thiol end groups with α,β-unsaturated carbonyl derivatives, via Michael addition, yielded telechelic polymers bearing isobutylsulfanylthiocarbonylsulfanyl end groups [72].

### 12.4.3
### Oxidation-Induced End-Group Removal

The oxidation of thiocarbonylthio group is reported to give sulfines that decompose to thioesters and elemental sulfur [81, 82]. Vana et al. used this approach to modify the dithiobenzoate end group of a PMA by reacting it with *tert*-butyl hydroperoxide and transforming the C=S group to C=O [83].

### 12.4.4
### Radical-Induced End-Group Removal

Free radicals can also be used to modify the thiocarbonylthio end group of a RAFT-synthesized polymer. The generic approach consists in reacting polymeric chains end-capped with a thiocarbonylthio group with an excess of radicals. The in situ addition of a radical to the reactive C=S bond of the thiocarbonylthio polymer end group leads to the formation of an intermediate radical, which can then fragment back either to the original attacking radical or toward the polymeric chain radical. The polymeric radical can then react with a trapping group and terminate.

The process has been applied to free radical reducing agents, consisting of a free radical source and a hydrogen atom donor, to replace the thiocarbonylthio group with hydrogen. To avoid side reactions, such as bimolecular terminations between polymeric radicals, an efficient hydrogen atom donor is required (Scheme 12.13). Tri-*n*-butylstannane [16, 60] is one of the most efficient of such reactants, but unfortunately generates toxic by-products during reaction. Tris(trimethylsilyl)silane was tested as an alternative source of hydrogen atom in the case of PSt, but was less efficient and bimolecular terminations were observed [16]. Hypophosphite salts were also shown to be very efficient free radical reducing agents for polymers of acrylate and St derivatives [84]. An alternative approach is to couple a source of radical with a strong hydrogen donor, such as a peroxide initiator and a secondary alcohol. Peroxides can induce radical cleavage of the C—S bond of a thiocarbonylthio group to generate a radical intermediate, which can undergo either intramolecular cyclization or intermolecular addition, depending on the conditions [85]. By performing the reaction in the presence of secondary alcohol, the radical formed from the fragmentation of the polymeric chains from the RAFT intermediate can be trapped by a hydrogen atom from the alcohol. The principle was demonstrated

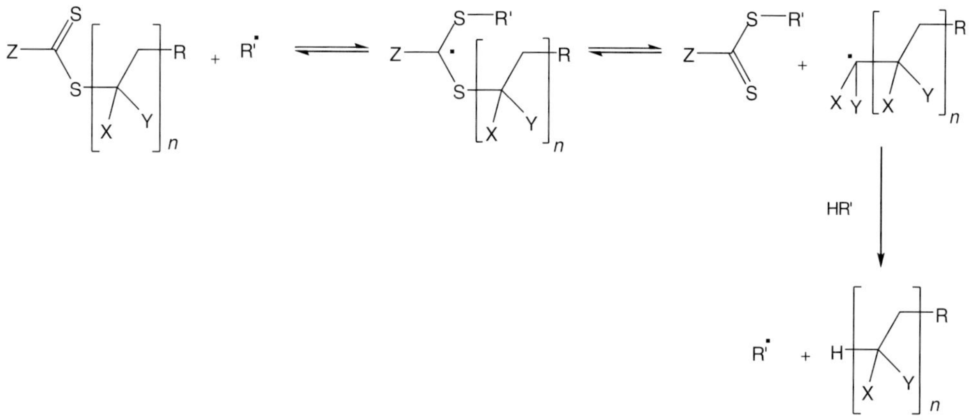


**Scheme 12.13** Substitution of the thiocarbonylthio end group by hydrogen via a radical process.

with xanthate-terminated poly(dimethyl sulfoxide) and PAA synthesized by MADIX (i.e. *ma*cromolecular *d*esign via *i*nterchange of *x*anthates) and further reacted with peroxide (di-*tert*-butyl peroxide (at 175 °C) or dilauryl peroxide (at 80 °C)) and percarbonate (butylcyclohexyl percarbonate (at 60 °C)) initiators in secondary alcohols as solvent (2-octanol and 2-propanol) [86].

Alternatively, the same process can be applied by utilizing an excess of free radical species. During the process of addition–fragmentation, and in the presence of an excess of free radicals, the RAFT equilibrium is displaced toward the formation of the polymeric chain radical, which can then recombine irreversibly with one of the free radicals present in excess in solution, thus forming a dead polymeric chain. This method not only allows for elimination of the thiocarbonylthio polymeric end group, but also introduces a new functionality at the end of the polymeric chain, provided by the free radical introduced in the system. Furthermore, the same process permits the recovery of the CTA during the same process, as shown in Scheme 12.14 [51].

### 12.4.5
### Radiation-Induced End-Group Removal

RAFT end groups can also be removed by UV irradiation, but their ability to decompose depends on their structures [87–89]. Lu et al. showed that trithiocarbonates such as *S*,*S*′-bis(α,α′-dimethyl-α″-acetic acid) trithiocarbonate (**1**) and *S*-1-dodecyl-*S*′-(α,α′-dimethyl-α″-acetic acid) trithiocarbonate (**2**) are relatively stable under UV irradiation, while dithioesters such as cumyl dithiobenzoate and 2-cyanoprop-2-yl(4-fluoro) dithiobenzoate are strongly sensitive to UV radiation at wave ranges around their characteristic absorption wavelength [87–89]. Among dithioesters, it seems that dithiobenzoates are the most sensitive to UV, as illustrated in a study

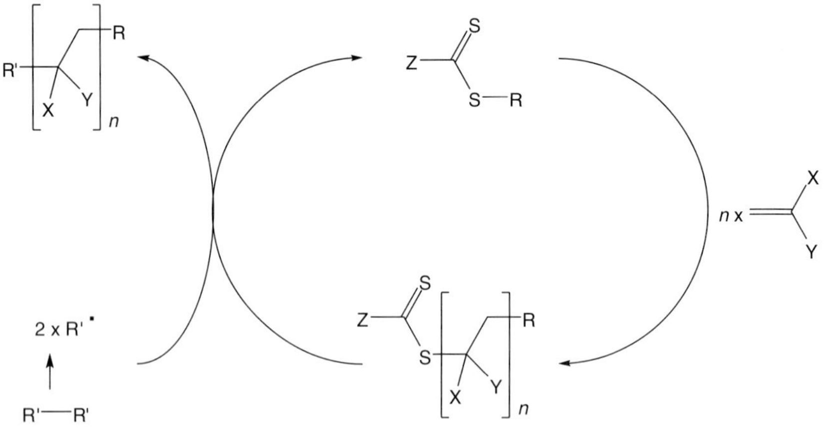


**Scheme 12.14** Reaction cycle to produce chain-end functional polymers via RAFT/MADIX, with recovery of the CTA.

by Quinn et al. who showed that 1-phenylethyl dithiobenzoate decomposes much faster than 1-phenylethyl phenyldithioactetate during polymerization. A proposed mechanism for their decomposition involves the formation of phenyl ethyl and benzyl radicals, which may initiate polymerization [87–89]. To date, decomposition of RAFT end groups under UV radiation have only been reported during polymerization, and led to a loss of control over molecular weight distribution of the polymer.

## 12.5 Conclusion

The RAFT process allows the easy introduction of end-group functionalities on polymeric chains, either by modifying the CTA which mediates the polymerization (via the R and/or the Z group) or by modifying the thiocarbonylthio chain-end group *post*polymerization. The versatility of RAFT polymerization toward functionalities makes it possible to introduce an almost infinite number of $\alpha$- and $\omega$-functionalities on polymeric chains. Functional (RAFT) polymers also open the possibility to synthesize complex molecular architectures by the combination of RAFT polymerization with other polymerization techniques.

## 12.6 Abbreviations

| | |
|---|---|
| AA | Acrylic acid |
| AMB | Anthracene-9-yl methyl benzodithioate |
| ATRP | Atom transfer radical polymerization |
| BA | Butyl acrylate |

| | |
|---|---|
| BHBT | *S,S'*-Bis(2-hydroxyethyl-2'-butyrate) trithiocarbonate |
| CTA | Chain transfer agent |
| CTTC | Cyclic trithiocarbonate |
| DMA | *N,N*-Dimethylacrylamide |
| DMAEMA | *N,N*-Dimethylethylaminoethyl methacrylate |
| EA | Ethyl acrylate |
| GMA | Glycidyl methacrylate |
| HEA | 2-Hydroxyethyl acrylate |
| HEMA | 2-Hydroxyethylmethacrylate |
| HPMA | *N*-(2-Hydroxypropyl) methacrylamide |
| MA | Methyl acrylate |
| MADIX | *Ma*cromolecular *d*esign via *i*nterchange of *x*anthates |
| MMA | Methyl methacrylate |
| NAM | *N*-Acryloylmorpholine |
| NIPAM | *N*-Isopropyl acrylamide |
| NMP | Nitroxide-mediated polymerization |
| NVP | *N*-Vinyl pyrrolidone |
| RAFT | Reversible addition–fragmentation chain transfer |
| ROP | Ring-opening polymerization |
| PEG | Poly(ethylene glycol) |
| PEG-A | Oligo(ethylene glycol) acrylate |
| PEGMA | Poly(ethylene glycol) methyl ether methacrylate |
| PLLA | Poly(L-lactide) |
| PMMA | Poly(methyl methacrylate) |
| PSt | Polystyrene |
| St | Styrene |
| tBAm | *tert*-Butyl acrylamide |
| TCEP | Tris(2-carboxyethyl) phosphine |
| VAc | Vinyl acetate |
| VBC | 4-Vinylbenzyl chloride |

## References


1. J. T., Lai, D., Filla, R., Shea *Macromolecules* **2002**, *35*, 6754–6756.
2. C. J., Ferguson, R. J., Hughes, B. T. T., Pham, B. S., Hawkett, R. G., Gilbert, A. K., Serelis, C. H., Such *Macromolecules* **2002**, *35*, 9243–9245.
3. G., Moad, Y. K., Chong, A., Postma, E., Rizzardo, S. H., Thang *Polymer* **2005**, *46*, 8458–8468.
4. M. H., Stenzel, T. P., Davis, A. G., Fane *J. Mater. Chem.* **2003**, *13*, 2090–2097.
5. R., Wang, C. L., McCormick, A. B., Lowe *Macromolecules* **2005**, *38*, 9518–9525.
6. M. H., Stenzel, C., Barner-Kowollik, T. P., Davis, H. M., Dalton *Macromol. Biosci.* **2004**, *4*, 445–453.
7. P.-E., Millard, L., Barner, M. H., Stenzel, T. P., Davis, C., Barner-Kowollik, A. H. E., Müller *Macromol. Rapid Commun.* **2006**, *27*, 821–828.
8. J., Liu, C., Hong, C.-Y., Pan *Polymer* **2004**, *45*, 4413–4421.
9. Y., You, C., Hong, W., Wang, W., Lu, C.-Y., Pan *Macromolecules* **2004**, *37*, 9761–9767.

**10.** F., D'Agosto, R., Hughes, M.-T., Charreyre, C., Pichot, R. G., Gilbert *Macromolecules* **2003**, *36*, 621–629.

**11.** S. E., Shim, Y., Shin, J. W., Jun, K., Lee, H., Jung, S., Choe *Macromolecules* **2003**, *36*, 7994–8000.

**12.** S. E., Shim, H., Lee, S., Choe *Macromolecules* **2004**, *37*, 5565–5571.

**13.** H., Lee, J. M., Lee, S. E., Shim, B., Hyung, S., Choe *Polymer* **2005**, *46*, 3661–3668.

**14.** V., Lima, X. L., Jiang, J., Brokken-Zijp, P. J., Schoenmakers, B., Klumperman, R., Van Der Linde *J. Polym. Sci., Part A: Polym. Chem.* **2005**, *43*, 959–973.

**15.** P., Takolpuckdee, C. A., Mars, S., Perrier, S. J., Archibald *Macromolecules* **2005**, *38*, 1057–1060.

**16.** A., Postma, T. P., Davis, R. A., Evans, G., Li, G., Moad, M. S., O'Shea *Macromolecules* **2006**, *39*, 5293–5306.

**17.** A., Postma, T. P., Davis, G., Li, G., Moad, M. S., O'Shea *Macromolecules* **2006**, *39*, 5307–5318.

**18.** N., Zhou, L., Lu, X., Zhu, X., Yang, X., Wang, J., Zhu, D., Zhou *Polym. Bull.* **2006**, *57*, 491–498.

**19.** N., Zhou, L., Lu, X., Zhu, X., Yang, X., Wang, J., Zhu, Z., Cheng *J. Appl. Polym. Sci.* **2006**, *99*, 3535–3539.

**20.** N., Zhou, L., Lu, J., Zhu, X., Yang, X., Wang, X., Zhu, Z., Zhang *Polymer* **2007**, *48*, 1255–1260.

**21.** L., Zhang, Y., Zhang, Y., Chen *Eur. Polym. J.* **2006**, *42*, 2398–2406.

**22.** L., Zhang, Y., Chen *Polymer* **2006**, *47*, 5259–5266.

**23.** D. L., Patton, R. C., Advincula *Macromolecules* **2006**, *39*, 8674–8683.

**24.** M., Bathfield, F., D'Agosto, R., Spitz, M.-T., Charreyre, T., Delair *J. Am. Chem. Soc.* **2006**, *128*, 2546–2547.

**25.** C., Hong, C.-Y., Pan *Macromolecules* **2006**, *39*, 3517–3524.

**26.** J., Liu, V., Bulmus, D. L., Herlambang, C., Barner-Kowollik, M. H., Stenzel, T. P., Davis *Angew. Chem. Int. Ed.* **2007**, *46*, 3099–3103.

**27.** J., Liu, V., Bulmus, C., Barner-Kowollik, M. H., Stenzel, T. P., Davis *Macromol. Rapid Commun.* **2007**, *28*, 305–314.

**28.** J. T., Lai, R., Shea *J. Polym. Sci., Part A: Polym. Chem.* **2006**, *46*, 4298–4316.

**29.** G., Zhou, I. I., Harruna *Macromolecules* **2005**, *38*, 4114–4123.

**30.** U., Schulze, T., Fónagy, H., Komber, G., Pompe, J., Pionteck, B., Iván *Macromolecules* **2003**, *36*, 4719–4726.

**31.** Y.-G., Li, Y.-M., Wang, C.-Y., Pan *J. Polym. Sci., Part A: Polym. Chem.* **2003**, *41*, 1243–1250.

**32.** P.-J., Shi, Y.-G., Li, C.-Y., Pan *Eur. Polym. J.* **2004**, *40*, 1283–1290.

**33.** J., Liu, C.-Y., Pan *Polymer* **2005**, *46*, 11133–11141.

**34.** D.-H., Han, C.-Y., Pan *J. Polym. Sci., Part A: Polym. Chem.* **2007**, *45*, 789–799.

**35.** X., Xu, J., Huang *J. Polym. Sci., Part A: Polym. Chem.* **2004**, *42*, 5523–5529.

**36.** X., Xu, J., Huang *J. Polym. Sci., Part A: Polym. Chem.* **2006**, *44*, 467–476.

**37.** X., Xu, J., Huang *J. Polym. Sci., Part A: Polym. Chem.* **2006**, *44*, 1048–1048.

**38.** X., Xu, Z., Jia, R., Sun, J., Huang *J. Polym. Sci., Part A: Polym. Chem.* **2006**, *44*, 4396–4408.

**39.** J., Rzayev, M. A., Hillmyer *Macromolecules* **2005**, *38*, 3–5.

**40.** M. K., Mahanthappa, F. S., Bates, M. A., Hillmyer *Macromolecules* **2005**, *38*, 7890–7894.

**41.** M., Hales, C., Barner-Kowollik, T. P., Davis, M. H., Stenzel *Langmuir* **2004**, *20*, 10809–10817.

**42.** J., Hong, Q., Wang, Y., Lin, Z., Fan *Macromolecules* **2005**, *38*, 2691–2695.

**43.** J., Hong, Q., Wang, Z., Fan *Macromol. Rapid Commun.* **2006**, *27*, 57–62.

**44.** B., Luan, B.-Q., Zhang, C.-Y., Pan *J. Polym. Sci., Part A: Polym. Chem.* **2005**, *44*, 549–560.

**45.** Z., Cheng, X., Zhu, G. D., Fu, E. T., Kang, K. G., Neoh *Macromolecules* **2005**, *38*, 7187–7192.

**46.** F. J., Xu, E. T., Kang, K. G., Neoh *J. Mater. Chem.* **2006**, *16*, 2948–2952.

**47.** C. M., Wager, D. M., Haddelton, S. A. F., Bon *Eur. Polym. J.* **2004**, *40*, 641–645.

**48.** H. C., Kolb, M. G., Finn, K. B., Sharpless *Angew. Chem. Int. Ed.* **2001**, *40*, 2004–2021.

**49.** D., Quémener, T. P., Davis, C., Barner-Kowollik, M. H., Stenzel *Chem. Commun.* **2006**, 5051–5053.

**50.** S. R., Gondi, A. P., Vogt, B. S., Sumerlin *Macromolecules* **2007**, *40*, 474–481.

**51.** S., Perrier, P., Takolpuckdee, C. A., Mars *Macromolecules* **2005**, *38*, 2033–2036.

**52.** Y., Liu, J., He, J., Xu, D., Fan, W., Tang, Y., Yang *Macromolecules* **2005**, *38*, 10332–10335.

**53.** J., Xu, J., He, D., Fan, W., Tang, Y., Yang *Macromolecules* **2006**, *39*, 3753–3759.

**54.** T., Arita, M., Buback, P., Vana *Macromolecules* **2005**, *38*, 7935–7943.

**55.** T. L., Legge, A., Slark, T.S., Perrier *J. Polym. Sci., Part A: Polym. Chem.* **2006**. *44*, 6980–6987.

**56.** B., Chong, G., Moad, E., Rizzardo, M., Skidmore, S. H., Thang *Aust. J. Chem.* **2006**, *59*, 755–762.

**57.** D. L., Patton, M., Mullings, T., Fulghum, R. C., Advincula *Macromolecules* **2005**, *38*, 8597–8602.

**58.** C. H., Depuy, R. W., King *Chem. Rev.* **1960**, *60*, 444.

**59.** K. S., Mori, T.S., Masuda *Tetrahedron Lett.* **1978**, *37*, 3447.

**60.** G., Moad, Y. K., Chong, A., Postma, E., Rizzardo, S. H., Thang *Polymer* **2005**, *46*, 8458–8468.

**61.** A., Postma, T. P., Davis, G., Moad, M. S., O'Shea *Macromolecules* **2005**, *38*, 5371–5374.

**62.** M., Destarac, C., Kalai, A. Z., Wilczewska, G., Mignani, S. Z., Zard *Polym. Prepr. (Am. Chem. Soc., Div. Polym. Chem.)* **2005**, *46*, 213–214.

**63.** D. B., Thomas, A. J., Convertine, R. D., Hester, A. B., Lowe, C. L., McCormick *Macromolecules* **2004**, *37*, 1735.

**64.** M. H., Stenzel, T. P., Davis, C., Barner-Kowollik *Chem. Commun.* **2004**, 1546–1547.

**65.** M. L., Llauro, J. F., Boisson, F., Delolme, C., Ladaviere, J., Claverie *J. Polym. Sci., Part A: Polym. Chem.* **2004**, *42*, 5439.

**66.** C., Schilli, M. G., Lanzendörfer, A. H. E., Müller *Macromolecules* **2002**, *35*, 6819–6827.

**67.** M., Hruby, V., Korostyatynets, M. J., Benes, Z., Matejka *Collect. Czech. Chem. Commun.* **2003**, *68*, 2159–2170.

**68.** R. T. A., Mayadunne, J., Jeffery, G., Moad, E., Rizzardo *Macromolecules* **2003**, *36*, 1505–1513.

**69.** R. T. A., Mayadunne, E., Rizzardo, J., Chiefari, J., Krstina, G., Moad, A., Postma, S. H., Thang *Macromolecules* **2000**, *33*, 243–245.

**70.** E. C. J., Rizzardo, B. Y. K., Chong, F., Ercole, J., Krstina, J., Jeffery, T. P. T., Le, R. T. A., Mayadunne, G. F., Meijs, C. L., Moad, G., Moad, S. H., Thang *Macromol. Symp.* **1999**, *143*, 291.

**71.** Z. M., Wang, J. P., He, Y. F., Tao, L., Yang, H. J., Jiang, Y. L., Yang *Macromolecules* **2003**, *36*, 7446–7452.

**72.** X.-P., Qiu, F. M., Winnik *Macromol. Rapid Commun.* **2006**, *27*, 1648–1653.

**73.** X.-P., Qiu, F. M., Winnik *Macromolecules* **2007**, *40*, 872–878.

**74.** M. R., Whittaker, Y. K., Goh, H., Gemici, T. M., Legge, S., Perrier, M. J., Monteiro *Macromolecules* **2006**, *39*, 9028–9034.

**75.** H., Gemici T. M., Legge, M., Whittaker, M. J., Monteiro, S., Perrier *J. Polym. Sci., Part A: Polym. Chem.*, **2007**, *45*, 2334–2340.

**76.** J. T., Xu, J. P., He, D. Q., Fan, X. J., Wang, Y. L., Yang *Macromolecules* **2006**, *39*, 8616–8624.

**77.** C. W., Scales, A. J., Convertine, C. L., McCormick *Biomacromolecules* **2006**, *7*, 1389–1392.

**78.** A. B., Lowe, B. S., Sumerlin, M. S., Donovan, C. L., McCormick *J. Am. Chem. Soc.* **2002**, *124*, 11562–11563.

**79.** S. G., Spain, L., Albertin, N. R., Cameron *Chem. Commun.* **2006**, 4198–4200.

**80.** B. S., Sumerlin, A. B., Lowe, P. A., Stroud, P., Zhang, M. W., Urban, C. L., McCormick *Langmuir* **2003**, *19*, 5559–5562.

**81.** F., Cerreta, A. M., Lenocher, C., Leriverend, P., Metzner, T. N., Pham *Bull. Soc. Chim. Fr.* **1995**, *132*, 67–74.

**82.** H., Alper, C., Kwiatkowska, J. F., Petrignani, F., Sibtain *Tetrahedron Lett.* **1986**, *27*, 5449–5450.

**83.** P., Vana, L., Albertin, L., Barner, T. P., Davis, C., Barner-Kowollik *J. Polym. Sci., Part A: Polym. Chem.* **2002**, *40*, 4032–4037.

**84.** W. B., Farnham, M., Fryd, G., Moad, S. H., Thang, E., Rizzardo WP 2005 113612 A1 [*Chem. Abstr.* **2005**, *144*, 23268].

**85.** L., Boiteau, J., Boivin, A., Liard, B., Quiclet-Sire, S. Z., Zard *Angew. Chem. Int. Ed* **1998**, *37*, 1128–1131.

**86.** M., Destarac, C., Kalai, L., Petit, A. Z., Wilczewska, G., Mignani, S. Z., Zard *Polym. Prepr. (Am. Chem. Soc., Div. Polym. Chem.)* **2005**, *46*, 372–373.

**87.** L., Lu, H. J., Zhang, N. F., Yang, Y. L., Cai *Macromolecules* **2006**, *39*, 3770–3776.

**88.** L. C., Lu, N. F., Yang, Y. L., Cai *Chem. Commun.* **2005**, 5287–5288.

**89.** J. F., Quinn, L., Barner, C., Barner-Kowollik, E., Rizzardo, T. P., Davis *Macromolecules* **2002**, *35*, 7620–7627.

# 13
# Toward New Materials Prepared via the RAFT Process: From Drug Delivery to Optoelectronics?

*Arnaud Favier, Bertrand de Lambert, and Marie-Thérèse Charreyre*

## 13.1
## Introduction

The development of controlled/living ionic and radical polymerization techniques during the last decades represents a major breakthrough in polymer chemistry and polymer science. Since these techniques lead to the synthesis of a wide range of tailor-made macromolecular architectures, significant improvements are expected in the current or future application fields of polymers. The industrial impact will however depend on the versatility and applicability of each technique, and of course on the extra cost for the final product.

In this context, controlled radical polymerizations (CRP) and especially the reversible addition–fragmentation chain transfer (RAFT) process [1–3] have attracted a lot of attention since they combine the advantages of conventional radical polymerization and living ionic polymerizations. Conventional radical polymerizations are cost-effective techniques, easy to process with a low sensitivity to water and oxygen and applicable to a wide range of monomers. In addition, CRP techniques result in a very efficient control over molecular weight (MW), molecular-weight distribution (MWD), microstructure, chain-end functionality and macromolecular architecture.

As shown in the previous chapters of this handbook, the RAFT process appears as one of the most interesting CRP techniques. First, RAFT polymerization is very similar to conventional radical polymerization since it only requires the addition of a chain-transfer agent (CTA) in the medium. Second, RAFT is a very versatile process able to control the (co)polymerization of a large variety of monomers leading to a virtually unlimited macromolecular design library.

As a consequence, RAFT polymerization is a well-suited and promising technique to prepare high-performance polymers for a wide range of applications. Indeed, an efficient control at the macromolecular level is a very important step to control and improve the macroscopic properties of the final materials (Fig. 13.1).

*Handbook of RAFT Polymerization.* Edited by Christopher Barner-Kowollik

ISBN: 978-3-527-31924-4

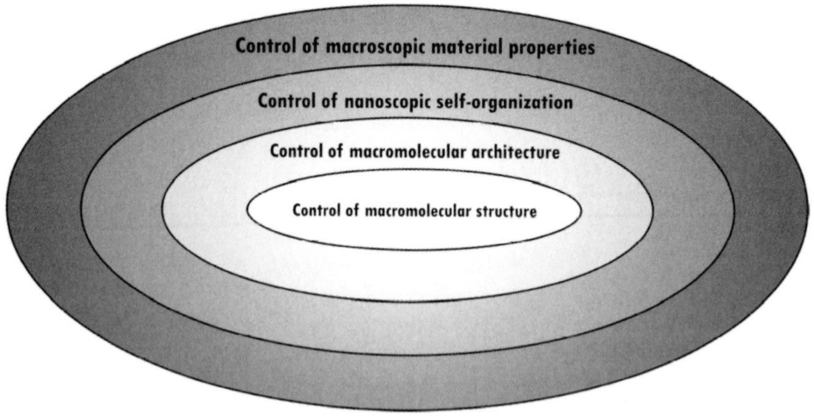


**Fig. 13.1** The various levels of control from the macromolecule to the material.

To be more explicit, it is necessary to briefly summarize without entering into the details of the polymerization mechanism, the parameters of the RAFT process that can be tuned to access to the desired well-defined macromolecules (Fig. 13.2).

- The initial concentrations in CTA, monomer and initiator control the MW and MWD of each polymer segment synthesized via the RAFT process.
- The nature of the CTA (and possible subsequent postfunctionalization) controls the chain-end functionality ($\alpha$- and $\omega$ chain ends bearing reactive functions or specific entities).
- The nature of the monomer and comonomer controls the microstructure (homopolymer, random or gradient copolymers), the lateral functionalities (reactive functions and specific entities) and the physicochemical properties (hydrophilic/hydrophobic, stimuli responsive and $T_g$) of each polymer segment.
- The CTA structure, the different steps of polymerization and the combination of RAFT process with other techniques control the macromolecular architecture (block copolymers, graft copolymers, star copolymers, branched or hyperbranched/dendritic copolymers).

Considering that all these building possibilities can be combined, it becomes obvious that RAFT is a very powerful tool for macromolecular design expected to be applied to various application fields. Nevertheless, a critical step lies between the synthesis of the macromolecules and the final application. Indeed, the processing of the polymer chains has a great influence on the properties and performance of the resulting product. Since RAFT technique is very versatile (nature, microstructure and architecture of the macromolecules), it provides great opportunities to control the self-assembly or self-organization of the chains in solution or in the solid state. Then, optimization of the processing technique could lead to materials with enhanced chemical, physical, mechanical and surface properties.

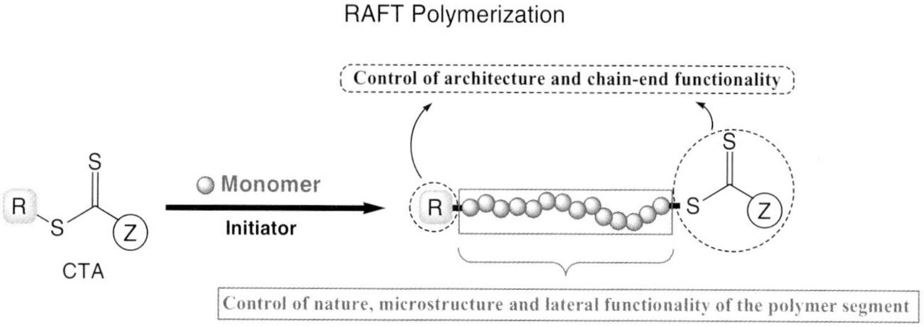


**Fig. 13.2** Formation of polymer building blocks via RAFT polymerization.

In this chapter, it is our objective to present an overview of the applications of polymers synthesized via the RAFT process in various fields, from biology (diagnostics, drug and gene delivery and tissue engineering) to optoelectronics, including organic/inorganic hydrid materials. However, to date still few articles describe the final application. Consequently, the various sections of this chapter will be presented either from the angle of the application or from the angle of the material *potentially* useful for such-and-such fields. We will emphasize on the advantages of the RAFT process and on the main limitations that remain. Finally, we will try to point the challenges for the future since the exponential interest in this technique will push forward the limits of polymer science and polymer-based applications.

## 13.2 Bio-Related Applications

Before to describe the applications of polymers synthesized via the RAFT process in precise bio-related fields (diagnostics, drug and gene delivery and tissue engineering), it is worth reporting the synthesis of polymer precursors that enable an easy binding of various biomolecules as well as the synthesis of biomolecule/polymer conjugates whether they have been or not yet used in a particular application.

### 13.2.1 Polymer Precursors for an Easy Binding of Biomolecules

#### 13.2.1.1 Polymer Precursors with Reactive Side Groups

Water-soluble polymers with reactive functions as side groups are commonly used as backbones for the covalent immobilization of biomolecules [4]. Since most biomolecules bear amino groups, polymers with anhydride or activated ester side groups are among the favorite reactive polymers (Fig. 13.3). Such polymers have long been synthesized by conventional free-radical polymerization [5]. The lack of MW homogeneity inherent to this technique was more or less problematic according to the application. However, a better reproducibility of the final biological result

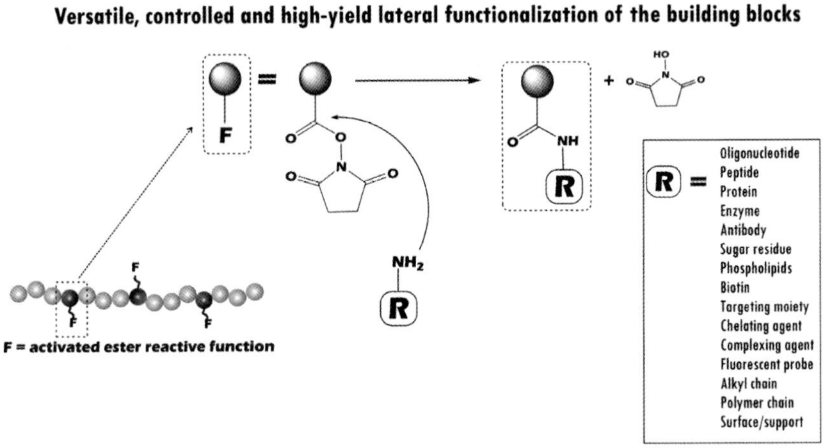


**Fig. 13.3** Functionalization of macromolecules using succinimidyl-activated ester reactive groups.

is usually expected when chains bear a similar number of biomolecules, resulting from a similar structure of the chains (microstructure and MW).

With the discovery of RAFT polymerization, it became possible to synthesize reactive polymers homogeneous in both size and interchain composition. As our group already had some experience with activated ester-based polymers especially with *N*-acryloxysuccinimide (NAS) [6, 7], it appeared challenging to polymerize NAS by the RAFT process. Water-soluble random copolymers with *N*-acryloylmorpholine (NAM), poly(NAM-*co*-NAS), and amphiphilic block copolymers with a hydrophobic block of *tert*-butyl acrylate (*t*BA), P[tBA-*b*-(NAM-*co*-NAS)], were prepared using *tert*-butyl dithiobenzoate as CTA [8]. A careful optimization of the experimental conditions (temperature, monomer concentration and dithioester/initator ratio) led to an improved control of poly(NAM-*co*-NAS) chains [9].

Moreover, an in-depth study of the NAM/NAS pair copolymerization kinetics revealed that the apparent reactivity ratios determined in a RAFT copolymerization were similar to that determined in a conventional one [10]. Then, at the azeotropic composition (60/40:NAM/NAS), it was possible to get polymer chains without composition drift, that is with a homogeneous microstructure (constant intrachain composition). Finally, extremely well controlled material (size, polydispersity, interchain composition and microstructure) could be synthesized.

RAFT copolymerization of NAS has been extended to other monomers and architectures, leading to random and block copolymers with dimethylacrylamide (DMA), poly(DMA-*co*-NAS) and poly(DMA-*b*-NAS) [11], double hydrophilic block copolymers P[NAM-*b*-(NAM-*co*-NAS)] [10] and amphiphilic block copolymers based on *t*BA, P[tBA-*b*-(NAM-*co*-NAS)] [8], or on two acrylamide derivatives, *N*-*tert*-butyl acrylamide (*t*BAM) as the hydrophobic monomer and NAM as the hydrophilic one, P[*t*BAM-*b*-(NAM-*co*-NAS)] (Figs. 13.4 and 13.5) [12].

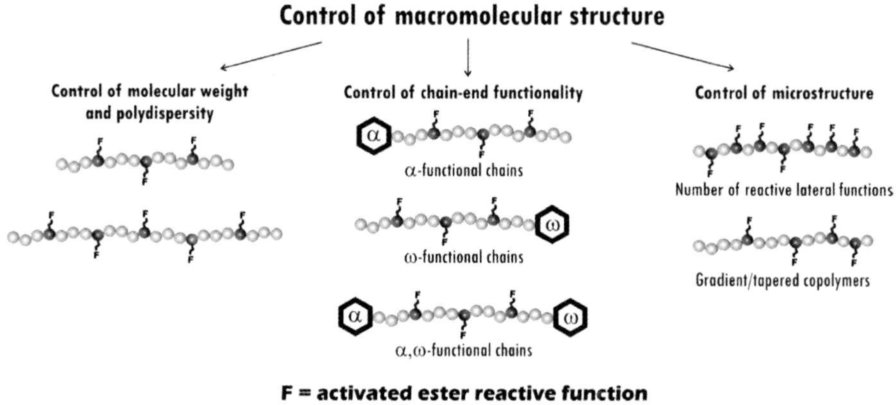


**Fig. 13.4** Control of the macromolecular structure and functionality of NAS-containing building blocks.

These various polymer precursors have been used for several applications, such as growing of nucleic acid probes for diagnostic tests on DNA chips (Section 13.2.4) [12, 13], electrostatic complexation of plasmid DNA for transfection of cells (Section 13.2.5) [14], as well as covalent binding of fluorescent dyes to prepare highly fluorescent polymers (Section 13.3.1) [15].

Recently, McCormick and coworkers used a poly(ethylene oxide) (PEO)-based macro-CTA (dithiobenzoate) to prepare PEO-*b*-P(DMA-*co*-NAS) diblock copolymers that could be extended to triblock copolymers, including a thermally responsive poly(*N*-isopropylacrylamide), poly(NIPAm), block [16]. These PEO-*b*-P(DMA-*co*-NAS)-*b*-P(NIPAm) triblock copolymers self-assemble into micelles in water at

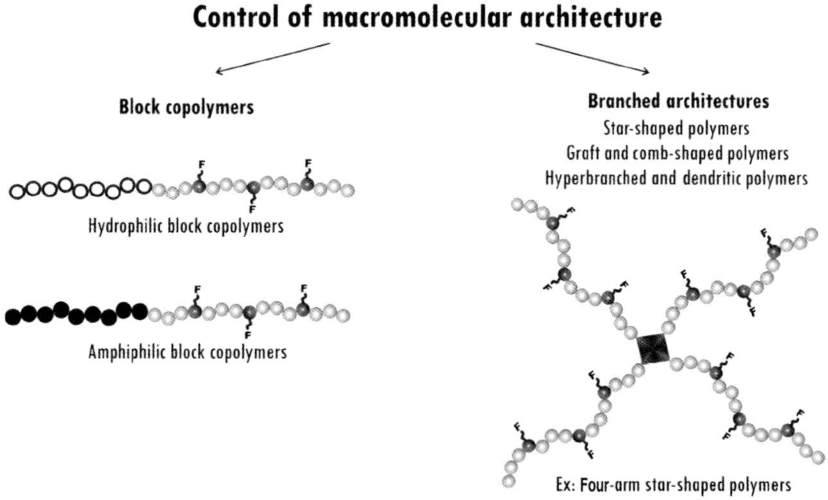


**Fig. 13.5** Examples of macromolecular architectures that may be obtained from NAS-containing building blocks using the RAFT process.

temperature above the lower critical solution temperature (LCST) of the poly(NIPAm) block. Moreover, after cross-linking of the intermediate layer with a diamine, shell cross-linked micelles were obtained with a PEO hairy layer at the periphery and a swollen core, used as vehicle for drug delivery (Section 13.2.5).

Alternatively, the methacrylate analog of NAS, *N*-methacryloxysuccinimide (*N*-MAS), was also polymerized by the RAFT process. It was rather difficult to get a good control of *N*-MAS homopolymerization and polydispersities around 1.5 were obtained using cyanopropyl dithiobenzoate [17]. In the presence of the same dithioester, *N*-MAS was also copolymerized with NIPAm [18], and with hydroxypropylmethacrylamide (HPMA) [19]. Both random copolymers were obtained with low polydispersities, poly(NIPAm-*co*-*N*-MAS) including 2–30% of *N*-MAS in the MW range of 14 000–72 000 $g \cdot mol^{-1}$, and poly(HPMA-*co*-*N*-MAS) including 19–28% of *N*-MAS in the MW range of 4000–54 000 $g \cdot mol^{-1}$. The latter copolymer was used for the covalent binding of a peptide leading to a multivalent inhibitor of the assembly of anthrax toxin.

Another kind of activated ester, pentafluorophenyl methacrylate (PFMA), has been homopolymerized by RAFT with cumyldithiobenzoate and 4-cyanopentanoic acid-4-cyanobenzoate, with a better MW control using the latter dithiobenzoate. The poly(PFMA) block mediated the polymerization of MMA, NAM and diethylacrylamide (DEAAm), leading to reactive diblock copolymers, poly(PFMA-*b*-MMA), poly(PFMA-*b*-NAM) and poly(PFMA-*b*-DEAAm) [20].

#### 13.2.1.2 **Polymer Precursors with Reactive End Groups**

A particular class of polymer precursors for the binding of various species (including biomolecules) are the reactive end-functionalized polymers. For instance, carboxylic end-functionalized polymers are easily obtained when using a CTA bearing a carboxylic function in the R group [21–24]. Then, amino derivatives can in principle be covalently bound to the chain. However, as amino compounds are known to aminolyse the thiocarbonylthio function, side products are usually formed in parallel. A smart way to avoid such reactions relies on the use of a CTA bearing an activated ester instead of a carboxylic function in the R group [25, 26]. In turn, to get an amino end-functionalized polymer, it is necessary to protect the amino function, for instance with a phtalimide group [22]. Finally, the possibility to bind thiol derivatives to the $\omega$ end of a polymer chain was recently given by the design of trithiocarbonates bearing a pyridyl-disulfide function [27].

#### 13.2.1.3 **Conclusion on Polymer Precursors Synthesized by the RAFT Process**

Finally, various water-soluble polymer architectures with activated ester side groups have been synthesized by RAFT polymerization with controlled MW and MWD. Even if *N*-MAS can also be polymerized by atom-transfer radical polymerization (ATRP) [28], the versatility of the RAFT process enables the use of a great number of possible comonomers, including acrylamide derivatives. The resulting copolymers can be used to prepare families of copolymers by covalently grafting increasing density of various amino derivatives via the reactive activated ester units. The potential applications of such well-controlled reactive copolymers (with reactive side- or end groups) are extremely promising in the biological field and other fields, and several examples will be described in the following sections.

### 13.2.2
### Biomolecule/Polymer Conjugates

In this section, various polymers involving bio-related species and synthesized by RAFT are reported. The bio-related species is, for instance, a small molecule able to interact with a protein either indirectly (an imidazole function) or directly (a biotin), a protein component such as an amino acid or a peptide, a cell membrane component such as a phospholipid. The bio-related molecule is introduced in the polymer structure either as *a monomer derivative* leading to a polymer with multiple biomolecules as side groups or as *a chain end* leading to a polymer with a precisely located unique biomolecule ($\alpha$ chain end if introduced via the R group of the CTA, or $\omega$ chain end if introduced via the Z group of the CTA or via a covalent reaction on the terminal —SH group released by hydrolysis of the thiocarbonylthio polymer chain end).

As very few articles have been published about each of these bio-related species and as some similarities may be encountered concerning the strategy, these examples are presented together in this section (Table 13.1). Articles related to polymers involving carbohydrate derivatives – another family of bioconjugates – are presented in the next section since they are more numerous.

#### 13.2.2.1 **Imidazole/Polymer Conjugates**
Highly branched polymers with multiple imidazole end groups have been synthesized for protein purification [29, 30]. Imidazole groups have the property to bind to transition metal ions ($Zn^{II}$, $Cu^{II}$ and $Ni^{II}$) in a similar way than do histidine groups. When using imidazole-functionalized polymers, it is possible to purify histidine-tagged proteins by forming polymer/metal ions/protein complexes. The advantage of introducing the imidazole groups as end groups of a branched polymer synthesized via the RAFT process relies on a better accessibility of these groups compared to a linear random copolymer incorporating imidazole units as side groups. Rimmer and coworkers used such highly branched polymer to extract histidine-tagged breast cancer susceptibility proteins from a crude cell lysate [30]. The polymer was based on P(NIPAm) to benefit from the thermally induced phase separation that favors extraction of the polymer/protein complex.

#### 13.2.2.2 **Biotin/Polymer Conjugates**
Another interesting application related to proteins consists in the control of the bioactivity of, for instance, an enzyme via the binding of a thermosensitive polymer close to the active site. Then, the reversible collapse of the polymer chain would result in a temporary blocking of the active site of the enzyme. This concept has been demonstrated by Hoffman and coworkers, using biotin $\omega$-end-labeled P(NIPAm) synthesized by RAFT and streptavidin protein [31]. Introduction of the biotin was

**Table 13.1** Bio-related species conjugated to polymer chains synthesized by RAFT

| Bio-related species | Location on the polymer chain | Nature of the polymer | Ref. |
|---|---|---|---|
| Imidazole group | ω end group | Highly branched polymer | [29] |
| Binding group for protein purification via transition metal ions | In the Z group of the CTA | P(NIPAm-*co*-S) backbone | [30] |
| | Polymerizable RAFT agent 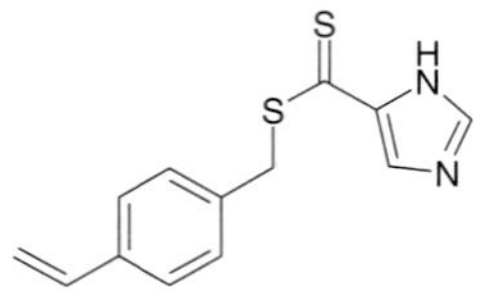 | P(NIPAm) and p(NIPAm-*b*-GMA) branches | |
| Biotin | ω end group | Stimuli-sensitive polymers | [31] |
| Strong specific interaction with streptavidin protein | Maleimide-biotin 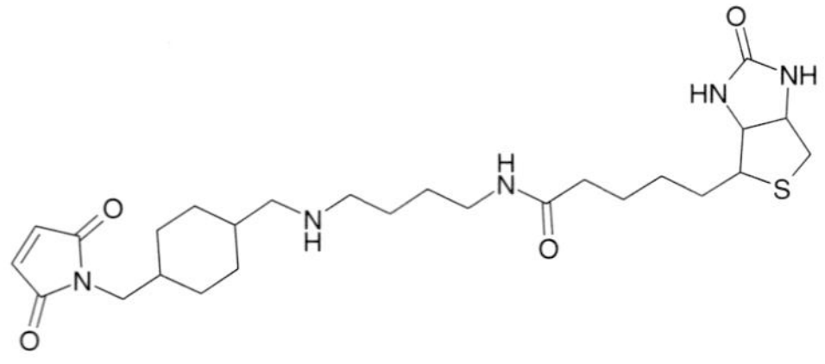 | P(NIPAm)-SH | [32] |
| | | P(AA-*b*-NIPAm)-SH | [33] |
| | | Y-shaped block copolymer | |
| | | PEG-lysine-*b*-P(NIPAm)-SH | |

| | | | |
|---|---|---|---|
| | α end group<br>Biotin dithioester<br>Amide link<br>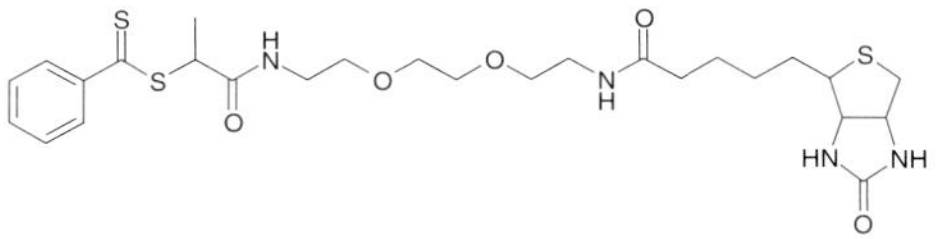 | Biotin-P(NAM)<br>Biotin-P(NAM-*co*-AmGal) | [25]<br>[34] |
| | α end group<br>Biotin trithiocarbonate<br>Ester link<br>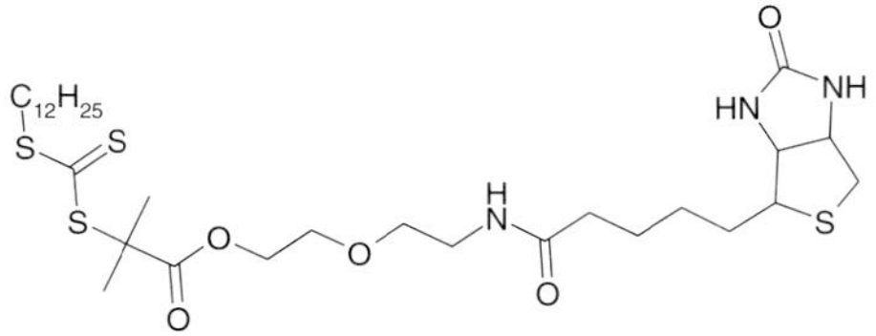 | Biotin-P(HPMA-*b*-NIPAm) | [35] |
| Amino acid<br>Ability to produce highly ordered structures | Side group (L-phenylalanine)<br>Acrylamide derivative<br>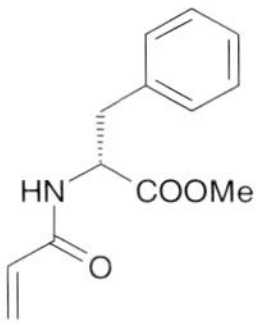 | Homopolymer in the presence of dithiocarbamate and dithioester | [36] |

(*Continued*)

Table 13.1 (*Continued*)

| Bio-related species | Location on the polymer chain | Nature of the polymer | Ref. |
|---|---|---|---|
| | Side group (L-proline)<br>Acrylamide derivative<br>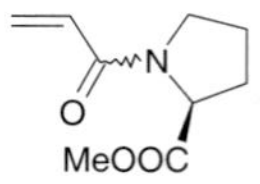 | Homopolymer<br>Random copolymer with DMA in the presence of dithioester | [37] |
| Peptide<br>Sequence-dependent bioactivity | α end group<br>Coupling of a carboxy-functionalized RAFT agent to a resin-bound peptide<br>Amide link<br>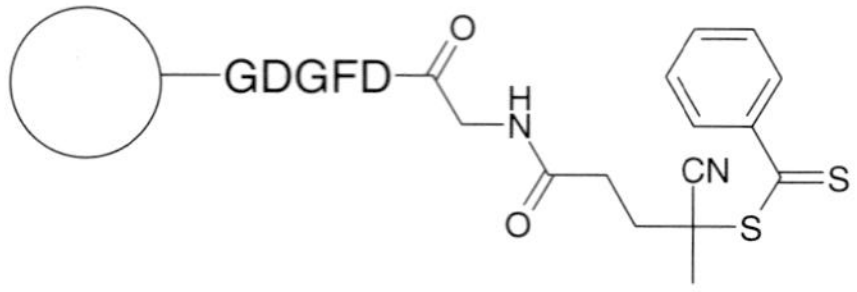 | Peptide/polymer conjugate from combination of solid-phase-supported synthesis and RAFT polymerization:<br>Peptide-P(BA) | [38] |
| Phospholipids<br>Ability for insertion in lipid membranes | Side group<br>Phosphorylcholine<br>Acrylate derivative (PCA)<br>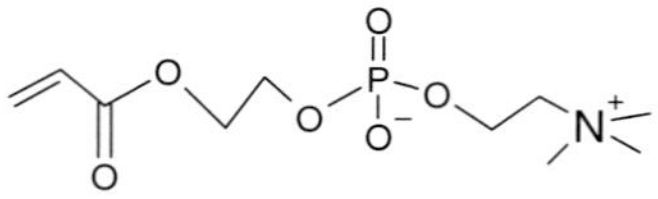 | Block copolymer<br>P(BA-*b*-PCA) in the presence of a carboxylic trithiocarbonate | [39] |

| | | |
|---|---|---|
| Side group<br>Phosphorylcholine<br>Methacrylate derivative (PCM)<br>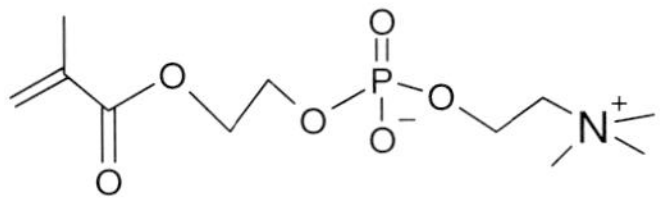 | Block copolymer<br>P(BMA-*b*-PCM) in the presence of 4-cyanopentanoic acid dithiobenzoate | [40] |
| α end group<br>Phospholipid dithioester<br>Amide link<br>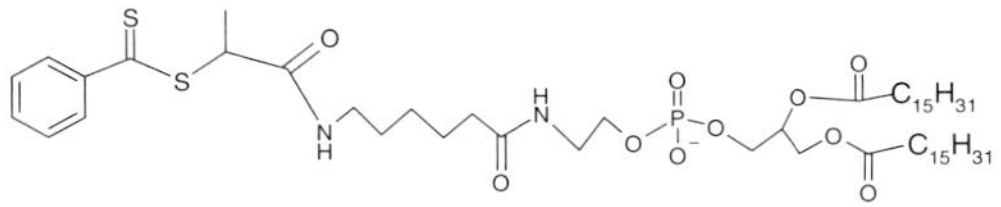 | Phospholipid-P(NAM) | [26, 41] |

NIPAm = *N*-isopropylacrylamide; S = styrene; GMA = glycidyl methacrylate; AA = acrylic acid; PEG = poly(ethylene glycol); NAM = *N*-acryloylmorpholine; AmGal = 6-*O*-acryloylamino-6-desoxy-1,2:3,4-di-*O*-isopropylidène-α-D-galactopyranose; HPMA = hydroxypropyl methacrylate; DMA = dimethyl acrylamide; BA = butyl acrylate; PCA = phosphorylcholine acrylate; BMA = butyl methacrylate; PCM = phosphorylcholine methacrylate.

carried out by a postpolymerization reaction on the terminal —SH group released after hydrolysis of the dithiocarbamate chain end (60% coupling yield). As biotin molecule binds to streptavidin with a very high association constant, it is then easy to prepare P(NIPAm)–streptavidin conjugates. However, upon increase of the temperature above LCST, the collapse of the poly(NIPAm) chain on the surface of streptavidin induces an aggregation behavior of the conjugates (formation of nanoparticles). Recently, the use of an ionizable block copolymer, poly(acrylic acid-*b*-NIPAm)-biotin, increased the electrostatic stability that avoided intermolecular aggregation of the conjugates at pH values where a sufficient number of carboxylic groups were ionized [32].

In a similar way, a biotin $\omega$-end-labeled Y-shaped block copolymer was prepared, with a poly(ethylene glycol) (PEG) block, a poly(NIPAm) block, a carboxylic group from L-lysine at the focal point and a biotin molecule introduced after aminolysis of the chain end of the poly(NIPAm) block (52% coupling yield) [33]. Such Y-shaped block copolymers were designed for surface modification with a potential thermally controlled presentation of the ligands.

An improved approach to significantly increase the percentage of end labeling of polymer chains by biotin consists in using biotinylated RAFT agents. This was recently achieved with biotin-dithioesters [25, 26] and with a biotin-trithiocarbonate [35]. In the first case, the biotin-dithioester (with an amide link) was synthesized with an almost quantitative yield from a commercial amino-PEO-biotin derivative, and was used to copolymerize NAM either with an activated ester, NAS, or with carbohydrate derivatives, respectively leading to biotin-$\alpha$-end-labeled reactive copolymers [26] and to biotin-$\alpha$-end-labeled glycopolymers based on galactose and *N*-acetyl glucosamine [34]. In the second case, the biotin-trithiocarbonate (with an ester link) was synthesized with 28% yield from biotin, and was used to synthesize poly(HPMA-*b*-NIPAm) block copolymers able to reversibly form core-shell nanostructures with biotin-$\alpha$-end-labeled poly(HPMA) arms. In both cases, a judicious choice of the RAFT agent/initiator ratio led to more than 90% of biotin-end-labeled chains.

#### 13.2.2.3 **Amino-Acid and Peptide/Polymer Conjugates**

Incorporation of amino-acid moieties in a synthetic polymer leads to macromolecules with biomimetic structures and properties. Homopolymers of L-phenylalanine- and L-proline-based acrylamide derivatives were synthesized by RAFT in the presence of dithiocarbamate or dithioester CTA [36, 37]. Well-controlled polymers were obtained in the range 4000–30 000 g·mol$^{-1}$ with polydispersity index of 1.2–1.5. Addition of Lewis acid led to improved tacticity of the homopolymer. The homopolymer bearing L-proline moieties showed a thermosensitive behavior with an LCST around 15–20 °C. This value was increased when the monomer was copolymerized with less than 50% of DMA. None application was described with these polymers that were designed for potential uses as controlled release systems, biochemical sensing and biocompatible materials.

Another approach consists in preparing peptide/polymer conjugates by combining solid-phase synthesis and RAFT process. A peptide-based RAFT agent was prepared either by coupling a carboxy-functionalized RAFT agent to the amino end of a resin-bound peptide, or by switching a solid-phase-supported ATRP macroinitiator into a peptide RAFT agent [38]. These two strategies avoided the usual chromatographic purification procedures. However, in the first case, some side product resulting from the competitive thioamidation reaction was also formed. The peptide RAFT agent mediated the polymerization of *n*-butyl acrylate (*n*BA) at 60 °C. Circular dichroism analysis of the peptide–poly(*n*BA) conjugate confirmed that the chirality of the peptide segment was preserved.

#### 13.2.2.4 **Phospholipid/Polymer Conjugates**

Random copolymers having phosphorylcholine side groups have interest as drug carriers since they are highly water soluble, biocompatible and antithrombogenic and since they form hydrophobic microdomains in aqueous solutions. Using RAFT polymerization, block copolymers of an acrylate (or methacrylate) derivative of phosphorylcholine with *n*-butylacrylate (or *n*-butyl methacrylate) were synthesized in the presence of a trithiocarbonate [39] or a dithiobenzoate [40]. The resulting amphiphilic block copolymers self-assembled into micelles in water. In the case of the methacrylate derivative, a solubilization test of a poorly water-soluble anticancer drug, paclitaxel, was carried out. The solubilized amount was significantly increased with block copolymers in comparison with random copolymers.

Alternatively, a phospholipid-end-labeled polymer was synthesized in the presence of a phospholipid RAFT agent [26, 41] obtained from a precursor RAFT agent bearing an activated ester function [25] and a dipalmitoyl-type phospholipids. Polymerization of NAM led to controlled phospholipid-end-labeled polymer chains that showed an amphiphilic behavior even with poly(NAM) sequences of 35 000 $g{\cdot}mol^{-1}$. Its ability to act as steric stabilizer of lipoparticles was investigated in aqueous solutions of increasing ionic strength.

#### 13.2.2.5 **Conclusion on Biomolecule/Polymer Conjugates**

The synthesis of this large variety of biomolecule/polymer conjugates by RAFT polymerization relies on the versatility of the RAFT process to polymerize monomers with polar or ionic groups, as well as on the possibility to use mild experimental conditions such that the integrity of the biomolecule is retained. This is especially important when the biomolecule is introduced as side group of a monomer or in the RAFT agent. The examples presented in this section illustrate the numerous and more complex biomolecule/polymer architectures that can be envisioned in a near future.

### 13.2.3 Polymers Involving Carbohydrate Derivatives

#### 13.2.3.1 **Glycopolymers**

Glycopolymers are synthetic polymers bearing saccharidic residues as side groups introduced either by polymerization of a carbohydrate derivative or by postmodification of a preformed polymer. Glycopolymers find numerous applications in many fields and especially in the biological field, for instance as biocompatible polymer backbone for immobilization of nucleic acid sequences [42] or as multivalent ligands favoring recognition processes with lectins and glycoproteins [43, 44].

Since the last decades, glycopolymers have been prepared by 'living' techniques to benefit from the MW control, including living anionic and cationic processes, ring-opening polymerization, methathesis polymerization and more recently, controlled radical polymerization like NMP, ATRP and RAFT, as reported by several reviews [45, 46].

Concerning RAFT polymerization, it was first applied to homopolymerize a glucose derivative (methacrylate group in the C-1 position) in water in the presence of cyanopentanoic acid dithiobenzoate as CTA [47]. Well-controlled homopolymers were prepared in the MW range 7000–27 000 g·mol$^{-1}$, while the synthesis of block copolymers with 3-sulfopropyl methacrylate appeared more favorable when beginning with a sulfopropyl methacrylate block.

Several other carbohydrate derivatives have been polymerized by RAFT so far, monosaccharides such as glucose, galactose, mannose and *N*-acetyl glucosamine, as well as a disaccharide, lactose, mostly with a methacrylate or an acrylamide function. Contrary to the case of ionic polymerization, when using a CRP process it is possible to polymerize carbohydrate derivatives in their deprotected form in water or in a water/ethanol or water/DMSO mixture. RAFT polymerization in water phase leads to fast kinetics and to MW up to 100 000 g·mol$^{-1}$ [48] (Table 13.2). Some carbohydrates derivatives have also been polymerized in a protected form, (isopropylidene or acetate groups) [34, 53, 55], which makes copolymerization possible with a wider range of comonomers. Concerning the kind of CTA, most of them were dithioesters (including a biotin-linked dithiobenzoate) [25, 34] and trithiocarbonates (including a grafted trithiocarbonate) [54], leading to well-controlled polymers in aqueous or organic medium.

Various polymer architectures have been designed, gradient copolymers, block copolymers, three-arm stars and copolymer brushes on a silicon wafer, among which some original diblock copolymers including two different carbohydrates, glucose and mannose, or two glucose linked by a different position to the polymerizable function [49].

It is noteworthy that considering carbohydrates, the position of the cycle where the polymerizable function is introduced is of much importance since it determines the type of application of the resulting glycopolymer. If the polymerizable function is at the C-6 position and if the C-1 is free, then it will be possible to use the masked aldehyde function at the C-1 to further bind amino-bearing compounds onto the glycopolymer [42, 53]. On the contrary, to favor a better recognition of the glycopolymer by a lectin, the polymerizable function should be introduced at the C-1 position.

#### 13.2.3.2 Sugar End-Functionalized Polymers

To get a carbohydrate derivative at a polymer chain end, the best solution consists in synthesizing an initiator or a CTA including a carbohydrate moiety. To date, two examples can be found in the literature concerning carbohydrate-derived CTA,

**Table 13.2** Carbohydrate derivatives that have been polymerized by the RAFT process

| Sugar | C-x | Monomer derivative | CTA/initiator/solvent | Polymer architecture Range of $M_n$ ($g \cdot mol^{-1}$) (PDI) | Ref. |
|---|---|---|---|---|---|
| Glucose | C-1 | methacrylate | CPADB/ACPA/water | Homopolymers<br>7000–27 000 (<1.1)<br>Block copolymers<br>P(MAGlu-*b*-SPMA) (1.6)<br>P(SPMA-*b*-MAGlu) (1.2) | [47] |
| | | | CPADB/ACPA/water/EtOH:9/1 | Homopolymers<br>25 000–52 000<br>Block copolymers<br>P(MAGlu-*b*-MAMan) (1.2) | [49] |
| | C-6 | methacrylate | CPADB/ACPA/water/EtOH:9/1 | Homopolymers<br>20 000–70 000 (<1.2) | [50] |
| | | | Water/EtOH:1/1 | Block copolymers<br>P(MAGlu-*b*-HEMA) (1.2)<br>P(MAGlu6-*b*-MAGlu) (1.2) | [49] |
| | C-6 | vinyl ester | Carbamate or xanthate/ACPA/water | Homopolymers (1.2) | [51] |
| Galactose | C-1 | methacrylate | CPADB/ACPA/Water/EtOH:9/1 | Homopolymers<br>21 000 (1.1) | [52] |
| protected | C-6 | acrylate<br>acrylamide<br>acrylamide + spacer arm | *t*-BDB<br>or BEDBA/AIBN/dioxane | Gradient copolymers<br>P(NAM-*co*-AGal)<br>P(NAM-*co*-AmGal)<br>5000–50 000 (<1.3) | <br>[53]<br>[34] |

*(Continued)*

**Table 13.2** (*Continued*)

| Sugar | C-*x* | Monomer derivative | CTA/initiator/solvent | Polymer architecture Range of $M_n$ ($g \cdot mol^{-1}$) (PDI) | Ref. |
|---|---|---|---|---|---|
| Mannose | C-6 | methacrylate | CPADB/ACPA/water/EtOH:9/1 | Block copolymers P(MAGlu-*b*-MAMan) (1.2) | [49] |
| protected | C-1 | acrylamide | *t*-BDB/AIBN/dioxane | Gradient copolymers P(NAM-*co*-AmMan) 5000–50 000 (<1.3) | [53] |
| Glucosamine | C-2 | acrylamide | BTCPA/ACPA/water/MeOH:5/1 | Homopolymers 6000–100 000 (<1.3) | [48] |
| | | | Water/DMSO:1/1 | Block copolymers P(NAGlu-*b*-NIPAm) (1.3) | |
| | | | Water/EtOH:5/1 | 3-Arm star copolymers P(HEA-*b*-NAGlu) (1.4) | |
| | | | Grafted BTCPA/water/EtOH:5/1 | P(NAGlu) brushes on Si wafer | [54] |
| | | | Water/DMSO:1/1 | 5000–40 000 P(NAGlu-*b*-NIPAm) brushes on Si wafer | |
| *N*-acetyl glucosamine protected | C-1 | acrylamide | *t*-BDB/AIBN/dioxane | Gradient copolymers P(NAM-*co*-AmNAcGlu) 5000–50 000 (< 1.3) | [53] |
| Lactose protected | C-1 | methacrylate | CDB/AIBN/$CHCl_3$ | Homopolymers 3000–22 000 (<1.3) | [55] |

CPADB: cyanopentanoic acid dithiobenzoate; *t*-BDB: *tert*-butyl dithiobenzoate; BEDBA: biotin ethylamide dithiobenzoate; BTCPA: 3-benzylsulfanylthiocarbonyl sulfanylpropionic acid; CDB: cumyl dithiobenzoate; MAGlu: methacryloxyethyl glucoside; SPMA: 3-sulfopropyl methacrylate; MAMan: methyl 6-*O*-methacryloyl-α-D-mannoside; HEMA: hydroxyethyl methacrylate; MAGlu6: methyl 6-*O*-methacryloyl-α-D-glucoside; NAM: *N*-acryloyl morpholine; AGal: 6-*O*-acryloyl-1,2:3,4-di-*O*-isopropylidène-α-D-galactopyranose; AmGal: 6-*O*-acryloylamino-6-désoxy-1,2:3,4-di-*O*-isopropylidène-α-D-galactopyranose; AmMan: 2-[2-(2-*N*-acryloyl-aminoethoxy)ethoxy]ethoxyl 2,3,4,6-tetra-*O*-acetyl-α-D-mannopyranoside; NAGlu: *N*-acryloyl glucosamine; NIPAm: *N*-isopropyl acrylamide; HEA: hydroxyethyl acrylamide; AmNAcGlu: 2-[2-(2-acrylamidoethoxy)ethoxy]ethyl 3,4,6-tri-*O*-acetyl-2-*N*-acetamido-2-desoxy-β-D-glucopyranoside.

a trithiocarbonate bearing a glucose moiety in the Z equivalent group [56] and a dithioester bearing a protected galactose moiety in the R group [25]. In the first case, the glucose is linked to the thiocarbonylthio group by the C-1 position via an ester function, whereas in the latter, the galactose is linked by the C-6 position via an amide function.

The glucose-derived trithiocarbonate mediated styrene polymerization in *N*-methyl pyrrolidone. The ω-functionalized chains were used to prepare porous films, with pore size depending on the MW of the chains [56]. A rearrangement of the carbohydrate moieties at the periphery of the pores was expected with the aim to elaborate new kinds of support for cell cultures. By a similar method, a multifunctional RAFT agent was obtained from β-cyclodextrine. However, the control of the arm growth was made difficult due to steric crowding.

The galactose-derived dithioester mediated NAM polymerization in dioxane. Well-controlled α-functionalized chains were obtained with narrow polydispersity (1.05) in the range 2000–40 000 $g{\cdot}mol^{-1}$. A matrix-assisted laser desorption ionization – time of flight mass spectrometry analysis confirmed the integrity of the carbohydrate chain end after polymerization [25]. These chains were further successfully used as stabilizer in a dispersion polymerization of *n*BA, leading to hairy latex particles of submicron size (150-nm diameter) that bear an average of 2400 sugar moieties per particle [57]. Such functionalized hairy particles are model particles potentially useful for biological diagnostic devices.

#### 13.2.3.3 **Polysaccharides**

Natural polysaccharides such as pullulan and cellulose have also been used as substrates for RAFT polymerization, either after binding vinylic bonds to the polysaccharide [58], or after binding a CTA derivative (a trithiocarbonate [56, 59] or a dithioester [60]) to some OH groups of the cellulose backbone. In the former case, addition of initiator and cumyldithiobenzoate led to reticulation of the pullulan backbone as a hydrogel (Section 13.2.5). In the latter case, grafting from polymerization of styrene was carried out in order to prepare honeycomb-structured films (Section 13.3.2) and composite materials (Section 13.3.3).

#### 13.2.3.4 **Conclusion on Carbohydrate-Containing Polymers Synthesized by RAFT**

In comparison with the other CRP techniques, NMP and ATRP, RAFT process also affords polymerization of protected or unprotected sugar-based monomers, however over a wider range, including acrylamide derivatives. Polymerization can be performed at low temperature in water or water/alcohol mixtures and the resulting glycopolymers do not contain any residual species disadvantageous for biomedical uses.

Until now, very few carbohydrate-containing polymers synthesized via the RAFT process have been used in the context of a biological application. However, most of them could be favorably evaluated in biological applications implying recognition events such as diagnostics, targeted drug or gene delivery, as well as in the study of pathogen–host cell interactions. They could also be used to bring biocompatibility to the surface of biomaterials, and to promote cell adhesion/proliferation in regenerative medicine. These various application fields are described in the next sections.

## 13.2.4
## Diagnostic Applications

The aim of biological diagnostic tests is to detect (i.e. capture and quantify) the presence of a biomolecule (antigen, antibody, DNA, RNA and whole pathogen) in a complex sample (blood, urine and other kinds of samples). Diagnostic tests are also used to classify different genes belonging to a same family or to a large panel. For instance, DNA biochip technology on microsystems has been intensively developed to answer the increasing need of DNA sequence multianalyses in reduced time and small volumes.

To reach such goals, several strategies have been developed. One of them relies on the immobilization of single-stranded DNA sequences or *oligonucleotides* (ODN) on synthetic polymer chains. The resulting polymer/ODN conjugates are a smart alternative to free ODN probes in order to capture the DNA target. First assays based on random copolymers led to improved results of the test in terms of signal intensity and sensitivity [4, 61]. To control the orientation of the conjugate on the solid support (to favor ODN probe accessibility), it appeared challenging to develop new kinds of conjugates based on graft and block copolymers. To synthesize such well-defined architectures, the RAFT process was chosen since it can be applied to a wide range of monomers (especially acrylamide derivatives) under smooth experimental conditions.

Favier developed the synthesis of graft copolymers via the RAFT process, with a backbone based on NAM and NAS monomers, respectively, bringing hydrophilicity and reactive groups. Glycopolymer grafts of poly(galactovinylether) were grafted onto these reactive units. Since the side chains provided aldehyde groups from the galactose units, it was possible to further bind $NH_2$-ODN. The resulting graft copolymer/ODN conjugate was evaluated in an ELOSA test for hepatitis B virus using the VIDAS platform developed by bioMérieux Company. The sensitivity limit was significantly improved in comparison with free ODN probes [62].

More recently, De Lambert et al. carried out the synthesis of amphiphilic block copolymers via the RAFT process, based on acrylamide derivatives [63]. The hydrophobic block consisted of *t*BAM and was designed to adsorb preferentially on the solid support, inducing a favorable orientation of the hydrophilic block in aqueous solution. The hydrophilic block consisted of a reactive random copolymer having some NAS units able to bind starters for ODN direct synthesis. The resulting amphiphilic block copolymer P[*t*BAM-*b*-(NAM-*co*-NAS)] was used to elaborate a block copolymer/ODN conjugate corresponding to polyd$T_{25}$ model sequence. A parallel conjugate synthesis was performed on a poly(NAM-*co*-NAS) random copolymer.

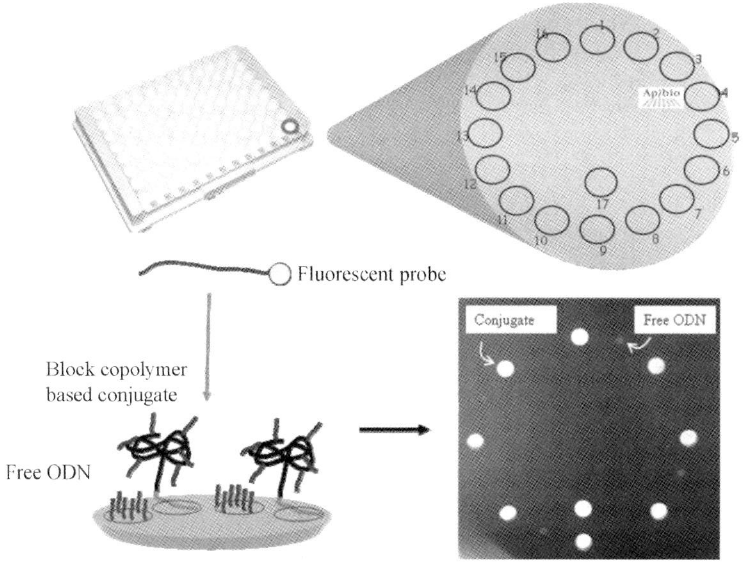


**Fig. 13.6** Hybridization on DNA biochip microarray using free ODN probes and polymer/ODN conjugate based on a block copolymer synthesized by the RAFT process.

The microarray system was based on a 96-well microplate format and the conjugates were spotted using a nanodroplet inkjet technique. When using a standard aqueous buffer, both conjugates – elaborated from random and block copolymers – similarly enhanced the signal corresponding to polydA$_{25}$ target hybridization in comparison with free ODN probes (Fig. 13.6). Moreover, when conjugates were spotted in a mixed DMF/$H_2O$ solvent, the conjugate based on the block copolymer reached a double intensity signal than that corresponding to the random copolymer that confirms the improved orientation mediated by the amphiphilic diblock [12].

The same strategy was then applied to two biological models especially difficult to study with classical systems. The first model, concerning the detection of alleles associated with insulin-dependent diabetes, was investigated to estimate the sensitivity threshold and the specificity. The second biological model, concerning blood platelet polymorphism determination, was studied to detect a single nucleotide mismatch. For both models, the polymer/ODN conjugates led to an increase in signal intensity in comparison with free ODN probes and to a better specificity at low DNA target concentration [13]. These successful results show that block copolymer/ODN conjugates can be useful tools for a wide range of biological models.

In conclusion, RAFT polymerization has been successfully applied to in vitro diagnostics and is currently explored for in vivo diagnostics, especially for magnetic resonance imaging (MRI) as described in a subsequent section (Section 13.3.3.2).

## 13.2.5
## Drug and Gene Delivery Applications

### 13.2.5.1 Drug Delivery

Considering therapeutic applications of polymers, controlled drug delivery is the most active area. Polymer systems are used as carrier or *vector* to safely transport drugs to appropriate body sites and to control the release rate [64]. The role of the polymer is multiple. First, it should promote the protection of the drug from any degradation or unwanted uptake from the body's immune system (notion of furtivity or shielding). Second, it may include specific moieties (sugars, peptides and antibodies) on the surface of the drug-loaded vector able to drive it to targeted cells or tissues (notion of targeting). Third, it should favor the release of a sufficient amount of the drug when required or over a desired timescale. Finally and most importantly, the polymer (and its possible degradation products) should not induce any toxic/inflammatory response and should be excreted out of the body. The toxicity and the resorbability are issues that need to be addressed. All these parameters as well as those relative to the drug characteristics (nature, hydrophilicity/hydrophobicity and sensitivity to degradation) and action mode (appropriate locus of release and optimal release profile) have to be considered when designing the polymers [65].

Generally, the polymer systems used for drug delivery are micelles/vesicles/nano- or microparticles [66–70], (hydro)gels [71] and branched macromolecules, as recently reviewed by Kumar et al. [72].

Due to its versatility toward polar and/or charged monomers and site-specific functionalization, RAFT polymerization has rapidly been considered to improve the properties of various kinds of polymer vectors. Controlled drug delivery is often mentioned as a potential application; however, only very few studies have reported the loading capacities and release properties.

**Micelles, Vesicles and Nanoparticles** Micelles, vesicles and particles (generally nanosized) are prepared via self-assembly or aggregation of amphiphilic block copolymers in water. Since a wide range of well-defined block copolymers is available, micelles/vesicles and nanoparticles are to date the most explored carriers. The drug can indeed be embedded in the core during the micellization process, while the hydrophilic outer shell can promote furtivity and dispersion of the drug/polymer system in water. Furthermore, since reactive functions or specific moieties can be introduced along [10] or at the chain end [25] of the hydrophilic blocks, it is possible to tailor the surface properties of the nano-objects (Fig. 13.7).

The block copolymers may include more than two blocks, blocks with branched architectures and stimuli-responsive blocks depending on conditions such as temperature, pH, ionic strength, light, electric or magnetic field. These block copolymers can be fully synthesized by the RAFT process or may possess segments differently produced, such as poly(ethylene oxide) (PEO) [73] or poly(lactic acid) (PLA) [74, 75]. Finally, one or several blocks can be cross-linked after self-assembly in order to form well-defined nanogel domains and 'freeze' the structure of the nano-objects.

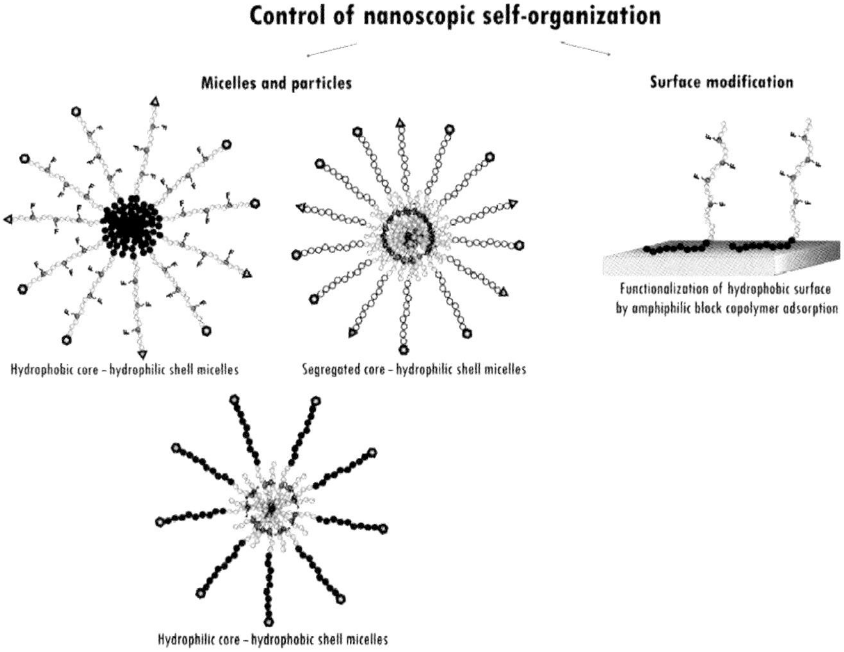


**Fig. 13.7** Examples of self-organization of NAS-containing block copolymers in solution or at the surface of hydrophobic supports.

Consequently, the number of combination to elaborate the carriers is very large and several have already been explored:

- The self-assembly in water of amphiphilic diblocks bearing a neutral hydrophilic block and a classical hydrophobic block has been studied by different groups [40, 63, 76, 77]. Yusa et al. demonstrated that aqueous solution of an amphiphilic poly(2-methacryloyloxyethylphosphorylcholine-*b*-butyl methacrylate) copolymer was able to solubilize a significant amount of paclitaxel, a poorly water-soluble anticancer drug [40]. The stability of the micelles/aggregates could be increased by cross-linking the hydrophobic cores [73, 78].
- Stimuli-responsive micelles have also been widely studied. The various synthetic routes to obtain these 'smart' materials have been recently compiled by McCormick et al. and will not be detailed here [79]. The different kinds of stimuli-responsive block copolymers can be classified according to their nature:
  - hydrophilic block copolymers that bear only one stimuli-responsive segment, either temperature responsive [80–85] or pH- and ionic strength responsive [83, 86–92],
  - hydrophilic block copolymers that bear two stimuli-responsive segments or 'schizophrenic' block copolymers [83, 85, 93–97] and
  - amphiphilic block copolymers that bear one or several stimuli-responsive segments [39, 74–76, 83, 98–102].

Compared to other polymerization techniques, RAFT appears as a very powerful technique to synthesize well-defined block copolymers with thermoresponsive segments, generally poly(*N*-isopropyl acrylamide) (PNIPAm), and/or pH-sensitive segments, such as poly(acrylic acid) (PAA) [89]. In water, the stimuli-responsive segments are able to reversibly switch from fully soluble to insoluble after a sharp transition in temperature or in pH/ionic strength, inducing the reversible formation of micelles. It is noteworthy that the modification of the terminal thiocarbonylthio RAFT function can be used to introduce a fluorescent probe at the end of the thermoresponsive poly(NIPAm) chains to study their conformation at the surface of poly(benzyl methacrylate) nanoparticles [101].

- Finally, the cross-linking of the micelle/nanoparticle shell can also be used to prevent dissociation [103], the cross-links being created by chemical bonding [16, 104, 105] or electrostatic interactions [106]. Clickable reactive functions coming from the initial RAFT agent can also be introduced in the cross-linked PAA outer corona [105]. Then, various entities such fluorescent probes or targeting moieties can be bound at the surface of the nanoparticles.

**Hydrogels** Hydrogels are another type of interesting carriers for drug delivery [71]. They consist of a cross-linked hydrophilic network that swells in aqueous solutions and then progressively releases the drug. The chemical or physical cross-links (or nodes) of the network are, respectively, obtained by covalent bonding and electrostatic or hydrophobic interactions.

The RAFT process can be used to improve the control of the physicochemical properties of the gels since they depend on the gel structure and degree of cross-linking (number and repartition of the cross-links), and thus on the architecture of the connecting chains. For instance, Crescenzi et al. have shown that gels obtained via RAFT polymerization from methacrylated pullulans are more regular and homogeneous and exhibit higher swelling properties than corresponding gels prepared by conventional polymerization [58].

In addition, it appears highly desirable to control the synthesis of polymers known to form physical gels, such as poly(vinylalcohol) obtained after hydrolysis of poly(vinyl acetate) [107–109], poly(acrylamide) [86, 110], as well as well-defined random copolymers of hydrophilic and hydrophobic monomers [99, 111].

Finally, since the RAFT process is particularly suited to produce well-defined stimuli-responsive architectures, 'smart' gels can be elaborated with improved properties [112–115]. They are commonly used for drug delivery due to their ability to expand or shrink, depending on the external conditions of pH and/or temperature.

**Branched Macromolecules** Branched macromolecules like star, comb, hyperbranched polymers or dendrimers [116] (Fig. 13.5) are also very interesting architectures (easily synthesized by RAFT polymerization) for the preparation of nanocarriers and hydrogels for drug delivery. One advantage of branched macromolecules relies on their ability to form unimolecular or multimolecular aggregates in diluted aqueous solutions [117–119] or highly functionalized gels in concentrated media [120–122].

#### 13.2.5.2 **Gene Delivery**

The objective of gene therapy is to introduce genetic material into specific cells (transfection) in order to replace deficient genes that are source of diseases [123]. Moreover, gene transfection can be used for the expression of biologically active proteins of interest. Besides therapeutics, applications also enter the field of vaccination.

Among the potential carriers for gene delivery, liposomes and cationic polyelectrolytes have been considered as an alternative to inactivated viral vectors for biosafety reasons despite their recent significant improvements. Cationic polymers like poly(ethylenimine), PEI [124, 125], are indeed able to complex DNA or RNA (negatively charged) to form interpolyelectrolyte aggregates sometimes called polyplexes [126]. This electrostatic complexation of the gene provides protection against enzymatic degradation and ensures the necessary compaction to enter the targeted cells.

In this context, the RAFT process opens new possibilities in terms of vector design since it does control the polymerization of polar cationic or cationizable monomers. For instance, well-defined homopolymers of P(dimethylaminoethyl methacrylate), PDMAEMA, were prepared [127], since the corresponding polymers prepared via conventional radical polymerization had led to very significant transfection results [128]. Moreover, reactive poly(NAM-*co*-NAS) building blocks were modified to form cationizable blocks. The NAS units were reacted with derivatives such as spermine, ethylene diamine or *N,N*-dimethyl ethylene diamine [62]. This is a smart strategy to obtain RAFT polymers bearing primary or secondary amine functions known to induce aminolysis of the RAFT agents. In addition, the grafting reaction of aminated side chains on the preformed polymer enables aminolysis of the thiocarbonylthio chain end that may induce toxicity issues.

The resulting well-defined polymers that carry primary, secondary or tertiary amine side groups were cationized over a large range of pH and exhibited low toxicity in 3-(4,5-dimethylthiazol-2-yl)-2,5-diphenyltetrazolium bromide (MTT) cell assays. All were shown to efficiently complex plasmid DNA as evidenced by agarose gel electrophoresis and picogreen displacement assays and to efficiently protect the plasmid from DNAses. Preliminary in vitro transfection tests of a plasmid coding for the expression of the fluorescent luciferase protein were performed with BHK-21 (baby hamster kidney) cells. The production of luciferase was low, indicating a too strong complexation of DNA that prevented its release. In further assays, a varying density of alkyl chains (dodecylamine) were bound to the polymer chain in addition to the *N,N*-dimethyl ethylenediamine side chains in order to try and decrease the strength of the plasmid complexation to favor its subsequent release. The resulting complexes were stable in the presence of a serum despite an incomplete compaction of the plasmid (depending on the charge density). However, the transfection results were not improved, indicating that the presence of hydrophobic side chains was not a preponderant parameter on the release ability [14].

The use of hydrophilic block copolymers bearing one neutral block and one cationic block is also a very attractive strategy since it enables the preparation of micellar aggregates with a polyplex core surrounded by water-soluble blocks (Fig. 13.7). The cationic blocks ensure the compaction of the plasmid, while the hydrophilic neutral blocks provide a corona with furtive properties toward the body's natural defences. Moreover, the hydrophilic blocks may also carry specific entities like sugars and/or peptides in order to improve the targeting properties of the vector and to facilitate entrance through the cell membrane.

This kind of core-shell polyplexes can be obtained from P[NAM-*b*-(NAM-*co*-NAS)] precursors synthesized using either a classical two-step method or a more convenient one-pot procedure by monomer sequential addition [10]. Another possibility developed by McCormick et al. for small interfering ribonucleic acid (siRNA) delivery applications is the synthesis of polymers consisting of one block of P(hydroxypropyl methacrylate), HPMA and one block of *N*-[3-(dimethylamino)propyl]methacrylamide, DMAPMA [129]. The cationic segments bearing tertiary amine groups provided a strong complexation of siRNA and the poly(HPMA) hydrophilic neutral blocks ensured protection of the nucleic acids from enzymatic degradation. In addition, the observed slow dissociation of the complexes suggested that siRNA may be released after entering the cells.

Another class of polyplex vectors can be obtained from grafted polymers. Bisht et al. synthesized carboxylic acid $\alpha$-functionalized poly(NIPAm) chains, either by RAFT ($M_w = 17\,000$ g·mol$^{-1}$; PDI = 1.02) or by conventional radical polymerization in the presence of a mercaptan irreversible transfer agent ($M_w$ = 4100 g·mol$^{-1}$; PDI = 2.1). The chains were subsequently bound to amino side groups of PEI [130]. The resulting graft copolymer with a cationic backbone and thermoresponsive side chains was studied as a potential candidate for thermoactivated DNA delivery. In this study, the controlled length of the poly(NIPAm) side chains seemed to have only a weak influence on the thermoresponsive and DNA complexation properties.

In conclusion, the contribution of RAFT polymerization to controlled gene delivery is potentially very important to improve both vector properties and understanding of the polyplex behavior in vitro and in vivo. Very few examples have been reported to date; however, various cationic architectures can be designed to tailor the complexation, compaction, protection, furtivity, targeting and release properties of the vectors. In addition, the possibility to introduce a function at the $\alpha$- or $\omega$ end of the polymer chain may be used to tag the vector in order to track its localization and fate in/out the cells or in the body. This could provide valuable information about the transfection mechanism and the optimal design of the vectors.

#### 13.2.5.3 **Outlook on Drug and Gene Delivery**

CRP techniques enable the access to more sophisticated controlled release systems via the control of the macromolecular architecture. The RAFT process is especially well suited due to its tolerance toward polar hydrophilic monomers. Well-defined stimuli-responsive and charged blocks can be directly synthesized in aqueous media and used to tailor the release profile. In addition, the large number of possibilities in terms of site-specific functionalization will transform the usual polymer carriers into bioactive vectors able to target specific cells and control the locus of release.

Nevertheless, important issues have still to be addressed, concerning the loading of the vectors and their in vivo behavior. For instance, the vectors (as well as their embedded material) have to remain intact during transportation to the locus of release and should not induce unacceptable inflammatory responses and toxicity, before, during and after the release. Then, the polymer materials should be removed from the body using the natural excretion pathways. All these critical parameters need to be considered when designing the polymers.

Finally, the new generation of delivery devices now accessible using techniques like the RAFT process should also provide new opportunities in various other fields, such as agriculture, cosmetics and personal care.

### 13.2.6
### Tissue Engineering and Regenerative Medicine

Tissue engineering applications lie at the interface of material and life science. The objective is to develop organ or tissue substitutes using living cells such as stem cells in order to replace, restore, maintain or improve deficient biological functions [131]. Whereas the first generation of materials in contact with or implanted inside of the human body was mainly biologically inert such as metallic hip substitutes, the introduction of bioactive glasses and ceramics and bioresorbable polymer materials have opened the way for advanced therapeutic strategies [132]. The new generation of bioactive materials is now designed to help the body repair itself (bones, cartilage, skin, vascular grafts, nerves, heart valves, liver or kidney).

Among these materials, one- (fibers), two- (films and membranes) or three-dimensional (gels and porous matrices) polymer materials are used to build scaffolds/templates that support the cells and guide a biomimetic tissue-regeneration process [133]. They are mainly based on biodegradable synthetic polymers such as polyesters (poly(lactic-*co*-glycolic acid), PLGA) or natural biopolymers such as chitosan, hyaluronic acid or collagen. The recourse to C—C backbone synthetic polymers such as poly(vinylalcohol) or poly(methyl methacrylate) is less widespread for in vivo applications due to their low biodegradability. Nevertheless, they can be used for the preparation of permanent material for orthopedic or dental applications and for external devices such as artificial liver or kidney and wound-healing applications.

From an engineering point of view, the most important challenge is to develop biocompatible, biodegradable or bioresorbable scaffolds with appropriate chemical, mechanical and biological properties for an optimal regeneration process [134]. In this context, RAFT polymerization can provide precious tools to tailor both the mechanical properties of the scaffold and the interface between the scaffold and the biological environment. First, the versatility toward a large variety of monomers and polymer architectures enables synthesis of adaptable ground for cell adhesion/proliferation and for the establishment of important biological processes like angiogenesis. Moreover, it is possible to produce low-molecular-weight building blocks more easily excreted out of the body and site-specific functionalized polymer chains like well-defined bioconjugates (for instance RGD peptide/polymer conjugates [135]).

Polymer materials for tissue engineering are usually based on hydrogels [136] or biodegradable porous matrices [133], which may contain controlled release systems (similar to those used for drug delivery) in order to progressively release bioactive molecules such as growth factors (proteins) that favor the regeneration process. Although the direct use of RAFT process for tissue engineering applications is still not widespread, it can provide interesting alternatives. On the one hand, RAFT polymerization can be used to produce highly functional hydrogels with improved physicochemical properties. RAFT-modified hydrogels based on natural biopolymers seem particularly promising as reported by Crescenzi et al. about reticulation of pullulan [58]. On the other hand, polymers obtained from the combination of RAFT with another polymerization technique such as ring-opening polymerization [74, 75, 137], or polymers obtained by grafting RAFT blocks onto natural biopolymers [56, 60], are also expected to give biodegradable materials with improved properties. Finally, Suzuki et al. have synthesized phosphate polymers via the RAFT process that can be mineralized by hydroxyapatite (calcium phosphate, CaP) [138]. Since CaP is the main component of bone matrices, this kind of inorganic/organic hybrid composites can find potential application in bone regeneration.

## 13.3
## Polymer-Based Materials for Various Applications

### 13.3.1
### Polymers with Fluorescent or Optoelectronic Properties

Since about 20 years, it has been demonstrated that polymer films may be used for a new generation of optoelectronic devices, such as organic light-emitting devices and organic solar cells [139, 140]. Moreover, fluorescent polymers may find numerous applications as biological or chemical sensors in military, biomedical and industrial fields [141].

Quite a large number of articles describe the use of RAFT polymerization to synthesize polymers bearing fluorescent or luminescent compounds, either as side groups along the chain or as end groups (introduced via the R or Z group of the CTA) (Table 13.3).

The introduction of several fluorescent molecules along the chain can result from the polymerization of a fluorescent monomer or from the covalent binding of a fluorescent derivative onto a precursor copolymer. The first strategy requires an easy synthesis pathway able to provide grams of the fluorescent monomer. Moreover, in case of an ionic derivative, the choice of the polymerization solvent may be limited.

**Table 13.3** Fluorescent or luminescent polymer chains synthesized by the RAFT process

| Fluorophore | Polymer | Properties and/or potential applications | Ref. |
|---|---|---|---|
| **Main chain fluorescent monomer** | | | |
| Anthracene | Alternated copolymer (ROP+RAFT) | Blue emission (365 nm) | [142] |
| An | poly(anthracene-*alt*-styrene) | | |
| **Side group fluorescent monomer** | | | |
| Carbazole | P(*N*-vinyl-carbazole) | Excimer emission (454 nm) | [143] |
| Vinyl derivative | P(*N*-ethyl-3-vinylcarbazole) | Polymeric LED | [144] |
| | P(S-*b*-*N*-ethyl-3-vinylcarbazole) | Photovoltaic devices | |
| | | Photorefractive materials | |
| Carbazole | Functional polymers | Hole transfer ability | [145] |
| Methacrylate derivative | Homopolymer | Electron transfer ability | |
| Oxadiazole | Homopolymer | Photoluminescence | |
| Methacrylate derivative | | Organic electronics | |
| **Side group covalent binding** | | | |
| | Postmodification of a polymer: | | |
| Coumarin 3 and 343 | P(4-vinylbenzylchloride) | Multilayered films for LED | [146] |
| Diphenylquinoline | Linear and star P(styrene) | Blue light emitting (560 nm) porous polymer films (electronic devices) | [147] |
| Lucifer yellow | P(NAM-*co*-NAS) | Highly fluorescent polymer (516 nm) for signal amplification (diagnostic tool) | [15] |
| **$\omega$ end group postreaction on polymer —SH** | | | |
| Pyrene maleimide | P[BzMA-*b*-(NIPAm-*co*-DMA)]-SH | Stimuli-responsive nanoparticles | [101] |
| | | Supramacromolecular sensors | |
| | P(NIPAm)-SH | Potential drug-delivery applications | [148] |

*(Continued)*

Table 13.3 (*Continued*)

| Fluorophore | Polymer | Properties and/or potential applications | Ref. |
|---|---|---|---|
| **$\omega$ end group Z group of the CTA** | | | |
| Carbazole | P(MA) | No optical properties studied | [149] |
| | Carbazyl = efficient stabilizing group for RAFT process | | |
| Naphthalene | P(GMA) | Precursor polymer with epoxy side groups | [150] |
| | P(NaphA) | UV absorption (270 nm) | [151] |
| | P(NaphA-*b*-MMA) | Fluorescence emission (444 nm) increases with $M_n$ | |
| | P(NaphA-*b*-S) | | |
| **$\alpha$ end group R group of the CTA** | | | |
| Anthracene | An-P(AcNaph) | Light harvesting polymers | [152] |
| An-dithiobenzoate (ester link) | An-P(AcNaph-*b*-MA) | Intrachain energy transfer | |
| An-dithiobenzoate (C—C link) | An-P(S) | | [153] |
| | An-P(S-*b*-MA) | | |
| 9,10-diphenyl anthryl | dpAn-P(AcNaph-*alt*-MalAnh) | Light harvesting polymers | [154] |
| dpAn-Dithiobenzoate (C—C link) | | | |
| Naphthalene | RAFT polymerization of various derivatives: | No properties studied | [155] |
| Naph-dithiobenzoate (amide link) | | | |
| Water soluble | Ionic and nonionic acrylate | | |
| | Anionic and zwitterionic styrene | | |
| | Cationic acrylamide | | |
| Pyrene | Py-P(S) | | [156] |
| Py-dithiobenzoate (C—C link) | Py-P(S-*b*-MA) | | |
| Phenanthrene | Phe-P(DcAAm) | Light scattering and fluorescence studies | [26, 157] |
| Phe-dithiobenzoate (amide link) | Phe-P(DcAAm-*b*-DEAAm) | Micelles in water | |
| Coumarin | Coum-P(AcNaph) | Light harvesting polymers | [158] |
| Coum-dithiobenzoate (C—C link) | Coum-P(AA-*b*-AcNaph) | | |

| | | | |
|---|---|---|---|
| Spiro-oxazine | SpOx-P(BA) | Photochromic transitions; | [159] |
| SpOx-trithiocarbonate (ester link) | SpOx-P(S) | Influence of the low $T_g$ block on the photochromic switching rates | |
| | SpOx-P(BA-*b*-S) | | |
| | SpOx-P(S-*b*-BA) | | |
| Quantum dots (Cd–Se) | Qd-P(S) | UV and fluorescence properties | [160] |
| Qd-trithiocarbonate | Qd-P(MA) | LED | |
| | Qd-P(BA) | Tunable lasers | |
| | Qd-P(S-*co*-MA) | Photovoltaic cells | |
| | Qd-P(S-*co*-AA) | Biological tags | |
| | Qd-P(S-*co*-isoprene) | | |
| | Qd-P(S-*b*-MA) | | |
| | Qd-P(S-*b*-BA) | | |
| **$\alpha$ end group R group of the CTA** | | | |
| Ruthenium/Ligand complex | | | |
| $Ru^{II}$-bipyridine-dithiobenzoate (ester link) | $Ru^{II}$-bipy-P(SCoum) | Light harvesting from Coumarin to $Ru^{II}$ bipyridine energy trap core | [161] |
| | Linear and six-arm homopolymer stars | | |
| | Six-arm block copolymer stars: | | |
| | $Ru^{II}$-bipy-P(SCoum-*b*-AcNaph) | EET from AcNaph to $Ru^{II}$-bipy core through Coumarin sequences | [162] |
| | $Ru^{II}$-bipy-P(SCoum-*b*-AcNaph-*b*-NIPAm) | | |
| | $Ru^{II}$-bipy-P(SCoum-*b*-NIPAm) | | [163] |
| Bipyridine-dithiobenzoate (C—C link) | P(S-bipy$Ru^{II}$-S) | Metallopolymers | [164] |
| | | Fluorescence (625 nm) | |
| | P(NIPAm-bipy$Ru^{II}$-NIPAm) | Thermosensitive metallopolymers | [165] |
| Terpyridine-dithiobenzoate (C—C link) | P(S-terpy-$Ru^{II}$-terpy-S) | (Sensors and luminescent films) | [166] |
| | P(S-terpy-$Ru^{II}$-terpy-NIPAm) | | |
| **$\alpha$ end group R group of the CTA** | | | |
| Europium/Ligand complex | | | |
| $Eu^{III}$Lig-dithiobenzoate (C—C link) | $Eu^{III}$Lig-P(MMA) | Luminescence titration (ppb range) | [167] |
| | $Eu^{III}$Lig-P(S) | Molecularly imprinted polymers | |

S = styrene; NAM = *N*-acryloylmorpholine; NAS = *N*-acryloxysuccinimide; BzMA = benzyl methacrylate; NIPAm = *N*-isopropylacrylamide; DMA = dimethyl acrylamide; MA = methyl acrylate; GMA = glycidyl methacrylate; NaphA = 2-naphtyl acrylate; MMA = methyl methacrylate; AcNaph = acenaphtylene; MalAnh = maleic anhydride; DcAAm = decyl acrylamide; DEAAm = diethyl acrylamide; AA = acrylic acid; BA = butyl acrylate, SCoum = styrene-coumarin.

This strategy has mainly been used for carbazole derivatives in order to prepare organic electronic materials [143–145]. The second strategy is favored when the precursor polymer is bearing highly reactive functions, for instance activated ester ones, which can bind amino- or hydroxyl fluorescent derivatives [15].

Alternatively, the introduction of an unique fluorescent label at one chain end has also been investigated. Pyrene, naphthalene and carbazole moieties have been introduced as $\omega$ end groups, whereas anthracene, phenanthrene, naphthalene, pyrene and coumarin have been introduced as $\alpha$ end groups via the R group of the CTA (a dithiobenzoate in all cases), with either an ester, an amide or a C—C bond between the fluorescent moiety and the thiocarbonylthio group (Table 13.3).

An original study reports RAFT polymerization from cadmium–selenium quantum dots, semiconductor nanoparticles with narrow fluorescence emission profiles and discrete energy bands. The strategy relies on the anchoring of a functional trithiocarbonate onto the quantum dots via di-*n*-octylphosphine-oxide ligand, a derivative of the usual tri-*n*-octylphosphine-oxide ligand. Various random and block copolymers were polymerized from the surface in order to improve the stabilization and the dispersion of the quantum dots in polymer matrices. The tailored quantum dot nanoparticles retained their structural and optical properties [160].

Ligands of ruthenium and europium have also been introduced in the R group of dithiobenzoates. After RAFT polymerization, the complexation of $Ru^{II}$ or $Eu^{III}$ leads to metallopolymers with luminescent properties [161, 162, 164–167]. A special application of the $Ru^{II}$-bypyridine-based dithioester consists in elaborating light harvesting polymers with a styrene derivative of coumarin. Indeed, excitation energy transfer from coumarin to the $Ru^{II}$ bipyridine energy trap core is observed, especially in the case of star architectures [161]. Moreover, introduction of a poly(NIPAm) second block induces a decrease of the competing quenching effect of the dithioester chain end.

Such quenching effect of the dithioester moiety has been reported in most of the articles about fluorescent polymers synthesized by RAFT. Recent fluorescence investigations using coumarin 343 demonstrated that static and dynamic quenching result from the reversible formation of an exciplex between coumarin and the dithiobenzoate chain end. This fluorescence quenching could be totally suppressed by aminolysis into a thiol end group [168].

### 13.3.2
### Thin Films and Membranes on Organic Substrates

Polymer grafting techniques provide a versatile tool to tailor surfaces of solid substrates. Two strategies have been explored. The first one consists in a reversible coating via the adsorption of polymer chains using electrostatic or hydrophobic interactions. The second approach consists in an irreversible grafting via covalent attachment of polymer chains to the surface, either by 'grafting to' or by 'grafting from'. The 'grafting-to' technique involves a reaction between one or several appropriate functional groups on the polymer chain and the surface, to chemically tether the polymer chains. However, steric hindrance near the surface (due to the already-grafted chains) limits the diffusion of incoming polymer chains, which usually results in low grafting density. The 'grafting-from' technique is based on the anchoring of initiators to the surface, followed by polymerization from the surface. The diffusion of small-sized monomers is little affected by the existing grafted chains, which significantly increases the grafting density. Finally, polymer chains can be immobilized on a surface as (i) thin films, either by adsorption or using a multiple point covalent anchoring ('grafting to'), or as (ii) polymer brushes, via a unique point covalent anchoring ('grafting to' and 'grafting from').

The CRP techniques have been intensively used for surface-initiated polymerization in order to modify the surface properties of various substrates. Homopolymers, diblocks, triblocks and star architectures can be easily tethered to the surface to produce brushes, multilayers and patterned surfaces. Among the various CRP techniques, the RAFT process is compatible with a wide range of monomers and requires easy experimental conditions. It has been applied to the synthesis of thin films, polymer brushes, honeycomb films and membranes on various organic substrates.

#### 13.3.2.1 **Thin Films and Polymer Brushes on Organic Substrates**

Polymer brushes resulting from surface-initiated polymerization are of high density that induces highly anisotropic conformation of the swollen brushes in suitable solvent. Such coatings bring strong resistance against compression and interesting size-exclusion properties. One of the most important requirements is a suitable polymerization solvent preventing the surface from swelling and leading to well-defined graft systems. RAFT polymerization has been used to improve the properties of different organic substrates, such as fluoropolymers, cellulose, poly(dimethylsiloxane) (PDMS), poly(propylene) (PP) and poly(styrene) (PS) (Table 13.4). The last part of the table reports a class of polymers defined as adhesives synthesized via RAFT polymerization in solution.

**Fluoropolymer Substrates** Among the different polymer substrates, fluoropolymers have often been studied due to their chemical and physical resistance. Concerning their biomedical applications as biomaterials, the key parameter is to improve the surface hydrophilicity of fluoropolymers to avoid contaminations. To reach this objective, Yoshikawa et al. suggested to graft a biocompatible polar polymer, poly(hydroxyethyl methacrylate) (PHEMA), from a nonpolar substrate, poly(tetrafluoroethylene-*co*-hexafluoropropylene) (FEP). They successfully synthesized PHEMA brushes via the RAFT process from the pretreated substrate ($O_2$ plasma) and claimed that these biointerfaces are very promising [169]. A similar 'grafting-from' strategy was used to tether poly(methyl methacrylate) (PMMA) and poly(ethyelene glycol) monomethacrylate (PEGMA) brushes via the RAFT process on azo-functionalized poly(vinylidene fluoride) (PVDF) surfaces [170]. Another strategy was proposed by Grasselli and Betz using an electron-beam-induced RAFT-mediated graft polymerization. Acrylic acid monomer was polymerized from

**Table 13.4** Thin films and polymer brushes synthesized via the RAFT process on organic substrates

| Substrate | Polymer brushes | $M_n$ (g·mol$^{-1}$) (PDI) | Ref. |
|---|---|---|---|
| Fluoropolymer | | | |
| FEP | PHEMA | 20 000–80 000 (PDI < 1.5) | [169] |
| PVDF | PMMA, PPEGMA | 2000–10 000 (PDI < 1.4) | [170] |
| PVDF | PAA | | [171] |
| Cellulose | PS | 21 000 (1.11) | [60] |
| | PS | 20 000–40 000 | [59] |
| PDMS | PDMA, PVBTAC, PSS | 40 000–160 000 (PDI < 1.4) | [172] |
| PP | PMA | | [173, 174] |
| | PS, P(S-*co*-TMI) | <20 000 | [175, 176] |
| PS | PNAM | 20 000 (PDI < 1.2) | [177] |
| Synthesis of adhesives for adsorption on metal or polymer substrates | P(MMA-*b*-SAN), P(GMA-*b*-SAN), P(HEMA-*b*-SAN), P(DMAEMA-*b*-SAN), P(S-*co*-MAh-*b*-SAN) | 10 000–40 000 (PDI < 1.5) | [178] |
| | P(VC2-*co*-MA-*co*-HEA), P(VC2-*co*-MA-*co*-MAUPHOS) | 7000 (1.5) | [179] |
| | PETMA | 3000–16 000 (PDI < 1.3) | [180] |

HEMA = hydroxyethyl methacrylate; MMA = methyl methacrylate; PEGMA = (polyethyelene glycol) monomethacrylate; AA = acrylic acid; S = styrene; DMA = dimethylacrylamide; VBTAC = (*ar*-vinylbenzyl)-trimethylammonium chloride; SS = 4-styrene sulfonate; MA = methyl acrylate; TMI = *m*-isopropenyl-$\alpha,\alpha'$-dimethylbenzyl isocyanate; NAM = *N*-acryloylmorpholine; AN = acrylonitrile; GMA = glycidyl methacrylate; DMAEMA = 2-(dimethylamino)ethyl methacrylate; MAh = maleic anhydride; VC2 = vinylidene chloride; MA = methyl acrylate; HEA = hydroxyethyl acrylate; MAUPHOS = phosphonated methacrylate; ETMA = 2,3-epithiopropyl methacrylate.

irradiated PVDF in order to improve its surface properties without loosing its excellent mechanical properties [171].

**Cellulose Fiber Substrates** Another often studied substrate is natural cellulose, which presents several advantages: It is a renewable natural resource, recyclable, of low cost and has good mechanical properties. However, cellulose fibers are very absorbent due to their high hydrophilicity, which can lead to composites failure. To reduce this hydrophilicity and improve adhesion to more hydrophobic materials, a strategy relies on the surface modification of cellulose fibers by grafting hydrophobic polymer chains onto their surface. Perrier and coworkers defined a two-step method to anchor RAFT agents on cellulose and to grow PS brushes from the surface. Although the density of polymer chains has to be improved, the reduced hydrophilicity of the cellulose fibers is promising for future hydrophobic composites [60]. In the same time, Hernandez-Guerrero et al. directly grafted RAFT agents based on trithiocarbonate containing carboxylic groups onto cellulose. Then, PS brushes were successfully synthesized from cellulose, in order to prepare highly regular honeycomb-structured porous films [59].

**poly(dimethylsiloxane) Substrates** Surface-initiated polymerization has been performed on PDMS substrates, often used for inertness and structural properties. Bae et al. treated PDMS under microwave plasma in the presence of maleic anhydride and followed by hydrolysis, leading to the appearance of carboxylic groups on the surface. In the presence of surface-tethered 2,2′-azobis(2-methylpropioamidine) dihydrochloride initiator, neutral DMA, anionic 4-styrene sulfonate (SS) and cationic (vinylbenzyl)-trimethylammonium chloride (p-VBAC) monomers were successfully homopolymerized via RAFT from the surface. The polyelectrolyte layers are planned to be involved in a layer-by-layer process in order to generate stable hydrophilic surfaces, whereas poly(DMA) provides a biocompatible surface [172].

**poly(propylene) Substrates** Solid phases for organic compound and peptide synthesis have been studied since the first supported reaction proposed by Merrifield on PS resins. A large variety of solid phases were then proposed based on different monomers. As the RAFT process is versatile, the use of Merrifield resins was considered to remove all impurities of the polymerization medium (dead chains, monomers and residual RAFT agent) to make this process industrially viable. Perrier and coworkers suggested to bind the RAFT agent via the Z group, polymerized methyl acrylate, and claimed that such technique could lead to well-defined block copolymers [173, 174].

More recently, a new support was developed where PS is grafted from a more rigid PP scaffold. Polymer was grafted from the surface via $\gamma$-radiation-initiated RAFT polymerization at low temperature [175]. Recently, the same group adapted this method to copolymerize styrene and *m*-isopropenyl-$\alpha,\alpha'$-dimethylbenzyl isocyanate (TMI), as TMI is able to scavenge primary and secondary amine [176].

**poly(styrene) Substrates** Polymer colloids with biointeractive functional groups on the surface have many applications in biological diagnostics [181]. D'Agosto et al. described the synthesis of P(NAM) chains in the presence of a RAFT agent containing a propionic acid group, and their subsequent grafting onto amino-functionalized polystyrene latex particles in order to get hydrophilic hairy particles. This strategy is promising to introduce biologically active species in the hairy layer via the copolymerization of suitable functional monomers [177].

**Adhesives on Metal or Polymer Substrates** Promoters for adhesion on metal or polymer substrates have been synthesized in solution via the RAFT process. These well-defined polymers exhibit functional side chains from acrylonitrile and phosphorus- or sulfur-containing monomers.

Fan et al. carried out the copolymerization of styrene and acrylonitrile via the RAFT process, leading to poly(styrene-*co*-acrylonitrile) (PSAN) copolymer, well-known thermoplastic for its excellent properties, such as solvent resistance, thermal stability, transparency and processability. To improve the interfacial adhesion of different phases, block copolymers containing PSAN and a block of PMMA, PHEMA, PGMA or PDMAEMA were successively prepared [178].

Phosphorus-containing polymers are of great interest, as they can be used for different applications like flame retardant, ion-exchange resins, dental adhesives, adhesion promoters on metal substrates and biotechnology. Rixens et al. reported the controlled synthesis of random and block copolymers from halogenated and phosphonated monomers using the RAFT process. The phosphonated groups were introduced either by chemical modification of hydroxyethylacrylate copolymers or by copolymerization with a phosphonated methacrylate. As these copolymers have barrier properties due to the vinylidene chloride monomer as well as adhesion and anticorrosion properties due to the phosphonated monomer, potential applications in paints and surface treatments seem suitable [179].

Tebaldi de Sordi et al. focused on 2,3-epithiopropyl methacrylate (ETMA) monomer that presents two reactive centers. It can be polymerized by ring-opening polymerization through episulfide group or by radical polymerization through the double bond of methacrylate group. This polymer is of great interest since its sulfur-containing ring pendants can promote chemical adhesion to metals and polar surfaces, and it was already used as dental adhesive. Well-defined PETMA homopolymers and block copolymers were synthesized by RAFT polymerization [180].

#### 13.3.2.2 Honeycomb Films and Membranes

A wide range of applications have been suggested for honeycomb-structured porous films, like separation membranes, photonic devices, chemical sensors or cell growth substrates. These porous structures are obtained from casting solutions of star or block copolymers in an organic solvent (e.g. carbon disulfide) onto a glass substrate under a humid atmosphere. To get a high regularity, polymer chains should exhibit an amphiphilic nature. Stenzel et al. elaborated porous films based on various polymer architectures synthesized via RAFT polymerization [182]. For instance, polystyrene homopolymers, graft copolymers and star polymers were prepared in the presence of trithiocarbonate RAFT agents previously synthesized from α-D-glucose, cellulose and β-cyclodextrin, respectively [56, 183].

Various other copolymers synthesized by the RAFT process were casted as honeycomb-structured porous films: (i) a PS comb polymer after introduction of trithiocarbonate groups on a poly(styrene-*co*-hydroxyethylmethacrylate) random copolymer backbone [59], (ii) an amphiphilic block copolymer composed of a poly(styrene) block and a poly(acryloyl phosphorylcholine) block, playing on the length of each block in order to change the pore size [184] and (iii) a poly(styrene-*b*-acrylic acid) amphiphilic block copolymer customized by pyrrole template as this monomer coordinates with acid groups via hydrogen bonds, in order to stimulate biological systems with electric current, beneficial for the cell growth [185].

Polyimides (PI) have often been studied as dielectric and packaging materials in microelectronic industry because of thermal stability, chemical resistance, mechanical strength and good adhesion to semiconductor and metals. The use of materials of ultralow dielectric constant is required to reduce the power dissipation of integrated circuits. One interesting approach consists of preparing nanoporous materials since the incorporated air (dielectric constant of 1) greatly reduces the dielectric constant of the resulting structure. Fu et al. used this strategy and prepared a RAFT-mediated graft copolymer composed of a poly(amic acid) backbone (PamA) and PMMA side chains grown from RAFT agents bound on the pendant carboxylic groups. Nanoporous PI films were obtained after thermal imidization of PamA backbone under inert atmosphere, followed by thermal decomposition of PMMA side chains. As RAFT polymerization provides well-defined side chains, nanoporous films exhibited more uniform and smaller pores – leading to lower dielectric constants – in comparison with films prepared via conventional radical polymerization [186].

From similar materials, Chen et al. described the synthesis of a graft copolymer via the RAFT process using an ozone-activated fluorinated polyimide (FPI). Microfiltration membranes were formed from the amphiphilic poly(FPI-g-PEGMA) graft copolymer by phase inversion in aqueous media [187]. Another pathway was explored by Rzayev and Hillmyer in order to prepare nanoporous PS-based materials containing hydrophilic pores for water purification or biomolecule separation. poly(LA-*b*-DMA-*b*-S) triblock and poly(LA-*b*-S) diblock copolymers were synthesized by combining ring-opening polymerization and RAFT. In the case of the ABC triblock structure only, nanoporous PS with hydrophilic pores was obtained using a selective etching protocol to remove the PLA block [188].

A last application reports a simple way to develop nanoporous films for lithography (fabrication of microelectronic devices), using the self-assembly behavior of block copolymers. Bang et al. synthesized poly(S-*b*-MMA-*b*-EO) triblock copolymer via the RAFT process and obtained nanoporous arrays promising for lithography, taking advantage of PMMA block degradability and PEO block long-range ordering [189].

#### 13.3.2.3 **Conclusion on Thin Films and Membranes on Organic Substrates**

RAFT polymerization is a very suitable method to immobilize polymer architectures on a wide range of organic substrates, using the classical strategies, namely adsorption, 'grafting-to' and 'grafting-from' approaches. The latter still suffer from limitations since the anchoring of RAFT agents on various substrates remains difficult. The resulting thin films, nanoporous films and polymer brushes find applications in many fields, such as elaboration of microelectronic devices, ultrafiltration membranes and biomaterials.

### 13.3.3 Organic/Inorganic Hybrid Materials

Hybrids materials composed of synthetic polymer and inorganic substrate exhibit complex nanostructures, combine the inherent properties of both phases and reveal original characteristics due to the interaction between the phases. The different coating pathways described in Section 13.3.2 (adsorption, 'grafting to' and 'grafting from') can be adapted to inorganic surfaces. Here again, the RAFT process is promising due to the versatility and the simplicity of the method. In comparison with other CRP techniques, only few articles describe the application of RAFT polymerization to grow polymer brushes from inorganic substrates due to the difficulty to tether RAFT agents on the surface. In this section, we report the applications of the RAFT process to immobilize polymer chains on inorganic substrates of different shapes, such as flat inorganic surfaces, inorganic nanoparticles and carbon nanotubes. Applications involving silane-based polymer chains are also described in a last part.

#### 13.3.3.1 Flat Inorganic Substrates

RAFT polymerization has been performed on different flat inorganic substrates, such as silica wafers, oriented single-crystal silicon substrates and clays (Table 13.5).

**Silica Wafers** Baum and Brittain were pioneers to use RAFT polymerization to tether homopolymer, diblock or triblock copolymers on silica wafers [190]. Using an azo-immobilized surface, they grow polymer brushes of PS-*b*-PDMA and PDMA-*b*-PMMA [191]. More recently, Stenzel et al. carried out the anchoring of the RAFT agent (via the Z group) on the surface of aminated silica wafers and obtained stimuli-responsive glycopolymer brushes of PNIPAm-*b*-PAGA. Such brushes are promising because of the significant role of glycopolymers in specific recognition events [54].

Suzuki et al. performed the controlled polymerization of phosphate-derived monomers, monoacryloxyethyl phosphate (MAEP) and 2(methacryloyloxy)ethyl phosphate (MOEP), via the RAFT process because of their potential use in biomedical applications, especially for regeneration of bones. Indeed, polymer materials containing phosphate groups are able to initiate events that lead to CaP mineral nucleation and biomineralization. Homopolymers of PMAEP and PMOEP were synthesized with MWs below 20 000 $g{\cdot}mol^{-1}$, to avoid any cross-linking reaction due to residual diene impurities from monomers. Block copolymers including a poly[(2-acetoacetoxy)ethyl methacrylate] (PAAEMA) block were also prepared and immobilized on aminated silica wafers through the ketone side groups. Then, evaluation of the calcification behavior of these polymers was done by simulated body

**Table 13.5** Polymer brushes synthesized via the RAFT process on flat inorganic substrates

| Substrate | Polymer brushes | $M_n$ (g·mol$^{-1}$) (PDI) | Ref. |
|---|---|---|---|
| Silica wafer | P(S-*b*-DMA), P(DMA-*b*-MMA) | | [190, 191] |
| | P(NIPAm-*b*-AGA) | <40 000 (PDI < 1.25) | [54] |
| | PMOEP, P(MOEP-*b*-AAEMA), PMAEP, P(MAEP-*b*-AAEMA) | 5000–20 000 (PDI < 1.5) | [138] |
| | PAMPS, PAMBA, P(AMPS-*b*-AMBA), PDMAEA | 10 000–40 000 (PDI < 1.2) | [192] |
| Silicon crystal | PDMAPS, P(DMSAPS-*b*-SS) | | [193] |
| | PVBC, P(VBC-*b*-FS) | 30 000 (PDI < 1.4) | [194] |
| | PMMA, PHEMA, PDMAEMA, P(DMAEMA-*b*-MMA), P(DMAEMA-*b*-HEMA) | 5000–20 000 (PDI < 1.3) | [195] |
| Clay | PS, PBA, PMMA | <30 000 (PDI < 1.5) | [196] |
| | PS | 40 000 (1.3) | [197] |

S = styrene; MMA = methyl methacrylate; BA = butyl acrylate; DMA = dimethylacrylamide; NIPAM = *N*-isopropylacrylamide; AGA = *N*-acryloylglucosamine; MAEP = monoacryloxyethyl phosphate; MOEP = 2(methacryloyloxy)ethyl phosphate; AAEMA = 2-(acetoacetoxy)ethyl methacrylate; SS = 4-styrene sulfonate; DMSAPS = *N,N'*-dimethyl(methyl-methacryloyl ethyl) ammoniumpropane sulfonate; VBC = 4-vinylbenzyl chloride; FS = pentafluorostyrene; HEMA = hydroxyethyl methacrylate; DMAEMA = 2-(dimethylamino)ethyl methacrylate; AMPS = sodium-2-acrylamido-2-methylpropane sulfonate; AMBA = sodium-3-acrylamido-3-methyl butonate; DMAEA = *N*-(dimethylamino)ethylacrylamide.

fluid measurements. The number of phosphate group and their accessibility are key parameters, concerning the amount and type of formed mineral. Block copolymers brought a significant improvement in comparison with random copolymers, as they provide accessible ionic phosphate groups from the surface for calcium chelation and CaP nucleation [138].

Layer by layer is an important method for the creation of structured and functional thin films deposited on solid surfaces (elaboration of multilayers via electrostatic interactions between opposite charged polyelectrolytes). Potential applications are for instance optical and electronical devices, separation membranes, biosensors, cell-repellent surfaces and catalyst systems. CRP techniques and especially the RAFT process enable preparation of well-defined polyelectrolyte or polyzwitterionic chains as random, graft and block copolymers. Morgan et al. synthesized poly(sodium-2-acrylamido-2-methylpropane sulfonate) (PAMPS) and poly(sodium-3-acrylamido-3-methylbutonate) (PAMBA) homopolymers as well as PAMPS-*b*-PAMBA block copolymers. The multilayer film was created using cleaned silica wafers first in contact with the cationic polyelectrolyte solution. Authors studied the correlation between the polyelectrolyte architecture and the resulting film in terms of morphology, dimension and stimuli-responsive behavior [192].

**Oriented Single-Crystal Silicon Substrates** Oriented single-crystal silicon substrates are central to the semiconductor and microelectronics industry. To better design and manipulate the physicochemical properties of the substrate, self-assembled monolayers based on well-defined functional polymers have been grafted onto or from the surface, using controlled radical polymerization. Applications concern passivation layers in microfluidic and microelectromechanical systems, recognition layers in sensors or adhesion promoters for metal on silicon substrates. Baum and Brittain suggested first to use the RAFT polymerization to form polymer brushes on AIBN-anchored silicon surfaces with free RAFT agents [191]. Using the same way, Zhai et al. reported the synthesis of polybetaine brushes on a silicon surface. This surface-initiated RAFT polymerization was carried out with immobilized azo initiator [193]. At the same period, Yu et al. used an azo-immobilized surface to perform RAFT polymerization of vinylbenzyl chloride (with free RAFT agent in solution) in order to get well-defined PVBC brushes on oriented crystal silicon substrate. The benzyl chloride groups of the PVBC brushes were then derivatized into viologen groups that confer redox-responsive properties [194]. Recently, the same group described another way to functionalize silicon substrates by directly immobilizing RAFT agent on the surface via the Z group. This approach was validated by the synthesis of PHEMA homopolymer and PMMA-*b*-PDMAEMA block copolymer brushes [195].

**Clays** Nanocomposites based on clays are of interest due to the improvement of mechanical, thermal and gas barrier properties in comparison with more classical polymer blends. To produce these composite materials, a preliminary step is required, using an intercalation agent to enable diffusion of monomers between clay layers. Salem and Shipp treated montmorillonite with cationic vinyl monomers as intercalation agents. The RAFT polymerization of P(styrene) and P(butyl acrylate) was carried out with classical dithiobenzoates, but no significative contribution was observed on the properties of this nanocomposite [196]. More recently, Zhang et al. described a similar protocol based on an intercalation agent containing a dithiobenzoate group and cationic species. RAFT polymerization of styrene was successful and resulted in a higher thermal stability of this nanocomposite in comparison with PS [197].

#### 13.3.3.2 **Inorganic Nanoparticles**

Composites nanoparticles are of great interest because of the potential properties of nanoparticles in the magnetic, optical and electronic fields. Polymer grafting is an interesting technique to tailor the surface of nanoparticles and consequently the interface between nanoparticles and the polymer matrix. This tailoring leads from simple particle dispersions in polymer matrices to highly ordered structures by self-assembly. RAFT polymerization was successfully used to tether polymer chains on various inorganic nanoparticles, such as silica nanoparticles, gold nanoparticles, magnetic nanoparticles and colloidal calcite dispersion (Table 13.6).

**Silica Nanoparticles** Silica nanoparticles are solid supports with good chemical resistance, mechanical stability and reasonable costs. Tsujii et al. first synthesize oligomeric PS brushes on silica beads via ATRP techniques. In a second step, they convert the halogen chain end into a thiocarbonylthio function to perform RAFT polymerization of styrene from the particles surface [198]. As the method requires the presence of RAFT agents in solution, a large amount of free polymer chains were produced detrimental to tethered chains. Benicewicz and coworker suggested solving this disadvantage by anchoring the RAFT agent directly on the silica nanoparticle surface using a RAFT-silane coupling agent. Well-defined polymer brushes of PS, PnBA and PS-*b*-PnBA were synthesized [199]. The same group developed another method, the anchoring of a carboxylic group containing RAFT agent onto aminated silica particles, to form controlled PMMA brushes [200]. In order to improve the distribution of the polymer brushes, Zhao and Perrier tethered the RAFT agent via its Z group on silica particles. A wide range of homopolymers and block copolymers were successfully grafted on the surface [204]. Zhang et al. used a different pathway to functionalize silica particles with stimuli-responsive block copolymer. They first synthesized in solution block copolymer composed of a stimuli-responsive P(NIPAm) block and a P($\gamma$-methacryloxypropyltrimethoxysilane) functional block. The specific reaction between the surface and the trimethoxysilyl side groups leads to the formation of stimuli-responsive core-shell nanoparticles [201]. A third pathway was explored by Guo et al. to tether lactose-containing polymer on silica beads. First, well-defined glycopolymers were synthesized by RAFT process in solution. In a second step, these polymer chains were tethered on particles functionalized with vinyl-silane. The nanoparticles exhibit sugar groups on the surface, with potential applications for separation materials and for analysis of substances with biological activity [55].

**Table 13.6** Polymer brushes synthesized via the RAFT process on inorganic nanoparticles

| Substrate | Polymer brushes | $M_n$(g·mol$^{-1}$) (PDI) | Ref. |
|---|---|---|---|
| Silica nanoparticles | PS | 10 000–50 000 (PDI < 1.4) | [198] |
| | PS, P*n*BuA, P(*n*BuA-*b*-S) | 10 000–35 000 (PDI < 1.2) | [199] |
| | PMAEL | 3000–30 000 (PDI < 1.4) | [55] |
| | PS, PMMA | <40 000 (PDI < 1.2) | [200] |
| | P(NIPAm-*b*-MPS) | 20 000–80 000 (PDI < 1.2) | [201] |
| | P(MAA-*co*-EDMA) | | [202] |
| | PS, PMMA | | [203] |
| | PMA, PMMA, PBA, PS, P(MA-*b*-BA), P(MMA-*b*-MA), P(MMA-*b*-S) | <20 000 (PDI < 1.2) | [204] |
| Gold nanoparticles | PAMPS, PVBTAC, PDMA, PNaPSS, P(MAEDAPS-*b*-DMA) | 10 000–60 000 (PDI < 1.3) | [205, 206] |
| | | | [206] |
| | PNIPAm | 21 000 (1.2) | [207] |
| | PEO-*b*-PDMAEMA | 45 000 (1.15) | [208] |
| | PAA | 15 000 (1.3) | [209] |
| | PS, PDMAEMA, PAA | 10 000 (1.15) | [210] |
| | PEG-A-*b*-PBA | 35 000 (1.4) | [27] |
| Magnetic nanoparticles | P(TEA-*b*-AM) | 5000–30 000 11 000–30 000 | [211] |
| | PS, PAA | 15 000–25 000 (1.2–1.4) | [212] |
| Calcite nanoparticles | PAA | <5000 (1.3) | [213, 214] |

S = styrene; *n*BuA = *n*-butyl acrylate; MAEL = 2-*O*-methacryloyloxyethoxyl-(2,3,4,6-tetra-*O*-acetyl-β-D-galactopyranosyl)-(1,4)-(2,3,6)-tri-*O*-acetyl-β-D-glucopyranoside; MA = methyl acrylate; BA = butyl acrylate; AA = acrylic acid; I = isoprene; AM = acrylamide; TEA = trimethylammonium ethylacrylate methyl sulfate; AMPS = sodium 2-acrylamido-2-methylpropane sulfonate; VBTAC = (*ar*-vinylbenzyl)-trimethylammonium chloride; DMA = dimethylacrylamide; MAEDAPS = 3-(2-*N*-methylacrylamido)ethyl dimethyl ammonio propane sulfonate; DMAEMA = 2-(dimethylamino)ethyl methacrylate; EO = ethylene oxide; NIPAM = *N*-isopropylacrylamide; EG-A = ethylene glycol acrylate; NaPSS = sodium-4-styrene sulfonate; MMA = methyl methacrylate; MPS = $\gamma$-methacryloxypropyltrimethoxysilane; MAA = methacrylic acid; EDMA = ethylene glycol dimethacrylate.

More recently, Titirici and Sellergren used the molecular imprinting concept that relies on generation of polymer-based elements designed for specific recognition of a given target. The range of applications is very wide, including chemical sensing, drug delivery, catalysis, solid-phase extraction and chiral separation. Usually, polymer particles can constitute the molecular imprinted polymer, as they provide good affinity and specificity. However, they exhibit low capacity and poor accessibility for the target due to an irregular shape. Composites nanoparticles afford one solution, as it distinguishes the regular morphology (inorganic part) from the imprinting ability (polymer). They immobilized azo initiator on mesoporous silica nanoparticles. Then, graft copolymerization of methacrylic acid and ethylene glycol dimethacrylate in the presence of L-phenylalanine anilide led to imprinted thin film composite beads that proved to be a highly selective chiral phase, resulting in separation of the template racemate [202].

In order to develop industrial applications for RAFT polymerization, the first step may require the removal of the sulfur-containing RAFT end groups that may be a limitation for potential applications. Immobilizing the RAFT agent appears as a suitable approach to reach this goal. The use of silica nanoparticles as solid support seems well adapted. According to Nguyen and Vana, immobilizing the RAFT agent via its Z group presents several advantages: (i) to obtain a narrow distribution of polymer chains without dead chain population, (ii) to produce sulfur-free polymer chains by adding AIBN on particles to cleave the polymer brushes and (iii) to recover the initial RAFT agent immobilized on silica particles [203].

**Gold Nanoparticles** Gold nanoparticles are of particular interest, as they provide optical, magnetic and electronic properties to hybrid materials. Consequently, numerous applications are possible in various fields like chemical separation, sensing, catalysis or biotechnology. The colloidal stability of gold nanoparticles appears as a crucial parameter. To facilitate the self-assembly of nanostructures, stabilizing agents like polymers have been tethered on the surface of gold nanoparticles. RAFT polymerization is one of the most suitable techniques to do so. Moreover, the reduction of the RAFT thiocarbonylthio end group in a thiol function is especially adapted to the modification of gold surfaces. Lowe et al. first used this strategy with different water-soluble homopolymers or block copolymers with anionic, cationic, neutral or zwitterionic blocks [205]. These polymers were prepared in the presence of 4,4′-azobis(4-cyanopentanoic acid) and 4-cyanopentanoic acid dithiobenzoate as initiator and RAFT agent, respectively, providing free carboxylic groups on the surface, useful for biochip and high-throughput screening applications [206]. Luo et al. adapted this method to tether responsive double hydrophilic block copolymer PEO-*b*-PDMA on gold surface [208]. More recently, Fustin et al. suggested a method to transfer the gold particles from organic solvent to aqueous environment. Nanoparticles first stabilized with PEG-*b*-PCL block copolymer in organic solvent were grafted with PAA chains synthesized by RAFT polymerization. The transfer of these particles in water was successful due to the specific chemisorption of the trithiocarbonate end group on gold surface [209].

In the same time, Hotchkiss et al. described the modification of gold nanorods using hydrophilic (PAA or PDMAEMA) or hydrophobic polymer (PS) with or without reducing agent. Interaction of the dithiobenzoate or trithiocarbonate chain ends with gold surface was sufficient to stabilize the particles in suitable solvent (water for PAA and PDMAEMA brushes and DMF for PS brushes) [210]. Stenzel and coworkers proposed a similar approach to modify the surface of gold nanoparticles with polymer chains that bear pyridyl disulfide end group. The use of a mild reducing agent is sufficient to obtain thiol-terminated polymers and stabilize the gold nanoparticles [27]. In contrast, the 'grafting-from' pathway was used by Tenhu and coworkers in order to control the thickness of the coating and the density of the chains at the surface of the hybrid nanoparticles. NIPAm monomer was polymerized to vary the surface properties, altering the particle solubility in water due to the thermoresponsive behavior [207].

**Magnetic Nanoparticles** Magnetic nanoparticles have attracted considerable interest because of their potential biomedical applications like MRI, targeted drug delivery or rapid biological separation. To avoid aggregation or precipitation of the nanoparticles, several methods to prepare polymer-coated nanoparticles have been described. Wang et al. proposed the 'grafting from' of PS or PAA brushes on ozone preactivated nanoparticles. The composite particles exhibit an excellent dispersability and stability in organic solvent [212]. In the same time, Berret et al. prepared iron-based contrast agents for MRI. The stabilization of the magnetic nanoparticles was performed using electrostatic interactions. Two cationic-neutral block copolymers PTEA-*b*-PAM were synthesized via the macromolecular design by interchange of xanthates process with the same MW for neutral segment. The size of the magnetic clusters and their contrast properties increased with the MW of the cationic block. This easy technique to prepare stable magnetic nanoparticles is very promising for biomedical applications [211].

**Calcite Nanoparticles** Calcite is a natural crystal of calcium carbonate, which can form colloidal calcite dispersion in presence of water and a small amount of dispersant. This colloidal dispersion is used in paper industry as coating. P(sodium acrylate) prepared via the RAFT process was used as dispersant. Since only chains of a given MW (5000 $g{\cdot}mol^{-1}$) adsorb on the surface, generating an electrostatic barrier against aggregation, the synthesis of well-defined polymer chains is crucial, justifying the choice of the RAFT polymerization [213, 214].

#### 13.3.3.3 **Carbon Nanotubes**

Carbon nanotubes possess unique structures, thermal stability and mechanical and electrical properties [215, 216]. Then, potential applications are numerous in fields like molecular electronics, sensors, high-strength fibers and biological electronic devices. The main problem of carbon nanotubes is their poor solubility in solvent due to the strong intertube Van der Waals interactions. The challenge consists of elaborating carbon nanotubes soluble in solvents (especially in water) and biologically compatible.

Although the tethering of polymer chains on the surface of carbon nanotubes remains difficult, it offers potential solubility in various solvents without altering the physical properties. Cui et al. first immobilized RAFT agents on the surface of carbon nanotubes and polymerized styrene via a 'grafting-from' approach leading to core-shell nanocomposites [217]. Using the same strategy, Hong et al. covered carbon nanotubes with thermoresponsive PNIPAm chains [218], with hydrophilic and biocompatible PHPMA chains [219] and with reactive P(S-*alt*-MAh) chains [220]. In the latter case, hydrophilic polymer chains (PEO—OH or glycopolymer) were further indirectly tethered to the surface via the anhydride groups, avoiding deterioration of the surface and consequently keeping the properties of the nanotubes.

At the same time, this group developed another way to functionalize the surface via a 'grafting-to' approach. They first anchored thiopyridine, highly reactive with thiol groups, and bound hydrophilic PHEMA and PHPMA polymer chains synthesized by the RAFT process. The resulting polymer brushes brought water solubility to the carbon nanotubes [221]. More recently, Wang et al. achieved the same objective using an acrylamide-based coating [222]. At the opposite, Xu et al. used a tethered RAFT agent to synthesize P(MMA-*b*-S) block copolymers from the surface, exhibiting dispersibility in various organic solvents [223].

#### 13.3.3.4 **Silane-Based Polymer Chains**

RAFT polymerization has also been used to synthesize polymer architectures based on silyl-containing methacrylate monomers, immobilized on flat inorganic substrates as well as on inorganic nanoparticles. Saricilar et al. applied RAFT polymerization to 3-[tris(trimethylsilyloxy)silyl]propyl methacrylate (TRIS) monomer, a precious monomer for contact lens industry due to its high Si—O content that brings high permeability to oxygen. Optimizing oxygen permeability without losing transparency, mechanical properties and hydrophilicity requires well-defined polymer structures. TRIS was successfully polymerized in solution with low dispersity (10 000–80 000 $g{\cdot}mol^{-1}$; PDI $< 1.2$), a promising result for future block copolymer synthesis [224].

Mellon et al. focused on the RAFT polymerization of $\gamma$-methacryloxypropyl trimethoxysilane (MPS) monomer, since the alkoxysilane reactive function can be used as coupling agent to design nanocomposites for applications in optics, coatings and catalysis. PMPS homopolymer and P(MMA-*b*-MPS) block copolymers were successfully synthesized with MWs up to 40 000 $g{\cdot}mol^{-1}$ and a low polydispersity (PDI $< 1.3$) [225].

Nguyen et al. polymerized *tert*-butyldimethylsilyl methacrylate monomer via the RAFT process, for the potential applications of the polymer in photoresists, dry etch resistance and antifouling paints. The presence of trialkylsilyl esters as pendant groups, subject to hydrolysis, led to water-soluble polymers and induced the decrease of the coating thickness. As this erosion was dependent on the number of silylated groups, RAFT polymerization was used to obtain well-defined structures with controlled MW and low polydispersity (30 000 g·mol$^{-1}$; PDI < 1.3) [226].

#### 13.3.3.5 Conclusion on Organic/Inorganic Hybrid Materials

In the recent years, RAFT polymerization has provided numerous hybrid materials, as this process is suitable for a large variety of monomers. Different approaches like adsorption, grafting to and grafting from have been performed on various inorganic substrates (silica wafers, clays, gold, magnetic, cadmium–selenium, calcite nanoparticles and carbon nanotubes) to get polymer architectures on the surface. New applications are expected since the polymer coating brings a new interfacial behavior (e.g. stabilized nanoparticles and carbon nanotubes in water), without impacting on the inorganic substrate own properties.

## 13.4 Conclusions

The large increase of publications related to polymers synthesized via the RAFT process clearly demonstrates its significant impact on polymer science and polymer-related applications, as reported in the various sections of this chapter. Like other polymerization techniques that enable the control of macromolecular architecture, the RAFT process appears as a powerful tool for the design of a large array of polymers. Its versatility, high tolerance to polar groups and ease to process make it a very attractive technique, especially for the development of functional polymers.

The characteristic of getting chains of predetermined MW and low polydispersity is a first advantage, considering size and composition homogeneity. This structural homogeneity is expected to lead to materials of homogeneous properties and consequently to more reproducible results in the application where that material is used. However, that last expectation remains to be proved in the next years in order that RAFT polymerization may reach an industrial level. In addition, for some kinds of application, the size homogeneity of the polymer chains may appear as a disadvantage.

In fact, the best advantage of RAFT polymerization relies on the ability to prepare well-defined block, graft and star copolymers with sequences of different nature. Moreover, if the RAFT process is combined with another polymerization technique (which is relatively easy for a number of processes, such as the other CRP techniques, catalytic polymerization, ring-opening (metathesis) polymerization and polycondensation, Table 13.7), or with a grafting technique coming from organic/inorganic chemistry and biochemistry, the variety of combination of the blocks is even increased. The resulting polymers and new generations of hybrid/composite materials are expected to have a very large range of tailored properties and most probably some unprecedented properties. The field of future applications is then widely opened.

**Table 13.7** Polymers obtained by combination of RAFT with another polymerization technique

| Polymer | RAFT segments | Other segments | Ref. |
|---|---|---|---|
| Block copolymers | Dithiobenzoate chain ends | PMMA, PDMAEMA and PEG ATRP macroinitiators | [227] |
| | Dithiobenzoate and trithiocarbonate chain ends | P(styrene) produced by NMP | [228] |
| | P(styrene) | PEG | [73] |
| | P(styrene) multiblocks | PEG multiblocks | [229] |
| | P(NIPAm) | PLA produced by ROP | [74, 75] |
| | Xanthate, dithiocarbamate trithiocarbonate chain ends | P(ethylene) produced by catalytic polymerization | [230] |
| | P(styrene) and P(*t*BA) | P(butadiene) produced by ROMP | [231] |
| Comb-shaped polymers | P(BIEM)-based methacrylic backbones | P(MA-*co*-1-Oct) produced by ATRP and PEG side chains | [232] |
| | P(NPMI-*co*-pCMS) backbone | P(styrene) side chains produced by ATRP | [233] |
| | P(styrene-*co*-pVBSC) backbone | PMMA side chains produced by ATRP | [234] |
| | P(HEMA-*co*-styrene) backbone | PCL side chains produced by ROP | [235] |
| | Methacrylic backbone | PEG side chains | [236] |
| | Methacrylic and PGMA backbone blocks | PEG side chains | [77] |
| | Methacrylic and PNIPAm backbone blocks | PEG side chains | [84] |
| | PMMA backbone | PDMS and PLA side chains | [237] |
| | P(MA-*b*-styrene-*co*-pCMS) backbone | P(THF) side chains produced by cationic ROP | [238] |
| | Methacrylic backbone | Lateral dendrons | [239] |

(*Continued*)

**Table 13.7** (*Continued*)

| Polymer | RAFT segments | Other segments | Ref. |
|---|---|---|---|
| Stars, dendritic and hyperbranched polymers | P(styrene) and PMMA arms | P(DOP) arm produced by cationic ROP | [240] |
| | P(styrene) arm | PLA produced by ROP and PEG arms | [241] |
| | P(styrene) arms | PLA arms produced by ROP and PEG core | [242] |
| | P(styrene) and P(BA) branches Trithiocarbonate chain ends | Polyester hyperbranched core | [243] |
| | P(styrene) and P(BA) branches Trithiocarbonate chain ends | Six- and twelve-arm polyester dendritic core | [244] |
| | P(styrene) branches Dithiobenzoate chain ends | Twelve-arm phosphorus-containing dendritic core | [245] |
| | P(styrene) branches Dithiobenzoate chain ends | Sixteen-arm polyester dendritic core | [246] |
| | P(styrene) and P(MA) branch blocks Dithiobenzoate chain ends | Sixteen-arm P(PPI) dendritic core | [247] |

ATRP = atom-transfer radical polymerization; NMP = nitroxide-mediated polymerization; ROP = ring-opening polymerization; ROMP = ring-opening metathesis polymerization; PMMA = P(methyl methacrylate); PDMAEMA = P(*N*,*N*-dimethylaminoethyl methacrylate); PEG = P(ethylene glycol); NIPAm = *N*-isopropylacrylamide; PLA = P(lactic acid); *t*-BA = *tert*-butyl acrylate; BIEM = 2-(2-bromoisobutyryloxy)ethyl methacrylate; MA = methyl acrylate; 1-Oct = 1-octene; NPMI = *N*-phenylmaleimide; PCMS = P(chloromethyl styrene); PVBSC = P(vinyl benzene sulfonyl chloride); HEMA = hydroxyethyl methacrylate; PCL = P($\varepsilon$-caprolactone); GMA = glycidyl methacrylate; PDMS = P(dimethyl siloxane); THF = tetrahydrofuran; DOP = 1,3-dioxepane; BA = *n*-butyl acrylate; PPI = P(propylene imine).

## References


1. J. Chiefari, B. Y. K. Chong, F. Ercole, J. Krstina, J. Jeffery, T. P. T. Le, R. T. A. Mayadunne, G. F. Meijs, C. L. Moad, G. Moad, E. Rizzardo, S. H. Thang, *Macromolecules* **1998**, *31*, 5559–5562.
2. T. P. T. Le, G. Moad, E. Rizzardo, S. H. Thang, E. I. Dupont deNemours and Company, WO 98/01478, **1998**.
3. P. Corpart, D. Charmot, T. Biadatti, S. Z. Zard, D. Michelet, Rhodia Chimie, WO 98/58974, **1998**.
4. T. Delair, M.-H. Charles, P. Cros, A. Laayounn, B. Mandrand, C. Pichot, *Polym. Adv. Technol.* **1998**, *9*, 349–361.
5. R. Arshady, *Adv. Polym. Sci.* **1994**, *111*, 1–41.
6. M.-N. Erout, A. Elaissari, C. Pichot, M.-F. Llauro, *Polymer* **1996**, *37*, 1157–1165.
7. F. D'Agosto, M.-T. Charreyre, L. Véron, M.-F. Llauro, C. Pichot, *Macromol. Chem. Phys.* **2001**, *202*, 1689–1699.
8. M.-T. Charreyre, F. D'Agosto, A. Favier, C. Pichot, B. Mandrand, bioMérieux S. A., WO 01/92361, **2001**.
9. A. Favier, M. T. Charreyre, C. Pichot, *Polymer* **2004**, *45*, 8661–8674.
10. A. Favier, F. D'Agosto, M. T. Charreyre, C. Pichot, *Polymer* **2004**, *45*, 7821–7830.
11. P. Relogio, M. T. Charreyre, J. P. S. Farinha, J. M. G. Martinho, C. Pichot, *Polymer* **2004**, *45*, 8639–8649.
12. B. de Lambert, C. Chaix, M. T. Charreyre, A. Laurent, A. Aigoui, A. Perrin-Rubens, C. Pichot, *Bioconjug. Chem.* **2005**, *16*, 265–274.
13. B. de Lambert, C. Chaix, M.-T. Charreyre, T. Martin, A. Aigoui, A. Perrin-Rubens, B. Mandrand, C. Pichot, *Anal. Biochem.* **2007**, doi:10.1016/j.ab.2007.09.008
14. A. Favier, F. Robin, A. Ganée, L. Veron, M. T. Charreyre, T. Delair, C. Pichot, Manuscript in preparation.
15. M.-T. Charreyre, P. Relogio, J. P. S. Farinha, J. M. G. Martinho, B. Mandrand, bioMérieux, S. A., CNRS, Instituto Superior Tecnico, WO 07/003781, **2007**.
16. Y. Li, B. S. Lokitz, C. L. McCormick, *Macromolecules* **2006**, *39*, 81–89.
17. C. M. Schilli, A. H. E. Müller, E. Rizzardo, S. H. Thang, B. Y. K. Chong, In *Advances in Controlled/Living Radical Polymerization*, ACS Symposium Series 854, K. Matyjaszewski, Ed., American Chemical Society, Washington, DC, **2003**, pp. 603–618.
18. E. N. Savariar, S. Thayumanavan, *J. Polym. Sci., Part A: Polym. Chem.* **2004**, *42*, 6340–6345.
19. M.-J. Yanjarappa, K. V. Gujraty, A. Joshi, A. Saraph, R. S. Kane, *Biomacromolecules* **2006**, *7*, 1665–1670.
20. M. Eberhardt, P. Theato, *Macromol. Rapid Commun.* **2005**, *26*, 1488–1493.
21. F. D'Agosto, R. Hughes, M.-T. Charreyre, C. Pichot, R. G. Gilbert, *Macromolecules* **2003**, *36*, 621–629.
22. G. Moad, Y. K. Chong, A. Postma, E. Rizzardo, S. H. Thang, *Polymer* **2005**, *46*, 8458–8468.
23. R. Wang, C. L. McCormick, A. B. Lowe, *Macromolecules* **2005**, *38*, 9518–9525.
24. J. T. Lai, R. Shea, *J. Polym. Sci., Part A: Polym. Chem.* **2006**, *44*, 4298–4316.
25. M. Bathfield, F. D'Agosto, R. Spitz, M.-T. Charreyre, T. Delair, *J. Am. Chem. Soc.* **2006**, *128*, 2546–2547.
26. F. D'Agosto, M. Bathfield, M.-T. Charreyre, bioMérieux S. A., CNRS, ESCPE LYON, WO 07/003782, **2007**.
27. J. Liu, V. Bulmus, C. Barner-Kowollik, M. H. Stenzel, T. P. Davis, *Macromol. Rapid Commun.* **2007**, *28*, 305–314.
28. A. Godwin, M. Hartenstein, A. H. E. Müller, S. Brocchini, *Angew. Chem. Int. Ed.* **2001**, *40*, 594–595.
29. S. Carter, S. Rimmer, A. Sturdy, M. Webb, *Macromol. Biosci.* **2005**, *5*, 373–378.
30. S. Carter, S. Rimmer, R. Rutkaite, L. Swanson, J. P. A. Fairclough, A. Sturdy, M. Webb, *Biomacromolecules* **2006**, *7*, 1124–1130.
31. S. Kulkarni, C. Schilli, A. H. E. Müller, A. S. Hoffman, P. S. Stayton, *Bioconjug. Chem.* **2004**, *15*, 747–753.
32. S. Kulkarni, C. Schilli, B. Grin, A. H. E. Muller, A. S. Hoffman, P. S. Stayton, *Biomacromolecules* **2006**, *7*, 2736–2741.

**33.** Y.-Z. You, D. Oupicky, *Biomacromolecules* **2007**, *8*, 98–105.
**34.** G. Gody, P. Boullanger, C. Ladavière, M. T. Charreyre, T. Delair, *Macromol. Rapid Commum.*, submitted.
**35.** C.-Y. Hong, C.-Y. Pan, *Macromolecules* **2006**, *39*, 3517–3524.
**36.** H. Mori, K. Sutoh, T. Endo, *Macromolecules* **2005**, *38*, 9055–9065.
**37.** H. Mori, H. Iwaya, A. Nagai, T. Endo, *Chem. Commun.* **2005**, 4872–4874.
**38.** M. G. J. ten Cate, H. Rettig, K. Bernhardt, H. G. Börner, *Macromolecules* **2005**, *38*, 10643–10649.
**39.** M. H. Stenzel, C. Barner-Kowollik, T. P. Davis, H. M. Dalton, *Macromol. Biosci.* **2004**, *4*, 445–453.
**40.** S.-I. Yusa, K. Fukuda, T. Yamamoto, K. Ishihara, Y. Morishima, *Biomacromolecules* **2005**, *6*, 663–670.
**41.** M. Bathfield, F. D'Agosto, C. Ladavière, M. T. Charreyre, T. Delair, Manuscript in preparation.
**42.** T. Delair, B. Badey, A. Domard, C. Pichot, B. Mandrand, *Polym. Adv. Technol.* **1997**, *8*, 297–304.
**43.** R. Roy, F. Tropper, *Glycoconj. J.* **1988**, *5*, 203.
**44.** J. E. Gestwicki, C. W. Cairo, L. E. Strong, K. A. Oetjen, L. L. Kiessling, *J. Am. Chem. Soc.* **2002**, *124*, 14922–14933.
**45.** M. Okada, *Prog. Polym. Sci.* **2001**, *26*, 67–104.
**46.** V. Ladmiral, E. Melia, D. M. Haddleton, *Eur. Polym. J.* **2004**, *40*, 431–449.
**47.** A. B. Lowe, B. S. Sumerlin, C. L. McCormick, *Polymer* **2003**, *44*, 6761–6755.
**48.** J. Bernard, X. Hao, T. P. Davis, C. Barner-Kowollik, M. H. Stenzel, *Biomacromolecules* **2006**, *7*, 232–238.
**49.** L. Albertin, M. H. Stenzel, C. Barner-Kowollik, L. J. R. Foster, T. P. Davis, *Macromolecules* **2005**, *38*, 9075–9084.
**50.** L. Albertin, M. Stenzel, C. Barner-Kowollik, L. J. R. Foster, T. P. Davis, *Macromolecules* **2004**, *37*, 7530–7537.
**51.** L. Albertin, C. Kohlert, M. H. Stenzel, L. J. R. Foster, T. P. Davis, *Biomacromolecules* **2004**, *5*, 255–260.
**52.** S. G. Spain, L. Albertin, N. R. Cameron, *Chem. Commun.* **2006**, 4198–4200.
**53.** G. Gody, P. Boullanger, M. T. Charreyre, T. Delair, Manuscript in preparation.
**54.** M. H. Stenzel, L. Zhang, W. T. S. Huck, *Macromol. Rapid Commun.* **2006**, *27*, 1121–1126.
**55.** T.-Y. Guo, P. Liu, J.-W. Zhu, M.-D. Song, B. H. Zhang, *Biomacromolecules* **2006**, *7*, 1196–1202.
**56.** M. H. Stenzel, T. P. Davis, A. G. Fane, *J. Mater. Chem.* **2003**, *13*, 2090–2097.
**57.** M. Bathfield, F. D'Agosto, R. Spitz, M.-T. Charreyre, C. Pichot, T. Delair, *Macromol. Rapid Commun.*, **2007**, *28*, 1540–1545.
**58.** V. Crescenzi, M. Dentini, D. Bontempo, G. Masci, *Macromol. Chem. Phys.* **2002**, *203*, 1285–1291.
**59.** M. Hernandez-Guerrero, T. P. Davis, C. Barner-Kowollik, M. H. Stenzel, *Eur. Polym. J.* **2005**, *41*, 2264–2277.
**60.** D. Roy, J. T. Guthrie, S. Perrier, *Macromolecules* **2005**, *38*, 10363–10372.
**61.** C. Minard-Basquin, C. Chaix, C. Pichot, B. Mandrand, *Bioconjug. Chem.* **2000**, *11*, 795–804.
**62.** A. Favier, PhD thesis, Synthesis of controlled (co)polymer architectures by the RAFT process for biological diagnostic and gene delivery applications, University Claude Bernard Lyon I, **2002**.
**63.** B. de Lambert, M.-T. Charreyre, C. Chaix, C. Pichot, *Polymer* **2007**, *48*, 437–447.
**64.** R. Langer, *Nature* **1998**, *392*, 5–10.
**65.** K. E. Uhrich, S. M. Cannizzarro, R. S. Langer, K. M. Shakesheff, *Chem. Rev.* **1999**, *99*, 3181–3198.
**66.** K. Kataoka, A. Harada, Y. Nagasaki, *Adv. Drug Deliv. Rev.* **2001**, *47*, 113–131.
**67.** A. V. Kabanov, E. V. Batrakova, V. Y. Alakhov, *J. Control Release* **2002**, *82*, 189–212.
**68.** A. Lavasanifar, J. Samuel, G. S. Kwon, *Adv. Drug Deliv. Rev.* **2002**, *54*, 169–190.
**69.** M. F. Francis, M. Cristea, F. M. Winnik, *Pure Appl. Chem.* **2004**, *76*, 1321–1335.
**70.** E. R. Gillies, J. M. J. Fréchet, *Pure Appl. Chem.* **2004**, *76*, 1295–1307.

71. C. Wang, R. J. Stuart, J. Kopecek, *Nature* **1999**, *397*, 418–420.
72. M. N. V. Ravi Kumar, N. Kumar, A. J. Domb, M. Arora, *Adv. Polym. Sci.* **2002**, *160*, 45–117.
73. Q. Zheng, G.-H. Zheng, C.-Y. Pan, *Polym. Int.* **2006**, *55*, 1114–1123.
74. M. Hales, C. Barner-Kowollik, T. P. Davis, M. H. Stenzel, *Langmuir* **2004**, *20*, 10809–10817.
75. Y. You, C. Hong, W. Wang, W. Lu, C. Pan, *Macromolecules* **2004**, *37*, 9761–9767.
76. S. Garnier, A. Laschewsky, *Macromolecules* **2005**, *38*, 7580–7592.
77. Z. Cheng, X. Zhu, E. T. Kang, K. G. Neoh, *Langmuir* **2005**, *21*, 7180–7185.
78. L. Zhang, K. Katapodi, T. P. Davis, C. Barner-Kowollik, M. H. Stenzel, *J. Polym. Sci., Part A: Polym. Chem.* **2006**, *44*, 2177–2194.
79. C. L. McCormick, S. E. Kirkland, A. W. York, *J. Macromol. Sci., Part C: Polym. Rev.* **2006**, *46*, 421–443.
80. S.-I. Yusa, Y. Shimada, Y. Mitsukami, T. Yamamoto, Y. Morishima, *Macromolecules* **2004**, *37*, 7507–7513.
81. A. J. Convertine, B. S. Lokitz, Y. Vasileva, L. J. Myrick, C. W. Scales, A. B. Lowe, C. L. McCormick, *Macromolecules* **2006**, *39*, 1724–1730.
82. B. Liu, S. Perrier, *J. Polym. Sci., Part A: Polym. Chem.* **2005**, *43*, 3643–3654.
83. M. Mertoglu, S. Garnier, A. Laschewsky, K. Skrabania, J. Storsberg, *Polymer* **2005**, *46*, 7726–7740.
84. D.-H. Han, C.-Y. Pan, *Macromol. Chem. Phys.* **2006**, *207*, 836–843.
85. K. Skrabania, J. Kristen, A. Laschewsky, O. Akdemir, A. Hoth, J.-F. Lutz, *Langmuir* **2007**, *23*, 84–93.
86. D. Taton, A.-Z. Wilczewska, M. Destarac, *Macromol. Rapid Commun.* **2001**, *22*, 1497–1503.
87. B. S. Sumerlin, A. B. Lowe, D. B. Thomas, A. J. Convertine, M. S. Donovan, C. L. McCormick, *J. Polym. Sci., Part A: Polym. Chem.* **2004**, *42*, 1724–1734.
88. Y. Mitsukami, A. Hashidzume, S.-I. Yusa, Y. Morishima, A. B. Lowe, C. L. McCormick, *Polymer* **2006**, *47*, 4333–4340.
89. S.-I. Yusa, Y. Shimada, Y. Mitsukami, T. Yamamoto, Y. Morishima, *Macromolecules* **2003**, *36*, 4208–4215.
90. M. S. Donovan, A. B. Lowe, T. A. Sandford, C. L. McCormick, *J. Polym. Sci., Part A: Polym. Chem.* **2003**, *41*, 1262–1281.
91. A. Laschewsky, M. Mertoglu, S. Kubowicz, A. F. Thünemann, *Macromolecules* **2006**, *39*, 9337–9345.
92. E. Khousakoun, J.-F. Gohy, R. Jérôme, *Polymer* **2004**, *45*, 8303–8310.
93. M. Arotçaréna, B. Heise, S. Ishaya, A. Laschewsky, *J. Am. Chem. Soc.* **2002**, *124*, 3787–3793.
94. C. M. Schilli, M. Zhang, E. Rizzardo, S. H. Thang, B. Y. K. Chong, K. Edwards, G. Karlsson, A. H. E. Müller, *Macromolecules* **2004**, *37*, 7861–7866.
95. X. Xin, Y. Wang, W. Liu, *Eur. Polym. J.* **2005**, *41*, 1539–1545.
96. P.-E. Millard, L. Barner, M. H. Stenzel, T. P. Davis, C. Barner-Kowollik, A. H. E. Müller, *Macromol. Rapid Commun.* **2006**, *27*, 821–828.
97. Z. Ge, Y. Cai, J. Yin, Z. Zhu, J. Rao, S. Liu, *Langmuir* **2007**, *23*, 1114–1122.
98. A. T. Nikova, V. D. Gordon, G. Cristobal, M. R. Talingting, D. C. Bell, C. Evans, M. Joanicot, J. A. Zasadzinski, D. A. Weitz, *Macromolecules* **2005**, *37*, 2215–2218.
99. D. Bendejacq, V. Ponsinet, M. Joanicot, Y.-L. Loo, R. A. Register, *Macromolecules* **2002**, *35*, 6645–6649.
100. M. Nuopponen, J. Ojala, H. Tenhu, *Polymer* **2004**, *45*, 3643–3650.
101. M. Nakayama, T. Okano, *Biomacromolecules* **2005**, *6*, 2320–2327.
102. P. Zhang, Q. Liu, A. Qing, J. Shi, M. Lu, *J. Polym. Sci., Part A: Polym. Chem.* **2006**, *44*, 3312–3320.
103. H. Huang, T. Kowalewski, E. E. Remsen, R. Gertzmann, K. L. Wooley, *J. Am. Chem. Soc.* **1996**, *118*, 7239–7240.
104. D. Wan, Q. Fu, J. Huang, *J. Polym. Sci., Part A: Polym. Chem.* **2005**, *43*, 5652–5660.
105. R. K. O'Reilly, M. J. Joralemon, C. J. Hawker, K. L. Wooley, *J. Polym. Sci., Part A: Polym. Chem.* **2006**, *44*, 5203–5217.

**106.** B. S. Lokitz, A. J. Convertine, R. G. Ezell, A. Heidenreich, Y. Li, C. L. McCormick, *Macromolecules* **2006**, *39*, 8594–8602.

**107.** M. Destarac, D. Charmot, X. Franck, S. Z. Zard, *Macromol. Rapid Commun.* **2000**, *21*, 1035–1039.

**108.** M. H. Stenzel, L. Cummins, G. E. Roberts, T. P. Davis, P. Van, C. Barner-Kowollik, *Macromol. Chem. Phys.* **2003**, *204*, 1160–1168.

**109.** A. Favier, C. Barner-Kowollik, T. P. Davis, M. H. Stenzel, *Macromol. Chem. Phys.* **2004**, *2005*, 925–936.

**110.** D. B. Thomas, B. S. Sumerlin, A. B. Lowe, C. L. McCormick, *Macromolecules* **2003**, *36*, 1436–1439.

**111.** T. C. Krasia, C. S. Patrickios, *Macromolecules* **2006**, *39*, 2467–2473.

**112.** X. Yin, A. S. Hoffman, P. S. Stayton, *Biomacromolecules* **2006**, *7*, 1381–1385.

**113.** P. Kujawa, H. Watanabe, F. Tanaka, F. M. Winnik, *Eur. Phys. J. E* **2005**, *17*, 129–137.

**114.** Q. Liu, P. Zhang, A. Qing, Y. Lan, J. Shi, M. Lu, *Polymer* **2006**, *47*, 6963–6969.

**115.** D. Fournier, R. Hoogenboom, H. M. L. Thijs, R. M. Paulus, U. S. Schubert, *Macromolecules* **2007**, *40*, 915–920.

**116.** J. M. J. Frechet, *J. Polym. Sci., Part A: Polym. Chem.* **2003**, *41*, 3713–3725.

**117.** R. H. R. S. Lambeth, R. Mueller, J. P. Poziemski, G. S. Miguel, L. J. Markoski, C. F. Zukoski, J. S. Moore, *Langmuir* **2006**, *22*, 6352–6360.

**118.** M. R. Whittaker, M. J. Monteiro, *Langmuir* **2006**, *22*, 9746–9752.

**119.** R. Plummer, D. J. T. Hill, A. K. Whittaker, *Macromolecules* **2006**, *39*, 8379–8388.

**120.** M. H. Stenzel, T. P. Davis, C. Barner-Kowollik, *Chem. Commun.* **2004**, 1546–1547.

**121.** J. Bernard, A. Favier, L. Zhang, A. Nilasaroya, T. P. Davis, C. Barner-Kowollik, M. H. Stenzel, *Macromolecules* **2005**, *38*, 5475–5484.

**122.** J. Bernard, A. Favier, T. P. Davis, C. Barner-kowollik, M. H. Stenzel, *Polymer* **2006**, *47*, 1073–1080.

**123.** D. W. Pack, A. S. Hoffman, S. Pun, P. S. Stayton, *Nature* **2005**, *4*, 581–593.

**124.** O. Boussif, F. Lezoualc'h, M. A. Zanta, M. D. Mergny, D. Scherman, B. Demeneix, J.-P. Behr, *Proc. Natl. Acad. Sci. USA* **1995**, *92*, 7297–7301.

**125.** R. Kircheis, L. Wightman, E. Wagner, *Adv. Drug Deliv. Rev.* **2001**, *53*, 341–358.

**126.** C. L. Gebhart, A. V. Kabanov, *J. Control Release* **2001**, *73*, 401–416.

**127.** M. Sahnoun, M. T. Charreyre, L. Veron, T. Delair, F. D'Agosto, *J. Polym. Sci., Part A: Polym. Chem.* **2005**, *43*, 3551–3565.

**128.** L. Veron, A. Ganée, M.-T. Charreyre, C. Pichot, T. Delair, *Macromol. Biosci.* **2004**, *4*, 431–444.

**129.** C. W. Scales, F. Huang, N. Li, Y. A. Vasilieva, J. Ray, A. J. Convertine, C. L. McCormick, *Macromolecules* **2006**, *39*, 6871–6881.

**130.** H. S. Bisht, D. S. Manickam, Y. You, D. Oupicky, *Biomacromolecules* **2006**, *7*, 1169–1178.

**131.** J. P. Vacanti, R. Langer, *Lancet* **1999**, *354*, 32–34.

**132.** L. L. Hench, J. M. Polak, *Science* **2002**, *295*, 1014–1017.

**133.** B. L. Seal, T. C. Otero, A. Panitch, *Mater. Sci. Eng. R* **2001**, *34*, 147–230.

**134.** L. G. Griffith, G. Naughton, *Science* **2002**, *295*, 1009–1014.

**135.** S. Patel, R. G. Thakar, J. Wong, S. D. McLeod, S. Li, *Biomaterials* **2006**, *27*, 2890–2897.

**136.** K. Y. Lee, D. J. Mooney, *Chem. Rev.* **2001**, *101*, 1869–1879.

**137.** S. Perrier, P. Takolpuckdee, J. Westwood, D. M. Lewis, *Macromolecules* **2004**, *37*, 2709–2717.

**138.** S. Suzuki, M. R. Whittaker, L. Grondahl, M. J. Monteiro, E. Wentrup-Byrne, *Biomacromolecules* **2006**, *7*, 3178–3187.

**139.** S. R. Forrest, M. E. Thompson, *Chem. Rev.* **2007**, *107*, 923–925.

**140.** S.-C. Lo, P. L. Burn, *Chem. Rev.* **2007**, *107*, 1097–1116.

**141.** S. W. Thomas III, G. D. Joly, T. M. Swager, *Chem. Rev.* **2007**, *107*, 1339–1386.

**142.** H. Mori, S. Masuda, T. Endo, *Macromolecules* **2006**, *39*, 5976–5978.

**143.** H. Mori, H. Ookuma, S. Nakano, T. Endo, *Macromol. Chem. Phys.* **2006**, *207*, 1005–1017.

**144.** H. Mori, S. Nakano, T. Endo, *Macromolecules* **2005**, *38*, 8192–8201.

**145.** P. Zhao, Q. D. Ling, W. Z. Wang, J. Ru, S.-B. Li, W. Huang, *J. Polym. Sci., Part A: Polym. Chem.* **2007**, *45*, 242–252.

**146.** J.-F. Baussard, J.-L. Habib-Jiwan, A. Laschewski, *Langmuir* **2003**, *19*, 7963–7969.

**147.** C. Barner-Kowollik, H. Dalton, T. P. Davis, M. H. Stenzel, *Angew. Chem. Int. Ed.* **2003**, *42*, 3664–3668.

**148.** C. W. Scales, A. J. Convertine, C. L. McCormick, *Biomacromolecules* **2006**, *7*, 1389–1392.

**149.** D. Hua, W. Sun, R. Bai, W. Lu, C. Pan, *Eur. Polym. J.* **2005**, *41*, 1674–1680.

**150.** J. Zhu, D. Zhou, X. Zhu, G. Chen, *J. Polym. Sci., Part A: Polym. Chem.* **2004**, *42*, 2558–2565.

**151.** W. Zhang, X. Zhu, X. Wang, J. Zhu, *J. Polym. Sci., Part A: Polym. Chem.* **2005**, *43*, 2632–2642.

**152.** M. Chen, K. P. Ghiggino, A. W. Mau, E. Rizzardo, S. H. Thang, G. J. Wilson, *Chem. Commun.* **2002**, 2276–2277.

**153.** N. Zhou, L. Lu, X. Zhu, X. Yang, X. Wang, J. Zhu, D. Zhou, *Polym. Bull.* **2006**, *57*, 491–498.

**154.** M. Chen, K. P. Ghiggino, A. W. H. Mau, W. H. F. Sasse, S. H. Thang, G. J. Wilson, *Macromolecules* **2005**, *38*, 3475–3481.

**155.** M. Mertoglu, A. Laschewsky, K. Skrabania, C. Wieland, *Macromolecules* **2005**, *38*, 3601–3614.

**156.** N. Zhou, L. Lu, J. Zhu, X. Yang, X. Wang, X. Zhu, Z. Zhang, *Polymer* **2007**, *48*, 1255–1260.

**157.** T. Prazeres, M. T. Charreyre, J. P. Farinha, J. M. G. Martinho, Manuscript in preparation.

**158.** M. Chen, K. P. Ghiggino, A. W. H. Mau, E. Rizzardo, W. H. F. Sasse, S. H. Thang, G. Wilson, *Macromolecules* **2004**, *37*, 5479–5481.

**159.** G. K. Such, R. A. Evans, T. P. Davis, *Macromolecules* **2006**, *39*, 9562–9570.

**160.** H. Skaff, T. Emrick, *Angew. Chem. Int. Ed.* **2004**, *43*, 5383–5386.

**161.** M. Chen, K. P. Ghiggino, A. Launikonis, A. W. H. Mau, E. Rizzardo, W. H. F. Sasse, S. H. Thang, G. J. Wilson, *J. Mater. Chem.* **2003**, *13*, 2696–2700.

**162.** M. Chen, K. P. Ghiggino, S. H. Thang, G. J. Wilson, *Angew. Chem. Int. Ed.* **2005**, *44*, 4368–4372.

**163.** M. Chen, K. P. Ghiggino, S. H. Thang, G. J. Wilson, *Polym. Int.* **2006**, *55*, 757–763.

**164.** G. Zhou, I. I. Harruna, *Macromolecules* **2004**, *37*, 7132–7139.

**165.** G. Zhou, I. I. Harruna, C. W. Ingram, *Polymer* **2005**, *46*, 10672–10677.

**166.** G. Zhou, I. I. Harruna, *Macromolecules* **2005**, *38*, 4114–4123.

**167.** G. E. Southard, K. A. van Houten, E. W. Ott, G. M. Murray, *Anal. Chim. Acta* **2007**, *58*, 202–207.

**168.** J. P. S. Farinha, P. Relogio, T. Prazeres, M. T. Charreyre, J. M. G. Martinho, *Macromolecules*, **2007**, *40*, 4680–4690.

**169.** C. Yoshikawa, A. Goto, Y. Tsujii, T. Fukuda, K. Yamamoto, A. Kishida, *Macromolecules* **2005**, *38*, 4604–4610.

**170.** Y. Chen, W. Sun, Q. Deng, L. Chen, *J. Polym. Sci., Part A: Polym. Chem.* **2006**, *44*, 3071–3082.

**171.** M. Grasselli, N. Betz, *Nucl. Instrum. Meth. B* **2005**, *236*, 201–207.

**172.** W. S. Bae, A. J. Convertine, C. L. McCormick, M. W. Urban, *Langmuir* **2007**, *23*, 667–672.

**173.** P. Takolpuckdee, C. A. Mars, S. Perrier, *Org. Lett.* **2005**, *7*, 3449–3452.

**174.** S. Perrier, P. Takolpuckdee, C. A. Mars, *Macromolecules* **2005**, *38*, 2033–2036.

**175.** L. Barner, N. Zwaneveld, S. Perera, Y. Pham, T. P. Davis, *J. Polym. Sci., Part A: Polym. Chem.* **2002**, *40*, 4180–4192.

**176.** L. Barner, S. Perera, S. Sandanayake, T. P. Davis, *J. Polym. Sci., Part A: Polym. Chem.* **2005**, *44*, 857–864.

**177.** F. D'Agosto, M. T. Charreyre, C. Pichot, R. G. Gilbert, *J. Polym. Sci., Part A: Polym. Chem.* **2003**, *41*, 1188–1195.

**178.** D. Fan, J. He, J. Xu, W. Tang, Y. Liu, Y. Yang, *J. Polym. Sci., Part A: Polym. Chem.* **2006**, *44*, 2260–2269.

**179.** B. Rixens, R. Severac, B. Boutevin, P. Lacroix-Desmazes, *Polymer* **2005**, *46*, 3579–3587.

**180.** M. L. Tebaldi deSordi, M. A. Ceschi, C. L. Petzhold, A. H. E. Müller, *Macromol. Rapid Commun.* **2007**, *28*, 63–71.

**181.** C. Pichot, B. Charleux, M.-T. Charreyre, J. Revilla, *Makromol. Chem., Macromol. Symp.* **1994**, *88*, 71–86.

**182.** M. H. Stenzel, C. Barner-Kowollik, T. P. Davis, *J. Polym. Sci., Part A: Polym. Chem.* **2006**, *44*, 2363–2375.

**183.** M. H. Stenzel-Rosenbaum, T. P. Davis, A. G. Fane, V. Chen, *Angew. Chem. Int. Ed.* **2001**, *40*, 3428–3432.

**184.** M. H. Stenzel, T. P. Davis, *Aust. J. Chem.* **2003**, *56*, 1035–1038.

**185.** D. Beattie, K. H. Wong, C. Williams, L. A. Poole-Warren, T. P. Davis, C. Barner-Kowollik, M. H. Stenzel, *Biomacromolecules* **2006**, *7*, 1072–1082.

**186.** G. D. Fu, B. Y. Zong, E. T. Kang, K. G. Neoh, C. C. Lin, D. J. Liaw, *Ind. Eng. Chem. Res.* **2004**, *43*, 6723–6730.

**187.** Y. Chen, L. Chen, H. Nie, E. T. Kang, R. H. Vora, *Mater. Chem. Phys.* **2005**, *94*, 195–201.

**188.** J. Rzayev, M. A. Hillmyer, *Macromolecules* **2005**, *38*, 3–5.

**189.** J. Bang, S. H. Kim, E. Drockenmuller, M. J. Misner, T. P. Russell, C. J. Hawker, *J. Am. Chem. Soc.* **2006**, *128*, 7622–7629.

**190.** S. G. Boyes, A. Granville, M. Baum, B. Akgun, B. K. Mirous, W. J. Brittain, *Surf. Sci.* **2004**, *570*, 1–12.

**191.** M. Baum, W. J. Brittain, *Macromolecules* **2002**, *35*, 610–615.

**192.** S. E. Morgan, P. Jones, A. S. Lamont, A. Heidenreich, C. L. McCormick, *Langmuir* **2007**, *23*, 230–240.

**193.** G. Zhai, W. H. Yu, E. T. Kang, K. G. Neoh, C. C. Huang, D. J. Liaw, *Ind. Eng. Chem. Res.* **2004**, *43*, 1673–1680.

**194.** W. H. Yu, E. T. Kang, K. G. Neoh, *Ind. Eng. Chem. Res.* **2004**, *43*, 5194–5202.

**195.** Q. Peng, D. M. Y. Lai, E. T. Kang, K. G. Neoh, *Macromolecules* **2006**, *39*, 5577–5582.

**196.** N. Salem, D. A. Shipp, *Polymer* **2005**, *46*, 8573–8581.

**197.** B. Q. Zhang, C.-Y. Pan, C. Y. Hong, B. Luan, P. J. Shi, *Macromol. Rapid Commun.* **2006**, *27*, 97–102.

**198.** Y. Tsujii, M. Ejaz, K. Sato, A. Goto, T. Fukuda, *Macromolecules* **2001**, *34*, 8872–8878.

**199.** C. Li, B. Benicewicz, *J. Polym. Sci., Part A: Polym. Chem.* **2005**, *43*, 1535–1543.

**200.** C. Li, J. Han, C. Y. Riu, B. C. Benicewicz, *Macromolecules* **2006**, *39*, 3175–3183.

**201.** Y. Zhang, S. Luo, S. Liu, *Macromolecules* **2005**, *38*, 9813–9820.

**202.** M. M. Titirici, B. Sellergren, *Chem. Mater.* **2006**, *18*, 1773–1779.

**203.** D. H. Nguyen, P. Vana, *Polym. Adv. Technol.* **2006**, *17*, 625–633.

**204.** Y. Zhao, S. Perrier, *Macromolecules* **2006**, *39*, 8603–8608.

**205.** A. B. Lowe, B. S. Sumerlin, M. S. Donovan, C. L. McCormick, *J. Am. Chem. Soc.* **2002**, *124*, 11562–11563.

**206.** B. S. Sumerlin, A. B. Lowe, P. A. Stroud, P. Zhang, M. W. Urban, C. L. McCormick, *Langmuir* **2003**, *19*, 5559–5562.

**207.** J. Raula, J. Shan, M. Nuopponen, A. Niskanen, H. Jiang, E. I. Kauppinen, H. Tenhu, *Langmuir* **2003**, *19*, 3499–3504.

**208.** S. Luo, J. Xu, Y. Zhang, S. Liu, C. Wu, *J. Phys. Chem. B* **2005**, *109*, 22159–22166.

**209.** C. A. Fustin, C. Colard, M. Filali, P. Guillet, A. S. Duwez, M. A. R. Meier, U. S. Schubert, J. F. Gohy, *Langmuir* **2006**, *22*, 6690–6695.

**210.** J. W. Hotchkiss, A. B. Lowe, S. G. Boyes, *Chem. Mater.* **2007**, *19*, 6–13.

**211.** J. F. Berret, N. Schonbeck, F. Gazeau, D. El Kharrat, O. Sandre, A. Vacher, M. Airiau, *J. Am. Chem. Soc.* **2006**, *128*, 1755–1761.

**212.** W. C. Wang, K. G. Neoh, E. T. Kang, *Macromol. Rapid Commun.* **2006**, *27*, 1665–1669.

**213.** J. Loiseau, N. Doërr, J. M. Suau, J. B. Egraz, M.-F. Llauro, C. Ladaviere, J. Claverie, *Macromolecules* **2003**, *36*, 3066–3077.

**214.** J. Loiseau, C. Ladaviere, J. M. Suau, J. Claverie, *Polymer* **2005**, *46*, 8565–8572.

**215.** P. M. Ajayan, *Chem. Rev.* **1999**, *99*, 1787–1799.

**216.** D. Tasis, M. Tagmatarchis, A. Bianco, M. Prato, *Chem. Rev.* **2006**, *106*, 1105–1136.

**217.** J. Cui, W. Wang, Y. You, C. Liu, P. Wang, *Polymer* **2004**, *45*, 8717–8721.

**218.** C.-Y. Hong, Y.-Z. You, C.-Y. Pan, *Chem. Mater.* **2005**, *17*, 2247–2254.

**219.** C. Y. Hong, Y. Z. You, C. Y. Pan, *J. Polym. Sci., Part A: Polym. Chem.* **2006**, *44*, 2419–2427.
**220.** C. Y. Hong, Y. Z. You, C. Y. Pan, *Polymer* **2006**, *47*, 4300–4309.
**221.** Y. Z. You, C. Y. Hong, C. Y. Pan, *Macromol. Rapid Commun.* **2006**, *27*, 2001–2006.
**222.** G. J. Wang, S. Z. Huang, Y. Wang, L. Liu, J. Qiu, Y. Li, *Polymer* **2007**, *48*, 728–733.
**223.** G. Xu, W. T. Wu, Y. Wang, W. Pang, Q. Zhu, P. Wang, Y. You, *Polymer* **2006**, *47*, 5909–5918.
**224.** S. Saricilar, R. Knott, C. Barner-Kowollik, T. P. Davis, J. P. A. Heuts, *Polymer* **2003**, *44*, 5169–5176.
**225.** V. Mellon, D. Rinaldi, E. Bourgeat-Lami, F. D'Agosto, *Macromolecules* **2005**, *38*, 1591–1598.
**226.** M. N. Nguyen, C. Bressy, A. Margaillan, *J. Polym. Sci., Part A: Polym. Chem.* **2005**, *43*, 5680–5689.
**227.** C. Wager, D. M. Haddleton, S. A. F. Bon, *Eur. Polym. J.* **2004**, *40*, 641–645.
**228.** A. Favier, B. Luneau, J. Vinas, D. Gigmes, D. Bertin, Manuscript in preparation.
**229.** Z. Jia, X. Xu, Q. Fu, J. Huang, *J. Polym. Sci., Part A: Polym. Chem.* **2006**, *44*, 6071–6082.
**230.** R. Godoy Lopez, C. Boisson, F. D'Agosto, R. Spitz, F. Boisson, D. Gigmes, D. Bertin, *Macromol. Rapid Commun.* **2006**, *27*, 173–181.
**231.** M. K. Mahanthappa, F. S. Bates, M. A. Hillmyer, *Macromolecules* **2005**, *38*, 7890–7894.
**232.** R. Venkatesh, L. Yajjou, C. E. Koning, B. Klumperman, *Macromol. Chem. Phys.* **2004**, *205*, 2161–2168.
**233.** Y. Shi, Z. Fu, W. Yang, *J. Polym. Sci., Part A: Polym. Chem.* **2006**, *44*, 2069–2075.
**234.** C. Li, Y. Shi, Z. Fu, *Polym. Int.* **2006**, *55*, 25–30.
**235.** X. Xu, J. Huang, *J. Polym. Sci., Part A: Polym. Chem.* **2004**, *42*, 5523–5529.
**236.** Y. Chen, W. Wang, W. Yu, Z. Yuan, E.-T. Kang, K.-G. Neoh, B. Krauter, A. Greiner, *Adv. Funct. Mater.* **2004**, *14*, 471–478.
**237.** J. F. Lutz, N. Jahed, K. Matyjaszewski, *J. Polym. Sci., Part A: Polym. Chem.* **2004**, *42*, 1939–1952.
**238.** W.-P. Wang, Y.-Z. You, C.-H. Hong, J. Xu, C.-Y. Pan, *Polymer* **2005**, *46*, 9489–9494.
**239.** A. Zhang, L. Wei, A. D. Schlüter, *Macromol. Rapid Commun.* **2004**, *25*, 799–803.
**240.** Y. G. Li, Y. M. Wang, C. Y. Pan, *J. Polym. Sci., Part A: Polym. Chem.* **2003**, *41*, 1243–1250.
**241.** P.-J. Shi, Y.-G. Li, C.-Y. Pan, *Eur. Polym. J.* **2004**, *40*, 1283–1290.
**242.** D.-H. Han, C.-Y. Pan, *J. Polym. Sci., Part A: Polym. Chem.* **2007**, *45*, 789–799.
**243.** M. Jesberger, L. Barner, M. H. Stenzel, E. Malmström, T. P. Davis, C. Barner-Kowollik, *J. Polym. Sci., Part A: Polym. Chem.* **2003**, *41*, 3847–3861.
**244.** X. Hao, C. Nilsson, M. Jesberger, M. H. Stenzel, E. Malmström, T. P. Davis, E. Ostmark, C. Barner-Kowollik, *J. Polym. Sci., Part A: Polym. Chem.* **2004**, *42*, 5877–5890.
**245.** V. Darcos, A. Dureault, D. Taton, Y. Gnanou, P. Marchand, A.-M. Caminade, J.-P. Majoral, M. Destarac, F. Leising, *Chem. Commun.* **2004**, 2110–2111.
**246.** C.-Y. Hong, Y.-Z. You, C.-Y. Pan, *J. Polym. Sci., Part A: Polym. Chem.* **2005**, *43*, 6379–6393.
**247.** Q. Zheng, C.-Y. Pan, *Macromolecules* **2005**, *38*, 6841–6848.

# Subject Index



*Handbook of RAFT Polymerization.* Edited by Christopher Barner-Kowollik

ISBN: 978-3-527-31924-4